ADVANCED MATERIALS FOR CLEAN ENERGY

ADVANCED MATERIALS FOR CLEAN ENERGY

EDITED BY
Qiang Xu
Tetsuhiko Kobayashi

CRC Press
Taylor & Francis Group
Boca Raton London New York

CRC Press is an imprint of the
Taylor & Francis Group, an **informa** business

CRC Press
Taylor & Francis Group
6000 Broken Sound Parkway NW, Suite 300
Boca Raton, FL 33487-2742

First issued in paperback 2020

© 2015 by Taylor & Francis Group, LLC
CRC Press is an imprint of Taylor & Francis Group, an Informa business

No claim to original U.S. Government works

Version Date: 20150622

ISBN 13: 978-0-367-57581-6 (pbk)
ISBN 13: 978-1-4822-0578-7 (hbk)

Visit the Taylor & Francis Web site at
http://www.taylorandfrancis.com

and the CRC Press Web site at
http://www.crcpress.com

Contents

Preface

Research for clean energy is in an explosive phase, driven by the rapid depletion of fossil fuels and growing environmental concerns as well as the increasing growth of mobile electronic devices. It has been the focus of a wide range of research fields to develop high-performance materials for alternative energy technologies and develop a fundamental understanding of their structure–property–performance relationship, which include materials for photovoltaics, solar energy conversion, thermoelectrics, piezoelectrics, supercapacitors, rechargeable batteries, hydrogen production and storage, and fuel cells.

After successfully organizing an international symposium on advanced materials for clean energy in 2011, an idea of publishing a book to clearly demonstrate the profound progress and provide a comprehensive recognition of advanced materials for clean energy to the researchers in this promising field was naturally conceived. It is our honor, as organizers, to invite leading scientists in this field to survey the key developments of the materials in a broad range and the important advances in their applications to date, which are outlined in the chapters as follows.

The chapters start with materials for photovoltaics. In Chapter 1, Ho and Wong provide a survey on arylamine-based photosensitizing metal complexes for dye-sensitized solar cells; in Chapter 2, Ning and Tian provide a review on p-type small electron-donating molecules for organic heterojunction solar cells; and in Chapter 3, Toyoshima gives a wide-range overview about inorganic materials for solar cell applications. In Chapter 4, Funahashi and coworkers demonstrate the development of thermoelectric technology from materials to generators. Piezoelectric materials for energy harvesting from a wide range of renewable energy sources are introduced by Maurya, Yan, and Priya in Chapter 5. In Chapter 6, Kang and Chen review the synthesis and application of various electrode materials for electrochemical capacitors focusing on the nanoarchitecture design of advanced electrodes. Being extensively investigated thus far, materials for batteries are dealt with in chapters from different directions. In Chapter 7, Inoue and Higuchi describe electrode materials for nickel/metal hydride (Ni/MH) rechargeable batteries; in Chapter 8, Casas-Cabanas and Palacín document the fabrication of electrode materials for lithium-ion rechargeable batteries; and in Chapter 9, Munakata and Kanamura review the materials for all-solid-state rechargeable batteries. A new trend in liquid electrolytes for electrochemical energy devices is highlighted by Matsumoto in Chapter 10. Recent development and future prospects of organic electrode active materials for rechargeable batteries are illustrated by Yao and Kobayashi in Chapter 11. Neburchilov and Wang present in Chapter 12 the synthesis and application of materials for metal–air batteries. Two chapters deal with solar energy/material conversion: Zhu introduces photocatalytic hydrogen production with semiconductor photocatalysts in Chapter 13, and Primo, Neațu, and García address photocatalytic CO_2 reduction with a focus on the fundamentals behind the reaction and the photocatalysts in Chapter 14. Two chapters are dedicated to hydrogen storage, the key technology for hydrogen economy: Zhu, Ouyang, and Wang review Mg-based hydrogen storage alloys, complex hydrides, metal amides, and imides for reversible high-capacity hydrogen storage in Chapter 15, and Kojima, Miyaoka, and Ichikawa provide an overview on ammonia and ammonia borane as chemical hydrogen storage materials in Chapter 16. Finally, three chapters are presented on fuel cells: In Chapter 17, Daimon describes cathode catalysts for polymer electrolyte fuel cell with an emphasis on the strategies for decreasing the use of the precious metal Pt; in Chapter 18, Sahai and Ma address the fundamentals and materials aspects of fuel cells directly using organic and inorganic liquids as fuels; and in Chapter 19, Merino-Jiménez, Ponce de León, and Walsh focus on the developments in electrodes, membranes, and electrolytes for direct borohydride fuel cells (DBFCs).

It is obvious that the accomplishments in materials for clean energy applications to date are exciting and the potential appears to be even greater. We are sure that this field will sustain its growth in the future.

We thank all the authors for their great contributions to this book as well as to this field. Sincere thanks to Barbara Glunn, David Fausel, and Cheryl Wolf (Taylor & Francis Group/CRC Press) for their conscientious cooperation in the editorial process.

Qiang Xu
Tetsu Kiyobayashi

Editors

Qiang Xu received his PhD in physical chemistry in 1994 from Osaka University, Japan. After working as a postdoctoral fellow at Osaka University for a year, he started his career as a research scientist in Osaka National Research Institute (ONRI) in 1995. Currently, he is a chief senior researcher at the National Institute of Advanced Industrial Science and Technology (AIST) and adjunct professor at Kobe University, Japan. His research interests include porous materials and nanostructured materials and related functional applications, especially for clean energy.

Tetsu Kiyobayashi received his PhD in electrochemistry in 1983 from Osaka University, Japan. After working as a Japan Society for the Promotion of Science postdoctoral fellow at Osaka University for a year, he started his career as a research scientist in the Government Industrial Research Institute of Osaka (GIRIO) in 1984. He worked as group leader in Osaka National Research Institute (ONRI) and as director of the Special Division for Green Life Technology, Research Institute for Ubiquitous Energy Devices, and Kansai Research Center in the National Institute of Advanced Industrial Science and Technology (AIST) and worked as guest professor in Kyoto University, Kobe University, and Osaka University. Currently, he is director-general for Environment and Energy Research, AIST. His research interests include electrochemistry, catalytic chemistry, and materials chemistry, especially for clean energy applications.

Contributors

Hiroaki Anno
Department of Electrical Engineering
Tokyo University of Science,
  Yamaguchi
Sanyo Onoda, Japan

Montse Casas-Cabanas
CIC energiGUNE
Miñano, Spain

Mingwei Chen
Advanced Institute for Materials
  Research
Tohoku University
Sendai, Japan

Hideo Daimon
Office for Advanced Research and
  Education
Doshisha University
Kyoto, Japan

Feng Dang
Key Laboratory for Liquid-Solid
  Structural Evolution and Processing
  of Materials
Shandong University
Jinan, Shandong, People's Republic of
  China

Takeyuki Fujisaka
Nippon Steel & Sumitomo Metal Co.
Tokyo, Japan

Ryoji Funahashi
Research Institute for Ubiquitous
  Energy Devices
National Institute of Advanced
  Industrial Science and Technology
Osaka, Japan

Hermenegildo García
Instituto Universitario de Tecnología
  Química
Universidad Politécnica de Valencia
Valencia, Spain
and
Center for Excellence in Advanced
  Materials Research
King Abdulaziz University
Jeddah, Saudi Arabia

Eiji Higuchi
Department of Applied Chemistry
Osaka Prefecture University
Osaka, Japan

Cheuk-Lam Ho
Department of Chemistry
Hong Kong Baptist University
Kowloon Tong, Hong Kong

Takayuki Ichikawa
Institute for Advanced Materials
  Research
Hiroshima University
Hiroshima, Japan

Hiroshi Inoue
Department of Applied Chemistry
Osaka Prefecture University
Osaka, Japan

Kiyoshi Kanamura
Graduate School of Urban
  Environmental Sciences
Tokyo Metropolitan University
Tokyo, Japan

Masaru Yao
Research Institute for Ubiquitous
  Energy Devices
National Institute of Advanced
  Industrial Science and Technology
Osaka, Japan

Jiefang Zhu
Department of Chemistry
Uppsala University
Uppsala, Sweden

Min Zhu
School of Materials Science and
  Engineering
South China University of Technology
Guangzhou, Guangdong, People's
  Republic of China

1 Arylamine-Based Photosensitizing Metal Complexes for Dye-Sensitized Solar Cells

Cheuk-Lam Ho and Wai-Yeung Wong

CONTENTS

1.1 INTRODUCTION

The continuous growth of energy demand around the world and the environmental pollution resulting from global warming have triggered global research toward the exploitation of clean and renewable energy sources over the past decades as the resources of coal, oil, and natural gas are dwindling rapidly.[1] Solar energy is generally considered as the most promising way to solve the global energy crisis as it is available profusely in most of the world's regions, is inexhaustible, and can be used for direct electricity production by means of photovoltaic and photoelectrochemical cells. In this context, dye-sensitized solar cells (DSSCs) have attracted tremendous attention as an economical solar energy conversion device that offers several advantages over traditional silicon-based p–n junction devices such as low cost and simpler fabrication procedures. Furthermore, DSSCs display higher efficiencies at low light levels and the angle of the incident light has no great effect on the performance.

In DSSC, the photosensitizer plays a crucial role as it is responsible for light absorption and electron injection into the conduction band of the TiO_2 after solar light excitation to generate the photocurrent via a redox mediator. Enormous effort is being dedicated to develop efficient dyes that are suitably characterized for their modest cost, facile synthesis and modification, bulky rigid conjugation structure, large molar extinction coefficients (ε), and outstanding thermal and chemical stability. Triphenylamine (TPA) and carbazole (Cz) and their related moieties have been widely employed as active components in optoelectric devices, such as light-emitting[2] and photovoltaic devices,[3–5] because of their desirable electron-donating and hole-transport capabilities.[6,7] In recent years, a substantial number of dyes with triarylamine (e.g., TPA) or Cz as electron donor were developed for DSSCs because of their good electron-donating and electron-transporting properties, as well as their special propeller starburst molecular structure.[8] Furthermore, it was found that functional triarylamine and Cz can suppress surface dye aggregation, retard interfacial charge recombination between the electron in TiO_2 electrode and dye center, and increase the stability of dyes when exposed to light and high temperature, thereby generating high power conversion efficiency (η) and stability. Among the metal-free organic dyes, the arylamine-based organic dyes, holding the record for validated efficiency of over 10.3%, are promising candidates for highly efficient DSSCs.[9] On the other hand, the cost issue concerning the use of metalated organic photosensitizers always puts an obstacle on their applications, despite the fact that the contribution of the sensitizer to the total cell cost is indeed limited, as efficient light harvesting requires a mono-layer of sensitizer molecules. Triarylamine or Cz units embedded in the metal complex can serve as the secondary electron donor moiety to enhance the light-harvesting ability, and metal complex dyes usually have higher thermal and chemical stability as compared with organic dyes. In this chapter, we present the major achievements that have been made in this field. The aim here is to provide promising design principles by using triarylamine and Cz chromophores in developing novel organometallic photosensitizing dyes for the new generation of functional molecules for DSSCs.

1.2 RUTHENIUM-CONTAINING PHOTOSENSITIZERS

A variety of transition-metal complexes and organic dyes have been successfully employed as sensitizers in DSSCs thus far; however, in terms of photovoltaic performance and long-term stability, Ru(II) polypyridyl complexes comprise the most successful family of DSSC sensitizers. They harvest visible light very efficiently with their absorption threshold being at about 800 nm. The well-known and easily tunable photophysical, photochemical, and electrochemical properties of these dyes make them excellent candidates for light-harvesting systems in energy conversion devices.[10] Since the remarkable pioneering invention of Ru(II)-based dyes, with an efficiency of 7% by O'Regan and Grätzel in 1991,[11] tremendous efforts have been expended by scientists concerning the development of DSSC technology based on Ru(II) photosensitizers. While **N3**,[11,12] **N719**,[13] and black dye[14] afford excellent DSSC performance, many research groups have attempted to modify their structures with the goal of improving photovoltaic performance that largely depends on the charge generation step (Scheme 1.1). The structure of the polypyridyl ligand determines the

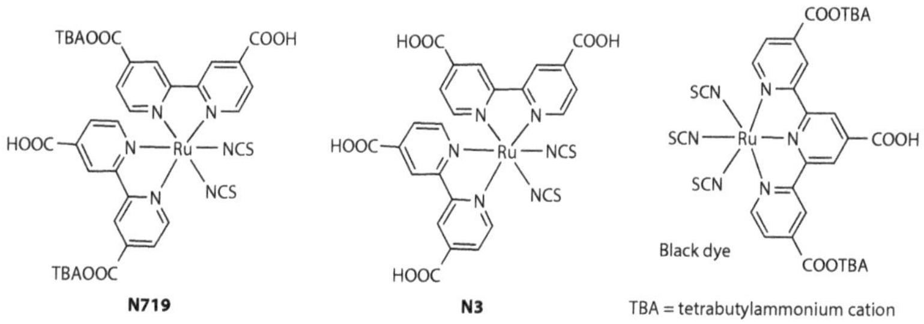

SCHEME 1.1 Structures of Ru(II) complexes **N719**,[11,12] **N3**,[13] and black dye.[14]

redox and spectroscopic properties of the Ru(II) complex, which can be modified by introduction of the chelating ligands.[15] The introduction of hole-transporting and electron-donating TPA or Cz unit involves destabilization of the t_{2g} metal orbital, that is, the highest occupied molecular orbital (HOMO) energy level, in order to tune the energy levels, which has a direct effect over the metal-to-ligand charge transfer (MLCT) transitions of the resulting Ru(II) complex and, consequently, induces a change in the absorption spectra of the molecule. Therefore, controllable adjustments of the excited-state energy levels can be performed with an appropriate selection of the ligand involved in the MLCT. We will discuss the TPA- and Cz-containing Ru(II) dyes based on their molecular engineering.

1.2.1 Ru(II)-Based Photosensitizers with 2,2′-Bipyridine-Based Ancillary Ligands

With regard to the development of Ru(II)-based sensitizers, efforts to enhance the light-harvesting efficiency and the photo- and thermal stability of the sensitizers are of paramount importance. To extend the absorption wavelength and increase the absorption intensity of ruthenium dyes, a popular strategy is to use ruthenium(II) heteroleptic complexes in which the carboxy units in one of the 4,4′-dicarboxy-2,2′-bypyridine ligands of **N3** are replaced by electron-rich conjugated segments.[16–21] By using the electron-rich heteroaromatic rings, such as TPA or Cz, the problem of lower dye density adsorbed on TiO_2 particles due to elongation of the conjugation length and an increase in molecular size can be lessened.[22–24] By using this concept, Lin and coworkers reported the synthesis of **Ru-1** with 9-hexylcarbazole substituted at the 4,4′-position of bipyridine (bpy) (Scheme 1.2).[25] An absorption spectrum typical of Ru(II) complex was measured for **Ru-1** with three absorption bands in the region within 300–800 nm, and the one below 350 nm is attributed to the intraligand π–π* transition of 4,4′-dicarboxylic acid-2,2′-bipyridine (dcbpy) and the Cz-based ancillary bpy ligand. The one from 380 to 410 nm is attributed to the π–π* transition of the ancillary bpy ligand and higher-energy MLCT transition, and the band located at 535 nm is characteristic of a lower-energy MLCT band from the Ru(II) d orbital to dcbpy π* orbital. This MLCT band is redshifted by about 5 nm and broader relative to that of **N3** and **N719** with comparable absorptivities (**Ru-1**, 1.42×10^4 M^{-1} cm^{-1}; **N3**, 1.45×10^4 M^{-1} cm^{-1}; **N719**, 1.40×10^4 M^{-1} cm^{-1}).[14,26] The electron-donating power the dyes, thus enhancing charge recombination. The enhancement of another charge recombination between injected electrons and oxidized electrolyte and quenching of the excited dye all contribute to poor device performance.

To date, the majority of work has proliferated exclusively to the use of Cz- or TPA-based 4,4′-disubstituted bpy derivatives as the ancillary ligands, while research work of the corresponding 5,5′-disubstituted bpy congeners was not well elucidated. Wong and coworkers reported three heteroleptic Ru(II) complexes (**Ru-9–Ru-11**) with uncommon TPA-based 5,5′-disubstituted-2,2-bipyridine chromophores for DSSC application.[33] These complexes are more accessible and less costly with higher synthetic yield from the starting 5,5′-dibromo-2,2′-bipyridine as compared to 4,4′-dibromo-2,2′-bipyridine. Their absorption wavelength (λ_{abs}) in the low-energy region is chemically tunable by structural design and follows the order **Ru-9** (535 nm) < **Ru-11** (553 nm) < **Ru-10** (585 nm), which is in line with the increasing conjugation length of the 5,5′-disubstituted ancillary ligands. While the intensities of their MLCT bands were also affected by the nature of substituents, their ε values are in the order of **Ru-9** (6400 M^{-1} cm^{-1}) < **Ru-11** (5900 M^{-1} cm^{-1}) < **Ru-10** (3200 M^{-1} cm^{-1}). These results indicate that increasing the conjugation length of the ancillary ligand can lower the MLCT energy but also decrease the absorption intensity of the MLCT transition. A good balance between these two factors should be taken into account when designing a new ancillary ligand for Ru(II) complexes. The dye **Ru-11**, containing an electron-releasing octyl chain in the thiophene, showed a more positive oxidation potential than **Ru-10** due to the electron richness of the bipyridyl ligand in **Ru-11**, rendering its Ru(II) center less susceptible to oxidation. The cell η values of **Ru-9–Ru-11** were diminished as compared to the 4,4′-disubstituted bpy derivatives. This is probably due to their weaker light-harvesting capacity and bulky molecular structures both leading to lower J_{sc} values. Although the η values of **Ru-9–Ru-11**-sensitized solar cells are moderate (2.51%, 2.00%, and 2.03%, respectively for **Ru-9–Ru-11**), these findings suggest that bpy chromophores substituted at the synthetically more accessible 5,5′-positions open an alternative class of ancillary ligands over the traditional 4,4′-disubstituted ones for use in DSSC research.

Two Ru(II) photosensitizers, **Ru-12**[34] and **Ru-13**,[35] containing substituted TPA moiety in connection with bpy by double bonds were developed by Yum and coworkers (Scheme 1.5). The insertion of TPA donor groups in **Ru-12** improved the absorption behavior relative to the unsubstituted one (**Ru-12**: λ_{abs} = 536 nm, ε = 19,060 M^{-1} cm^{-1}), leading to highly efficient visible light-harvesting capacity, with very high J_{sc} of 19.2 mA cm^{-2}, η of 10.3%, and IPCE of 87%. By replacing the methyl group by triethylene glycol methyl ether on the TPA group in **Ru-13**, the ε was increased but the lowest-energy MLCT band was blueshifted (λ_{abs} = 524 nm, ε = 30861 M^{-1} cm^{-1}). Although higher light-harvesting yield in the lowest-energy MLCT band was detected for **Ru-13**, slightly poorer η and J_{sc} values were recorded. The highest η of **Ru-13**-based devices is 9.25%. This may probably be due to its narrower absorption range as compared to **Ru-12**. The solid-state solar cells showed J_{sc} of 6.75 mA cm^{-1}, V_{oc} of 864 mV, and FF of 57%, leading to a respectable η of 3.30%. **Ru-13** dye exhibited an excellent stability in combination with an ionic liquid electrolyte. For more than 1000 h of light soaking at 60°C, the device lost only 46 mV of photovoltage

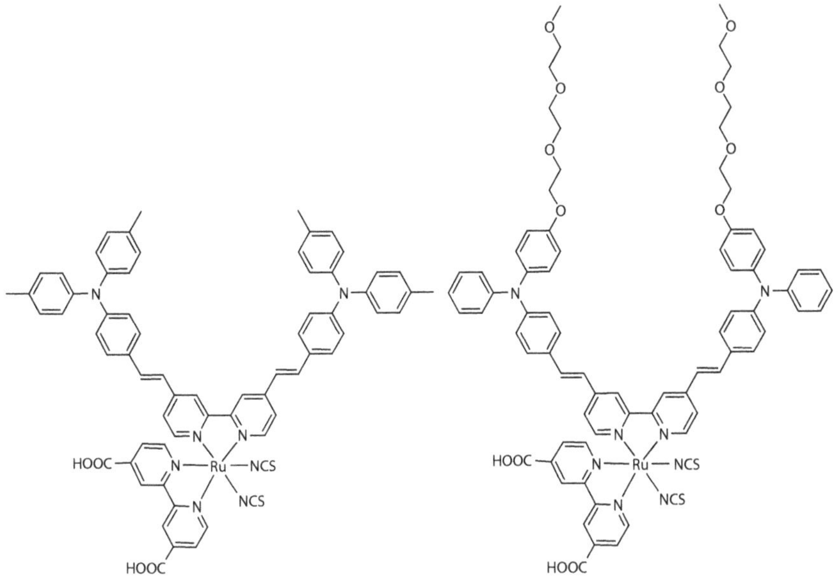

SCHEME 1.5 Structures of Ru(II) complexes **Ru-12**[34] and **Ru-13**.[35]

with essentially no change in the photocurrent. These findings will encourage the inclusion of the TPA unit in the molecular dye structure to further increase the photovoltaic performance and the stability of DSSCs.

Modification of the dye structure is one of the essential strategies to improve the performance of DSSCs. Ko and Nazeeruddin and coworkers introduced the *bis*-dimethylfluorenyl amino benzo[*b*]thiophene donor unit as an antenna into the Ru(II) complexes (**Ru-14** and **Ru-15**) that showed strong absorption with enhanced ε and panchromatic response (Scheme 1.6).[36] Both photosensitizers showed similar absorption bands and range with the lowest-energy MLCT bands located at 539 and 537 nm, respectively, but **Ru-14** with an extra antenna exhibited a higher ε of 22.8×10^3 M^{-1} cm^{-1}. Due to the presence of a bulky structure, there is less dye coverage of **Ru-14** that gives rise to a largely unoccupied area on the TiO$_2$ surface, and hence higher dark current and lower η were observed as compared with **Ru-15** dye (8.20% and 9.16% for **Ru-14** and **Ru-15**, respectively). **Ru-15** exhibited a high plateau in a broad spectral range from 450 to 600 nm, reaching a maximum of 83% at 550 nm. Therefore, the performance of a sensitizer does not depend entirely on the absorptivity values, but the structure and molecular size of the dye also play pivotal roles.

The presence of *N*,*N*′-dibutylamino or *N*,*N*′-dimethylaniline moieties as electron-donating groups on the periphery of 2,2′-bipyridine by utilizing styryl or methane groups in **Ru-16**–**Ru-21** has a great influence on η (Scheme 1.7).[16] The enhancement in J_{sc} of **Ru-16** and **Ru-17** is correlated with the increased ε over the entire visible spectral region as compared with **N3**. This implies that the *N*,*N*′-dibutylamino

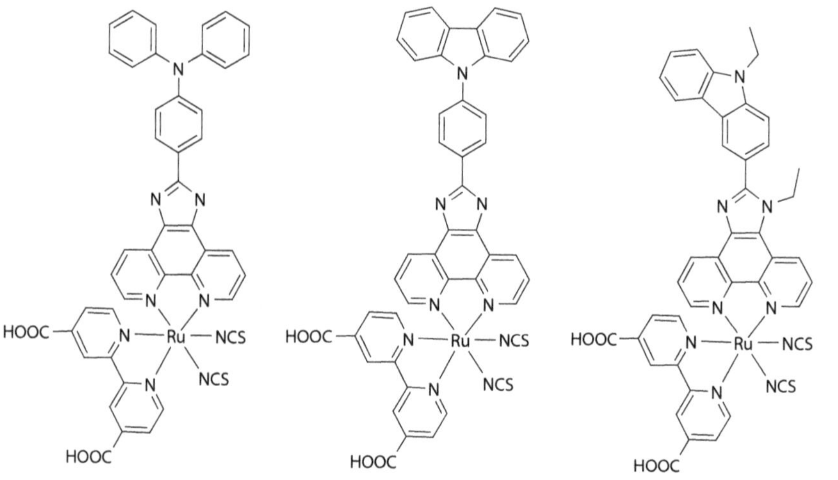

SCHEME 1.8   Structures of Ru(II) complexes **Ru-22–Ru-24**.[38,40,41]

Ru-24 was 10.9 mA cm^{-2}, while the V_{oc} was 620 mV, corresponding to an FF of 0.59 and an overall η of 5.3%. The overall η of **Ru-22–Ru-24** is limited by the larger steric congestion of their structures, resulting in less efficient adsorption by TiO$_2$. But the enhanced photoelectric performance of **Ru-23** and **Ru-24** is attributable to the reduced electron recombination rate of TiO$_2$ conduction electron with oxidized Ru(II) dyes and hence the elongated charge separated-state lifetime. These characteristics make **Ru-22** and **Ru-23** attractive for further photoelectric applications.

1.2.2   Ru(II) Photosensitizers with C^N-Based Ancillary Ligands

Much effort has been made to increase the photovoltaic performance and stability of a device. A way to improve the stability is the development of a dye without NCS donor ligands because monodentate NCS is believed to provide the weakest dative bonding within the metal complexes, making the sensitizer unstable. Accordingly, cyclometalated NCS-free Ru(II) complexes have been developed to extend the lifespan of practical solar cells.[42–45] This kind of Ru(II) complexes is worthy of further development because it not only improves the stability of the solar cell but also broadens the absorption coverage when compared to the standard dye **N719**.[46,47] The majority of work has proliferated exclusively to the use of 2-phenylpyridine derivatives as the cyclometalating ligands.[42–45,48–50] Actually, the common starting precursor 4,4′-dibromo-2,2′-bipyridine for modifying the N^N ancillary ligands of the Ru(II) dyes is relatively costly with complicated synthetic procedures.[33] On the other hand, the C^N cyclometalating ligands are much easier to be modified by simple synthetic procedures and inexpensive starting precursor. Four novel Ru(II) photosensitizers (**Ru-25–Ru-28**) based on thiazole and benzothiazole C^N ligands were reported (Scheme 1.9).[51] The ε values of benzothiazole-containing dyes **Ru-27** and **Ru-28** are higher than that of thiazole-containing dyes **Ru-25** and

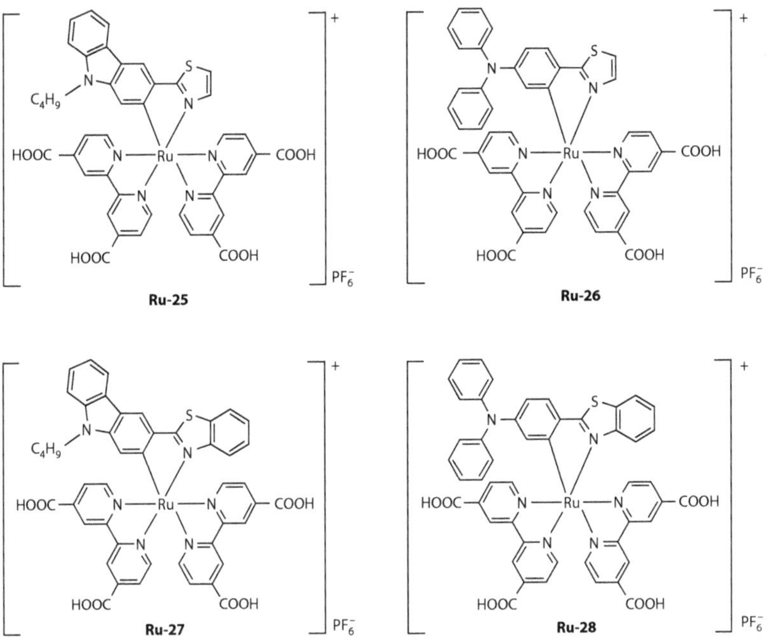

SCHEME 1.9 Structures of Ru(II) complexes **Ru-25–Ru-28**.[51]

Ru-26 due to the higher π-conjugation length of the benzothiazole ligands. **Ru-27** has the strongest light-harvesting ability as the ε values of the lowest-energy MLCT bands for **Ru-25–Ru-28** decrease in the order of **Ru-27** > **Ru-28** > **Ru-25** ~ **Ru-26**. Remarkably, upon the introduction of Cz and TPA groups to the cyclometalating ligands, the light-harvesting ability can be improved, which can be reflected by the presence of one new band at about 370 nm, more redshifted absorption bands, and higher ε as compared to **N719**. All of these dyes can convert visible light to photocurrent in the absorption region from 310 to 750 nm efficiently and **Ru-27** governs the highest IPCE values at all coverage wavelengths among **Ru-25–Ru-28**, which is in agreement with its highest ε. All the dyes exhibited favorable HOMO and LUMO energy levels for electron injection and regeneration. **Ru-27** showed the highest J_{sc} value of 6.25 mA cm^{-2} and the value is almost double of those for the other three dyes. These findings imply that benzothiazole chromophore opens up an alternative route to afford a good class of cyclometalating ligands for DSSC research based on Ru(II) metal.

In line with the continuation of the efforts for improving the solar energy conversion efficiency, Berlinguette and coworkers reported a bulky trichromic *tris*(heteroleptic) cycloruthenated dye **Ru-29** where one of the bidentate bpy ligands bears two TPA groups and devoid of the use of the labile NCS ligand.[52] As compared to **Ru-25–Ru-28**, this dye is insoluble in most of the polar organic solvents and readily soluble in dimethylformamide (DMF) only. The UV–Vis spectrum of **Ru-29** reveals a relatively intense MLCT band centered at 580 nm with ε of 3.3×10^4 M^{-1} cm^{-1}, complementing with the large intensity of the intraligand

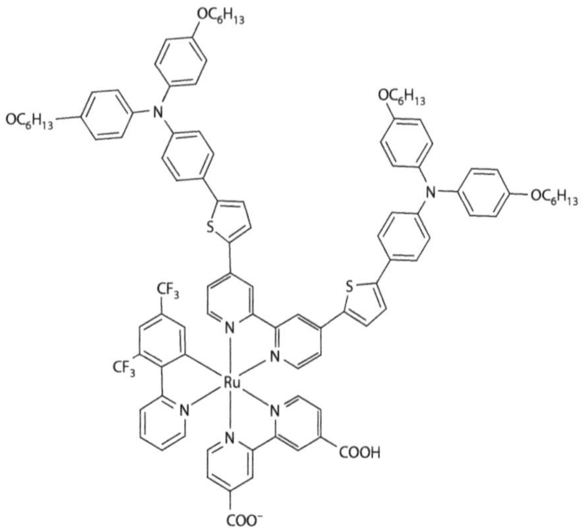

Ru-29

SCHEME 1.10　Structure of Ru(II) complex **Ru-29**.[52]

charge transfer (ILCT) band centered at 465 nm ($\varepsilon = 6.7 \times 10^4$ M^{-1} cm^{-1}) emanating from the TPA groups (Scheme 1.10). The overall DSSC performances with and without coadsorbent chenodeoxycholic acid are 5.9% and 6.2%, respectively (mask size = 0.28 cm^2). The IPCE trace for **Ru-29** showed an onset of current at nearly 790 nm and the curve plateaus within 65%–70% in the region of 450–600 nm. An η value of 7.3% with a reduced mask size (0.13 cm^2) was recorded. As compared to **Ru-12** with two NCS ligands in place of the cyclometalating ligand, the performance of **Ru-29** is relatively inferior. This may be due to the lower driving force for intramolecular electron transfer (i.e., reduction of the photooxidized Ru site by the TPA) relative to **Ru-12** because the two NCS groups do not raise the HOMO energy to the same extent as the phenylpyridine ligand. Further studies of these intramolecular and interfacial electron transfer processes are needed.

Wu and coworkers attempted the introduction of the TPA group to cyclometalated Ru(II) photosensitizers as the bridge between the Ru(II) chromophore and the NiO surface.[53] Besides TPA-derived bpy ligand, the other two N^N ligands were systematically tuned from 2,2′-bipyridine (**Ru-30**) to 1,10-phenanthroline (**Ru-31**) and to bathophenanthroline (**Ru-32**) (Scheme 1.11). Following the series, due to the incorporation of the expanded π-conjugation ligand, the ε slightly increased from **Ru-30** to **Ru-32**. The spectrum of **Ru-32** was redshifted by around 20 nm as compared to those of **Ru-30** and **Ru-31**. From calculations, the light-harvesting ability of **Ru-31** and **Ru-32** increases by 5.6% and 7.8% compared to that of **Ru-30**. In addition, their reduction potentials were affected by the extension of conjugation in the ligands but the oxidation potentials remain relatively unchanged. The first reduction potential exhibited a positive shift away from **Ru-30** to **Ru-32** that is consistent with their absorption spectra. Because of the fact that N^N and N^C

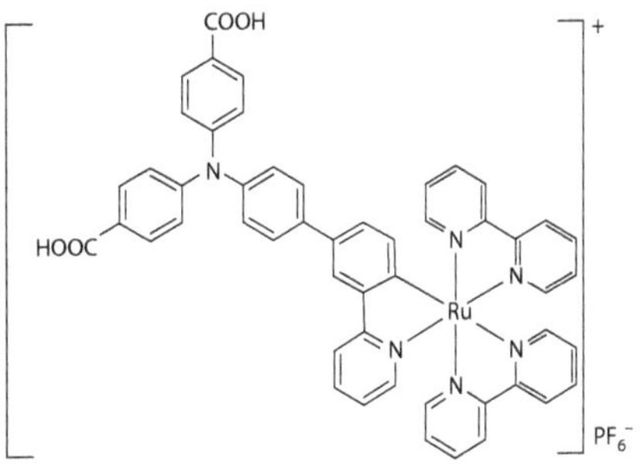

SCHEME 1.11 Structures of Ru(II) complexes **Ru-30–Ru-32**.[53]

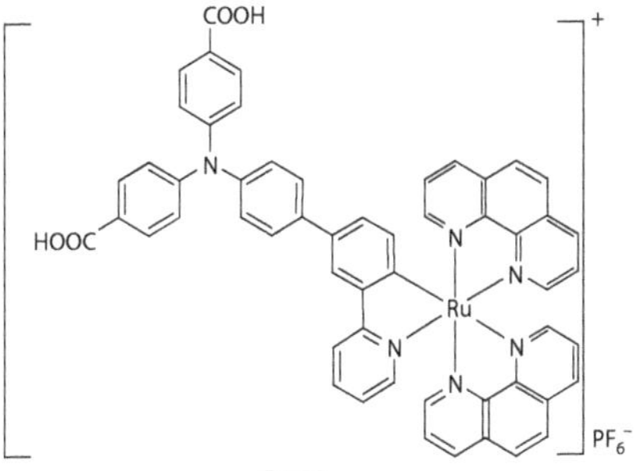

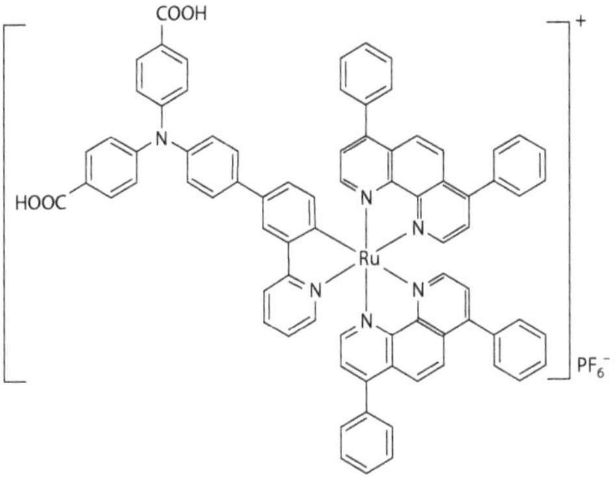

ligands are close in energy, which have been proven by DFT calculations, this resulted in multiple transitions that account for the broad absorption spectra of these cyclometalated Ru(II) complexes. The principal contribution to the lower-energy band is mainly charge transfer transitions and metal-perturbed intraligand transitions from the ruthenium metal center with the contribution of the N^C ligand to either the N^N or N^C ligand. Although **Ru-30–Ru-32** exhibited a broad absorption, suitable HOMO and LUMO levels for efficient electron injection, and easy dye regeneration, unexpected poor DSSC performances were examined with nanoporous NiO films sensitized with the dyes. The highest η was only 0.099% for **Ru-30**. The J_{sc} value of 3.04 mA cm^{-2} for **Ru-30** is likely caused by the slow geminate charge recombination and efficient dye regeneration. The results imply that the ancillary N^N ligands play a very important role in the electron–hole recombination kinetics and thus the device performance. The reasons behind the poor efficiencies of these three devices are not yet known.

1.2.3 Ru(II) Photosensitizers with Terpyridine-Based Ancillary Ligands

Besides bipyridine-based Ru(II) complexes, Ru(II) terpyridine photosensitizers were also developed for DSSC application. In this regard, black dye, [Ru(Htctpy) the inferior light-harvesting capacity of **Ru-35** in the visible region that is manifested in the anionic ring that increases the energies of the ILCT transitions. Generally, complexes containing methyl-substituted TPA donor units have a tendency to generate lower performance in the DSSC than other TPA-bearing substituents. The highest η value of 8.02% was obtained for **Ru-37** as its IPCE showed a maximum of about 83% and appreciable response down to 650 nm. It is also noteworthy that **Ru-37** showed an IPCE of 10% at 900 nm, whereas the IPCE for **N3** is negligible at wavelengths as short as 750 nm, which demonstrate the majority and wide range of light harvested by both the metal chelate and the TPA unit that is being converted into useful electrical energy. One of the drawbacks from the interaction of the TPA unit with the electrolyte may adversely affect the V_{oc}, resulting in 30 mV lower in V_{oc} value for **Ru-37** as compared to **N3**. Therefore, the combination of the high directionality and dipole influence of proper design on Ru(II) photosensitizers is important for highly efficient DSSCs.

In conjunction with the current endeavors, Chi and coworkers presented a breakthrough design employing TPA-functionalized 2,6-*bis*(5-pyrazolyl)pyridine chelate to synthesize the *bis*-tridentate Ru(II) complex **Ru-38** (Scheme 1.13).[58] This dye showed intense visible absorption bands down to 510 nm, which are assigned as MLCT bands to the 4,4′,4″-tricarboxy-2,2′:6′,2″-terpyridine (H_3tctpy) chelate, together with contributions from ligand-to-ligand charge transfer (LLCT) band originating from the pyrazolate groups. Relative to black dye, **Ru-38** exhibited a broad absorption at the longer wavelength region, with two weak peak maxima centered at about 650 and 720 nm and with a shoulder extending to 800 nm; these features are attributed to mixed MLCT and LLCT transitions to H_3tctpy. Due to the lack of the three NCS anions, the intense MLCT band at 606 nm in black dye is substantially suppressed in **Ru-38**. **Ru-38** showed an attractive IPCE value of 92% at 515 nm and maintained a high value of approximately 75% in the longer wavelength region up to 685 nm. For comparison, DSSC fabricated using black dye showed inferior IPCE below 560 nm in the region between 640 and 710 nm and increased IPCE in the regions between 560 and 640 nm and beyond 730 nm. The replacement of the three NCS ligands with the extremely bulky 2,6-*bis*(5-pyrazolyl)pyridine ligand may allow better packing upon adsorption on the TiO_2 surface, which could prevent interfacial charge recombination, giving a higher V_{oc} of 770 mV in **Ru-38**-based device than black dye one. The neutral pyridyl group flanked by two 2,6-*bis*(5-pyrazolyl) termini may also be intrinsically more capable of exerting a dipole moment pointing toward the TiO_2 surface that results in the higher value of V_{oc}. This is in agreement with the electrochemical impedance spectroscopy (EIS) that reveals that the recombination process from **Ru-38** in DSSCs is slower than that of black dye. A higher current density of 20.27 mA cm^{-2} for **Ru-38** than that of 19.49 mA cm^{-2} for black dye is a manifestation of the increase of ε upon extending the π conjugation of the auxochrome. This molecular structure also benefits the charge transport efficiency in the DSSC device. All the factors gave an overall η of 10.5% for **Ru-38**. This result offers an insight into the importance of the structural design of efficient Ru(II) photosensitizer that may allow the negative dipole of sensitizers to reside closer to the TiO_2 surface, which is the prerequisite for raising the TiO_2 conduction band energy level to achieve high V_{oc} and decent overall η.

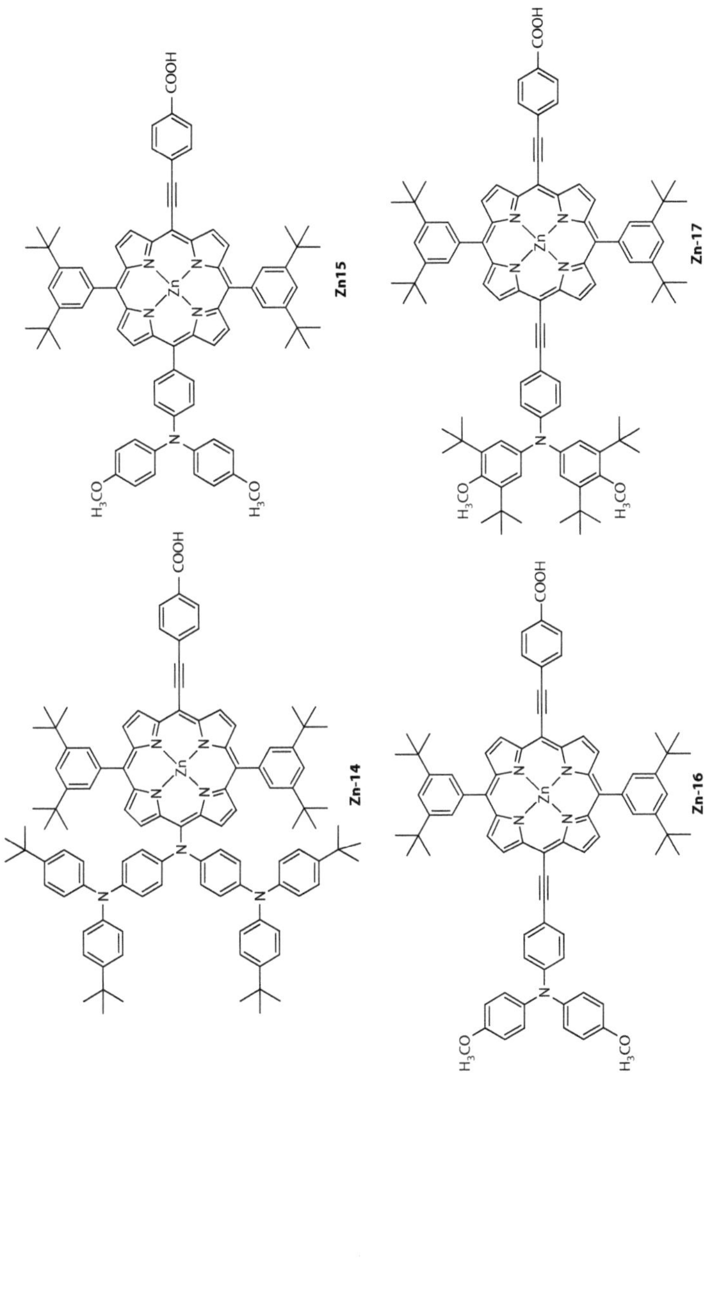

SCHEME 1.17 Structures of Zn(II) complexes **Zn-14–Zn-17**.[68]

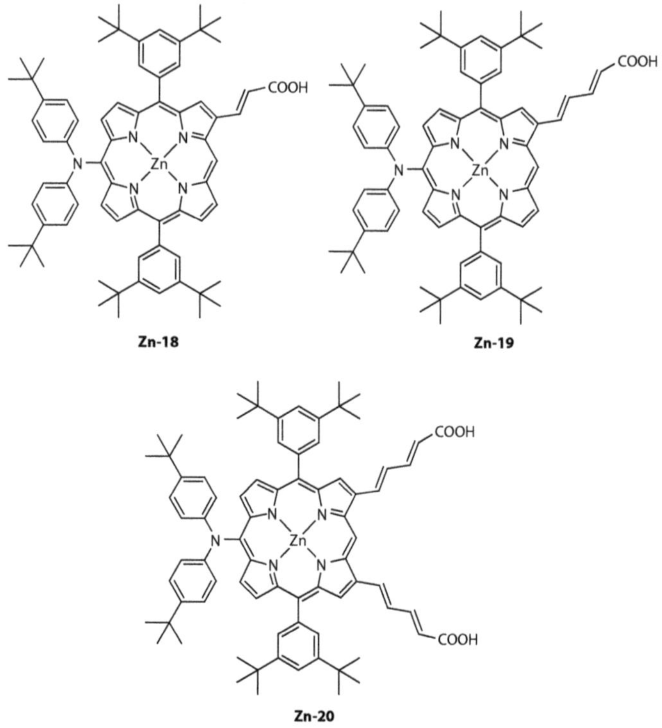

SCHEME 1.18   Structures of Zn(II) complexes **Zn-18–Zn-20**.[71]

spectra. The photovoltaic performance is also strongly correlated to the size of the molecules. The maximal η values were found to increase as the size and complexity of the β-substituents increased. The η values of **Zn-18–Zn-20** were found to be 3.20%, 4.22%, and 4.36%, respectively, under a conventional DSSC device structure. After treatment of the TiO$_2$ films in a dilute aqueous TiCl$_4$ solution, an η of 6.35% was achieved for **Zn-20** and the difference in η is originated primarily from the non-parallel values of J_{sc}. This surface modification technique provides a larger surface coverage of **Zn-20** on the TiO$_2$, leading to an increase on the light-harvesting efficiency. Applying a wider optical bandgap anatase TiO$_2$, **Zn-20** produced cells with the highest η of 7.47% due to the enhancement in IPCE values (max. IPCE = 77.3% at the Soret band) over a wide range of absorption wavelengths. These values were considered to reflect both a favorable light-harvesting capability and efficient electron injection ability. These results suggest that an optimization of both the donor–π–acceptor structure of the porphyrin dye and the morphology of nanostructured TiO$_2$ in which it is supported could lead to DSSCs with yet improved efficiencies.

In principle, tuning of the HOMO and LUMO levels could be achieved through the choice of suitable donor groups and their positions. The effects of the meso donor groups on cell performance were investigated by introducing various numbers of TPA and trimethoxyphenyl groups as the electron donors in **Zn-21–Zn-25** (Scheme 1.19).[72,73] Interestingly, a systematic tuning of the absorption bands and

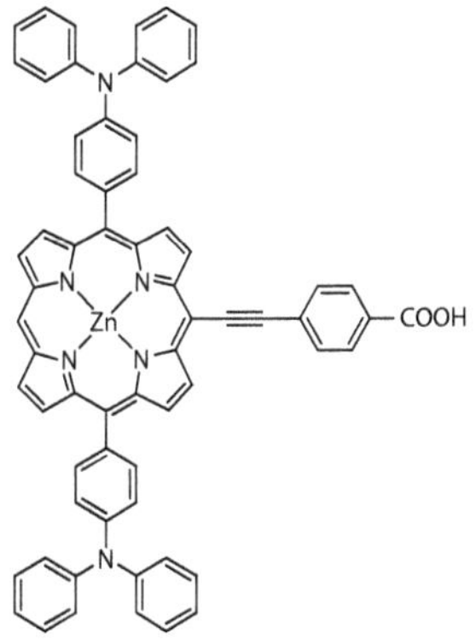

Zn-21

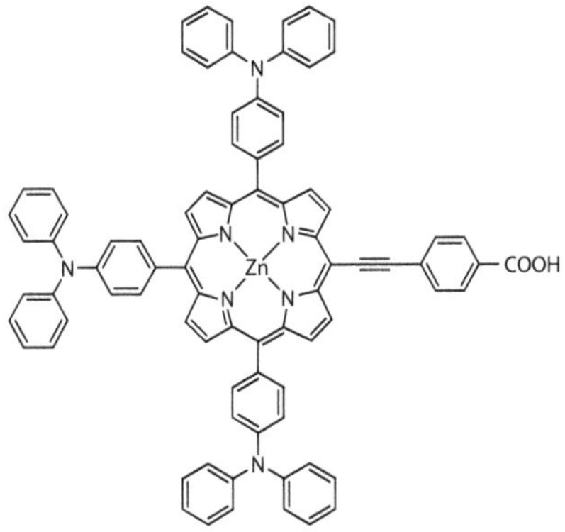

Zn-22

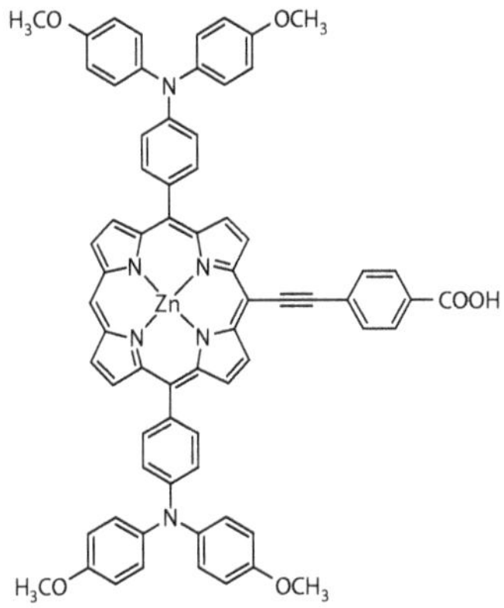

Zn-23

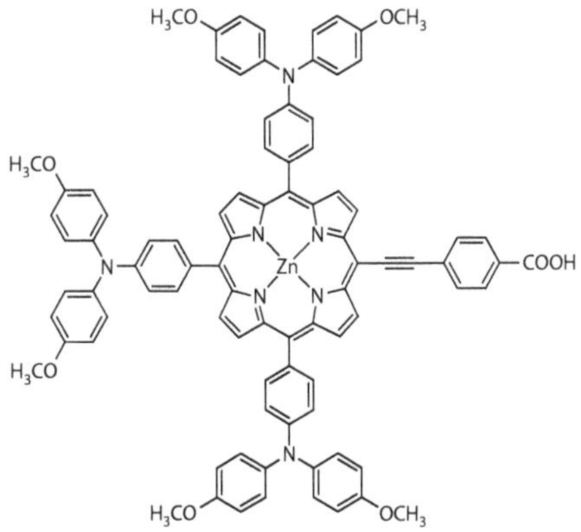

Zn-24

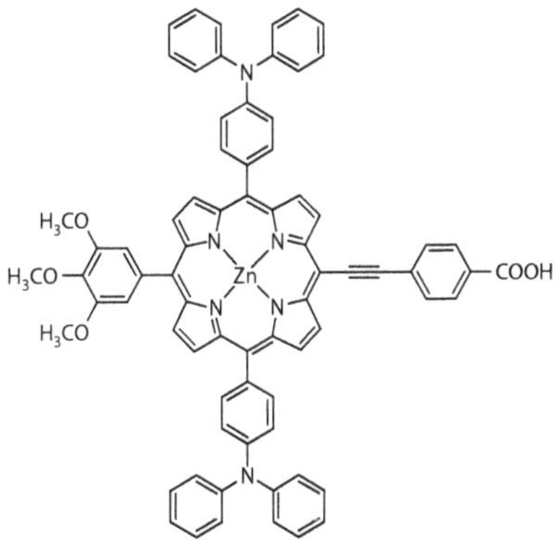

Zn-25

SCHEME 1.19 Structures of Zn(II) complexes **Zn-21–Zn-25**.[72,73]

to 600 nm. The absorption maxima for photosensitizers **Pt-5–Pt-8** are located at 461, 511, 461, and 508 nm, respectively. The thiophene π-bridges did not influence the oxidation and reduction potentials of the dyes but the nature of donor group did. Owing to its rigidifying structure, which restricts the rotation of σ-bonds, the dimethylfluorene unit exhibits a stronger electron-donating ability, higher conjugation efficiency, and suitable HOMO/LUMO energy levels as compared to the TPA unit. Thus, **Pt-5** and **Pt-6** have higher IPCE values and a broader light response region than **Pt-7** and **Pt-8**. The device with **Pt-6** gave the highest η of 4.09% among these four dyes, which can be attributed to the highest V_{oc} of 640 mV. **Pt-5** has a relatively high V_{oc} of 620 mV; therefore, sensitization with dimethylfluorene as an electron donor would result in efficient intramolecular charge transfer in the rigidifying structure, owing to the effect of coplanarity in impeding the electron recombination between the aromatic molecules in the excited and ground states. EIS analysis also indicated that the resistance to charge recombination and the electron lifetime increased from **Pt-7** to **Pt-8**, **Pt-5**, and **Pt-6**; thus, the replacement of the triarylamine group with a dimethylfluorene group is effective in modifying the TiO$_2$/dye/electrolyte interface and increasing V_{oc} or J_{sc}.

Three unsymmetric Pt(II) photosensitizers **Pt-9–Pt-11** (Scheme 1.22) with TPA and benzothiadiazole acceptor were reported by Wong and coworkers.[78] These dyes were found to have strong electron-donating strength due to TPA units that resulted in a higher degree of electronic delocalization and hence a stronger ICT effect in the molecule. Because of the longer electronic conjugation length in **Pt-10**, it exhibited the strongest ICT transition at 547 nm among these dyes. The time-dependent density functional theory (TD-DFT) calculations indicate that the HOMOs are highly delocalized over the aryleneethynylene donor moieties and the LUMOs are mainly localized on the benzothiadiazole units and another aryleneethynylene acceptor and anchoring unit. The absorption abilities can be reasonably enhanced by reducing the energy gap for the HOMO/LUMO transitions using a stronger electron donor. The devices based on **Pt-9–Pt-11** showed photovoltaic performance with V_{oc} of 0.58–0.62 V, J_{sc} of 3.35–3.63 mA cm^{-2}, and FF of 0.70–0.73. The higher J_{sc} value may be caused by the stronger ICT interaction, which leads to a more efficient absorption of the solar spectrum. The η values of **Pt-9–Pt-11**-sensitized solar cells range from 1.42% to 1.57%. Poor results of **Pt-9–Pt-11**-based DSSCs may be attributed to the less efficient electron collection than that based on the conventional Ru(II) dye **N719**. But all the previously mentioned results suggest that platinum(II)-acetylide small molecules have great

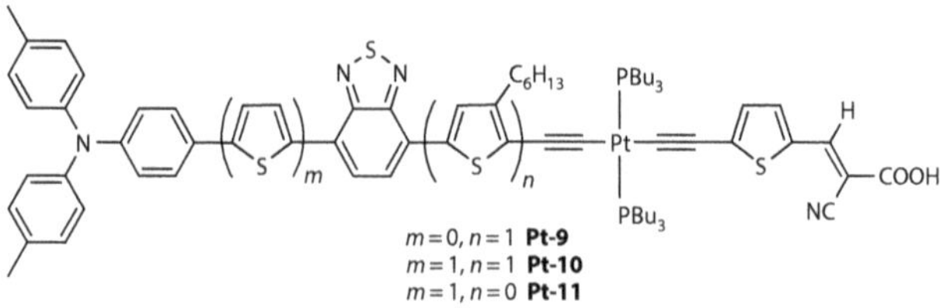

$$m=0, n=1 \ \textbf{Pt-9}$$
$$m=1, n=1 \ \textbf{Pt-10}$$
$$m=1, n=0 \ \textbf{Pt-11}$$

SCHEME 1.22   Structures of Pt(II) complexes **Pt-9–Pt-11**.[78]

potential for use as effective photosensitizers in DSSC applications, and further improvements in device efficiency can be achieved by tuning the ICT absorption and energy levels of Pt(II)-containing small molecules.

1.5 CONCLUDING REMARKS

Energy crisis is one of the most challenging problems confronting mankind today. While the efficiencies of DSSCs based on metal-containing photosensitizing dyes are still lower than those of the best performing silicon-based devices, intensive research is currently performed in order to increase the photoconversion efficiency in DSSCs. Therefore, a deeper understanding of the roles that the molecular dye structures play over the interfacial charge transfer processes, both recombination and regeneration, is a must. From the results presented, triarylamine- and Cz-containing metal-based photosensitizers should be considered as a promising class of dyes for new-generation high-performance DSSCs. We are optimistic that more encouraging data should come up in the near future along this line of research.

ACKNOWLEDGMENTS

W-YW thanks the Science, Technology and Innovation Committee of Shenzhen Municipality (JCYJ20120829154440583), the Hong Kong Baptist University (FRG2/12-13/083), the Hong Kong Research Grants Council (HKBU203011), and the Areas of Excellence Scheme, University Grants Committee of HKSAR, China (project No. AoE/P-03/08), for financial support.

REFERENCES

1. N. Armaroli and V. Balzani, *Angew. Chem. Int. Ed.*, 2007, **46**, 52.
2. U. Mitschke and P. Bauerle, *J. Mater. Chem.*, 2000, **10**, 1471.
3. S. Gunes, H. Neugebauer and N. S. Sariciftci, *Chem. Rev.*, 2007, **107**, 1324.
4. T. Kitamura, M. Ikeda, K. Shigaki, T. Inoue, N. A. Anderson, X. Ai, T. Q. Lian and S. Yanagida, *Chem. Mater.*, 2004, **16**, 1806.
5. D. P. Hagberg, T. Edvinsson, T. Marinado, G. Boschloo, A. Hagfeldt and L. C. Sun, *Chem. Commun.*, 2006, 2245.
6. S. K. Lee, T. Ahn, N. S. Cho, J. I. Lee, Y. K. Jung, J. Lee and H. K. Shim, *J. Polym. Sci. A: Polym. Chem.*, 2007, **45**, 1199.
7. H. Choi, C. Baik, S. O. Kang, J. Ko, M. S. Kang, K. Md. Nazeeruddin and M. Grätzel, *Angew. Chem. Int. Ed.*, 2008, **120**, 333.
8. Z. Ning and H. Tian, *Chem. Commun.*, 2009, 5483.
9. W. Zeng, Y. Cao, Y. Bai, Y. Wang, Y. Shi, M. Zhang, F. Wang and Y. Pan, *Chem. Mater.*, 2010, **22**, 1915.
10. S. Campagna, F. Puntoriero, F. Nastasi, G. Bergamini and V. Balzani, *Top. Curr. Chem.*, 2007, **280**, 117.
11. B. O'Regan and M. Grätzel, *Nature*, 1991, **353**, 737.
12. M. K. Nazeeruddin, A. Key, L. Rodicio, R. Humphry-Baker, E. Muller, P. Liska, N. Vlachopoulos and M. Grätzel, *J. Am. Chem. Soc.*, 1993, **115**, 6382.
13. M. K. Nazeeruddin, S. M. Zakeeruddin, R. Humphry-Baker, M. Jirousek, P. Liska, N. Vlachopoulos, V. Shklover, C.-H. Fischer and M. Grätzel, *Inorg. Chem.*, 1999, **38**, 6298.

14. M. K. Nazeeruddin, P. Pechy, T. Renouard, S. M. Zakeeruddin, R. Humphry-Baker, P. Comte, P. Liska, L. Cevey, E. Costa, V. Shklover, L. Spiccia, G. B. Deacon, C. A. Bignozzi and M. Grätzel, *J. Am. Chem. Soc.*, 2001, **123**, 1613.

15. A. Islam, H. Sugihara and H. Arakawa, *J. Photochem. Photobiol. A: Chem.*, 2003, **158**, 131.

16. S.-R. Jang, C. Lee, H. Choi, J. Ko, J. Lee, R. Vittal and K.-J. Kim, *Chem. Mater.*, 2006, **18**, 5604.

17. C. Lee, J.-H. Yum, H. Choi, S. O. Kang, J. Ko, R. Humphry-Baker, M. Grätzel and M. K. Nazeeruddin, *Inorg. Chem.*, 2008, **47**, 2267.

18. C.-Y. Chen, S.-J. Wu, C.-G. Wu, J.-G. Chen and K.-C. Ho, *Angew. Chem. Int. Ed.*, 2006, **45**, 5882.

19. C. Karthikeyan, H. Wietasch and M. Thelakkat, *Adv. Mater.*, 2007, **19**, 1091.

20. C.-Y. Chen, J.-G. Chen, S.-J. Wu, J.-Y. Li, C.-G. Wu and K.-C. Ho, *Angew. Chem. Int. Ed.*, 2008, **47**, 7342.

21. B. C. O'Ragen, K. Walley, M. Juozapavicius, A. Anderson, F. Matar, T. Ghaddar, S. M. Zakeeruddin, C. Klein and J. R. Durrant, *J. Am. Chem. Soc.*, 2009, **131**, 3541.

22. P. Wang, S. M. Zakeeruddin, J.-E. Moser, R. Humphry-Baker, P. Comte, V. Aranyos, A. Hagfeldt, M. K. Nazeeruddin and M. Grätzel, *Adv. Mater.*, 2004, **16**, 1806.

23. P. Wang, C. Klein, R. Humphry-Baker, S. M. Zakeeruddin and M. Grätzel, *J. Am. Chem. Soc.*, 2005, **127**, 808.

24. D. Kuang, C. Klein, S. Ito, J.-E. Moser, R. Humphry-Baker, N. Evans, F. Duriaux, C. Grätzel, S. M. Zakeeruddin and M. Grätzel, *Adv. Mater.*, 2007, **19**, 1133.

25. Y.-S. Yen, Y.-C. Chen, Y.-C. Hsu, H.-H. Chou, J. T. Lin and D.-J. Yin, *Chem. Eur. J.*, 2011, **17**, 6781.

26. H. M. Nguyen, D. N. Nguyen and N. Kim, *Adv. Nat. Sci.: Nanosci. Nanotechnol.*, 2010, **1**, 025001.

27. C. Y. Chen, S. J. Wu, C. G. Wu, J. G. Chen and K. C. Ho, *Angew. Chem. Int. Ed.*, 2006, **45**, 5822.

28. C. Y. Chen, S. J. Wu, C. G. Wu, J. G. Chen and K. C. Ho, *Adv. Mater.*, 2007, **19**, 3888.

29. S.-Q. Fan, C. Kim, B. Fang, K.-X. Liao, G.-J. Yang, C.-J. Li, J.-J. Kim and J. Ko, *J. Phys. Chem. C*, 2011, **115**, 7747.

30. C.-Y. Chen, N. Pootrakulchote, S.-J. Wu, M. Wang, J.-Y. Li, J.-H. Tsai, C.-G. Wu, S. M. Zakeeruddin and M. Grätzel, *J. Phys. Chem. C*, 2009, **113**, 20752.

31. C.-Y. Chen, N. Pootrakulchote, T.-H. Hung, C.-J. Tan, H.-H. Tsai, S. M. Zakeeruddin, C.-G. Wu and M. Grätzel, *J. Phys. Chem. C*, 2011, **115**, 20043.

32. W.-S. Han, J.-K. Han, H.-Y. Kim, M. J. Choi, Y.-S. Kang, C. Pac and S. O. Kang, *Inorg. Chem.*, 2011, **50**, 3271.

33. F.-R. Dai, W.-J. Wu, Q.-W. Wang, H. Tian and W.-Y. Wong, *Dalton Trans.*, 2011, **40**, 2314.

34. J.-H. Yum, I. Jung, C. Baik, J. Ko, M. K. Nazeeruddin and M. Grätzel, *Energy Environ. Sci.*, 2009, **2**, 100.

35. J.-H. Yum, S.-J. Moon, C. S. Karthikeyan, H. Wietasch, M. Thelakkat, S. M. Zakeeruddin, M. K. Nazeeruddin and M. Grätzel, *Nano Energy*, 2012, **1**, 6.

36. H. Choi, C. Baik, S. Kim, M.-S. Kang, X. Xu, H. S. Kang, S. O. Kang, J. Ko, M. K. Nazeeruddin and M. Grätzel, *New J. Chem.*, 2008, **32**, 2233.

37. N. Onozawa-Komatsuzaki, O. Kitao, M. Yanagida, Y. Himeda, H. Sugihara and K. Kasuga, *New J. Chem.*, 2006, **30**, 689.

38. X. H. Li, J. Gui, H. Yang, W. J. Wu, F. Y. Li, H. Tian and C. H. Huang, *Inorg. Chim. Acta*, 2008, **361**, 2835.

39. S. H. Fan, K. Z. Wang and W. C. Yang, *Eur. J. Inorg. Chem.*, 2009, 508.

40. S. H. Fan, A.-G. Zhang, C.-C. Ju, L.-H. Gao and K.-Z. Wang, *Inorg. Chem.*, 2010, **49**, 3752.

41. S. H. Fan, A.-G. Zhang, C.-C. Ju and K.-Z. Wang, *Solar Energy*, 2011, **85**, 2497.
42. C. Dragonetti, A. Valore, A. Colombo, D. Roberto, V. Trifiletti, N. Manfredi, M. M. Salamone, R. Ruffo and A. Abbotto, *J. Organomet. Chem.*, 2012, **714**, 88.
43. A. Abbotto, C. Coluccini, E. Dell'Orto, N. Manfredi, V. Trifiletti, M. M. Salamone, R. Ruffo, M. Acciarri, A. Colombo, C. Dragonetti, S. Ordanini, D. Roberto and A. Valore, *Dalton Trans.*, 2012, **41**, 11731.
44. P. G. Bomben, K. D. Thériaultm and C. P. Berlinguette, *Eur. J. Inorg. Chem.*, 2011, 1806.
45. P. G. Bomben, K. D. Robson and C. P. Berlinguette, *Inorg. Chem.*, 2009, **48**, 9631.
46. M. K. Nazeeruddin, F. De Angelis, S. Fantacci, A. Selloni, G. Viscardi, P. Liska, S. Ito, B. Takeru and M. Grätzel, *J. Am. Chem. Soc.*, 2005, **127**, 16835.
47. M. K. Nazeeruddin, R. Splivallo, P. Liska, P. Comte and M. Grätzel, *Chem. Commun.*, 2003, 1456.
48. P. G. Bomben, T. J. Gordon, E. Schott and C. P. Berlinguette, *Angew. Chem. Int. Ed.*, 2011, **123**, 10870.
49. S. H. Wadman, J. M. Kroon, K. Bakker, M. Lutz, A. L. Spek, G. P. M. van Klink and G. van Koten, *Chem. Commun.*, 2007, 1907.
50. S. H. Wadman, M. Lutz, D. M. Tooke, A. L. Spek, F. Hartl, R. W. A. Havenlth, G. P. M. van Kllnk and G. van Koten, *Inorg. Chem.*, 2009, **48**, 1887.
51. C.-H. Siu, C.-L. Ho, J. He, T. Chen, X. Cui, J. Zhao and W.-Y. Wong, *J. Organomet. Chem.*, 2013, **748**, 75.
52. P. G. Bomben, J. Borau-Garcia and C. P. Berlinguette, *Chem. Commun.*, 2012, **48**, 5599.
53. Z. Ji, G. Natu and Y. Wu, *ACS Appl. Mater. Interfaces*, 2013, **5**, 8641.
54. M. K. Nazeeruddin, P. Pechy and M. Grätzel, *Chem. Commun.*, 1997, 1705.
55. H.-W. Lin, Y.-S. Wang, Z.-Y. Huang, Y.-M. Lin, C.-W. Chen, S.-H. Yang, K.-L. Wu, Y. Chi, S.-H. Liu and P.-T. Chou, *Phys. Chem. Chem. Phys.*, 2012, **4**, 14190.
56. M. K. Nazeeruddin and M. Grätzel, *J. Photochem. Photobiol. A: Chem.*, 2001, **145**, 79.
57. K. C. D. Robson, B. D. Koivisto, A. Yella, B. Sporinova, M. K. Nazeeruddin, T. Baugartner, M. Grätzel and C. P. Berlinguette, *Inorg. Chem.*, 2011, **50**, 5494.
58. C.-C. Chou, K.-L. Wu, Y. Chi, W.-P. Hu, S. J. Yu, G.-H. Lee, C.-L. Lin and P.-T. Chou, *Angew. Chem. Int. Ed.*, 2011, **50**, 2054.
59. M. V. Martínez-Díaz, G. D. L. Torre and T. Torres, *Chem. Commun.*, 2010, **46**, 7090.
60. H. Imahori, T. Umeyama and S. Ito, *Acc. Chem. Res.*, 2009, **42**, 1809.
61. H. Imahori, Y. Matsubara, H. Iijima, T. Umeyama, Y. Matano, S. Ito, M. Niemi, N. V. Tkachenko and H. Lemmetyinen, *J. Phys. Chem. C*, 2010, **114**, 10656.
62. S. Eu, S. Hayashi, T. Umeyama, A. Oguro, M. Kawasaki, N. Kadota, Y. Matano and H. Imahori, *J. Phys. Chem. C*, 2007, **111**, 3528.
63. Y.-C. Chang, C.-L. Wang, T.-Y. Pan, S.-H. Hong, C.-M. Lan, H.-H. Kuo, C.-F. Lo, H.-Y. Hsu, C.-Y. Lin and E. W.-G. Diau, *Chem. Commun.*, 2011, **47**, 4010.
64. P. Wang, S. M. Zakeeruddin, J. E. Moser, M. K. Nazzeruddin, T. Sekiguchi and M. Grätzel, *Nat. Mater.*, 2003, **2**, 402.
65. K. D. Seo, M. J. Lee, H. M. Song, H. S. Kang and H. K. Kim, *Dyes Pigments*, 2012, **94**, 143.
66. M. J. Lee, K. D. Seo, H. M. Song, M. S. Kang, Y. K. Eom, H. S. Kang and H. K. Kim, *Tetrahedron Lett.*, 2011, **52**, 3879.
67. T. Bessho, S. M. Zakeeruddin, C.-Y. Yeh, E. W.-G. Diau and M. Grätzel, *Angew. Chem. Int. Ed.*, 2010, **49**, 6646.
68. C.-P. Hsieh, H.-P. Lu, C.-L. Chiu, C.-W. Lee, S.-H. Chuang, C.-L. Mai, W.-N. Yen, S.-J. Hsu, E. W.-G. Diau and C.-U. Yeh, *J. Mater. Chem.*, 2010, **20**, 1127.
69. A. Yella, H.-W. Lee, H. N. Tsao, C. Yi, A. K. Chandiran, Md. K. Nazeeruddin, E. W.-G. Diau, C.-Y. Yeh, S. M. Zakeeruddin and M. Grätzel, *Science*, 2011, **334**, 29.
70. H.-P. Lu, C.-L. Mai, C.-Y. Tsia, S.-J. Hsu, C.-P. Hsieh, C.-L. Chiu, C.-Y. Yeh and E. W.-G. Diau, *Phys. Chem. Chem. Phys.*, 2009, **11**, 10270.

wavelength, while without strong electron donors, several weak electron donor units are needed. The D–A structure dyes are generally used for small-molecule solar cells.

2.5.2 LUMO Energy Level

To achieve high-efficiency devices, quick separation of excitons generated is a critical requirement. To fulfill this purpose, the LUMO level of the molecule should be at least 0.3 eV higher than that of the electron-accepting molecule to ensure efficient electron transfer from the donor to the acceptor, that is, the LUMO energy level should be higher than −4.3 eV.[8] For dyes with a bandgap over 2 eV, the LUMO energy level is usually higher than the value required. However, for dyes with a bandgap smaller than 2 eV, judicious molecular design is needed to fulfill the LUMO level requirement. A general method to adjust the energy level of the molecules is to change the electron-donating or electron-accepting strength. Generally the LUMO energy can be uplifted by decreasing the electron-accepting strength or increasing the electron-donating strength.

2.5.3 HOMO Energy Level

The HOMO energy level is another critical parameter for dyes. It was normally accepted that the voltage of the device is related to the difference between the LUMO energy level of the electron-accepting dyes and the HOMO energy level of the electron-donating dyes.[9] Uplifting the HOMO energy level can reduce the voltage of the device, and the downshift of the LUMO level increases the voltage. However, it was realized that other factors such as the charge recombination degree also play an important role to the voltage of the device.[9] In addition, it is also important to match the HOMO energy level of the electron-donating molecules to the energy level of the hole injection layer. The existence of the energy barrier between the hole injection layer and the active layer can reduce the current of the device. Similar to the LUMO energy level, the HOMO energy level can be adjusted by changing the donor and acceptor units also. The increase of the electron-donating strength can effectively uplift the HOMO level, while its decrease downshifts the HOMO level.

2.5.4 Molecular Ordering and Film Morphology

Through judicious molecular engineering, most molecules can fulfill the absorbance range and energy level requirement; however, among the big amount of dyes developed so far, only a small part shows great device performance. One important reason is the low hole mobility caused by unideal molecular ordering and film morphology, which limited the current and the fill factor of the device.[8] To increase device performance, much effort was made to improve the molecular ordering and morphology of the electron-donating molecule, and it was found that the long crystal domain and small intermolecular distance tend to bring a high hole mobility of the film.

Both the vacuum evaporation and solution process are used for the fabrication of small molecule–based solar cells. Vacuum evaporation was firstly introduced for the fabrication of planar structure small molecules due to its wide application in small-molecule organic–light emitting diode.[8] The vacuum process tends to bring a crystal-like structure and close molecular packing state in the film. The solution process was introduced for the fabrication of small molecule–based solar cells due to easy fabrication and low cost. However, it tends to form an amorphous film morphology and the intermolecular distance is usually larger than that by vacuum evaporation. For the solution process, as the development of the molecular design and the understanding of the relationship between the film morphology and molecular structure, the molecular ordering and crystal domain size have been largely improved.

Except for molecular ordering, quick exciton separation is also important for device performance. A big interface between the electron donor and acceptor is critical to ensure quick exciton separation. To achieve quick electron separation, good miscibility of the donor and acceptors is important, and big molecular aggregates need to be avoided.

Film morphology and molecular ordering are affected by many factors such as the molecular shape, substitute, molecular length, electron donor or acceptor units, as well as the film fabrication process. In the following, we will analyze and summarize the general rules for the relationship between the molecular shape and film ordering.

2.6 SPECIFIC EXAMPLES OF SMALL MOLECULES FOR PHOTOVOLTAIC DEVICES

Here, the description of specific molecules will be presented. Meanwhile, the relationship between the molecule's structure and overall device performance will be analyzed. Due to the significant difference of the vacuum evaporation process and solution film fabrication process, the discussion will be separated mainly into two parts, that is, vacuum-processed film and solution phase–processed film. Under each kind of process, the molecules will be categorized and analyzed based on the molecular shape and the substituent units.

2.6.1 Vacuum-Processed Molecules

2.6.1.1 Short-Length Dyes with Strong Electron Donors

As discussed in Section 2.5, for strong D–A dyes, even the short length of the molecules can bring broad absorption in the visible region. Vacuum evaporation was generally used for the small-length molecules, and good film morphology and molecular ordering can be achieved. The triphenylamine donor is the most frequently used electron donor for this kind of D–A dyes. For bulk heterojunction solar cells, it is expected that the use of triarylamine can improve hole mobility and carrier transport balance of the film.[11] Roncali and coworkers synthesized symmetrical and asymmetrical starburst dye based on the triphenylamine core and different acceptor unit numbers.[12] As the electron acceptors increase, the bandgap of the dye is reduced from 1.91 eV for one dicyanovinyl (DCN) to 1.78 eV for three DCN units. By using the bilayer planar heterojunction structure, a device based on dye **1** (Figure 2.7) shows a power conversion efficiency (PCE) over 1%. Shirota and coworkers synthesized a star-shaped, hole transporting, amorphous material **2** (Figure 2.7) and used

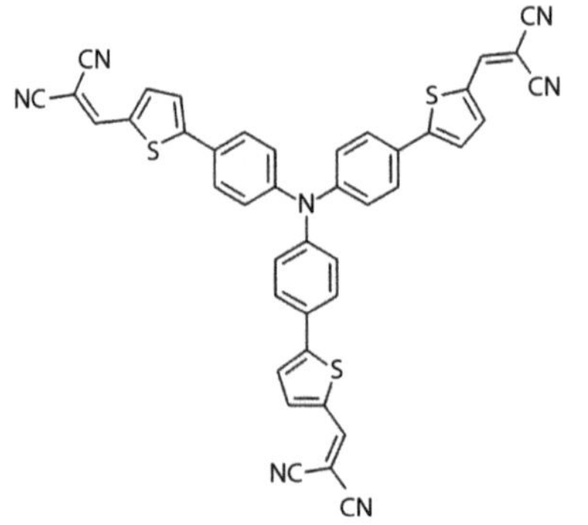

FIGURE 2.7 Strong electron donor triphenylamine-based starburst shape molecules.

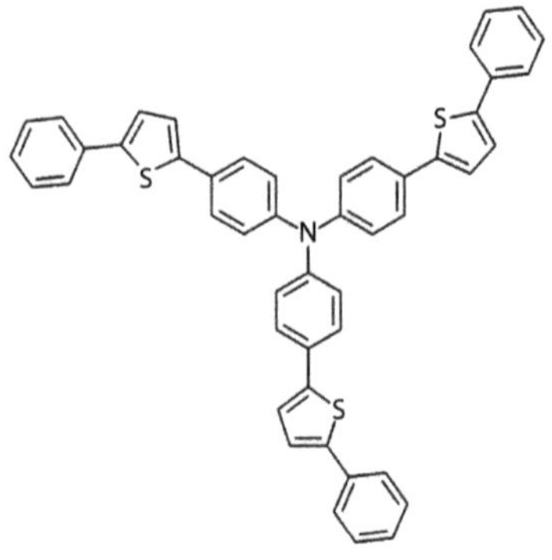

it in a bulk heterojunction solar cell as an electron donor.[13] The cell based on the bilayer heterojunction structure exhibited PCEs of 2.2%, with high ff values of 0.66. By using the time of flight (TOF) method, the hole mobility of the film is tested to be as high as 1.0×10^{-2} cm^2 V^{-1} s^{-1}, which is among the highest hole mobility achieved so far. The high hole mobility of the film explained the high fill factor of the device. The high V_{oc} value of 0.92 V obtained can be ascribed to the large bandgap of the dye, while the low current is due to the short absorption wavelength in the visible region. The film shows an amorphous character after annealing.

To extend the absorption wavelength of the dyes, thiophene units were added between the donor and acceptor to extend the absorption wavelength. Dye **3** (Figure 2.8) shows strong absorption at the visible region and absorption peak around 515 nm.[14] The HOMO and LUMO energy levels are −5.1 and −3.4 eV, respectively, which align well with the C60 electron acceptor. By using the planar heterojunction structure, devices based on these dyes showed the highest efficiency of 2.65% with current 6.5 mA cm^{-2} and V_{oc} 0.91 V. It was found that the increase in thiophene unit increases device performance. To further increase the light absorption region, a stronger electron-accepting unit, tricyanovinylene, was used, and dye **4** (Figure 2.8), based on tricyanovinylene, gives a much longer absorption peak of 640 nm.[15] But the device performance is poor, probably due to the low LUMO energy level, which affected the electron transferring from the dye to electron acceptor.

To reduce the molecular length and increase the absorption wavelength at the same time, Wong's group tried to add the second electron acceptor between the original D–A pair. Dye **5** (Figure 2.9) with benzothiazole and DCN as the double acceptors

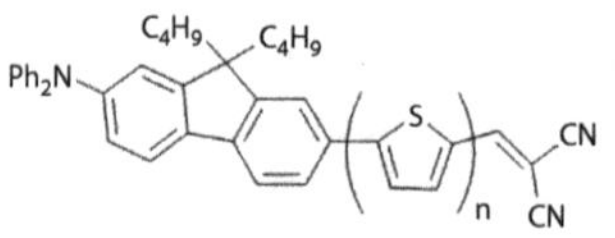

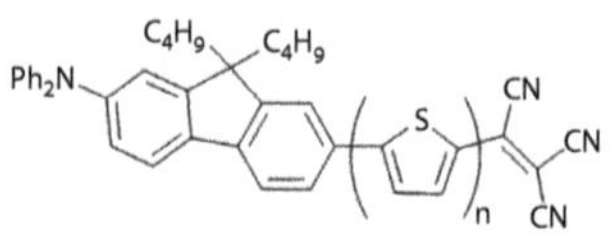

FIGURE 2.8 Triphenylamine-based D–A dyes with weak electron donors in the middle.

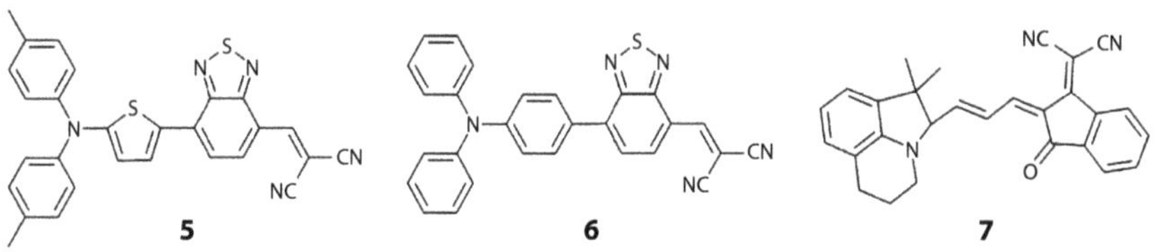

FIGURE 2.9 Strong donor-based dyes with double electron acceptors.

shows an absorption peak at 670 nm, and the HOMO and LUMO energy level are −5.15 and −3.71 eV, respectively.[16] By using C70 as the electron acceptor, the device showed a record device performance of 5.81% with V_{oc} 0.79 V, J_{sc} 14.68 mA cm^{-2}, and ff 0.50. The low voltage of the device was speculated to be caused by the relatively small bandgap and high HOMO energy level. To increase the voltage of the device, triphenylamine was used as the electron donor to replace ditolylaminothienyl.[17] The HOMO and LUMO energy levels of dye **6** (Figure 2.9) are −5.30 and −3.44 eV, respectively. And the device based on it shows a voltage of 0.93 V. The film shows good hole mobility of 4.32×10^{-4} cm^2 V^{-1} s^{-1}. Although the current of the device is a bit lower than that by dye **5**, the device performance is 1% higher, and a record performance of 6.8% was achieved. The increase in the device performance of the double acceptor unit dye was ascribed to the better molecular packing, which brings high hole mobility and fill factor of the device.

Except triarylamine, a strong electron donor unit, merocyanine, was also used for bulk heterojunction solar cell dyes by the vacuum process. The HOMO and LUMO energy levels of dye 7 are −5.75 and −3.65 eV, respectively.[18] Based on dye 7, bulk heterojunction solar cell devices were fabricated with the typical architecture ITO/PEDOT:PSS (40 nm)/dye: phenyl-C61-butyric acid methyl ester (PCBM)/Al (120 nm). By using PEDOT:PSS as the hole injection layer, a device performance of 4.9% was achieved, with V_{oc} 0.77 V and J_{sc} 12.1 mA cm^{-2}. By using deep work function MoO$_3$ to replace PEDOT as the hole transporting layer, the voltage of the device was effectively increased to 0.97 V. The increase in the voltage of the device brings a conversion efficiency over 6.0%. The high voltage of the device by using MoO$_3$ as the electrode indicates that its voltage is related to the hole injection layer.

Based on the results earlier, it can be seen that the molecular structure is critical to the device performance. It can be summarized that for vacuum-processed molecules with strong electron donors, a small molecular size is important for good molecular packing and device performance.

2.6.1.2 Long-Length Weak Electron Donor Dyes

Except for strong electron donors, weak electron donors and a long molecular size can also bring a long absorption wavelength. Oligothiophene-based molecules were widely explored for small molecule–based bulk heterojunction solar cells. Pure oligothiophene molecules tend to aggregate quickly in solution and the miscibility with PCBM is affected. Bäuerle and coworkers found that by adding DCN as the electron acceptor in the end of the thiophene chain, the aggregation can be effectively controlled, and the miscibility with PCBM was significantly improved.[19] Bäuerle et al.

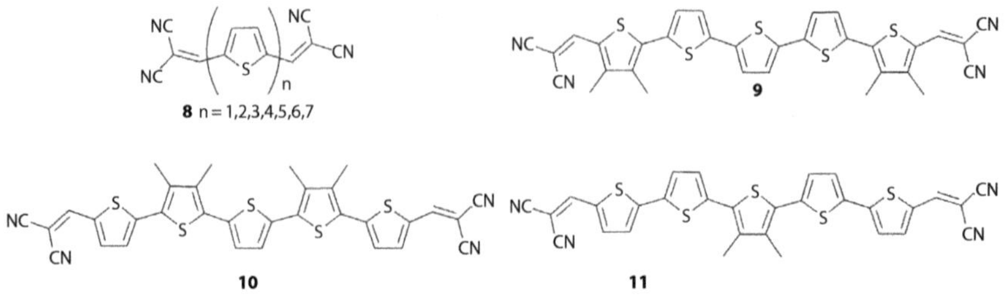

FIGURE 2.10 Oligothiophene-based weak electron donor molecules.

compared a series of oligothiophene dyes **8** (Figure 2.10) with different thiophene unit numbers. The absorption wavelength can be modulated from 406 (one thiophene unit) to 518 nm (four thiophene units), and the molar extinction coefficient is also increased with the increase in the thiophene unit number. There is a small increase in the absorption wavelength when the unit number is over six, which is 530 nm for six units and 532 nm for seven units. On the other hand, the solubility becomes worse as the thiophene units increase. The HOMO energy level is upshifted from two thiophene units −6.50 eV to six units −5.43 eV, and the LUMO level is upshifted from one unit −4.30 eV to six units −3.87 eV. By using a planar heterojunction structure, the efficiency of the devices with four and six thiophene units are 1%–3%, respectively. By using a bulk heterojunction structure, the device performance was significantly improved, and the best device performance of 5.2% was achieved with a five thiophene unit dye.[20]

Except for the effect of the thiophene unit number, the side alkyl chain is also important to device performance. Bäuerle group explored for five thiophene unit dyes **9–11** (Figure 2.10) with a methyl group on different thiophene units.[21] Dye **11** with a methyl substitute on the middle thiophene unit shows a much higher performance than dye **9** with methyl on the ending thiophene units. The conversion efficiency of dyes **9** and **10** are both 4.8%, while dye **11** shows a much higher current with efficiency as high as 6.9% (J_{sc} 11.5 mA cm^{-2}, V_{oc} 0.95 V, ff 0.63), which is the highest reported efficiency achieved for small-molecule solar cells utilizing the vacuum evaporation process. It can be seen that the side chain is important to the device performance. To clarify the mechanism behind it, Bäuerle and coworkers compared the crystal structure of the four thiophene unit dyes **12–14** (Figure 2.11) with different alkyl substitutes.[22] In the single-crystal structure, dye **12** with methyl substitute exhibited the highest number of nonbonding short contact, leading to a perfect coplanar layer structure with strong π–π interaction and electronic coupling between molecules. From the x-ray diffraction (XRD) measurement, dye **12** with a methyl unit shows a smaller distance between the molecules as well. Dye **12** shows a hole mobility of 2×10^{-4} cm^2 V^{-1} s^{-1}, much higher than dyes **13** 9.4×10^{-5} cm^2 V^{-1} s^{-1} and **14** 2.7×10^{-5} cm^2 V^{-1} s^{-1}. This is consistent with the efficiency of the device, which is 3.8% for dye **12** and 1.5% for dye **13** and 2.2% for dye **14**. This finding confirmed that the intermolecular distance is important to the hole mobility and device performance.

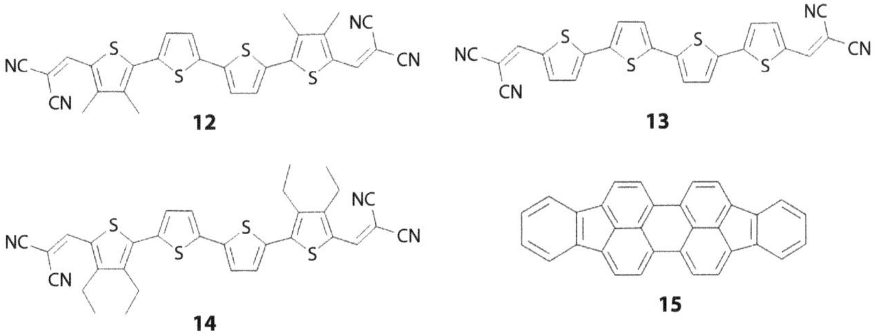

FIGURE 2.11 Long conjugating length dyes with weak electron donors.

Acene derivatives were also explored for organic photovoltaics based on the vacuum process. Its great planar structure allows the formation of a great crystal structure and close molecular packing distance, which brings significantly high mobility of 1.0 cm^2 V^{-1} s^{-1}. One of the best acene molecules reported so far is the diindenoperylene dye **15** (Figure 2.11).[23] For a planar structural device, the fill factor of the film can be as high as 0.75, indicating the high hole mobility and balanced carrier transport of the film. A bulk heterojunction device with C60 as the acceptor shows a voltage of 1 V and efficiency over 4%.

2.6.2 Solution-Processed Small Molecules for Organic Photovoltaics

Although great success has been made for vacuum-processed small-molecule solar cells, the high cost caused by the high vacuum evaporation process still somehow limited its development. The solution-phase film fabrication process, due to its easy and convenient fabrication, has received more and more attention in recent years. Importantly, the solution-phase printing technique can significantly facilitate the fabrication of a large area device. However, compared with vacuum evaporation, the morphology and molecular ordering are much harder to control for solution-phase fabrication. In the following, we will summarize the recent development of solution-processed small-molecule solar cells. To understand the relationship between the molecular structure and device performance better, the molecules are separated into four categories, that is, donor (D)–acceptor (A), A–D–A, D–A–D, and D–A–D–A–D.

2.6.2.1 D–A Structure Dyes

With the great success of the vacuum process for triarylamine-based molecules, much effort was made to use the solution process for these molecules as well. By using the solution process, starburst triarylamine molecules **16** (Figure 2.12) show efficiency around 1%, which is lower than its counterpart that is fabricated by vacuum evaporation.[12] Similar device performance was observed for the D–A triarylamine-based molecules. The solution-processed bulk heterojunction device based on dye **17** (Figure 2.12) shows a conversion efficiency of only 1.72%, much lower than that by vacuum evaporation as well.[24] The solution process was also used for

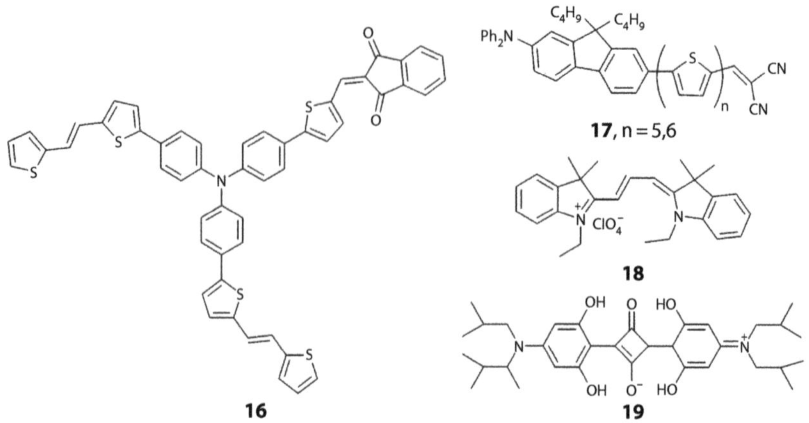

FIGURE 2.12   D–A shape molecules for solution-processed solar cells.

merocyanine dye **7**. The hole mobility of the film is only 5.0×10^{-5} cm² V⁻¹ s⁻¹, much lower than the film made by the vacuum process either, which was ascribed to the bad molecular packing. The device fabricated by the solution process with dye **7** shows a much lower photocurrent compared with that by the vacuum process. All these results indicate that it is hard to form a close molecular packing with the short D–A dyes utilizing the solution process.

Due to its strong absorbance in the visible and infrared region, cyanine and squaraine dyes were widely used for solar cells. Tian and coworkers prepared a planar heterojunction device by using squaraine dye **18** (Figure 2.12) as the electron donor and C60 as the electron acceptor.[25] The device showed the highest incident photon to current efficiency above 3%. Squaraine dye was utilized by Forrest and coworkers for bulk heterojunction solar cells using the solution process.[26] The HOMO and LUMO energy levels of dye **19** (Figure 2.12) are around −5.3 and −3.4 eV, respectively, and the absorption maximum wavelength is 652 nm in solution. The highest performance of dye **19** can reach 5.2%. The high device performance of dye **19** was ascribed to the improved good molecular ordering, which might be caused by the good planar structure of the dyes, as well as the ionic compound character.

2.6.2.2   A–D–A Structure Dyes

Except for asymmetrical donor acceptor dyes, symmetrical molecules were also explored to reduce the intermolecular distance and improve the intermolecular packing. By putting the electron acceptors in the end, A–D–A structure dyes can possibly increase the electron transport between molecules as well.

Sharma et al. introduced the cyanovinylene-4-nitrophenyl acceptor as the electron acceptor for A–D–A structure dyes. Dye **20** (Figure 2.13) with alkoxyl-substituted phenyl in the middle shows maximum absorption at 630 nm, and the HOMO and LUMO energy levels are −5.0 and −3.2 eV, respectively.[27] Bulk heterojunction solar cells show a conversion efficiency of 1.4%. With the annealing process,

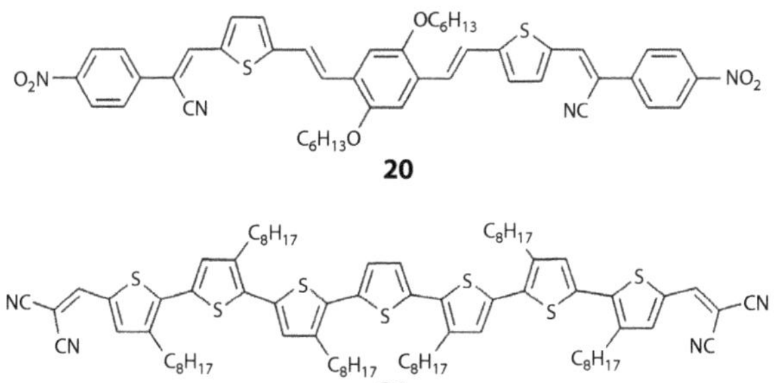

FIGURE 2.13 A–D–A shape molecules for solution-processed photovoltaics.

the conversion efficiency was improved to 2.3%. The change of the thiophene unit into anthracene brings a higher efficiency of 2.5%.[27]

Chen and coworkers firstly explored long oligothiophene dyes for solution-processed devices. By adding long alkyl chains on each thiophene unit, the solubility of the seven thiophene unit–based dye was significantly improved.[28] The optical bandgap of dye **21** (Figure 2.13) is 1.71 eV, and the absorption peak is over 600 nm. By using the bulk heterojunction structure, the highest efficiency of 3.7% was achieved with J_{sc} 12.4 mA cm^{-2} and V_{oc} 0.88 V under AM 1.5G illumination.[29]

Based on the initial success of oligothiophene dye **21**, Chen et al. further introduced a new electron donor benzodithiophene in the center. Meanwhile, acceptors octyl cyanoacetate and 3-ethylrhodanine were introduced to replace the original DCN unit.[30] Compared with DCN acceptor dyes, the bandgap of the new dyes, **22** and **23** (Figure 2.14), are increased due to the weaker electron-accepting character of the new electron acceptor. The HOMO and LUMO energy levels are around −5.0 and −3.2 eV, respectively. Although dye **22** has a higher HOMO energy level than **21**, the voltage of the device was not reduced. Meanwhile, the current and voltage of the devices are much improved, leading to a much improved efficiency of 7.3%. The high performance of the device was ascribed to better molecular ordering by changing the ending unit. Meanwhile, although dye **23** has smaller bandgap compared with dye **22**, the voltage of dye **23** is even higher. This is consistent with the dark current measurement, which is much lower for dye **23** than dye **22**. This result indicated that the voltage is not only related to the HOMO energy level and the bandgap of the molecules, but the charge recombination also plays an important role. It was speculated that the space hindrance between 3-ethylrhodanine and PCBM can reduce the charge recombination from the LUMO of the acceptor to the HOMO of the donor. In addition, it is worth to note that in this work, polydimethylsiloxane (PDMS) was added in the film fabrication, and improved film morphology was observed.

Based on dye **23**, by adding thiophene side chains on the central thiophene unit, the film morphology was further improved.[31] By using grazing-incidence wide-angle x-ray scattering measurement, the domain size of dye **24** (Figure 2.14) is around 10–15 nm, which is close to the best organic polymer solar cells. The hole mobility of the film is 3.29×10^{-4} cm^2 V^{-1} s^{-1}, close to the electron mobility of the film, 4.19×10^{-4} cm^2 V^{-1} s^{-1}. Based on dye **24**, a recorded device performance over 8% was achieved.

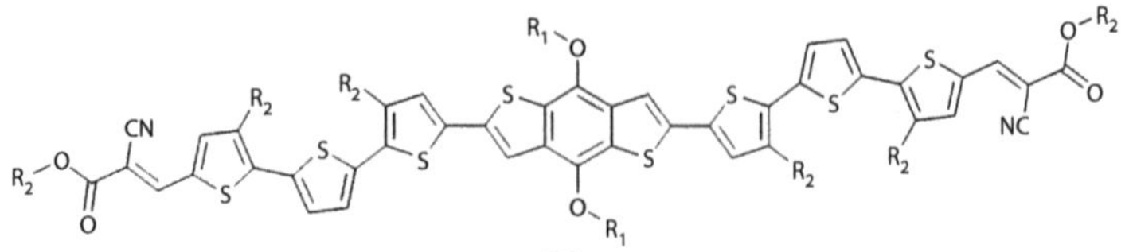

FIGURE 2.14 A–D–A shape molecules with benzodithiophene donor as the central donor unit.

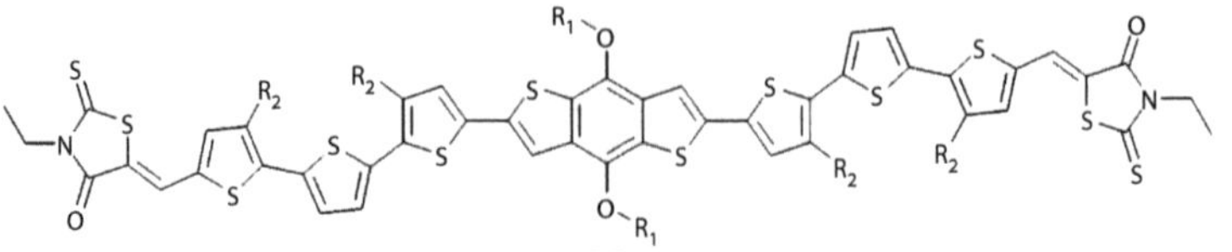

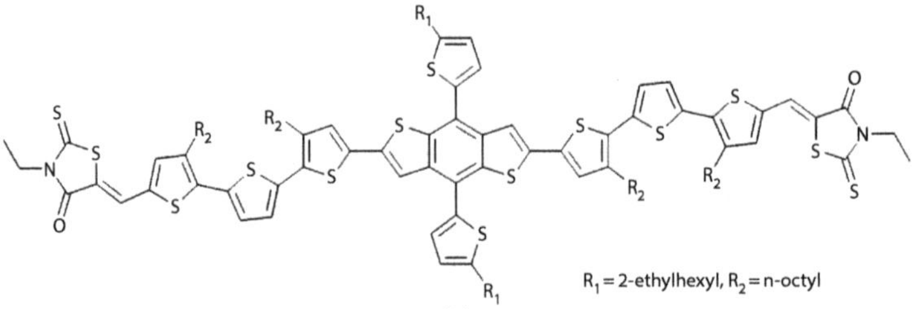

2.6.2.3 D–A–D Structure Dyes

Due to the high hole mobility of dyes with symmetrical multi-triphenylamine-based small molecules, much effort was made to develop similar molecules for solution-processed small-molecule solar cells. Thiadiazoloquinoxaline was used as an electron acceptor for this kind of D–A–D style molecule with triphenylamine as the ending electron donor.[32] Dye **25** (Figure 2.15) shows a charge transfer band at 698 nm in solution with the absorption edge extended to 880 nm, corresponding to a bandgap of 1.41 eV. The current of the device is only 1.9 mA cm^{-2}, which is probably caused by the low LUMO energy level. A weaker electron acceptor dibenzo[f,h]thieno-[3,4-b]quinoxaline was used to increase the bandgap.[33] The HOMO and LUMO energy levels of dye **26** (Figure 2.15) are −5.3 and −3.3 eV, respectively. A bulk heterojunction device based on this dye shows a conversion efficiency of 1.7%.

Benzothiadizole acceptor was also explored for triarylamine-based D–A–D molecules; dye **27** (Figure 2.15) shows a bandgap around 1.8 eV and device performance of 0.2%.[34] Dyes **28** and **29** with the 2-pyran-4-ylidenemalononitrile unit as the electron acceptor were developed for solution-processed small-molecule solar cells as well, and different conjugating lengths were compared.[35,36] Although dyes **28** and **29** (Figure 2.15) show similar bandgap and HOMO and LUMO energy levels, dye **29** shows a much higher efficiency of 1.5% compared with 0.8% of dye **28**. The higher performance of dye **29** was ascribed to the better molecular packing. The device

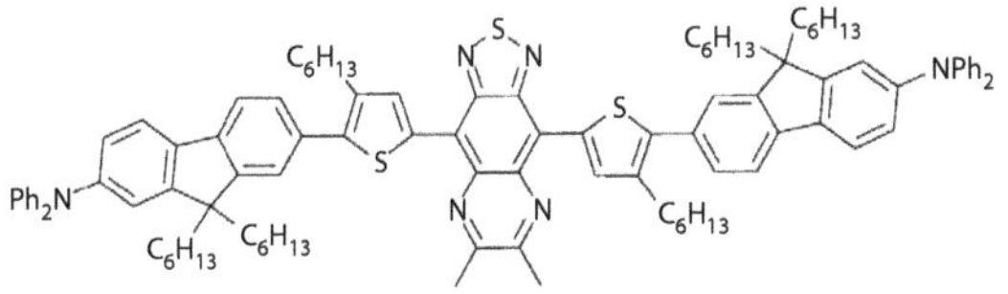

FIGURE 2.15 D–A–D dye with triarylamine as the ending units.

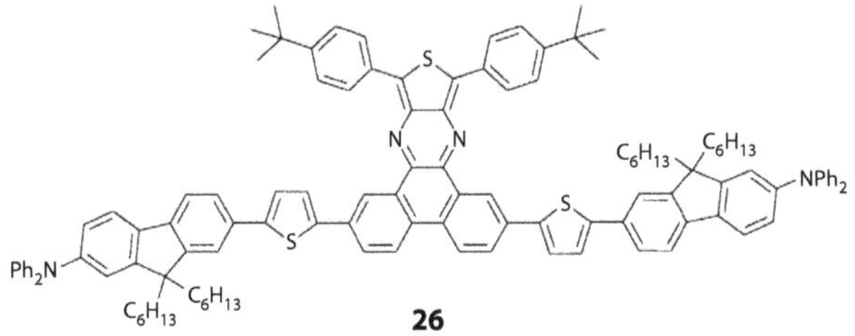

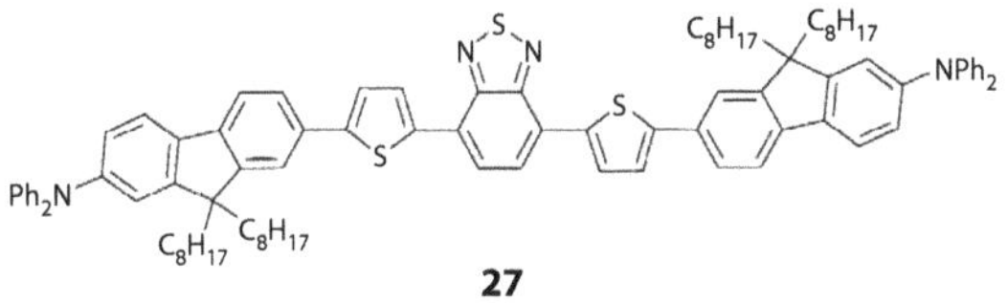

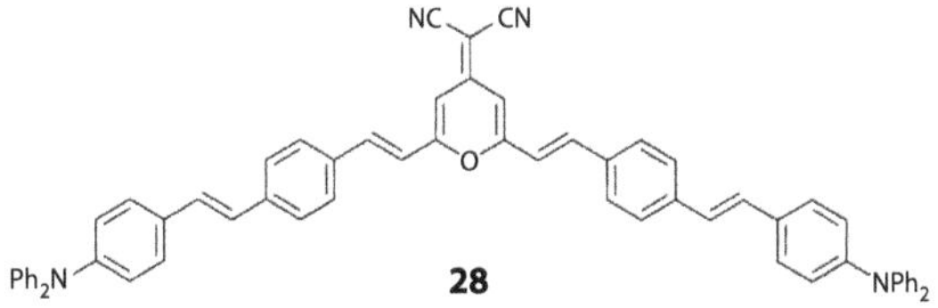

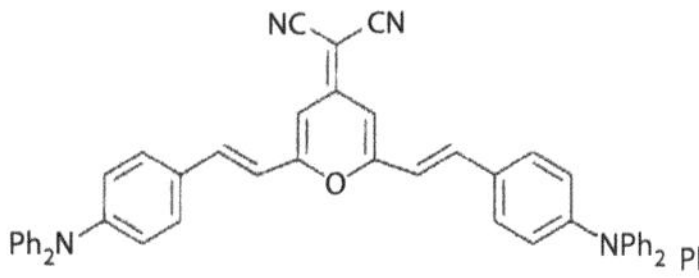

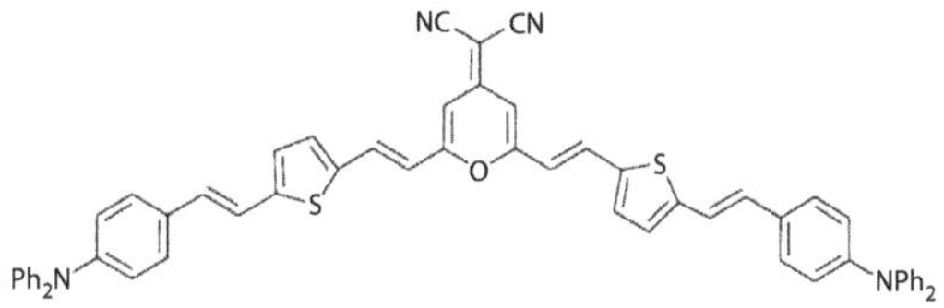

performance was further increased when the conjugating unit benzene was replaced by a thiophene unit, and dye **30** (Figure 2.15) shows a conversion efficiency of 2.1%.[37]

For all the previously mentioned dyes with triphenylamine as the ending electron donor unit, the fill factors of the device are all between 0.3 and 0.5, which can be ascribed to the low hole mobility of the film. Due to the nonplanar structure of triarylamine and a weak intermolecular interaction, the solution process probably cannot bring a close molecular distance compared with the vacuum process. The addition of other additives in the film or posttreatment to the film can possibly solve this problem in the future.

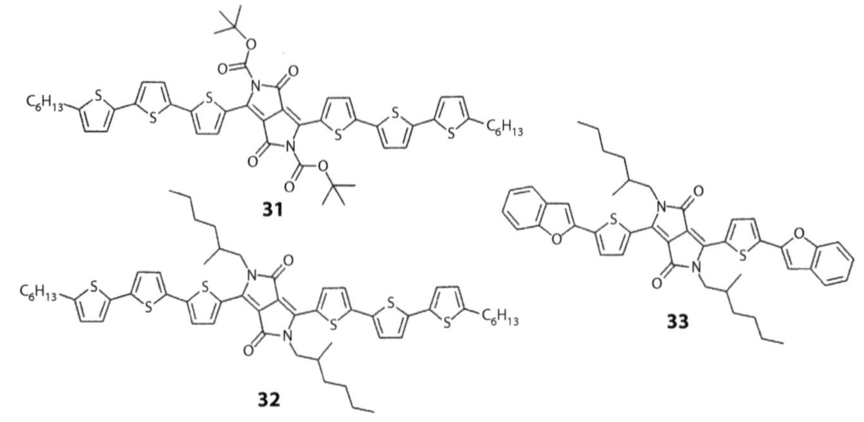

FIGURE 2.16 D–A–D structure molecules with thiophene or benzofuran as the ending donor units.

Except for triarylamine, thiophene units were also explored as the electron donor unit for D–A–D structure dyes. Nguyen and coworkers developed a new class of D–A–D dye **31** (Figure 2.16) with diketopyrrolopyrrole (DPP) as the central electron acceptor and thiophene as the donor unit.[38] To increase the solubility of the compound, the nitrogen atoms of the DPP unit were protected by tert-butyloxycarbonyl (Boc) groups. Compound **31** shows absorption band edge at 700 nm (1.77 eV) in solution and at 810 nm (1.53 eV) in thin films, and the HOMO energy value is about 4.9 eV. Bulk heterojunction device based on dye **31** shows efficiency of 2.3%. To improve thermal stability and solubility, ethylhexyl chains were used to replace the Boc groups.[39] The HOMO energy level is downshifted to −5.2 eV. In addition, film morphology was much improved, and the hole mobility of dye **32** (Figure 2.16) is two orders of magnitude larger than **31**. Bulk heterojunction device by compound **32** shows performance of 2.9%. Following this work, benzofuran terminal groups were used to replace the oligothiophene units as the electron donor, as shown in Figure 2.16 (**33**). The efficiency was dramatically increased to 4.4%.[40] It is worth to note that the device performance before the annealing process is only 0.3%, indicating that the annealing process plays a critical role in adjusting film morphology.

2.6.2.4 D–A–D–A–D Structure Dyes

To further improve the dye performance, a new kind of dye structure, that is, by combining the D–A–D and A–D–A structures, was introduced. Dye **34** (Figure 2.17) with triarylamine in the center, benzothiazole electron acceptor in the middle, and thiophene donor unit in the end was developed by Zhang and coworkers.[41] Dye **34** shows a bandgap about 1.9 eV and a conversion efficiency of 1.96%. By increasing the length of the outside thiophene chain and using the starburst dye structure, dye **35** (Figure 2.17) shows a recorded conversion efficiency of 4.3% among all triarylamine-based molecules.[42] The maximum absorption wavelength of dye **35** is 538 nm, the bandgap is 1.9 eV, and the hole mobility of the film is 4.9×10^{-4} cm^2 V^{-1} s^{-1}.

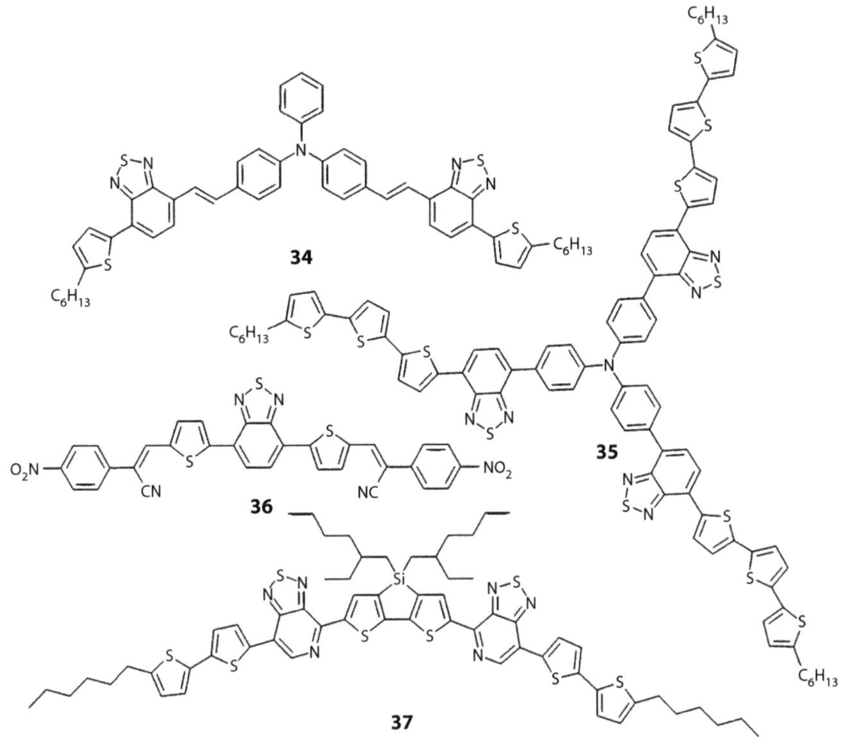

FIGURE 2.17   The molecular structure of D–A–D–A–D dyes.

Cyanovinylene-4-nitrophenyl acceptor was also used for D–A–D–A–D structure dyes. Dye **36** (Figure 2.17) shows an optical bandgap of 1.67 eV and absorption peak over 600 nm.[43] The LUMO and HOMO energy levels are around −5.25 and −3.55 eV, respectively. Bulk heterojunction device based on dye **36** shows a conversion efficiency of 1.3%.

Bazan and coworkers synthesized a new kind of D–A–D–A–D dye (**37**) (Figure 2.17) with a dithienosilole as the central core and hexylbithienyl-thiadiazolopyridine as the acceptor.[44] The dye showed an absorption maximum at 625 nm in solution and at 720 nm in thin films and an optical bandgap of 1.51 eV. The device based on the as-cast film gave a PCE of 0.7%, which was improved to 3.2% by thermal annealing at 110°C for 2 min. The increased molecular ordering after annealing was confirmed by conductive and photoconductive atomic force microscopy, dynamic secondary ion mass spectrometry, and grazing incident wide-angle x-ray scattering measurement.

To further improve the morphology of the film, Heeger et al. used a diiodooctane as the additive for film fabrication.[45] A significantly reduced domain size was observed for dye **37**, which is close to the ideal domain size of 10 nm for organic bulk heterojunction solar cells. Improved molecular ordering and morphology bring a much increased fill factor of the device, and the efficiency of the device was improved from 4.5% to 6.7%.

2.7 CONCLUSION AND PERSPECTIVE

The fast development of the efficiency of small molecule–based solar cells has attracted more and more interest nowadays. The vacuum process brings the highest efficiency of organic solar cells by using the tandem device structure. For solution-processed solar cells, the morphology problem of small molecules has been largely solved by judicious molecular design and the film treatment process. Although polymer solar cells still show the highest efficiency of single-junction device nowadays, small-molecule solar cells can effectively compensate the shortcoming of the polymer solar cells such as the random molecular weight and impurities. As molecule design and mechanism understanding are developed, it can be expected that the efficiency of small molecules can be largely improved in the near future.

However, there are still some big challenges ahead to improve efficiency. To further push forward performance enhancement, the following factors need to be further addressed: (1) the low hole mobility of the film (low hole mobility can reduce carrier collection in the device side that is far away from the cathode and cause low photocurrent; on the other hand, the unbalanced carrier transport will bring a low fill factor of the device), (2) quick excitons separation (compared with polymers, small molecules are easier to aggregate in the film fabrication process, which will bring a reduction of the interface between the electron donor and acceptor and delay exciton separation), (3) judicious control of the molecular domain size (a big enough domain size is important to maintain the carrier transport between molecules, while too long domain can affect the electron transport from the electron-donating dye to the fullerene acceptor; it is critical to find the optimized domain size for each kind of molecule; the addition of additives post–film treatment may play an important role in solving this problem), and (4) mechanism understanding (up to now, the relationship between the molecular structure and carrier transport in small molecules is still far from clear; the combination of theoretic modeling and high-resolution spectroscopy might be helpful in clarifying the correlation between the molecular structure and hole transporting). Except for the factors listed earlier, other factors such as the improvement of the electron acceptor and electrode material and the use of plasmonic-enhanced absorbance also have the potential to significantly increase device performance.

REFERENCES

1. N. S. Lewis, *Science*, 2007, **315**, 798.
2. Z. Ning, Y. Fu and H. Tian, *Energy Environ. Sci.*, 2010, **3**, 1170.
3. G. A. Chamberlain, *Sol. Cells*, 1983, **8**, 47.
4. C. W. Tang, *Appl. Phys. Lett.*, 1986, **48**, 183.
5. M. Hiramoto, H. Fujiwara and M. Yokoyama, *Appl. Phys. Lett.*, 1991, **58**, 1062.
6. N. S. Sariciftci, L. Smilowitz, A. J. Heeger and F. Wudl, *Science*, 1992, **258**, 1474.
7. S. Guünes, H. Neugebauer and N. S. Sariciftci, *Chem. Rev.*, 2007, **107**, 1324.

8. A. Mishra and P. Bäuerle, *Angew. Chem. Int. Ed.*, 2012, **51**, 2020–2067.
9. A. W. Hains, Z. Liang, M. A. Woodhouse, and B. A. Gregg, *Chem. Rev.*, 2010, **110**, 6689–6735.
10. Y. Li, Q. Guo, Z. Li, J. Pei and W. Tian, *Energy Environ. Sci.*, 2010, **3**, 1427–1436.
11. Z. Ning and H. Tian, *Chem. Commun.*, 2009, **37**, 5483.
12. S. Roquet, A. Cravino, P. Leriche, O. Alévêque, P. Frère and J. Roncali, *J. Am. Chem. Soc.*, 2006, **128**, 3459.
13. H. Kageyama, H. Ohishi, M. Tanaka, Y. Ohmori and Y. Shirota, *Adv. Funct. Mater.*, 2009, **19**, 3948.
14. P. F. Xia, X. J. Feng, J. Lu, S.-W. Tsang, R. Movileanu, Y. Tao and M. S. Wong, *Adv. Mater.*, 2008, **20**, 4810.
15. P. F. Xia, X. J. Feng, J. Lu, R. Movileanu, Y. Tao, J.-M. Baribeau and M. S. Wong, *J. Phys. Chem. C*, 2008, **112**, 16714.
16. L. Lin, Y. Chen, Z. Huang, H. Lin, S. Chou, F. Lin, C. Chen, Y. Liu and K. Wong, *J. Am. Chem. Soc.*, 2011, **133**, 15822–15825.
17. Y. Chen, L. Lin, C. Lu, F. Lin, Z. Huang, H. Lin, P. Wang, Y. Liu, K. Wong, J. Wen, D. J. Miller and S. B. Darling, *J. Am. Chem. Soc.*, 2012, **134**, 13616–13623.
18. N. M. Kronenberg, V. Steinmann, H. Bürckstümmer, J. Hwang, D. Hertel, F. Würthner and K. Meerholz, *Adv. Mater.*, 2010, **22**, 4193.
19. S. Haid, A. Mishra, M. Weil, C. Uhrich, M. Pfeiffer and P. Bäuerle, *Adv. Funct. Mater.*, 2012, **22**, 4322–4333.
20. R. Fitzner, E. Reinold, A. Mishra, E. Mena-Osteritz, H. Ziehlke, C. Kcrner, K. Leo, M. Riede, M. Weil, O. Tsaryova, A. Weis, C. Uhrich, M. Pfeiffer and P. Bäuerle, *Adv. Funct. Mater.*, 2011, **21**, 897.
21. R. Fitzner, E. Mena-Osteritz, A. Mishra, G. Schulz, E. Reinold, M. Weil, C. Körner, H. Ziehlke, C. Elschner, K. Leo, M. Riede, M. Pfeiffer, C. Uhrich and P. Bäuerle, *J. Am. Chem. Soc.*, 2012, **134**, 11064–11067.
22. R. Fitzner, C. Elschner, M. Weil, C. Uhrich, C. Körner, M. Riede, K. Leo, M. Pfeiffer, E. Reinold, E. Mena-Osteritz and P. Bäuerle, *Adv. Mater.*, 2012, **24**, 675–680.
23. J. Wagner, M. Gruber, A. Hinderhofer, A. Wilke, B. Bröker, J. Frisch, P. Amsalem, A. Vollmer, A. Opitz, N. Koch, F. Schreiber and W. Brütting, *Adv. Funct. Mater.*, 2010, **20**, 4295.
24. W. Zhang, S. C. Tse, J. Lu, Y. Tao and M. S. Wong, *J. Mater. Chem.*, 2010, **20**, 2182.
25. F. Meng, K. Chen, H. Tian, L. Zuppiroli and F. Nuesch, *Appl. Phys. Lett.*, 2003, **82**, 3788.
26. G. Wei, R. R. Lunt, K. Sun, S. Wang, M. E. Thompson and S. R. Forrest, *Nano Lett.*, 2010, **10**, 3555.
27. J. A. Mikroyannidis, M. M. Stylianakis, P. Balraju, P. Suresh and G. D. Sharma, *ACS Appl. Mater. Interfaces*, 2009, **1**, 1711.
28. Y. Liu, J. Zhou, X. Wan and Y. Chen, *Tetrahedron*, 2009, **65**, 5209.
29. B. Yin, L. Yang, Y. Liu, Y. Chen, Q. Qi, F. Zhang and S. Yin, *Appl. Phys. Lett.*, 2010, **97**, 023303.
30. J. Zhou, X. Wan, Y. Liu, Y. Zuo, Z. Li, G. He, G. Long, W. Ni, C. Li, X. Su and Y. Chen, *J. Am. Chem. Soc.*, 2012, **134**, 16345–16351.
31. J. Zhou, Y. Zuo, X. Wan, G. Long, Q. Zhang, W. Ni, Y. Liu, Z. Li, G. He, C. Li, B. Kan, M. Li and Y. Chen, *J. Am. Chem. Soc.*, 2013, **135**, 8484–8487.
32. M. Sun, L. Wang, X. Zhu, B. Du, R. Liu, W. Yang and Y. Cao, *Sol. Energy Mater. Sol. Cells*, 2007, **91**, 1681.
33. M. Velusamy, J.-H. Huang, Y.-C. Hsu, H.-H. Chou, K.-C. Ho, P.-L. Wu, W.-H. Chang, J. T. Lin and C.-W. Chu, *Org. Lett.*, 2009, **11**, 4898.
34. C. He, Q. He, Y. He, Y. Li, F. Bai, C. Yang, Y. Ding, L. Wang and J. Ye, *Sol. Energy Mater. Sol. Cells*, 2006, **90**, 1815.

35. C. He, Q. He, X. Yang, G. Wu, C. Yang, F. Bai, Z. Shuai, L. Wang and Y. Li, *J. Phys. Chem. C*, 2007, **111**, 8661.
36. L. Xue, J. He, X. Gu, Z. Yang, B. Xu and W. Tian, *J. Phys. Chem. C*, 2009, **113**, 12911.
37. J. Zhang, G. Wu, C. He, D. Deng and Y. Li, *J. Mater. Chem.*, 2011, **21**, 3768.
38. A. B. Tamayo, B. Walker and T.-Q. Nguyen, *J. Phys. Chem. C*, 2008, **112**, 11545.
39. A. Tamayo, T. Kent, M. Tantitiwat, M. A. Dante, J. Rogers and T.-Q. Nguyen, *Energy Environ. Sci.*, 2009, **2**, 1180.
40. B. Walker, A. B. Tamayo, X.-D. Dang, P. Zalar, J. H. Seo, A. Garcia, M. Tantiwiwat and T.-Q. Nguyen, *Adv. Funct. Mater.*, 2009, **19**, 3063.
41. Y. Yang, J. Zhang, Y. Zhou, G. Zhao, C. He, Y. Li, M. Andersson, O. Inganäs and F. Zhang, *J. Phys. Chem. C*, 2010, **114**, 3701.
42. H. Shang, H. Fan, Y. Liu, W. Hu, Y. Li and X. Zhan, *Adv. Mater.*, 2011, **23**, 1554.
43. P. Suresh, P. Balraju, G. D. Sharma, J. A. Mikroyannidis and M. M. Stylianakis, *ACS Appl. Mater. Interfaces*, 2009, **1**, 1370.
44. G. C. Welch, L. A. Perez, C. V. Hoven, Y. Zhang, X.-D. Dang, A. Sharenko, M. F. Toney, E. J. Kramer, T.-Q. Nguyen and G. C. Bazan, *J. Mater. Chem.*, 2011, **21**, 12700.
45. Y. Sun, G. C. Welch, W. Leong, C. J. Takacs, G. C. Bazan and A. J. Heeger, *Nat. Mater.*, 2012, **11**, 44–48.

3 Inorganic Materials for Solar Cell Applications

Yasutake Toyoshima

CONTENTS

3.1 INTRODUCTION

It has been more than half a century since the first solar cell was invented,[1] which was made of single-crystal Si. Up to now, Si is still the major semiconductor material that has been used in solar cell for photovoltaics (PV). In this chapter, a wide-range overview about the inorganic materials that have been used for solar cells will be presented.

Before going into the variety of materials, the fundamental aspects of PV are described. First, the sunlight spectra are shown in Figure 3.1. Due to the atmospheric absorption (mainly by ozone molecules in the UV region and water and carbon dioxide molecules in the infrared region) and scattering, these spectra have somewhat complicated features on the earth ground. *AM* is abbreviation of air mass and the number means the *thickness* of the atmosphere (unity corresponds to vertical thickness), while the letter "D" or "G" means that each spectrum is directly from the sun or from the entire sky including blue light scattered by the atmosphere. In Figure 3.1, photon number–based spectrum (on the basis of the Si bandgap) is also presented since a PV cell current is proportional to the total number of absorbed photons, not to the total absorbed energy. In Figure 3.2, photoabsorption characteristics of various semiconductor materials are summarized. Recently reported corrections[2] for CuInSe$_2$ (CIS) and CdTe profiles are compiled in Figure 3.2 along with the previous ones that are shown in broken curves. In addition, a commonly used dye molecule called N719 is also plotted in this figure by simply assuming the molecular volume

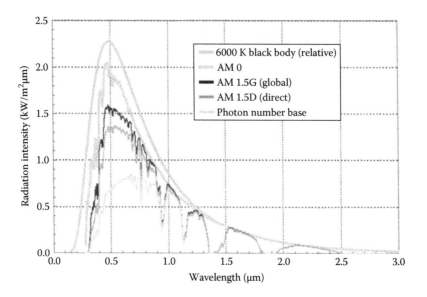

FIGURE 3.1   Standard solar spectra.

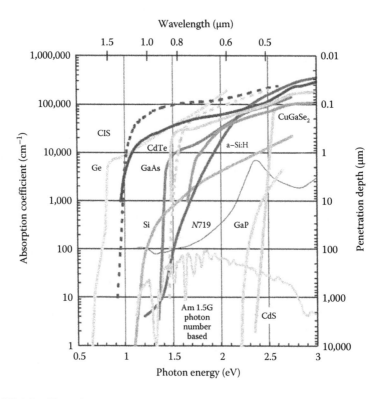

FIGURE 3.2   Photoabsorption characteristics of semiconductors.

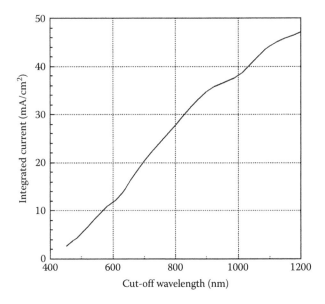

FIGURE 3.3 Relationship between the integrated current and cutoff wavelength.

to be 1 nm³. In Figure 3.3, the maximum current corresponding to the bandgap cut-off is plotted. This is obtained simply by integrating the number of higher-energy photons and is basically the same as the one reported by Henry,[3] except for two corrections of total intensity from 844 to 1000 W/m² and the revision of standard solar spectra.[4] This figure is useful for a brief estimation of current collecting efficiency for a semiconductor of a given bandgap. When it comes to the voltage, the situation is rather complicated. Generally speaking, half (or 2/3 at best) of the bandgap is usually obtained for open-circuit voltage. Refer to the Shockley–Queisser limit for estimating the upper limit of energy conversion for a single-junction solar cell.[5] Two major origins to limit the energy conversion efficiency are shown in Figure 3.4. One is

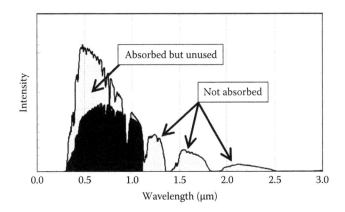

FIGURE 3.4 Two kinds of major losses.

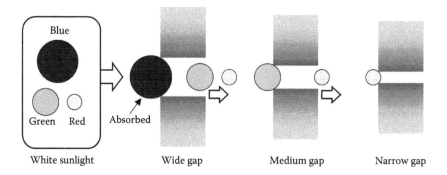

| White sunlight | Wide gap | Medium gap | Narrow gap |

FIGURE 3.5   How tandem structure works efficiently to absorb white light.

longer-wavelength light that cannot be absorbed, and the other is shorter-wavelength light that has excess energy above the bandgap that cannot be utilized for conversion although absorbed. To partly improve this situation, a multi-bandgap PV cell, commonly called as *tandem cell*, is employed. The basic concept of a tandem cell is schematically shown in Figure 3.5. It should be noted that a semiconductor will not absorb the light whose energy is below the bandgap. Practical implementation of this tandem concept into PV cells is discussed in Section 3.6.

3.2   CRYSTAL SILICON SOLAR CELLS

Currently, crystal silicon–based PV cells are still in major use in the field of solar cell application. First, the purification process is shown in Figure 3.6, since purity is the most important factor in semiconductor materials. These processes are commonly called *Siemens method* as a whole. The key feature of this method is purification by multiple distillations up to 300 times (or more) of trichlorosilane ($SiHCl_3$) gas, whose boiling point is close to ambient temperature. Ultimate purity of 11 nine (99.999999999%) can be achieved by this method, which is good (and necessary) for semiconductor use. For solar cells, however, such an ultimate purity is sometimes said to be unnecessary to achieve practical conversion efficiency, from the viewpoint of production cost. Those moderate-purity materials of five to seven nine (99.999-99.99999%) level are called *solar grade*. The problematic feature in this ultimate purification is the low conversion efficiency in the solidification of $SiHCl_3$ gas by electric current heating in a bell jar–shaped reactor. As shown in Figure 3.6, the maximum (theoretical) conversion is 25% at best. In addition, the heat loss is quite significant since higher pressure is preferred in this solidification reaction. Attempts to improve this reactor are still ongoing.[6] These solidified polysilicon rods are broken into small pieces and then melted and made to a single-crystal rod by the Czochralski (CZ) method as shown in Figure 3.7. For solar cells, polycrystalline Si made by cast methods is also used.[7] Since the cuboid made by casting would be usually larger than a meter, it is suitable for mass productions. These crystal Si ingots are sliced into wafers by a multiwire saw as shown in Figure 3.8.

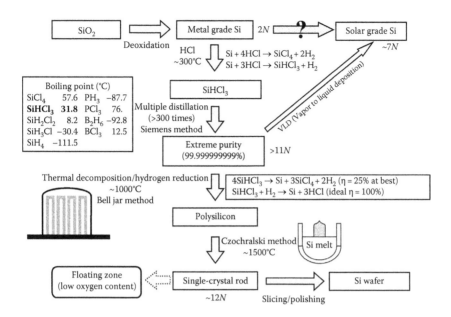

FIGURE 3.6 Purification process of silicon.

A typical sequence for solar cell production from p-type polycrystalline wafer is shown in Figure 3.9. The silicon nitride (SiN_x) layer at the light incident side produced by plasma-enhanced chemical vapor deposition (PECVD) will play multiple roles, namely, (a) an antireflecting layer for light incidence, (b) interface passivation, (c) carrier collecting effect by positive fixed charges in the layer, and (d) hydrogen reservoir to passivate point defects in polycrystalline Si, mainly at the grain boundaries. Passivation is carried out not only by placing an insulating layer but also by minority (positive) carrier repulsion due to the fixed charge in this layer. Since this material is insulating, some trick is necessary for front electrode contact. This can be achieved by so-called *glass frit* incorporation into the Ag paste for front grid electrodes. This glass frit is mainly made of lead borosilicate[8] and/or other metal oxide mixtures[9] that melt at relatively low temperature. During the firing process, this glass frit melt penetrates through the nitride layer, forming the Ag wire contact to the underlying Si emitter. The counterdoped layer at the light incident side is usually called *emitter* for convention. At the same time of this firing, rear-side Al electrode metallization is also performed, which also includes glass frit mainly made of silica particles to reduce the mechanical stress (which causes wafer bowing) due to a large thermal expansion coefficient in Al. This Al electrode works as a backside reflector (BSR) for transmitted (unabsorbed) light back into the active layer. In addition, a small amount of Al diffused into the rear end of the Si layer will work as a p-type dopant, forming an enforced p-type-doped part to enhance positive carrier collection. This effect is called backside field (BSF). It is proposed that Al oxide produced by atomic layer deposition can also be employed to enhance positive carrier collection, due to the negatively charged defects in this layer.[10] A heavily doped layer

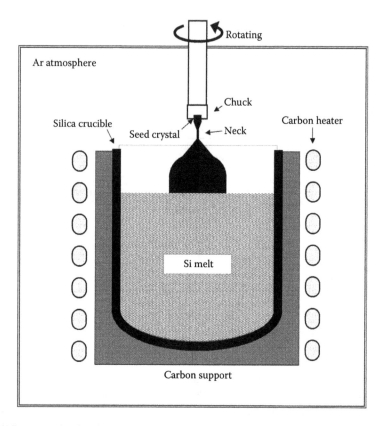

FIGURE 3.7   CZ method.

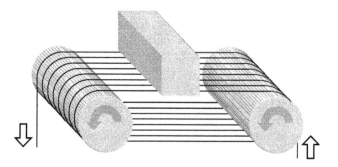

FIGURE 3.8   Multiwire saw.

also has an effect on interfacial passivation. Since carrier recombination requires coexistence of both positive and negative carriers, it can be reduced by minimizing the minority carrier density in heavily doped layers. In Figures 3.10 and 3.11, typical appearances of Si cell and such cell-based PV module are shown, respectively, where PVF, PET, PEN, and EVA stand for polyvinyl fluoride, polyethylene terephthalate, polyethylene naphthalate, and ethylene-vinyl acetate, respectively.

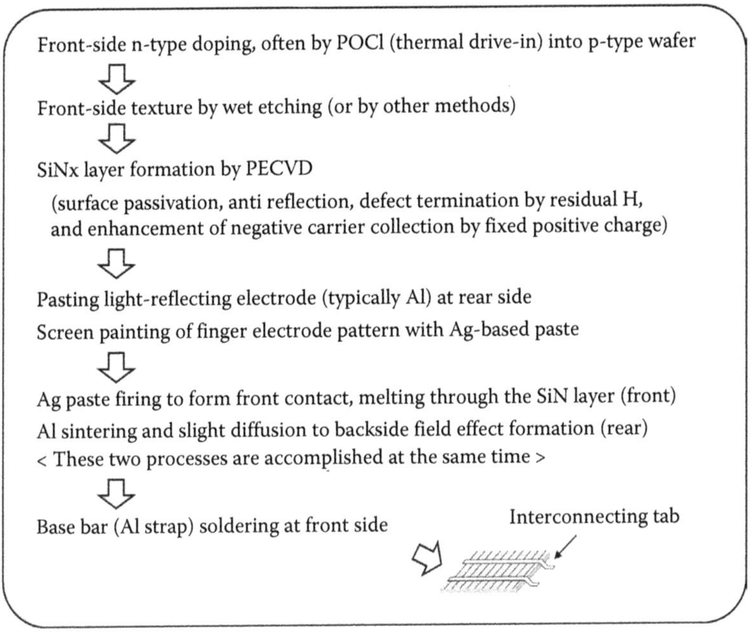

Front-side n-type doping, often by POCl (thermal drive-in) into p-type wafer

⇩

Front-side texture by wet etching (or by other methods)

⇩

SiNx layer formation by PECVD

(surface passivation, anti reflection, defect termination by residual H, and enhancement of negative carrier collection by fixed positive charge)

⇩

Pasting light-reflecting electrode (typically Al) at rear side

Screen painting of finger electrode pattern with Ag-based paste

⇩

Ag paste firing to form front contact, melting through the SiN layer (front)

Al sintering and slight diffusion to backside field effect formation (rear)

< These two processes are accomplished at the same time >

⇩

Base bar (Al strap) soldering at front side   Interconnecting tab

FIGURE 3.9 Typical procedure for PV cell formation (multicrystal Si).

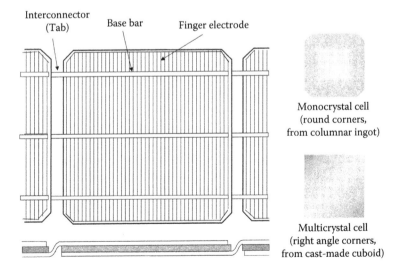

Interconnector (Tab)   Base bar   Finger electrode

Monocrystal cell (round corners, from columnar ingot)

Multicrystal cell (right angle corners, from cast-made cuboid)

FIGURE 3.10 Typical appearance of Si-based PV cell.

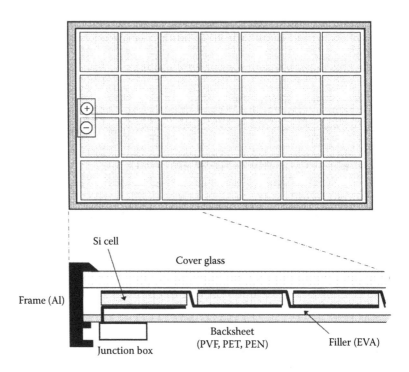

FIGURE 3.11   Typical structure of Si-based PV module.

Hereafter, several types of high-efficiency cells are presented. One of them shows the highest efficiency* among the Si-based cells, called passivated emitter, rear locally diffused (PERL) cell[11] (Figure 3.12). Among all, this cell has a quite unique structure of *inverted pyramids* for antireflection at the front side, although the cell area is considerably small.

The next one is the heterojunction with intrinsic thin layer (HIT) cell[12] (Figure 3.13), which has already been applied for commercial production of PV modules. The key feature in this PV cell is p–n junction formation not by thermal drive-in of impurity but by doped amorphous layer formation, which will result in less heat stress in the single-crystal Si wafers. In addition, hydrogen contained in the intrinsic intermediate layer plays efficient passivating roles, resulting in a quite high open-circuit voltage. However, since the light absorbed in the amorphous layer cannot contribute to the photocurrent generation, the short-circuit current is somewhat lower (below 40 mA/cm^2). It should be noted that in this HIT structure, a transparent electrode such as transparent conductive oxide (TCO) that can be produced without heating up is necessary since the top amorphous layer is low conducting and will crystallize with heating up. Room temperature sputtered indium tin oxide (ITO) layers will match this requirement. Care should be taken in producing such TCO layer so as not to disturb the light absorption close to the absorption threshold as described in Figure 3.14.

---

* The most efficient Si-based cell is now the HIT type.[51]

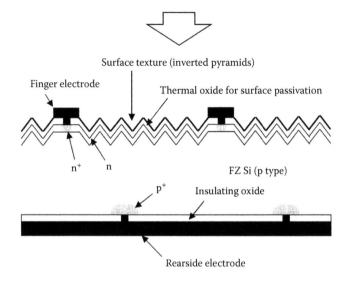

FIGURE 3.12   PERL cell.

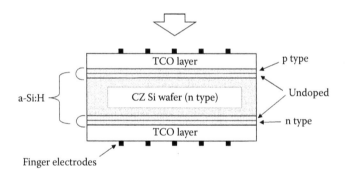

FIGURE 3.13   HIT cell (surface textures are omitted for simplicity).

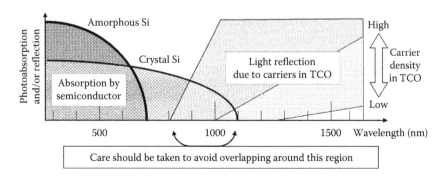

FIGURE 3.14   Consideration on interference between c-Si absorption and TCO transmission.

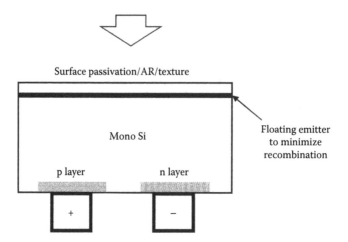

FIGURE 3.15 Back contact cell.

All these cell structures have front-side electrodes of base bars and fingers that will hinder the light incidence to some extent. Increase in the photocurrent, resulting in elevated conversion efficiency, is expected upon removal of these front-side electrodes. Such a concept for PV cell is called back contact,[13] as shown in Figure 3.15. To avoid shadowing would be one reason for high efficiency. However, removing the

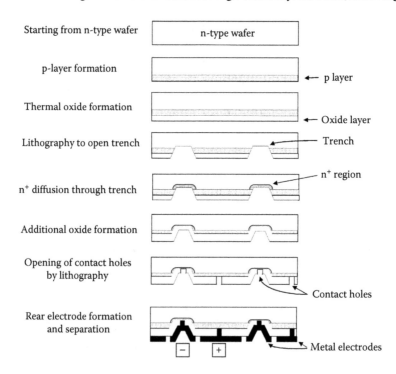

FIGURE 3.16 Rear contact fabrication for back contact cell.

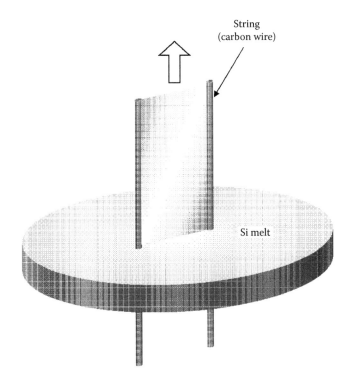

FIGURE 3.17   String ribbon method.

metal contacts from the front side, where photocarrier generation is most abundant, should be the most important feature in achieving high efficiency in this type of PV cell since the metal/semiconductor interface usually works to enhance carrier recombination. In this sense, front-side passivation by heavy doping, usually called floating emitter, is another key issue to achieve a higher efficiency. The somewhat complicated production procedure of a back contact cell is schematically presented in Figure 3.16.[14]

There used to be thin-wafer production techniques called *ribbon methods* in which very thin Si plates are directly produced.[15] These methods were meaningful to avoid kerf loss at wafer slicing especially when the Si materials are quite expensive. However, it has a certain disadvantage in mass production. That is, producing each single wafer one by one takes a lot of time. So the solidification process should be done quite quickly, resulting in a not so good enough crystal quality. In Figure 3.17, the last-survived ribbon method, called string ribbon, is displayed, which had ended commercial production a few years ago.

3.3 THIN-FILM SILICON SOLAR CELLS

In 1975, an epoch-making report was published, which showed that the semiconducting type of n and p in a silicon-based amorphous material can be controlled by impurity doping.[16] Soon after this report, the amorphous material was employed for

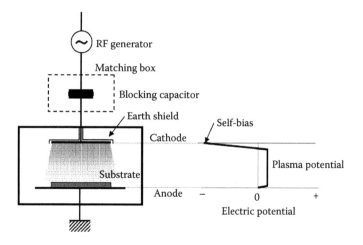

FIGURE 3.18   PECVD reactor. RF—radio frequency.

solar cell application.[17] In spite of large expectations to the potential of this material, light-induced degradation effect (called Staebler–Wronski effect after the reporter) is found,[18] which seems fatal for PV use. After some time, these features are regarded to be originating from hydrogen incorporation in this material. So this material is formally called hydrogenated amorphous silicon, or a-Si:H, although a simple way of amorphous silicon or a-Si is most commonly used alternatively.

Preparation of a-Si is done in the PECVD reactor from monosilane (SiH_4) gas. A schematic representation of a PECVD reactor is shown in Figure 3.18. The thin-film growth is usually performed at the anode side to minimize ion impingement–induced damages. The cathode electrode is self-biased negatively by the presence of a blocking capacitor in the impedance matching circuit attached between the cathode electrode and the power supply unit that is generating a radiofrequency of 13.56 MHz. A drawback in this growth technique is a quite low utilization of start-ing gas material (<5%), which is necessary to keep the quality of film reasonable and uniform over the entire deposited area.

Although both the electrons and holes can transport in this material, their mobil-ity is quite different. Electrons can move quicker than holes by about 1000 times. This feature puts a stiff restriction on the solar cell structure, as schematically explained in Figure 3.19. Carrier generation is most abundant at the light incident side. Meanwhile, one of the carriers only has to move a short distance to the front-side electrode, and the other has to move all the way to the rear electrode. If the latter carrier moves slowly, it may not reach the electrode. This is why a-Si solar cells must have a p-type layer at the light incident side. If only this is reverse, tandem-type cell of quite high efficiency can be achieved with the CIS bottom layer. In addition, poor conductivity in the lateral direction of the top p layer makes it necessary to cover the entire top surface by TCO, commonly made of F-doped tin oxide (FTO). It should also be noted that since doped n and p layers include too much defects, photogen-erated carriers can only survive in the undoped layer. As a result, the p–i–n structure is a must for a-Si-based solar cells.

FIGURE 3.19 Selection of p–i–n or n–i–p upon carrier mobility.

Still, the tandem structure is meaningful to reduce photodegradation (Figure 3.20). Since photodegradation is caused by carrier recombination, a steeper band profile produced by the thinner i-layer, which separates the carriers more quickly, will be useful to reduce photodegradation, although light absorption is also decreased. Texture formations of TCO at the incident side have a complementary effect on light absorption. A tandem structure can also complement this decrease. However, the bottom cell must be thicker than the top cell in order to match the photocurrent, a prerequisite of tandem structures. Instead of simply increasing the thickness of the bottom cell, several attempts to reduce the bandgap of the bottom cell is performed. One way to do this is by adding the Ge (by adding the GeH$_4$ gas into the SiH$_4$ gas) to make the bandgap of the bottom cell narrower. An example is shown in Figure 3.21, which is called a series connection through apertures formed in film (SCAF) module.[19] The other way to avoid the rather expensive GeH$_4$ use is by employing the microcrystal Si (μc-Si) material, as shown in Figure 3.22, which is called a hybrid cell.[20] The μc-Si materials can be made basically by employing

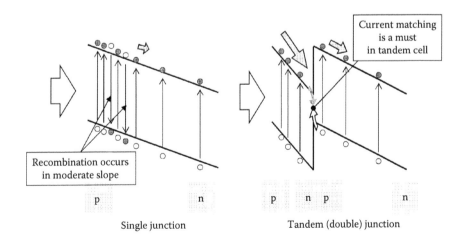

FIGURE 3.20 How tandem structure works to reduce photodegradation.

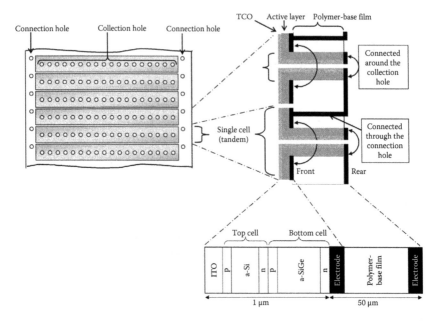

FIGURE 3.21   Details of substrate-type SCAF module structure.

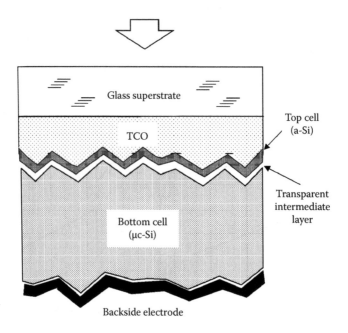

FIGURE 3.22   Amorphous and microcrystalline hybrid cell.

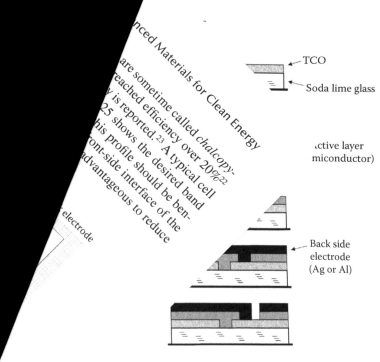

are sometime called chalcopy-
reached efficiency over 20%.
y is reported.[23] A typical cell
25 shows the desired band
his profile should be desired band
ont-side interface of the
advantageous to reduce

...electrode

...TCO

...Soda lime glass

...ctive layer
...miconductor)

Back side
electrode
(Ag or Al)

...monolithic structure formation for thin-film PV module.

...on conditions in the PECVD reactors. In the latter case, an inter-
...low refractive index is inserted between the top and bottom cells
...ct the top-cell transmitted light back. By employing this interlayer,
...ess of the top cell can be further reduced, preferred for the reduction of
...egradation. Such an interlayer should be desired to be resistive to reduction
...oxidation) by hydrogen atoms that are necessary for microcrystal formation. It
...s known that ZnO-based TCO is more resistive to such reduction than FTO. The
doping element Al or Ga is mostly used to increase the conductivity of ZnO, called
AZO or GZO, respectively.

In Figure 3.23, the preparation sequence for the so-called monolithic structure
of a thin-film-type PV module is presented.[21] Among these processes, laser scribing
for cell separation would be most time consuming in mass production, in addition to
the microcrystal growth for the bottom cell of the hybrid type. Since the glass plate
initially employed as the substrate for the film growth will be at the front side of light
incidence in practical use, this module structure is called *superstrate type*. The other
one is called substrate type, as usual.

3.4 CIGS SOLAR CELLS

Because copper indium diselenide ($CuInSe_2$ or CIS) shows strong photoabsorption
due to its direct transition gap nature, it is expected to be a candidate for highly
efficient thin-film solar cells, although the bandgap is almost as short as Si. So it is
quite important to increase the bandgap by substitutional doping of Ga in the lattice
position of In. Such materials are called CIGS. Based on the crystal structure, these

materials (and also this type of PV cells and modules)
rites. Currently, only the CIGS-based solar cells have
in the thin-film category. Quite recently, 20.8% efficienc
structure is shown in Figure 3.24. In addition, Figure 3.
profile in the CIGS layer. Since CIGS is natively p-type,
eficial for the minority carrier (electron) transport. At the f
CIGS layer, however, a slight increase of bandgap would be

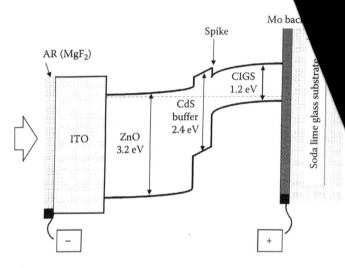

FIGURE 3.24 Basic structure of CIGS cell.

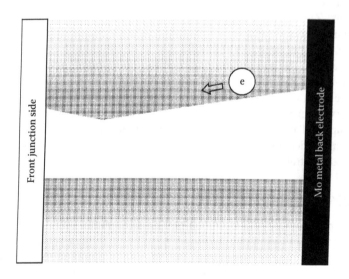

FIGURE 3.25 Desired band profile for the CIGS layer.

carrier recombination. This is also the origin of placing a buffer layer of a wider gap with a spike at the conduction band side.

As these cells are substrate type, a Mo back electrode is sputtered on the soda lime glass substrate at first, as shown in Figure 3.26. It is somewhat mysterious that Na diffusion from the soda lime glass into the CIGS layer through this Mo layer is a must for a high-efficiency cell. Next, the way to fabricate the CIGS layer is categorized into two cases, of small cells and of large modules. For small cell fabrication, a multisource coevaporation technique is employed, and selenization (and sulfuration) is done at the same time, since Se (and possibly S) is evaporated simultaneously. For large modules, on the other hand, metal components of desired compositions are deposited mainly by sputtering, and selenization (and sulfuration) is done afterwards in the ambient of H_2Se and H_2S at elevated temperature. So a precise control of compositions to achieve a desired bandgap profile, such as shown in Figure 3.25, can only be accomplished easily by coevaporation, leading to much higher efficiencies in small cells. This, in return, would be a big issue for large-area module productions. The buffer layer production is usually performed by chemical bath deposition (CBD), which uses a chemical solution to form the layer of desired composition. Although the best performance cell is achieved using a Cd-based buffer layer, attempts to realize a Cd-free structure of high efficiency is a matter of great interest from an environmental viewpoint.

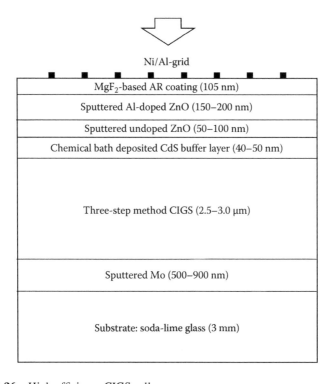

Ni/Al-grid

MgF$_2$-based AR coating (105 nm)
Sputtered Al-doped ZnO (150–200 nm)
Sputtered undoped ZnO (50–100 nm)
Chemical bath deposited CdS buffer layer (40–50 nm)
Three-step method CIGS (2.5–3.0 μm)
Sputtered Mo (500–900 nm)
Substrate: soda-lime glass (3 mm)

FIGURE 3.26 High-efficiency CIGS cell.

3.5 CdTe SOLAR CELLS

It has been quite some time now since CdTe is said to be the best candidate for solar cells due to its best matched bandgap of about 1.5 eV. However, the toxic Cd is problematic to use for solar cells, especially in Japan. The improvement of conversion efficiency had not been so prominent, except for the recent few years. It is quite rapidly increasing now, reaching 19.0%.[24] In addition, its production cost is probably most inexpensive among all the commercialized PV modules. Unfortunately, since the high-efficiency cell structure is not disclosed, a somewhat old one[25] is shown in Figure 3.27. The light-absorbing CdTe layer seems rather too thick in this case, so it would be much suitable to make it thinner for commercial production. The CdTe layer is produced by close-spaced sublimation (CSS), which is quite rapid (takes only a few minutes for a 10 μm absorber growth). In addition, high vacuum is not necessary for CSS. Instead, oxygen-based gas ambient is usually employed. These features are quite advantageous for low-cost production. The window layer of CdS is, like the CIGS case, produced by the CBD method. However, this CBD layer is grown prior to the CdTe layer because this cell structure is the superstrate type.

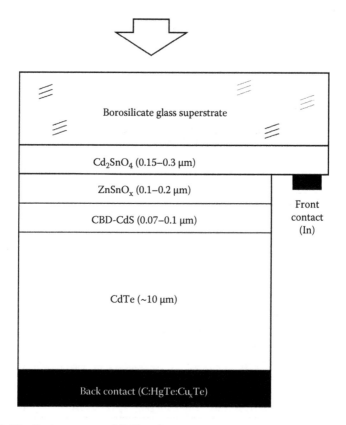

FIGURE 3.27 Basic structure of CdTe cell.

3.6   III–V COMPOUND SEMICONDUCTOR TANDEM SOLAR CELLS

The III–V compound semiconductors generally have superior properties in carrier transport over Si and are a better match to PV applications. A comprehensive diagram[26] of the relationship between the lattice constant and the bandgap is shown in Figure 3.28. Among them, GaAs is the best match to the solar spectrum due to its adequate bandgap of 1.4 eV for single-junction use. Actually, the best conversion efficiency of GaAs single-junction cell is higher than that of Si.[27] Because of its direct transition nature, the thickness necessary (and thus adequate) for a solar cell is much thinner than the thickness commonly used for wafers. So a lift-off technology is employed to fabricate the thin-film-type GaAs cells,[28] which is also beneficial for space use. In addition, the open-circuit voltage is also increased by thinning, as is commonly expected.

Although the conversion efficiency of a GaAs single-junction cell is already superior than the Si cells, the key issue using the III–V compound semiconductors for solar cells is their versatility in composition and thus the band gap, as schematically shown in Figure 3.29. This feature may be quite desirable in the multijunction (tandem) cell design. Note that a semiconductor cannot absorb the light whose energy is below the bandgap. Utilizing this nature, a triple-junction cell made of high-, medium-, and low-energy absorbers is expected to perform a quite high efficiency, as previously shown in Figure 3.5. A well-designed combination of bandgap for these three absorbers is essential for high performance. As shown in Figure 3.30, there are several ways for such choice. Because there is a large *valley* at around 1.4 μm in the

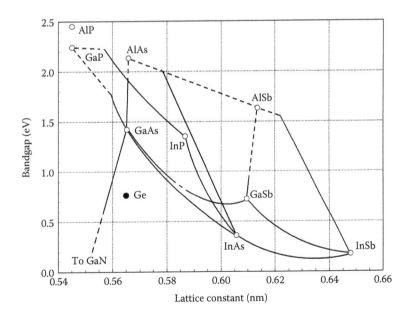

FIGURE 3.28 Relationship between the lattice constant and the bandgap for III–V compound semiconductors.

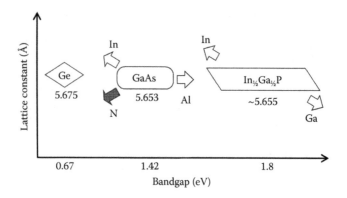

FIGURE 3.29 Schematics on how compositional change affects the bandgap and lattice constant.

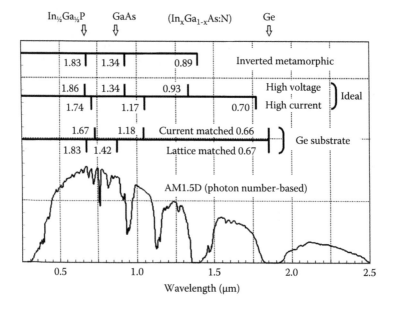

FIGURE 3.30 Bandgap design for triple-junction cells: ideal and practical.

sunlight spectra due to the strong absorption of water molecules in the atmosphere, two ways are available for an ideal combination of these three bandgaps.

One of them utilizes the longer-wavelength region beyond this valley by using the low bandgap material. This choice seems most practical from the viewpoint of fabrication, which employs the Ge wafer for the bottom junction. By chance, it is lucky that the lattice constant of Ge and GaAs is quite similar and thus rather easy to perform heteroepitaxial growth between these two semiconductors. In order to further continue the lattice-matched heteroepitaxy, the top layer composition is selected to be $In_{1/2}Ga_{1/2}P$, whose lattice constant is also close to the underlying two layers (Figure 3.29). However, this choice, denoted as *lattice matched/Ge substrate* in Figure 3.30, has a problematic nature. That is, the number of photons absorbed in the middle layer is far smaller than the other two layers. So the current matching, one of the most important prerequisites in the multijunction solar cell, will not be satisfied. To overcome this problem, the bandgap of the middle layer should be reduced to increase the current generation therein. For this purpose, a metamorphic technology using the buffer layer is necessary since the bandgap–lowered middle layer is no longer lattice matched to the base Ge layer. Figure 3.31 schematically describes how the buffer layer works in the lattice-mismatched growth. Although the thickness necessary for this buffer layer is considerable, it works quite well for high-efficiency tandem cells.[29]

For an ideal bandgap combination, another way, denoted as *high voltage* in Figure 3.30, also requires this buffer technology to achieve the so-called *inverted metamorphic* structure.[30] In this choice, only the sunlight whose wavelength is shorter than the valley is designed to be absorbed. Although this will cause a decrease in the generated current, the bandgap of the bottom junction layer, and thus the total voltage, will be increased in return. So the Ge substrate is not adequate to fabricate this structure. Instead, a GaAs wafer is employed for the starting substrate. At first, the lattice-matched InGaP layer, or the top layer, is grown on the substrate. This is why it is called *inverted*. Next, a little bit modified GaAs layer is grown as the middle junction. Then, the lattice-mismatched bottom layer, whose bandgap is larger than that of Ge, is grown with the aid of the previously mentioned buffering technology.

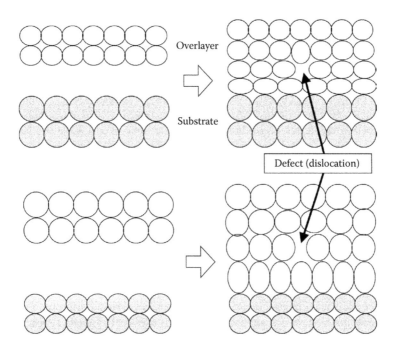

FIGURE 3.31 Schematics on how the buffer layer works in lattice-mismatched heteroepitaxy.

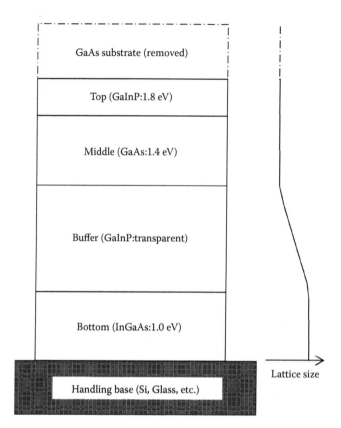

FIGURE 3.32 Schematics of inverted metamorphic cell.

Again in this case, the buffer layer is considerably thick as shown in Figure 3.32. Finally, the starting base of the GaAs substrate is removed (etched out), and the triple-junction cell is transferred onto a holding baseplate, such as a Si wafer or glass plate. Generally speaking, a higher voltage with lower current will be welcomed from the viewpoint of large-scale integration, since Joule loss is reduced by lowering the current. However, this inverted metamorphic process totally sacrifices the valuable GaAs wafer, and it needs another breakthrough to be utilized for practical productions.

The behavior of nitrogen addition to GaAs, as shown in Figure 3.30, is quite exceptional since both the bandgap and lattice constant are decreasing. This is due to a quite large band bowing occurring between GaAs and GaN, wherein the intermediate bandgap profile does not follow the linear interpolation but shifts downwards.[31] A similar behavior, although the shift is smaller, can be seen in other binary combinations shown in Figure 3.29.

Since these triple-junction cells are quite expensive, installation with the optical concentrator system is a must in a terrestrial use. It also needs a quite precise mechanical tracking system to follow the moving sun, called a heliostat. Although such systems are already installed in the world, their outdoor performance is not as good as the previously mentioned cell efficiency.[32] In addition, one should note that a tracking system has a low ground cover ratio, to avoid shadowing, and thus has a lower performance per unit land area compared to nonconcentrating systems.[33]

3.7 EMERGING MATERIALS FOR SOLAR CELLS

3.7.1 METAL SILICIDES

There has been some attention to use metal silicides for solar cells. Typically, β-FeSi$_2$ used to gather much attention because of its strong and wide-range photoabsorption.[34] However, its bandgap, which is narrower than Si, seems too small for a single-junction solar cell, resulting in a not so high efficiency.[35]

Instead, wide-gap silicides, such as BaSi$_2$, are of interest recently. The band gap of BaSi$_2$, which is not larger than that of Si, can be increased by the addition of Sr,[36] although the bandgap of pure SrSi$_2$ is quite small (about 43 meV[37]). Attempts to fabricate high-efficiency solar cells are underway.[38]

3.7.2 PEROVSKITES

Quite recently, there have been intensive works on this new category of solar cell materials, which is called *perovskites*, due to their crystal structure (see Figure 3.33). The key material is CH$_3$NH$_3$PbI$_3$. This inorganic–organic hybrid compound is first employed[39] as a sensitizer in a dye-sensitized solar cell (DSC), focusing on its unique optical properties.[40] That is, this iodide,[41] along with its bromide counterpart,[42] shows strong excitonic absorption near the band edge. The bandgap of about 1.5 eV[43] is also suitable for a solar cell material candidate. Recent reports have prevailed that the TiO$_2$-based DSC structure is no longer necessary for high-efficiency cells exceeding 15%.[44,45] Further increase in conversion efficiency is greatly expected utilizing these new materials. If the excitonic behavior is ruling the conduction mechanism, the so-called bulk heterostructure[46] might be useful, as for the organic semiconductor–based solar cells. Since these compounds are soluble in water, long-term stability against humidity will be a big issue for solar module applications.

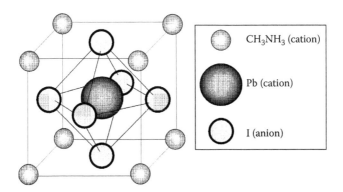

FIGURE 3.33 Crystal structure of CH$_3$NH$_3$PbI$_3$.

3.8 INORGANIC MATERIALS EMPLOYED FOR DYE-SENSITIZED AND ORGANIC SOLAR CELLS

A dye-sensitized cell (Figure 3.34) has three major components, in addition to the common thin-film solar cells: porous TiO_2 layer for electron transport, sensitizer dye for optical absorber, and iodine-based organic redox solution.[47] The use of ZnO instead of TiO_2 seems to be not working suitably in this DSC application probably because of the hole transporting behavior. TiO_2 is said to have a hole blocking nature, while ZnO can easily transport holes judging from its bipolar nature (both n- and p-type conduction are possible).[48] The Ru-based complex dyes, such as N3, N719, and N749, are most commonly used for sensitizers. The redox reaction of $3I^- \leftrightarrow I^- + I_2 + 2e^-$ is carrying the electrons from the back electrode (typically Pt) to the photo-excited and electron-removed dye molecules.

For solar cells made of organic semiconductors, TiO_2 is also used for the electron transporting layer. It is quite interesting that MoO_3 and Cs_2CO_3 are used as the doping material for organic semiconductors to make them p and n types, respectively.[49] Note that in this field of organic solar cells, a hole conductor is called donor and an electron conductor is called acceptor. Other inorganic compounds, such as LiF, are used to improve the transport at the electrode interface.[50]

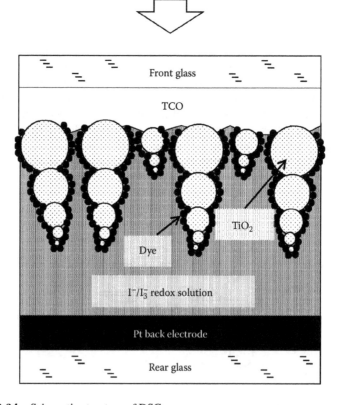

FIGURE 3.34 Schematic structure of DSC.

REFERENCES

1. D. M. Chapin, C. S. Fuller and G. L. Pearson, *J. Appl. Phys.*, 1954, **25**, 676–677.
2. S. Minoura, K. Kodera, T. Maekawa, K. Miyazaki, S. Niki and H. Fujiwara, *J. Appl. Phys.*, 2013, **113**, 063505.
3. C. H. Henry, *J. Appl. Phys.*, 1980, **51**, 4494–4500.
4. M. A. Green, K. Emery, Y. Hishikawa and W. Warta, *Prog. Photovolt.: Res. Appl.*, 2009, **17**, 85–94.
5. W. Shockley and H. J. Queisser, *J. Appl. Phys.*, 1961, **32**, 510–519.
6. Wacker Chemie AG, *US Pat. Appl.*, 2013/0089488, 2013.
7. M. A. Green, *Phil. Trans. R. Soc. A*, 2013, **371**, 20110413.
8. M. M. Hilali, S. Sridharan, C. Khadilkar, A. Shaikh, A. Rohatgi and S. Kim, *J. Electron. Mater.*, 2006, **35**, 2041–2047.
9. S. J. Jeon, S. M. Koo and S. A. Hwang, *Sol. Energy Mater. Sol. Cells*, 2009, **93**, 1103–1109.
10. R. Har-Lavin and D. Cahen, *IEEE J. Photovolt.*, 2013, **3**, 1443–1458.
11. J. Zhao, A. Wang and M. A. Green, *Prog. Photovolt.: Res. Appl.*, 1999, **7**, 471–474.
12. M. Taguchi, A. Yano, S. Tohoda, K. Matsuyama, Y. Nakamura, T. Nishiwaki, K. Fujita and E. Maruyama, *IEEE J. Photovolt.*, 2014, **4**, 96–99.
13. E. V. Kerschaver and G. Beaucarne, *Prog. Photovolt.: Res. Appl.*, 2006, **14**, 107–123.
14. Sun Power Corp., *US Pat.*, 8399287, 2013.
15. M. A. Green, *Solar Energy*, 2003, **74**, 181–192.
16. W. E. Spear and P. G. LeComber, *Solid State Commun.*, 1975, **17**, 1193–1196.
17. D. E. Carlson and C. R. Wronski, *Appl. Phys. Lett.*, 1976, **28**, 671–673.
18. D. L. Staebler and C. R. Wronski, *Appl. Phys. Lett.*, 1977, **31**, 292–294.
19. Fuji Electric Co., *US Pat.*, 623 5982, 2001.
20. Y. Tawada, *Philos. Mag.*, 2009, **89**, 2677–2685.
21. RCA Corp., *US Pat.*, 429 2092, 1981.
22. P. Jackson, D. Hariskos, E. Lotter, S. Paetel, R. Wuerz, R. Menner, W. Wischmann and M. Powalla, *Prog. Photovolt.: Res. Appl.*, 2011, **19**, 894–897.
23. Zentrum für Sonnenenergie und Wasserstoff-Forschung Baden-Württemberg (ZSW), press release, October 24, 2013.
24. M. Gloeckler, I. Sankin and Z. Zhao, *IEEE J. Photovolt.*, 2013, **3**, 1389–1393.
25. X. Wu, *Solar Energy*, 2004, **77**, 803–814.
26. A. Y. Cho, *Thin Solid Films*, 1983, **100**, 291–317.
27. M. A. Green, K. Emery, Y. Hishikawa, W. Warta and E. D. Dunlop, *Prog. Photovolt.: Res. Appl.*, 2013, **21**, 827–837.
28. G. J. Bauhuis, P. Mulder, E. J. Haverkamp, J. C. C. M. Huijben and J. J. Schermer, *Sol. Energy Mater. Sol. Cells*, 2009, **93**, 1488–1491.
29. W. Guter, J. Schöne, S. P. Philipps, M. Steiner, G. Siefer, A. Wekkeli, E. Welser, E. Oliva, A. W. Bett and F. Dimroth, *Appl. Phys. Lett.*, 2009, **94**, 223504.
30. J. F. Geisz, S. Kurtz, M. W. Wanlass, J. S. Ward, A. Duda, D. J. Friedman, J. M. Olson, W. E. McMahon, T. E. Moriarty and J. T. Kiehl, *Appl. Phys. Lett.*, 2007, **91**, 023502.
31. V. Virkkala, V. Havu, F. Tuomisto and M. J. Puska, *Phys. Rev. B*, 2013, **88**, 035204.
32. E. F. Fernández, P. Pérez-Higueras, A. J. G. Loureiro and P. G. Vidal, *Prog. Photovolt.: Res. Appl.*, 2013, **21**, 693–701.
33. S. A. Halasaha, D. Pearlmutter and D. Feuermann, *Energy Policy*, 2013, **52**, 462–471.
34. K. Lefki and P. Muret, *J. Appl. Phys.*, 1993, **74**, 1138–1142.
35. Z. Liu, S. Wang, N. Otogawa, Y. Suzuki, M. Osamura, Y. Fukuzawa, T. Ootsuka, Y. Nakayama, H. Tanoue and Y. Makita, *Sol. Energy Mater. Sol. Cells*, 2006, **90**, 276–282.
36. K. Morita, M. Kobayashi and T. Suemasu, *Jpn. J. Appl. Phys.*, 2006, **45**, L390–L392.

37. M. Imai, A. Sato, T. Kimura and T. Aoyagi, *Thin Solid Films*, 2011, **519**, 8496–8500.
38. T. Suemasu, T. Saito, K. Toh, A. Okada and M. A. Khan, *Thin Solid Films*, 2011, **519**, 8501–8504.
39. A. Kojima, K. Teshima, Y. Shirai and T. Miyasaka, *J. Am. Chem. Soc.*, 2009, **131**, 6050–6051.
40. G. C. Papavassiliou, G. A. Mousdis and I. B. Koutselas, *Adv. Mater. Opt. Electron.*, 1999, **9**, 265–271.
41. M. Hirasawa, T. Ishihara and T. Goto, *J. Phys. Soc. Jpn.*, 1994, **63**, 3870–3879.
42. K. Tanaka, T. Takahashi, T. Ban, T. Kondo, K. Uchida and N. Miura, *Solid State Commun.*, 2003, **127**, 619–623.
43. J. H. Noh, N. J. Jeon, Y. C. Choi, Md. K. Nazeeruddin, M. Graetzel and S. I. Seok, *J. Mater. Chem. A*, 2013, **1**, 11842–11847.
44. J. Burschka, N. Pellet, S.-J. Moon, R. Humphry-Baker, P. Gao, M. K. Nazeeruddin and M. Graetzel, *Nature*, 2013, **499**, 316–319.
45. M. Liu, M. B. Johnston and H. J. Snaith, *Nature*, 2013, **501**, 395–398.
46. G. Yu, J. Gao, J. C. Hummelen, F. Wudl and A. J. Heeger, *Science*, 1995, **270**, 1789–1791.
47. M. Graetzel, *Inorg. Chem.*, 2005, **44**, 6841–6851.
48. J. C. Fan, K. M. Sreekanth, Z. Xie, S. L. Chang and K. V. Rao, *Prog. Mater. Sci.*, 2013, **58**, 874–985.
49. M. Kubo, Y. Shinmura, N. Ishiyama, T. Kaji and M. Hiramoto, *Appl. Phys. Express*, 2012, **5**, 092302.
50. C. J. Brabec, S. E. Shaheen, C. Winder, N. S. Sariciftci and P. Denk, *Appl. Phys. Lett.*, 2002, **80**, 1288–1290.
51. M. A. Green, K. Emery, Y. Hishikawa, W. Warta and E. D. Dunlop, *Prog. Photovolt.: Res. Appl.*, 2014, **22**, 701–710.

4 Development of Thermoelectric Technology from Materials to Generators

Ryoji Funahashi, Chunlei Wan,
Feng Dang, Hiroaki Anno, Ryosuke O. Suzuki,
Takeyuki Fujisaka, and Kunihito Koumoto

CONTENTS

4.1   INTRODUCTION

4.1.1   Energy and Environment Crisis

We have been relishing a lot of affluence thanks to energy. Fossil energy provides us fun to drive, warmth to escape from cold, brightness of illumination, etc. However, drying up of oil is feared recently. Moreover, consumption of fossil fuel produces carbon dioxide. Carbon dioxide is considered as a greenhouse gas and emitted from many places, for example, power generation, industry, transport, agriculture, and waste disposal. The amount of carbon dioxide emission will increase with increasing consumption of fossil energy, gas, oil, and coal year by year (Figure 4.1).[1] It is predicted that the amount of CO_2 emission in 2030 will reach almost twice as much as in 1990.

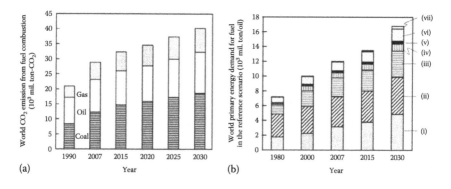

FIGURE 4.1   World CO_2 emissions from fuel combustion by sector in the reference scenario (a) and world primary energy demand in the reference scenario (b) for coal (i), oil (ii), natural gas (iii), nuclear (iv), hydro (v), biomass and waste (vi), and other renewable sources (vii).

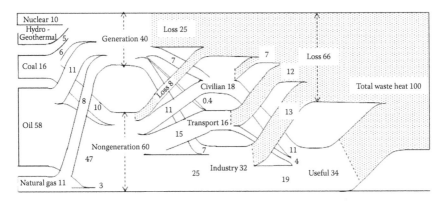

FIGURE 4.2   Energy flow in Japan in 1992.

We have been using a huge amount of energy and then are facing serious problems in energy and the environment. The demand of primary energy in the world was as much as 12,013 million tons of oil per year in 2007 (Figure 4.2).[1] It has spent a long time to appeal using up the oil. All oil deposits are estimated to last for about 40 years in the present consumption rate. But the amount of oil consumption is increasing.

4.1.2   Unutilized Heat Energy

The primary energy demand in the world reached about 12,730 Mtoe (million ton of oil equivalent) in 2010.[2] The average of total utilizing primary energy efficiency is as low as 30%, with 70% exhausted to the air as waste heat,[3] as shown in Figure 4.2. It is clear that improved efficiencies of energy conversion systems could have a significant impact on energy consumption and carbon dioxide emission rate. Where a large sum of heat is localized, mechanical conversion systems, such as steam turbine or Stirling engine, can be used to generate electricity. However, most sources of waste heat are widely dispersed. Although technologies of storage and transport of dilute heat energy have been developed, most waste heat cannot be used effectively. Electricity is a convenient form of energy that is easily transported, redirected, and stored, and thus there are a number of advantages to the conversion of waste heat emitted from our living and industrial activities to electricity. Thermoelectric (TE) conversion is regarded as the strongest candidate to generate electricity from dilute waste heat.

Waste heat exists as a solid or fluid (exhaust gas and cooling water) with a widely spread temperature region of about 320–1300 K. The industry-classified amounts of waste energy from gas, water, and solid are shown for each temperature region in Figure 4.3.[4] The gaseous waste heat exhausted from mainly chimneys has a temperature region from 373 to 773 K. And the amount of energy is larger than from water and solid. The electric, chemical, and ceramic industries are producing a large amount of waste heat energy. The second-largest shape for waste heat is water. The maximum temperature is limited up to 373 K because of the boiling of water.

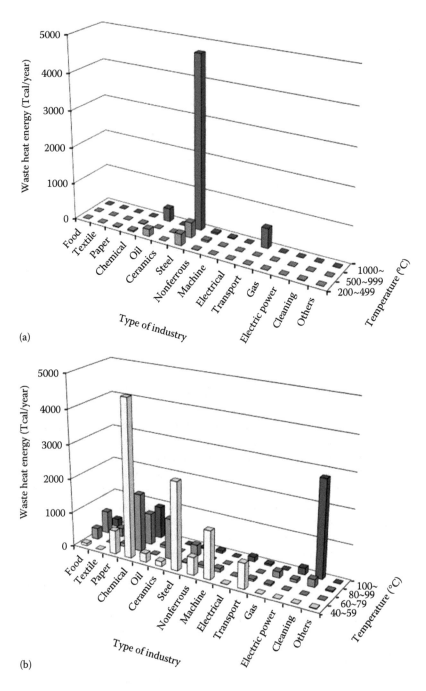

FIGURE 4.3 Industry-classified amounts of waste heat energy in (a) solid and (b) water.

(*Continued*)

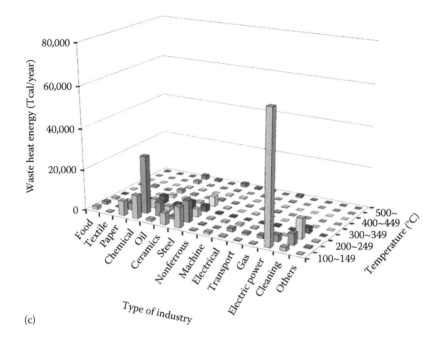

(c)

FIGURE 4.3 (Continued) Industry-classified amounts of waste heat energy in (c) gas forms for each temperature.

The amount of energy with about 343 K is especially outstanding. Although almost all types of industries are producing hot water waste, the chemical, iron, and machine industries are exhausting the largest amount of waste heat energy from the coolant. Heat energy with a temperature higher than 773 K is radiated from the molten iron. As mentioned earlier, the waste heat covers the wide temperature region with different shapes and is dispersed extensively. In other words, the amount of waste heat energy produced at one place is not large. Electrical generation by turbines using such a small amount of waste heat cannot be carried out or can be carried out with low conversion efficiency. Moreover, waste heat from automobiles is increasing by the diffusion of the automobiles. The automobile entry exceeded a billion in 2010.[5] Since the amount of waste heat from each automobile is small, a huge amount of thinly dispersed waste heat energy by automobiles is not utilized.

In a few decades, photovoltaic (PV) generation has been developed and is a widely spread application. The amount of solar energy is divided equally into two kinds: ultraviolet (UV) and visible light and infrared light. PV generation utilizes UV and visible light. Infrared is not used widely for electrical power generation, but only thermal exchanging to produce hot water. In order to develop technology for thermal condensation and storage, many people started to pay attention in using solar heat energy. The so-called *sunbelt* regions, which exist in North Africa, western United States, Central Australia, and Central Asia, get a large amount of infrared energy. It is rewarding to use solar heat for thermal exchange only but to *heat for electricity*. Electricity can be transported easily and can be stored for a long time. Large-scale power generation using steam turbines has been operating in Mojave Desert, United States. The cost of maintenance of the infrastructure and transportation of electricity is high for centralized systems at a large scale.

For widespread *sustainable* solar energy application, the distributable systems for power generation at a small scale and low cost are indispensable. So far the technology for solar heat condensation can be developed to enhance the temperature up to 873 K. TE generation using the Seebeck effect is expected for solar heat application; independent of the energy scale, this conversion method should be suitable to solar heat power generation.

4.1.3 Thermoelectric Generation

TE generation uses the Seebeck effect in which a voltage is induced between the ends of an electrical conducting material that has a temperature gradient. The voltage is caused by the diffusion of electrons away from the hot end of the material toward the cold end. As a result the distribution of electrons varies along the length of the material, thereby generating a voltage between the ends of such material. At present, the average absolute values of the Seebeck coefficient of good TE materials are in the range of 150–250 μV/K. Therefore, even when a temperature difference of 500 K is applied between the ends of one TE material, it can generate about 0.15 V at most. In order to get a high output voltage, many TE legs must be connected electrically series and thermally in parallel. In both p- and n-type devices, the former has a higher voltage at the hot side and the latter at the low-temperature side. The integrated objects composed of the p- and n-type devices are called modules.

Strong points and advantages of TE generation are shown in Table 4.1. The conversion efficiency is dependent on the temperature gradient but independent on the scale of heat energy. No carbon dioxide or isotope substances are emitted. Moreover, maintenance is not necessary because there are no moving parts in this generator. This was successfully demonstrated in NASA's Mars Science Laboratory mission.[6]

TABLE 4.1
Strong Points and Advantages of Thermoelectric Conversion

Strong Points	Advantages
Independence of heat amount	Recovery of dilute waste heat
No moving parts	No maintenance, long lifetime
	Low cost, less noise, no vibration
Power generation by temperature difference	Solar energy, geothermal energy, body heat, no emission, simple structure
High power density (W/cm², W/kg)	Automobile, ship, portable devices

A dimensionless TE figure of merit ZT is used for the assessment of TE materials. ZT is calculated as follows:

$$ZT = \frac{S^2 T}{\rho\kappa} \tag{4.1}$$

Here, S, T, ρ, and κ indicate the Seebeck coefficient, absolute temperature, electrical resistivity, and thermal conductivity, respectively. Better materials have higher ZT values. That is, good materials show large S, low ρ, and low κ. However, these three factors are dominated by charge carrier density and three-way deadlock.

Because the TE conversion efficiency for materials reaches about 8%–15% at an average $ZT = 1$, this value is an immediate goal for power application. Temperature dependence of ZT values are indicated in Figure 4.4 for several TE materials.[7–10] ZT values for all materials greatly depend on the temperature. The materials should be used in the temperature region, in which ZT values reach maximum. Application in a wide temperature region demands multiple TE materials. Bi_2Te_3 is an excellent material from room temperature to about 473 K. For high-temperature applications, oxides are promising materials because of their oxidation resistance and high melting temperature. On the other hand, almost all materials shown in Figure 4.5 are oxidized at temperatures of maximum ZT values. New materials with oxidation resistance at the middle-range temperature should be developed for unutilized heat

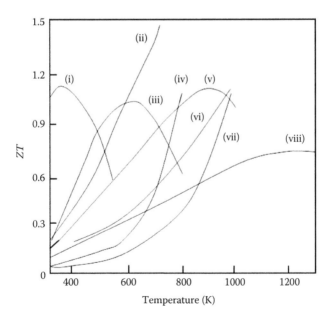

FIGURE 4.4 Temperature dependence of dimensionless TE figure of merit ZT for $Bi_{0.5}Sb_{1.5}Te_3$ (i), Zn_4Sb_3 (ii), PbTe alloy (iii), $NaCo_2O_4$ (iv), $CeFe_{3.5}Co_{0.5}Sb_{12}$ (v), $Ca_3Co_4O_9$ (vi), $Bi_2Sr_2Co_2O_9$ (vii), and SiGe alloy (viii).

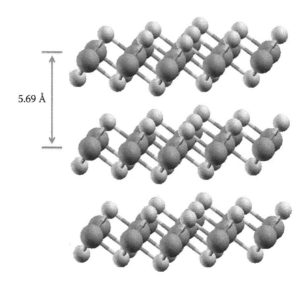

5.69 Å

FIGURE 4.5   Crystal structure of layered TiS$_2$. (Reproduced from Zhang, R.Z. et al., *Phys. Chem. Chem. Phys.*, 14, 15641, 2012. With permission from the PCCP Owner Societies.)

energy as mentioned earlier. Research on new TE materials is aimed at high *ZT* composed of abundant elements, development of TE modules using such materials, and design of TE cascade units, which are carried out by the Core Research for Evolution Science and Technology (CREST) project of the Japanese Science and Technology Agency (JST), *Development of Highly Efficient Thermoelectric Materials and Systems* from 2008 to 2014.

4.2   THERMOELECTRIC PROPERTIES OF 2D TITANIUM SULPHENE

4.2.1   INTRODUCTION

Dimension reduction has become a novel route for enhancing the TE performance of structures such as nanowire[11] or even nanosheets.[12] Environment-friendly TiS$_2$, consisting of abundant and safe chemical elements, has been shown to be a promising TE material[13,14] with a power factor as high as the conventional TE material, Bi$_2$Te$_3$. The two-dimensionality of the layered TiS$_2$ is of most interest in further enhancing the TE performance.

4.2.2   THEORETICAL CALCULATION

Dimension reduction has become a novel route for enhancing the TE performance of structures such as nanowire[11] or even nanosheets.[12] Environment-friendly TiS$_2$, consisting of abundant and safe chemical elements, has been shown to be a promising TE material[13,14] with a power factor as high as the conventional TE material, Bi$_2$Te$_3$. The two-dimensionality of the layered TiS$_2$ is of most interest in further enhancing the TE performance.

TiS$_2$ crystallizes in the trigonal CdI$_2$ structure (space group $P\bar{3}m1$) and consists of S–Ti–S sandwiches, separated in the z direction by the van der Waals gap, as illustrated in Figure 4.5. In order to model the TiS$_2$ monolayer, we used a super-cell consisting of two parts: a nanosheet (monolayered or multilayered) that has the same crystal structure as the bulk and a vacuum layer with a thickness of 22.76 Å. This 22.76 Å thickness was well tested to make sure that the interaction between two neighboring nanosheets can be neglected. Next, the atomic positions were fully relaxed. The bond lengths between Ti and S atoms in the monolayer were found to be nearly the same as those of the bulk and no surface reconstruction was found, indicating a smooth surface. Density functional theory (DFT) calculations were then performed using the package Quantum ESPRESSO.[15]

Figure 4.6 shows the calculated Seebeck coefficient of a monolayer TiS$_2$ nanosheet (S_{2D}) and a bulk single crystal (S_{3D}) at room temperature. A large increase in S_{2D} compared with S_{3D} can be seen. The experimental data of single crystals are also shown for comparison. The discrepancy between the calculated results and the experimental data is small (<5%), which verified our calculation methods and indicated that our calculated results are reliable. When the carrier concentration is as high as 10^{21} cm^{-3}, S_{3D} is fairly large, about −120 μV/K, which is beneficial for good TE performance, whereas S_{2D} is even larger, about −180 μV/K, corresponding to 40% enhancement compared with S_{3D}. Also, both S_{2D} and S_{3D} decrease with increasing carrier concentration and S_{2D} is always about 40% larger than S_{3D}.

The enhancement in S_{2D} can be understood by analyzing the density of states (DOS) of monolayer and bulk TiS$_2$, both plotted in Figure 4.7. Near the conduction band minimum (CBM), the DOS of monolayer TiS$_2$ is much larger than that of the bulk. This is because electronic quantum confinement occurs in monolayer TiS$_2$,

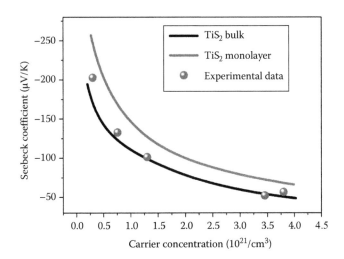

FIGURE 4.6 Calculated Seebeck coefficients of TiS$_2$ bulk and monolayer nanosheets as a function of carrier concentration at room temperature. (Reproduced from Zhang, R.Z. et al., *Phys. Chem. Chem. Phys.*, 14, 15641, 2012. With permission from the PCCP Owner Societies.)

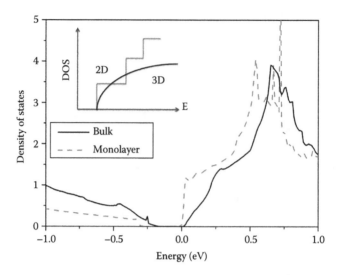

FIGURE 4.7 DOS of TiS$_2$ bulk and monolayer nanosheets. The inset shows the DOS for an ideal 2D system and an ordinary 3D system. (Reproduced from Zhang, R.Z. et al., *Phys. Chem. Chem. Phys.*, 14, 15641, 2012. With permission from the PCCP Owner Societies.)

that is, electrons are confined in the 2D layer and are unable to move along the third direction. Consequently, the electronic structure is very different from that of the bulk, where electrons can move freely in all three directions.[16] To clarify this, the DOS of an ideal 2D system is also shown in the inset of Figure 4.7.

Near the CBM, the step in the DOS of monolayer TiS$_2$ is quite similar to that of an ideal 2D system, which is expected for electronic confinement and indicates the 2D nature of the monolayer. The carrier effective mass of monolayer TiS$_2$ (m_{2D}^*) was also calculated from DOS. The value of m_{2D}^* is $5.5m_0$ (m_0 is the mass of a free electron), much larger than that of the bulk ($2.45m_0$); this favors a large Seebeck coefficient. Also from Figure 4.7, the bandgap of the monolayer (0.3 eV) is found to be larger than that of the bulk (0.18 eV). Furthermore, the calculated bandgap 0.18 eV is smaller than the experimental value (0.3 eV) of the TiS$_2$ single crystal, because the exchange–correlation functional used in DFT calculation tends to underestimate the bandgap.

4.2.3 EXPERIMENTAL RESULTS

TiS$_2$ nanosheets were prepared through a chemical exfoliation method (Figure 4.8). Firstly, TiS$_2$ single crystal was electrochemically intercalated with organic cations, such as hexylammonium. TiS$_2$ single crystal was set as the cathode and the platinum sheet was set as the anode. For the electrolyte, organic cations such as hexylammonium chloride were dissolved in an organic solvent such as dimethyl sulfoxide (DMSO). When an electrical potential is applied, part of the Ti^{4+} in TiS$_2$ is reduced to Ti^{3+} and the TiS$_2$ layers are negatively charged. The organic cations in the electrolyte

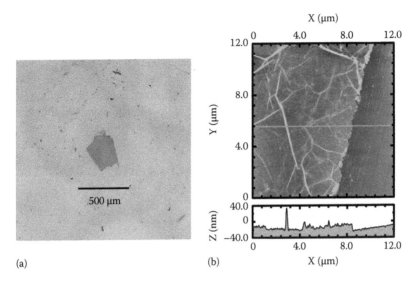

FIGURE 4.8   (a) Optical microscope and (b) AFM images of a TiS$_2$ nanosheet.

solution were then intercalated into the van der Waals gap, driven by the Coulomb force. In the electrolyte solution the organic cations were surrounded by polar solvent molecules (DMSO) as a result of the solvation effect. They are then co-intercalated into the van der Waals gap of the TiS$_2$, together with the organic cations. The hybrid TiS$_2$/organic material was then treated with highly polar organic molecules, such as formamide, which can combine with the organic cations through the cation–dipole effect. The hybrid material then swelled significantly, compared with the original TiS$_2$ single crystal and was finally exfoliated into blue nanosheets with a typical size of 500 μm. Atomic force microscope (AFM) is used to confirm that the TiS$_2$ nanosheet has a thickness of 15.4 nm, about 27 monolayers (Figure 4.8).

The Seebeck coefficient was measured to be −74.9 μV/K, which was reduced compared with that of the pristine TiS$_2$ single crystal (−166 μV/K). Hall measurement showed that the carrier concentration increased from 3.68×10^{20} to 2.75×10^{21} cm^{-3} as a result of the reduction of TiS$_2$ (from Ti^{4+} to Ti^{3+}) during the electrochemical reaction process. As is known, the Seebeck coefficient S is determined by the following equation[17]:

$$S = \frac{8\pi^2 k_B^2}{3eh^2} m^* T \left(\frac{\pi}{3n} \right)^{2/3} \tag{4.2}$$

The effective mass m^* is then calculated to be $7.31 m_0$, which is improved compared with that of TiS$_2$ single crystal ($4.18 m_0$). Although there are differences between the experimental values and the calculation values in the effective mass, this result confirms that 2D gas tends to form in the thin TiS$_2$ nanosheets and enhance the DOS and the effective mass.

4.2.4 SUMMARY

The electronic structure of the TiS$_2$ nanosheet was investigated through DFT calculation. With decreasing thickness of the TiS$_2$ nanosheet, 2D behavior was observed, which can increase the DOS at the CBMs. The Seebeck coefficient is therefore enhanced due to its 2D nature. We also used the chemical exfoliation method to prepare large TiS$_2$ nanosheets with several hundred micrometer size, and the effective mass was enhanced, which can contribute to a higher Seebeck coefficient.

4.3 THERMOELECTRIC PERFORMANCE OF SrTiO$_3$ ENHANCED FROM PARTICULATE FILM OF NANOCUBES

4.3.1 INTRODUCTION

Strontium titanate (SrTiO$_3$, abbreviated as STO) is a perovskite-type material with a cubic crystal structure. Its basic band structure consists of highly populated O-p- and Ti-e$_{2g}$-states with a bandgap Eg ~ 3.2 eV. STO is an important electrical material showing various properties, such as dielectricity, piezoelectricity, and ferroelectricity, and has been investigated as a photocatalyst, such as in the oxidation of hydrocarbons and H$_2$ generation and reduction. As a TE material, STO exhibits a high Seebeck coefficient (S) owing to a large DOS effective mass ($m^* = 6 - 15m_0$).[18,19] On the other hand, STO has a high thermal conductivity at low temperatures (7–11 W/m K at 300 K) owning to its crystal structure with strong bonded lightweight atoms in highly symmetric structure.

Heavily doped STO is considered a promising n-type TE oxide. The doping of heavy elements in the STO structure can increase the phonon scattering effect to decrease the thermal conductivity. STO is a band insulator, and the heavy elements can be introduced in the Sr site for trivalent cations A^{3+} (La, etc.) or the Ti site for pentavalent cations (B^{5+}, Nb, etc.) precisely under reducing atmosphere with controlled density of free carriers. Now the highest ZT value of Nb-doped STO is 0.37 at 1000 K for an epitaxial thin film of 20 at.%.[19] Another way to decrease the thermal conductivity was to prepare the nanoceramics. A lower thermal conductivity in bulk polycrystalline STO with a smaller grain size was identified in our work (Figure 4.9).[20] Through enhanced phonon scattering at grain boundaries, the thermal conductivity would decrease to 2 W/m K when the grain size decreases to 10 nm.

4.3.2 2D ELECTRON GAS

In 1993, Hicks and Dresselhaus theoretically predicted the quantum well effect to enhance the Seebeck coefficient of TE materials.[21] The Seebeck coefficient of TE materials will be enhanced significantly by an increase in DOS near the conduction band edge when the carrier electrons are confined from 3D to 2D space.

We prepared Nb-STO epitaxial film with 2D superlattice to obtain high TE performance. The 2D superlattice consisted of a 20% Nb-doped layer with several unit cells thick sandwiched in pure STO barrier layers. A drastic increase in |S| was identified when the thickness of the Nb-doped layer is less than four unit cells of STO

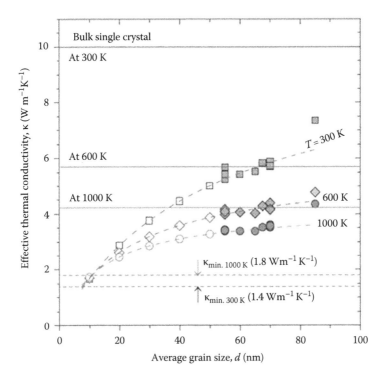

FIGURE 4.9 Variation of effective thermal conductivity of the nanograined STO ceramics with the average grain size at different temperatures.

as shown in Figure 4.10.[22] The $|S|$ at room temperature reached ~480 μV/K, which is ~4.4 times larger than that of $SrTi_{0.8}Nb_{0.2}O_3$ bulk materials. As a result, the ZT value of the Nb-doped STO layer reached an extremely large value of 2.4 when using the k of STO single crystal. On the other hand, the ZT value of the superlattice composed of the Nb-doped layers and STO barrier layers was only 0.1.

Motivated by the excellent properties of 2D electron gas (2DEG) in the STO super-lattice, we proposed a 3D superlattice model to enhance the ZT value of STO bulk materials. The ZT value of bulk STO is enhanced by changing the 2DEG phase into a 3D network. The 3D superlattice model consisted of periodically close-packed STO nanocubes as shown in Figure 4.11.[23] The STO nanocube is designed as a core–shell structure. The core is an La-doped STO phase and the shell is an Nb-doped STO surface layer. Through the close-packed structure of STO nanocubes, a large amount of interface between the nanocubes is introduced into the 3D superlattice ceramics. Three effects can be predicted from the interface. The first is the Nb-doped surface acting as the 2DEG phase with a lowered CBM compared to that of the La-doped phase. The second is the energy filtering effect between the nanocubes. The third is the phonon scattering effect to reduce thermal conductivity due to lattice (κ_{latt}). Theoretical simulation has shown that the ZT value increases when the thickness of the Nb-doped layer decreases. The ZT value, ~1.2, at 300 K was predicted for the 3D superlattice bulk STO in the best case.

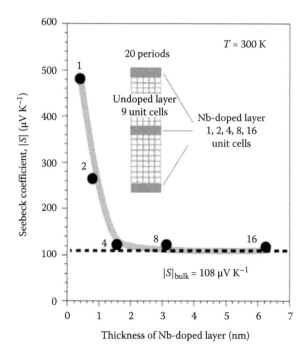

FIGURE 4.10 Seebeck coefficient ($|S|$) at 300 K of $SrTiO_3/SrTi_{0.8}Nb_{0.2}O_3/SrTiO_3$ superlattice with different thicknesses of the Nb-doped layer.

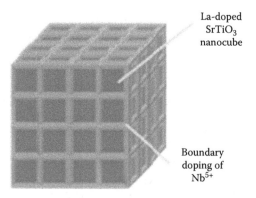

FIGURE 4.11 3D superlattice structure of STO ceramics.

The 3D superlattice STO ceramic is thought to be a revolutionary effort for the development of practical TE materials. To realize this model, a synthesis technique of nanocubes, a self-assembled process, is needed. Now we have successfully prepared La-doped STO nanocubes with Nb surface doped and developed a self-assembled process to prepare a particulate film.

4.3.3 Synthesis of Nanocubes and Their Self-Assembly

La-STO nanocubes were synthesized by using a six-coordinated Ti precursor (TALH).[24,25] Oleic acid (OLA) was used to control the morphology of STO nanocubes and hydrazine was used to accelerate the formation of the STO phase. The selection of TALH encouraged OLA to selectively adsorb on the Sr-rich (100) face of STO leading to the formation of nanocubes. Monodispersed La-STO nanocubes (La, 5%) with a size of ca. 15 nm were obtained as shown in Figure 4.12. A Nb precursor was added to the intermediate phase composed of a nanocube and unreacted Ti-based gel. The Nb precursor dissolved under high pH conditions and entered the Ti-based gel uniformly. The unreacted Ti-based gel transformed into the Nb-doped surfaced of La-STO nanocubes finally. The thickness of the Nb-doped surface was identified to be 1.7 nm as shown in Figure 4.13.[26]

The particulate film was prepared by evaporating toluene of the dispersion on the Si substrate under UV irradiation. The particulate films of La-STO nanocubes appeared yellow, blue, and purple, corresponding to different thicknesses. The UV irradiation caused decomposition of the organic phases (OLA and residual toluene), facilitating assembly of the nanocubes into a self-assembled film. The in situ decomposition of the organic phase resulted in good contacts between nanocubes and a mechanically robust self-assembled particulate film. A smooth surface with a roughness smaller than 50 nm was obtained for the particulate film as shown in Figure 4.14.[27] Scanning electron microscope (SEM) images (Figure 4.14) of the particulate film also indicated a highly uniform structure with no cracks. All the cube faces were 100 planes. Only 200 peaks were identified in the x-ray diffraction (XRD) pattern of the self-assembled film. It firmly indicates that the nanocubes aggregated epitaxially with each other. Supposing that the nanocubes have uniform size and

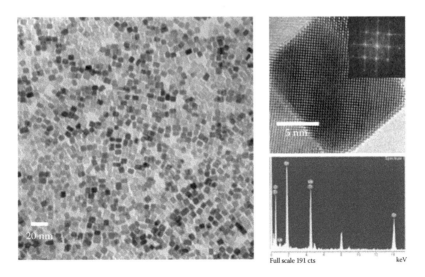

Full scale 191 cts   keV

FIGURE 4.12   TEM and high resolution transmission electron microscopy images (La 5%) and EDX analysis of the La-SrTiO₃ nanocubes.

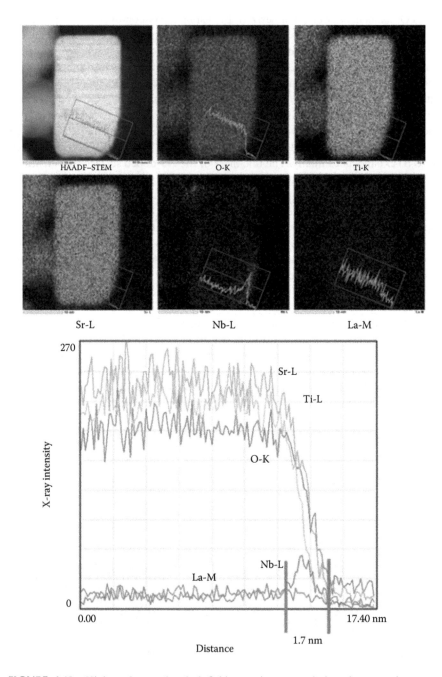

FIGURE 4.13 High-angle annular dark-field scanning transmission electron microscopy image and EDX analysis of Nb/La-STO nanocubes.

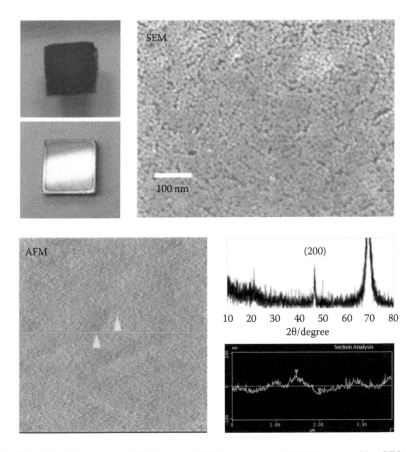

FIGURE 4.14   Photograph, SEM images, XRD pattern, and AFM images of La-STO particulate film.

the self-assembled particulate film has the perfect layer structure, the relative density of the self-assembled particulate film was estimated by counting the number of nanocubes in a given area (SEM image in Figure 4.14). The relative density (packing ratio) of the self-assembled particulate film was estimated to be 78%. The packing ratio (78%) of nanocubes in the self-assembled particulate film exceeded that (68%) of a body-centered cubic and slightly larger than that (74%) of a face-centered cubic (FCC) structure composed of spherical particles.

4.3.4   HUGE THERMOELECTRIC POWER FACTOR

To obtain TE properties, the Nb/La-STO particulate films were calcined under a reducing atmosphere to generate conduction electrons through oxygen release. A drastic increase of Seebeck coefficient was obtained for the Nb/La-STO particulate film. The largest $|S|$ is ~470 μV/K at 380 K. The electrical conductivity of particulate film also increased significantly through Nb surface doping as shown in Figure 4.15. Electrical conductivity decreases and the absolute Seebeck coefficient

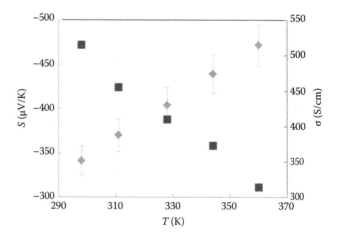

FIGURE 4.15 Seebeck coefficient and electrical conductivity of Nb/La-STO particulate film.

increases with increasing temperature from 293 to 353 K, which indicates that this particulate film is a typical n-type degenerate semiconductor. A rather high electrical conductivity observed might have been due to the presence of 2DEG grain boundary layers. A giant Seebeck coefficient may have been generated in two ways: 2DEG, which played a role in the electronic transport along the grain boundaries,[22] and the energy filtering effect across the grain boundaries.[28]

4.3.5 SUMMARY

Although thermal conductivity of a particulate film has not been measured precisely because the measuring technique is not yet well established, the present results firmly indicate that the 3D superlattice with nano length-scale structures that can give rise to quantum effects, such as quantum confinement effect and energy filtering effect, is an effective structure to generate high TE performance. In order to realize the goal of 3D superlattice ceramics, the self-assembly process to obtain a truly close-packed structure of nanocubes should be further developed.

4.4 THERMOELECTRIC SILICON CLATHRATES

4.4.1 INTRODUCTION

Group 14 (Si, Ge, and Sn)-based compounds with a type-I clathrate structure,[29] cubic unit cell, and space group $Pm\overline{3}n$, as shown in Figure 4.16,[30,31] have received considerable attention as novel materials[31–68] because of the potential applicability of the phonon glass and electron crystal (PGEC) concept.[69,70] In the type-I clathrate crystal structure, there are 54 atoms per primitive unit cell. The framework of the structure is formed by covalent tetrahedrally bonded atoms, Si, Ge, Sn, Al, Ga, or In, or some of the transition elements (6c, 16i, and 24k sites). The electronic structure

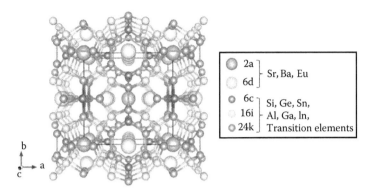

	2a ⎤	
	6d ⎦	Sr, Ba, Eu
	6c ⎤	Si, Ge, Sn,
	16i ⎥	Al, Ga, In,
	24k ⎦	Transition elements

FIGURE 4.16   Type-I clathrate crystal structure.

and the carrier transport properties are governed mainly by the framework.[35,37,43] Sr, Ba, or Eu guest atoms (2a and 6d sites) that loosely bond to the host structure are incorporated into the voids inside two different polyhedra that are connected to each other by these shared faces: two pentagonal dodecahedra and six tetrakaideca-hedra (12 pentagonal and 2 hexagonal faces) per cubic unit cell. The thermal vibra-tion *rattling* of guest atoms strongly scatters heat-carrying phonons, leading to the extremely low lattice thermal conductivity.[32,33,38,45,50]

Some properties of semiconductor clathrates are summarized in Table 4.2.[29] Among clathrates, silicon-based clathrates, especially the $Ba_8Al_xSi_{46-x}$ system, have received considerable attention because of significant advantages from the practi-cal points of view. Aluminum and silicon are nontoxic and quite abundant and thus less expensive than gallium, germanium, or tin. For example, world quantities of production for Ba, Al, and Si are larger than those for Sr, Ga, Ge, and Sn, as shown in Table 4.3.[71] Furthermore, high thermal stability and wide operation temperature are expected due to the high melting point in silicon-based clathrates, as noted in Table 4.2. Thus, crystallographic and TE properties have been extensively investi-gated for silicon-based clathrates.[31,39,40,49,61–63,66,67,72–75]

TABLE 4.2
Some Properties of Semiconductor Clathrates

Compound	Type	p, n	Lattice Constant, a (nm)	Density, ρ (g/cm³)	Melting Point, T_m (K)
$Sr_8Ga_{16}Si_{30}$	I	n	1.0460	3.86	1311
$Ba_8Al_{16}Si_{30}$	I	n	1.0607	3.30	1357
$Ba_8Ga_{16}Si_{30}$	I	n	1.0530	4.34	1384
$Sr_8Ga_{16}Ge_{30}$	I	n	1.0734	5.36	1073
$Ba_8Al_{16}Ge_{30}$	I	n	1.0835	4.84	1244
$Ba_8Ga_{16}Ge_{30}$	I	p, n	1.0767	5.84	1230
$Ba_8Ga_{16}Sn_{30}$	VIII	p, n	1.1595	6.15	723

TABLE 4.3

World Quantities of Production in 2010 for Mineral Resources for Clathrate Constituents

Element	Quantity of Production (10^3 ton)
Aluminum, Al	40,811
Silicon, Si	6,900
Barium, Ba	6,900
Tin, Sn	676
Strontium, Sr	405
Gallium, Ga	0.161
Germanium, Ge	0.118

Source: Mineral Resources Report in 2011, Japan Oil, Gas and Metals National Corporation (JOGMEC), http://mric.jogmec.go.jp/mric_search/ (in Japanese).

In this study, we report the TE properties of some polycrystalline silicon clathrates, $Ba_8Al_xSi_{46-x}$ and $Ba_8Ga_xSi_{46-x}$ systems, prepared by combining arc melting and spark plasma sintering (SPS) techniques. In addition, the effect of guest substitution on the TE properties is discussed on polycrystalline $Ba_{8-y}Eu_yGa_xSi_{46-x}$ system.

4.4.2 THERMOELECTRIC PROPERTIES OF SILICON CLATHRATES

We prepared systematically polycrystalline silicon clathrate samples of $Ba_8Al_xSi_{46-x}$ and $Ba_8Ga_xSi_{46-x}$ by combining arc melting and SPS techniques. The samples were characterized by powder XRD and field-emission SEM combined with energy-dispersive x-ray spectroscopy (EDX). Measurements of electrical conductivity, Seebeck coefficient, thermal conductivity, and Hall coefficient were performed to assess the TE properties of the materials. The TE properties depend on the contents of Al or Ga, which is a factor to tune the carrier concentration to improve the dimensionless TE figure of merit ZT. In this section, we focus mainly on two silicon clathrate samples, EDX chemical compositions $Ba_{7.77}Al_{15.16}Si_{31.07}$ and $Ba_{7.79}Ga_{14.81}Si_{31.40}$, because they exhibit the maximum ZT values in our $Ba_8Al_xSi_{46-x}$ and $Ba_8Ga_xSi_{46-x}$ systems, respectively.

Figure 4.17 shows the temperature dependence of electrical conductivity for silicon clathrates $Ba_{7.77}Al_{15.16}Si_{31.07}$[67] and $Ba_{7.79}Ga_{14.81}Si_{31.40}$.[31] The temperature dependence indicates that these silicon clathrates have attained a saturation region and that the electrical conductivity decreases with increasing temperature according to the scattering mechanism of carriers. The temperature dependence of carrier mobility μ follows a relation $\mu \propto T^r$, where r is $-1/2$ for alloy disorder scattering and $-3/2$ for acoustic phonon scattering.[76,77] The solid curves in Figure 4.17 are fits to the experimental data by alloy disorder scattering by assuming that the carrier concentration is constant due to the saturation region, and the dashed line for acoustic phonon scattering is presented for comparison. Because the temperature

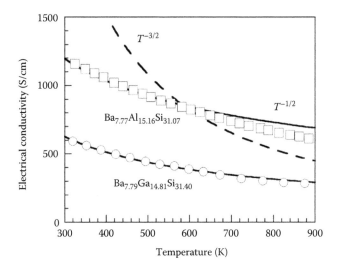

FIGURE 4.17   Temperature dependence of electrical conductivity for silicon clathrates.

behavior for both clathrates agrees well with the relation $T^{-1/2}$, the alloy disorder scattering dominates the transport properties at the measured temperature range. It is likely that the disorder scattering is due to a disordered occupation of Al or Ga atoms substituted for Si atoms in the host lattice of the clathrate crystal structure. For Ge and Si clathrates such as $Ba_8Cu_xGa_yGe_{46-x-y}$,[47] $Sr_8Ga_xGe_{46-x}$,[51] and $Ba_8Ga_xSi_{46-x}$,[31] it was also reported that the alloy disorder scattering was the dominant scattering mechanism in the above room temperature range. Rietveld analysis on the $Ba_8Al_xSi_{46-x}$ system performed by Tsujii et al.[61] showed good results of fitting based on a crystal structure model in which Al and Si atoms were assumed to occupy the host sites at random. However, thorough investigation on the $Ba_8Al_xSi_{46-x}$ system indicated that the Al and Si occupations have site preference depending on the Al content.[78]

Some properties of polycrystalline samples of silicon clathrates $Ba_{7.77}Al_{15.16}Si_{31.07}$[67] and $Ba_{7.79}Ga_{14.81}Si_{31.40}$[31] at room temperature are presented in Table 4.4. The Hall carrier concentration value is higher in $Ba_{7.77}Al_{15.16}Si_{31.07}$ than $Ba_{7.79}Ga_{14.81}Si_{31.40}$, while the Hall mobility value is lower in $Ba_{7.77}Al_{15.16}Si_{31.07}$ than $Ba_{7.79}Ga_{14.81}Si_{31.40}$. The difference in the Hall mobility between two silicon clathrates is attributed not only to the carrier concentration dependence of carrier mobility but also to the difference in the effective mass, as mentioned later. As observed generally in semiconductors, in some clathrate systems such as n-type $Ba_8Ga_xGe_{46-x}$,[44] $Ba_8Al_xSi_{46-x}$,[75] and $Ba_8Ga_xSi_{46-x}$,[31] the Hall mobility increases gradually with decreasing carrier concentration. In $Ba_8Al_xSi_{46-x}$ and $Ba_8Ga_xSi_{46-x}$ clathrates, a Hall mobility value as high as 10 cm^2/(V s) would be expected at the carrier concentration range of 10^{20} cm^{-3}, as is the case of $Ba_8Ga_xGe_{46-x}$ clathrates.

Figure 4.18 shows the temperature dependence of the Seebeck coefficient for silicon clathrates $Ba_{7.77}Al_{15.16}Si_{31.07}$[67] and $Ba_{7.79}Ga_{14.81}Si_{31.40}$.[31] The difference in the absolute values of the Seebeck coefficient between $Ba_{7.77}Al_{15.16}Si_{31.07}$ and $Ba_{7.79}Ga_{14.81}Si_{31.40}$

TABLE 4.4

Chemical Composition from EDX, Lattice Constant *a*, Density *d*, Hall Carrier Concentration *n*, and Hall Mobility μ of Silicon Clathrates

Compound	EDX	*a* (Å)	*d* (g/cm³)	*n* (cm⁻³)	μ (cm²/[V s])
$Ba_8Al_{15}Si_{31}$	Ba: 7.77	10.6400	3.20	9.7×10^{20}	7.4
	Al: 15.16				
	Si: 31.07				
$Ba_8Ga_{15}Si_{31}$	Ba: 7.79	10.5454	4.34	4.4×10^{20}	8.4
	Ga: 14.81				
	Si: 31.40				

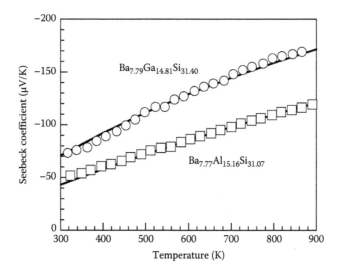

FIGURE 4.18 Temperature dependence of the Seebeck coefficient for silicon clathrates.

comes mainly from the difference in the carrier concentration between them. The Seebeck coefficient increases with increasing temperature below the intrinsic region, which is a typical behavior for a degenerate semiconductor. The DOS effective mass *m** was estimated from the carrier concentration dependence of the Seebeck coefficient at room temperature by assuming a single parabolic band model with alloy disorder scattering as the dominant scattering mechanism (energy dependence of relaxation time: $E^{-1/2}$) on Boltzmann's transport equation.[79] The Seebeck coefficient *S* is given by

$$S = -\frac{k_B}{e} \left[2\frac{F_1(\eta)}{F_0(\eta)} - \eta \right] \quad (4.3)$$

where

k_B is the Boltzmann constant

η is the reduced Fermi energy

F_n is the Fermi–Dirac integral of the nth order

The carrier concentration n is given by

$$n = \frac{4}{\sqrt{\pi}} \left(\frac{2\pi m^* k_B T}{h^2} \right)^{3/2} \cdot F_{1/2}(\eta) \tag{4.4}$$

where h is the Planck constant. The effective mass m^* values for $Ba_{7.77}Al_{15.16}Si_{31.07}$ and $Ba_{7.79}Ga_{14.81}Si_{31.40}$ were estimated to be about $2.3m_0$ and $2.0m_0$, respectively, from the values of the Seebeck coefficient and the Hall carrier concentration at room temperature from Equations 4.3 and 4.4. The curves in Figure 4.18 are calculated using Equations 4.3 and 4.4 as functions of η by using $m^* = 2.3m_0$ and $n = 1.2 \times 10^{21}$ cm^{-3} for $Ba_{7.77}Al_{15.16}Si_{31.07}$ and $m^* = 2.0m_0$ and $n = 4.36 \times 10^{20}$ cm^{-3} for $Ba_{7.79}Ga_{14.81}Si_{31.40}$. The calculated curves agree well with the experimental data. The effective mass m^* values of $Ba_{7.79}Ga_{14.81}Si_{31.40}$ are comparable to that of $Ba_8Ga_{16}Si_{30}$ reported by Kuznetsov et al. ($m^* = 2.2m_0$).[36] The effective mass m^* value of $Ba_{7.77}Al_{15.16}Si_{31.07}$ is slightly smaller than that of nominal $Ba_8Al_{15}Si_{31}$ (EDX chemical composition: $Ba_{8.02}Al_{14.74}Si_{31.26}$) ($m^* = 2.7m_0$) reported by Tsujii et al.[61] The effective mass m^* of polycrystalline $Ba_8Ga_xGe_{46-x}$ clathrate was reported to be $m^* = 1.3m_0$,[44] and $Ba_8Ga_xGe_{46-x}$ clathrate single crystal exhibits high $ZT = 1.35$ at 900 K.[48] Thus, it can be deduced from the effective mass value that the Seebeck coefficient for both silicon clathrates should be essentially as large as that of n-type $Ba_8Ga_xGe_{46-x}$ if the carrier concentration is further reduced to the optimum value.

Figure 4.19 shows the temperature dependence of the lattice thermal conductivity κ_L of silicon clathrates $Ba_{7.77}Al_{15.16}Si_{31.07}$[67] and $Ba_{7.79}Ga_{14.81}Si_{31.40}$.[31] The lattice thermal conductivity κ_L was estimated by subtracting the electronic thermal conductivity κ_e from the measured thermal conductivity κ. The electronic thermal conductivity κ_e was estimated by the Wiedemann–Franz relation $\kappa_e = LT\sigma$, where L is the Lorenz number. L was calculated using the reduced Fermi energy determined by the Seebeck coefficient on Boltzmann's transport equation[79]:

$$L = \left(\frac{k_B}{e} \right)^2 \frac{3F_0(\eta)F_2(\eta) - 4F_1(\eta)^2}{F_0(\eta)^2} \tag{4.5}$$

The values of lattice thermal conductivity κ_L for $Ba_{7.77}Al_{15.16}Si_{31.07}$ and $Ba_{7.79}Ga_{14.81}Si_{31.40}$ are approximately 1.2 and 1.1 W/(m K) at room temperature, respectively. The lattice thermal conductivity for silicon clathrates is approximately two orders of magnitude lower than that for silicon with a diamond-type crystal structure. It is also noted that the lattice thermal conductivity for silicon clathrates is almost equivalent to that for amorphous SiO_2.[32,33] The extremely low values of κ_L for clathrates are essentially attributed to the strong scattering of phonons by the thermal

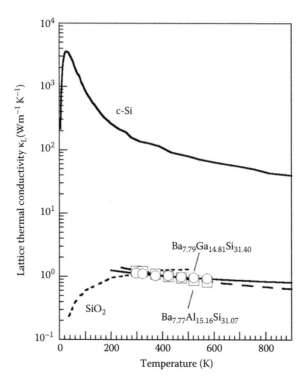

FIGURE 4.19   Temperature dependence of lattice thermal conductivity κ_L for silicon clathrates. Literature data: single-crystal silicon with diamond-type crystal structure (c-Si) and amorphous SiO_2.[32,33]

vibration *rattling* of barium guests in the voids of the framework. The lattice thermal conductivity κ_L decreases as the temperature increases from room temperature, according to the typical behavior at high temperatures due to the domination of phonon–phonon scattering (Umklapp process).[80–83] The curves shown in Figure 4.19 are fits to the relation $\kappa_L \propto T^{-b}$ based on the Umklapp process, where the values of b are 0.59 and 0.29 for $Ba_{7.77}Al_{15.16}Si_{31.07}$ and $Ba_{7.79}Ga_{14.81}Si_{31.40}$, respectively.

Figure 4.20 shows the temperature dependence of the dimensionless TE figure of merit ZT for silicon clathrates $Ba_{7.77}Al_{15.16}Si_{31.07}$[67] and $Ba_{7.79}Ga_{14.81}Si_{31.40}$.[31] ZT was estimated from the temperature dependence of the Seebeck coefficient, the electrical conductivity, and the thermal conductivity. As shown in Figure 4.20, $Ba_{7.77}Al_{15.16}Si_{31.07}$ and $Ba_{7.79}Ga_{14.81}Si_{31.40}$ show relatively high ZT of about 0.4 and 0.55, respectively, at 900 K.

Figure 4.21 shows the carrier concentration dependence of ZT for silicon clathrates $Ba_{7.77}Al_{15.16}Si_{31.07}$[67] and $Ba_{7.79}Ga_{14.81}Si_{31.40}$.[31] ZT was calculated by a modified model proposed originally by Slack[69] on Boltzmann's transport equation using the experimental data sets of the effective mass m^*, Hall mobility μ_0, and the lattice thermal conductivity κ_{L0} at room temperature: $2.3m_0$, 7.4 cm²/(V s), and 1.2 W/(m K) for $Ba_{7.77}Al_{15.16}Si_{31.07}$ and $2.0m_0$, 10 cm²/(V s), and 1.1 W/(m K) for

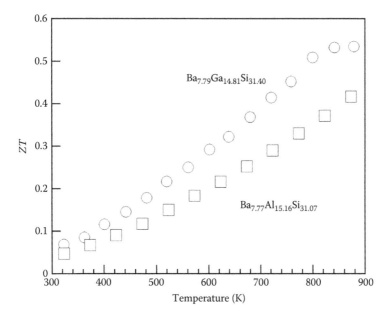

FIGURE 4.20   Temperature dependence of dimensionless TE figure of merit ZT for silicon clathrates.

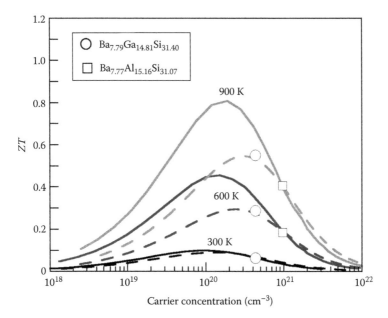

FIGURE 4.21   Carrier concentration dependence of dimensionless TE figure of merit ZT for silicon clathrates. Dashed and solid curves are calculations based on a model described in the text for $Ba_{7.77}Al_{15.16}Si_{31.07}$ and $Ba_{7.79}Ga_{14.81}Si_{31.40}$, respectively.

$Ba_{7.79}Ga_{14.81}Si_{31.40}$. The model calculation takes into account the temperature dependence as $\mu = \mu_0(T/T_0)^{-1/2}$ in alloy disorder scattering and $\kappa_L = \kappa_{L0}(T/T_0)^{-b}$, assuming constant $L = 2.44 \times 10^{-8}$ V^2/K^2. If the Lorenz number calculated from Equation 4.4 is used in ZT estimation, the values of ZT will be larger than the values shown in Figure 4.21. The calculation explains well the carrier concentration dependence of ZT for both silicon clathrates. Further, the calculation predicts that the maximum ZT at 900 K is approximately 0.5 and 0.8 at the optimum carrier concentration of approximately 3×10^{20} and 2×10^{20} cm^{-3} for $Ba_{7.77}Al_{15.16}Si_{31.07}$ and $Ba_{7.79}Ga_{14.81}Si_{31.40}$, respectively. We have proposed that ZT values will be further improved if the carrier concentration is lower in $Ba_8Al_xSi_{46-x}$ and $Ba_8Ga_xSi_{46-x}$ systems[31,67] by the substitution of transition-metal elements such as gold for the host atoms, for example, the $Ba_8Au_xGa_ySi_{46-x-y}$ system.[84] In addition, we have proposed that the ZT value will be improved if the carrier mobility is enhanced by the optimization of the synthesis process such as annealing.

4.4.3 Effects of Rare-Earth Substitution on Guest Site: $Ba_{8-y}Eu_yGa_xSi_{46-x}$

It is of great importance to investigate the effect of guests not only on the phonon properties but also on the transport properties in TE silicon clathrates. The effect of rare-earth substitution on the TE properties was reported on several clathrates: $Ba_6Eu_2Cu_4Si_{42}$,[39] $Ba_{8-y}Eu_yAu_6Ge_{40}$,[85] $Ba_7Eu_2Al_{13}Si_{33}$,[72] $Ba_{8-y}Eu_yAl_{14}Si_{31}$,[86] $Ba_{8-y}Yb_y$ $Ga_{16}Ge_{30}$,[54] $Ba_6Eu_2AuGa_xSi_{45-x}$,[58] and so on. In some clathrates including rare-earth guests, $Ba_{8-x}Yb_xGa_{16}Ge_{30}$ and $Ba_{8-x}Eu_xAu_6Ge_{40}$, rare-earth guests play an important role in the reduction of the lattice thermal conductivity.[54,85] On the other hand, Akai et al.[87] and Koga et al.[88] made a theoretical investigation on the effect of rare-earth guests on the electronic structure and carrier transport properties. They suggested that the conduction band edges are modified significantly by Eu or Yb substitution on the 2a site in the smaller cage, leading to an enhanced effective mass. Recent experimental results on $Ba_{8-x}Eu_xAu_6Ge_{40}$[85] and $Ba_6A_2AuGa_xSi_{45-x}$ (A = Sr, Eu)[58] supported their theoretical predictions. We then discuss the effect of Eu substitution for Ba guests in the smaller cage on the TE properties of polycrystalline $Ba_{8-y}Eu_yGa_xSi_{46-x}$ (nominal $y = 0, 1, 2$) clathrate system.[89]

The Rietveld structure refinements were performed on XRD data sets using Cr Kα radiation for $Ba_{8-y}Eu_yGa_xSi_{46-x}$ ($y = 0, 1, 2$) clathrate samples prepared by combining arc melting and SPS techniques. The structure refinement results suggested that the Eu guest atoms preferentially occupied the 2a site in the smaller cage. A similar trend was found in $Ba_6Eu_2Cu_4Si_{42}$[39] and $Ba_{8-y}Eu_yAl_{14}Si_{31}$.[86] The structure refinement results also showed that the atomic displacement parameters (ADPs) on the 2a site for $Ba_{8-y}Eu_yGa_xSi_{46-x}$ clathrates were significantly increased by a factor more than two relative to $Ba_8Ga_xSi_{46-x}$ clathrates. The thermal motion of the guest atoms is often associated with that of an Einstein oscillator.[90] In the Einstein model, the mean square displacement $\langle u^2 \rangle$ of the guest atoms at high temperature ($h\nu < k_BT$), assuming that the value of $\langle u^2 \rangle$ is all due to the dynamic motion of the guest atom, is given by $U_{iso} = \langle u^2 \rangle = h^2T/(4\pi^2mk_B\Theta_E^2)$, where Θ_E is the Einstein temperature, m is the reduced mass, and h and k_B are the Planck and Boltzmann constants, respectively. On the other hand, the host structure's ADPs can be modeled with the

TABLE 4.5

Debye Temperature Θ_D and Einstein Temperature Θ_E for $Ba_8Ga_xSi_{46-x}$ and $Ba_{8-y}Eu_yGa_xSi_{46-x}$ Clathrates

Sample	Θ_D (K) U_{iso} (Frame)	Θ_E (K)		
		$U_{11} = U_{22} = U_{33}$ (M1)	U_{11} (M2)	$U_{22} = U_{33}$ (M2)
$Ba_{7.79}Ga_{14.81}Si_{31.40}$	317	97	89	53
$Ba_{6.21}Eu_{1.62}Ga_{13.67}Si_{32.50}$	381	79	74	55
$Ba_8Ga_{16}Si_{30}$ neutron[a]	387	124	98	69
$Ba_8Ga_{16}Si_{30}$ x-ray[a]	416	127	101	77
$Ba_8Ga_{16}Ge_{30}$[a]	312	124	101	73
$Sr_8Ga_{16}Ge_{30}$[a]	313	151	104	163

[a] Data from Bentien, A., et al., *Phys. Rev. B*, 71, 144107, 2005.

Debye expression. At high temperatures $(T > \Theta_D)$, U_{iso} is linear in T and is given by $U_{iso} = 3h^2T/(4\pi^2 Mk_B\Theta_D^2)$, where M is the reduced mass and Θ_D is the Debye temperature.[90] By using the ADP (U_{ij}) data, we estimated Θ_E for Ba/Eu on the 2a site, and the results are summarized in Table 4.5. The estimation of Θ_E for $Ba_8Ga_xSi_{46-x}$ with nominal $x = 15$ ($Ba_{7.79}Ga_{14.81}Si_{31.40}$ clathrate) reasonably agrees with the reported values for $Ba_8Ga_{16}Si_{30}$ single crystal by Bentien et al.[46] and DFT calculations.[91] It is noted in $Ba_{8-y}Eu_yGa_xSi_{46-x}$ (EDX chemical composition: $Ba_{6.21}Eu_{1.62}Ga_{13.67}Si_{32.50}$) that the Θ_E for guest atoms on the 2a site in the smaller cage is significantly decreased by Eu substitution.

Figure 4.22 shows the temperature dependence of the lattice thermal conductivity κ_L for $Ba_{8-y}Eu_yGa_xSi_{46-x}$ clathrates.[89] The lattice thermal conductivity for $Ba_8Ga_xSi_{46-x}$ clathrates is as low as approximately 1.1 W/(m K) at room temperature, which is in good agreement with the reported value.[33] It is obviously found that the lattice thermal conductivity of $Ba_{8-y}Eu_yGa_xSi_{46-x}$ (EDX: $Ba_{6.21}Eu_{1.62}Ga_{13.67}Si_{32.50}$) clathrate is extremely low compared to $Ba_8Ga_xSi_{46-x}$ clathrates. The value of κ_L for $Ba_{8-y}Eu_yGa_xSi_{46-x}$ almost corresponds to that for type-I $Eu_8Ga_{16}Ge_{30}$.[33] Structure refinement showed that the atomic displacement parameter U for guest atoms on the 2a site in $Ba_{8-y}Eu_yGa_xSi_{46-x}$ was more than two times larger than that for Ba atoms on the 2a site in $Ba_8Ga_{16}Si_{30}$, as described already. As a result, the Einstein temperature Θ_E for atoms on the 2a site is greatly decreased in $Ba_{8-y}Eu_yGa_xSi_{46-x}$, as shown in Table 4.5. The phonon mean free path was estimated to be approximately 2.7 Å, which is nearly the distance between guest atoms. Thus, it is strongly suggested that the large reduction in the lattice thermal conductivity for $Ba_{8-y}Eu_yGa_xSi_{46-x}$ can be attributed to the increase in the scattering of phonons by the thermal *rattling* motion of guest Eu atoms that interact with heat-carrying phonons of the framework with lower frequency.

Figure 4.23 shows the carrier concentration dependence of the Seebeck coefficient for $Ba_{8-y}Eu_yGa_xSi_{46-x}$ clathrates.[89] The Seebeck coefficient for $Ba_{8-y}Eu_yGa_xSi_{46-x}$ clathrates is smaller than that for $Ba_8Ga_xSi_{46-x}$ clathrates due to the high carrier

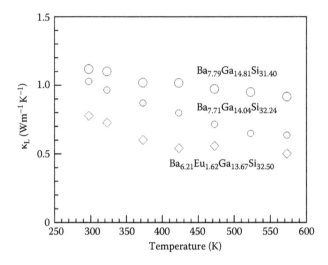

FIGURE 4.22 Temperature dependence of lattice thermal conductivity κ_L for $Ba_{8-y}Eu_yGa_xSi_{46-x}$ clathrates.

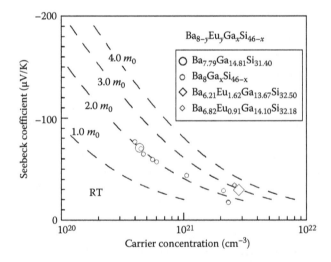

FIGURE 4.23 Carrier concentration dependence of the Seebeck coefficient at room temperature for $Ba_{8-y}Eu_yGa_xSi_{46-x}$ clathrate. Dashed lines are calculations based on Boltzmann's transport equation.

concentration. However, at the same carrier concentration, the Seebeck coefficient for $Ba_{8-y}Eu_yGa_xSi_{46-x}$ clathrates is larger than that for $Ba_8Ga_xSi_{46-x}$ clathrates. To discuss the effect of Eu substitution on the Seebeck coefficient, we estimated the DOS effective mass m^* from the analysis on a single parabolic band model with alloy disorder scattering as the dominant scattering mechanism, as already mentioned. The effective mass for $Ba_8Ga_xSi_{46-x}$ clathrates with different Ga compositions was estimated to be approximately $2m_0$ independent of the carrier concentration. On the other hand, the enhanced effective mass of approximately $3m_0$ was estimated for $Ba_{8-y}Eu_yGa_xSi_{46-x}$ clathrates. According to the calculations based on DFT by Akai et al.[87] and Koga et al.,[88] the substitutions by rare-earth elements on the 2a site in the smaller cage for Ba atoms affect significantly the conduction band bottom. This effect is naturally explained in terms of the change in the degree of hybridization between guests and framework due to the difference in the ionic radii of guest atoms and the cage size. Recently, we have experimentally confirmed the increases in the conduction band effective mass by guest substitution from a similar analysis of the carrier concentration of the Seebeck coefficient in some clathrate systems, $Ba_{8-x}Eu_xAu_6Ge_{40}$,[85] $Ba_6A_2AuGa_xSi_{45-x}$ (A = Sr, Eu),[58] and $Ba_6Eu_2Cu_6Si_{40}$.[92] Very recently, Prokofiev et al.[93] reported the thermopower enhancement in clathrate including Ce rare-earth guest.

4.4.4 SUMMARY

High-temperature TE properties of polycrystalline silicon clathrates $Ba_8Al_xSi_{46-x}$ and $Ba_8Ga_xSi_{46-x}$ prepared by arc melting and spark plasma sintering techniques were investigated to assess the materials for TE energy conversion applications. Both $Ba_8Al_xSi_{46-x}$ and $Ba_8Ga_xSi_{46-x}$ systems have an advantage of extremely low lattice thermal conductivity, which stems from the strong interaction of thermal vibration *rattling* of guest atoms with the heat-carrying phonons of the framework of clathrates. The electrical conductivity is good and the Seebeck coefficient is relatively high, resulting in a relatively high TE figure of merit at high temperatures in these silicon clathrates. The transport properties are well explained in terms of Boltzmann's transport theory, which predicts the direction of material design for further improvement. The effects of rare-earth substitution on the transport properties were investigated on the $Ba_{8-y}Eu_yGa_xSi_{46-x}$ system. Eu guest substitutions give rise to the increase of the electron effective mass through the modification of the electronic structure, which reasonably agrees with recent calculations on DFT. In addition, Eu substitution causes a significant decrease in the lattice thermal conductivity. The analysis of the transport properties combined with the structural properties indicates the importance of the interactions of guests in the smaller cage with the framework not only in the thermal properties but also in the electronic properties of clathrates.

4.5 NEW N-TYPE SILICIDE THERMOELECTRIC MATERIAL WITH HIGH OXIDATION RESISTANCE

4.5.1 INTRODUCTION

Good TE materials are indispensable to TE power generation. However, these materials should have not only a high dimensionless TE figure of merit ZT but also high chemical stability and should not contain harmful elements.

The temperature range for waste heat for industries, cars, and incinerators or condensed solar heat is spread from 350 to 1000 K. Almost all metallic TE compounds as shown in Figure 4.4 have a problem of oxidation at such temperatures. Oxide TE

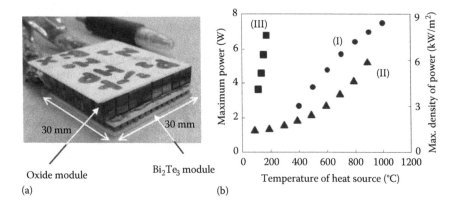

FIGURE 4.24 A photograph of a cascade module composed of oxide and Bi_2Te_3 modules (a) and temperature dependence of maximum power P_{max} and density of power for cascade (I), oxide (II), and Bi_2Te_3 (III) modules (b).

materials are considered to be promising ones because of their durability against high temperature, cost, no content of toxic elements, and so on. Some layered CoO_2 compounds show high ZT values in air.[9,10,94] Fabrication and properties of modules using oxide materials have been produced.[95–98] Though ZT values around unity are obtained at temperatures higher than 900 K, enhancement of ZT is necessary at temperatures lower than 900 K for waste heat application. It is not efficient that only single oxide devices are used over the wide temperature range. Cascade modules accumulated oxide and Bi_2Te_3 modules have been developed to obtain a high generating power over a wide temperature range (Figure 4.24a). Actually, cascading is effective to enhance the power density at the hot side temperature higher than 873 K (Figure 4.24b).[99] However, ZT values of oxide materials are good at 800 K or higher, and common Bi_2Te_3 modules can be used at temperatures lower than 473 K or lower because of the degradation of junctions between TE legs and electrodes. In other words, there is a *blank area* at the middle temperature range (473–900 K).

The development of not only TE materials with high ZT values and oxidation resistance but also modules with high generating power density and durability is an important task of pressing urgency. Silicide compounds are one of the strongest candidates for the middle temperature range because of the formation of a passive layer on the surface. In fact, $MnSi_{1.75}$ and $FeSi_2$ are well known as TE materials usable at high temperatures even in air.[100–103] TE properties of a new n-type silicide material have been discovered and are studied in detail. $Mn_3Si_4Al_2$ silicide shows a good oxidation resistance. Enhancement of the ZT values of this silicide has been succeeded by elemental substitution. In order to achieve TE solar power generation, the module composed of the silicide materials has been developed.

4.5.2 NEW n-TYPE $Mn_3Si_4Al_2$

A precursor ingot was prepared by arc melting Si (99.999%), Mn (99.9%), and Al (99.99%) metallic chips with an atomic ratio of Mn:Si:Al = 3:4:3. After removing oxide on the surface of the ingot by polishing, the melted ingot was ground using an agate mortar and pestle and packed into disk-shaped pellets. The precursor pellets were sintered by pulse electric current sintering at 1023 K for 15 min under a uniaxial pressure of 30 MPa.

XRD of the powder samples after sintering was used to investigate the crystallographic structure and purity. The XRD patterns were analyzed using the Rietveld method with the help of the Jana 2006 software to calculate cell parameters. The microstructure of the sintered samples was observed using a SEM in backscattering mode. The analysis of elemental composition was performed by EDX analysis. Transmission electron microscope (TEM) observation and electron diffraction (ED) measurement were carried out to assign the crystallographic structure.

The samples were cut into rectangular bars, about 3×3 mm^2 in cross section and 5–10 mm long. The Seebeck coefficient (S) values were calculated from a plot of TE voltage against the temperature differential as measured at 373–973 K in air using an instrument designed by our laboratory. Two Pt-Pt/Rh (R-type) thermocouples were adhered to both ends of the samples using silver paste. The Pt wires of the thermocouples were used for voltage terminals. The measured thermovoltage was plotted against the temperature difference. The slope corresponds to the apparatus' Seebeck coefficient. The actual S values were obtained by correction using the S values of the Pt wire. Electrical resistivity (ρ) was measured using the standard dc four-probe method in air from room temperature to 1000 K. Silver paste was used for the connections between the samples and both the current and voltage lead wires. Thermal conductivity (κ) measurement was carried out using the laser flash method in a direction parallel to the pressing axis during pulse electric current sintering.

The XRD pattern for the $Mn_3Si_4Al_2$ sample is shown in Figure 4.25. The crystallographic structure of all the samples can be assigned to a hexagonal close-packed $CrSi_2$ structure.[104] The a- and c-cell parameters for this sample are 0.447 and 0.644 nm, respectively. Some weak diffraction peaks due to secondary phases are observed in the XRD pattern. Microscopic observation using the SEM indicates results that are consistent with the XRD data. A large area due to the secondary phase is observed. The atomic ratio was measured by EDX analysis. Two phases, $Mn_{3.0}Si_{4.0}Al_{2.3}$ (bright portion A) and $MnSi_{2.9}Al_{2.1}$ (dark portion B), are formed (Figure 4.26). The bright portion corresponds to the main phase possessing $Mn_3Si_4Al_2$. Many starting compositions were tried to prepare the $Mn_3Si_4Al_2$ (MSA-342) phase, though the ratio of Mn:Si:Al = 3:4:3 described earlier is close to the optimum one to obtain good ZT values. The starting composition is not coincident with the composition of the main phase. One of the presumable reasons is oxidation of Al during arc melting. Such oxide is located around the surface and is removed by polishing. Completely pure samples have never been prepared yet, but we believe almost all physical properties can be discussed based on the MA-342 phase.

A detailed structure is also investigated by TEM and ED measurement from some incident directions (Figure 4.27). Strong fundamental diffraction spots observed were confirmed to originate from the $CrSi_2$ structure as indexed in the figure. In addition, our careful analysis revealed the presence of weak diffraction spots as indicated by arrows. These spots were present at 1/3 1/3 0 positions in the reciprocal space, which suggest a superlattice structure. The diffraction peaks indexed by

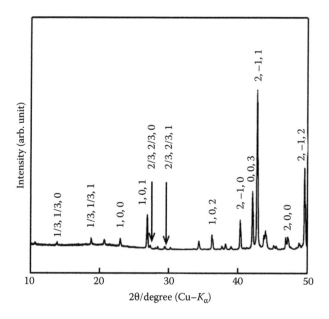

FIGURE 4.25 XRD pattern for the sample with $Mn_3Si_4Al_2$ after SPS. Closed circles and arrows correspond to the diffraction peaks due to secondary phases and superlattice structure as mentioned later, respectively.

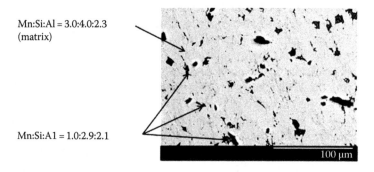

FIGURE 4.26 Scanning electron microscopic image for the sample with $Mn_3Si_4Al_3$ of the starting composition after SPS.

1/3 1/3 0, 1/3 1/3 1, 2/3 2/3 0, and 2/3 2/3 1 are also observed in XRD patterns for all samples. This means that there is a structural modulation with a threefold period in the [110] direction with respect to the $CrSi_2$ structure. The relationship between the superlattice structure and TE or other properties is an open question.

Temperature dependence of the S values is shown in Figure 4.28a. S values for the $Mn_3Si_4Al_2$ samples increase up to 773 K and then suddenly decrease at 973 K. The maximum value of S is 93 $\mu V/K$ at 773 K. The ρ value increases with increasing

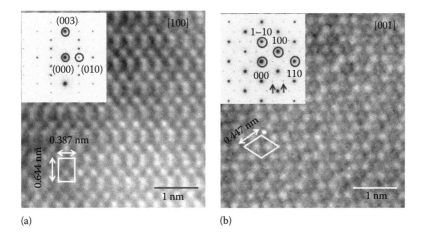

FIGURE 4.27   TEM images and ED patterns for the $Mn_3Si_4Al_2$ phase for the direction of incidence of [100] (a) and [001] (b). Arrows in the ED patterns indicate the existence of a superlattice structure.

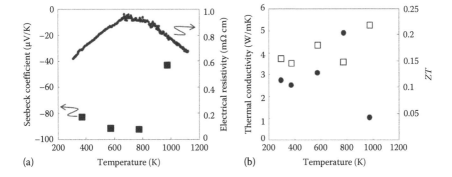

FIGURE 4.28   Temperature dependence of the Seebeck coefficient and electrical resistivity (a) and thermal conductivity and dimensionless figure of merit ZT (b) for the $Mn_3Si_4Al_2$ sample of the starting composition after pulse electron.

temperature up to about 800 K showing a metallic behavior. This character changes to an insulator-like one around 800 K. This transition temperature almost corresponds to the temperature of the sudden decrease in S. We believe that this is due to the decrease in the Seebeck coefficient because of the transition to the intrinsic range. The power factor $(=S^2/\rho)$ reaches 0.11 mW/(m K^2) at 773 K in air.

Figure 4.28b shows the temperature dependence of the thermal conductivity κ. Basically the κ values increase with increasing temperature and reach as high as 5 W/(m K). In order to obtain high ZT, the κ values should be reduced. Though the ZT for the nonsubstituted MSA-342 sample increases with increasing temperature up to 773 K, it decreases suddenly at 973 K. This is due to the decrease in the Seebeck coefficient because of the transition to the intrinsic range. The maximum ZT reaches

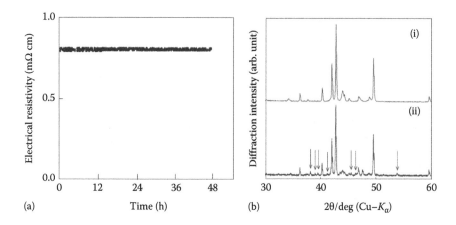

(a)   Time (h)   (b) $2\theta/deg$ (Cu–K_α)

FIGURE 4.29 Electrical resistivity measured continuously at 873 K in air (a) and XRD patterns before (i) and after (ii) heat treatment at 873 K for 48 h in air (b) for the $Mn_3Si_4Al_2$ sample.

0.2 at 773 K. In order to enhance the *ZT* values, controlling the carrier density by elemental substitution for elevation of the transition temperature into the intrinsic range and optimizing the microstructure including the secondary phases as scattering sites of phonons for low κ values are vital.

Figure 4.29a shows the ρ values measured continuously at 873 K in air. Though the measurement was performed for 48 h, the ρ values were maintained constant. Additional diffraction peaks which can be assigned to alumina, silica, or complex oxides of the constituent elements are detected in the XRD patterns for the heated sample (Figure 4.29b). The color of the surface of the sample after heat treatment at 873 K turns from light gray to dark gray (Figure 4.30).

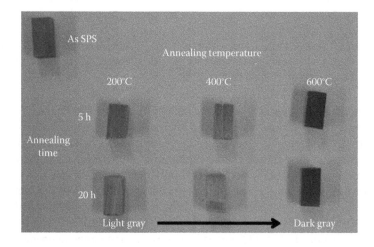

FIGURE 4.30 Photograph of $Mn_3Si_4Al_2$ samples after SPS processing and heat treatment in 473–873 K for 5 or 20 h in air.

The brown layer can be removed easily by polishing with a sheet of emery paper and the light gray surface appears under the brown layer. These results indicate that the MSA-342 phase is oxidized but occurs only on the surface. Because the formed oxide layer acts as a passive layer to protect the inside of the MSA-342 bulk from oxidation, the ρ values are kept constant at high temperatures even in air. In order to make the MSA-342 material into a good TE device in the middle temperature range, high sintered density is necessary to prevent oxygen penetration into the inside of the bulk.

4.5.3 ELEMENTAL SUBSTITUTION

Since $Mn_3Si_4Al_2$ shows good oxidation resistance, this silicide is a promising n-type TE material for waste heat recovery or solar heat application. But ZT values are limited as low as 0.2. Enhancement of ZT values can be expected by improvement of the Seebeck coefficient and suppression of thermal conductivity. The optimization of the electronic structure is necessary for the former and microscopic control is effective for the latter. To enhance the Seebeck coefficient, elemental substitution of Mn with 3D transition metals has been tried. As a result, Cr substitution elevates the Seebeck coefficient. The ZT value reaches 0.3 at 573 K in air for the $Mn_{2.7}Cr_{0.3}Si_4Al_2$ sample.

Precursor ingots were prepared by arc melting Si, Mn, Ti, V, Cr, Fe, Co, Ni, Cu, and Al metallic chips with an atomic ratio of $Mn:M:Si:Al = 3.0 - x:x:4.0:3.0$, where M indicates Ti, V, Cr, Fe, Co, Ni, Cu, and $x = 0.3$ or 1.0. Precursor ingots were prepared by arc melting. After removing oxides on the surface by polishing, the melted ingots were ground using an agate mortar and pestle and packed into disk-shaped pellets. The precursor pellets were sintered by SPS at 1023–1073 K for 15 min under a uniaxial pressure of 30 MPa. Measurements of TE properties have been carried out as the same procedures as mentioned earlier.

XRD patterns of the $Mn_{2.7}M_{0.3}Si_4Al_2$ powder indicate several weak diffraction peaks due to the secondary phases; the MSA-342 phase, however, is formed as a main phase. But intensity of the secondary phases are stronger in the $x = 1.0$ samples than the $x = 0.3$ ones.

Figure 4.31 indicates the temperature dependence of the Seebeck coefficient for all the samples. Though the starting composition was $Mn_{3-x}M_xSi_4Al_3$ as mentioned earlier, the final compositions for the matrix main phase of all the samples were close to $(Mn+M):Si:Al = 3:4:2$. An outstanding increase in the S values is observed only for the Cr-substituted sample with $x = 0.3$. The S values decrease in all other samples with both $x = 0.3$ and $x = 1.0$. The reduction of the S values tends to be small in the samples substituted with the prior elements compared with the posterior ones than Mn in the periodic table (Figure 4.32). This result indicates that not only carrier density but ionic radii should affect the S values.

The temperature dependence of the ρ value and the power factor ($= S^2/\rho$) for the samples substituted by $x = 0.3$ is shown in Figure 4.33. The ρ values tend to be lower in the samples substituted by the posterior elements than Mn because of the high level of electron doping. This result is consistent with the low S values. The substitution with Cr is effective to enhance the power factor. In order to optimize the Cr

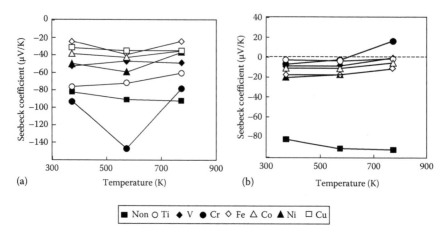

FIGURE 4.31 Temperature dependence of the Seebeck coefficient for the samples with $Mn_{3-x}M_xSi_4Al_2$ ($x = 0.3$ (a) and 1.0 (b)). M is Ti, V, Cr, Fe, Co, Ni, or Cu.

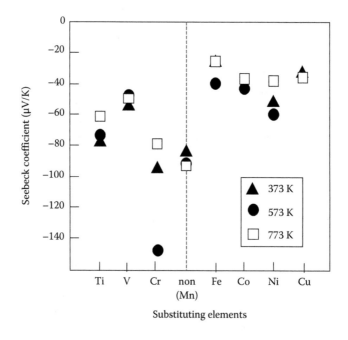

FIGURE 4.32 Seebeck coefficient of $Mn_3Si_4Al_2$ (non) and $Mn_{2.7}M_{0.3}Si_4Al_2$ samples at 373–773 K.

composition, the MSA-342 samples with different x values of $Mn_{3-x}Cr_xSi_4Al_2$ have been prepared by the same conditions as described earlier.

Several samples with Cr substitution possessing different contents of Cr were prepared as the same procedures as mentioned earlier. Figure 4.34 indicates the powder XRD patterns for $Mn_{3-x}Cr_xSi_4Al_2$ ($x = 0-0.7$). Some weak diffraction peaks

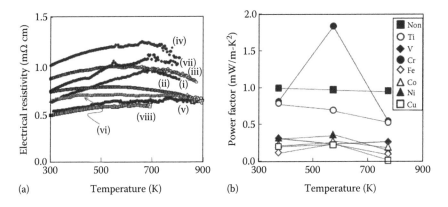

FIGURE 4.33 Temperature dependence of electrical resistivity (a) and (b) power factor ($=S^2/\rho$) for the samples of $Mn_3Si_4Al_2$ (i) and $Mn_{2.7}M_{0.3}Si_4Al_2$. M: Ti (ii), V (iii), Cr (iv), Fe (v), Co (vi), Ni (vii), and Cu (viii).

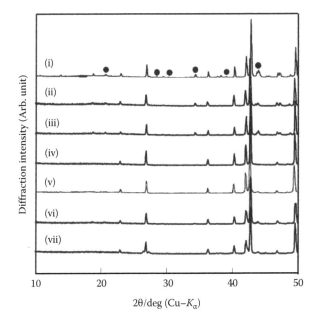

FIGURE 4.34 XRD patterns for $Mn_{3-x}Cr_xSi_4Al_2$ ($x = 0–0.7$). Diffraction peaks for the secondary phases are pointed out by closed circles.

due to the secondary phases are observed at $x \leq 0.1$ in XRD patterns. The a- and c-cell parameters calculated from the XRD patterns increase once at $x=0.05$ and decrease with increasing Cr substitution (Figure 4.35a). The decrease in the parameters becomes gentle at x higher than 0.3. The average Cr content measured by EDX analysis in the main phase increases with increasing x values up to $x=0.2$,

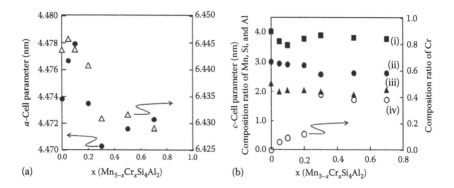

FIGURE 4.35   *a*- and *c*-cell parameters for $Mn_{3-x}Cr_xSi_4Al_2$ (a). Average composition of Mn, Cr, Si, and Al in MSA-342 grains measured by EDX analysis (b).

and a discontinuous increase is observed at 0.3 as shown in Figure 4.35b. The Cr composition in the main phase is constant at *x* values greater than 0.3. Namely, the solid solution of Cr into the Mn sites is saturated around $x=0.3$. The average compositions of Si and Al are almost independent of the *x* values in the Cr-substituted samples.

The temperature dependence of the *S* values is shown in Figure 4.36a. The *S* values for all samples increase up to 573 K at $x=0–0.3$ then suddenly decrease at 773 K in the Cr-substituted samples with $x=0.05–0.3$. Such a decrease in the *S* value is also observed in the non-Cr-substituted sample at 973 K. On the other hand, the *S* values decrease with temperature from 373 K at greater *x* values than 0.5. The Cr substitution effectively enhances the *S* values at 573 K. The maximum value of *S* reaches 160 μV/K at $x=0.3$, which is more than 1.6 times greater than for the non-Cr-substituted sample. The ρ values are elevated by the Cr substitution (Figure 4.36b). Electrical conduction behavior of all samples changes from a metallike temperature dependence to an insulator-like one at 623–723 K. These transition temperatures almost correspond to the temperature of the decrease in *S*. The Cr substitution seems to decrease the carrier density in the MSA-342 at temperatures lower than 573 K. Of course the secondary phases and microstructure should be considered as the reasons for the change in electrical properties. Because the amount of the secondary phases are small considering the weak XRD peaks in all samples, the change of the Seebeck coefficient and electrical resistivity of the samples is mainly dominated by the Cr content. Considering the behaviors of *S* and ρ, the MSA-342 phase could be in an intrinsic range at temperatures higher than the transition temperature. The transition temperature from metallike to insulator-like behavior is lowered by the Cr substitution.

Figure 4.36c indicates the temperature dependence of the thermal conductivity κ. there is decrease at 773 K for the non-Cr-substituted sample, the κ values increase with temperature. The increase in κ after a decrease at 773 K indicates also that the MSA-342 phase is in the intrinsic range at 973 K in the non-Cr-substituted sample. In order to obtain high *ZT*, the reduction of κ values is indispensable. The decrease

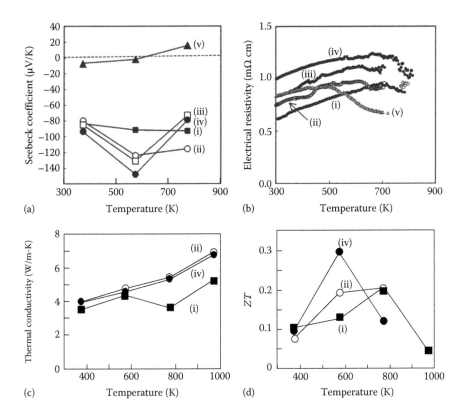

FIGURE 4.36 Temperature dependence of electrical resistivity (a), Seebeck coefficient (b), thermal conductivity (c), and dimensionless figure of merit ZT (d) for the $Mn_{3-x}Cr_x$-Si_4Al_3 samples. $x = 0$ (i), 0.05 (ii), 0.1 (iii), 0.3 (iv), and 1.0 (v).

in κ between 573 and 773 K for the non-Cr-substituted sample might be caused by existence of the secondary phase; however, it is an open question. The high transition temperature into the intrinsic range could be another reason for the decrease in κ. Namely, if the transition temperature in the Cr-substituted samples can be risen, the κ values at temperatures higher than 773 K are lower than measured ones as shown in Figure 4.36c. Though the dimensionless TE figure of merit ZT for the samples with $x=0$ and 0.05 increases with increasing temperature up to 773 K, the Cr-substituted sample with $x=0.3$ has a peak at 573 K (Figure 4.36d). ZT reaches 0.3 at 573 K for the sample with $x=0.3$ and 0.21 at 773 K for the $x=0$ and 0.05, respectively. These values are lower than the n-type Mg_2Si compound[105,106]; however, it is comparable to the p-type $MnSi_{1.7}$ compound[109] or higher than the n-type $FeSi_2$ compound.[110] Because of the small S and high κ values, ZT values markedly decrease in the intrinsic range. In order to enhance the ZT values, controlling the carrier density by atomic substitution for elevation of the transition temperature into the intrinsic range and optimizing the microstructure including the secondary phases for low κ values are vital.

4.5.4 THERMOELECTRIC MODULE COMPOSED OF SILICIDE MATERIALS

A TE module composed of 64 pairs of n- and p-type legs was fabricated. The compositions of the p- and n-type devices are $MnSi_{1.7}$[111] and the non-Cr-substituted $Mn_3Si_4Al_2$, respectively. The p- and n-type disks were sintered at 1173 and 973 K, respectively, for 15 min under a uniaxial pressure of 30 MPa by pulse electric current sintering. The sintered disks were cut into a cross section of 3.5×7.5 mm^2 and a length of 5 mm. After coating by Ni, the devices were connected using Ag paste and sheets. The module had an Al_2O_3 substrate with 64.5×64.5 mm^2 on one surface, namely, a module with a half skeleton structure. Power generation was measured using an electrical plate shape furnace in air. The Al_2O_3 substrate was heated at a furnace temperature of 373–873 K and the other surface of the module was cooled by water circulation at a temperature of 293 K.

The photograph of thermoelectric module is show in the inset of Figure 4. 37. The maximum output power increased with the hot side temperature and reached 9.4 W at 873 K of the furnace temperature (Figure 4.37). This wattage corresponds to 2.3 kW/m^2 against the surface area of the substrate. After the measurement of power generation, no damage and deterioration of internal resistance of the module were observed. Though durability and lifetime testing will be carried out more minutely, the silicide module is a promising device for the recovery of waste heat in the middle range temperature.

The long lifetime tests of the silicide module have been carried out under air and vacuum atmospheres. Between the measurements of power generation, the external

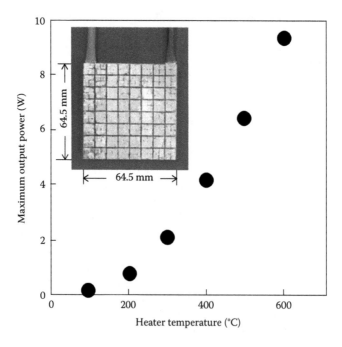

FIGURE 4.37 A photograph of the $Mn_3Si_4Al_2/MnSi_{1.7}$ module (inset) and temperature dependence of the output power measured in air.

load was disconnected from the module. In the air atmosphere, the normalized power of the module decreases at temperatures higher than 773 K. This is due to an increase in the internal resistance of the module. No damages of the module were observed clearly after the test. However, the color of the surface of the Ag electrodes of the n-type devices turned to black. Though the reason of this change in color has never been understood, Al was detected at the surface of the Ag electrodes by EDX analysis. On the other hand, no deterioration of the TE output power has been observed in the test in vacuum even at 973 K for 3 days. Some reactions at the junction of the n-type devices are the main reasons for the degradation of the silicide module in air.

4.5.5 SUMMARY

The TE properties of $Mn_{3-x}Cr_xSi_4Al_2$ (MSA-342) with a $CrSi_2$ structure were investigated. This material shows an n-type character and ZT of 0.21 at 773 K and ZT of 0.3 at 573 K for the non-Cr-substituted and the Cr-substituted sample with $x=0.3$, respectively. Since MSA-342 is in the intrinsic range at temperatures higher than 573–773 K, the ZT values markedly decrease. Electrical resistivity measured at 873 K was constant for 2 days in air. This indicates that MSA-342 devices have good oxidation resistance at high temperatures in air. This is caused by the formation of the passive oxide layer around the surface. The MSA-342 is a promising n-type material for power generation applications in the middle temperature range from 473 to 773 K. A TE module consisting of 64 pairs of legs was fabricated using $Mn_3Si_4Al_2$ and $MnSi_{1.7}$ devices as n- and p-type devices, respectively. Output power reached 9.4 W at a furnace temperature of 873 K in air.

4.6 OPTIMAL DESIGN OF CASCADE-TYPE THERMOELECTRIC MODULES

4.6.1 NOMENCLATURE

T Temperature (K)
Q_h Heat transfer rate from the heat source to the hot surface of a cascade module (W)
Q_c Heat transfer rate from the cold surface of the cascade module to the heat sink (W)
P Output power of the cascade module (W)
η Conversion efficiency; $=P/Q_h$
n Number of p–n pairs in a single-stage module
m Number of stages in a multistage cascade module
d Leg length of the TE element (m)
a Cross-sectional area of the TE element (m^2)
l Ratio of a to d (m); $=a/d$
S Relative Seebeck coefficient (V/K)
ρ Electrical resistivity (Ω m)
λ Thermal conductivity (W/[m K])
K Thermal conductance (W/K)

R Electrical resistance (Ω)

R_L External load (Ω)

I Current (A)

E Electromotive force (V)

V Electrical potential (V)

J Current density (A/m)

Subscript

i ith stage from the hot side in the multistage cascade module

Superscripts

p   p-Type material

n   n-Type material

4.6.2 INTRODUCTION

TE generation based on the Seebeck effect can directly convert heat into electricity. A TE generation system has the advantage of not requiring a large-scale system and has been studied as a way to recover unused heat, such as waste heat from automobiles,[107] fuel cells,[108] and marine engines,[109] as well as solar heat.[110–112] However, the conversion efficiency of TE generation systems generally remains low.

It has been believed that TE generation efficiency is determined only by the working temperature and material properties, because the TE power is generally proportional to the square of the temperature difference. However, conventional single-stage TE modules operating under large temperature difference conditions cannot effectively convert heat to power. This is mainly because there are no suitable TE materials that maintain a good performance over a wide temperature range. For example, a good performance of BiTe material at room temperature cannot be extended to the higher temperature range because of strong temperature dependencies of thermal properties. Generally an excellent property of a TE material is limited at a certain temperature range, and it is easily lost outside a good temperature range.

The performance can be enhanced by stacking several TE modules vertically (so-called cascade modules).[99,113,114] For example, a module consists of two stacking layers where their materials are selected judged by their performance suitable for the operating temperature ranges. The TE generation efficiency of the cascade module increases if the thermal connection between the materials is ideal.

Another method to improve the efficiency is to optimize the parameters of the module structure, such as the TE element size and number of p–n pairs. Harman[113] derived a general expression for the overall efficiency of multistage cascade modules and the optimal ratio of p–n pair number. Figure 4.38 illustrates the cascade module. A 1D heat balance model was used in a two-stage module by setting a given intermediate junction temperature. In a practical module construction, Zhang et al.[114] indicated that cascade TE generators using TE oxides have a high potential for heat recovery from high-temperature waste. Funahashi[99] fabricated the cascade modules consisting of oxide and Bi_2Te_3 modules, and he indicated experimentally that the cascade structures are very effective for high efficiency. Kaibe et al.[115] showed that the efficiency is as high as 10% using the stacking of TE silicide and Bi_2Te_3.

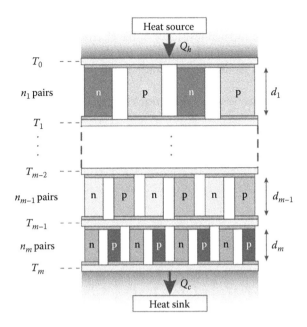

FIGURE 4.38   Illustration of a cascade-type TE module.

This value reported at 2005 is still a world record up to now. Although the cascade design has been developed to improve the TE performance, and several commercial modules with multistacking are listed in the market, the optimal structure and design remain elusive and understudied.

The purpose of this section is to introduce the optimization technique in the cascade module structure. When the module structure can be approximated in a 1D system, the heat balance can be analyzed in an analytical way. However, the TE behaviors do not hold the linear dependencies on temperature, and they are neglected at the primitive steps for simplicity of calculation. A rough design of the TE module can be mathematically estimated in an analytical solution from 1D analysis. When we need the more precise temperature distribution for the optimal design of a realistic module, 2D and 3D numerical calculations should be applied to solve the heat transfer equations in the module with the consideration of TE effects.

This section shows as the example of 1D analysis how the TE element dimensions and the number of p–n pairs can be optimized. This analytical approach can roughly predict the maximum power output by neglecting the Peltier heat. This estimation is then followed by numerical calculations based on the finite volume method to optimize the TE element sizes and to evaluate the efficiencies.

4.6.3   MODELING

4.6.3.1   1D Analytical Model and Evaluation

A cascade module consisting of m stages is analyzed as illustrated in Figure 4.38. It is assumed that the input heat on the top surface, Q_h, is transferred to a heat sink without any energy loss, such as that caused by thermal or electrical contact resistance, and that the thermal resistance of the electric insulators can be ignored. These assumptions are certainly not realistic, but these effects are not so serious in the actual modules. As the first assumption, we will set these assumptions for analytical exressions.

The heat balance equations[113,116] are written as

$$
\begin{cases}
Q_h = \left[K_1(T_0 - T_1) + S_1 T_0 I - \dfrac{1}{2} R_1 I^2 \right] n_1 \\[2mm]
\vdots \\[2mm]
\left[K_{m-1}(T_{m-2} - T_{m-1}) + S_{m-1} T_{m-1} I + \dfrac{1}{2} R_{m-1} I^2 \right] n_{m-1} \\[2mm]
= \left[K_m(T_{m-1} - T_m) + S_m T_{m-1} I - \dfrac{1}{2} R_m I^2 \right] n_m \\[2mm]
Q_c = \left[K_m(T_{m-1} - T_m) + S_m T_m I + \dfrac{1}{2} R_m I^2 \right] n_m
\end{cases}
\tag{4.6}
$$

where $K_i(T_{i-1} - T_i)$, $S_i T_i I$, and $(1/2)R_i I^2$ represent the heat conduction, Peltier heat, and Joule heat, respectively, of one p–n pair. The terms K_i and R_i are written as

$$
K_i = \lambda_i^p \frac{a_i^p}{d_i} + \lambda_i^n \frac{a_i^n}{d_i} = \lambda_i^p l_i^p + \lambda_i^n l_i^n
\tag{4.7}
$$

$$
R_i = \rho_i^p \frac{d_i}{a_i^p} + \rho_i^n \frac{d_i}{a_i^n} = \frac{\rho_i^p}{l_i^p} + \frac{\rho_i^n}{l_i^n}
\tag{4.8}
$$

In order to maximize the conversion efficiency under constant temperature conditions, the optimal relationship between l_i^p and l_i^n is well studied[113,116] and given by

$$
\frac{l_i^n}{l_i^p} = \sqrt{\frac{\rho_i^n \lambda_i^p}{\rho_i^p \lambda_i^n}}
\tag{4.9}
$$

This optimization comes only from the intrinsic material properties, and it does not relate with the cell dimensions. Using the optimized element design, Equations 4.7 through 4.9 can be rewritten as

$$K_i = \left(\lambda_i^p + \lambda_i^n \sqrt{\frac{\rho_i^n \lambda_i^p}{\rho_i^p \lambda_i^n}} \right) l_i^p \tag{4.10}$$

$$R_i = \left(\rho_i^p + \rho_i^n \sqrt{\frac{\rho_i^p \lambda_i^n}{\rho_i^n \lambda_i^p}} \right) \frac{1}{l_i^p} \tag{4.11}$$

It is noteworthy that the variable parameter related to the TE element sizes is only l_i^p, and l_i^n is automatically determined by Equation 4.9 under the optimized TE element conditions.

A practical assumption is taken here that all modules are connected electrically in series. This is a reasonable assumption because a series circuit has the advantage that the number of output electric leads is reduced to only two terminals. By this merit, the heat loss from the leads can be minimized compared with parallel and individual connections. The current and electromotive force are then given by

$$I = \frac{\sum_{i=1}^{m} n_i S_i (T_{i-1} - T_i)}{R_L + \sum_{i=1}^{m} n_i R_i} \tag{4.12}$$

$$E = \sum_{i=1}^{m} n_i S_i (T_{i-1} - T_i) - \sum_{i=1}^{m} n_i R_i I \tag{4.13}$$

It is assumed that Q_h, Q_c, $T_0,\ldots, T_m$, $l_1^p \supset, l_m^p$, and $n_1,\ldots, n_m$ are variables in the m-stage cascade model. The number of free variables then totals $3m+3$, while the number of parameters is fixed at $m+1$ in Equation 4.6. Therefore, $2m+2$ of these $3m+3$ variables can be freely set as target values.

Although the junction temperatures $T_0,\ldots, T_m$ are generally unknown, they were fixed as target temperature assumptions in this work. In addition, $n_1,\ldots, n_m$ and Q_h were here fixed as external variables. In these initial settings, there are many variations for future optimization and its analysis.

Under these conditions, $l_1^p \supset, l_m^p$ and Q_c were solved by applying Equation 4.6. When each stage of a cascade module is connected electrically in series, the number of p–n pairs is needed to optimize the relation of electromotive force and internal resistance of a stage to those of another stage. This optimization can enhance the conversion efficiency as the total module. It is natural that the optimal dimensions of TE elements depend on the number of p–n pairs. These two parameters, dimensions and pair number, are requested to be optimized simultaneously.

Generally the TE output power from the module is defined as

$$P = Q_h - Q_c = IE \tag{4.14}$$

The maximum power P_{max} is then calculated as the mathematical product of I and E.[116] P_{max} is given by

$$P_{max} = \frac{\left[\sum_{i=1}^{m} n_i S_i (T_{i-1} - T_i) \right]^2}{4 \sum_{i=1}^{m} n_i R_i} \tag{4.15}$$

where the external resistance is tacitly optimized. This assumption means that the external load should be equivalent with the internal resistance in any case, although this is a general description in any circuit of direct current.

In order to find P_{max}, the analytical solution was taken. The power was calculated by repeatedly varying $n_1,..., n_m$ in order to find optimal values. In a practical module, the variation of the number of pairs causes some complicated issues. For example, the arrangement of TE elements should be modified due to the requested number of pairs, and the size of empty space between the TE elements should be changed. These factors slightly affect the heat transfer rate Q_h or Q_c. However, these complicated issues are here simplified because the 1D heat transfer model is here considered.

4.6.3.2 Numerical Model and Evaluation Procedure

In the 1D model, only the heat balance at the junctions is considered, and we have no knowledge about the temperature or current density distribution inside the TE materials because a differential equation for the heat conduction is not included. For simplicity, the homogeneous TE properties in the TE elements are assumed. Taking energy conservation inside a control volume into consideration, the heat conduction equation under steady-state conditions[117,118] is taken and written as

$$\nabla \cdot (\lambda \nabla T) + \rho \, | \, \mathbf{J} \, |^2 - T \mathbf{J} \cdot \nabla S = 0 \tag{4.16}$$

where the first, second, and third terms represent the heat conduction, Joule heat generated by the current along the control volume, and heating or cooling generated by the Thomson effect, respectively. The current density $\mathbf{J}$ is related to the electrical potential and temperature as[117,118]

$$\nabla V = -\rho \mathbf{J} - S \nabla T \tag{4.17}$$

where the first term on the right-hand side is the voltage drop due to Ohm's law and the second term is the increase in voltage generated by the Seebeck effect. The following differential equation can be derived from Equation 4.17 by applying charge conservation under steady-state conditions ($\nabla \cdot \mathbf{J} = 0$):

$$\nabla \cdot \left(-\frac{1}{\rho} \nabla V \right) = \nabla \cdot \left(\frac{S}{\rho} \nabla T \right) \tag{4.18}$$

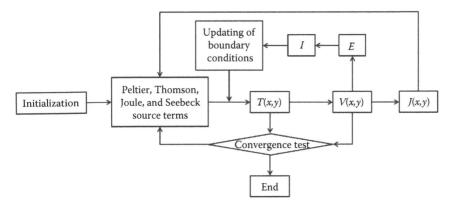

FIGURE 4.39 Workflow of the numerical calculation.

The temperature and electrical potential distributions can be obtained by solving the simultaneous differential equations of Equations 4.16 and 4.18.

Generally, it is difficult to find an analytical solution, except for the simplest initial conditions. These two simultaneous equations can be solved numerically based on the finite volume method. The detailed algorithm and method for applying the TE numerical model to FLUENT were reported in Refs. [109,117–121]. Due to the development of mechanical simulators in compact computers, the transport equations can be numerically solved using the commercial software. ANSYS FLUENT is one of them, although their original TE functions are not in a level for realistic analysis at present. Here some examples will be introduced, where the handmade functions[109,117–121] are coded and installed in FLUENT to solve the two terms on the right-hand side of Equation 4.17.

Figure 4.39 shows the workflow of the numerical calculation. After giving the initial values such as the temperature and electrical potential distributions and the boundary conditions, the terms induced by TE phenomena are evaluated. Subsequently, Equations 4.15 and 4.18 are solved, and temperature and electrical potential distributions are calculated. By updating the source terms and boundary conditions with the calculated results, FLUENT iterates the procedure until the solution converges.

4.6.4 1D ANALYTICAL EVALUATION

Here we focus on three-stage cascade modules; m should be 3 in Figure 4.39. Table 4.6 shows the TE material combinations which will be used here. These materials[61,99,122–124] are extensively examined as the TE future materials applicable over the wide temperature range by the national project team, JST-CREST conducted by Prof. Koumoto.

Cases 1 and 2 are the experimental data recently obtained, and Case 3 is the ideal target data of their team. For 1D analytical calculations, some variables should be fixed. Here the temperature dependencies of the TE properties are firstly ignored, although the detailed properties are reported.[61,99,116,122–124] Using these

increases. In order to prevent heat transfer, it is clear that the cross-sectional area of TE elements should be smaller or the TE elements should be longer.

Figure 4.41 shows an example calculated for Case 1. The current is calculated by Equation 4.12 when the external load R_L is equal to the internal resistance. The electrical resistance in the first stage is constant because n_1 is fixed, and those at the second and third stages are not constant but vary correspondingly to the shape of the TE elements. These behaviors of electrical resistance reflect the characteristic behavior as shown in Figure 4.41.

Figure 4.42a through c shows contour plots of the efficiency against n_2 and n_3 for Cases 1, 2, and 3, respectively. There is a clear optimal set of n_2 and n_3 for the corresponding n_1 value. Cases 1-A, 2-A, and 3-A indicate the traditional cascade module design ($n_1 = n_2 = n_3$), and Cases 1-B, 2-B, and 3-B are the optimized conditions found by 1D analysis.[119,121] The improvement of efficiency is significant by selecting a new design. It is noteworthy that the highest efficiency such as 19.6% can be achieved by setting the optimal design, even if the materials with $ZT < 1.0$ are used.

Comparing Cases 1-B, 2-B, and 3-B, it is obvious that the optimal condition depends on the TE material combination. When the materials with a higher performance (higher ZT value in Table 4.1) are used, the optimization procedure becomes crucial and is effective in enhancing the conversion efficiency. Additionally, optimization of the TE element dimensions (l_1^p, l_2^p, and l_3^p) is needed to obtain the target temperature distribution. Therefore, in designing a cascade module, it is not satisfactory

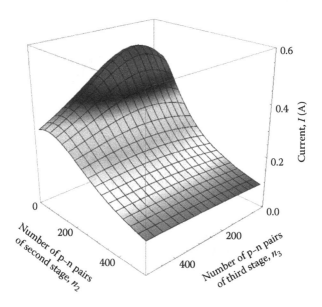

FIGURE 4.41 Calculated current in the module of Case 1 as a function of number of pairs at the second and third stages, n_2 and n_3, where $n_1 = 100$. (From Suzuki, R.O. and Fujisaka, T., *Proceedings of Powder Metallurgy World Congress & Exhibition [PM2012 Yokohama]*, Yokohama, October 14–18, 2012, 16E-S1-17, 0463_201211060004.pdf, 2012.)

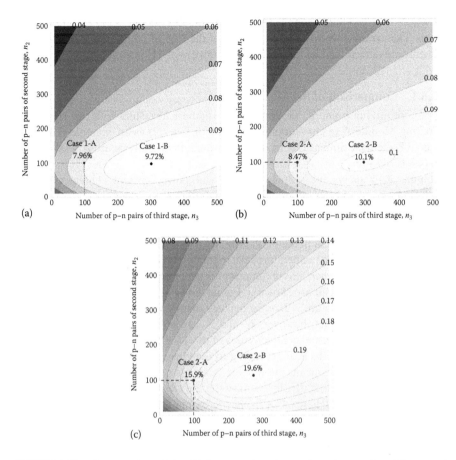

FIGURE 4.42   Contour plots of the efficiency against the number of p–n pairs of the second and third stages in a three-stage module at $n_1 = 100$: Case 1 (a), Case 2 (b), and Case 3 (c).

to simply select high-performance TE materials; the structure should also be designed with the optimal length, cross-sectional area, and number of p–n pairs.

4.6.5   2D NUMERICAL EVALUATION

In the 1D analytical procedure, both the number of p–n pairs and the TE element dimensions are optimized. However, dimensional optimization is not completed because, although l is optimized, both a and d are not individually determined. In order to optimize these parameters, a 2D or 3D analysis is effective.

Here we focus on two-stage cascade modules ($m = 2$) to simplify the problem. A temperature range of 300–800 K is assumed for comparison with the 1D results, and Case 1 is used. In numerical analysis, we also include electrodes and insulators made of copper (0.2 mm thick) and alumina (0.6 mm thick), respectively. Their properties are taken from Ref. [125]. Both thermal and electrical contact resistances are ignored for simplicity, although they are critically important for practical fabrication. The arrangement of p–n elements and the size of empty space between the elements are determined by reference to a commercial TE module.

4.6.5.1 Comparison of the Models

Before optimizing a and d, we compare the numerical results with the 1D results by taking Module A (a conventional structure; $n_1=n_2=2$) as an example. The optimized TE element dimensions from the 1D analytical analysis are $l_1^p = 0.450$ mm and $l_2^p = 0.744$ mm.

Figure 4.43 shows the voltage–current and power–current characteristics of Module A calculated using both the 1D analytical model and the 2D numerical model. The maximum power results of the analytical and numerical models are 73.5 and 68.1 mW, respectively. To discuss this difference, we examine the calculated temperature and electrical potential distributions in both models.

Figure 4.44 shows the temperature, electrical potential, and current density distributions for Module A when maximum power is generated. The surface temperatures T_{11} and T_{12} of the insulator are shown in Figure 4.44a as a function of the horizontal position x. The temperature is homogeneously distributed along the x direction, as shown in the contour plot; however, T_{11} and T_{12} are not completely constant due to the difference in the thermal conductivities of the TE elements and the Peltier effect that generates and absorbs heat according to the current at the junctions. The voltage drop caused by the current through electrodes is negligible, as shown in Figure 4.44b, although the current density increases at the electrodes.

The temperature difference, $T_{11}-T_{12}$, is approximately 5 K, which is not identical to that obtained in the 1D model: The thermal resistance in the insulators is ignored there, and it is assumed that no temperature difference exists inside the insulators. Although the insulators are thin and have a thermal conductivity that is 10 times higher than the TE materials, the TE power is sensitive to the temperature difference and drops drastically even if the temperature difference between both sides of the TE

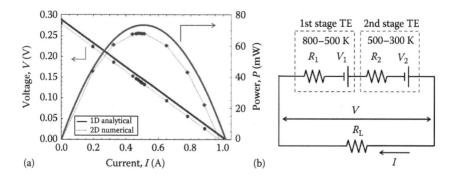

FIGURE 4.43 Voltage–current and power–current characteristics of Module A calculated using a 1D analytical model and 2D numerical model (a) (From Fujisaka, T. et al., *J. Electron. Mater.*, 42, 1688, 2013) and the equivalent electric circuit is inserted (b).

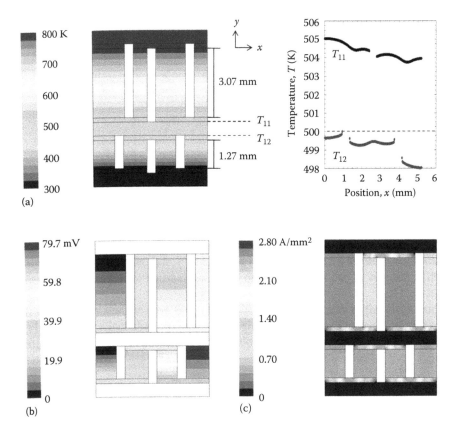

FIGURE 4.44 Temperature profile (a), electrical potential profile (b), and current density profile of Module A (c).

element decreases slightly due to the thermal resistance of the insulators. Therefore, it is important to minimize the thermal resistance by reducing the insulator thickness and using insulators with a higher thermal conductivity.

4.6.5.2 Optimization of the Cross-Sectional Area and Leg Length

Figure 4.45 shows the power obtained at $m=2$ and $n_1=2$ as a function of n_2; Module A in the previous section corresponds to $n_2=2$. The maximum power is obtained at $n_2=6$, where $l_1^p = 0.462$ mm and $l_2^p = 0.220$ mm. Based on these 1D results, two possible structures are reported,[119] which are equivalent in the 1D analysis. Here one model, as referred to Module B, is here introduced from the 2D numerical analysis.

Figure 4.46 shows the temperature, electrical potential, and current density distributions in Module B. Table 4.7 lists the dimensions of these modules and their performance.

The temperature difference is approximately 5 K in Module B, as in the case of Module A. From this, we can conclude that the 1D heat transfer model was too highly simplified but partly reasonable.

REFERENCES

1. International Energy Agency, *World Energy Outlook 2009 Edition*, (IEA PUBLICATIONS, Paris), p. 623, 2009.
2. International Energy Agency, *World Energy Outlook 2012 Edition*, (IEA PUBLICATIONS, Paris), p. 51, 2012.
3. M. Hirata, *Syou Enerugi Ron* (Ohmu Sha, Tokyo), p. 37, 1994 [in Japanese].
4. The Energy Conservation Center, *Hesei 16 nendo Shou Enerugi Gijutu Fukyu-sokushinjigyou Houkokusyo*, pp. 74–75, 2005 [in Japanese].
5. Homepage of Japan Automobile Manufactures Association, Inc., http://www.jama.or.jp/index.html, 2012 [in Japanese].
6. Homepage of Idaho National Laboratory, http://www.inl.gov/marsrover/, 2012.
7. G. Chen, M. S. Dresselhaus, G. Dresselhaus, J. P. Fleurial and T. Caillat, *Int. Mater. Rev.*, 2003, **48**, 45–66.
8. K. Fujita, T. Mochida and K. Nakamura, *Jpn. J. Appl. Phys.*, 2001, **40**, 4644–4647.
9. R. Funahashi, I. Matsubara, H. Ikuta, T. Takeuchi, U. Mizutani and S. Sodeoka., *Jpn. J. Appl. Phys.*, 2000, **39**, L1127–L1129.
10. R. Funahashi and M. Shikano, *Appl. Phys. Lett.*, 2002, **81**, 1459–1461.
11. A. I. Hochbaum, R. Chen, R. D. Delgado, W. Liang, E. C. Garnett, M. Najarian, A. Majumdar and P. Yang, *Nature*, 2008, **451**, 163–167.
12. D. Teweldebrhan, V. Goyal and A. A. Balandin, *Nano Lett.*, 2010, **10**, 1209–1218.
13. H. Imai, Y. Shimakawa and Y. Kubo, *Phys. Rev. B: Condens. Matter Mater. Phys.*, 2001, **64**, 241104.
14. C. Wan, Y. Wang, W. Norimatsu, M. Kusunoki and K. Koumoto, *Appl. Phys. Lett.*, 2012, **100**, 101913.
15. R. Z. Zhang, C. L. Wan, Y. F. Wang and K. Koumoto, *Phys. Chem. Chem. Phys.*, 2012, **14**, 15641–15644.
16. M. Dresselhaus, G. Chen, M. Tang, R. G. Yang, H. Lee, D.-Z. Wang, Z.-F. Ren, J. P. Fleurial and P. Gogna, *Adv. Mater.*, 2007, **19**, 1043–1053.
17. G. J. Snyder and E. S. Toberer, *Nat. Mater.*, 2008, **7**, 105–114.
18. S. Ohta, T. Nomura, H. Ohta and K. Koumoto, *J. Appl. Phys.*, 2005, **97**, 034106.
19. S. Ohta, T. Nomura, H Ohta and K. Koumoto, *Appl. Phys. Lett.*, 2005, **87**, 092108.
20. Y. F. Wang, K. Fujinami, R. Z. Zhang, C. L. Wan, N. Wang, Y. S. Ba and K. Koumoto, *Appl. Phys. Express*, 2010, **3**, 031101.
21. L. D. Hicks and M. S. Dresselhaus, *Phys. Rev. B*, 1993, **47**, 12727–12731.
22. H. Ohta, S.-W. Kim, Y. Mune, T. Mizoguchi, K. Nomura, S. Ohta, T. Nomura, Y. Nakanishi, Y. Ikuhara, M. Hirano, H. Hosono and K. Koumoto, *Nat. Mater.*, 2007, **6**, 129–134.
23. R. Z. Zhang, C. L. Wang, J. C. Li, and K. Koumoto, *J. Am. Ceram. Soc.*, 2010, **93**, 1677; R. Z. Zhang and K. Koumoto, *J. Electron. Mater.*, 2013, **42**, 1568–1572.
24. F. Dang, K. Mimura, K. Kato, H. Imai, S. Wada, H. Haneda and M. Kuwabara, *CrystEngComm*, 2011, **13**, 3878–3883.
25. F. Dang, K. Mimura, K. Kato, H. Imai, S. Wada, H. Haneda and M. Kuwabara, *Nanoscale*, 2012, **4**, 1344–1349.
26. F. Dang, K. Turuda, C. L. Wan and K. Koumoto, *International Conference on Thermoelectrics (ICT 2013)*, Kobe, 2013.
27. F. Dang, C. L. Wan, K. Turuda, N. Park, W. Seo and K. Koumoto, *ACS Appl. Mater. Interfaces*, 2013, **5**, 10933–10937.
28. Y. F. Wang, X. Y. Zhang, L. M. Shen, N. Z. Bao, C. L. Wan, N. H. Park, K. Koumoto, A. Gupta, *J. Power Sources*, 2013, **241**, 255–258.
29. B. Eisenmann, H. Schäfer, and R. Zagler, *J. Less-Common Metals*, 1986, **118**, 43–55.

30. K. Momma and F. Izumi, *J. Appl. Cryst.*, 2011, **44**, 1272–1276 (Crystal structure was drown on data of Ref. [31] by using a program VESTA).
31. H. Anno, H. Yamada, T. Nakabayashi, M. Hokazono and R. Shirataki, *J. Solid State Chem.*, 2012, **193**, 94–104.
32. G. Nolas, J. L. Cohn, G. A. Slack and S. B. Schujman, *Appl. Phys. Lett.*, 1998, **73**, 178–180.
33. J. L. Cohn, G. S. Nolas, V. Fessatidis, T. H. Metcalf and G. A. Slack, *Phys. Rev. Lett.*, 1999, **82**, 779–782.
34. G. S. Nolas, G. A. Slack and S. B. Schujman, *Recent Trends in Thermoelectric Materials Research I, Semiconductors and Semimetals*, Volume ed. T. M. Tritt, Academic Press, San Diego, CA, 2001, vol. 69, ch. 6, pp. 255–300.
35. N. P. Blake, L. Møllnitz, G. Kresse and H. Metiu, *J. Chem. Phys.*, 1999, **111**, 3133–3144.
36. V. L. Kuznetsov, L. A. Kuznetsova, A. E. Kaliazin and D. M. Rowe, *J. Appl. Phys.*, 2000, **87**, 7871–7875.
37. N. P. Blake, S. Latturner, J. D. Bryan, G. D. Stucky and H. Metiu, *J. Chem. Phys.*, 2001, **115**, 8060–8073.
38. B. C. Sales, B. C. Chakoumakos, R. Jin, J. R. Thompson and D. Mandrus, *Phys. Rev. B*, 2001, **63**, 245113.
39. Y. Mudryk, P. Rogl, C. Paul, S. Berger, E. Bauer, G. Hilsher, C. Godart and H. Noël, *J. Phys.: Condens. Matter*, 2002, **14**, 7991–8004.
40. D. Nataraj, J. Nagao, M. Ferhat and T. Ebinuma, *J. Appl. Phys.*, 2003, **93**, 2424–2428.
41. H. Anno, M. Hokazono, M. Kawamura, J. Nagao and K. Matsubara, *IEEE Proceedings of XXI International Conference on Thermoelectrics (ICT 2002)*, IEEE, Piscataway, NJ, 2002, pp. 77–80.
42. A. Bentien, B. B. Iversen, J. D. Bryan, G. D. Stucky, A. E. C. Palmqvist, A. J. Schultz and R. W. Henning, *J. Appl. Phys.*, 2002, **91**, 5694–5699.
43. G. K. H. Madsen, K. Schwarz, P. Blaha and D. J. Singh, *Phys. Rev. B*, 2003, **68**, 125212.
44. H. Anno, M. Hokazono, M. Kawamura and K. Matsubara, *IEEE Proceedings of XXII International Conference on Thermoelectrics (ICT 2003)*, IEEE, Piscataway, NJ, 2003, pp. 121–126.
45. F. Bridges and L. Downward, *Phys. Rev. B*, 2004, **70**, 140201(R).
46. A. Bentien, E. Nishibori, S. Paschen and B. B. Iversen, *Phys. Rev. B*, 2005, **71**, 144107.
47. M. Hokazono, H. Anno and K. Matsubara, *Mater. Trans.*, 2005, **46**, 1485–1489.
48. A. Saramat, G. Svensson, A. E. C. Palmqvist, C. Stiewe, E. Mueller, D. M. Rowe, J. D. Bryan and G. D. Stucky, *J. Appl. Phys.*, 2006, **99**, 023708.
49. C. L. Condron, J. Martin, G. S. Nolas, P. M. B. Piccoli, A. J. Schultz and S. M. Kauzlarich, *Inorg. Chem.*, 2006, **45**, 9381–9386.
50. M. A. Avila, K. Suekuni, K. Umeo, H. Fukuoka, S. Yamanaka and T. Takabatake, *Phys. Rev. B*, 2006, **74**, 125109.
51. I. Fujita, K. Kishimoto, M. Sato, H. Anno and T. Koyanagi, *J. Appl. Phys.*, 2006, **99**, 093707.
52. J.-H. Kim, N. L. Okamoto, K. Kishida, K. Tanaka and H. Inui, *Acta Materialia*, 2006, **54**, 2057–2062.
53. N. L. Okamoto, K. Kishida, K. Tanaka and H. Inui, *J. Appl. Phys.*, 2007, **101**, 113525.
54. X. Tang, P. Li, S. Deng and Q. Zhang, *J. Appl. Phys.*, 2008, **104**, 013706.
55. K. Suekuni, M. A. Avila, K. Umeo, H. Fukuoka, S. Yamanaka, T. Nakagawa and T. Takabatake, *Phys. Rev. B*, 2008, **77**, 235119.
56. E. S. Toberer, M. Christensen, B. B. Iversen and G. J. Snyder, *Phys. Rev. B*, 2008, **77**, 075203.
57. S. Deng, X. Tang and R. Tang, *Chin. Phys. B*, 2009, **18**, 3084–3089.
58. K. Koga, K. Suzuki, M. Fukamoto, H. Anno, T. Tanaka and S. Yamamoto, *J. Electron. Mater.*, 2009, **38**, 1427–1432.

59. S. Deng, Y. Saiga, K. Suekuni and T. Takabatake, *J. Appl. Phys.*, 2010, **108**, 073705.
60. S. Deng, Y. Saiga, K. Kajisa and T. Takabatake, *J. Appl. Phys.*, 2011, **109**, 103704.
61. N. Tsujii, J. H. Roudebush, A. Zevalkink, C. A. Cox-Uvarov, G. J. Snyder and S. M. Kauzlarich, *J. Solid State Chem.*, 2011, **184**, 1293–1303.
62. T. Nakabayashi, M. Hokazono, H. Anno, Y. Ba and K. Koumoto, *Mater. Sci. Eng.*, 2011, **18**, 142008.
63. U. Aydemir, C. Candolfi, A. Ormeci, Y. Oztan, M. Baitinger, N. Oeschler, F. Steglich and Yu. Grin, *Phys. Rev. B*, 2011, **84**, 195137.
64. I. Zeiringer, M. X. Chen, I. Bednar, E. Royanian, E. Bauer, R. Podloucky, A. Grytsiv, P. Rogl and H. Effenberger, *Acta Materialia*, 2011, **59**, 2368–2384.
65. Y. Saiga, B. Du, S. K. Deng, K. Kajisa and T. Takabatake, *J. Alloys Compd.*, 2012, **537**, 303–307.
66. I. Zeiringer, M. X. Chen, A. Grytsiv, E. Bauer, R. Podloucky, H. Effenberger and P. Rogl, *Acta Materialia*, 2012, **60**, 2324–2336.
67. H. Anno, M. Hokazono, R. Shirataki and Y. Nagami, *J. Mater. Sci.*, 2013, 48, 2846–2854.
68. X. Yan, E. Bauer, P. Rogl and S. Paschen, *Phys. Rev. B*, 2013, **87**, 115206.
69. G. A. Slack, *CRC Handbook of Thermoelectrics*, ed. D. M. Rowe, CRC Press, Boca Raton, FL, 1995, pp. 407–440.
70. G. A. Slack, *Materials Research Society Symposium Proceedings*, ed. T. M. Tritt, M. G. Kanatzidis, H. B. Lyon, Jr. and G. D. Mahan, MRS Press, Warrendale, PA, 1997, vol. 478, pp. 47–54.
71. Mineral Resources Report in 2011, Japan Oil, Gas and Metals National Corporation (JOGMEC), http://mric.jogmec.go.jp/mric_search/ [in Japanese].
72. C. L. Condron, S. M. Kauzlarich, F. Gascoin and G. J. Snyder, *Chem. Mater.*, 2006, **18**, 4939–4945.
73. C. L. Condron, S. M. Kauzlarich, T. Ikeda, G. J. Snyder, F. Haarmann and P. Jeglic, *Inorg. Chem.*, 2008, **47**, 8204–8212.
74. N. Mugita, Y. Nakakohara, T. Motooka, R. Teranishi and S. Munetoh, *Mater. Sci. Eng.*, 2011, **18**, 142007.
75. H. Anno, M. Hokazono, R. Shirataki and Y. Nagami, *J. Electron. Mater.*, 2013, **47**, 2326–2336.
76. S. Krishnamurthy, A. Sher and A. Chen, *Appl. Phys. Lett.*, 1985, **47**, 160–162.
77. C. B. Vining, *J. Appl. Phys.*, 1991, **69**, 331–341.
78. J. H. Roudebush, C. de la Cruz, B. C. Chakoumakos and S. M. Kauzlarich, *Inorg. Chem.*, 2012, **51**, 1805–1812.
79. H. J. Goldsmid, *Electronic Refrigeration*, Pion Limited, London, 1986, ch. 3, pp. 58–64.
80. P. G. Klemens, *Phys. Rev.*, 1960, **119**, 507–509.
81. J. Callaway and H. C. Baeyer, *Phys. Rev.*, 1960, **120**, 1149–1154.
82. B. Abeles, *Phys. Rev.*, 1963, **131**, 1906–1911.
83. J. Yang, *Thermal Conductivity—Theory, Properties, and Applications—Physics of Solids and Liquids*, ed. T. M. Tritt, Kluwer Academic/Plenum Publishers, New York, 2004, ch. 1.1, pp. 1–20.
84. H. Anno, K. Suzuki, K. Koga and K. Matsubara, *IEEE Proceedings of XXVI International Conference on Thermoelectrics (ICT 2007)*, IEEE, Piscataway, NJ, 2007, pp. 226–229.
85. H. Anno, H. Fukushima, K. Koga, K. Okita and K. Matsubara, *IEEE Proceedings of XXV International Conference on Thermoelectrics (ICT 2006)*, IEEE, Piscataway, NJ, 2006, pp. 36–39.
86. C. L. Condron, S. M. Kauzlarich and G. S. Nolas, *Inorg. Chem.*, 2007, **46**, 2556–2562.
87. K. Akai, G. Zhao, K. Koga, K. Oshiro and M. Matsuura, *IEEE Proceedings of XXIV International Conference on Thermoelectrics (ICT 2005)*, IEEE, Piscataway, NJ, 2005, pp. 230–233.

88. K. Koga, H. Anno, K. Akai, M. Matsuura and K. Matsubara, *Mater. Trans.*, 2007, **48**, 2108–2113.
89. H. Anno, T. Nakabayashi and M. Hokazono, *Adv. Sci. Technol.*, 2010, **74**, 26–31.
90. B. C. Sales, D. G. Mandrus and B. C. Chakoumakos, *Recent Trends in Thermoelectric Materials Research II, Semiconductors and Semimetals*, Volume ed. T. M. Tritt, Academic Press, San Diego, CA, 2001, vol. 70, ch. 1, pp. 1–36.
91. N. P. Blake, D. Bryan, S. Latturner, L. Møllnitz, G. D. Stucky and H. Metiu, *J. Chem. Phys.*, 2001, **114**, 10063–10074.
92. H. Anno, K. Okita, K. Koga, S. Harima, T. Nakabayashi, M. Hokazono and K. Akai, *Mater. Trans.*, 2012, **53**, 1220–1225.
93. A. Prokofiev, A. Sidorenko, K. Hradil, M. Ikeda, R. Svagera, M. Waas, H. Winkler, K. Neumaier and S. Paschen, *Nat. Mater.*, 2013, **12**, 1096–1101.
94. I. Terasaki, Y. Sasago and K. Uchinokura, *Phys. Rev. B*, 1997, **56**, R12685–R12687.
95. R. Funahashi, M. Mikami, T. Mihara, S. Urata and N. Ando, *J. Appl. Phys.*, 2006, **99**, 066117.
96. S. Urata, R. Funahashi, T. Mihara, A. Kosuga, S. Sodeoka and T. Tanaka, *Int. J. Appl. Ceram. Technol.*, 2007, **4**, 535–540.
97. P. Tomes, C. Suter, M. Trottmann, A. Steinfeld and A. Weidenkaff, *J. Mater. Res.*, 2011, **26**, 1975–1982.
98. A. Inagoya, D. Sawaki, Y. Horiuchi, S. Urata, R. Funahashi and I. Terasaki, *J. Appl. Phys.*, 2011, **110**, 123712.
99. R. Funahashi, *Sci. Adv. Mater.*, 2011, **3**, 682–686.
100. T. Yamada, Y. Miyazaki and H. Yamane, *Thin Solid Films*, 2011, **519**, 8524–8527.
101. I. Aoyama, M. I. Fedorov, V. K. Zaitsev, F. Y. Solomkin, I. S. Eremin, A. Y. Samunin, M. Mukoujima, S. Sano and T. Tsuji, *Jpn. J. Appl. Phys.*, 2005, **44**, 8562–8570.
102. R. Wolfe, J. H. Wernick and S. E. Haszko, *Phys. Lett.*, 1965, **19**, 449–454.
103. K. Morikawa, H. Chikauchi, H. Mizoguchi and S. Sugihara, *Mater. Trans.*, 2007, **48**, 2100–2103.
104. JCPDS Card No. 35-0781.
105. V. K. Zaitsev, M. I. Fedorov, E. A. Gurieva, I. S. Eremin, P. P. Konstantinov, A. Y. Samunin and M. V. Vedernikov, *Phys. Rev. B*, 2006, **74**, 045207.
106. J. Tani and H. Kido, *Physica B: Condens. Matter*, 2005, **364**, 218–224.
107. D. M. Rowe, J. Smith, G. Thomas and G. Min, *J. Electron. Mater.*, 2011, **40**, 784–788.
108. L. A. Rosendahl, P. V. Mortensen and A. A. Enkeshafi, *J. Electron. Mater.*, 2011, **40**, 1111–1114.
109. M. Chen, Y. Sasaki and R. O. Suzuki, *Mater. Trans.*, 2011, **52**, 1549–1552.
110. N. Wang, L. Han, H. He, N. Park and K. Koumoto, *Energy Environ. Sci.*, 2011, **4**, 3676–3679.
111. D. Kraemer, B. Poudel, H. P. Feng, J. C. Caylor, B. Yu, X. Yan, Y. Ma, X. Wang, D. Wang, A. Muto, K. McEnaney, M. Chiesa, Z. Ren and G. Chen, *Nat. Mater.*, 2011, **10**, 532–538.
112. R. O. Suzuki, A. Nakagawa, H. Sui and T. Fujisaka, *J. Electron. Mater.*, 2013, **42**, 1960–1965.
113. T. C. Harman, *J. Appl. Phys.*, 1958, **29**, 1471–1473.
114. L. Zhang, T. Tosho, N. Okinaka and T. Akiyama, *Mater. Trans.*, 2008, **49**, 1675–1680.
115. H. Kaibe, I. Aoyama, M. Mukoujima, T. Kanda, S. Fujimoto, T. Kurosawa, H. Ishimabushi, K. Ishida, L. Rauscher, Y. Hata and S. Sano, *Proceedings of 24th International Conference on Thermoelectrics 2005 (ICT 2005)*, IEEE, Piscataway, NJ, p. 242.
116. D. M. Rowe, *CRC Handbook of Thermoelectrics*, CRC Press, Boca Raton, FL, 1995.
117. M. Chen, L. A. Rosendahl and T. J. Condra, *Int. J. Heat Mass Transf.*, 2011, **54**, 345–355.
118. M. Chen, S. J. Andreasen, L. A. Rosendahl, S. K. Kær and T. J. Condra, *J. Electron. Mater.*, 2010, **39**, 1593–1600.
119. T. Fujisaka, H. Sui, R. O. Suzuki, *J. Electron. Mater.*, 2013, **42**, 1688–1696.

120. R. O. Suzuki, Y. Sasaki, T. Fujisaka and M. Chen, *J. Electron. Mater.*, 2012, **41**, 1766–1700.
121. R. O. Suzuki and T. Fujisaka, *Proceedings of Powder Metallurgy World Congress & Exhibition (PM2012 Yokohama)*, Yokohama, October 14–18, 2012, 16E-S1-17, 0463_201211060004.pdf, 2012.
122. W. Luo, H. Li, F. Fu, W. Hao and X. Tang, *J. Electron. Mater.*, 2011, **40**, 1233–1237.
123. C. Wan, Y. Wang, N. Wang, W. Norimatsu, M. Kusunoki and K. Koumoto, *J. Electron. Mater.*, 2011, **40**, 1271–1280.
124. R. Zhang, C. Wang, J. Li and K. Koumoto, *J. Am. Ceram. Soc.*, 2010, **93**, 1677–1681.
125. Japan Society of Mechanical Engineering, *JSME Data Book Heat Transfer*, Japan Society of Mechanical Engineering, Maruzen, Tokyo, 4th edn., 1986.

5 Piezoelectric Materials for Energy Harvesting

Deepam Maurya, Yongke Yan, and Shashank Priya

CONTENTS

5.1 INTRODUCTION

Recent emphasis on wireless sensor nodes and distributed low-power electronic network has placed demand on developing energy harvesters that can locally harvest the electric energy from environmental sources such as solar, wind, water flow, heat, and mechanical vibrations. Figure 5.1 shows the compatibility of several commonly used electronic devices with the harvester output power.

Technological advances in improving the efficiency of low-power electronic devices coupled with the developments of energy harvesting solutions are expected to provide a significant reduction in the cost of maintenance and simplify the installation. In this chapter, first, we briefly discuss various environmental sources and then focus on vibration energy harvesting specifically using piezoelectric materials. A comprehensive overview of the technological advances in design of piezoelectric materials including lead-free piezoelectric materials is provided.

5.2 SOLAR ENERGY HARVESTING

Solar energy has attracted significant attention in recent years due to its abundance. As is well known, the total amount of energy received by the earth from the sun in

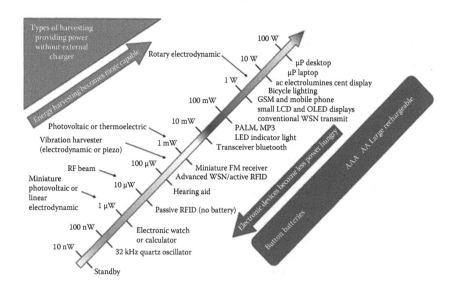

FIGURE 5.1 Energy harvesting device becoming increasingly efficient along with shrinking integrated circuit (IC) chip line geometries and lower consumption levels. (Adapted with permission from IDTechEx, Cambridge, MA.)

one day is sufficient to satisfy the energy requirements for the whole year. The variety and variation in solar cell efficiency demonstrated over the years are depicted in Figure 5.2 (source NREL). This plot describes the advances made in enhancing the performance of various configurations of the solar cell. The power conversion efficiency of a solar cell can be given as[1]

$$\eta = \frac{P_{out}}{P_{in}} = \frac{J_{sc}V_{oc}FF}{P_{in}} \tag{5.1}$$

where
P_{out} is the electrical output
P_{in} is the absorbed solar energy
J_{sc} is the short circuit current density
V_{oc} is the open-circuit voltage
FF is the fill factor (ratio of maximum obtainable power to the product of the V_{oc}
   and J_{sc}[1]

High-efficiency multijunction cells require expensive fabrication processes and thereby find limited applications like deep space explorations. With respect to energy harvesting, flexible technologies such as dye-sensitized solar cells (DSSCs) and hybrid solar cells are gaining a lot of attention as they can provide decent efficiencies (>10%) in a conformal package.[2] Recent research on solid-state DSSCs utilizing perovskite-based electrolyte has provided further boost to the field. In addition to being an excellent fundamental materials research topic, a solid-state DSSC overcomes the problem of packaging and provides a pathway for the integration with multiple platforms.[2]

Recently, ABO_3-type ferroelectrics have attracted great attention as photovoltaic materials.[3] Although photovoltaic properties of ferroelectric materials are known for the past 50 years, they did not receive much attention due to their reported low power conversion efficiency.[3] Interestingly, recent findings have demonstrated that the low conversion efficiency can be improved by exploiting the physical characteristics of the perovskites.[4–6] Some of the interesting features of ferroelectric photovoltaics include extremely large and above bandgap open-circuit voltage.[4] This is fundamentally different from conventional semiconductor solar cells.[3] The ferroelectric photovoltaic effect and conventional p–n junction photovoltaic effects are schematically shown in Figure 5.3a and b.[1] The absorbed photons, in the semiconductor of conventional p–n junction solar cells, can pump the electrons from the valence band to the conduction band leaving holes in the valence band.[7] Subsequently, the built-in electric field in the p–n junction separates the electrons and holes, which are collected by the respective electrodes.[7] The value of the open-circuit voltage (V_{oc}) in p–n junction solar cells can be theoretically estimated by the quasi-fermi energy difference of photogenerated electrons and holes.[7] This is further controlled by the bandgap of the light-absorbing semiconductors.[7] However, the experimental results on ferroelectric photovoltaic materials suggested that the output photovoltage is proportional to the electrode spacing and magnitude of the electric polarization.[4,8,9] Thus, the output V_{oc}

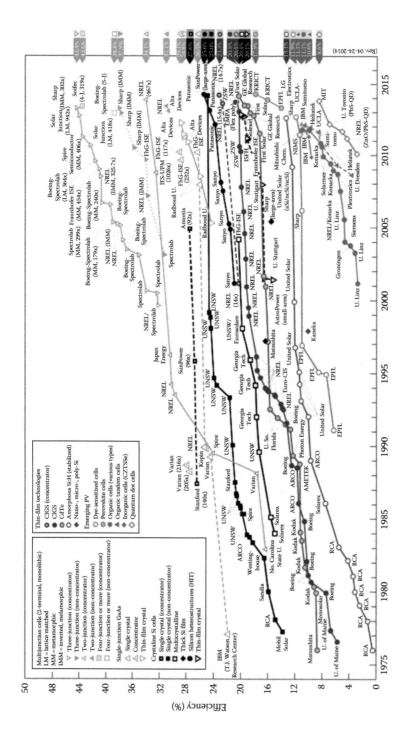

FIGURE 5.2 Increase in efficiency as a function of time for different kinds of solar cells. (Adapted from Kurtz, S., Opportunities and challenges for development of a mature concentrating photovoltaic power industry, TP-520-43208, National Renewable Energy Laboratory, Golden, CO, 2009.)

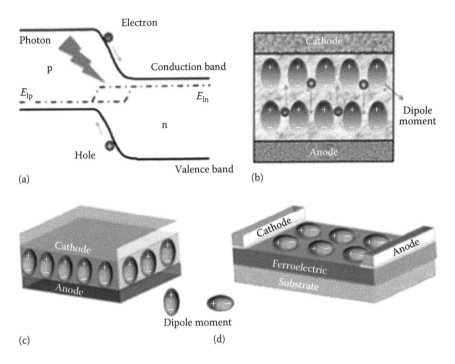

FIGURE 5.3 The working principle of (a) p–n junction solar cells and (b) ferroelectric photovoltaic devices. Ferroelectric photovoltaic device architectures, (c) vertical and (d) lateral, in which a large photovoltage proportional to the electrode spacing can be measured along the polarization direction (P). (Yuan, Y. et al., *J. Mater. Chem. A*, **2**, 6027–6041, Copyright 2014. Adapted by permission of The Royal Society of Chemistry.)

can be several orders of magnitude larger than the bandgap of the ferroelectric materials.[4,10,11] The configurations of typical thin-film ferroelectric photovoltaic devices are shown in Figure 5.3c and d.[1]

In a recent work, Cao et al. have reported a 72-fold increase in the efficiency of Pb(Zr, Ti)O$_3$ (PZT) based ferroelectric thin-film solar cells by using an n-type Cu$_2$O cathode buffer layer.[12] They reported that an ohmic contact on Pt/Cu$_2$O, an n$^+$–n$^-$ heterojunction on Cu$_2$O/PZT, and a Schottky barrier on PZT/indium tin oxide (ITO) provided a favorable energy-level alignment for efficient electron extraction on the cathode.[12] In another interesting recent work, Grinberg et al. reported high photocurrent density in [KNbO$_3$]$_{1-x}$[BaNi$_{1/2}$Nb$_{1/2}$O$_{3-\delta}$]$_x$ (KBNNO)-based compositions.[13] The photocurrent density was found to be 50 times higher than that of archetype (Pb,La) (ZrTi)O$_3$ materials.[13] This material was also found to absorb up to six times more solar energy than other existing ferroelectric materials.[13] These interesting results on the ferroelectric photovoltaic effect clearly demonstrate the potential of piezoelectric materials in solar energy harvesting. Domain engineering and chemical modifications could be successfully employed to enhance the efficiency of ferroelectric-based photovoltaic devices.

5.3 WIND ENERGY HARVESTING

Wind forms the major source of mechanical energy and is readily available in most places thereby providing an excellent opportunity to tap and power electronic components.[14] Commonly, in order to harvest wind energy, wind turbines are used.[14] There are broadly three ways to classify wind turbines: (1) on the basis of the orientation of axis of rotation (vertical or horizontal), (2) on the basis of the component of aerodynamic forces (lift or drag) that power the wind turbine, and (3) on the basis of the energy generating capacity (micro, small, medium, or large). Kishore and Priya[15] have defined the nomenclature of horizontal-axis wind turbines based on the size of the wind turbine rotor as follows:

1. Microscale wind turbine (μSWT): rotor diameter ≤ 10 cm
2. Small-scale wind turbine (SSWT): 10 cm < rotor diameter ≤ 100 cm
3. Mid-scale wind turbine (MSWT): 1 m < rotor diameter ≤ 5 m
4. Large-scale wind turbine (LSWT): rotor diameter > 5 m

For energy harvesting applications, SSWT becomes the category of choice.[15] Recently, Kishore et al. have demonstrated small-scale wind energy portable turbine (SWEPT, Figure 5.4) that operates near ground level and provides a significant magnitude of power at quite low wind speeds. SWEPT is a three-bladed, 40 cm rotor diameter, direct-drive, horizontal-axis wind turbine that operates in a wide range of

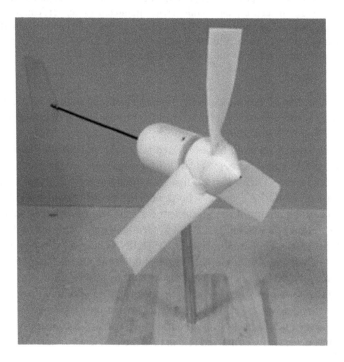

FIGURE 5.4 An experimental prototype of "SWEPT."

wind speeds between 1.7 and 10 m/s and produces a rated power output of 1 W at a wind speed of 4.0 m/s.[15] This power is sufficient for many of the wireless sensing applications as can be seen in Figure 5.4.

Kishore et al.[16] have also reported an ultralow start-up speed piezoelectric windmill that consisted of a 72 mm diameter horizontal-axis wind turbine rotor with 12 alternating polarity magnets around its periphery and a 60 mm × 20 mm × 0.7 mm piezoelectric bimorph element having a magnet at its tip. This wind turbine was found to produce a peak electric power of 450 μW at a wind speed of 4.2 mph.[16] An extremely low cut-in wind speed of 1.9 mph was achieved by operating the bimorph in the actuator mode for <4 s.[16] These results are quite interesting from the perspective that a finite amount of electrical power can be produced continuously at ultralow wind speeds of 2–3 mph.

5.4 WAVE AND FLOW ENERGY HARVESTING

Ocean waves are generated by winds that are the result of nonuniform heating of the atmosphere. Consequently, water particles adopt a circular motion as a part of the waves carrying kinetic energy (KE).[17,18] The total estimated power of waves breaking on the world's coastline is around 2–3 million megawatts.[17] The sheer power of the wave motion can easily exceed 50 kW/m of wave front, thereby making wave energy harvesting attractive.[19] The wave energy harvesting system converts the KE and potential energy (PE) of wave oscillations into electricity.[17] A typical ocean wave energy harvesting system may consist of a turbine, generator, wave power absorber, and power electronic interfaces.

Recently, there has been some progress in the related area of harvesting continuous water flow using vortex-induced vibrations where water current flowing around the cylinders results in a linear motion that can be coupled with the electrical generators to produce electricity.[20–22] Research in this area is focused on understanding the vortex–vortex interaction, fluid–structure interaction, and fluid resonance and in the design of the mechanical systems that can effectively convert the linear motion into rotational motion.[23–25] Electrical generators that can optimally operate at low torque magnitudes and rpm will be of tremendous importance. Piezoelectrics may also play some role in ultralow-speed water flow energy harvesting.[26,27] Various research groups are investigating the coupling of the bimorph-type transducer with the oscillating body in the water flow.[26,28] There is also interest in the design of hair-type oscillating piezoelectric structures that respond to very-small-magnitude waves. Materials challenges in this area are related to finding high-performance piezoelectric polymers and flexible piezoelectric structures that can operate under high-pressure conditions.

5.5 THERMAL ENERGY HARVESTING

5.5.1 THERMOELECTRIC-BASED ENERGY HARVESTING

Traditionally, the Seebeck effect is utilized in thermoelectric (TE) energy conversion as commonly seen in thermocouples having a negative and a positive thermoelement. Since a single thermocouple is capable of handling only a small amount of power at very low voltage, practical TE devices contain modules having many thermocouples.[29] The TE conversion can be utilized in various scenarios such as hot effluent of power-plant smokestacks, heat generated in photovoltaic panels, and automobiles. TE material design relies on discovering strategies to improve the figure of merit ZT, where Z is directly proportional to S^2 (Seebeck coefficient) and inversely proportional to the electrical resistivity and thermal conductivity.[30] Traditional TE materials rely on telluride, silicide, and skutterudite systems.[31] Recently, there has been interesting development in oxide materials where improved performance has been obtained through texturing.[32–36] The results show that by incorporating single crystals in the bulk materials, there is significant improvement in the texture strength, which can be used to modulate the transport behavior.[37] Using a texturing approach, there are several possible oxide candidates that can be considered for high-temperature TE materials for power generation because of their high thermal stability, oxidation resistance, and reduced toxicity, such as $(ZnO)_2In_2O_3$, $NaCo_2O_4$, $Ca_3Co_4O_9$, $(Zn_{1-x}Al_x)O$, $(Ba,Sr)PbO_3$, $(Ca,M)MnO_3$ (M = Bi, In), $(La,Sr)CrO_3$, Li-doped NiO, $Pr_{0.9}Ca_{0.1}CoO_3$, and $Ho_{0.9}Ca_{0.1}CoO_3$.[38] Some of these materials do not utilize expensive rare-earth elements and are cost-effective in large-scale manufacturing. To exemplify, Tani et al. have synthesized textured bulk ceramics of the layer-structured homologous compound $(ZnO)_5In_2O_3$ by using reactive templated grain growth (RTGG) through the conventional tape casting process.[39] They utilized platelike $ZnSO_4·3Zn(OH)_2$ particles to form a single phase of $(ZnO)_5In_2O_3$ with the c-axis perpendicular to the sheet surface. Snyder and Toberer have recently reviewed these materials and discussed the importance of morphology of microstructural features with the spatial resolution approaching the nanoscale regime.[40]

A major advancement in harvesting low-temperature gradients has been obtained through the breakthroughs in the design of shape memory engines.[41] Shape memory alloys (SMAs) exhibit temperature-dependent cyclic deformation. SMAs undergo reversible phase transformation when cycled through the martensite-to-austenite transformation temperature that generates strain that is used to develop heat engine. An SMA heat engine first converts environmental heat into mechanical energy through SMA deformation and then into electric energy using a microturbine. The results reported in literature show that 0.12 g of SMA wire can produce 2.6 mW of mechanical power that is sufficient to drive a miniature electromagnetic generator that can produce 1.7 mW of electrical power.[41] This generated electric energy was sufficient to power a wireless sensor node.

5.5.2 PYROELECTRIC-BASED THERMAL ENERGY HARVESTING

Pyroelectric energy harvesting is another interesting technique to harvest heat energy. Pyroelectric energy harvesting generates power from temperature fluctuations in contrast to TE energy harvesting, which utilizes the temperature gradient.[1,42] The pyroelectric effect is generally observed in ferroelectric materials. These materials exhibit a change in spontaneous polarization as a function of temperature. The pyro-electric coefficient λ can be given as[42]

$$\lambda = \frac{dP_s}{dT} \tag{5.2}$$

where
P_s is the spontaneous polarization
T is the temperature

The pyroelectric current can be given as[42]

$$I = \frac{dQ}{dt} = S\lambda \frac{dT}{dt} \tag{5.3}$$

where
Q is the induced charge
S is the electrode surface of the sample

Pyroelectric materials are used in metal–insulator–metal configuration with polar-ization in the direction perpendicular to the electrodes. Pyroelectric materials have been considered interesting due to their high thermodynamic efficiency.[42] The main technical challenge in pyroelectric materials is to have temperature oscillations for harvesting energy. Some researchers have used cyclic pumping in order to transform the temperature gradient into a time variable temperature.[42] In order to make the process feasible, the consumption of power in cyclic pumping should be negligible in comparison to generated power.[1] For higher current, pyroelectric materials need to have a larger surface area and enhanced pyroelectric coefficients coupled with a higher rate of change in temperature.

5.6   VIBRATION ENERGY HARVESTING

Mechanical energy harvesting for powering the distributed sensor nodes and struc-tural health monitoring has received global attention. Vibrations can be found in most of the places such as human motions, machines, pumps, vehicles, railway tracks and wagons, floors and walls, and other civil structures. Research reported in the past two decades has shown significant improvement in enhancing the power density and bandwidth of vibration energy harvesters.[43] In parallel, the power con-sumption of microelectronic components has been decreasing, and thus there is con-siderable opportunity in developing "self-powered devices" by meeting all the power needs through environmental sources. Here we explain the nature of various vibra-tion sources and materials used for vibration energy harvesting.[44]

5.6.1 Automobile Vibrations (Suspension Energy Harvesting)

Figure 5.5 shows the energy distribution of a 2.5L 2005 Camry indicating that one-fifth of the fuel energy is lost into mechanical energy.[45] It has been estimated that only less than half of that mechanical energy is transferred to the driving wheel.[43] Generally, vehicles utilize only 14%–30% of the available fuel energy in driving and overcoming resistance from the road friction and drag from the air according to the U.S. Department of Energy.[46] Opportunities to improve fuel efficiency include recovery of waste heat, regenerative braking, tire energy harvesting, and regenerative shock absorbers.[45,47,48]

An integrated model has been reported to assess the energy harvesting potential of vehicle suspension systems taking into account the road–vehicle–harvester system, and this model has been validated through road tests (Figure 5.6).[43,47,48] In this model, the excitations from road nonuniformity were modeled

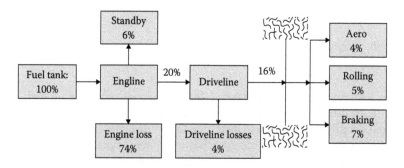

FIGURE 5.5 Energy distribution and consumption in various sections of the 2.5L Camry. (Adapted from Bandivadekar, L. et al., Laboratory for Energy and the Environment, Report No. LFEE 2008-05 RP. Copyright 2008, Massachusetts Institute of Technology, Cambridge, MA. With permission.)

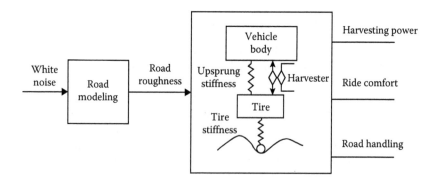

FIGURE 5.6 Integrated road–vehicle–harvester model. (Adapted from Zuo, L. and Tang, X., *J. Intel. Mater. Syst. Struct.*, 24, 1405, 2013. Copyright 2013, The Author(s). With permission.)

as a stationary random process. For different road conditions, the displacement power spectral densities are taken from the international standard organization (ISO 2631-1:1997, 1997). This effort was reported to increase the fuel efficiency by 2.4%–9%.

5.6.2 VIBRATIONS IN CIVIL STRUCTURES

Vibrations in civil structures have traditionally been a concern for reliability and sustainability of these structures as they suffer huge dynamic loadings exerted during earthquakes, tsunamis, wind storms, traffic flow, and human motions.[45,49] Vibrations in civil structures can be used to harvest energy that at the same time dampens the excess mechanical energy. The harvested energy can be used to meet the power requirements of various on-board structural health monitoring systems and active tuned mass dampers (TMDs).[49]

5.6.3 VIBRATION ENERGY HARVESTING FROM RAILWAY TRACKS

A typical freight train exerts a load of 20,000–30,000 pounds inducing 1/8–3/8 in. of deflection on the track.[50,51] The deflection frequency depends on the distance between the two train wagons. For a train moving 25–75 mph, the frequency of track deflection was estimated to be 0.6–1.8 Hz. The average power available from the railway track can be estimated as[45,50,51]

$$P_{avg} = \frac{NFD}{T} = \frac{FD}{\Delta T} \tag{5.4}$$

where
  N is the number of wheels passing through
  F is the normal force exerted by the wheel
  D is the track deflection
  T is the total time taken by the train to pass
  ΔT is the average time taken by each wheel to pass

For a four-wheel train wagon moving at a speed of 40 mph and having a weight of 100 tons and length of 80 ft, the four wheels were expected to pass over in 1.36 s ($\Delta T = 0.34$).[45,50,51] In this case, the average power potential (from the earlier equation) was estimated to be about 2 kW with ¼ in. track deflection according to Equation 5.4. By providing 5% of the support force to a harvester, the harvestable energy was estimated to be 200 W.[45,50,51] This energy could be used in powering trackside health monitoring systems, LED signal lights, etc.

5.6.4 ENERGY HARVESTING FROM HUMAN MOTION

Various human activities can be used for energy harvesting purposes as reviewed in detail by Starner and Paradiso (Figure 5.7).[52] Human body energy harvesting is

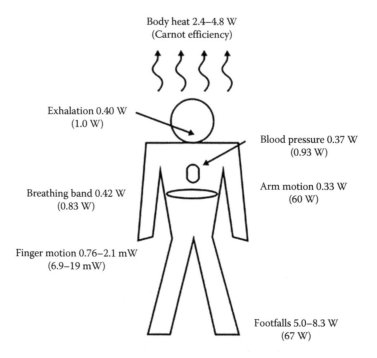

Body heat 2.4–4.8 W
(Carnot efficiency)

Exhalation 0.40 W
(1.0 W)

Blood pressure 0.37 W
(0.93 W)

Breathing band 0.42 W
(0.83 W)

Arm motion 0.33 W
(60 W)

Finger motion 0.76–2.1 mW
(6.9–19 mW)

Footfalls 5.0–8.3 W
(67 W)

FIGURE 5.7 Energy harvesting potential from various human activities. (Adapted from Starner, T. and Paradiso, J., in: C. Piguet (ed.), *Low Power Electronics Design*, CRC Press, Boca Raton, FL, 2004, p. 912. With permission.)

especially interesting for powering portable electronic devices and personal health-care systems. Figure 5.7 shows various human activities like walking, arm motion, finger motion, and breathing and blood pressure, which can be tapped to generate small magnitudes of electric energy.[45,52] Several efforts in this area have been made using piezoelectric and electromagnetic harvesting mechanisms. Piezoelectric uni-morph and bimorphs, cymbals, microfiber composites, and thunders are the common transducers utilized for this purpose.[45,53–56] In the case of electromagnetics, motor-ized system with gears and pulleys, magnetic levitation, and simple coil and moving magnet assembly have been utilized. Current focus in further improving the perfor-mance is toward developing low-resonance frequency structures and integration of the harvester with normal human kinetics.

5.7 MATERIALS FOR VIBRATION ENERGY HARVESTING

Mechanical energy harvesting has been mainly investigated for meeting the low power sensing needs and there is continuous ongoing effort in improving their power density. Mechanical energy can be converted into electric energy through various possible transduction methods utilizing electromagnetic, electrostatic, electroactive, and piezoelectric polymers. In this section, we will briefly discuss the various types of approaches and materials followed by a detailed description of piezoelectric materials, which are the main focus of this chapter.

5.7.1 ELECTROMAGNETIC INDUCTION

Electromagnetic induction commonly utilizes Faraday's principle to harvest vibration energy. The generation of electricity is due to the relative motion between a conductor (e.g., coil) and a magnetic field (due to a magnet), and system performance is dependent upon the dynamics of the interaction between the coil and magnetic flux.[57] The power output of an inductive harvester depends on the number of turns of the coil, the strength of the magnetic field, and the relative velocity between the coil and the magnet. At resonance, the output power is significantly affected by its quality factor and the internal resistance of the generator coil.[57] Contrary to piezoelectric energy harvesting, electromagnetic energy harvesting is characterized by low voltage and high current output.

5.7.2 ELECTROSTATIC ENERGY HARVESTING

Electrostatic (capacitive) energy harvesting of mechanical vibrations is based on the movement of charged capacitor plates of a variable capacitor against the electrostatic forces between the electrodes.[58] The capacitors are generally used in a parallel plate capacitor configuration having vacuum, air, or dielectric material (high $-k$ dielectric) in between electrodes. However, an electrostatic energy harvester requires a dc voltage to oppositely charge the capacitor plates.[58,59] The vibration changes stimulate the change in capacitance resulting in energy transfer per cycle, which further depends on the capacitive characteristics of filler material.

5.7.3 ELECTROACTIVE POLYMERS

The dielectric elastomers (DEs) and ionic polymer–metal composites (IPMCs) exhibit coupling between the mechanical stress (or strain) and electrical potential (or charge).[60] Electromechanical coupling in DEs is due to polarization-based or electrostatic mechanisms.[61] However, coupling in IPMCs is due to diffusion or conduction of charged species in the polymer network.[62] IPMCs are polymeric membranes possessing a molecule that can be easily ionized consisting of a fixed part with a negative charge and usually a cation/cationic group free to move in the polymer net.[62] On hydration, solvent (water) molecules get connected with the free cations in IPMCs.[62] When an electric field is applied, the free cations migrate in the direction of the E-field parasitically carrying water molecules along with them.[62] Consequently, the cathode region expands while the region near the anode contracts.[63] Furthermore, metal particles have Coulomb interactions (attractive on the anode and repulsive on the cathode) with the fixed negative charges. This consequently results in bending.[62,63] However, during the mechanical bending of an IPMC strip, the gradient in charge concentration develops. Consequently, the cations migrate from a high-density region toward a lower-density region generating a differential voltage between the electrodes, which depends on the mechanical excitation.[64] Therefore, IPMCs were found to be capable of vibration energy harvesting.[65]

5.7.4 PIEZOELECTRIC ENERGY HARVESTING

Piezoelectricity is found in crystallized materials that possess noncentrosymmetry. This effect induces an electric polarization proportional to an applied mechanical stress (direct piezoelectric effect) or a mechanical strain proportional to an applied electric field (converse piezoelectric effect). Piezoelectric materials are utilized in a multitude of applications such as sensors, actuators, and ultrasonic transducers that are important in various industrial and scientific areas. Piezoelectric materials used in vibration energy harvesting primarily utilize direct piezoelectric effect.[44] The energy harvesting performance highly depends on the coupling between the mechanical source and piezoelectric materials.[44] Another factor contributing the energy output is the ability of the piezoelectric material to withstand an applied force and the ability to repeatedly undergo a recoverable strain.[66] In piezoelectric vibration energy harvesters, the strain in the piezoelectric material is generated through a tip mass going through acceleration.[44,66] The most common piezoelectric cantilever configuration for energy harvesting is depicted in Figure 5.8.[66] In this configuration, one end of the cantilever is clamped to the vibration source while the other end has an attached tip mass.[66] In case of only a piezoelectric cantilever, the equal and opposite strain generated gets canceled and no current generation is achieved. In order to achieve energy generation through bending, the piezoelectric layer is moved away from the neutral axis by attaching an elastic layer (generally nonpiezoelectric) or joining two piezoelectric layers poled in the opposite direction.[66]

The primary factor for the selection of piezoelectric materials is the transduction rate whose magnitude is governed by the product of the piezoelectric strain constant, d, and the effective piezoelectric voltage constant, g, since the electric energy available under an alternating stress excitation is given as follows[67]:

$$P = U_e = \frac{1}{2}CV^2 = \frac{1}{2}\varepsilon_0\varepsilon_r\frac{A}{t}\left(-\frac{g \cdot F \cdot t}{A}\right)^2 = \frac{1}{2}(d \cdot g)\cdot\left(\frac{F}{A}\right)^2 \cdot (At)$$

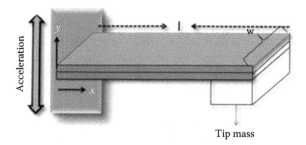

Tip mass

FIGURE 5.8 Schematic representation of a typical piezoelectric cantilever vibration energy harvester.

or energy density (U_e),

$$U_e = \frac{1}{2}(d \cdot g) \cdot \left(\frac{F}{A}\right)^2 \tag{5.5}$$

where
  F is the applied force
  A is the area
  t is the thickness of the active piezoelectric material

Equation 5.5 shows that under given experimental conditions, a material with a higher ($d \cdot g$) product will generate higher power.[67] If the energy harvester is working at resonance, the consideration of electromechanical coupling and mechanical quality factor becomes important. Here, we review various forms of piezoelectric materials for energy harvesting purposes: polycrystalline, single crystals, and textured piezoelectric ceramics. We will further discuss the progress made in the synthesis of environment-friendly lead-free piezoelectric materials for energy harvesting applications.[67] We also briefly review piezoelectric materials for energy harvesting in extremely high-temperature environment and flexible energy harvesters.

5.8   PIEZOELECTRIC MATERIALS

5.8.1   Polycrystalline Piezoelectric Ceramics

PZT solid solutions with morphotropic phase boundary (MPB) have been dominating the piezoelectric industry since their discovery in 1950. MPB is an intrinsic region of a phase diagram where two or more phases coexist. The solid solution of piezoelectric materials having MPB shows enhanced piezoelectric and dielectric responses at this phase boundary. For example, the phase diagram of PZT has $PbTiO_3$ and $PbZrO_3$ as end members.[68] Traditionally, the high piezoelectric response has been attributed to the enhanced number of polar axis in MPB composition. In a particular crystallographic symmetry, there are always a fixed number of equivalent polar axes in which the dipoles can be switched. The Ti-rich side of MPB in PZT possesses the tetragonal crystal symmetry with six equivalent [001] directions as polar axes. However on the Zr-rich side of the MPB, PZT with rhombohedral symmetry possesses eight equivalent [111] directions as polar axes. Because at MPB two different phases coexist, the switching of polar axes in PZT has 14 available directions under applied E-field. This phenomenon was considered to be responsible for giving rise to high piezoelectric response at MPB. Noheda et al.[69] discovered the monoclinic phase between tetragonal and rhombohedral phases around MPB of PZT. Some well-known piezoelectric materials having MPB are $(1-x)Pb(Mn_{1/3}Nb_{2/3})O_3-xPbTiO_3$, $(1-x)Pb(Zn_{1/3}Nb_{2/3})O_3-xPbTiO_3$, and $(Na_{0.5}Bi_{0.5})TiO_3-BaTiO_3$. High-resolution x-ray diffraction (XRD) investigations suggested that the major intrinsic contribution from both rhombohedral and tetragonal phases comes from the tilting of the P_s vectors.[70] The monoclinic phase with 24 permitted orientation states is expected to provide bridging polarization continuity leading to enhanced piezoelectric response.[70] Finding MPB in a solid solution of piezoelectric materials is one of the important methods to achieve a high piezoelectric response.

In addition, it has been shown that enhanced piezoelectric response can be related to the presence of miniaturized ferroelectric domains (<1 μm) with high mobility.[71] Wada et al. were able to synthesize a miniaturized domain structure in $BaTiO_3$ single crystal by applying E-field along nonpolar $[111]_C$ direction through a poling process conducted above T_c.[71] The high piezoelectric response ($d_{31} = -230$ pC/N) observed in this system was attributed to the high density of domain walls.[71] Rao and Wang have performed theoretical calculations and have confirmed that the enhanced piezoelectric response in the specimen with miniaturized domains was correlated with domain wall broadening under applied E-field.[72] The enhanced piezoelectric response in nonpolar directions was believed to originate from the broadening of domain walls.

Furthermore, the doping (soft and hard) of various elements in PZT has been effectively used to tune the piezoelectric response. The doping of elements like Nb^{5+} or Ta^{5+} (donor) results in "soft" PZTs (e.g., PZT-5). The enhanced piezoelectric response in soft PZT has been attributed to ease the domain wall motion due to doping-induced Pb vacancies.[73] The acceptor-type dopant (e.g., Fe^{3+} or Sc^{3+}) gives rise to "hard" characteristics as exemplified by PZT-8.[73] The hard PZTs have low piezoelectric response but high mechanical quality factor. This hard PZT behavior has been attributed to the pinning of domain wall motion due to the presence of doping-induced oxygen vacancies.[73,74] Table 5.1 summarizes the piezoelectric, dielectric, and elastic properties of typical PZTs: soft PZT-5H, semihard PZT-4, and hard PZT-8. Note that soft PZTs exhibit high electromechanical coupling (k), high piezoelectric coefficient (d), and high relative permittivity, in comparison with hard PZTs, while Q_M is quite high in hard PZTs. Thus, soft PZTs should be used for off-resonance applications, while hard PZTs are suitable for on-resonance applications.[73]

5.8.2 Single-Crystal Piezoelectric Ceramics

Relaxor-based ferroelectric single crystals, such as $Pb(Mg_{1/3}Nb_{2/3})O_3$–$PbTiO_3$ (PMN–PT) and $Pb(Zn_{1/3}Nb_{2/3})O_3$–$PbTiO_3$ (PZN–PT), have attracted extensive attention over the last 15 years due to their ultrahigh piezoelectric properties.[75] The longitudinal piezoelectric strain coefficient d_{33} and electromechanical coupling factor k_{33}, being on the order of >1500 pC/N and 0.9, far outperform the state-of-art polycrystalline ceramics $Pb(Zr,Ti)O_3$ (PZT). The properties of relaxor-PT single crystals are summarized in Table 5.2.[75] These properties are highly promising for various electromechanical applications.[75] The piezoelectric response is highly dependent on the crystallographic orientation of the single crystals, and it can be further improved by selecting MPB composition and substituting suitable elements on the desired lattice sites.

TABLE 5.1

Piezoelectric and Dielectric Properties of Various PZT Ceramics

Properties	Soft PZT-5H	Semihard PZT-4	Hard PZT-8
EM coupling factor			
k_p	0.65	0.58	0.51
k_{31}	0.39	0.33	0.30
k_{33}	0.75	0.70	0.64
k_{15}	0.68	0.71	0.55
Piezoelectric coefficient			
d_{31} ($\times 10^{-12}$ m/V)	−274	−122	−97
d_{33}	593	285	225
d_{15}	741	495	330
g_{31} ($\times 10^{-3}$ Vm/N)	−9.1	−10.6	−11.0
g_{33}	19.7	24.9	25.4
g_{15}	26.8	38.0	28.9
Permittivity			
$\varepsilon^X_{33/\varepsilon_0}$	3400	1300	1000
$\varepsilon^X_{11/\varepsilon_0}$	3130	1475	1290
Dielectric loss (tan δ) (%)	2.00	0.40	0.40
Mechanical quality factor (Q_M)	65	500	1000

Source: Adapted from *Advanced Piezoelectric Materials: Science and Technology*, Uchino, K./Uchino, K. (ed.), p. 696, Copyright 2010, with permission from Elsevier.

5.8.3 TEXTURED PIEZOELECTRIC CERAMICS

Despite the promising future single crystals for energy harvesting and many other applications, these single crystals have been facing challenges in their deployment mainly due to the high cost of fabrication. Moreover, obtaining large-size single crystals with required chemical homogeneity has been quite challenging. The use of expensive single crystals is generally limited to medical imaging applications where performance considerations outweigh the materials cost. The synthesis of high-performance piezoelectric ceramics can be categorized into two approaches: one is to explore a new phase diagram and another is to tailor the microstructures of known materials. The synthesis of grain-oriented (textured) materials (or microstructure tailoring) has been proven to be an effective method to obtain the best performance from the material with anisotropic properties. The method of synthesizing textured materials has been successfully applied to various systems including alumina structural ceramics,[76,77] TE oxide ceramics,[32,35] and piezoelectric ceramics.[78]

A schematic diagram of the texturing process and resulting microstructural development is shown in Figure 5.9. In order to obtain textured ceramics, microsized template crystals are aligned in the ceramic matrix powder using the tape casting

TABLE 5.2

Comparison of Longitudinal Piezoelectric Coefficients for Relaxor-PT Crystals

Relaxor-PT Crystal	[001] Poled				[011] Poled				[111] Poled			
	$\varepsilon^x_{33/\varepsilon_0}$	s^E_{33} (pm²/N)	d_{33} (pC/N)	k_{33}	$\varepsilon^x_{33/\varepsilon_0}$	s^E_{33} (pm²/N)	d_{33} (pC/N)	k_{33}	$\varepsilon^x_{33/\varepsilon_0}$	s^E_{33} (pm²/N)	d_{33} (pC/N)	k_{33}
PIN–PMN–PT(R)	7200	77.8	2740	0.95	4400	52	1300	0.92	700	6.8	74	0.36
PMN–PT(R)	8200	120	2820	0.95	3800	70	1350	0.88	640	13.3	190	0.69
PZN–7PT(R)	5622	142	2455	0.92	3180	62	1150	0.87	—	—	—	—

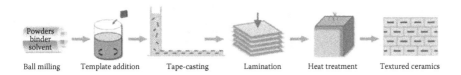

Powders
binder
solvent

Ball milling   Template addition   Tape-casting   Lamination   Heat treatment   Textured ceramics

FIGURE 5.9   Schematic diagram of the TGG process.

method. During high-temperature processing, epitaxial growth occurs on the template seeds giving rise to textured grains.[78] Anisotropically shaped (whisker or platelet) template seeds should have a smaller (<15%) lattice mismatch with the matrix phase to get a high degree of texturing.[78] Anisotropically shaped $BaTiO_3$ (BT) templates were synthesized using the topochemical microcrystal conversion process.[79–81] These templates were used to fabricate various highly textured piezoelectric materials with excellent properties.

5.8.3.1 Textured PMN–PT Ceramics

The oriented single crystal of $0.675\ Pb(Mg_{1/3}Nb_{2/3})O_3$–$0.325PbTiO_3$ (PMN–32.5PT) had high piezoelectric properties with $d_{33} > 2000$ pC/N and $k_{33} > 0.9$. In order to synthesize cost-effective $\langle 001 \rangle$-oriented PMN–PT ceramics with high performance comparable to single crystals, the TGG process was employed.

Figure 5.10a shows the XRD patterns of PMN–PT–xBT samples. Here, PMN–PT–0BT represents a random ceramic without BT seeds ($x=0$), and PMN–PT–xBT ($x=1, 3, 5$) represents textured ceramics with different BT template contents.[82] XRD patterns confirmed the formation of a perovskite structure without any noticeable secondary phase. Domination of (00l) peaks in PMN–PT–xBT ($x=1, 3, 5$) samples indicated high $\langle 001 \rangle$ orientation. The Lotgering factor[83] calculated for these samples was found to be almost the same ($f > 98\%$). Figure 5.10b displays the scanning electron microscopy (SEM) images of random and textured cross-sectional samples. Compared to the equiaxed grains in random ceramics (left), all the matrix grains in the textured sample (right) were well aligned with the brick-wall microstructure. Another feature of textured ceramics is the existence of an aligned platelike BT template inside the oriented PMN–PT matrix. The BT template microcrystals had length of 5–10 μm and thickness of 0.5–1 μm. Therefore, besides a large grain size, the textured ceramics had two other unique characteristics: $\langle 001 \rangle$ grain orientation and the existence of a heterogeneous BT "core." Figure 5.10c shows the electron backscatter diffraction (EBSD) inverse pole figures of the random ceramic and textured ceramic clearly illustrating that the textured grains exhibit strong $\langle 001 \rangle$-preferred orientation.

The piezoelectric response and dielectric loss tangent factor (tan δ) for PMN–PT–xBT are shown in Figure 5.11. The d_{33} value was found to increase with increasing BT content to achieve a maximum value of 1000 pC/N at $x=1$.[80,84] The enhanced piezoelectric response was attributed to the high degree of texturing and domain structure similar to that of single crystals.[84] The trends of variation in the value of d_{31} were similar to that of d_{33}. However, the value of tan δ showed a reverse trend as

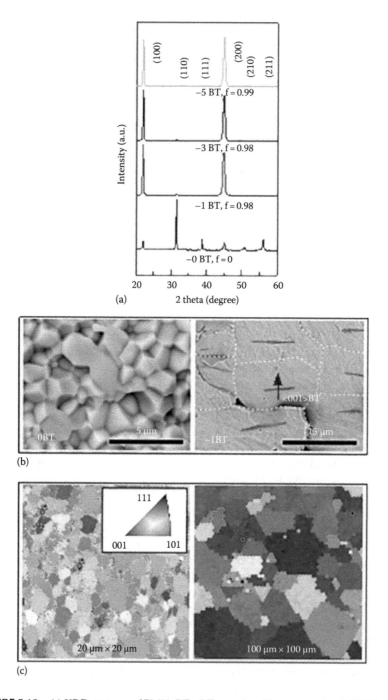

FIGURE 5.10 (a) XRD patterns of PMN–PT–xBT samples; (b) cross-sectional SEM images of PMN–PT–0BT (left) and PMN–PT–1BT (right); and (c) EBSD images of PMN–PT–0BT (left) and PMN–PT–1BT (right) surfaces. (Adapted with permission from Yan, Y. et al., *Appl. Phys. Lett.*, 103, 082906. Copyright 2013, American Institute of Physics.)

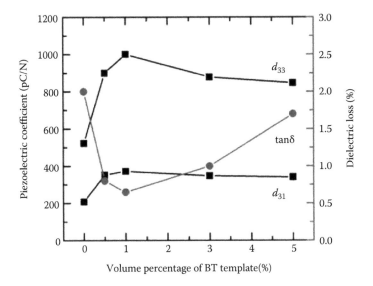

FIGURE 5.11 Piezoelectric properties and dielectric loss of PMN–PT–xBT ceramics. (Adapted with permission from Yan, Y. et al., *Appl. Phys. Lett.*, 100, 192905. Copyright 2012, American Institute of Physics.)

compared to the d_{33} and d_{31} values. Interestingly, the textured PMN–PT–1BT ceramics were found to exhibit loss tangent value as low as 0.6%, which is about 1/3 of the magnitude obtained for most of the soft piezoelectric ceramics (>2.0%). Therefore, grain-oriented PMN–PT–1BT ceramics, with high piezoelectric response and low dielectric loss, are ideal substitutes for currently used soft piezoelectrics in piezoelectric energy harvesting technologies.

5.8.3.2 [001] Textured Pb(Mg$_{1/3}$Nb$_{2/3}$)O$_3$–PbZrO$_3$–PbTiO$_3$ Ceramics

The domain engineered PMN–PT relaxor piezoelectric single crystals have been characterized by high functional response. However, low ferroelectric rhombohedral to tetragonal phase transition temperature ($T_{R-T} \sim 60°C–90°C$) and low coercive fields ($E_c \sim 2–3$ kV/cm) were found to limit the stability of these relaxor-based single crystals.[75] Many attempts have been made to grow PZT-based single crystals near MPB to achieve high and stable piezoelectric properties, but so far there has been only limited success in growing PZT-based single crystals[85] due to inherent technical challenges.

Yan et al. synthesized textured PMN–PZT ceramics with enhanced piezoelectric properties ($d_{33} > 1000$ pC/N, $g_{33} > 50 \times 10^{-3}$ Vm/N, and tan δ < 1.2%).[86] Generally, high d_{33} piezoelectric materials possess low g_{33} value and vice versa (Figure 5.12a) because the d_{33} is mostly proportional to the square root of the dielectric constant of the piezoelectric materials. Interestingly, the textured PMN–PZT ceramics exhibited both high d_{33} and high g_{33} values due to the template-controlled dielectric characteristics.[86] Table 5.3 summarizes the piezoelectric and dielectric properties of randomly

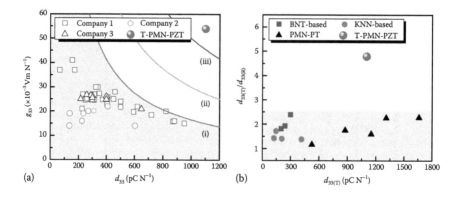

FIGURE 5.12 Comparison of (a) g_{33} and (b) d_{33} values of various piezoelectric ceramics. Different colored lines (i), (ii), and (iii) indicate the plots of the $g_{33} \cdot d_{33} = c$ function: (i) $c = 16{,}500 \times 10^{-15}$ m^2 N^{-1}, (ii) $c = 30{,}000 \times 10^{-15}$ m^2 N^{-1}, and (iii) $c = 50{,}000 \times 10^{-15}$ m^2 N^{-1}. (Adapted with permission from Yan, Y. et al., *Appl. Phys. Lett.*, 102, 042903. Copyright 2013, American Institute of Physics.)

TABLE 5.3

Piezoelectric and Dielectric Properties of PMN–PZT Piezoelectrics: Randomly Oriented Ceramic (R-Ceramic), ⟨001⟩ Textured Ceramic (T-Ceramic), and ⟨001⟩ Single Crystal (S-Crystal)

Properties	R-Ceramic	T-Ceramic	S-Crystal[87]
Piezoelectric charge constant, d_{33} (pC/N)	230	1100	1530
Electromechanical coupling constant, k	0.4 (k_p)	0.84 (k_p)	0.93 (k_{33})
Relative dielectric permittivity, $\varepsilon_{33}/\varepsilon_0$	915	2310	4850
Piezoelectric voltage constant, g_{33} ($\times 10^{-3}$ Vm/N)	28.4	53.8	35.6
$d_{33} \cdot g_{33}$ ($\times 10^{-15}$ m^2/N)	6,532	59,180	54,468
Remnant polarization, P_r (μC/cm^2)	30	36	29
Coercive field, E_c (kV/cm)	8.2	7.4	4.5
Curie temperature, T_c (°C)	233	204	211

Source: Adapted with permission from Yan, Y. et al., *Appl. Phys. Lett.*, 102, 042903. Copyright 2013, American Institute of Physics.

oriented ceramic (R-ceramic), ⟨001⟩ textured ceramic (T-ceramic), and ⟨001⟩ single crystal (S-crystal) of the PMN–PZT composition. The textured ceramic exhibited giant d_{33} of 1100 pC/N, which was 4.8 times higher than that of its randomly oriented counterpart having $d_{33} \sim 230$ pC/N.[86] This increased ratio of the d_{33} value between R- and T-ceramic (4.8) was much higher than that of other textured piezoelectric ceramics (usually <2.5) as shown in Figure 5.12b. These textured piezoelectric materials are very promising in piezoelectric energy harvesting.

5.8.3.3 Cofired Textured Piezoelectric for Energy Harvesting

As a transduction material, the piezoelectric material can generate a very large voltage but at low current amplitude. However, to charge a battery or supercapacitor, a large current is needed to shorten the charging time. Also, the matching impedance for piezoelectrics is of the order of ~10 kΩ, while the sensor nodes have impedance on the order of 100 Ω. For example, the capacitance and output charge of the multilayer capacitor can be dramatically increased by tuning the number of layers and materials properties[88]:

$$C_n = \varepsilon_0 \varepsilon_r \frac{A}{t/n} \times n = n^2 C,$$

$$R_{\text{match}} = \frac{1}{2\pi f C} \cdot \frac{2\zeta}{\sqrt{4\zeta^2 + k^4}},$$

$$V_{\text{open}} = E \cdot t = -g \cdot X \cdot t = -\frac{g \cdot F \cdot t}{A},$$

$$R_{\text{match},n} = \frac{1}{n^2} \times R_{\text{match}}; \quad V_{\text{open},n} = \frac{1}{n} \times V_{\text{open}}; \quad I_n = n \times I; \quad P_n = P \qquad (5.6)$$

where
  C, R_{match}, I, and V_{open} are the capacitance, matching resistance, current, and open voltage of a single-layer structure, respectively
  C_n, $R_{\text{match},n}$, I_n, and $V_{\text{open},n}$ are the capacitance, matching resistance, current, and open voltage of the n-layer structure, respectively
  f is the operating frequency
  ζ and k are the damping ratio and piezoelectric coupling coefficient, respectively
  X is the stress applied on the piezoelectric materials[88]

It can be seen that through a multilayer structure, the matching resistance and voltage can be decreased and the current can be increased. To fabricate multilayer piezoelectric ceramics, a cost-effective low-temperature cofired ceramics (LTCC) process method is utilized.[89] This processing method has been widely used for the synthesis of a multilayer ceramic capacitor (MLCC) and multilayer actuator (MLA).[89] The TGG process was used in combination with the LTCC process to successfully achieve a high-performance low-cost multilayer textured piezoelectric bender. The synthesis of cofired multilayers requires precise control of the processing temperature and time due to the difference in temperature stability of different materials and its stability on the materials properties. The cofiring temperature was selected such that it was lower than the melting point of the Ag electrode (961°C). The processing conditions were optimized carefully to avoid low-temperature-sintering-induced defect (such as low density, poor mechanical adhesion, and electrical contact of the piezoelectric–electrode interface) in the piezoelectric ceramic body, which could have been detrimental to the functional properties. Therefore, there was a trade-off between the electrode composition and texture degree. The Lotgering factor was found to increase rapidly at 800°C and then saturated after 850°C. The samples

TABLE 5.4

Piezoelectric and Energy Harvesting Properties of Random Single-Layer, Textured Single Layer, Bilayer, and Trilayer Samples Tested at the Acceleration of 0.1 g

Properties	R-1	T-1	T-2	T-3
d_{31}^{*} (pC/N)	78	366	780	1048
g_{31}^{*} ($\times 10^{-3}$ Vm/N)	9.6	14.7	7.3	5.0
$d_{31}^{*}g_{31}^{*}$ ($\times 10^{-15}$ m^2/N)	749	5380	5694	5438
R_{match} (kΩ)	600	500	120	60
U (V)	3.83	7.22	3.92	2.87
I (μA)	6.4	14.4	32.7	47.8
P_{max} (μW)	24.4	104.3	128.1	137.2

Source: Yan, Y., et al., *Energy Harvesting and Systems*, 1, 189–195, 2014.

cofired at 925°C exhibited 90% degree of texture. The piezoelectric properties of these materials are listed in Table 5.4.[90]

Table 5.4 lists the piezoelectric and energy harvesting properties of random and textured samples with different layers. The piezoelectric properties of the textured sample were significantly higher than that of their random counterpart. The $d \cdot g$ values of textured samples were about 800% higher than that of the random counterpart, which could be the principal reason for 500%–600% increase in output power from the textured samples. For comparative study, the dimension and test configuration for energy harvesting were kept similar.

The open-circuit voltage of the single-layer textured sample (T-1) was almost twice as that of the single-layer random sample (R-1).[90] The generated voltage of the piezoelectric ceramic is determined by the piezoelectric voltage coefficient (g). The g value for T-1 (14.7×10^{-3} Vm/N) is much higher than that of R-1 (9.6×10^{-3} Vm/N). For cofired textured sample (T-1, T-2, and T-3), the generated voltage was found to decrease with the increase in the number of layers. The optimum resistive load was also found to decrease with the increase in the number of layers. In an ideal condition, open-circuit voltage of an n-layer generator should be n times smaller than that of a single-layer generator, and the optimum resistive load of an n-layer generator is required to be n^2 times smaller than that of a single-layer generator. This was found to be in agreement with the experimental results listed in Table 5.4. The output power of cofired multilayer textured samples was five to six times higher than that of random sample, which further suggests a high potential of textured piezoelectric ceramics in energy harvesting applications.

5.8.3.4 Lead-Free Piezoelectric Materials for Environment- Friendly Energy Harvesters

Conventional energy harvesting devices have been using lead-based piezoelectric materials due to their superior functional response. However, various regulating agencies across the globe are imposing restrictions on devices containing lead-based materials. The electromechanical properties (piezoelectric constant, coupling factor) for known lead-free compositions were far inferior to those of lead-based systems.[91,92] Different methods have been used to improve the functional response of lead-free piezoelectric systems.[93] Out of these possibilities, improvement in the properties of familiar lead-free piezoelectric materials has been quite interesting.[94,95]

Recently, the $Na_{0.5}Bi_{0.5}TiO_3–BaTiO_3$ (NBT–BT) system has been considered as a potential candidate for the replacement of lead-based piezoelectric materials.[94] However, efforts on compositional modification for improving the piezoelectric response in this material system were not successful as it reduced the depoling temperature that affected their temperature stability (Figure 5.13a).[96] Using the microstructural control technique, Maurya et al. demonstrated enhanced piezoelectric response in grain-oriented lead-free piezoelectric NBT–BT ceramics at MPB.[94]

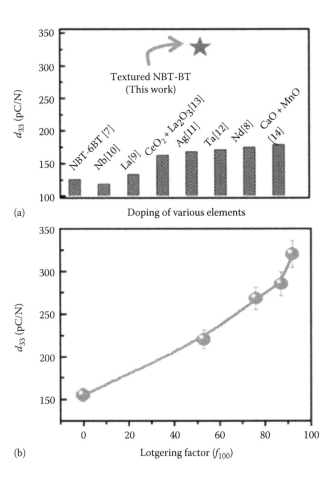

FIGURE 5.13   (a) d_{33} versus various dopants and (b) d_{33} versus the Lotgering factor. (Maurya, D. et al., *J. Mater. Chem. C*, 1, 2102. Copyright 2013. Adapted by permission of The Royal Society of Chemistry.)

The grain-oriented NBT–BT textured sample, with 90% degree of texturing, was found to exhibit 200% improvement in the piezoelectric properties in comparison to its randomly oriented counterpart (Figure 5.13b).[94] The advantage of microstructure control over substitution of various elements in terms of high d_{33} is very clear in Figure 5.13a and b. The grain-oriented NBT–BT sample had d_{33} about 322 pC/N in comparison to 160 pC/N in the randomly oriented sample. The value of d_{33} for the textured sample was almost comparable to that of NBT-based single crystals having a similar composition. The grain-oriented NBT–BT sample had d_{33} of 322pC/N in comparison to 160 pC/N in the randomly oriented sample. The value of d_{33} for the textured sample was almost comparable to that of NBT-based single crystals[97,98] having a similar composition and hard PZT (Ferroperm) ceramics (Table 5.5). As an application, the textured specimen was found to exhibit 300% improvement in magnetoelectric response (Figure 5.14) that is vital for dual-phase energy harvesters.

Figure 5.15a shows relative permittivity and loss tangent at various frequencies as a function of temperature for poled grain-oriented NBT–BT sample. The value of depoling temperature (T_d) and the maximum of relative permittivity (T_m) for textured samples were found to be ~90°C and 300°C, respectively. This composition was at MPB and it could be noted that the value of T_d increases on either side of MPB. Thus, a compositional shift toward the rhombohedral or tetragonal regime in the phase diagram could enhance the temperature of the operation. These single-phase textured lead-free piezoelectric materials are much simpler to synthesize and are cost-effective. The textured and randomly oriented specimens were found to exhibit a characteristic diffuse nature of the phase transition, which suggests the presence of small-sized ferroelectric nanodomains. Figure 5.15b shows the P–E hysteresis loops at 1 kHz for NBT–BT ceramics with different crystallographic textures. The randomly oriented specimen exhibited a square loop with high coercive field, while textured specimens were found to exhibit a pinched hysteresis loop. The specimen with 92% texturing was found to show $2P_r$ ~ 65 pC/N.

TABLE 5.5
Piezoelectric and Ferroelectric Properties of PZ27 (PZT-Based Polycrystalline, Ferroperm) Ceramics and NBT–BT Single Crystal, Mn: NBT–BT (Pt) (Single Crystal Grown Using Pt Seed), and Textured NBT–BT Ceramics

Sample	d_{33} (pC/N)	k_t (%)	k_{31} (%)	P_r (μC/cm^2)	E_c (kV/mm)
PZ27 ceramics (Ferroperm)	425	47	33	—	—
NBT–BT single crystal[20]	280	56	—	16.44	2.67
Mn: NBT–BT single crystal(Pt)[21]	287	55.6	39.7	35	2.91
Textured NBT–BT ceramics	322	57.3	—	35	1.8

Source: Maurya, D. et al., *J. Mater. Chem. C*, 1, 2102. Copyright 2013. Adapted by permission of The Royal Society of Chemistry.

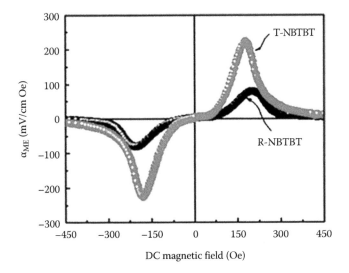

FIGURE 5.14 ME voltage coefficients for trilayer laminates of Metglas/NBT–BT/Metglas for randomly oriented (R) and textured NBT–BT (T) systems. (Maurya, D. et al., *J. Mater. Chem. C*, 1, 2102. Copyright 2013. Adapted by permission of The Royal Society of Chemistry.)

The grain-oriented sample had a sharp increase in the values of remnant polarization (P_r) at a threshold E-field (Figure 5.15c and d). On the other hand, the randomly oriented specimen had an almost monotonous increase in the remnant polarization with increasing field. In order to understand the domain structure and its influence on the piezoelectric properties, piezoresponse force microscopy (PFM) was performed (Figure 5.16a and b). The grain-oriented sample had smaller-sized domains with rather homogenous arrangement resulting in coherent switching of the highly mobile domain. However, the randomly oriented NBT–BT specimen had a wide distribution of domain sizes as observed in Figure 5.16a and b, which further explained the unique piezoelectric and ferroelectric properties of textured NBT–BT samples.

5.8.3.5 Piezoelectric Materials for High-Temperature Energy Harvesting

The high-temperature stability of energy harvesters becomes inevitable while working in hot environments like near-engine sensors, remote sensors for geothermal explorations, and deep bore drilling.[99] LiNbO$_3$ single crystals have been found to be stable up to 460°C.[100] Bedekar et al.[99] reported comparative results on the high-temperature stability and energy harvesting capability of single crystals of yttrium calcium oxyborate (YCa$_4$O(BO$_3$)$_3$ [YCOB]), lanthanum gallium silicate (La$_3$Ga$_5$SiO$_{14}$ [LGS]), and lithium niobate (LiNbO$_3$ [LN]). The YCOB and LGS crystals were found to be promising for generating electricity up to 1000°C and therefore suitable for deploying in extremely hot environments. The temperature-dependent dielectric and piezoelectric responses of these crystals are shown in Figure 5.17a and b.[99] The YCOB crystal had a room temperature value of d_{33} ~ 6.3 pC/N and g_{33} ~ 0.065 Vm/N. However, the g_{33} value for LGS crystals was found to be 0.04 Vm/N. The YBCO crystal was found to have almost constant values of g_{33}

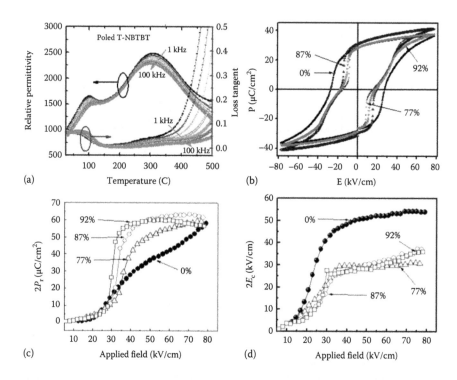

FIGURE 5.15 (a) Temperature dependence of relative permittivity and loss tangent of poled textured NBT–BT specimen, (b) $P–E$ hysteresis of various NBT–BT textured ceramics with different degrees of orientation, (c) remanent polarization ($2P_r$) versus E-field for textured and randomly oriented specimen, and (d) coercive field ($2E_c$) versus E-field for textured and randomly oriented specimen. (Maurya, D. et al., *J. Mater. Chem. C*, 1, 2102. Copyright 2013. Adapted by permission of The Royal Society of Chemistry.)

and d_{33} up to 1000°C. LGS crystals were also found to exhibit a similar behavior up to 1000°C indicating high-temperature stability for energy harvesting devices. Furthermore, one can also achieve high-temperature stability (up to 400°C) in PZT-based solid solutions by moving away from the phase boundary, however at the expense of losing sensitivity.

Moreover, wide bandgap materials having a wurtzite structure have recently been considered as promising candidates for harvesting energy in hostile high-temperature environment.[66] However, these materials have been found to have a relatively lower piezoelectric response. GaN, SiC, and AlN are the potential candidates from this category for high-temperature applications.[101–103]

5.8.3.6 Piezoelectric Polymer Materials for Flexible Energy Harvesting

Increasing demand of flexible electronics has inspired many researchers to develop efficient flexible energy harvesting materials.[104,105] Moreover, flexible harvesters can be mounted on a curved surface with ease. Most of the materials utilized in classical piezoelectric energy harvesting are ceramics, which are hard and brittle. Therefore,

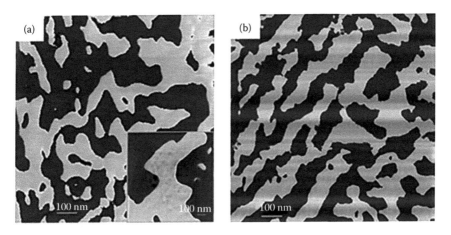

FIGURE 5.16 (a) PFM image of domains in randomly oriented NBT–BT (inset is close-up view) and (b) PFM phase image depicting domains in textured NBT–BT. (Maurya, D. et al., *J. Mater. Chem. C*, 1, 2102. Copyright 2013. Adapted by permission of The Royal Society of Chemistry.)

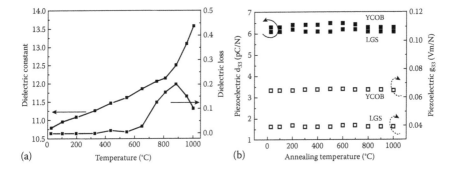

FIGURE 5.17 (a) Dielectric behavior as a function of temperature measured at 100 kHz of YCOB crystal and (b) piezoelectric coefficients, d_{33} and g_{33}, as a function of the annealing temperature for YCOB and LGS crystals. (Adapted from Bedekar, V. et al., *Jpn. J. Appl. Phys.*, 48, 091406. Copyright 2009, The Japan Society of Applied Physics. With permission.)

polymer-based piezoelectric materials have been of great interest for fabricating flexible piezoelectric energy harvesters due to their low density, biocompatibility, flexibility, toughness, and low cost.[104] The flexible technology further becomes important for the devices that are prone to bend and stretch such as the devices mounted on human body.[104,106] Though various kinds of piezoelectric polymers are available, polyvinylidence fluoride (PVDF) has been the dominant piezoelectric polymer.[66] The failure strain of PVDF has been more than 2%.[104] However, PVDF had relatively low d_{33} ~ 33 pC/N but high g_{33} ~ 339 × 10^{-3} Vm/N.[66] The high d_{33} value makes PVDF a suitable candidate for responding to the low value of applied stress. In order to improve the piezoelectric response of polymer-based flexible piezoelectric systems, various polymer–ceramic composites (0–0, 0–1, 0–2, 0–3, 1–1, 1–2, 1–3, 2–2, 2–3, and 3–3) could be synthesized.[107] The advantage of mixing piezoelectric ceramics with polymers is to combine the functional responses of both the materials to achieve a synergic effect.[108] Recently, Baur et al.[109] fabricated PVDF composites containing carbon fullerenes (C_{60}) and single-walled carbon nanotubes (SWNTs) over a range of compositions and reported two times improvement in d_{31} values of the PVDF–SWNT composite. Some of the polymer-based piezoelectric composites are listed in Table 5.6.[110] These materials could be used to harvest energy from respiration, rainfall, footsteps, wind, etc. In order to make an efficient environment-friendly energy harvester, the polymer–ceramic composites could be blended with lead-free piezoelectric materials.

Recently, Jeong et al.[115] fabricated a flexible lead-free nanocomposite generator (NCG) device for high-output energy harvesting. In order to make a piezoelectric composite, the lead-free piezoelectric $0.942(K_{0.480}Na_{0.535})NbO_3$–$0.058LiNbO_3$ (KNLN) particles were mixed with the Cu nanorods (NRs) in a polydimethylsiloxane (PDMS) matrix. The copper NR filler in the composite was found to play the role of nanoelectrical bridges that lead to short voltage lifetimes for higher output performance.[115]

A schematic of the NCG containing KNLN particles and Cu NRs is depicted in Figure 5.18a. The ITO electrode on the flexible polyethylene terephthalate (PET) substrate was used as current collectors in the device. In order to prevent electrical breakdown during high–voltage poling process and extreme mechanical deformation, pure PDMS dielectric layers are used. The flexibility of this energy harvester allowed the researchers to mount this energy harvester on any curved surface as shown in Figure 5.18b. The SEM image of the bent KNLN-based flexible device with composite thickness (~250 μm) is shown in Figure 5.18c. The piezoelectric composite was sandwiched between two flexible substrates.[115] The plastic substrates and composite were intact even after mechanical deformation due to stickiness of the PDMS. The KNLN particles in the composite were used to harvest the mechanical energy during the bending process. In order to optimize the composition of the nanocomposite for energy harvesting, different amounts of KNLN particles and Cu NRs were used.[115] With the increase in the amount of Cu NRs up to 1 g at the fixed weight

TABLE 5.6
Piezoelectric Polymer Composites and Their Functional Responses

Composition	Dielectric Constant	d_{33}	$d_{33} \times g_{33}$
PZT–PVDF (vol., 65%)[111]	45	33	2.73
PMN–PT–P(VDF–TrFE) (vol., 40%)[112]	37.3	31	2.91
PZT–polyester resin (vol., 65%)[113]	88	29	1.08
PZT–P(VDF–HFP) (vol., 48%)[114]	62.6	11.3	0.23
PZT–P(VDF–CTFE) (vol., 60%)[110]	118	87	7.27

Source: Adapted with kind permission from Springer Science+Business Media: *J. Electroceram.*, 30, 2013, 30, Choi, Y., Yoo, M.-J., Kang, H.-W., Lee, H.-G., Han, S., and Nahm, S.

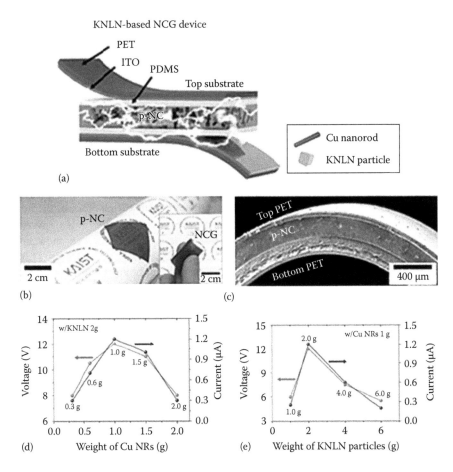

FIGURE 5.18 (a) Schematic of an NCG device using KNLN particles and Cu NRs. (b) Photograph of the flexible p-NC layer attached to a rolled paper. The inset shows the final NCG device bent by fingers. (c) Cross-sectional SEM image of a bent KNLN-based NCG. (d) Variation of output voltage and current produced from nanocomposite devices with the different weights of Cu NRs at the fixed weight of KNLN particles. (e) Generated output signals of NCG devices with various weights of KNLN particles at the fixed amount of Cu NRs. (Jeong, C.K., Park, K.-I., Ryu, J., Hwang, G.-T., and Lee, K.J.: *Adv. Funct. Mater.* 2014. 24, 2620–2629. Copyright Wiley-VCH Verlag GmbH & Co. KGaA. Adapted with permission.)

(2 g) of KNLN particles, the output voltage and current signals from the deformed nanocomposite device were found to rise (Figure 5.18d). Now, the weight of KNLN particles was increased while keeping the amount of Cu NRs fixed at 1 g (Figure 5.18e). The generated output of the nanocomposite device was found to be optimum with 2 g of KNLN particles due to the Maxwell–Wagner–Sillars polarization. This flexible large-area nanocomposite-based energy harvester was found to generate ~mW power (≈140 V and ≈8 µA) from biomechanical energy.[115] This research suggests a possibility to design a cost-effective environment-friendly nanocomposite to fabricate flexible energy harvesting devices.

5.9 CONCLUSIONS

In this chapter we first provide a brief review of various renewable sources of energy harvesting including solar, wind, wave and water flow, thermal, and vibration energy. We further review state-of-the-art materials and approaches to harvest these energy sources with emphasis on piezoelectric vibration energy harvesting. Mostly piezoelectric materials used in energy harvesting devices are polycrystalline in nature due to their cost-effectiveness. On the other hand, single crystals exhibit a high piezo-electric response in certain crystallographic orientations resulting in enhanced energy output. However, the synthesis of large-size and high-quality single crystals has been quite challenging and expensive. High-performance cost-effective textured piezoelec-tric materials were developed with a piezoelectric response similar to that of their single-crystal counterparts. Furthermore, using the unique texturing process, the fun-damental barrier of obtaining the high energy density coefficient ($d \cdot g$) was overcome to achieve the highest reported ($d \cdot g$) coefficient. With superior properties, these tex-tured materials can be used to fabricate efficient piezoelectric energy harvesters for piezoelectric vibration energy harvesting. Moreover, by combining texturing and low-temperature cofiring ceramics processes, high-performance and cost-effective multi-layer textured piezoelectric ceramic systems were fabricated, which exhibited six times increase in power output from vibration energy harvesting. These multilayer energy harvesters could be directly used to power electrical devices without the additional electrical circuit. Furthermore, environmentally benign textured lead-free piezoelectric materials with high piezoelectric response were synthesized. These high-performance lead-free piezoelectric materials can be used to design environment-friendly piezoelec-tric energy harvesters. A brief review about the piezoelectric materials, for harvesting energy in the high-temperature hostile environment up to 1000°C, is provided.

In addition, due to recent demand of flexible electrons, polymer-based compos-ites have also attracted great attention. We also provide a review on materials for efficient flexible energy harvesters. Though the piezoelectric materials due to their unique characteristics are increasingly becoming popular in harvesting various energy sources as mentioned earlier, recently, they have attracted great attention for harvesting solar and thermal energy.

ACKNOWLEDGMENTS

The authors gratefully acknowledge the financial support from the Office of Naval Research (ONR) through the Center for Energy Harvesting Materials and Systems (CEHMS). D. Maurya would like to acknowledge the financial support from the Office of Basic Energy Sciences, Department of Energy, through grant number DE-FG02-07ER46480.

REFERENCES

1. Y. Yuan, Z. Xiao, B. Yang and J. Huang, *J. Mater. Chem. A*, 2014, **2**, 6027–6041.
2. S. Thomas, T. G. Deepak, G. S. Anjusree, T. A. Arun, S. V. Nair and A. S. Nair, *J. Mater. Chem. A*, 2014, **2**, 4474–4490.
3. J. Kreisel, M. Alexe and P. A. Thomas, *Nat. Mater.*, 2012, **11**, 260.

4. S. Y. Yang, J. Seidel, S. J. Byrnes, P. Shafer, C. H. Yang, M. D. Rossell, P. Yu, Y. H. Chu, J. F. Scott, J. W. Ager, L. W. Martin and R. Ramesh, *Nat. Nano*, 2010, **5**, 143–147.
5. M. Alexe and D. Hesse, *Nat. Commun.*, 2011, **2**, 256.
6. J. Seidel, D. Fu, S.-Y. Yang, E. Alarcón-Lladó, J. Wu, R. Ramesh and J. W. Ager, *Phys. Rev. Lett.*, 2011, **107**, 126805(1–4).
7. M. A. Green, *Solar Cells: Operating Principles, Technology, and System Applications*, Prentice-Hall, New Jersey, 1982.
8. V. M. Fridkin, *Crystallogr. Rep.*, 2001, **46**, 654–658.
9. L. Pintilie, I. Vrejoiu, G. Le Rhun and M. Alexe, *J. Appl. Phys.*, 2007, **101**, 064109(1–8).
10. A. M. Glass, D. von der Linde and T. J. Negran, *Appl. Phys. Lett.*, 1974, **25**, 233–235.
11. B. I. Sturman and V. M. Fridkin, *Photovoltaic and Photo-refractive Effects in Noncentrosymmetric Materials*, Gordon and Breach, Philadelphia, 1992.
12. D. Cao, C. Wang, F. Zheng, W. Dong, L. Fang and M. Shen, *Nano Lett.*, 2012, **12**, 2803–2809.
13. I. Grinberg, D. V. West, M. Torres, G. Gou, D. M. Stein, L. Wu, G. Chen, E. M. Gallo, A. R. Akbashev, P. K. Davies, J. E. Spanier and A. M. Rappe, *Nature*, 2013, **503**, 509–512.
14. E. Hau, *Wind Turbines: Fundamentals, Technologies, Application, Economics*, Springer, Berlin Heidelberg, Softcover reprint of hardcover 2nd edn. (October 14, 2010), 2006.
15. R. A. Kishore and S. Priya, *J. Wind Eng. Ind. Aerod.*, 2013, **118**, 12–19.
16. R. A. Kishore, D. Vučković and S. Priya, *Ferroelectrics*, 2014, **460**, 98–107.
17. O. C. O. Alireza Khaligh, *Energy Harvesting*, CRC Press is an Imprint of Taylor & Francis Group, New York, 2010.
18. O. S. A. R. Bedard, *Proceedings of the IEEE Power Engineering Society 2005 Meeting Panel Session*, EPR I solutions, San Rafael, CA, June 2005.
19. J. Rastegar and R. Murray, *Active and Passive Smart Structures and Integrated Systems 2010, Pts 1 and 2*, Bellingham, WA, 2010, vol. 7643.
20. D. Shiels, A. Leonard and A. Roshko, *J. Fluid Struct.*, 2001, **15**, 3–21.
21. E. Guilmineau and P. Queutey, *J. Fluid Struct.*, 2004, **19**, 449–466.
22. M. M. Bernitsas, K. Raghavan, Y. Ben-Simon and E. M. Garcia, *J. Offshore Mech. Arct. Eng.*, 2008, **130**, 041101(1–15).
23. A. Mehmood, A. Abdelkefi, M. R. Hajj, A. H. Nayfeh, I. Akhtar and A. O. Nuhait, *J. Sound Vibration*, 2013, **332**, 4656–4667.
24. M. S. Bhuyan, M. Othman, S. H. M. Ali, B. Y. Majlis and M. S. Islam, *2012 10th IEEE International Conference on Semiconductor Electronics (ICSE)*, Kuala Lumpur, 2012.
25. J. J. Allen and A. J. Smits, *J. Fluid Struct.*, 2001, **15**, 629–640.
26. X. D. Xie, Q. Wang and N. Wu, *J. Sound Vibration*, 2014, **333**, 1421–1429.
27. G. W. Taylor, J. R. Burns, S. A. Kammann, W. B. Powers and T. R. Welsh, *IEEE J. Oceanic Eng.*, 2001, **26**, 539–547.
28. C. Viñolo, D. Toma, A. Mànuel and J. Rio, *Eur. Phys. J. Spec. Top.*, 2013, **222**, 1685–1698.
29. H. J. Goldsmid, *Materials, Preparation, and Characterization in Thermoelectrics*, ed. D. M. Rowe, CRC Press, Taylor & Francis Group, Boca Raton, FL, 2012.
30. P. Ball, *MRS Bulletin*, 2013, **38**(06), 446–447.
31. M. Hamid Elsheikh, D. A. Shnawah, M. F. M. Sabri, S. B. M. Said, M. Haji Hassan, M. B. Ali Bashir and M. Mohamad, *Renew. Sust. Energy Rev.*, 2014, **30**, 337–355.
32. S. Tajima, T. Tani, S. Isobe and K. Koumoto, *Mater. Sci. Eng.: B*, 2001, **86**, 20–25.
33. L. D. Zhao, B. P. Zhang, J. F. Li, H. L. Zhang and W. S. Liu, *Solid State Sci.*, 2008, **10**, 651–658.
34. H. Itahara, C. Xia, Y. Seno, J. Sugiyama, T. Tani and K. Koumoto, *Twenty-Second International Conference on Thermoelectrics, 2003—ICT*, La Grande Motte, France, 2003.

35. H. Itahara, W. S. Seo, S. Lee, H. Nozaki, T. Tani and K. Koumoto, *J. Am. Chem. Soc.*, 2005, **127**, 6367–6373.
36. S. D. Bhame, D. Pravarthana, W. Prellier and J. G. Noudem, *Appl. Phys. Lett.*, 2013, **102**, 211901(1–5).
37. E. Guilmeau, R. Funahashi, M. Mikami, K. Chong and D. Chateigner, *Appl. Phys. Lett.*, 2004, **85**, 1490–1492.
38. Y. Masuda, D. Nagahama, H. Itahara, T. Tani, W. S. Seo and K. Koumoto, *J. Mater. Chem.*, 2003, **13**, 1094–1099.
39. T. Tani, S. Isobe, W. S. Seo and K. Koumoto, *J. Mater. Chem.*, 2001, **11**, 2324–2328.
40. G. J. Snyder and E. S. Toberer, *Nat. Mater.*, 2008, **7**, 105–114.
41. R. A. K. Dragan Avirovik, D. Vuckovic, and S. Priya, *Energy Harvest. Syst.*, 2014, **0**, 1–6.
42. A. Cuadras, M. Gasulla and V. Ferrari, *Sens. Actuat. A: Phys.*, 2010, **158**, 132–139.
43. L. Zuo and X. Tang, *J. Intel. Mater. Syst. Struct.*, 2013, **24**, 1405–1430.
44. H. Kim, Y. Tadesse, S. Priya, *Energy Harvesting Technologies*, eds. S. Priya and D. J. Inman, Springer, New York, 2009, pp. 3–39.
45. K. B. A. Bandivadekar, K. Bodek, L. Cheah, C. Evans, T. Groode, J. Heywood, E. Kasseris, M. Kromer and M. Weiss, *On the Road in 2035: Reducing Transportation's Petroleum Consumption and GHG Emissions LFEE 2008–05 RP*, Massachusetts Institute of Technology, Cambridge, MA, 2008.
46. U.S. Department of Energy, Fuel economy: Where the energy goes, Accessed on March 22, 2014, http://www.fueleconomy.gov/feg/atv.shtml.
47. L. Zuo and P.-S. Zhang, *ASME 2011 Dynamic Systems and Control Conference DSCC2011*, Arlington, VA, 2011.
48. L. Zuo and P.-S. Zhang, *J. Vib. Acoust.*, 2013, **135**, 011002–011002.
49. X. Tang and L. Zuo, *J. Intel. Mater. Syst. Struct.*, 2012, **23**, 2117–2127.
50. D. Bowness, A. C. Lock, W. Powrie, J. A. Priest and D. J. Richards, *Proc. Inst. Mech. Eng. F: J. Rail Rapid Transit*, 2007, **221**, 13–22.
51. K. J. Phillips, Master Thesis, University of Nebraska-Lincoln, 2011.
52. T. Starner and J. Paradiso, *Low Power Electronics Design*, ed. C. Piguet, CRC Press, Boca Raton, FL, 2004, pp. 1–35.
53. C. R. Saha, T. O'Donnell, N. Wang and P. McCloskey, *Sens. Actuat. A: Phys.*, 2008, **147**, 248–253.
54. P. D. Mitcheson, E. M. Yeatman, G. K. Rao, A. S. Holmes and T. C. Green, *Proc. IEEE*, 2008, **96**, 1457–1486.
55. A. Khaligh, Z. Peng and Z. Cong, *IEEE Trans. Ind. Electron.*, 2010, **57**, 850–860.
56. S. P. Beeby, M. J. Tudor and N. M. White, *Meas. Sci. Technol.*, 2006, **17**, R175–R195.
57. S. P. Beeby and T. O. Donnell, *Energy Harvesting Technologies*, ed. S. Priya and D. J. Inman, Springer, New York, 2009, pp. 129–161.
58. S. Meninger, J. O. Mur-Miranda, R. Amirtharajah, A. P. Chandrakasan and J. H. Lang, *IEEE Trans. Very Large Scale Integr. Syst.*, 2001, **9**, 64–76.
59. E. O. Torres and G. A. Rincon-Mora, *IEEE Trans. Circ. Syst. I: Reg. Pap.*, 2009, **56**, 1938–1948.
60. N. Elvin and A. Erturk, *Advances in Energy Harvesting Methods*, ed. N. Elvin and A. Erturk, Springer, New York, 2013, pp. 3–14.
61. S. Roundy, *J. Intel. Mater. Syst. Struct.*, 2005, **16**, 809–823.
62. J. Brufau-Penella, M. Puig-Vidal, P. Giannone, S. Graziani and S. Strazzeri, *Smart. Mater. Struct.*, 2008, **17**, 015009(1–15).
63. M. Shahinpoor, Y. Bar-Cohen, J. O. Simpson and J. Smith, *Smart. Mater. Struct.*, 1998, **7**, R15–R30.
64. K. Sadeghipour, R. Salomon and S. Neogi, *Smart. Mater. Struct.*, 1992, **1**, 172–179.
65. B. R. Martin, Master Thesis, Virginia Tech, 2005.

66. C. R. Bowen, H. A. Kim, P. M. Weaver and S. Dunn, *Energy Environ. Sci.*, 2014, **7**, 25–44.
67. S. Priya, *IEEE Trans. Ultrason. Ferroelectr. Freq. Contr.*, 2010, **57**, 2610–2612.
68. B. Jaffe, W. R. Cook and H. L. Jaffe, *Piezoelectric Ceramics*, Academic Press, New York, 1971.
69. B. Noheda, J. A. Gonzalo, L. E. Cross, R. Guo, S. E. Park, D. E. Cox and G. Shirane, *Phys. Rev. B*, 2000, **61**, 8687–8695.
70. M. J. Hoffmann, H. Kungl, J. T. Reszat and S. Wagner, *Polar Oxides*, Wiley-VCH Verlag GmbH & Co. KGaA, Weinheim, 2005, pp. 137–150.
71. S. Wada, K. Yako, H. Kakemoto, T. Tsurumi and T. Kiguchi, *J. Appl. Phys.*, 2005, **98**, 014109(1–7).
72. W.-F. Rao and Y. U. Wang, *Appl. Phys. Lett.*, 2007, **90**, 041915(1–3).
73. K. Uchino, in *Advanced Piezoelectric Materials: Science and Technology*, ed. K. Uchino, Woodhead Publishing, Philadelphia, 2010, pp. 1–82.
74. K. Uchino, *Ferroelectric Devices*, 2nd edn., CRC Press, 2009.
75. S. Zhang and F. Li, *J. Appl. Phys.*, 2012, **111**, 031301(1–50).
76. S.-H. Hong and G. L. Messing, *J. Am. Ceram. Soc.*, 1999, **82**, 867–872.
77. E. Suvaci, M. M. Seabaugh and G. L. Messing, *J. Eur. Ceram. Soc.*, 1999, **19**, 2465–2474.
78. G. L. Messing, S. Trolier-McKinstry, E. M. Sabolsky, C. Duran, S. Kwon, B. Brahmaroutu, P. Park, H. Yilmaz, P. W. Rehrig, K. B. Eitel, E. Suvaci, M. Seabaugh and K. S. Oh, *Crit. Rev. Solid State Mater. Sci.*, 2004, **29**, 45–96.
79. Y. Yan, K.-H. Cho and S. Priya, *Appl. Phys. Lett.*, 2012, **100**, 132908(1–5).
80. Y. Yan, Y. U. Wang and S. Priya, *Appl. Phys. Lett.*, 2012, **100**, 192905(1–4).
81. Y. Yan, Y. Zhou and S. Priya, *Appl. Phys. Lett.*, 2013, **102**, 052907(1–5).
82. Y. Yan, Y. Zhou, S. Gupta and S. Priya, *Appl. Phys. Lett.*, 2013, **103**, 082906(1–5).
83. F. K. Lotgering, *J. Inorg. Nucl. Chem.*, 1959, **9**, 113–123.
84. Y. Yan, K.-H. Cho and S. Priya, *J. Am. Ceram. Soc.*, 2011, **94**, 1784–1793.
85. A. A. Bokov, X. Long and Z.-G. Ye, *Phys. Rev. B*, 2010, **81**, 172103(1–4).
86. Y. Yan, K.-H. Cho, D. Maurya, A. Kumar, S. Kalinin, A. Khachaturyan and S. Priya, *Appl. Phys. Lett.*, 2013, **102**, 042903(1–5).
87. S. Zhang, S.-M. Lee, D.-H. Kim, H.-Y. Lee and T. R. Shrout, *Appl. Phys. Lett.*, 2008, **93**, 122908(1–3).
88. H.-C. Song, H.-C. Kim, C.-Y. Kang, H.-J. Kim, S.-J. Yoon and D.-Y. Jeong, *J. Electroceram.*, 2009, **23**, 301–304.
89. M. R. Gongora-Rubio, P. Espinoza-Vallejos, L. Sola-Laguna and J. J. Santiago-Avilés, *Sens. Actuat. A: Phys.*, 2001, **89**, 222–241.
90. Y. Yan, A. Marin, Y. Zhou and S. Priya, *Energy Harvesting and Systems*, 2014, **1**, 189–195.
91. D. Maurya, M. Murayama and S. Priya, *J. Am. Ceram. Soc.*, 2011, **94**, 2857–2871.
92. D. Maurya, V. Petkov, A. Kumar and S. Priya, *Dalton Trans.*, 2012, **41**, 5643–5652.
93. D. Maurya, A. Kumar, V. Petkov, J. E. Mahaney, R. S. Katiyar and S. Priya, *RSC Adv.*, 2014, **4**, 1283–1292.
94. D. Maurya, M. Murayama, A. Pramanick, W. T. Reynolds, K. An and S. Priya, *J. Appl. Phys.*, 2013, **113**, 114101(1–9).
95. D. Maurya, A. Pramanick, K. An and S. Priya, *Appl. Phys. Lett.*, 2012, **100**, 172906(1–5).
96. D. Maurya, Y. Zhou, Y. Yan and S. Priya, *J. Mater. Chem. C*, 2013, **1**, 2102–2111.
97. W. Ge, H. Liu, X. Zhao, B. Fang, X. Li, F. Wang, D. Zhou, P. Yu, X. Pan, D. Lin and H. Luo, *J. Phys. D: Appl. Phys.*, 2008, **41**, 115403(1–5).
98. Q. Zhang, Y. Zhang, F. Wang, Y. Wang, D. Lin, X. Zhao, H. Luo, W. Ge and D. Viehland, *Appl. Phys. Lett.*, 2009, **95**, 115403(1–5).
99. V. Bedekar, J. Oliver, S. Zhang and S. Priya, *Jpn. J. Appl. Phys.*, 2009, **48**, 091406(1–5).

100. Z. Chang, S. Sherrit, X. Bao and Y. Bar-Cohen, *Proceedings of SPIE Conference on Smart Structures and Materials*, San Diego, CA, 2004, SPIE vol. 5388, pp. 320–326.

101. M. Minary-Jolandan, R. A. Bernal, I. Kuljanishvili, V. Parpoil and H. D. Espinosa, *Nano Lett.*, 2011, **12**, 970–976.

102. E. D. Le Boulbar, M. J. Edwards, S. Vittoz, G. Vanko, K. Brinkfeldt, L. Rufer, P. Johander, T. Lalinský, C. R. Bowen and D. W. E. Allsopp, *Sens. Actuat. A: Phys.*, 2013, **194**, 247–251.

103. D. Damjanovic, *Curr. Opin. Solid State Mater. Sci.*, 1998, **3**, 469–473.

104. Y. Qi and M. C. McAlpine, *Energy Environ. Sci.*, 2010, **3**, 1275–1285.

105. J. Fang, H. Niu, H. Wang, X. Wang and T. Lin, *Energy Environ. Sci.*, 2013, **6**, 2196–2202.

106. J. Chang, M. Dommer, C. Chang and L. Lin, *Nano Energy*, 2012, **1**, 356–371.

107. R. E. Newnham, D. P. Skinner and L. E. Cross, *Mater. Res. Bull.*, 1978, **13**, 525–536.

108. S. R. Khaled, D. Sameoto and S. Evoy, *Smart Mater. Struct.*, 2014, **23**, 033001.

109. C. Baur, J. R. DiMaio, E. McAllister, R. Hossini, E. Wagener, J. Ballato, S. Priya, A. Ballato and D. W. Smith, *J. Appl. Phys.*, 2012, **112**, 033001(1–26).

110. Y. Choi, M.-J. Yoo, H.-W. Kang, H.-G. Lee, S. Han and S. Nahm, *J. Electroceram.*, 2013, **30**, 30–35.

111. X. Cai, C. Zhong, S. Zhang and H. Wang, *J. Mater. Sci. Lett.*, 1997, **16**, 253–254.

112. K. H. Lam and H. L. W. Chan, *Compos. Sci. Technol.*, 2005, **65**, 1107–1111.

113. W. Nhuapeng and T. Tunkasiri, *J. Am. Ceram. Soc.*, 2002, **85**, 700–702.

114. W. Michael and A. Kristin, *J. Phys. D: Appl. Phys.*, 2008, **41**, 165409(1–6).

115. C. K. Jeong, K.-I. Park, J. Ryu, G.-T. Hwang and K. J. Lee, *Adv. Funct. Mater.*, 2014, **24**, 2620–2629.

116. L. Zuo and X. Tang, *J. Intel. Mater. Syst. Struct.*, 2013, **24**, 1405–1430.

6 Advanced Electrode Materials for Electrochemical Capacitors

Jianli Kang and Mingwei Chen

CONTENTS

6.1 INTRODUCTION

Energy is one of the most important topics in the twenty-first century. Traditional energy produced from the combustion of fossil fuels has generated a huge adverse effect on the world economic activity and ecology. As a result, we are observing a rapid increase in the sustainable and renewable energy production from the wind, water, and sun, as well as the development of electric and hybrid electric vehicles with low CO_2 emissions. Because the wind, water, and sun cannot be constantly available nor have stable outputs on demand, electrochemical energy storage systems are becoming crucial. To meet the high-energy requirements of future applications from portable electronics to electric vehicles and large industrial equipments, the performance of the current electrochemical energy storage systems needs to be substantially improved by developing new electrode materials.[1]

At the forefront of the electrochemical energy storage systems are batteries and electrochemical capacitors (ECs). Batteries, including Li-ion batteries and other advanced secondary batteries, can store a high energy density. However, they suffer from a somewhat slow power delivery or uptake. Faster and higher-power energy storage systems are needed in a number of applications, which have been given to ECs.[2,3] ECs, also known as supercapacitors or ultracapacitors, are power devices that can be fully charged or discharged in seconds. Since the first commercial aqueous-electrolyte ECs were developed by NEC (Japan) under the energy company SOHIO's license for power-saving units in electronics in 1971,[4] new applications in mobile electronics, transportation vehicles, renewable energy production, and aerospace systems bolstered further research. A recent report by the U.S. Department of Energy[5] assigned equal importance to ECs and batteries for future energy storage systems. Articles on ECs in popular magazines also show increasing interest by the general public.

For the EC systems, the electrode materials are the most important factor to determine the capacitive properties of the devices. In this chapter, we will briefly review the energy storage principles, microstructure, and properties of advanced electrode materials, including carbon, transition metal oxides, conductive polymers, and their composites, which have been intensely studied in the last decade. The focus of this chapter will be placed on the recent progress in the nanoarchitecture design of advanced electrodes for next-generation ECs that have both high energy density, comparable to batteries, and high power density.

6.2 ENERGY STORAGE MECHANISMS OF ECs

Depending on the charge-storage mechanisms, ECs can be classified into two types: electric double-layer capacitor (EDLC) and faradaic pseudocapacitor. ECs that only involve physical adsorption of ions in the manner, without any chemical reaction, are called EDLCs, while ECs with additional or alternative pseudocapacitance belonged to faradaic pseudocapacitors, in which the valence electrons of electrochemical active materials are transferred across the electrode/electrolyte interface, resulting in a potential-dependent capacitance. Although the two mechanisms usually function together in one EC system, it is more convenient to discuss them separately.

6.2.1 EDLCs

EDLCs store the electric charge directly across the double layer of an electrode, which was initially introduced by von Helmholtz in the nineteenth century with a model that oversimplified the behavior representing capacitive charge.[6] The double-layer capacitance can be estimated according to the following equation:

$$C = \frac{\varepsilon_r \varepsilon_0 A}{d}$$

where
   ε_r is the electrolyte dielectric constant
   ε_0 is the dielectric constant of the vacuum
   d is the effective thickness of the double layer (charge separation distance)
   A is the electrode surface area

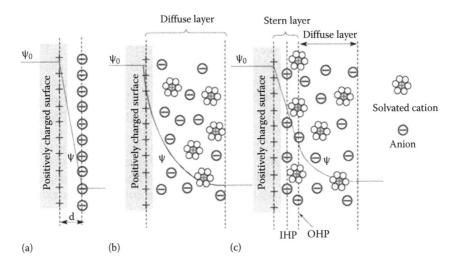

FIGURE 6.1 Progressive evolutions of the models describing the electric double layer at a positive electrode—(a) the initial Helmholtz model, (b) the Gouy–Chapman model, and (c) the Stern model—where the division of the inner Helmholtz plane (IHP) and outer Helmholtz plane (OHP) shows the specifically and nonspecifically absorbed ions. OHP is also the plane where the diffused layer begins. The distance, d, in the Helmholtz model describes the double-layer distance, and ψ_0 and ψ are the surface and electrode/electrolyte interface potentials, respectively. (Zhang, L. and Zhao, X.S., *Chem. Soc. Rev.*, 38, 2520, Copyright 2009. Adapted by permission of The Royal Society of Chemistry.)

The capacitance model was later refined by Gouy and Chapman and Stern and Geary, who suggested the presence of a diffuse layer in the electrolyte due to the accumulation of ions close to the electrode surface, as shown in Figure 6.1. The double-layer capacitance is typically in the range of 10–40 µF cm^{-2} for a smooth electrode, depending on the electrode materials and the nature of the electrolytes used.[7]

6.2.2 Pseudocapacitors

The mechanism for the charge storage by pseudocapacitive materials involves the electron transfer reactions and hence the reduction or oxidation (redox) changes in the electrode materials, which are also referred to as the faradaic reactions or processes. Several faradaic models in the pseudocapacitive electrodes have been proposed[8,9]: (1) reversible surface adsorption of proton or metal ions from the electrolyte, (2) redox reactions involving ions from the electrolyte, and (3) reversible doping–dedoping processes in conducting polymers. The first two processes are primarily surface reactions, which are highly dependent on the surface area of the electrode. The third process is more of a bulk process, and hence the specific capacitance of the electrode materials is less dependent on the surface area, although a relatively high surface area with small pores is desirable to rapidly transfer the ions efficiently to or from the electrode.

The electron transfer to or from a pseudocapacitive material brings a net charge into the material, which must be neutralized by the ingression or intercalation of the ions of the opposite charge from the electrolyte into a solid phase. Thus, the kinetic difficulties for the ion transport and substantial solid phase changes are common in pseudocapacitive materials, which results in an inferior performance, such as lower performance rate and poor cyclic stability, compared to the EDLC storage system.[10] However, unlike redox processes in a battery, the faradaic process in a pseudocapacitor is a thermodynamic change of potential during charge accumulation and has better reversibility. Consequently, the pseudocapacitors can be considered to possess both battery-like behavior with faradaic reactions occurring across the double-layer and electrostatic capacitor-like behaviors with high reversibility and high power. Generally, the capacitance can reach up to over $100~\mu F~cm^{-2}$ of the pseudocapacitive materials.[11]

6.3   ELECTRODE MATERIALS FOR ECs

In general, the electrode materials of ECs can be categorized into three types: (1) carbon-based materials with high specific surface area (SSA), such as activated carbon,[12–14] carbon nanotubes,[15–17] carbon nanofibers,[18–20] and graphene[21–23]; (2) transition metal oxides, such as RuO_2,[24–26] MnO_2,[27–29] NiO,[30–32] Co_xO_y,[33,34] $NiCo_2O_4$,[35–37] SnO_2,[38–40] V_2O_5,[41,42] and MoO_x,[43,44]; and (3) conducting polymers, such as polyaniline (PANI),[17,45,46] polypyrrol (PPy),[47–49] polythiophene (PTh),[50,51] and their corresponding derivatives.[52,53]

The advantages of carbon-based materials include abundance, low cost, high chemical stability, high SSA, nontoxic, easy processing, and good electronic conductivity. Carbon-based ECs are close to the EDLCs, inferring that their capacitance heavily depends on the SSA of the electrode materials. To develop high-performance EDLC electrode materials, many micro-, meso-, and macroporous carbon materials have been investigated.[54–56] It is found that the measured capacitance of the porous carbon materials with a high surface area does not linearly increase by increasing the SSA.[57] The pore size distribution of the carbon electrode plays an important role in realization of the effective surface area.[58,59] According to Largeot et al.,[60] the pore size of the electrode materials, close to the ion size of an electrolyte (with respect to an ionic liquid electrolyte), yields the maximum double-layer capacitance and both larger and smaller pores led to a significant drop in capacitance. Further, Kondrat et al.[61] indicated that the energy density of EC devices was a nonmonotonic function of the pore size of monodispersed porous electrodes. The *optimal* pore size, providing the maximal energy density, increases with the operating voltages and saturated at a high voltage. Graphene, as a 2D material, can be potentially processes and are fully accessible with a high SSA and high conductivity. The theoretical SSA of a single-layer graphene can reach up to $2670~m^2~g^{-1}$.[62] By chemical activation with KOH, the exfoliated graphene oxide (GO) can achieve SSA values up to $3100~m^2~g^{-1}$ with a high electrical conductivity and a low oxygen and hydrogen content.[63] Consequently, graphene has become one of the most promising candidates for EDLC applications and attracted great interest during the past several years. However, the gravimetric density of graphene is still low, especially in designing thick electrodes to assemble large cells of hundreds of farads.[62] Most graphene-derived carbons likely have a limited exploited surface area, presumably because they may form stacked agglomerates during processing and thus may have limited ion diffusion paths throughout some narrow channels.[64,65] The critical issue for graphene is therefore to avoid restacking of sheets during electrode preparation. To address this, hierarchically structured graphene networks have been developed. The effects of molecular structures (oxygen functionalities, heteroatom doping), deposited metal oxides, and conductive polymers on the performance of hierarchically structured graphene electrodes have been recently reviewed by Dong et al.[66]

Transition metal oxides are one of the best candidates as electrode materials for redox ECs because of their high theoretical capacitance originating from various oxidation states available for redox charge transfer. Among the transition metal oxides, RuO_x has been the most extensively studied candidate due to its ultrahigh theoretical capacitance of ca. 2000 F g^{-1} in a wide potential window, a nearly metallic electrical conductivity, excellent thermal stability, and long cycle life, as well as its high rate capability.[67,68] Despite these fascinating electrochemical characteristics, the commercial applicability of RuO_x is far hindered by the high material costs and lack of abundance. More cost-effective and abundant transition metal systems,[69] such as MnO_2, NiO, Co_2O_3, SnO_2, and MoO_x, have been investigated to replace RuO_x for pseudocapacitive electrodes. Although these oxides have high theoretical specific capacitance, most of them are semiconductors or insulators. The poor electrical conductivity of these materials restrains the realization of their high theoretical capacitance, especially at high charge/discharge rates in electrochemical supercapacitors. The ultrahigh specific capacitance was achieved only in the form of thin films or small nanoparticles with a very low loading amount (less than 10 µg cm^{-2}).[70] Increasing the oxide loading amount and film thickness leads to the dramatic loss of the specific capacitance due to the loss of electrical conductance with the increase of oxide film thickness.[71,72] As a result, the specific capacity of the device is too low for commercial applications. To overcome this deficiency, a composite design with conductive reinforcements, such as nanostructured carbon, conducting polymers, and metal nanoparticles, has been employed to enhance the conductivity of the pseudocapacitive oxide materials.[11,69,73,74]

Conducting polymers possess many advantages that make them promising materials of ECs, such as low cost, high conductivity in a doped state, high storage capacity, good reversibility, and adjustable redox activity through chemical modification. The pseudocapacitance of electronically conducting polymers comes from the fast and reversible oxidation and reduction throughout the entire bulk not just on the surface, related to the π-conjugated polymer chains. Over the last few years, a number of groups have reported promising researches based on polymer electrodes.[17,45–53] The specific capacitance of the conducting polymer electrodes can reach ultrahigh values by optimizing the process. However, their long-term stability during cycling is far from satisfactory. Swelling and shrinking of conducting polymers may occur during the doping/dedoping process, which often leads to mechanical degradation of the electrode and fading electrochemical performance during cycling, and thus compromise conducting polymers as electrode materials.[53] The charge-storage mechanism in polymer electrodes is still not completely understood.

6.4 ADVANCED ELECTRODE OPTIMIZATION FOR HIGH PERFORMANCE

Although nanotechnologies have greatly advanced high-performance EC electrodes, techniques to realize the full potential of electrode materials, especially for pseudocapacitive materials, by achieving simultaneously optimal electrode structures, conductivity, and crystallinity are still under development. Particularly, the problem of being difficult in scaling up the nanotechnology for industrial applications is far from resolved. Herein, we introduce some novel concepts recently proposed for electrode design, which may be implemented in next-generation ECs with both high energy density and high power density.

6.4.1 GRAPHENE ELECTRODES WITH A LARGE PACKING DENSITY

Graphene, as one of the most promising candidates as electrode materials for next-generation ECs, have attracted great attention and thousands of articles have been published every year. In principle, graphene-based ECs can achieve an EDL capacitance as high as 550 F g^{-1} if their entire surface area is used.[75] Some recent publications have used Ragone plots to argue that the energy density of graphene-based ECs can even reach to that of batteries.[76,77] However, as indicated by Gogotsi and Simon,[78] the Ragone plot on a gravimetric basis may not give a realistic picture of the true performance of the assembled device since the weight of other device components (current collectors, electrolyte, separator, and packaging) needs to be taken into account. Furthermore, since graphene-based nanomaterials usually have a low packing density to avoid restacking (<0.5 g cm^{-3}), all the empty space in the electrode will be flooded with electrolytes, increasing the weight without adding capacitance. Therefore, volumetric energy density is more important than the widely used gravimetric density. Especially for microdevices and thin-film ECs in which the weight of the active materials on a chip or a smart fabric is negligible, the gravimetric energy density is almost irrelevant compared to the areal or volumetric energy density. Thus, developing a new technique to prepare porous yet densely packed graphene electrodes by fully utilizing the potential capacitance at a high operating voltage is crucial for high-performance graphene-based ECs.

Based on the previously mentioned consideration, significant progress has recently been made to achieve high volumetric specific capacitance (C_{vol}) of graphene-based ECs. By simply compressing the activated microwave-expanded graphite oxide (a-MEGO)-based electrode material, a volumetric capacitance of up to 110 F cm^{-3} was achieved.[79] However, high compression force induced the mesopores of a-MEGO to collapse, which resulted in a higher effective series resistance and thus lower power density in the compressed samples. Kim et al.[80] recently reported a highly porous graphene-derived carbon with hierarchical pore structures in which mesopores are integrated into macroporous scaffolds (Figure 6.2). The unique 3D pore structures in the produced carbons gave rise to an SSA value of up to 3290 m^2 g^{-1} and provided an efficient pathway for electrolyte ions to diffuse into the interior surface of bulk electrode particles. Meanwhile, the electrode kept a high packing density of ~0.59 g cm^{-3}. Thus, these carbons exhibited both high gravimetric (174 F g^{-1}) and volumetric

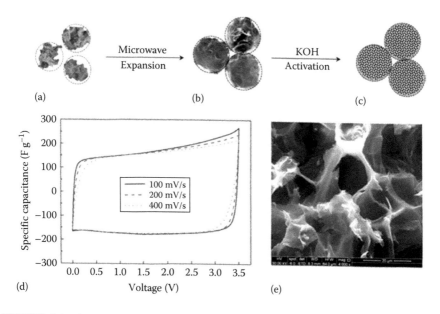

FIGURE 6.2 Schematic illustration of the fabrication of highly porous graphene-derived carbon with hierarchical pore structures. (a) Graphene oxide (GO) sheets are transformed into crumpled ball-like GO particles by aerosol spray drying technique; (b) under microwave irradiation, crumpled GO particles form hollow graphene-based spheres; (c) chemical activation with potassium hydroxide; (d) cyclic voltammetry curves of activated SMEGO (a-MEGO) in [EMIM][TFSI]/AN electrolyte; and (e) SEM micrographs of a-MEGO. (Adapted with permission from Kim, T. et al., *ACS Nano*, 7, 6899. Copyright 2013 American Chemical Society.)

(~100 F cm^{-3}) specific capacitance in an ionic liquid electrolyte. Due to the porous carbon electrodes without applying any further electrode densification process (such as applying high-force mechanical compression), the electrode exhibited both high volumetric and gravimetric power density, as well as high energy density.

More recently, the breakthrough in dense integration of graphene materials for compact capacitive energy storage was made by Yang et al.[81] The chemically converted graphene (CCG) hydrogel films, with a metastable and adaptive pore structure, were compressed irreversibly by capillary pressure to increase the packing density through controlled removal of volatile solvent trapped in the gel. The graphene sheets, stacking in nearly face-to-face fashion in the films, possessed an ultrahigh packing density of up to ~1.33 g cm^{-3} keeping a continuous ion transport network, which was nearly double that of traditional active porous carbon (0.5–0.7 g cm^{-3}).[82] When the packing density of the liquid electrolyte-mediated CCG (EM-CCG) was increased from 0.13 to 1.33 g cm^{-3} by changing the ratio of volatile and nonvolatile liquids, gravimetric specific capacitance (C_{wt-C}) only dropped from 203.2 to 191.7 F g^{-1} at a low current density of 0.1 A g^{-1} and keep over 100 F g^{-1} at a high current density of 100 A g^{-1}. In contrast, the C_{wt-C} of the completely dried CCG (1.49 g cm^{-3}) was only 155.2 F g^{-1} and dramatically dropped to 10.2 F g^{-1}. The volumetric specific capacitance (C_{vol}) of the highly compact EM-CCG films (1.25–1.33 g cm^{-3}) reached up to 255.5 F cm^{-3} in aqueous electrolyte and 261.3 F cm^{-3} in organic electrolyte at 0.1 A g^{-1}, which were much higher than those of existing porous carbon materials. Although the dried CCG film was able to deliver a high C_{vol} at low current density, the value decreased rapidly with the increase of operation rate, as shown in Figure 6.3e. The effect of areal mass loadings (or electrode thickness) on the electrochemical performance further verified the indispensable role of the preincorporated electrolyte in ion transport. Both C_{wt-C} and C_{vol} generally decreased with increasing thickness of the electrodes. However, the EM-CCG film displayed a much slower rate of decrease than the dried film (Figure 6.3f). As a result, the large-scale device assembled by the EM-CCG films, delivering a much higher energy density and power density with a high cyclic stability, is promising for real applications.

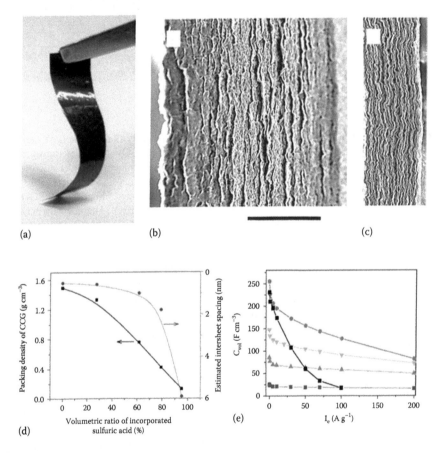

(a) (b) (c)

(d) (e)

FIGURE 6.3 Characterization of liquid EM-CCG films. (a) A photograph showing the flexibility of the film. (b and c) SEM images of cross sections of the obtained EM-CCG films containing 78.9 volumetric percent (vol.%) and 27.2 vol.% of H_2SO_4, respectively, corresponding to the packing density of 0.42 and 1.33 g cm^{-3}. (d) The relation between the volumetric ratio of incorporated electrolyte and packing density as well as the estimated intersheet spacing. (e) Volumetric capacitance of the EM-CCM films with varied charging/discharging current densities. *(Continued)*

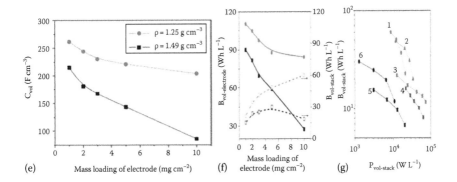

FIGURE 6.3 (Continued) Characterization of liquid EM-CCG films. (f) Volumetric capacitance as a function of the areal mass loading of the EM-CCG film (1.25 g cm^{-3}) and the dried CCG film (1.49 g cm^{-3}) at the current density of 0.1 A g^{-1}. (g) Energy density as a function of the areal mass loading of the EM-CCG film (1.25 g cm^{-3}) and the dried CCG film (1.49 g cm^{-3}) at the current density of 0.1 A g^{-1}. (h) Ragone plots of the ECs assembled by different EM-CCG films with a packing density and areal mass loading of CCG, (1) 1.25 g cm^{-3}, 10 mg cm^{-2}; (2) 1.25 g cm^{-3}, 5 mg cm^{-2}; (3) 1.25 g cm^{-3}, 1 mg cm^{-2}, and (4) 0.39 g cm^{-3}, 1.0 mg cm^{-2}, and the dried CCG films (1.49 g cm^{-3}) with the areal mass loading of (5) 1 mg cm^{-2} and (6) 5 mg cm^{-2}. The data were obtained from the prototype ECs with EMIMBF$_4$/AN as the electrolyte and an operation voltage of 3.5 V. (From Yang, X., Cheng, C., Wang, Y., Qiu L., and Li, D., *Science*, 341, 534, Copyright 2013. Adapted with permission from AAAS.)

6.4.2 Nanoporous Metal-Enhanced Pseudocapacitive Electrodes

As mentioned earlier, the crucial issues for pseudocapacitors are their lack of cyclic stability and rate capability because of the volumetric variation induced by the faradaic redox reactions and low conductivity of pseudocapacitive materials. Despite some significant progress achieved by introducing conductive reinforcements, the enhancement of oxide conductivity by external conductive reinforcements is often limited by weak oxide/conductor interfaces. Recently, Lang et al.[83] proposed that dealloyed nanoporous metals can be functioned as both current collector and conductive reinforcement, which provides an alternative approach to address these problems. Dealloyed nanoporous metals have been representing a new class of functional materials with larger surface areas and unique structural properties of mechanical rigidity, high corrosion resistance, and electrical conductivity, as well as a very flexible porous network structure with feature dimensions tunable within a wide range from a few nanometers to several microns.[84,85] Thus, nanoporous metals can be directly used as electrodes for EDLCs. The EDLC device, assembled by two nanoporous gold (NPG) sheets sandwiched with a piece of cotton paper,[86] can deliver a volumetric capacitance of ~10–20 F cm^{-3} in the room temperature ionic liquid (RTIL) electrolyte and ~17–24 F cm^{-3} in KOH solution. The energy and power densities of EDLCs based on NPG electrodes are ~16.5–21.7 mW h cm^{-3} and ~0.3–6 W cm^{-3}, respectively.

Obviously, the energy density of the nanoporous metal electrodes, similar to the carbon-based electrodes, is far from satisfactory due to the limited double-layer capacitance. Nevertheless, nanoporous metals can be an ideal candidate that functions as both a current collector and conductive reinforcement because of its high conductivity, high surface area, and self-supporting architecture. Lang et al.[83] fabricated 100 nm thick NPG/MnO_2 hybrid electrodes by incorporating nanocrystalline MnO_2 into the nanopores of NPG via a facile electroless plating process. Microstructure characterizations demonstrated that the nanocrystalline MnO_2 was uniformly plated into the nanopore channels while keeping open nanoporosity (Figure 6.4a). The nanocrystalline MnO_2 grew epitaxially on the gold ligament surfaces and formed a chemically bonded metal/oxide interface (Figure 6.4b and c), which was further confirmed by the electron energy-loss spectroscopy (EELS) spectra (Figure 6.4d). The noticeable shift of Mn $L_{2,3}$ edges to the low-energy side in the

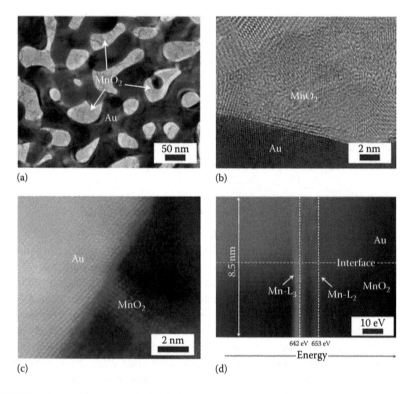

FIGURE 6.4 (a) Bright-field TEM image of the nanoporous gold/MnO_2 hybrid with an MnO_2 plating time of 20 min. (b) HRTEM image of the hybrid showing nanocrystalline MnO_2 with a grain size of about 5 nm. (c) High-angle annular dark-field STEM image taken from the gold/MnO_2 interface region. Nanocrystalline MnO_2 epitaxially grows on the gold surface, forming a chemically bonded metal/oxide interface. (d) EELS profile image from a region across the interface. The horizontal axis shows the electron energy loss in the range of 586–705 eV. EELS spectra were obtained from 30 points with an interval of 0.28 nm and an exposure time of 1.6 s for each spectrum acquisition. The Mn L_3 (642 eV) and L_2 (653 eV) edges shift by about 0.5–1 eV to lower energies near the interface between the MnO_2 and the gold. (Adapted by permission from Macmillan Publishers Ltd. *Nat. Nanotechnol.*, Lang, X. et al., 6, 232, copyright 2011.)

gold/MnO_2 interface region provided compelling evidence of charge transfer between the gold substrate and MnO_2 nanocrystals. The excellent contact between the MnO_2 and gold ligaments, as well as the continuous gold network with ultrahigh electrical conductivity, assists the fast electron transport in the electrodes. Furthermore, the well-interconnected porous channels in the electrode not only allowed rapid ion transfer but also provided extremely large SSA of the electrode/electrolyte interface. All these contribute to facilitating the full use of large pseudocapacitance of MnO_2. Electrochemical measurement indicates that the resultant NPG/MnO_2 hybrid electrodes exhibit an ultrahigh specific capacitance (1145 F g^{-1}), as well as high power density with low internal resistance. Even in organic electrolytes,[87] the NPG/MnO_2 films with excellent electrical conductivity, rich porous structure, and large surface area still give rise to low internal resistance, good ionic contact, and thus enhanced redox reaction of MnO_2, resulting in high specific capacitance with a large working potential window. Thus, the MnO_2/NPG-based supercapacitors possess competitive energy density (~40 W h kg^{-1}) and power density (~150 kW kg^{-1}). This strategy was also successfully applied in other pseudocapacitive material systems, such as RuO_2,[88] PPy,[89,90] and PANI.[91] By introducing NPG as both current collector and conductive reinforcement, these pseudocapacitive materials exhibit ultrahigh specific capacitances, almost approaching their theoretical capacitances, solidly confirming the indispensable role and universality of NPG in high-performance pseudocapacitive electrodes.

In order to load a large amount of active materials for high-capacity electrodes, Kang et al.[70] developed a novel sandwich electrode with retained high specific capacitance by utilizing the open porosity of nanoporous metals. Figure 6.5 showed the SEM images of MnO_2 films electroplated on NPG for different periods. As shown in Figure 6.5a, the nanostructured MnO_2 starts to grow within the nanopore channels of NPG. After completely covering the active internal surface of nanopores, MnO_2 sprouts out from the nanopore channels and symmetrically grows on the two outer surfaces of the NPG sheet, forming a sandwich structure with two thick MnO_2 coatings (Figure 6.5b through d). The high-magnification SEM image (Figure 6.5e) indicated that the thick MnO_2 layers are porous and continuously grow from the pore channels of NPG. The top-view SEM micrograph (Figure 6.5f) also proved the porous morphology of the thick MnO_2 coatings. The porous structure (tens of nanometer pore size) of the novel sandwich electrode ensures electrolyte transfer freely through the whole electrode from both sides, facilitating the full utilization of the pseudocapacitance of MnO_2. As shown in Figure 6.6a, the CV curves of the sandwich electrodes keep a symmetrical rectangular shape even with the MnO_2 loading amount of high as ~200 µg cm^{-2}, inferring fast and accessible electron/ion transport in the thick MnO_2 films. The highest specific capacitance value of ~916 F g^{-1}, ~70% of the theoretical capacitance of MnO_2, was achieved at the loading amount of 53 µg cm^{-2}. Further increasing the loading amounts of MnO_2 gives rise to the decrease of the capacitance. However, an ultrahigh capacitance of ~772 F g^{-1} was still retained for the thick film with a loading amount of ~128 µg cm^{-2}, which is about two to five times larger than that of MnO_2 deposited on flat Au films and literature data at a similar loading amount. Importantly, even the MnO_2 loading amount is as large as ~200 µg cm^{-2} (~2.1 µm thick), the specific capacitance of the

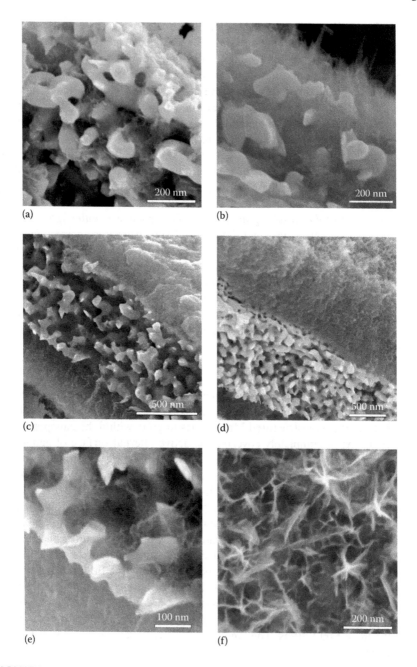

FIGURE 6.5   SEM micrographs and growth rate of electroplated MnO_2 on NPG. (a through e) Cross-sectional SEM images of $MnO_2/NPG/MnO_2$ electrodes with the MnO_2 plating time for (a) 10 min, (b) 20 min, (c) 30 min, (d) 40 min, and (e) zoom-in image of (d). (f) The top view of the 40 min electroplated MnO_2. (Kang, J., Chen, L., Hou, Y., Li, C., Fujita, T., Lang, X., Hirata, A., and Chen, M.: *Adv. Energy Mater.* 2013. 3. 857–863. Copyright Wiley-VCH Verlag GmbH & Co. KGaA. Adapted with permission.)

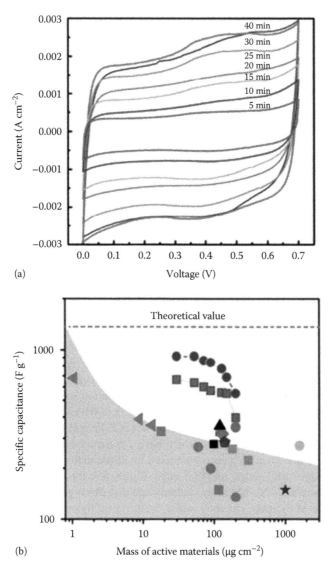

(a)

(b)

FIGURE 6.6 (a) CV curves of $MnO_2/NPG/MnO_2$ electrodes with different plating time at a scan rate of 50 mV s^{-1}. (b) The plot of specific capacitance and loading mass of MnO_2 in the NPG-based sandwich electrodes (● at a scan rate of 5 mV s^{-1}; ▪ at a scan rate of 50 mV s^{-1}). For comparison, the value of MnO_2 deposited on Au film (■ at a scan rate of 50 mV s^{-1}) and literature data of the MnO_2 electrodes are listed in the plot (●[92][at a scan rate of 20 mV s^{-1}]; ◆[93] [at a scan rate of 50 mV s^{-1}]; ▪[94][at a scan rate of 20 mV s^{-1}]; ●[71][at a scan rate of 5 mV s^{-1}]; ▲[95][at a scan rate of 2 mV s^{-1}]; ●[96][at a scan rate of 50 mV s^{-1}]; ▪[97][at a scan rate of 50 mV s^{-1}]; ◀[98] [at a scan rate of 50 mV s^{-1}]; ●[99][at a scan rate of 20 mV s^{-1}]; ★[100][at a scan rate of 20 mV s^{-1}]). (Kang, J., Chen, L., Hou, Y., Li, C., Fujita, T., Lang, X., Hirata, A., and Chen, M.: *Adv. Energy Mater.* 2013. 3. 857–863. Copyright Wiley-VCH Verlag GmbH & Co. KGaA. Adapted with permission.)

thick $MnO_2/NPG/MnO_2$ electrode still keeps as high as ~550 F g^{-1} at a scan rate of 5 mV s^{-1}. Moreover, the sandwiched electrodes do not show strong scanning rate dependence. At a high scan rate of 50 mV s^{-1}, there is only a relative small capacitance fading (e.g., from 665 to 557 F g^{-1} for 30 min in an MnO_2 plated sample). Due to the symmetrical architecture of the $MnO_2/NPG/MnO_2$ electrode and excellent interfacial contact between gold and MnO_2, the novel electrode processes ultrahigh cyclic stability. The capacitance remained ~97.1% of the maximum value after 3000 cycles at a scan rate of 50 mV s^{-1}.

Although the NPG-based electrodes exhibited excellent capacitive performance, gold, as the noble metal, is very expensive, which consequently narrows their practical applications. Moreover, the fabrication of oxide(polymer)/NPG composites requires to chemically or electrochemically deposit the oxides (polymers) into deep nanopores of NPG, which limits the loading amount of active materials and often forms an inhomogeneous structure with diminished capacitance. To address these problems, more recently, Kang et al.[101] developed a simple approach to fabricate low-cost transition metal–based oxyhydroxide/nanoporous metal electrodes by electrochemical polarization of a dealloyed nanoporous Ni–Mn alloy in an alkaline solution. The resultant Ni–Mn oxyhydroxide grew in nanopore channels by the paradigm of domain matching epitaxy, which was evidenced by high-resolution transition electron microscope (HRTEM) and electron diffraction patterns. As shown in Figure 6.7a, the nanoporous structure is still retained after polarization treatment although there is a layer of oxyhydroxide on the metallic ligaments within the pore channels. Electron diffraction (the inset of Figure 6.7a) indicates that the rock-salt-type oxyhydroxide epitaxially grows on the metal ligaments, consistent with the X-ray diffraction (XRD) results. Inverse fast Fourier transform (IFFT) of a high-resolution scanning transition electron microscope (STEM)-high-angle annular dark-field (HAADF) image clearly displays the distribution of the face-centered cubic (FCC) metal substrate and rock-salt oxyhydroxide phases (Figure 6.7b), which have a coherent crystallographic relation with a small lattice mismatch accommodated by dislocations. The EELS mappings of the Ni L-edge, Mn L-edge, and O K-edge (Figure 6.7c through e), taken from the polarized nanoporous Ni–Mn alloy, demonstrate that Mn together with O partitions into the oxide in the pore channels while the metallic ligaments with enriched Ni keep continuous in the nanoporous hybrid material. With XRD and X-ray photoelectron spectroscopy (XPS) analysis combined, it is concluded that the polarized oxyhydroxide is a single phase with multicomponent and mixed valence, instead of a mixture of individual oxides. It can be expressed as $(Ni_a{}^{II}Mn_b{}^{II}Mn_c{}^{II})O_d(OH)_e \cdot fH_2O$. The values of a, b, c, d, e, and f are approximately estimated by calculating the areas under each deconvulated peaks in the XPS spectra.

The typical CV curves of the self-grown oxyhydroxide/nanoporous metal, perfomed in 1 M KOH solution using a three-electrode system, consist of a pair of well-defined redox peaks at 0.2 and 0.4 V (vs. Ag/AgCl) and a rectangular-shape area without obvious redox peaks (Figure 6.8a). The pair of redox peaks is consistent with the reversible reactions between Ni^{2+}/Ni^{3+} and OH$^-$ anions. For the rectangular-shape area, it is found that the area size mainly depends on the amount of the residual Mn in the Mn–Ni oxyhydroxide, inferring that the rectangular-shape region results from the capacitive behavior of Mn cations. Based on the contributions of both Ni and Mn

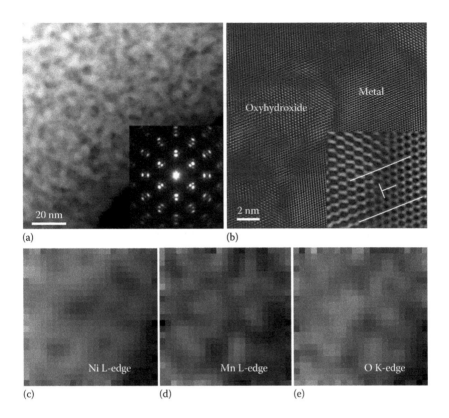

FIGURE 6.7 Microstructure and chemistry of the nanoporous self-grown Ni–Mn oxyhydroxide/NPM electrode. (a) STEM-HAADF image showing a nanoporous structure. The inset is the selected area electron diffraction pattern taken from this sample. (b) IFFT-HRSTEM atomic image displaying the distribution of the self-grown oxyhydroxide and metal substrate. The inset is the zoom-in micrograph showing the mismatch dislocations at the oxyhydroxide/metal interface. The EELS mappings of (c) Ni L-edge, (d) Mn L-edge, and (e) O K-edge. (Kang, J.L., Hirata, A., Qiu, H.-J., Chen, L.Y., Ge, X.B., Fujita, T., and Chen, M.W.: *Adv. Mater.* 2014. 26. 269–272. Copyright Wiley-VCH Verlag GmbH & Co. KGaA. Adapted with permission.)

cations, a maximum specific capacitance of up to 505 F cm^{-3} at 0.5 A cm^{-3} can be achieved. With the increase of current density from 0.5 to 10 A cm^{-3}, the specific capacitance still retains as high as ~339 F cm^{-3}, suggesting that the electrode has a high rate of capacitive performance. Combined with the advantages of low cost, easy operation, and environmentally friendly nature, as well as high cyclic stability, this novel oxyhydroxide/nanoporous metal electrode shows a great promise for applications in commercial high-capacity ECs.

6.4.3 Atomic Doping–Enhanced Pseudocapacitive Electrodes

The low electrical conductivity of transition metal oxides is known to be the key factor limiting the utilization of their theoretical capacitances in a thick film and

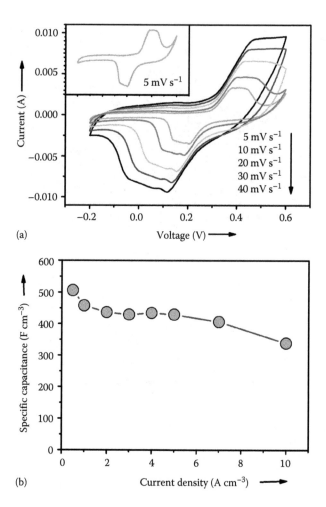

(a)

(b)

FIGURE 6.8 Capacitive performance of the self-grown oxyhydroxide/NPM electrode. (a) Cyclic voltammogram curves at different scan rates. The inset is a slow-scan curve with well-defined redox reaction peaks. (b) The specific capacitance vs. current densities. (Kang, J.L., Hirata, A., Qiu, H.-J., Chen, L.Y., Ge, X.B., Fujita, T., and Chen, M.W.: *Adv. Mater.* 2014. 26. 269–272. Copyright Wiley-VCH Verlag GmbH & Co. KGaA. Adapted with permission.)

large loading amount. Previous works are mainly focused on introducing conductive reinforcement to enhance electron transfer. Is it possible to directly improve the intrinsic conductivity of transition metal oxides themselves? Recently, Kang et al.[102] devised a strategy to address this: boosting the intrinsic conductivity of transition metal oxides by nonequilibrium doping of free electron metal atoms as electron donors, which are expected to change the electronic structure of transition metal oxides for better conductivity and high capacitive performance. They first electrochemically deposited a 350 nm thick porous MnO_2 layer on 100 nm thick NPG and then bombarded the MnO_2 electrode with gold atoms by physical vapor deposition

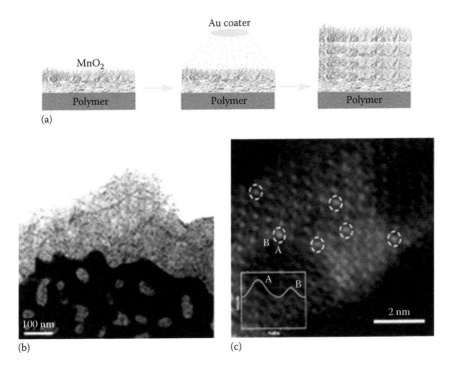

FIGURE 6.9 (a) Schematic illustration of the fabrication process of the Au-doped MnO_2 electrodes; (b and c) STEM images of Au-doped MnO_2. The inset of (c) indicates the intensity difference between Au-occupied columns and undoped Mn-O atomic columns. (Kang, J., Hirata, A., Kang, L., Zhang, X., Hou, Y., Chen, L., Li, C., Fujita, T., Akagi, K., and Chen, M.: *Angew. Chem. Int. Ed.* 2013. 52. 1664–1667. Copyright Wiley-VCH Verlag GmbH & Co. KGaA. Adapted with permission.)

using a sputtering coater. Repeating the sequence of MnO_2 growth followed by gold doping several times enabled the construction of micrometer-thick electrodes ready for use as supercapacitors (Figure 6.9a). Alternate layering of MnO_2 and gold atoms ensured the dopants were well integrated within the oxide's lattices, which was further evidenced by XRD, XPS, STEM, and EELS analysis. As shown in Figure 6.9b and c, most of the doped Au atoms homogeneously distribute in MnO_2 lattices in the form of individual atoms except for a small number of gold nanoparticles. The peak shifts in XRD, XPS, and EELS patterns solidly confirm the lattice distortion and electronic structure changes of MnO_2 by gold atom doping. The electronic resistance of thick MnO_2 with doping gold atoms of ~9.9 at.% can decrease to 11%, which dramatically enhances the electrochemical performance of the MnO_2 electrodes (Figure 6.10). The thick gold-doped MnO_2 films (~1.35 μm, equal to ~100 μg cm^{-2}) show a capacitance of 65% higher than that of the pure films—the best charge storage ever achieved from thick MnO_2 electrodes. More excitingly, the new material displayed outstanding stability: after a few hours' use, the capacitance even increased slightly due to the electrochemical dopant redistribution during voltammetric cycling. The concept developed in this study is not limited only to Au-doped MnO_2 but can be

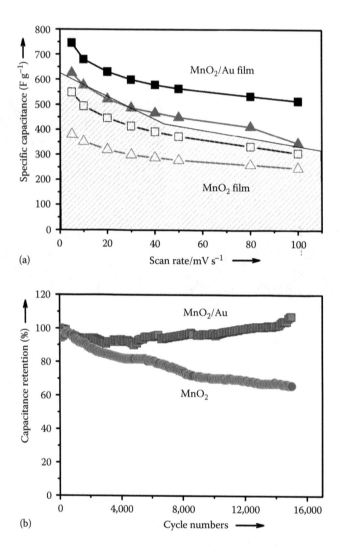

FIGURE 6.10 (a) The specific capacitance of pure MnO_2 and 9.9 at.% Au-doped MnO_2 with the thickness of ~0.7 μm (□■) and ~1.35 μm (△▲). The solid symbols (■▲) represent Au-doped MnO_2 and the open ones (□△) denote pure MnO_2. (b) Cycling stability of the pure MnO_2 and Au-doped MnO_2 electrodes with the same thickness of ~1.35 μm as a function of the cycle number. The measurement of the capacitance retention was carried out in the cyclic voltammetry at a scan rate of 50 mV s^{-1}. (Kang, J., Hirata, A., Kang, L., Zhang, X., Hou, Y., Chen, L., Li, C., Fujita, T., Akagi, K., and Chen, M.: *Angew. Chem. Int. Ed.* 2013. 52. 1664–1667. Copyright Wiley-VCH Verlag GmbH & Co. KGaA. Adapted with permission.)

applied to many oxide materials to improve their electrochemical capacitance by enhancing their conductivity using nonequilibrium doping of free electron metals. Even for conductive polymers, gold atom doping also dramatically improved their EC performance.[103] Therefore, this study may provide an alternative approach to improve the capacitive properties of transition metal oxides and conductive polymers for high-performance ECs and batteries.

6.5 SUMMARY AND PERSPECTIVES

With increasing demands for clean and sustainable energy, the advantages of high power density, high efficiency, and long life have made ECs as one of the major emerging devices for energy storage and power supply. To meet the key challenge of low-energy ECs, discovering new electrode materials with high capacitance and a wide potential window has been the recent focus of EC research. Although numerous active electrode materials have been developed in the past two decades, they usually have various deficiencies, which limit their commercial applications in real-world electronics. In this chapter, we briefly reviewed the research progress of the advanced electrode materials and the scientific and technical challenges in the practical implementation of these materials. Three new strategies, discussed in this chapter, may be available to promote the practical applications of ECs: (1) graphene electrodes with a large packing density, (2) nanoporous metal-enhanced pseudocapacitive electrodes, and (3) atomic doping–enhanced pseudocapacitive electrodes. These concepts may inspire new efforts for next-generation ECs with both high energy density, comparable to batteries, and high power density. Moreover, it is worth to point out that it is necessary to establish testing standards for ECs to compare the performance data taken from different research groups. The specific capacitance data, presented in most papers, were only taken from the active component. However, in practical application, it is important to provide the overall performance data of the composite electrodes or whole devices to assess the ECs.

REFERENCES

1. P. Simon and Y. Gogotsi, *Nat. Mater.*, 2008, **7**, 845.
2. C. Liu, F. Li, L. Ma and H. Cheng, *Adv. Energy Mater.*, 2010, **22**, E28.
3. J. R. Miller and P. Simon, *Science*, 2008, **321**, 651.
4. D. I. Boos, *US Pat.*, 3 536 963 (to Standard Oil, SOHIO), 1970.
5. U.S. Department of Energy, *Basic Research Needs for Electrical Energy Storage*, http://www.osti.gov/accomplishments/documents/fullText/ACC0330.pdf, accessed on April 4, 2007.
6. L. L. Zhang and X. S. Zhao, *Chem. Soc. Rev.*, 2009, **38**, 2520.
7. J. H. Chae, K. C. Ng, G. Z. Chen, *Proc. IMechE A: J. Power Energy*, 2010, **224**, 479.
8. A. Burke, *J. Power Sources*, 2000, **91**, 37.
9. Y. M. Vol'fkovich and T. M. Serdyuk, *Russ. J. Electrochem.*, 2002, **38**, 935.
10. B. E. Conway, *Electrochemical Supercapacitors: Scientific Fundamentals and Technological Applications*, Kluwer Academic/Plenum Publishers, New York, 1999.
11. X. Zhao, B. M. Sánchez, P. J. Dobson and P. S. Grant, *Nanoscale*, 2011, **3**, 839.

12. J. Gamby, P. L. Taberna, P. Simon, J. F. Fauvarque and M. Chesneau, *J. Power Sources*, 2011, **101**, 109.
13. L. Zhang, S. L. Candelaria, J. Tian, Y. Li, Y. X. Huang and G. Cao, *J. Power Sources*, 2013, **236**, 215.
14. B. Hsia, M. S. Kim, C. Carraro and R. Maboudian, *J. Mater. Chem. A*, 2013, **1**, 10518.
15. D. N. Futaba, K. Hata, T. Yamada, T. Hiraoka, Y. Hayamizu, Y. Kakudate, O. Tanaike, H. Hatori, M. Yumura and S. Iijima, *Nat. Mater.*, 2006, **5**, 987.
16. C. Du, J. Yeh and N. Pan, *Nanotechnology*, 2005, **16**, 350.
17. K. Wang, Q. Meng, Y. Zhang, Z. Wei and M. Miao, *Adv. Mater.*, 2013, **25**, 1494.
18. K. Wang, Y. Wang, Y. Wang, E. Hosono and H. Zhou, *J. Phys. Chem. C*, 2009, **113**, 1093.
19. G. Kim, B. T. N. Ngoc, K. S. Yang, M. Kojima, Y. A. Kim, Y. J. Kim, M. Endo and S. C. Yang, *Adv. Mater.*, 2007, **19**, 2341.
20. C. Ma, Y. Song, J. Shi, D. Zhang, X. Zhai, M. Zhong, Q. Guo and L. Liu, *Carbon*, 2013, **51**, 290.
21. Y. Wang, Z. Shi, Y. Huang, Y. Ma, C. Wang, M. Chen and Y. Chen, *J. Phys. Chem. C*, 2009, **113**, 13106.
22. Y. Meng, Y. Zhao, C. Hu, H. Cheng, Y. Hu, Z. Zhang, G. Shi and L. Qu, *Adv. Mater.*, 2013, **25**, 2326.
23. M. F. El-Kady and R. B. Kaner, *Nat. Commun.*, 2013, **4**, 1475.
24. J. H. Lim, D. J. Choi, H. K. Kim, W. I. Cho and Y. S. Yoon, *J. Electrochem. Soc.*, 2001, **148**, A275.
25. C. C. Hu, K. H. Chang, M. C. Lin and Y. T. Wu, *Nano Lett.*, 2006, **6**, 2690.
26. X. Wu, Y. Zeng, H. Gao, J. Su, J. Liu and Z. Zhu, *J. Mater. Chem. A*, 2013, **1**, 469.
27. H. Y. Lee and J. B. Goodenough, *J. Solid State Chem.*, 1999, **144**, 220.
28. P. Ragupathy, H. N. Vasan and N. Munichandraiah, *J. Electrochem. Soc.*, 2008, **155**, A34.
29. L. Peng, X. Peng, B. Liu, C. Wu, Y. Xie and G. Yu, *Nano Lett.*, 2013, **13**, 2151.
30. K. W. Nam and K. B. Kim, *J. Electrochem. Soc.*, 2002, **149**, A346.
31. J. W. Lee, T. Ahn, J. H. Kim, J. M. Ko and J. D. Kim, *Electrochim. Acta*, 2011, **56**, 4849.
32. B. Qu, L. Hu, Y. Chen, C. Li, Q. Li, Y. Wang, W. Wei, L. Chen and T. Wang, *J. Mater. Chem. A*, 2013, **1**, 7023.
33. S. K. Meher and G. R. Rao, *J. Phys. Chem. C*, 2011, **115**, 15646.
34. X. C. Dong, H. Xu, X. W. Wang, Y. X. Huang, M. B. Chan-Park, H. Zhang, J. H. Wang, W. Huang and P. Chen, *ACS Nano*, 2012, **6**, 3206.
35. H. Jiang, J. Ma and C. Li, *Chem. Commun.*, 2012, **48**, 4465.
36. G. Zhang and X. W. Lou, *Adv. Mater.*, 2013, **25**, 976.
37. Y. Chen, B. Qu, L. Hu, Z. Xu, Q. Li and T. Wang, *Nanoscale*, 2013, **5**, 9812.
38. G. Oldield, T. Ung and P. Mulvaney, *Adv. Mater.*, 2000, **12**, 1519.
39. N. L. Wu, *Mater. Chem. Phys.*, 2002, **75**, 6.
40. H. Cui, Y. Liu, W. Ren, M. Wang and Y. Zhao, *Nanotechnology*, 2013, **24**, 345602.
41. H. Y. Lee and J. B. Goodenough, *J. Solid State Chem.*, 1999, **148**, 81.
42. Z. Chen, V. Augustyn, J. Wen, Y. Zhang, M. Shen, B. Dunn and Y. Lu, *Adv. Mater.*, 2011, **23**, 791.
43. J. Rajeswari, P. S. Kishore, B. Viswanathan and T. K. Varadarajan, *Electrochem. Commun.*, 2009, **11**, 572.
44. Z. Cui, W. Yuan and C. M. Li, *J. Mater. Chem. A*, 2013, **1**, 12926.
45. F. Fusalba, P. Gouérec, D. Villers and D. Bélanger, *J. Electrochem. Soc.*, 2001, **148**, A1.
46. H. Zhou, H. Chen, S. Luo, G. Lu, W. Wei and Y. Kuang, *J. Solid State Electrochem.*, 2005, **9**, 574.
47. K. Jurewicz, S. Delpeux, V. Bertagna, F. Béguin and E. Frackowiak, *Chem. Phys. Lett.*, 2001, **347**, 36.
48. H. An, Y. Wang, X. Wang, L. Zheng, X. Wang, L. Yi, L. Bai and X. Zhang, *J. Power Sources*, 2010, **195**, 6964.

49. Y. Zhao, J. Liu, Y. Hu, H. Cheng, C. Hu, C. Jiang, L. Jiang, A. Cao and L. Qu, *Adv. Mater.*, 2013, **25**, 591.
50. A. Laforgue, P. Simon, C. Sarrazin and J. Fauvarque, *J. Power Sources*, 1999, **80**, 142.
51. R. B. Ambade, S. B. Ambade, N. K. Shrestha, Y. C. Nah, S. H. Han, W. Lee and S. H. Lee, *Chem. Commun.*, 2013, **49**, 2308.
52. P. A. Basnayaka, M. K. Ram, E. K. Stefanakos and A. Kumar, *Electrochim. Acta*, 2013, **92**, 376.
53. G. A. Snook, P. Kao, A. S. Best, *J. Power Sources*, 2011, **196**, 1.
54. X. Feng, Y. Liang, L. Zhi, A. Thomas, D. Wu, I. Lieberwirth, U. Kolb and K. Müllen, *Adv. Funct. Mater.*, 2009, **19**, 2125.
55. J. Kang, K. Qin, H. Zhang, A. Hirata, J. Wang, M. Chen, N. Zhao, R. Sun, T. Fujita, C. Shi and Z. Qiao, *Carbon*, 2012, **50**, 5162.
56. T. Kim, G. Jung, S. Yoo, K. S. Suh and R. S. Ruoff, *ACS Nano*, 2013, **7**, 6899.
57. J. Chmiola, G. Yushin, Y. Gogotsi, C. Portet, P. Simon and P. L. Taberna, *Science*, 2006, **313**, 1760.
58. G. Gryglewicz, J. Machnikowski, E. Lorenc-Grabowska, G. Lota and E. Frackowiak, *Electrochim. Acta*, 2005, **50**, 1197.
59. S. Pohlmann, B. Lobato, T. A. Centeno and A. Balducci, *Phys. Chem. Chem. Phys.*, 2013, **15**, 17287.
60. X. Largeot, C. Portet, J. Chmiola, P. Taberna, Y. Gogotsi and P. Simon, *J. Am. Chem. Soc.*, 2008, **130**, 2730.
61. S. Kondrat, C. R. Pérez, V. Presser, Y. Gogotsi and A. A. Kornyshev, *Energy Environ. Sci.*, 2012, **5**, 6474–6479.
62. P. Simon and Y. Gogotsi, *Acc. Chem. Res.*, 2013, **46**, 1094.
63. Y. Zhu, S. Murali, M. D. Stoller, K. J. Ganesh, W. Cai, P. J. Ferrira, A. Pikle, P. M. Wallace, K. A. Cychosz, M. Thommes, D. Su, E. A. Stach and R. S. Ruoff, *Science*, 2011, **332**, 1537.
64. Z. S. Wu, Y. S. Tan, S. B. Yang, X. L. Feng and K. Mullen, *J. Am. Chem. Soc.*, 2012, **134**, 19532.
65. B. G. Choi, M. Yang, W. H. Hong, J. W. Choi and Y. S. Hub, *ACS Nano*, 2012, **6**, 4020.
66. L. Dong, Z. Chen, D. Yang and H. Lu, *RSC Adv.*, 2013, **3**, 21183–21191.
67. S. Hadzi-Jordanov, H. Angersteir-Kozlowska, M. Vukovic and B. E. Conway, *J. Electrochem. Soc.*, 1978, **125**, 1471.
68. J. P. Zheng, P. J. Cygan and T. R. Jow, *J. Electrochem. Soc.*, 1995, **142**, 2699.
69. M. Zhi, C. Xiang, J. Li, M. Li and N. Wu, *Nanoscale*, 2013, **5**, 72.
70. J. Kang, L. Chen, Y. Hou, C. Li, T. Fujita, X. Lang, A. Hirata and M. Chen, *Adv. Energy Mater.*, 2013, **3**, 857.
71. P. Ragupathy, D. H. Park, G. Campet, H. V. Vasan, S. J. Hwang, J. H. Choy and N. Munichandraiah, *J. Phys. Chem. C*, 2009, **113**, 6303.
72. M. Xu, W. Jia, S. J. Bao, Z. Su and B. Dong, *Electrochim. Acta*, 2010, **55**, 5117.
73. M. B. Sassin, C. N. Chervin, D. R. Rolison and J. W. Long, *Acc. Chem. Res.*, 2013, **46**, 1062.
74. J. Jiang, Y. Li, J. Liu, X. Huang, C. Yuan and X. W. Lou, *Adv. Mater.*, 2012, **24**, 5166.
75. M. F. El-Kady, V. Strong, S. Dubin and R. B. Kana, *Science*, 2012, **335**, 1326.
76. X. Yang, J. Zhu, L. Qiu and D. Li, *Adv. Mater.*, 2011, **23**, 2833.
77. U. N. Maiti, J. Lim, K. E. Lee, W. J. Lee and S. O. Kim, *Adv. Mater.*, 2013, **26**, 615–619.
78. Y. Gogotsi and P. Simon, *Science*, 2011, **334**, 917.
79. S. Murali, N. Quarles, L. L. Zhang, J. R. Potts, Z. Tan, Y. Lu, Y. Zhu and R. S. Ruoff, *Nano Energy*, 2013, **2**, 764–768.
80. T. Y. Kim, G. Jung, S. Yoo, K. S. Suh and R. S. Ruoff, *ACS Nano*, 2013, **7**, 6899.
81. X. Yang, C. Cheng, Y. Wang, L. Qiu and D. Li, *Science*, 2013, **341**, 534.
82. A. Burke, *Electrochim. Acta*, 2007, **53**, 1083.

83. X. Lang, A. Hirata, T. Fujita and M. Chen, *Nat. Nanotech.*, 2011, **6**, 232–236.

84. L. Chen, X. Lang and M. Chen, *Nanoporous Materials: Synthesis and Applications*, ed. Q. Xu, CRC Press, Boca Raton, FL, 2013, p. 125.

85. T. Fujita, P. F. Guan, K. McKenna, X. Y. Lang, A. Hirata, L. Zhang, T. Tokunaga, S. Arai, Y. Yamamoto, N. Tanaka, Y. Ishikawa, N. Asao, Y. Yamamoto, J. Erlebacher and M. W. Chen, *Nat. Mater.*, 2012, **11**, 775.

86. X. Lang, H. Yuan, Y. Iwasa and M. Chen, *Scripta Mater.*, 2011, **64**, 923–926.

87. L. Y. Chen, J. L. Kang, Y. Hou, P. Liu, T. Fujita, A. Hirata and M. W. Chen, *J. Mater. Chem. A*, 2013, **1**, 9202.

88. L. Y. Chen, Y. Hou, J. L. Kang, A. Hirata, T. Fujita and M. W. Chen, *Adv. Energy Mater.*, 2013, **6**, 851.

89. F. Meng and Y. Ding, *Adv. Mater.*, 2011, **23**, 4098.

90. Y. Hou, L. Y. Chen, L. Zhang, J. L. Kang, T. Fujita, J. H. Jiang and M. W. Chen, *J. Power Sources*, 2013, **225**, 304.

91. X. Lang, L. Zhang, T. Fujita, Y. Ding and M. Chen, *J. Power Sources*, 2012, **197**, 325.

92. K. R. Prasad and N. Miura, *J. Power Sources*, 2004, **135**, 354.

93. T. Xue, C. L. Xu, D. D. Zhao, X. H. Li and H. L. Li, *J. Power Sources*, 2007, **164**, 953.

94. N. Nagarajan, H. Humadi and I. Zhitomirsky, *Electrochem. Acta*, 2006, **51**, 3039.

95. Y. Wang, H. Liu, X. Sun and I. Zhitomirsky, *Scripta Mater.*, 2009, **61**, 1079.

96. Y. Dai, K. Wang, J. Zhao and J. Xie, *J. Power Sources*, 2006, **161**, 737.

97. K. W. Nam and K. B. Kim, *J. Electrochem. Soc.*, 2006, **153**, A81.

98. S. C. Pang, M. A. Anderson and T. W. Chapman, *J. Electrochem. Soc.*, 2000, **147**, 444.

99. K. T. Lee, C. B. Tsai, W. H. Ho and N. L. Wu, *Electrochem. Commun.*, 2010, **12**, 686.

100. Z. P. Feng, G. R. Li, J. H. Zhong, Z. L. Wang, Y. N. Ou and Y. X. Tong, *Electrochem. Commun.*, 2009, **11**, 706.

101. J. L. Kang, A. Hirata, H.-J. Qiu, L. Y. Chen, X. B. Ge, T. Fujita and M. W. Chen, *Adv. Mater.*, 2014, **26**, 269.

102. J. Kang, A. Hirata, L. Kang, X. Zhang, Y. Hou, L. Chen, C. Li, T. Fujita, K. Akagi and M. Chen, *Angew. Chem. Int. Ed.*, 2013, **52**, 1664.

103. Y. Hou, Doctoral Dissertation. Tohoku University, 2013.

7 Electrode Materials for Nickel/Metal Hydride (Ni/MH) Rechargeable Batteries

Hiroshi Inoue and Eiji Higuchi

CONTENTS

7.1 INTRODUCTION

7.1.1 CELL REACTION

A nickel/metal hydride (Ni/MH) cell is basically composed of a hydrogen storage alloy/metal hydride (M/MH) negative electrode, a nickel oxyhydroxide/nickel hydroxide ($NiOOH/Ni(OH)_2$) positive electrode, an alkaline aqueous electrolyte solution, and a separator, as illustrated in Figure 7.1. Positive and negative electrode reactions and total cell reaction are represented as follows[1]:

$$\text{(Positive electrode)} \quad Ni(OH)_2 + OH^- = NiOOH + H_2O + e^- \quad (7.1)$$

$$\text{(Negative electrode)} \quad M + H_2O + e^- = MH + OH^- \quad (7.2)$$

$$\text{(Cell reaction)} \quad Ni(OH)_2 + M = NiOOH + MH \quad (7.3)$$

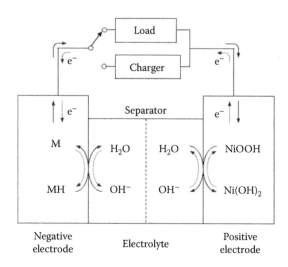

FIGURE 7.1 Construction of an Ni/MH cell.

During charging, at the positive electrode $Ni(OH)_2$ is oxidized to $NiOOH$ and at the negative electrode water is reduced to atomic hydrogen, followed by being absorbed in M to form MH. On the other hand, during discharging at the positive electrode $NiOOH$ is reduced to $Ni(OH)_2$ and at the negative electrode MH is oxidized to M. The cell reaction is apparently represented by a reversible transfer of hydrogen between both electrodes. It is notable that the total amount of water and hydroxide and the concentration of the electrolyte solution are unchangeable during charging and discharging because water does not apparently contribute to the cell reaction. This is different from the lead–acid cell and nickel–cadmium cell where the volume of water changes during charging or discharging. The unique cell reaction mechanism contributes to high capacity, high power, long cycle life, etc.

In commercial sealed Ni/MH batteries, the M/MH negative electrode is about 50% higher in capacity than the $NiOOH/Ni(OH)_2$ positive electrode. This means the capacity of the Ni/MH battery is controlled by that of the positive electrode. Moreover, in order to suppress gas evolution at the M/MH negative electrode during charging and discharging, charge and discharge reserves are set in the negative electrode, as shown in Figure 7.2. In overcharging, M is still converted to MH at the negative electrode along Equation 7.2 or the charge reserve is consumed while oxygen evolves at the positive electrode (Equation 7.4). The oxygen diffuses to the negative electrode through a separator and chemically reacts with MH to return to water (Equation 7.5). Conesquently, there is no rise of pressure in the cell:

$$\text{(Positive electrode)} \quad 2OH^- = H_2O + 1/2O_2 + 2e^- \tag{7.4}$$

$$\text{(Negative electrode)} \quad 2MH + 1/2O_2 = 2M + H_2O \tag{7.5}$$

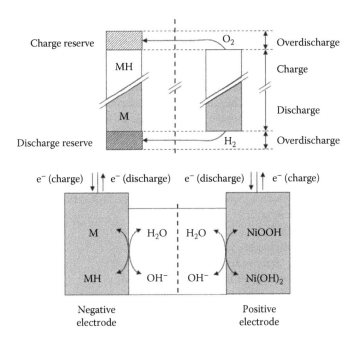

FIGURE 7.2 Cell reactions in charging and discharging and charge and discharge reserves.

On the other hand, in overdischarging, hydrogen evolves at the positive electrode (Equation 7.6). The hydrogen diffuses to the negative electrode through a separator and dissociates to atomic hydrogen on M (Equation 7.7), followed by a charge-transfer reaction (Equation 7.8). So there is also no rise of pressure in the cell:

$$\text{(Positive electrode)} \quad 2H_2O + 2e^- = H_2 + 2OH^- \tag{7.6}$$

$$\text{(Negative electrode)} \quad H_2 + 2M = 2MH \tag{7.7}$$

$$2OH^- + 2MH = 2M + 2H_2O + 2e^- \tag{7.8}$$

In this way the Ni/MH battery has not only an overcharge protection mechanism but also an overdischarge protection mechanism. The Ni/Cd battery also has a similar overcharge protection mechanism to the Ni/MH battery. But the overdischarge protection mechanism is characteristic of the Ni/MH battery, which is a great advantage for high-voltage batteries composed of some hundreds of cells for electric vehicle (EV) application.

7.1.2 History of Development of Ni/MH Batteries and Their Features

Since 1980, the progress of electronics led to a rapid popularization of light and compact cordless appliances such as video cameras, mobile phones, laptop computers, shavers, toys, and tools, which created a strong demand for compact, high energy density, long life, and maintenance-free rechargeable batteries. At that time sealed Ni/Cd batteries were commonly used, but further improvement of energy density was an urgent problem, leading to the development of the Ni/MH battery. The sealed Ni/MH battery was firstly commercialized in Japan in 1990 by virtue of the development of new metal hydrides as negative electrode active material. The capacity of an AA size Ni/MH battery was almost twice as large as that of a conventional Ni/Cd battery, leading to a long discharge period of time. Other advantages of the Ni/MH battery are summarized as follows:

1. Average discharge voltage of around 1.2 V (theoretical electromotive force is ca. 1.32 V), which is comparable to that of the Ni/Cd battery
2. Flat discharge curve
3. High energy density
4. Long charge–discharge cycle life
5. Self-overcharge and overdischarge protection
6. Good high-rate chargeability and dischargeability
7. Wide operation temperature
8. Ecofriendly active materials without any toxic compounds such as Cd and Pb
9. High safety

In the 1990s, the progress of the portable equipment market and environmental regulation for Ni/Cd batteries in Western countries caused a significant increase in the sales volume and price of Ni/MH batteries. In 2000, the sales volume of Ni/MH batteries in Japan was over a billion and the sales price was 116 billion JPY, as shown in Figure 7.3. However, the substitution of lithium-ion batteries (LIBs) for thinner and lighter portable equipment and the price down of Ni/MH batteries rapidly reduced

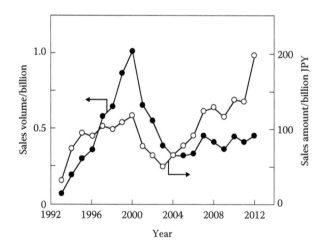

FIGURE 7.3 Annual change in sales volume and amount of Ni/MH batteries.

their sales volume and price. After that, new applications of Ni/MH batteries were explored. Large-sized Ni/MH batteries are earliest applied to hybrid electric vehicles (HEVs) such as Toyota *Prius* and Honda *Insight*. The HEVs with high-power Ni/MH batteries were firstly commercialized at the end of 1997. The production of HEV has been on the increase due to the emission regulation of greenhouse gases, and in 2012, the production of HEV in the world was about 1.6 million. In 2012, the sales volume is at the level of the previous year and the sales price is up about 50% from the previous year.

7.1.3 APPLICATIONS

An HEV combines a conventional internal combustion engine propulsion system with an electric propulsion system and chooses either propulsion system in accordance with a situation to keep high energy utilization efficiency. The energy utilization efficiency of HEV is represented as the product of energy efficiency from the oil well or natural gas field to the tank or battery (well-to-tank efficiency) and energy efficiency from the tank or battery to the wheel (tank-to-wheel efficiency), called well-to-wheel efficiency. The well-to-wheel efficiency of HEV is about 32%, which is higher than that of gasoline vehicles (ca. 14%) and fuel cell vehicles (ca. 29%). Batteries for HEV application are not only a component of vehicles but also a power source. The Ni/MH batteries for HEV need to have the following performances: (1) high power required for motor drive, (2) wide operation temperature range ($-40°C$ to $60°C$), (3) high energy density to make batteries compact, and (4) long life (more than 10 years). Cylindrical and square Ni/MH batteries are used for HEV application. To further improve the performances, the reduction in inner resistances with the modification of current collection mechanism, electrolyte composition, negative and positive electrode active materials, etc., is in progress.

A new large-sized Ni/MH battery module *GIGACELL®* was developed by Kawasaki Heavy Industries, Ltd. In the GIGACELL, negative and positive electrode sheets are alternately inserted in a strip-form separator in a square cell and each cell is stacked to make a bipolar structure. The bipolar design of the GIGACELL prevents overheating of the battery with a cooling fan effectively sending air through the structure, which holds down energy loss between cells, actualizing higher capacity. In addition, in the GIGACELL no welding is used in its construction, which facilitates disassembly and reassembly processes. The GIGACELL is applied to a power source for light rail vehicles *SWIMO®* and regenerative brakes and electric power storage systems that temporarily store electricity generated by solar and wind power.

For backup power supplies of telecommunication equipment during power failure, lead–acid batteries were often used before. However, their capacity is insufficient to backup current telecommunication equipment with large electricity consumption. The Ni/MH battery is around three times as large in gravimetric energy density as the lead–acid battery and has high safety. For backup application, good charging performance at high temperatures, low self-discharging, and long life are required, and large-sized Ni/MH batteries with these performances are developed.

Dry cells are used for various portable equipment and are also indispensable items in an emergency. However, the dry cells are discarded after use because they are not rechargeable. The substitution of rechargeable cells for dry cells must be desirable in terms of resource conservation and reduction of waste. To substitute the Ni/MH cells for dry cells, a serious problem was that self-discharge of the Ni/MH cells was too serious to use without recharging. The self-discharge was significantly improved by modification of active materials, in particular the development of new negative electrode active material. Nowadays the capacity loss after charging for the Ni/MH cells such as Panasonic *eneloop*® and *EVOLTA*® was only 15% in a year, and they exhibited charge/discharge cycle life more than 1000 cycles.

7.1.4 MATERIALS DEVELOPMENT FOR HIGH-PERFORMANCE NI/MH BATTERIES

For realizing higher capacity, higher power, longer cycle life, lower cost, etc., of Ni/MH batteries, battery components, in particular the negative and positive electrodes, have been improved. To improve the negative electrode performance, the following researches have been carried out: (1) the preparation of AB_5-type mischmetal (Mm)-based hydrogen storage alloys with multicomponents and nonstoichiometric compositions; (2) reduction or elimination of the Co component that is expensive; (3) surface modification with electroless plating Ni or Cu plating film, alkaline solutions, and reducing agents; (4) microcrystallization and homogenization of alloy composition with rapid solidification and annealing; and (5) mixing of additives to inhibit corrosion and enhance electrocatalytic activity. On the other hand, to improve the positive electrode performance, (1) preparation of spherical $Ni(OH)_2$ particles, (2) conductive CoOOH-coating, (3) mixing of additives such as zinc, and (4) pretreatment of Co species-coated $Ni(OH)_2$ have been investigated.

A technical breakthrough to realize the high capacity of the Ni/MH battery has been done for the negative electrode active material. In 2000, a new high capacity

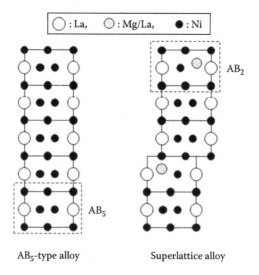

FIGURE 7.4   Unit cell structures of typical AB_5-type and superlattice alloys.

hydrogen storage alloy whose crystal structure was different from conventional AB_5-type Mm-based hydrogen storage alloys was proposed.[2] The new hydrogen storage alloy had a superlattice structure in which the AB_5 unit cell and AB_2 unit cell were regularly stacked as shown in Figure 7.4. New Ni/MH batteries with the superlattice negative electrodes were superior to the conventional ones on capacity, discharge at high current densities (e.g., 3 C-rate that is current required for discharging out the theoretical capacity of the battery for 1/3 h) and low temperatures (0°C), charge at high temperatures (40°C), storage performance at high temperatures (40°C), etc., and have been commercialized. But since then there have been no other technical breakthrough that led to commercial application on the development of new alloys for the negative electrode. In the next section, technical trends on the hydrogen storage alloys in the last 10 years are reviewed.

7.2 TECHNICAL TRENDS ON THE DEVELOPMENT OF HYDROGEN STORAGE ALLOYS FOR THE NEGATIVE ELECTRODE

Hydrogen storage alloys as the negative electrode active material play roles as hydrogen reservoirs and reaction fields where electrochemical hydrogen absorption/desorption reactions can smoothly proceed. New hydrogen storage alloys should have features such as high hydrogen storage capacity, high activity for the negative electrode reaction, and corrosion tolerance to alkaline solutions. From the viewpoint of high energy density, high hydrogen storage capacity is the most important motivation for the development of new alloys. In this section, the development of some kinds of alloys with high hydrogen storage capacity, magnesium-containing alloys and vanadium-containing alloys, is reviewed.

7.2.1 MAGNESIUM-CONTAINING HYDROGEN STORAGE ALLOYS

Magnesium-containing hydrogen storage alloys have attracted attention from the viewpoint of high hydrogen storage capacity and low cost of raw materials. The hydrogen storage capacity and discharge capacity of various Mg-containing alloys are summarized in Table 7.1. For example, Mg_2Ni has a high theoretical specific discharge capacity of ca. 1000 mA h g^{-1}, but its charge–discharge cycle performance is poor because of slow hydrogen desorption rate and passivation in alkaline solutions. The MgNi prepared from Mg and Ni powders by mechanical alloying exhibited over 500 mA h g^{-1},[3,4] but its cycle stability was not so better although it was a little improved by the addition of Ti and V constituents.[5]

In 1997, new ternary Mg-containing alloys, RMg_2Ni_9 (R=La, Ce, Pr, Nd, Sm, and Gd), with a $PuNi_3$-type structure were prepared by Kadir et al.[6] They also prepared various ternary Mg-containing alloys and evaluated their crystal structures, hydrogen absorption/desorption, and electrochemical properties.[7–11] $LaCaMgNi_9$ and $LaCaMgN_6Mn_3$ had a discharge capacity of 356 and 342 mA h g^{-1}, respectively, which were higher than that of typical AB_5 alloy. But their utilization efficiency, which was defined as the ratio of experimental discharge capacity to theoretical one, estimated from P–C–T curves was low due to the high content of the AB_2 structure with low utilization efficiency and inclined plateau of their P–C–T curves.[10]

TABLE 7.1
Hydrogen Storage Capacity and Discharge Capacity of Various Mg-Containing Alloys

Composition	Pretreatment	H_2 Pressure (MPa)	P–C–T Temperature (K)	Hydrogen Storage Capacity	Unit	H/M	Charge–Discharge Temperature (K)	C_{dis}(calc) (mA h g⁻¹)	C_{dis} (mA h g⁻¹)	Condition	References
$CaMg_2Ni_9$		3.3	273	1.48	wt.%		273	396	166	C: 100 mA g⁻¹, 5 h R: 0.5 h D: 40 mA g⁻¹, −0.87 V vs. MMO	[6]
$(Y_{0.5}Ca_{0.5})(MgCa)Ni_9$		3.3	263	1.98	wt.%						[7]
$(La_{0.65}Ca_{0.35})(Mg_{1.32}Ca_{0.68})Ni_9$		3.3	283	1.87	wt.%						[8]
$LaCaMgNi_9$		3.3	293	1.80	wt.%	1.10	293	484	356	C: 60 mA g⁻¹, 7 h D: 60 mA g⁻¹, 1 V	[9]
$CaTiMgNi_9$		3.3	293	1.87	wt.%	1.00	293	502	316	C: 60 mA g⁻¹, 7 h D: 60 mA g⁻¹, 1 V	[9]
$LaCaMgNi_6Al_3$		3.3	293	1.87	wt.%	0.99	293	501	284	C: 60 mA g⁻¹, 7 h D: 60 mA g⁻¹, 1 V	[9]
$LaCaMgNi_6Mn_3$		3.3	293	1.80	wt.%	1.08	293	484	342	C: 60 mA g⁻¹, 7 h D: 60 mA g⁻¹, 1 V	[9]
$La_{0.5}Ca_{1.5}MgNi_9$		3.3	293	1.8	wt.%	β-phase, 0.60 γ-phase, 1.08					[10]
$La_{0.67}Mg_{0.33}Ni_{2.5}Co_{0.5}$		1	333			1.1	298		387	C: 100 mA g⁻¹, 5 h R: 10 min D: 100 mA g⁻¹, −0.5 V vs. MMO	[2]
$La_{0.7}Mg_{0.3}Ni_{2.8}Co_{0.5}$							298		410	C: 100 mA g⁻¹, 5 h R: 10 min D: 100 mA g⁻¹, −0.5 V vs. MMO	[2]

(Continued)

TABLE 7.1 (Continued)
Hydrogen Storage Capacity and Discharge Capacity of Various Mg-Containing Alloys

Composition	Pretreatment	H_2 Pressure (MPa)	P–C–T Temperature (K)	Hydrogen Storage Capacity	Unit	H/M	Charge–Discharge Temperature (K)	C_{dis} (calc) (mA h g^{-1})	C_{dis} (mA h g^{-1})	Condition	References
$La_{0.75}Mg_{0.25}Ni_{3.0}Co_{0.5}$							298		390	C: 100 mA g^{-1}, 5 h R: 10 min D: 100 mA g^{-1}, −0.5 V vs. MMO	[2]
$La_{0.8}Mg_{0.2}Ni_{3.2}Co_{0.3}(MnAl)_{0.2}$									345	C: 0.2 C-rate D: 0.2 C-rate	[13]
$La_{0.6}Pr_{0.2}Mg_{0.2}Ni_{3.2}Co_{0.3}(MnAl)_{0.2}$									348	C: 0.2 C-rate D: 0.2 C-rate	[13]
$La_{0.6}Y_{0.2}Mg_{0.2}Ni_{3.2}Co_{0.3}(MnAl)_{0.2}$									349	C: 0.2 C-rate D: 0.2 C-rate	[13]
$La_{0.8}Mg_{0.2}Ni_{3.2}Co_{0.3}(MnAl)_{0.2}$							303		294	C: 300 mA g^{-1}, 3 h R: 5 min D: 100 mA g^{-1}, −0.6 V vs. MMO	[14]
$La_{0.67}Mg_{0.33}Ni_{2.5}Co_{0.5}$	Annealing (1123 K, vacuum)						303		398	C: 100 mA g^{-1}, 5 h R: 10 min D: 100 mA g^{-1}, −0.5 V vs. MMO	[15]
$La_{0.67}Mg_{0.33}Ni_{2.5}Co_{0.5}$	Annealing (1223 K, vacuum)						303		357	C: 100 mA g^{-1}, 5 h R: 10 min D: 100 mA g^{-1}, −0.5 V vs. MMO	[15]
La_2MgNi_9			298			1.05	298		403	C: 300 mA g^{-1}, 2 h D: 100 mA g^{-1}, −0.6 V vs. MMO	[16]

(Continued)

TABLE 7.1 (Continued)
Hydrogen Storage Capacity and Discharge Capacity of Various Mg-Containing Alloys

Composition	Pretreatment	H_2 Pressure (MPa)	P–C–T Temperature (K)	Hydrogen Storage Capacity	Unit	H/M	Charge–Discharge Temperature (K)	C_{dis} (calc) (mA h g⁻¹)	C_{dis} (mA h g⁻¹)	Condition	References
La₀.₇Mg₀.₃(Ni₀.₈₅Co₀.₁₅)₃.₅						1.05	303		396	C: 100 mA g⁻¹, 5 h; R: 10 min; D: 60 mA g⁻¹, −0.6 V vs. MMO	[18]
La₀.₇Mg₀.₃Ni₂.₆Mn₀.₁Co₀.₇₅							303		403	C: 100 mA g⁻¹, 5 h; R: 10 min; D: 60 mA g⁻¹, −0.6 V vs. MMO	[19]
LaMg₁₁Ni + 200 wt.% Ni	Ball-milling (400 rpm, 50 h)						303		1002ᵃ (334ᵇ)	C: 300 mA g⁻¹ (LaMg₁₁Ni), 5 h; R: 10 min; D: 180 mA g⁻¹ (LaMg₁₁Ni), −0.6 V vs. MMO	[20]
LaMgNi₄							298		390	C: 150 mA g⁻¹, 3 h; D: 30 mA g⁻¹, 0.5 V	[21]
La₀.₇₅Mg₀.₁₅Ni₃.₅							303		344	C: 60 mA g⁻¹ (alloy), 500 min; D: 60 mA g⁻¹ (alloy), −0.6 V vs. MMO	[23]
La₂MgNi₉		1	293	1.08							[24]
LaMg₁₀NiAl + 150 wt.% Ni	Ball-milling (580 rpm, 14 h)						RT		953ᵃ (381ᵇ)	C: 1000 mA g⁻¹ (alloy), 80 min; R: 5 min; D: 50 mA g⁻¹ (alloy), −0.6 V vs. MMO	[25]

(Continued)

TABLE 7.1 (*Continued*)
Hydrogen Storage Capacity and Discharge Capacity of Various Mg-Containing Alloys

Composition	Pretreatment	H_2 Pressure (MPa)	P-C-T Temperature (K)	Hydrogen Storage Capacity Unit	H/M	Charge–Discharge Temperature (K)	C_{dis}(calc) (mA h g^{-1})	C_{dis} (mA h g^{-1})	Condition	References
$LaMg_2TiNi_2$ + 150 wt.% Ni	Ball-milling (580 rpm, 12 h)					RT		910[a] (364[b])	C: 1000 mA g^{-1}(alloy), 80 min R: 5 min D: 50 mA g^{-1}(alloy), -0.6 V vs. MMO	[26]
$La_{0.67}Mg_{0.33}Ni_{2.5}Co_{0.5}$	Annealing (1173 K, Ar at 4 MPa, 24 h)		293		1.06	293		403	C: 100 mA g^{-1}, 5 h D: 100 mA g^{-1} -0.6 V vs. MMO	[27]
$La_{1.5}Mg_{0.5}Ni_7$	Annealing (1073 K, Ar at 4 MPa, 24 h)					298		391	C: 100 mA g^{-1}, 5 h D: 100 mA g^{-1} -0.6 V vs. MMO	[28]
$La_{1.8}Ti_{0.2}MgNi_{8.9}Al_{0.1}$	Annealing (1273 K, vacuum, 10 h)					RT		360	C: 100 mA g^{-1}, 5 h R: 10 min D: 80 mA g^{-1}, -0.5 V vs. MMO	[29]
$La_2Ca_3Mg_2Ni_{23}$						298		404	C: 100 mA g^{-1}, 6 h D: 50 mA g^{-1}, -0.6 V vs. MMO	[30]
$La_{0.7}Mg_{0.3}Ni_{3.5}$ + 40 wt.% $Ti_{0.17}Zr_{0.08}V_{0.35}Cr_{0.1}Ni_{0.3}$	Ball-milling (400 rpm, Ar at 0.2–0.3 MPa, 0.5 h)					303		335[a] (239[b])	C: 100 mA g^{-1}, 6 h D: 50 mA g^{-1}, -0.6 V vs. MMO	[31]

[a] mA h/g (alloy).
[b] mA h/g (composite).

Kohno et al. also prepared new ternary $La_xMg_{1-x}Ni_{y-0.5}Co_{0.5}$ ($x=0.67$–0.75, $y=3.0$–3.5) alloys and found that $La_{0.7}Mg_{0.3}Ni_{2.8}Co_{0.5}$ had a high discharge capacity of 410 mA h g^{-1} and good charge–discharge cycle durability for 30 cycles,[2] but the discharge capacity fell as the number of charge–discharge cycles exceeded about 100.[12] Yasuoka et al. modified the composition of $La_{0.7}Mg_{0.3}Ni_{2.8}Co_{0.5}$ for improving cycle durability and found that alloy oxidation was accelerated by Co and greatly suppressed by partial substitution of Ni with Al.[12] Moreover, they clarified that a 1700 mA h AA size cell using the $Mm_{0.83}Mg_{0.17}Ni_{3.1}Al_{0.2}$ negative electrode was 30% longer in cycle durability than that using the AB_5-type $MmNi_{3.3}Co_{0.8}Al_{0.2}Mn_{0.6}$ negative electrode.[12] The new Ni/MH batteries with superlattice alloys are commercialized as *eneloop* by Panasonic Co., Ltd., and *Twicell* by FDK Twicell Co., Ltd.

Ozaki et al. investigated the relationship among alloy composition, crystal structure (phase abundance), and electrode properties for low-cobalt AB_3–AB_4 rare earth–Mg–Ni alloys in detail.[13] In $La_{0.8}Mg_{0.2}Ni_{3.4-x}Co_{0.3}(MnAl)_x$ ($0<x<0.4$) alloys, five kinds of stacking-structured phases such as rhombohedral Gd_2Co_7 type, hexagonal Ce_2Ni_7 type, hexagonal Pr_5Co_{19} type, rhombohedral Ce_5Co_{19} type, and La_5MgNi_{24} type (Figure 7.5) were identified from Rietveld analysis of x-ray diffraction (XRD) patterns. The content of each phase was greatly influenced by the partial substitution of Ni with Mn and Al. In terms of the maximum discharge capacity and capacity retention for 50 cycles, the $La_{0.8}Mg_{0.2}Ni_{3.2}Co_{0.3}(MnAl)_{0.2}$ was the best, and $La_{0.6}Pr_{0.2}Mg_{0.2}Ni_{3.2}Co_{0.3}(MnAl)_{0.2}$ in which La was partially substituted by Pr still improved both properties. Chu et al. also took up $La_{0.7}Mg_{0.3}Ni_{3.5-x}$ $(MnAl)_x$ ($x=0$–0.2) alloys and found that all alloys consisted of $(LaMg)Ni_3$, $LaNi_5$, and impurity LaNi phases.[14] In terms of capacity retention for 100 cycles,

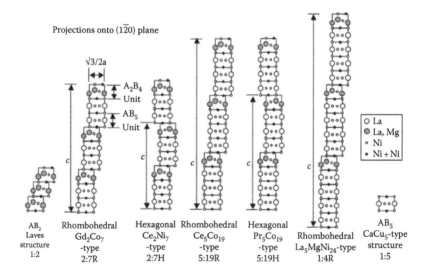

FIGURE 7.5 Projections onto $(1\bar{2}0)$ plane of crystal structures observed in $La_{0.8}Mg_{0.2}Ni_{3.4-x}$ $Co_{0.3}(MnAl)_x$ ($0<x<0.4$) alloys. (Reprinted from *J. Alloys. Compd.*, 446–447, Ozaki, T., Kanemoto, M., Kakeya, T., Kitano, Y., Kuzuhara, M., Watada, M., Tanase, S., and Sakai, T., pp. 620–624, Copyright 2007, with permission from Elsevier.)

the $La_{0.8}Mg_{0.2}Ni_{3.2}Co_{0.3}(MnAl)_{0.2}$ alloy was the best because of the formation of Al-containing passive film on the alloy surface.

Research and development of the rare earth–Mg–Ni-based alloys with superlattice structures as the negative electrode for Ni/MH batteries have been actively done in China during the 2000s. Wang and Lei et al. annealed $La_{0.67}Mg_{0.33}Ni_{2.8}Co_{0.5}$, which was the same composition as Kohno et al., at 1123 and 1223 K under vacuum due to composition homogenization. They found that the alloy annealed at 1123 K exhibited a higher discharge capacity, exchange current density, and limiting current density than that at 1223 K, while the cycle life of the latter was longer than that of the former.[15] They investigated that the effect of the La/Mg ratio on electrochemical properties of $La_xMg_{3-x}Ni_9$ ($x=1.0$–2.2) and La_2MgNi_9 ($x=2$) exhibited the largest discharge capacity (ca. 400 mA h g^{-1}), most rapid initial activation, and high-rate dischargeability.[16,17] Moreover, they investigated the charge–discharge properties of $La_{0.7}Mg_{0.3}(Ni_{0.85}Co_{0.15})_x$ ($x=2.5$–5.0),[18] $La_{0.7}Mg_{0.3}Ni_{3.4-x}Mn_{0.1}Co_x$ ($x=0$–1.6),[19] and ($LaMg_{11}Ni+200$ wt.% Ni) composite[20] prepared by ball-milling of $LaMg_{11}Ni$ with Ni.

Wang et al. prepared $LaMgNi_4$ with the cubic $SnMgCu_4$ structure by sintering tablets of mixed powders of Mg, La, and Ni.[21] The maximum discharge capacity of $LaMgNi_4$ was 305 mA h g^{-1}, but it was increased up to 390 mA h g^{-1} by ball-milling for 50 h. They also prepared $MMgNi_4$ (M=Ce, Pr, Nd) with the cubic single phase, but their maximum discharge capacity and cycle performance was lower than those of $LaMgNi_4$.[22] Dong et al. prepared $La_{0.75+x}Mg_{0.25-x}Ni_{3.5}$ ($x=0$–0.1) alloys that were composed of $LaNi_5$, $(La, Mg)_2Ni_7$, and a few $LaNi_2$ phases.[23] The LaMg phase appeared as La/Mg ratio was more than 3 and its abundance was increased with the La/Mg ratio. $La_{0.75}Mg_{0.15}Ni_{3.5}$ exhibited a maximum discharge capacity of 344 mA h g^{-1} and the best cycle stability. On the other hand, Denys and Yartys investigated the crystal structure of $La_{3-x}Mg_xNi_9$ ($x=0.5$–2.0) alloys by synchrotron XRD studies.[24] The main phase of each alloy was trigonal $PuNi_3$-type intermetallic that was a stacking structure of $CaCu_5$-type $LaNi_5$ and Laves $MgZn_2$-type $La_{2-x}Mg_xNi_4$ slabs along the trigonal (001) axis. When the x value was increased, the thermodynamic stability of the hydrides was decreased and reversible hydrogen storage capacity was increased.

Wang et al. prepared the composites of $LaMg_{10}Ni_{2-x}Al_x$ ($x=0$–1.5) and $LaMg_{10-x}Ti_xNi_2$ ($x=1$, 2) with carbonyl Ni powders by ball-milling in a cyclohexane solution.[25] The original $LaMg_{10}Ni_{2-x}Al_x$ alloys were composed of the body-centered orthorhombic $LaMg_{12}$ main phase and Mg_2Ni hexagonal phase and changed to amorphous or nanocrystalline structure after ball-milling with Ni. The ($LaMg_{10}NiAl+150$ wt.% Ni) composite exhibited the highest electrocatalytic activity and discharge capacity of 953 mA h g^{-1} (381 mA h g(composite)$^{-1}$) because high active Raney Ni would be formed at the electrode surface by dissolution of Al in a concentrated alkaline solution, but it was halved within 10 initial cycles.[25] In contrast, the ($LaMg_9TiNi_2+150$ wt.% Ni) composite had the first discharge capacity of 910 mA h g^{-1} (364 mA h g(composite)$^{-1}$), which was higher than that of the ($LaMg_8Ti_2Ni_2+150$ wt.% Ni) composite (750 mA h g^{-1} [300 mA h g(composite)$^{-1}$]).[26] The former was higher in electrocatalytic activity or lower in charge-transfer resistance than the latter because a small amount of TiNi phase was included with La_2Mg_{17} and Mg_2Ni phases in the former, leading to the improvement of cycle performance for the former.

Zhang et al. examined the effect of annealing on the negative electrode characteristics of some Li–Mg–Ni alloys. $La_{0.67}Mg_{0.33}Ni_{2.5}Co_{0.5}$ had $PuNi_3$-type (La, Mg)(Ni, Co)$_3$, Ce_2Ni_7-type (La, Mg)$_2$(Ni, Co)$_7$, $MgCu_4Sn$-type $LaMg(Ni, Co)_4$, and CaCu-type $La(Ni, Co)_5$ phases, and their abundance was different depending on the annealing temperature. The annealing at 1173 K gave the Ce_2Ni_7-type phase as the main phase and the highest discharge capacity of 403 mA h g^{-1} at third cycle.[27] In addition, the $La_{0.67}Mg_{0.33}Ni_{2.5}Co_{0.5}$ alloy annealed at 1173 K exhibited the highest cycle stability and highest high-rate dischargeability, and highest kinetic parameters such as exchange current density, limiting current density, and hydrogen diffusion coefficient. $La_{1.5}Mg_{0.5}Ni_7$, which was similar in composition to an alloy prepared by Kohno et al.,[2] was also annealed.[28] Annealing at 1073–1173 K converted multiphase to double-phase structure of Gd_2Co_7-type and Ce_2Ni_7-type phases. The maximum discharge capacity was decreased as the annealing temperature was increased, while exchange current density and diffusion coefficient of hydrogen were increased. Moreover, high-rate dischargeability and cycle performance were also improved as annealing temperature was increased. Jiang et al. also annealed $La_{1.8}Ti_{0.2}MgNi_{8.9}Al_{0.1}$ at 1073–1273 K to obtain a similar positive effect.[29] Annealing led to the disappearance of the $LaNi_2$ phase and improved the abundance of $La(Ni, Al)_5$ and $LaMg_2Ni_9$ phases. Consequently, the maximum discharge capacity decreased with increasing annealing temperature, but the annealing at 1073 and 1173 K was optimal in terms of cycle performance and high-rate dischargeability.

Si et al. investigated the structure and electrochemical properties of $La_{5-x}Ca_x$-Mg_2Ni_{23} ($x=0$–3).[30] $La_5Mg_2Ni_{23}$ consisted of Ce_2Ni_7-type, Gd_2Co_7-type, $LaNi_5$, Pr_5Co_{19}-type, Ce_5Co_{19}-type, and $LaMgNi_4$ phases, while $La_{5-x}Ca_xMg_2Ni_{23}$ consisted of $PuNi_3$-type, Gd_2Co_7-type, and $CaCu_5$-type phases except for $La_2Ca_3Mg_2Ni_{23}$ additionally containing 4 wt.% Ni. The $La_2Ca_3Mg_2Ni_{23}$ alloy had the highest discharge capacity (404 mA h g^{-1}) and high-rate dischargeability due to the optimum Ca content and the highest abundance of the $PuNi_3$-type and Gd_2Co_7-type phases. Cycle stability is lowered in the order of Gd_2Co_7-type $>PuNi_3$-type$>LaNi_5$, $LaMgNi_4$. The abundance of the Gd_2Co_7-type and $PuNi_3$-type phases influenced the cycle stability of $La_{5-x}Ca_xMg_2Ni_{23}$.

Chu et al. ball-milled $La_{0.7}Mg_{0.3}Ni_{3.5}$ with $Ti_{0.17}Zr_{0.08}V_{0.35}Cr_{0.1}Ni_{0.3}$ to prepare $La_{0.7}Mg_{0.3}Ni_{3.5}-x$ wt.% $Ti_{0.17}Zr_{0.08}V_{0.35}Cr_{0.1}Ni_{0.3}$ ($x=5$–40) composites and investigated negative electrode properties of the composites.[31] At $x=40$ wt.%, the maximum discharge capacity was the lowest, but the cycle stability was the best due to the improvement of anticorrosion performance. In addition, the kinetic parameters such as charge-transfer resistance, exchange current density, limiting current density, and hydrogen diffusion coefficient were also improved at $x=40$ wt.%.

7.2.2 VANADIUM-CONTAINING ALLOYS

Vanadium hydrides have been studied for practical use in chemical heat pumps, hydrogen compressors, and isotope separations because of their large hydrogen storage capacity (H/M=2). In addition, vanadium has advantages such as strong resistance to pulverization during cycling, high-rate diffusion, and rapid activation. Vanadium has a body-centered cubic (bcc) structure. When vanadium absorbs hydrogen, it is reversibly converted to VH_2 with a face-centered cubic (fcc) structure via VH with a body-centered tetragonal (bct) structure. However, since VH is generally too stable to desorb hydrogen, the usable hydrogen storage capacity is only half of the maximum capacity. Titanium–vanadium alloys, which form a single-phase solid solution over the whole fractions, were notable candidates of a new hydrogen reservoir because of their lower costs than V and their large hydrogen storage capacity.[32] However, owing to the slow desorption rate of hydrogen from Ti–V solid solution hydrides, some catalysts for hydrogen desorption are required. Ni is known to become the catalyst for facilitating the hydrogen absorption and desorption, and it has been substituted for constituent elements or added to the alloys using several methods. The hydrogen storage capacity and discharge capacity of various Ti–V–Ni alloys are shown in Table 7.2. Tsukahara et al. reported that the $TiV_3Ni_{0.56}$ alloy had a 3D network of two phases, the V-based bcc solid solution primary phase as a hydrogen reservoir and the TiNi-based secondary phase as an electrocatalyst and a current collector, and the formation of the network structure led to the improvement of charge–discharge characteristics.[33–36] In addition, they prepared $V_3TiNi_{0.56}M_{0.24}$ (M = Zr, Nb, Ta, Mo, Mn, Fe, Co, and Pd) alloys.[37] In the alloys with M = Zr and Hf, the Laves-type secondary phase was mostly precipitated, while in the other alloys, the TiNi-based secondary phase formed a network structure. The alloys with M = Hf, Nb, Ta, and Pd showed higher discharge capacity than 330 mA h g^{-1}. Zhang et al. annealed $V_3TiNi_{0.56}Co_{0.14}Nb_{0.047}Ta_{0.047}$ alloy at 1073 K for less than 24 h, leading to the grain growth of the bcc solid solution main phase and the decrease in the TiNi-based secondary phase.[38] Inoue et al. also prepared $TiV_{4-x}Ni_x$ (x = 0–1) alloys and the alloys with x = 0.4–0.6 showed the highest discharge capacity.[39]

$TiV_{1.4}$ alloy forms a hydride with a composition of $TiV_{1.4}H_{4.6}$[40] that corresponds to the calculated discharge capacity of 1036 mA h g^{-1} and is higher than the calculated discharge capacity (852 mA h g^{-1}) of TiV_4H_8. Therefore, the $TiV_{1.4}$ alloy has a potential as a new negative electrode material for use in Ni/MH batteries. Iwakura et al. prepared $TiV_{1.4-x}Ni_x$ (x = 0–1) alloys and investigated their electrochemical and structural characteristics.[41] They found that the $TiV_{0.9}Ni_{0.5}$ electrode showed the highest hydrogen storage capacity (hydrogen-to-metal ratio [H/M] = 1.69 at 0.5 MPa) and discharge capacity (390 mA h g^{-1}) among the $TiV_{1.4-x}Ni_x$ electrodes. In addition, the $TiV_{0.9}Ni_{0.5}$ alloy was composed of two phases; the primary phase $TiV_{2.1}Ni_{0.3}$ alloy with a bcc solid solution structure had a much higher hydrogen storage capacity (H/M = 1.99 at 0.5 MPa) and discharge capacity (470 mA h g^{-1}) than the original $TiV_{0.9}Ni_{0.5}$ alloy. However, the $TiV_{2.1}Ni_{0.3}$ alloy exhibited poor charge–discharge cycle durability because of the dissolution of the V constituent in the 6 M KOH electrolyte solution.

In order to suppress the deterioration of the $TiV_{2.1}Ni_{0.3}$ alloy, the surface of the $TiV_{2.1}Ni_{0.3}$ alloy was modified with an amorphous MgNi alloy by using a mechanochemical technique.[42,43] The spectral data on the modified $TiV_{2.1}Ni_{0.3}$ alloy showed that a mutual diffusion layer of the Mg, Ni, Ti, and V constituents was formed at the interface between the $TiV_{2.1}Ni_{0.3}$ and MgNi alloys and played an important role in suppressing the deterioration of the alloy. The surface modification of the $TiV_{2.1}Ni_{0.3}$ alloy with Ni or Raney Ni, which was an effective catalyst for hydriding and dehydriding, was also effective in improving the charge–discharge cycle

TABLE 7.2
Hydrogen Storage Capacity and Discharge Capacity of Various V-Containing Alloys

Composition	Pretreatment	H_2 Pressure (MPa)	P–C–T Temperature (K)	Hydrogen Storage Capacity		Charge–Discharge Temperature (K)	C_{dis} (calc) (mA h g^{-1})	C_{dis} (mA h g^{-1})	Condition	References
				Unit	H/M					
TiV$_3$Ni$_{0.56}$		3.3	333		0.59	293		420	C: 100 mA g^{-1} D: 25 mA g^{-1}, −0.7 V vs. MMO	[33]
TiV$_3$Ni$_{0.56}$Si$_{0.046}$		3.3	328		0.57	293		400	C: 100 mA g^{-1} D: 25 mA g^{-1}, −0.7 V vs. MMO	[37]
TiV$_3$Ni$_{0.56}$Co$_{0.047}$		3.3	318		0.55	293		400	C: 100 mA g^{-1} D: 25 mA g^{-1}, −0.7 V vs. MMO	[37]
V$_3$TiNi$_{0.56}$Co$_{0.14}$Nb$_{0.047}$Ta$_{0.047}$	Annealing (1073 K, vacuum, 24 h)							380	C: 100 mA g^{-1}, 5 h D: 10 mA g^{-1}, −0.7 V vs. MMO	[38]
TiV$_{3.4}$Ni$_{0.6}$						303	852	350	C: 50 mA g^{-1}, 12 h R: 10 min D: 50 mA g^{-1}, −0.7 V vs. MMO	[39]
TiV$_{2.1}$Ni$_{0.3}$		3.3	303		1.99	303	1055	470	C: 100 mA g^{-1}, 8 h R: 10 min D: 50 mA g^{-1}, −0.75 V vs. MMO	[41]
TiV$_{2.1}$Ni$_{0.3}$ + 5 wt.% MgNi	Ball-milling (180 rpm, Ar at 0.1 MPa, 3 h)					303		440[a] (419[b])	C: 100 mA g^{-1}, 8 h R: 10 min D: 50 mA g^{-1}, −0.75 V vs. MMO	[42]
TiV$_{2.1}$Ni$_{0.3}$ + 30 wt.% Raney Ni	Ball-milling (180 rpm, Ar at 0.1 MPa, 3 h)					303		495[a] (381[b])	C: 100 mA g^{-1}, 8 h R: 10 min D: 50 mA g^{-1}, −0.75 V vs. MMO	[44]
TiV$_{2.1}$Ni$_{0.3}$ + 30 wt.% Raney Ni	Ball-milling (180 rpm, 0.01 M NaH$_2$PO$_2$, Ar at 0.1 MPa, 3 h)					303		620[a] (477[b])	C: 100 mA g^{-1}, 8 h R: 10 min D: 50 mA g^{-1}, −0.75 V vs. MMO	[46]
TiV$_{2.1}$Ni$_{0.5}$						298		372	C: 100 mA g^{-1}, 6.5 h R: 10 min D: 50 mA g^{-1}, −0.7 V vs. MMO	[47]

(Continued)

TABLE 7.2 (*Continued*)
Hydrogen Storage Capacity and Discharge Capacity of Various V-Containing Alloys

Composition	Pretreatment	H_2 Pressure (MPa)	P–C–T Temperature (K)	Hydrogen Storage Capacity Unit	H/M	Charge–Discharge Temperature (K)	C_{dis}(calc) (mA h g^{-1})	C_{dis} (mA h g^{-1})	Condition	References
$TiV_{2.1}Ni_{0.3}Hf_{0.05}$						298		415	C: 100 mA g^{-1}, 6.5 h R: 10 min D: 50 mA g^{-1}, −0.7 V vs. MMO	[48]
$TiV_{1.7}Cr_{0.4}Ni_{0.3}$						303		440	C: 100 mA g^{-1}, 8 h R: 10 min D: 50 mA g^{-1}, −0.75 V vs. MMO	[49]
$Ti_{0.25}V_{0.35}Co_{0.1}Ni_{0.3}$						313		240	C: 60 mA g^{-1} D: 60 mA g^{-1}, 0.80 V	[51]
$Ti_{0.2}V_{0.68}Cr_{0.12}$ + 7 wt.% Ni	Ball-milling (500 rpm, Ar at 0.1 MPa, 25 min)					303		460[a] (430[b])	C: 100 mA g^{-1}, 6 h R: 5 min D: 30 mA g^{-1}, −0.75 V vs. MMO	[52]
$Ti_{0.8}Zr_{0.2}V_{3.2}$ $Mn_{0.64}Cr_{0.96}Ni_{1.2}$						303		320	C: 100 mA g^{-1}, 5 h R: 10 min D: 60 mA g^{-1}, −0.75 V vs. MMO	[54]
$Ti_{0.8}Zr_{0.2}V_{1.6}$ $Mn_{0.32}Cr_{0.48}Ni_{0.6}$						303		300	C: 60 mA g^{-1}, 10 h R: 10 min D: 60 mA g^{-1}, −0.6 V vs. MMO	[55]
$Ti_{0.17}Zr_{0.07}La_{0.01}$ $V_{0.35}Cr_{0.1}Ni_{0.3}$						303		338	C: 300 mA g^{-1}, 3 h R: 5 min D: 100 mA g^{-1}, −0.6 V vs. MMO	[56]
$Ti_{0.8}Zr_{0.2}V_{2.7}$ $Mn_{0.5}Cr_{0.6}Ni_{1.15}$ $Co_{0.1}Fe_{0.2}$						303		325	C: 300 mA g^{-1}, 2 h R: 10 min D: 300 mA g^{-1}, −0.6 V vs. MMO	[57]

[a] mA h/g (alloy).
[b] mA h/g (composite).

durability.[44,45] Composite particles prepared by ball-milling the $TiV_{2.1}Ni_{0.3}$ alloy with Raney Ni in the presence of a 0.01 M aqueous solution of sodium hypophosphite (NaH_2PO_2) as a reducing agent, showed a maximum discharge capacity of 620 mA h g^{-1} (477 mA h g(composite)$^{-1}$) although the cycle durability was not improved.[46] Guo et al. prepared $TiV_{2.1}Ni_x$ $(x=0.1–0.9)$ alloys and found that the maximum discharge capacity was the highest at $x=0.5$, but high-rate discharge-ability was not good.[47] The addition of Hf in $TiV_{2.1}Ni_{0.5}$ changed the secondary phase from the bcc TiNi-based phase to the C14 Laves phase.[48] The maximum discharge capacity was highest for $TiV_{2.1}Ni_{0.5}Hf_{0.05}$, while the cycle stability was decreased with increasing Hf content.

Bulk modification, such as partial substitution with foreign elements, is another important strategy for improving cycle durability. Chromium can be oxidized to form a passive layer, which confers high corrosion resistance against acidic and basic aqueous solutions. The effect of Cr substitution on negative electrode properties of $TiV_{2.1}Ni_{0.3}$ was investigated.[49,50] Consequently, $TiV_{2.1-x}Cr_xNi_{0.3}$ $(x=0.1–1.0)$ alloys were composed of two phases, the bcc solid solution primary phase and the TiNi-based secondary phase, like the $TiV_{2.1}Ni_{0.3}$ alloy. The V and Cr contents in the primary phase were higher than those in the secondary phase, although the Ti and Ni contents were higher in the secondary phase. The maximum discharge capacity of $TiV_{1.7}Cr_{0.4}Ni_{0.3}$ showed the highest, while the cycle stability of $TiV_{1.6}Cr_{0.5}Ni_{0.3}$ was the best due to the effective suppression of the dissolution of V. In particular, the loss of discharge capacity per cycle for $TiV_{1.6}Cr_{0.5}Ni_{0.3}$ was less than one-tenth than that for $TiV_{2.1}Ni_{0.3}$, which corresponded to a cycle life over 500 cycles. Hydrogen diffusion in the primary phase as the main hydrogen reservoir was the rate-controlling step in charge and discharge processes.

Chai and Zhao investigated the crystal structure and negative electrode properties of $Ti_{0.25}V_{0.35}Cr_{0.40-x}Ni_x$ $(x=0.05–0.40)$ alloys that were composed of a bcc solid solution primary phase and a little TiNi-based secondary phase.[51] The alloy with $x=0.30$ was the best in terms of the maximum discharge capacity and cycle stability. Kim et al. obtained a discharge capacity of 460 mA h g^{-1} (430 mA h g(composite)$^{-1}$) by ball-milling $Ti_{0.2}V_{0.68}Cr_{0.12}$ powder with 7 wt.% Ni powder for 25 min.[52] The ball-milling exhibited a positive effect for cycle stability due to the suppression of V dissolution.

Akiba and Iba proposed a new concept of hydrogen storage alloy, *Laves phase–related bcc solid solution*, that was formed in multicomponent nominal AB_2 alloys that consisted of Zr and Ti for the A metal site and group 5, 6, and 7 transition metals for the B metal sites.[53] They found an almost pure *Laves phase–related bcc solid solution* had a large hydrogen capacity (more than 2 mass%) and fast hydrogen absorption and desorption kinetics at ambient temperature and pressure. Zhu et al. prepared $(Ti_{0.8}Zr_{0.2})(V_{0.533}Mn_{0.107}Cr_{0.16}Ni_{0.2})_x$ $(x=6–7)$ and $Ti_{1-x}Zr_xV_{1.6}Mn_{0.32}Cr_{0.48}Ni_{0.6}$ $(x=0.2–0.5)$ alloys that were composed of a C14 Laves phase, V-based bcc phase, and a small amount of TiNi-based phase.[54,55] With increasing B-side elements and Zr constituent, the maximum discharge capacity was decreased and cycle stability was improved. High-rate dischargeability was improved by increasing the Zr constituent or decreasing the B-side elements. $Ti_{0.17}Zr_{0.08-x}La_xV_{0.35}Cr_{0.1}Ni_{0.3}$ $(x=0–0.04)$,[56] $Ti_{0.8}Zr_{0.2}V_{2.7}Mn_{0.5}Cr_{0.6}Ni_{1.15}Co_{0.1}Fe_{0.2}$,[57] etc., were also prepared and their structural and electrochemical properties were investigated.

7.3 TECHNICAL TRENDS ON THE DEVELOPMENT OF A POSITIVE ELECTRODE

At the end of the 1920s, high-power discharge performance of the nickel positive electrode was improved by the use of the sintered nickel electrode prepared by Ackermann. In the 1980s, a paste-type Ni positive electrode was developed, leading to the increase in energy density.[58,59] The paste-type Ni positive electrode was produced by packing spherical β-phase $Ni(OH)_2$ powders into pores of a porous nickel substrate. The coprecipitation with Zn^{60} and the coating with cobalt oxyhydroxide $(CoOOH)^{59,61}$ were effective for improving the discharge capacity and suppressing the formation of γ-phase nickel oxyhydroxide (γ-NiOOH) that caused electrode swelling and capacity decay. The Ni positive electrode is reversibly transformed between $Ni(OH)_2$ as an insulator and NiOOH as a semiconductor in charge/discharge processes as follows:

$$Ni(OH)_2 + OH^- \rightleftharpoons NiOOH + H_2O + e^- \tag{7.9}$$

The theoretical capacity of β-$Ni(OH)_2$ is 289 mA h g^{-1} based on Equation 7.9.[62]

$Ni(OH)_2$ has two polymorphic forms (α-phase and β-phase) and these phases are transformed into γ-NiOOH and β-NiOOH as oxidized forms, respectively. Bode et al. proposed the relationship between phase transformation and redox of α- and β-$Ni(OH)_2$ in charging and discharging as shown in Figure 7.6.[63] β-$Ni(OH)_2$ is often used as a positive electrode active material for Ni/MH batteries because α-$Ni(OH)_2$ is unstable in alkaline solutions. The crystal structure of β-$Ni(OH)_2$ consists of

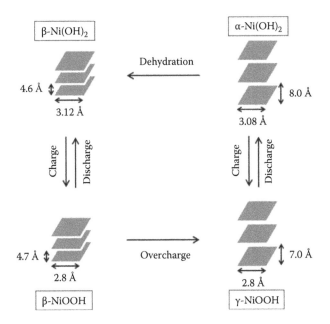

FIGURE 7.6 Bode's diagram for α-$Ni(OH)_2$, β-$Ni(OH)_2$, β-NiOOH, and γ-NiOOH.

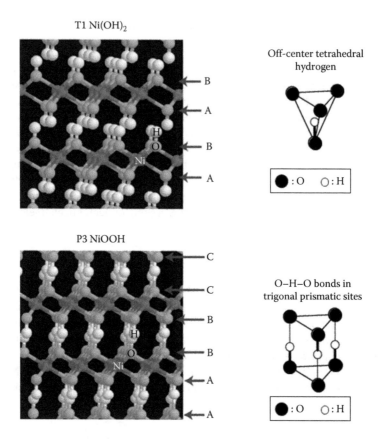

T1 Ni(OH)$_2$

Off-center tetrahedral
hydrogen

● : O ○ : H

P3 NiOOH

O–H–O bonds in
trigonal prismatic sites

● : O ○ : H

FIGURE 7.7 Crystal structures of β-Ni(OH)$_2$ having the T1 oxygen stacking sequence ABAB and β-NiOOH having the P3 oxygen stacking sequence AABBCC. Also shown right is local coordination of hydrogen small circles by oxygen large circles. (Reprinted from Van der Ven, A. et al., *J. Electrochem. Soc.*, 153, A210, Copyright 2006, with permission from The Electrochemical Society.)

close-packed oxygen planes with an ABAB stacking sequence as shown in Figure 7.7.[64] The Ni atoms occupy octahedral sites between alternating oxygen layers to form O–Ni–O bonds, while the H atoms reside in the tetrahedral sites of the remaining layers between oxygen to form O–H bonds and are bound to one of the four oxygen atoms surrounding the tetrahedral sites.[64,65] Since the chemical formula of β-Ni(OH)$_2$ can be also written as β-H$_2$NiO$_2$, a maximum of two H atoms can be removed during charging to become β-NiO$_2$. Thus β-Ni(OH)$_2$ consists of a rigid NiO$_2$ host (which can undergo structural transformations) and removable H atoms that occupy interstitial sites in the host.[64]

For β-Ni(OH)$_2$, the distance between the neighboring two Ni atoms (a) and that between the neighboring two NiO$_2$ layers (c) were 3.12 and 4.6 Å, respectively.[66,67] The a and c values for β-NiOOH formed in charging were 2.8 and 4.7 Å, respectively. γ-NiOOH is known to be formed in overcharging. The a and c values for

TABLE 7.3
Values of a, c and Density of β-Ni(OH)$_2$,
β-NiOOH, α-Ni(OH)$_2$, and γ-NiOOH

	Lattice Constant (Å)		
	a	c	Density (g cm^{-3})
β-Ni(OH)$_2$	3.12	4.6	3.97
β-NiOOH	2.8	4.7	4.68
α-Ni(OH)$_2$	3.08	8	2.82
γ-NiOOH	2.8	7	3.79

γ-NiOOH are 2.8 and 7.0 Å, respectively (Table 7.3). In this way the c values of α-Ni(OH)$_2$ and γ-NiOOH are larger than those of β-Ni(OH)$_2$ and β-NiOOH, respectively, because water molecules and electrolyte cations (K$^+$, Na$^+$, and Li$^+$) are intercalated between the NiO$_2$ layers for α-Ni(OH)$_2$ and γ-NiOOH. Moreover, the change in c value with the transformation of α-Ni(OH)$_2$/γ-NiOOH in charging/discharging is larger than that with the transformation of β-Ni(OH)$_2$/β-NiOOH, indicating that the former has a larger volume change than the latter. Such a large volume change can cause mechanical strains and rapid capacity decay during charge–discharge cycles.[66]

To realize the higher energy density of Ni/MH batteries, the increase in the specific capacity of the Ni positive electrode is an important issue. For this purpose, the development of α-Ni(OH)$_2$ has been tried.[68–70] γ-NiOOH partially contains tetravalent Ni ions (Ni^{4+}), and the average valence of Ni in γ-NiOOH is 3.3–3.7.[69,71] One electron moves in the transformation of β-Ni(OH)$_2$/β-NiOOH, while more than one electron can move in the transformation of α-Ni(OH)$_2$/γ-NiOOH, suggesting that the latter can give a higher specific capacity than the former. However, α-Ni(OH)$_2$ is unstable in alkaline solutions and it is steadily transformed to β-Ni(OH)$_2$ as a thermodynamically stable form upon aging. To stabilize α-Ni(OH)$_2$, partial substitution of Ni^{2+} in the Ni(OH)$_2$ with other divalent and trivalent ions such as Co^{2+}, Zn^{2+}, Fe^{3+}, Mn^{3+}, and Al^{3+} was tried.[69,72–77] Ni(OH)$_2$ containing 10 mol% Al^{3+} with a mixture structure of α- and β-phases was prepared by hydrothermal treatment and exhibited a maximum discharge capacity of 333 mA h g^{-1}, corresponding to 1.46 exchanged electrons per Ni atom in α-phase Ni(OH)$_2$.[77] Ni$_{0.75}$ Zn$_{0.5}$ (OH)$_2$ (CO$_3$)$_{0.25}$·yH$_2$O was rechargeable with a stabilized reversible discharge capacity of 410 ± 15 mA h g^{-1}.[69] In recent studies, several new Al-substituted materials such as α-Ni(OH)$_2$ thin film,[78] α-Ni(OH)$_2$/carbon composite,[70] and nickel–aluminum layered double hydroxide/carbon (Ni–Al LDH/C) composite[79] have been developed by a liquid-phase deposition method. The Ni–Al LDH/C (19.2 Al%) results in high-performance cathode materials, high discharge capacity (>390 mA h g(composite)$^{-1}$) at 1 C-rate, and cycle life over 300 cycles at 2 C-rate.[79]

It is known that additives such as CoO and Co(OH)$_2$ can improve the utilization of the positive electrode active material because conductive CoOOH formed during the first charging makes a conducting network on the active material surface to improve electric conductivity.[59,80,81] Oshitani et al. reported that Co(OH)$_2$, which was formed by the dissolution of a divalent cobalt-based precursor such as CoO into the electrolyte as a cobalt(II) complex ion like $Co(OH)_4^{2-}$ and then reprecipitation around the platelets of $Ni(OH)_2$, was oxidized to CoOOH during the first charging, leading to the formation of a conductive network.[59] Consequently, the utilization of the positive electrode active material was significantly improved. Recently, Inoue et al. found that a simple pretreatment, discharging before the first charging, of the positive electrode with commercial $Co(OH)_2$-coated $Ni(OH)_2$ particles improved battery performance such as discharge capacity, high-rate discharge, and cycle durability.[82,83]

Recently, the effect of the morphology of $Ni(OH)_2$ on its charge–discharge properties has been investigated. For this purpose, several new particle shapes, for instance, ribbon- and board-like,[84] tube-type,[85] fiber-type,[62,86] and large pseudo-single-crystal $Ni(OH)_2$,[87] were prepared chemically and electrochemically. In terms of the discharge capacity, the ribbon- and board-type $Ni(OH)_2$ electrodes exhibited 130 and 260 mA h g^{-1}, respectively,[84] while the discharge capacity of the tube-type $Ni(OH)_2$ electrode was 315 mA h g^{-1}, which was higher than the theoretical discharge capacity of β-$Ni(OH)_2$. This was ascribable to the effect of the hollow-inside tube structure.[85] Meanwhile, the discharge curves of the Ni/MH cell with the fiber-type positive electrode that was prepared utilizing the electrochemical technique exhibited a wide plateau region at a discharge voltage of 1.3 V and its discharge capacity was 305 mA h g^{-1} at 1 C-rate.[62,86] Even in charging at 200 C-rate, the discharge capacity was 220 mA h g^{-1}, indicating excellent high-rate chargeability of the Ni/MH cell. In addition the Ni/MH cell with the fiber-type positive electrode exhibited good voltage retention after 700 cycles. A approximate 80% of the α-$Ni(OH)_2$ phase was transformed into a γ-NiOOH phase to 318 mA h g^{-1} after charging.[62,86] Gourrier et al. prepared a single-crystal-type $Ni(OH)_2$ electrode with a micrometric hexagonal structure.[87] The discharge capacity of the single-crystal-type $Ni(OH)_2$ was 93 mA h g^{-1}, but it was electrochemically active and relatively stable during a charge–discharge test.

In an industrial process the Ni/MH battery is firstly charged at 40°C–80°C, and then a network of conductive cobalt phase that is similar to the Co_3O_4 spinel phase is formed within the $Ni(OH)_2$ electrode. In these conditions, the stability of the conductive network is significantly improved. Tronel et al. reported that when the CoO and $Co(OH)_2$ phases were oxidized at 90°C, the Co_3O_4 phase is primarily produced with CoOOH as a secondary product.[88,89] In addition, they showed that the Co_3O_4 phase was formed by the reaction of the CoOOH phase with Co^{2+} species in the electrolyte at 90°C. Moreover, when a Co_3O_4-type $H_xLi_yCo_{3-d}O_4$ spinel phase synthesized by electrooxidation of CoO powder in a mixed alkaline electrolyte (KOH, LiOH, NaOH) was heat-treated at 430°C, H_2O was removed and the Co/O ratio and the amount of Co^{2+} in tetrahedral sites were increased. Moreover, the Co^{4+}/Co^{3+} ratio in the (Co_2O_4) octahedral framework was increased and a large number of defects that tended to localize electrons were removed. Consequently, electrical conductivity was increased by three orders of magnitude.[90]

Pralong et al. reported that two kinds of CoOOH textures, mosaic and monolithic, were produced during the first charging at different rates and exhibited different electrochemical behaviors.[80,81] The mosaic texture was two digits higher in electric conductivity than the monolithic one. These textures had different formation mechanisms. At charge rates of C/2 or more, the mosaic texture was formed by electrochemical oxidation of $Co(OH)_2$ to nonstoichiometric $Co_x^{IV}Co_{1-x}^{III}OOH_{1-x}$ in the solid state, while at a low charge rate of C/100, the monolithic texture was formed by the oxidation of $Co(OH)_4^{2-}$ that partially dissolved in the electrolyte to less conductive CoOOH. Xia et al. studied the effects of metallic Co and CoO as additives in positive electrodes on the electrochemical performance of Ni/MH batteries.[91] When CoO are charged at 50°C in 6 M KOH solution, only Co_3O_4 was formed, while when metallic Co was used as a starting material, a CoOOH phase was formed with a Co_3O_4 phase that was produced by CoOOH and $Co(OH)_4^{2-}$. The final product was dependent on the solubility of cobalt and the kinetics of the reaction that consumes $Co(OH)_4^{2-}$. The CoOOH phase, which worked well between the Ni foam frame and $Ni(OH)_2$ particles, enhanced the rate capability of the Ni/MH battery. The Co_3O_4 phase, which worked well in connecting $Ni(OH)_2$ particles, improves the utilization of $Ni(OH)_2$.

For early commercial Ni–MH batteries, a loss of cell capacity was often observed with repeating shallow discharging and full charging or overcharging, which was called memory effect.[92–94] The effect is reversible by reconditioning, which is fully repeatedly discharging the battery after charging it. In addition to the loss of capacity, the so-called second plateau that corresponds to a partial transfer of the usual 1.2 V capacity to a 0.8 V potential plateau was another drawback.[66] Both the second plateau and the memory effect disappear if the cell is deeply discharged, indicating that the second plateau is related to the memory effect. The ultimate origin of these phenomena is unclear, but several explanations have been proposed in literatures: (1) an ohmic drop due to the formation of a barrier insulating layer at the active material/electron collector interface[95,96]; (2) the presence of another phase, such as γ-NiOOH, in the oxidized electrode[67,97]; or (3) the existence of an insulating almost stoichiometric phase, $Ni(OH)_{2-\varepsilon}$, in the vicinity of the current collector.[98] Morishita et al. described that under partial charge–discharge cycling conditions at a state of charge (SOC) of 50%–70%, a greater effect on the suppression of memory effect was observed using conductive materials such as nanosized $Co(OH)_2$ and CoO powders in the positive electrode.[99] Watanabe et al. suggested from Raman scattering that $Ni(OH)_2$ showing the memory effect involved more disorder and/or H^+ vacancy than the normal and reconditioned $Ni(OH)_2$, leading to the deterioration of the redox mechanism of the nickel oxide electrode based on the reversible insertion–reinsertion of H^+ ions and electrons.[66]

7.4 OUTLOOK FOR THE FUTURE

The Ni–MH batteries were commercialized in 1993, while LIBs were commercialized in the next year. Since LIBs have a much larger energy density than Ni–MH batteries, LIBs are currently used for almost all mobile applications. However, LIBs have a serious issue on safety, which has never been solved, and applying LIBs to large-scale applications like vehicles, aircrafts, and power storage is one of such issues. As is evident from no serious accidents of HEV, for example, Toyota Prius, in which the Ni–MH batteries are used, it is well known that they are highly reliable on safety compared to LIBs. Moreover, in the fall of 2013, Toyota Motor Corporation announced that their fuel cell vehicles that are going to be sold in 2015 adopted the Ni–MH batteries as a power source. The improvement of energy density is a never-ending assignment of Ni–MH batteries. For this purpose, some new breakthroughs of active materials will be necessary.

REFERENCES

1. T. Sakai, M. Matsuoka and C. Iwakura, Rare earth intermetallics for metal-hydrogen batteries, *Handbook on the Physics and Chemistry of Rare Earths* ed. K. A. Gschneidner, Jr. and L. Eyring, Elsevier Science B. V., Amsterdam, 1995, vol. 21, pp. 133–178.
2. T. Kohno, H. Yoshida, F. Kawashima, T. Inaba, I. Sakai, M. Yamamoto and M. Kanda, *J. Alloys Compd.*, 2000, **311**, L5.
3. Y. Q. Lei, Y. M. Wu, Q. M. Yang, J. Wu and Q. D. Wang, *Z. Phys. Chem.*, 1994, **183**, 379.
4. C. Iwakura, S. Nohara, H. Inoue and Y. Fukumoto, *Chem. Commun.*, 1996, **5**, 1831.
5. C. Iwakura, S. Nohara and H. Inoue, *Solid State Ionics*, 2002, **148**, 499.
6. K. Kadir, T. Sakai and I. Uehara, *J. Alloys Compd.*, 1997, **257**, 115.
7. K. Kadir, N. Kuriyama, T. Sakai, I. Uehara and L. Eriksson, *J. Alloys Compd.*, 1999, **284**, 145.
8. K. Kadir, T. Sakai and I. Uehara, *J. Alloys Compd.*, 1999, **287**, 264.
9. K. Kadir, T. Sakai and I. Uehara, *J. Alloys Compd.*, 2000, **302**, 112.
10. J. Chen, N. Kuriyama, H. T. Takeshita, H. Tanaka, T. Sakai and M. Haruta, *Electrochem. Solid-State Lett.*, 2000, **3**, 249.
11. J. Chen, H. T. Takeshita, H. Tanaka, N. Kuriyama, T. Sakai, I. Uehara and M. Haruta, *J. Alloys Compd.*, 2000, **302**, 304.
12. S. Yasuoka, Y. Magari, T. Murata, T. Tanaka, J. Ishida, H. Nakamura, T. Nohma, M. Kihara, Y. Baba and H. Teraoka, *J. Power Sources*, 2006, **156**, 662.
13. T. Ozaki, M. Kanemoto, T. Kakeya, Y. Kitano, M. Kuzuhara, M. Watada, S. Tanase and T. Sakai, *J. Alloys Compd.*, 2007, **446–447**, 620.
14. H. L. Chu, Y. Zhang, S. J. Qiu, Y. N. Qi, L. X. Sun, F. Xu, Q. Wang and C. A. Dong, *J. Alloys Compd.*, 2008, **457**, 90.
15. H. G. Pan, Y. F. Liu, M. G. Gao, Y. F. Zhu, Y. Q. Lei and Q. D. Wang, *Int. J. Hydrogen Energy*, 2003, **28**, 113.
16. B. Liao, Y. Q. Lei, G. L. Lu, L. X. Chen, H. G. Pan and Q. D. Wang, *J. Alloys Compd.*, 2003, **356–357**, 746.
17. B. Liao, Y. Q. Lei, L. X. Chen, G. L. Lu, H. G. Pan and Q. D. Wang, *J. Power Sources*, 2004, **129**, 358.
18. H. G. Pan, Y. F. Liu, M. X. Gao, Y. Q. Lei and Q. D. Wang, *J. Electrochem. Soc.*, 2003, **159**, A565.
19. Y. F. Liu, H. G. Pan, M. X. Gao and Y. Q. Lei, *J. Alloys Compd.*, 2004, **376**, 304.
20. L. Wang, X. H. Wang, L. X. Chen, C. P. Chen and Q. D. Wang, *J. Alloys Compd.*, 2005, **403**, 357.
21. Z. M. Wang, H. Y. Zhou, Z. F. Gu, G. Cheng and A. B. Yu, *J. Alloys Compd.*, 2004, **377**, L7.
22. Z. M. Wang, H. Y. Zhou, Z. F. Gu, G. Cheng and A. B. Yu, *J. Alloys Compd.*, 2004, **384**, 279.
23. X. P. Dong, F. X. Lu, Y. H. Zhang, L. Y. Yang and X. L. Wang, *Mater. Chem. Phys.*, 2008, **108**, 251.
24. R. V. Denys and V. A. Yartys, *J. Alloys Compd.*, 2011, **509S**, S540.
25. Y. Wang, Z. W. Lu, X. P. Gao, W. K. Hu, X. Y. Jiang, J. Q. Qu and P. W. Shen, *J. Alloys Compd.*, 2005, **389**, 290.
26. Y. Wang, X. Wang, X. P. Gao and P. W. Shen, *Int. J. Hydrogen Energy*, 2007, **32**, 4180.

27. F. L. Zhang, Y. C. Luo, J. P. Chen, R. X. Yan, L. Kang and J. H. Chen, *J. Power Sources*, 2005, **150**, 247.
28. F. L. Zhang, Y. C. Luo, J. P. Chen, R. X. Yan, L. Kang and J. H. Chen, *J. Alloys Compd.*, 2007, **430**, 302.
29. W. Q. Jiang, X. H. Mo, J. Guo and Y. Y. Wei, *J. Power Sources*, 2013, **221**, 84.
30. T. Z. Si, G. Pang, D. M. Liu, Q. A. Zhang and N. Liu, *J. Alloys Compd.*, 2009, **480**, 756.
31. H. L. Chu, S. J. Qiu, L. X. Sun, Y. Zhang, F. Xu, M. Zhu and W. Y. Hu, *Int. J. Hydrogen Energy*, 2008, **33**, 755.
32. S. Ono, K. Nomura and Y. Ikeya, *J. Less-Common Met.*, 1980, **72**, 159.
33. M. Tsukahara, K. Takahashi, T. Mishima, T. Sakai, H. Miyamura, N. Kuriyama and I. Uehara, *J. Alloys Compd.*, 1995, **226**, 203.
34. M. Tsukahara, K. Takahashi, T. Mishima, H. Miyamura, T. Sakai, N. Kuriyama and I. Uehara, *J. Alloys Compd.*, 1995, **231**, 616.
35. M. Tsukahara, K. Takahashi, T. Mishima, A. Isomura and T. Sakai, *J. Alloys Compd.*, 1996, **236**, 151.
36. M. Tsukahara, K. Takahashi, T. Mishima, A. Isomura and T. Sakai, *J. Alloys Compd.*, 1996, **243**, 151.
37. M. Tsukahara, K. Takahashi, T. Mishima, A. Isomura and T. Sakai, *J. Alloys Compd.*, 1996, **245**, 59.
38. Q. A. Zhang, Y. Q. Lei, L. X. Chen and Q. D. Wang, *Mater. Chem. Phys.*, 2001, **69**, 241.
39. H. Inoue, S. Arai and C. Iwakura, *Electrochim. Acta*, 1996, **41**, 937.
40. E. L. Muetterties, *The Hydrogen Series*, ed. E. L. Muetterties, Marcel Dekker, New York, 1971, vol. 1, p. 23.
41. C. Iwakura, W.-K. Choi, R. Miyauchi and H. Inoue, *J. Electrochem. Soc.*, 2000, **147**, 2503.
42. W.-K. Choi, T. Tanaka, R. Miyauchi, T. Morikawa, H. Inoue and C. Iwakura, *J. Alloys Compd.*, 2000, **299**, 141.
43. W.-K. Choi, T. Tanaka, T. Morikawa, H. Inoue and C. Iwakura, *J. Alloys Compd.*, 2000, **302**, 82.
44. H. Inoue, R. Miyauchi, R. Shin-Ya, W.-K. Choi and C. Iwakura, *J. Alloys Compd.*, 2002, **330–332**, 597.
45. H. Inoue, R. Miyauchi, T. Tanaka, W.-K. Choi, R. Shin-Ya, J. Murayama and C. Iwakura, *J. Alloys Compd.*, 2001, **325**, 299.
46. R. Shin-Ya, T. Tanaka, S. Nohara, H. Inoue and C. Iwakura, *J. Alloys Compd.*, 2004, **365**, 303.
47. R. Guo, L. X. Chen, Y. Q. Lei, B. L. Liao, T. Ying and Q. D. Wang, *Int. J. Hydrogen Energy*, 2003, **28**, 803.
48. R. Guo, L. X. Chen, Y. Q. Lei, B. L. Liao, Y. W. Zeng and Q. D. Wang, *J. Alloys Compd.*, 2003, **352**, 270.
49. H. Inoue, S. Koyama and E. Higuchi, *Electrochim. Acta*, 2012, **59**, 23.
50. H. Inoue, N. Kotani, M. Chiku and E. Higuchi, *ECS Trans.*, 2014, **58**, 19.
51. Y. J. Chai and M. S. Zhao, *Int. J. Hydrogen Energy*, 2005, **30**, 279.
52. J. H. Kim, H. Lee, P. S. Lee, C. Y. Seo and J. Y. Lee, *J. Alloys Compd.*, 2003, **348**, 293.
53. E. Akiba and H. Iba, *Intermetallics*, 1998, **6**, 461.
54. Y. F. Zhu, H. G. Pan, M. X. Gao, Y. F. Liu and Q. D. Wang, *J. Alloys Compd.*, 2002, **345**, 201.
55. Y. F. Zhu, H. G. Pan, M. X. Gao, Y. F. Liu and Q. D. Wang, *Int. J. Hydrogen Energy*, 2002, **27**, 287.
56. S. J. Qiu, H. L. Chu, J. Zhang, Y. Zhang, L. X. Sun, F. Xu, D. L. Sun, L. Z. Ouyang, M. Zhu, J. P. E. Grolier and M. Frenkel, *Int. J. Hydrogen Energy*, 2009, **34**, 7246.
57. H. Miao and W. G. Wang, *J. Alloys Compd.*, 2010, **508**, 592.

58. I. Matsumoto, M. Ikeyama, T. Iwaki and H. Ogawa, *Denki Kagaku oyobi Kogyo Buturi Kagaku*, 1986, **54**, 159.

59. M. Oshitani, H. Yufu, K. Takashima, S. Tsuji and Y. Matsumaru, *J. Electrochem. Soc.*, 1989, **136**, 1590.

60. M. Oshitani, M. Watada, H. Yufu and Y. Matsumaru, *Denki Kagaku oyobi Kogyo Buturi Kagaku*, 1989, **57**, 480.

61. M. Yano, T. Ogasawara, Y. Baba, M. Tadokoro and S. Nakahori, *Electrochemistry*, 2001, **69**, 858.

62. T. Takasaki, K. Nishimura, T. Mukai, T. Iwaki, K. Tsutsumi and T. Sakai, *J. Electrochem. Soc.*, 2012, **159**, A1891.

63. H. Bode, K. Dehmelk and J. Witte, *Electrochem. Acta*, 1966, **11**, 1079.

64. A. Van der Ven, D. Morgan, Y. S. Meng and G. Cederc, *J. Electrochem. Soc.*, 2006, **153**, A210.

65. P. Oliva, J. Leonardi, J. F. Laurent, C. Delmas, J. J. Braconnier, M. Figlarz, F. Fievet and A. de Guibert, *J. Power Sources*, 1982, **8**, 229.

66. N. Watanabe, T. Arakawa, Y. Sasaki, T. Yamashita and I. Koiwa, *J. Electrochem. Soc.*, 2012, **159**, A1949.

67. N. Sac-Epée, M. R. Palacín, B. Beaudoin, A. Delahaye-Vidal, T. Jamin, Y. Chabre and J.-M. Tarascon, *J. Electrochem. Soc.*, 1997, **144**, 3896.

68. N. Watanabe, H. Nakayama, K. Fukao and F. Munakata, *J. Appl. Phys.*, 2011, **110**, 023519.

69. M. Dixit, P. V. Kamath and J. Gopalakrishnan, *J. Electrochem. Soc.*, 1999, **146**, 72.

70. A. B. Béléké, A. Hosokawa, M. Mizuhata and S. Deki, *J. Ceram. Soc. Japan*, 2009, **117**, 392.

71. W. E. O'Grady, K. I. Pandya, K. E. Swider and D. A. Corrigan, *J. Electrochem. Soc.*, 1996, **143**, 1613.

72. C. Delmas, J. J. Braconnier, Y. Borthomieu and P. Hagenmuller, *Mat. Res. Bull.*, 1987, **22**, 741.

73. L. Demourgues-Guerlou and C. Delmas, *J. Power Sources*, 1993, **45**, 281.

74. L. Demourgues-Guerlou, C. Denage and C. Delmas, *J. Power Sources*, 1994, **52**, 269.

75. L. Demourgues-Guerlou and C. Delmas, *J. Power Sources*, 1994, **52**, 275.

76. L. Guerlou-Demourgues and C. Delmas, *J. Electrochem. Soc.*, 1996, **143**, 561.

77. P. V. Kamath, M. Dixit, L. Indira, A. K. Shukla, V. Ganesh Kumar and N. Munichandraiah, *J. Electrochem. Soc.*, 1994, **141**, 2956.

78. S. Deki, A. Hosokawa, A. B. Béléké and M. Mizuhata, *Thin Solid Films*, 2009, **517**, 1546.

79. A. B. Béléké, E. Higuchi, H. Inoue and M. Mizuhata, *J. Power Sources*, 2013, **225**, 215.

80. V. Pralong, A. Delahaye-Vidal, B. Beaudoin, B. Gérand and J.-M. Tarascon, *J. Mater. Chem.*, 1999, **9**, 955.

81. V. Pralong, A. Delahaye-Vidal, B. Beaudoin, J. B. Leriche and J.-M. Tarascon, *J. Electrochem. Soc.*, 2000, **147**, 1306.

82. H. Inoue, T. Mizuta and E. Higuchi, *ECS Trans.*, 2010, **25**, 113.

83. E. Higuchi, T. Mizuta and H. Inoue, *Electrochemistry*, 2010, **78**, 420.

84. D. Yang, R. Wang, M. He, J. Zhang and Z. Liu, *J. Phys. Chem. B*, 2005, **109**, 7654.

85. W. Li, S. Zhang and J. Chen, *J. Phys. Chem. B*, 2005, **109**, 14025.

86. T. Takasaki, K. Nishimura, T. Mukai, T. Iwaki, K. Tsutsumi and T. Sakai, *J. Electrochem. Soc.*, 2013, **160**, A564.

87. L. Gourrier, S. Deabate, T. Michel, M. Paillet, P. Hermet, J.-L. Bantignies and F. Henn, *J. Phys. Chem. C*, 2011, **115**, 15067.

88. F. Tronel, L. Guerlou-Demourgues, M. Ménétrier, L. Croguennec, L. Goubault, P. Bernard and C. Delmas, *Chem. Mater.*, 2006, **18**, 5840.

89. F. Tronel, L. Guerlou-Demourgues, L. Goubault, P. Bernard and C. Delmas, *J. Power Sources*, 2008, **179**, 837.

90. M. Douin, L. Guerlou-Demourgues, M. Ménétrier, E. Bekaert, L. Goubault, P. Bernard and C. Delmas, *Chem. Mater.*, 2008, **20**, 6880.
91. Y. Xia, Y. Yang and H. Shao, *J. Power Sources*, 2011, **196**, 495.
92. Y. Sato, K. Ito, T. Arakawa and K. Kobayakawa, *J. Electrochem. Soc.*, 1996, **143**, L225.
93. Z. X. Shu, P. A. Aiken and G. K. Maclean, *J. Power Sources*, 1992, **39**, 67.
94. Y. Sato, S. Takeuchi and K. Kobayakawa, *J. Power Sources*, 2001, **93**, 20.
95. R. Barnard, G. T. Crickmore, J. A. Lee and F. L. Tye, *J. Appl. Electrochem.*, 1980, **10**, 61.
96. R. Barnard, C. F. Randell and F. L. Tye, *J. Appl. Electrochem.*, 1980, **10**, 127.
97. R. A. Huggins, *Solid State Ionics*, 2006, **177**, 2643.
98. C. Léger, C. Tessier, M. Ménétrier, C. Denage and C. Delmas, *J. Electrochem. Soc.*, 1999, **146**, 924.
99. M. Morishita, Y. Shimizu, K. Kobayakawa and Y. Sato, *Electrochim. Acta*, 2008, **53**, 6651.

8 Electrode Materials for Lithium-Ion Rechargeable Batteries

Montse Casas-Cabanas and M. Rosa Palacín

CONTENTS

8.1 INTRODUCTION

Lithium-ion batteries were commercialized in 1991[1] as a natural result from the wide knowledge in intercalation chemistry developed by inorganic and solid-state chemists in the 1970s. The first generation of such batteries allowed storing more than twice the energy compared to nickel or lead batteries of the same size and mass. These typically consisted of Li_xCoO_2 at the positive electrode and a carbonaceous negative electrode, with an electrolyte containing a lithium salt such as $LiPF_6$ dissolved in an organic solvent or, most commonly, a mixture of them (see Figure 8.1).

During the charge process, lithium ions are extracted from $LiCoO_2$ and liberated to the electrolyte, with concomitant oxidation of Co^{3+} to Co^{4+}. At the same time, lithium-ion uptake takes place at the graphite electrode to form a graphite intercalation compound (Li_xC_6). The electrons liberated at the positive electrode pass through

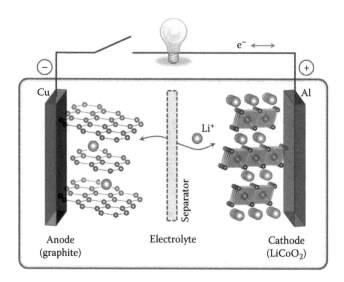

FIGURE 8.1 Schematic illustration of a Li-ion battery using $LiCoO_2$ and graphite as electrode materials. (Reprinted with permission from Goodenough, J.B. and Park, K.S., *J. Am. Chem. Soc.*, 135, 1167. Copyright 2013 American Chemical Society.)

the external circuit to reach the negative electrode. The corresponding reactions at both electrodes are

$$LiCoO_2 \leftrightarrow Li_{1-x}CoO_2 + xLi^+ + xe^-$$

$$C + xLi^+ + xe^- \leftrightarrow Li_xC \text{ (with a stoichiometry close to } LiC_6 \text{ for the fully charged state)}$$

The open-circuit voltage of the cell is the difference between the electrochemical potentials of the anode and the cathode. The stored energy is the product of the average voltage and the capacity of the cell. Upon battery discharge, the chemical potentials of each electrode will tend to equilibrate by approaching each other, thus reducing the cell voltage. The energy separation of the lowest unoccupied molecular orbital and the highest occupied molecular orbital of the electrolyte (LUMO and HOMO, respectively) determines its electrochemical stability window (see Figure 8.2). Exceeding this window results in electrolyte decomposition: An anode with a μ_A above the LUMO will reduce the electrolyte and a cathode with a μ_C below the HOMO will oxidize it.

Organic electrolytes used in Li-based batteries have a voltage stability window >4.5 V. These are most typically linear alkyl carbonates, such as dimethyl carbonate, diethyl carbonate, or ethyl methyl carbonate, that are unstable below ca. 0.8 V versus Li^+/Li^0 and do form a passivating solid layer at the surface negative electrode that allows successful operation outside their thermodynamic stability window. This layer is usually termed solid electrolyte interphase (SEI)[3] and is electronically insulating but permeable to lithium ions.[4] Although its composition will depend on the

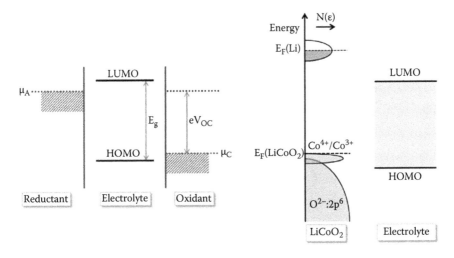

FIGURE 8.2 Relative energies of the electrolyte window Eg and the electrode electro-chemical potentials μA and μC with no electrode electrolyte reaction and schematic energy diagram of μA(Li) and μA(LiCoO₂) and their relative energy positions with respect to the HOMO and LUMO of a conventional carbonate-based electrolyte. (Adapted with permission from Goodenough, J.B. and Park, K.S., *J. Am. Chem. Soc.*, 135, 1167. Copyright 2013 American Chemical Society.)

specific electrolyte salt and solvents, its main components are lithium carbonate and lithium alkylcarbonates, as the most common electrolyte solvents are alkyl carbonates.[5] Similar *surface layers* at the positive electrode due to oxidative decomposition of the electrolyte were detected later and have thus been characterized to a lesser extent.[6,7]

Practical electrodes used in commercial cells are composed of a metal current collector (aluminum for the positive and copper for the negative) onto which is casted a suspension containing the active electrode material and additives that increase electronic conductivity (typically various types of carbon) together with a binder (usually polymeric, such as polyvinylidene fluoride [PVDF]) to improve adhesion, mechanical strength, and ease of processing. Both electrodes are separated by a microporous polyethylene or polypropylene separator film, the whole being impregnated with the electrolyte (see Figure 8.3). While voltage is governed by the properties of bulk electrode materials, the balance between energy and power densities is usually tuned through engineering aspects (such as electrode thickness and amount and type of additives).

Electrode formulation had been in the past largely based on empiric trends and kept as seldom disclosed industrial know-how. This tendency has started to change in recent years, with the topic capturing the interest of academic community. Appealing correlations have been established between the type of polymer used as binder, the suspension rheological properties, the morphology (thickness, tortuosity, etc.) and mechanical properties of the dried composite electrode, and the resulting electrochemical properties.[8–12]

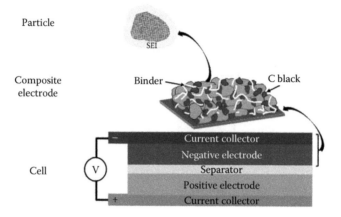

FIGURE 8.3 Schematic view of the full cell architecture and one of the composite electrodes and its components.

The choice of the separator is not as trivial as it may seem. Even if it is not directly involved in the electrochemical reaction itself, its function is critical in preventing internal short circuit while providing a path for ionic conduction in the liquid electrolyte throughout the interconnected porous structure.[13] In addition, many of them are designed with a shutdown feature to enhance battery safety (e.g., with one component having a low melting point than the other and filling the pores and stopping current flow in the cell if that temperature threshold is reached).[14]

All such technological issues are key in the electrochemical performance of the cells under diverse operation conditions but fall out of the scope of this chapter, which is focused on electrode materials. Thus, the reader is referred to the previously cited references for a deeper insight.

8.2 POSITIVE ELECTRODE MATERIALS

Several materials are currently commercialized as positive electrodes in batteries with competing chemistries intended for different applications. All these compounds react with lithium through intercalation reactions, in which lithium ions are topotactically introduced into the host material without major atomic rearrangement. Since in all commercial systems the positive side must act as a source of lithium ions, air-stable lithium-containing compounds are preferred to facilitate cell assembly.[15]

Current Li-ion batteries are assembled in the discharged state and must be charged before use.

Intercalation reactions can be homogeneous (single phase) or heterogeneous (two phase), as depicted in Figure 8.12 in Section 8.3. The former involve the formation of a solid solution whose composition continuously varies throughout the intercalation domain $x_{min} > x > x_{max}$. The host structure does not suffer from major structural changes other than a continuous variation in volume to accommodate the change in composition. Two-phase reactions involve the topotactic nucleation and growth, through interface motion, of the second phase within the pristine one, and it is generally assumed that such process involves extra kinetic barriers with respect to single-phase reactions.[16]

The different positive electrode materials are usually classified according to their crystal structure, and three main families can be found: layered, spinel, and polyanionic compounds.

8.2.1 LAYERED MATERIALS

8.2.1.1 Conventional Layered Materials

Layered lithium transition metal oxides represent the largest category of lithium positive electrode materials currently commercialized. These compounds, with general formula $LiMO_2$ (M = Co, Ni, Mn, or a combination of them), crystallize in a structure that can be viewed as consisting of alternating MO_2 and Li layers, providing 2D lithium diffusion pathways.[17] Depending on the arrangement of the oxygen sublattice, the structure type is named after a nomenclature developed by Delmas et al. in which the alkali ion environment is labeled with a capital letter (usually O for octahedral, P for prismatic, T for tetrahedral) followed by an integer that refers to the number of MO_2 layers included in the unit cell (see examples in Figure 8.4b).[18]

As mentioned earlier, the first positive electrode to reach the market was indeed a layered compound, $LiCoO_2$,[1] and it is still among the most frequently used. Reversible Li^+ removal from $LiCoO_2$ occurs within the range 4.5–3.0 V versus Li^+/Li^0 with a theoretical specific capacity of about 280 mA h g^{-1} (see Ref. [19]) for full delithiation. The maximum practical specific capacity is, however, limited to ca. 140 mA h g^{-1}, which corresponds to the reversible extraction of ≤0.55 Li^+ per formula unit. Higher capacities could be achieved by further Li^+ extraction, but it has detrimental effects upon cycle life that have been found to be rooted in several simultaneous phenomena, namely, structural transformations that might lead to structural collapse when almost all lithium is removed from the electrode,[20] electrolyte electrochemical decomposition at high voltages, and decomposition caused by the poor stability of the material at low lithium contents.[2] The latter represents a major safety issue since it is at the origin of the exothermic formation of oxygen, causing thermal runaway of the cells.[21] The different structural transformations occurring throughout Li^+ extraction are shown in Figure 8.4. The first biphasic domain is found for 0.75 ≤ x ≤ 0.94 in Li_xCoO_2 and corresponds to an insulator to metal transition.[22,23] A second biphasic domain is observed around $Li_{0.5}CoO_2$ and is due to an ordering of lithium ions and vacancies that allows minimizing electrostatic repulsions among lithium ions and leads to a monoclinic distortion of the structure.[22] At higher potential, two different phase transitions occur via layer gliding (see Figure 8.4b), with the formation of $Li_{≈0.15}CoO_2$ (with a structure denoted H1-3 exhibiting AB AB CA CA BC BC oxygen stacking, resulting in a hybrid framework with lithium ions occupying alternate layers: O3 environment for occupied layers and O1 environment for unoccupied layers)[24,25] and CoO_2 (O1 structure, with AB oxygen stacking).[26]

Most commercial materials actually contain a slight excess of lithium, resulting in $Li_{1+x}CoO_2$ compositions that are formed when an excess of Li_2CO_3 is used in the

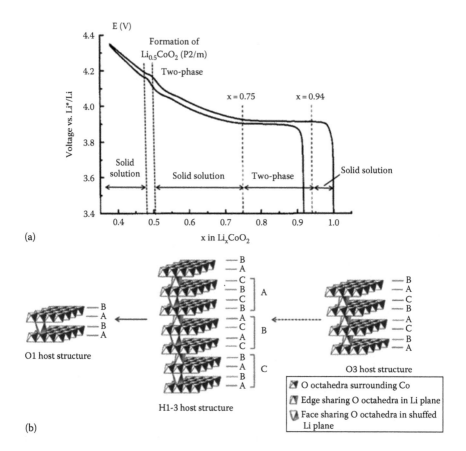

(a)

(b)

FIGURE 8.4 (a) Typical potential–composition curve for $LiCoO_2$. (Adapted from Shao-Horn, Y. et al., *J. Electrochem. Soc.*, 150, A366, Copyright 2003. With permission from The Electrochemical Society.) (b) Schematic illustration of parent $LiCoO_2$-O3, $Li_{\approx 0.15}CoO_2$-H1-3, and CoO_2-O1 host structures. (Adapted from Van der Ven, A. et al., *J. Electrochem. Soc.*, 145, 2149, Copyright 1998. With permission from The Electrochemical Society.)

synthesis. This excess Li_2CO_3 acts as a flux medium that promotes grain growth and pore removal resulting in a safer material in the charged state owing to its lower surface reactivity.[28] In addition, residual Li_2CO_3 in the electrode material provides a safety valve when the cell is overcharged above 4.8 V through the thermally induced formation of CO_2.[29]

The most recent developments on this material are related to the coating of particles with a thin protective layer of phosphate or oxides like Al_2O_3, which enable the use of electrodes consisting of nanoparticles with improved high rate performances, otherwise prevented by their high surface reactivity toward the electrolyte.[30–32]

In order to reduce the high cost of $LiCoO_2$, alternative layered compounds have been developed. The general approach was to replace cobalt with other transition metals like nickel or manganese. This resulted in several compositions such as $LiNi_{0.8}Co_{0.15}Al_{0.05}O_2$ (NCA)[33] and $LiNi_{1/3}Mn_{1/3}Co_{1/3}O_2$ (NMC)[34] that, albeit still containing reduced amounts of cobalt, are now a commercial success. Both materials operate at an average potential of 3.8 V and deliver capacities around 190 and 165 mA h g^{-1}, respectively.

Cobalt-free $LiNiO_2$ was potentially a good candidate to replace isostructural $LiCoO_2$ owing to its lower cost and higher practical capacity (0.7 Li$^+$ per formula unit can reversibly be extracted), but the difficulties to prepare stoichiometric $LiNiO_2$ and its low thermal stability have so far prevented its commercialization. Indeed, despite an almost stoichiometric material being achieved[35] $Li_{1-x}Ni_{1+x}O_2$ is usually obtained since there is a tendency toward loss of lithium and nickel reduction to Ni^{2+}. The excess nickel is located in the lithium layers due to the similarity in ionic radius with lithium, resulting in poor electrochemical performance.[36,37]

Conversely, the two polytypes of $LiMnO_2$ are thermodynamically unstable against transformation to a spinel framework during electrochemical cycling, a step that induces penalties in cycle life.[38–40] Such a transition can be prevented by substituting manganese by other redox-active transition metals such as nickel.[41] Following this approach, $LiNi_{0.5}Mn_{0.5}O_2$ has been widely investigated over the past years.[34,42–44] This material is also isostructural to $LiCoO_2$ with ordering of nickel, manganese, and lithium ions in the transition metal layers.[45,46] The advantage of $LiNi_{0.5}Mn_{0.5}O_2$ with respect to $LiCoO_2$ is that only nickel acts as a (double) redox center while manganese provides stability to the host structure.[34,42,47] The cell potential varies from 4.6 to 3.6 V versus Li$^+$/Li^0 and delivers a capacity around 180 mA h g^{-1}.[48] Structural changes in $LiNi_{0.5}Mn_{0.5}O_2$ upon lithium removal are complex and involve modifications in the cation arrangement and ordering of lithium and vacancies.[49] Li/Ni exchange also affects Li$^+$ mobility as in $LiNiO_2$ although the effects are less dramatic since it has been found that nickel ions are able to migrate upon Li$^+$ removal, facilitating Li diffusion in the interslab space.[49] A material with fewer defects can be obtained by an ion exchange reaction from $NaNi_{0.5}Mn_{0.5}O_2$, which has no structural defects due to the ionic radii mismatch between sodium and nickel ions.[43] The rate capability of $LiNi_{0.5}Mn_{0.5}O_2$ made by ion exchange was found to rival that of the cobalt-substituted materials[43] although the cycle life remains to be improved.

The most practical strategy to increase the layered character of the structure is, however, to maintain a certain amount of cobalt in the structure, as in $Li[Ni_{0.8}Co_{0.15}Al_{0.05}]O_2$ or $Li[Ni_yMn_yCo_{1-2y}]O_2$.[34,37,50–52] The former exhibits an improved thermal stability compared to $LiNiO_2$ with the help of aluminum doping, although it is significantly worse than that of the latter.[53,54] In these materials, the electrochemically active species are predominantly nickel with cobalt playing an active redox role in the later stages of lithium removal. The most studied material of this family is $LiNi_{1/3}Mn_{1/3}Co_{1/3}O_2$ for its excellent electrochemical performance (165 mA h g^{-1} between 3.5 and 4.3 V).[53,55] As in $LiNi_{0.5}Mn_{0.5}O_2$, the oxidation state of manganese ions remains unchanged providing stability. In addition, the partial substitution of cobalt by manganese results in lower cost of the material.

$LiNi_{1/3}Mn_{1/3}Co_{1/3}O_2$ is also isostructural to $LiCoO_2$ although it has been shown that nickel and manganese ions exhibit short-range ordering[56] while the possibility of long-range ordering is still under discussion.[57] The transition to the O1 structure type (as in $LiCoO_2$) has been reported for chemically oxidized samples[58,59]; however, electrochemical lithium extraction from $Li_{1-x}Ni_{1/3}Mn_{1/3}Co_{1/3}O_2$ has been shown to proceed homogeneously in the O3 phase over the entire range of x.[55]

A concentration-gradient material has been recently successfully prepared in order to enhance the capacity of $LiNi_{1/3}Mn_{1/3}Co_{1/3}O_2$ while maintaining a good thermal stability.[60] In this material, the core of the particles is rich in Ni, with a composition close to $Li(Ni_{0.8}Co_{0.1}Mn_{0.1})O_2$, while the concentration of manganese and cobalt increases when approaching the surface getting close to $Li(Ni_{0.46}Co_{0.23}Mn_{0.31})O_2$. With this imaginative design the capacity is increased through the nickel-rich regions (reaching reversible capacities above 200 mA h g^{-1}) while the concentration closer to the surface provides improved thermal stability.

8.2.1.2 Li-Rich Layered Oxides

In the recent years, research efforts have been driven towards a new family of materials derived from layered oxides.[61–63] These materials are usually formulated as solid solutions between two layered materials, $xLi_2MnO_3 \cdot (1-x)LiMO_2$, although they are also often described as Li-rich oxides of the following formula: $Li_{1+(x/(2+x))}M'_{(1-(x/(2+x)))}O_2 (M' = Mn + M)$.[64] These materials exhibit capacities exceeding 250 mA h g^{-1} while operating above 3 V and therefore have received particular attention. The structure and mechanism of these compounds are highly complex and are still the source of intense debate.

The structure of Li_2MnO_3, which can be formulated as $Li[Li_{1/2}Mn_{2/3}]O_2$, is composed of alternating lithium and $[Li_{1/2}Mn_{2/3}]O_2$ layers arranged to form a *honeycomb* ordering scheme (Figure 8.5a), while in $LiMO_2$ lithium layers alternate with MO_2 layers (which in turn can exhibit cationic ordering) as described earlier (Figure 8.5b). Li_2MnO_3 is nearly electrochemically inert since the compound is an insulator and all the manganese ions are in 4+ oxidation state and it is generally admitted that further oxidation is unachievable.[65] The first reports on Li-rich layered oxides described the composite materials as a real solid solution since both components are structurally compatible,[62] although a second model was proposed by Thackeray et al.[66] considering the structure as intergrowths of nanodomains of each of the two phases whose size will depend on the elemental ratio in the $LiMO_2$ component. Since further complexity is brought by the presence of cationic ordering and stacking disorder,[67] and preparation methods and calcination temperatures are probably important factors that influence the structure of the final material, defenders and detractors of each of the models have not yet found a unifying explanation.[64,68–71] Schematic illustrations of both models are shown in Figure 8.5c and d.[69,70]

The potential profile (see Figure 8.6)[72] exhibits a first charge cycle divided into two different regions, an initial sloping region from OCV (ca. 3.5 V) to 4.5 V in which lithium ions are extracted and transition metal ions as nickel and cobalt are oxidized to 4+ with manganese remaining as Mn^{4+}, and then a long plateau at 4.5 V whose length has been related to the amount of lithium in the transition metal layer.[73] After the first charge cycle, the plateau no longer appears and the potential profile becomes sloping throughout the full oxidation. This change seems to be driven by textural and structural modifications that are not yet fully understood.

Interestingly, the high capacities obtained require the participation of the lithium ions located in the transition metal layers (which would first migrate to the

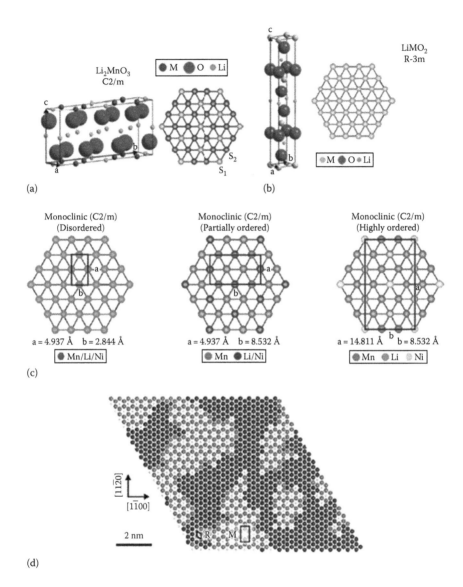

FIGURE 8.5 (a) Monoclinic (C2/m) Li_2MnO_3 unit cell and atomic arrangement in the transition metal layer. (b) Trigonal (R-3m) $LiMO_2$ unit cell and atomic arrangement in the transition metal layer. (c) Proposed atomic ordering in a single transition metal layer of $Li[Li_{0.2}Ni_{0.2}Mn_{0.6}]O_2$ considering a solid solution–type structure for different degrees of ordering of Li, Ni, and Mn atoms. (d) Proposed domain structure of a TM plane in $Li_{1.2}Co_{0.4}Mn_{0.4}O_2$, showing coexistence of Co and $LiMn_2$ domains. In-plane sections of the rhombohedral (R) and monoclinic (M) unit cells are indicated in panels (a through c). (Adapted with permission from Jarvis, K. A. et al., *Chem. Mater.*, 23, 3614, Copyright 2011 American Chemical Society.) (d: Reprinted with permission from Bareño, J. et al., *Chem. Mater.*, 23, 2039, Copyright 2011 American Chemical Society.)

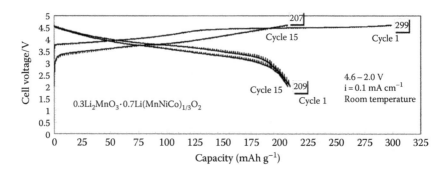

FIGURE 8.6 Electrochemical cycling profiles of the initial and 15th cycles of lithium half-cells, operated at room temperature between 4.6 and 2.0 V, containing a parent $0.3Li_2MnO_3 \cdot 0.7LiMn_{0.333}Ni_{0.333}Co_{0.333}O_2$ electrode. (Reprinted with permission from Johnson, C.S. et al., *Chem. Mater.*, 20, 6095. Copyright 2008 American Chemical Society.)

lithium layers first according to first-principles molecular dynamics calculations)[74] and result equally from cationic and anionic redox processes. Indeed, it has been shown that, concomitant to lithium extraction, O_2 is released during the high potential plateau[75] as previously proposed by Lu and Dahn.[76] The stabilization of the structure would then proceed through two proposed mechanisms. The first hypothesis would involve the diffusion of oxygen ions from the bulk to the surface to sustain the reaction resulting in a structure with oxygen vacancies. The second, and the most generally accepted, would involve a structural rearrangement involving transition metal diffusion from the surface into the bulk and entering the octahedral sites vacated by Li+ ions resulting in a densification of the host structure.[75,77]

Recent works have shed some more light into the mechanism and have shown that, concomitant to the irreversible removal of O_2 in the first cycle, reversible oxidation of O^{2-} occurs together with a partial participation of manganese in the redox process.[65,71,74,78] When Mn[4+] is replaced by a non-redox-active element like Sn[4+], the release of oxygen gas is minimized and the destabilization associated with the creation of O- anions leads to the condensation of O_2^{2-} (peroxo-like species) favored by the low bonding of Sn[4+].[79]

Upon cycling the average discharge potential progressively decreases, indicative of different local environments due to cation migration despite the oxygen substructure being preserved,[65] which could be related to a layered to spinel transformation.[80] Mn migration from the surface to the bulk occupying vacant lithium sites has also been observed after spinel formation at the surface, resulting in a nonoverlithiated layered oxide with cationic segregation.[81] It has been shown that the potential decay can be suppressed with substitution of large cations like Sn[4+] instead of Mn[4+].[79]

Other variations of the materials described earlier have also been investigated and include the replacement of Li_2MnO_3 by related compounds such as Li_2TiO_3 and Li_2ZrO_3, or the replacement of $LiMO_2$ by a spinel-type compound like $Li_{1+x}Mn_{2-x}O_4$[82–84] or even by another Li_2MnO_3-type compound like Li_2RuO_3 or Li_2SnO_3, in which case the composite material structurally organizes as a classical solid solution.[65,79,85]

This family of materials is still in a prospective research stage despite their attractive electrochemical capacities. Indeed, initial coulombic inefficiencies related to the amount of Li_2MnO_3 component,[79] stability upon cycling, and rate performance still need to be addressed to satisfy the market requirements. Voltage degradation represents a serious drawback since it can largely affect the energy output of the battery and cannot be mitigated with surface coatings.[64] Despite the promising results obtained in recent investigations, the chemical complexity of these composites still requires further understanding.

8.2.2 SPINEL-TYPE MATERIALS

As mentioned in the previous section, $LiMn_2O_4$ crystallizes in the spinel structure (Figure 8.7a). This compound (and its doped variants) is also a well-established commercial lithium-ion battery cathode material. Removal of lithium to form $Li_{1-x}MnO_2$ occurs at ≈4 V versus Li^+/Li^0 (Figure 8.7b), with a capacity of about 120 mA h g^{-1} considering the extraction/insertion of 0.8 Li^+ per formula unit. It has been shown that a reduction of the cell mismatch between the two end members through cationic substitution of manganese by nickel and lithium results in improved capacity retention.[86] The potential profile of $LiMn_2O_4$ exhibits two plateaux separated by a step of 100 mV at x ≈ 0.5. At this composition, an intermediate phase with charge ordering of the Mn ions along with long-range ordering of the lithium ions is formed.[87–89]

Below 3 V versus Li^+/Li^0 (Figure 8.7b), additional lithium can be inserted into interstitial octahedral vacancies in $LiMn_2O_4$ to form the Jahn–Teller distorted tetragonal phase $Li_2Mn_2O_4$, but the strong distortion induced by the formation of Mn^{3+} strongly damages the electrode material and the reaction is therefore avoided.[91] Despite working on the same Mn^{4+}/Mn^{3+} redox couple, this 1 V difference between lithium insertion and lithium extraction has its origin in the different sites occupied by lithium atoms in the structure (tetrahedral 8a sites in $Li_{1-x}Mn_2O_4$ and octahedral

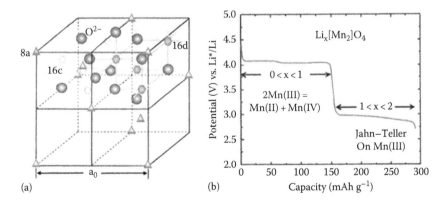

FIGURE 8.7 (a) Two quadrants of the spinel structure where the positions of Mn (16d) and Li^+ in $Li_{1-x}Mn_2O_4$ and $Li_{1+x}Mn_2O_4$ (8a and 16c, respectively) are indicated. (b) Potential profile of $Li_xMn_2O_4$. (Adapted with kind permission from Springer Science+Business Media: *J. Solid State Electr.*, 16, 2012, 2019, Goodenough, J., B. Copyright Springer-Verlag 2012.)

16c sites in $Li_{1+x}Mn_2O_4$). It has also been shown that close to room temperature, at 290 K, cubic $LiMn_2O_4$ transforms into an orthorhombic phase with partial charge ordering of Mn^{3+} and Mn^{4+} associated with a loss of conductivity.[92]

While the $LiMn_2O_4$ spinel has several advantages (such as an excellent power capability thanks to the 3D diffusion of lithium ions within the structure, the fact that manganese is inexpensive and environmentally benign and the better thermal stability), its weakness is capacity fading. Its origin is manganese dissolution in the electrolyte through the disproportionation of Mn^{3+} into Mn^{2+} and Mn^{4+} causing a progressive capacity fade as Mn^{2+} leaches out into the electrolyte in the presence of acidic species (i.e., trace amounts of HF formed by decomposition of $LiPF_6$ in the electrolyte).[93] Diffusion of Mn^{2+} toward the negative electrode might also result in safety hazards as the resistance of the SEI increases and causes a local increase in temperature that can induce thermal runaway. It has been shown that substitution of Mn by Li[94] or Al[95] in order to reduce the Mn^{3+}:Mn^{4+} ratio greatly improves the stability of the material, although at the expense of a decrease in capacity. In order to compensate for the lower capacity, double substitutions $LiAl_yMn_{2-y}O_{4-z}F_z$ emerged to achieve optimum electrochemical performance.[96] By finely tuning the substituent contents, both manganese dissolution and the formation of $Li_2Mn_2O_4$ on the surface of the material at the end of the 4 V plateau are minimized.[48] Another strategy to avoid manganese dissolution is to protect the material with a surface coating of inert oxides such as Al_2O_3 in order to avoid direct contact with the electrolyte.[97]

Mixed with layered oxides like $LiNi_{1/3}Mn_{1/3}Co_{1/3}O_2$ to increase the capacity, $LiMn_2O_4$ is now a leading candidate for automotive applications and already powers several models of hybrid and electric cars.[98] The approach of blending two cathode materials is currently a developing field since the resultant blend can be tailored to have a higher energy or power densities (provided, for instance, by a layered cathode material) coupled to a lower cost and enhanced stability (provided by 3D structures like $LiMn_2O_4$ or $LiFePO_4$; see Section 8.2.3).[99]

On the other hand, research attention has more recently turned into the substituted $LiMn_{1.5}Ni_{0.5}O_4$ spinel.[100,101] In this material, although all manganese ions are in 4+ oxidation state (and the couple Mn^{5+}/Mn^{4+} is inaccessible), nickel atoms, in 2+ oxidation state, can deliver two electrons and reach 4+ oxidation state when Li^+ ions are extracted from the structure. Hence, as $LiMn_2O_4$, $LiMn_{1.5}Ni_{0.5}O_4$ still provides one mobile Li^+ for two framework cations while the parasitic effects of Mn^{3+} are avoided. It exhibits a theoretical capacity of 147 mA h g^{-1} with both nickel oxidation processes occurring at 4.70–4.75 V versus Li^+/Li^0 and being separated by a small step (100 mV). The resulting energy density (690 W h kg^{-1}) is the highest among commercially available cathode materials (560 W h kg^{-1} for $LiCoO_2$, 480 W h kg^{-1} for $LiMn_2O_4$, 560 W h kg^{-1} for $LiFePO_4$, and 630 W h kg^{-1} for $LiNi_{1/3}Mn_{1/3}Co_{1/3}O_2$), and the cell voltage when coupled with a graphite anode almost reaches 5 V. Nonetheless, several drawbacks must still be overcome before commercialization of this material can be envisaged, namely, capacity fade during cycling and the formation of a rock salt impurity phase during synthesis.[98,102]

It is generally accepted that the main origin of capacity fading upon cycling is the formation of an unstable layer at the surface of the particles.[98] The operation potential of $LiMn_{1.5}Ni_{0.5}O_4$ lies above the HOMO of the electrolyte, and hence, it decomposes forming a surface layer that alters the normal operation of the material, especially at elevated temperatures (55°C).[98] Two different strategies can be followed to create a stable passivating solid layer on the surface of the cathode[2]: identify additives that assist the formation of an insoluble Li^+-conducting protective layer on the electrode surface or build a coating on the material with an oxide that is permeable to lithium ions. It has been shown that doping with trivalent ions like Fe^{3+} can be a low-cost alternative to surface coating, which besides adding processing costs might not be completely effective unless complete uniformity is achieved. Indeed, these cations self-segregate during the synthesis resulting in a material whose surface is iron enriched and nickel deficient.[103] The lower catalytic activity of iron toward electrolyte decomposition when compared to nickel suppresses the formation of thick passivating layers and enhances the electrochemical performance.[98]

8.2.3 Polyanionic Materials

$LiFePO_4$ is now considered one of the best available positive electrodes because of its cycling performance, safe operation, low cost, and nontoxicity. This material crystallizes in the olivine structure from which Li^+ ions can be reversibly extracted at 3.5 V versus Li^+/Li^0 yielding a capacity of 170 mA h g^{-1}.[104] Despite the intrinsic advantages of such a low cell voltage in terms of safety, this limits the energy density and is considered the main drawback of $LiFePO_4$ together with its low conductivity. In order to overcome the kinetic limitations that result from the poor transport of electrons and ions in $LiFePO_4$ two strategies are generally used in combination. The first one is coating the particles of $LiFePO_4$ with carbon in order to enhance electronic conduction.[105] The second one refers to the decrease in the particle size so as to reduce the diffusion lengths for Li ions, an approach that has a strong impact on this material since $LiFePO_4$ is a 1D Li-ion conductor.[106–108]

Although the Fe^{3+}/Fe^{2+} couple in a simple oxide would normally operate at a potential <2.5 V versus Li^+/Li^0, polyanion hosts of the XO_4^{n-} type (X = S, P, Mo, W) induce a lowering of the Fe^{3+}/Fe^{2+} redox energy and thus an increase in the potential versus Li^+/Li^0. This phenomenon, denoted inductive effect, is attributed to the influence of the strength of the X–O bonding on the covalency of the Fe–O bond (and thereby on the position of the Fermi level). When the X–O bond is the strongest, the cell voltage is higher.[104]

The lithium insertion/extraction mechanism in $LiFePO_4$ is without a doubt the most studied electrochemical process among Li-ion battery systems. The large amount of literature available on this field has been recently compiled by Malik et al.[109] in an excellent review that rationalizes the diversity of results encountered by different research groups. The ensemble of the works related to this active field represent a step forward toward the understanding of electrochemical systems based on intercalation reactions, and therefore, the most important aspects of $LiFePO_4$ reaction mechanism are herein summarized.

Since electrochemical activity in $LiFePO_4$ was first reported by Padhi et al.,[104] it has been shown that this material exhibits a complex phase transition mechanism that depends strongly on the microstructure (size, morphology, defects) and is still the source of exciting debate. In the first report, a two-phase process was proposed together with a shrinking core model for diffusion (lithium insertion proceeds from the surface of the particle moving inward behind a two-phase interface whose surface area shrinks upon lithiation and grows back upon extraction). $LiFePO_4$ represents thus an exception, since most commercialized cathode materials such as $LiCoO_2$, $LiMn_2O_4$, NMC, and NCA form solid solution domains over large concentration ranges. It was later shown that solid solutions of Li_xFePO_4 throughout the range $0 < x < 1$ could be driven by temperature ($350°C–400°C$)[110] and that, despite the extremely flat charge/discharge profile, two solid solution regions ($0 < x < \alpha$, $1 - \beta < x < 1$) existed outside the miscibility gap (i.e., the biphasic region) at room temperature.[111] Li_xFePO_4 compositions inside the miscibility gap should thus be described as a mixture of $Li_\alpha FePO_4$ and $Li_{1-\beta}FePO_4$ phases. Soon after it was shown that the miscibility gap is size dependent (reduced for nanoscale materials) and it was speculated that the miscibility gap could completely disappear below a critical size.[112] A full solid solution reaction was indeed achieved with nonstoichiometric particles of 40 nm with microstructural defects.[113] The enhanced solubility of lithium in smaller particles is explained by the different lattice parameters of the two end members $FePO_4$ and $LiFePO_4$. Indeed, when the ratio between the interfacial area and the particle volume increases, the interfacial energy penalty caused by lattice mismatch strains destabilizes the two-phase coexistence in smaller particles.[114] The average composition of the lithium-rich and lithium-poor phases has also been shown to vary within the miscibility gap, more pronouncedly in smaller particles.[115,116] This effect is attributed to the confinement of the concentration gradient that takes place at large fractions of the material for small particles, influencing the overall compositions of the two coexisting phases.[116]

Studies on $LiFePO_4$ have also demonstrated that particles of different sizes exhibit different chemical potentials due to the particle size dependence of the interfacial energy penalty.[117] This difference drives lithium exchange between particles of different size and composition in the absence of electrical current.[118,119] Since electrodes are multiparticle systems, lithium exchange allows avoiding the energy penalty of creating an interface and the system evolves toward a more stable state with fully intercalated and fully deintercalated $LiFePO_4$ particles[109,120] (see Figure 8.8) as has been experimentally shown and termed as *domino cascade* process.[121] The fact that a multiparticle system thus contains several equilibria also accounts for the occurrence of an additional potential hysteresis that does not vanish when the current is decreased.[120]

Lithium diffusion has been shown to be mostly 1D along [010][122–124] (see Figure 8.9a), and therefore, this is the direction along which size should be minimized in order to improve kinetics by reducing lithium diffusion lengths. Li–Fe antisite defects (i.e., when Li and Fe atoms exchange positions) are common in $LiFePO_4$[123] and nanosizing also allows minimizing the effects of channel blocking by two or more point defects (see Figure 8.9b).[125] The obstructing presence of antisite defects also enhances lithium crossover to adjacent channels changing the lithium migration from one to two or three dimensions (Figure 8.9c).[125] The shrinking core model for diffusion initially proposed by Padhi et al.[104] did not account for the anisotropy in lithium motion and a new core–shell model has been proposed. According to this

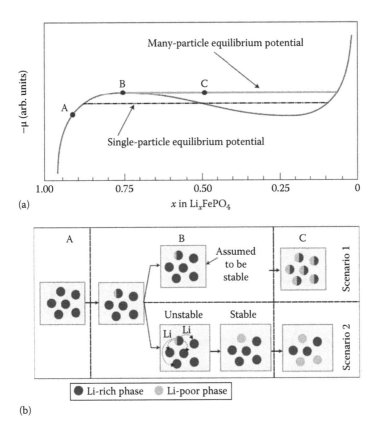

(a)

(b)

FIGURE 8.8 Two possible scenarios of new phase formation in a multiparticle system. (a) The main difference between the single- and multiparticle equilibrium potential curve. (b) Schemes A, B, and C describe roughly the possible situation at points A, B, and C in graph (a). Although Scenario 1 is widely accepted in the literature, Scenario 2 is much more likely. (Reprinted by permission from Macmillan Publishers Ltd. *Nat. Mater.*, Dreyer, W. et al., 9, 448, Copyright 2010.)

model, lithium insertion proceeds from the surface of the particle as in the shrinking core model, but lithium extraction originates in the center of the particles, with the interface moving outward. Hence, the particles always maintain a structure with a core of $LiFePO_4$ and a shell of $FePO_4$.[126]

The first report on $LiFePO_4$ already drew the attention to isostructural $LiFe_{1-x}Mn_xPO_4$ compositions, in which the Mn^{3+}/Mn^{2+} couple is located at a 4.1 V versus Li^+/Li^0[104] resulting in a significant increase in energy density. Further reports showed that increasing manganese substitution induces severe kinetic limitations that arise from a lower conductivity and lattice frustration effects due to the Jahn–Teller effect in charged $LiFe_{1-x}Mn_xPO_4$ electrodes.[127,128] Other reports proposed replacement of Fe by Co, resulting in a cell voltage of 4.8 V versus Li^+/Li^0.[129,130] While the energy density is significantly increased, the high operation potential results in severe electrolyte degradation, impeding practical application.

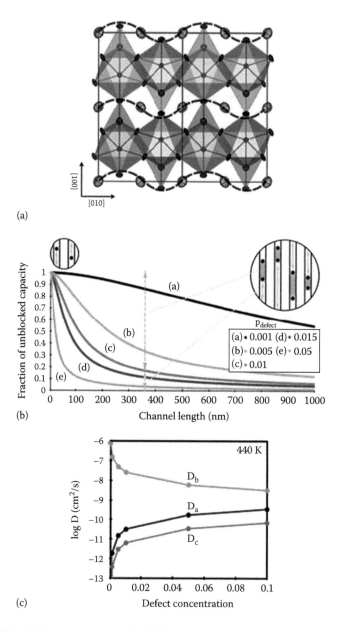

FIGURE 8.9 (a) Crystal structure of LiFePO$_4$ with Rietveld refined atomic displacement ellipsoids and expected curved 1D continuous chains of lithium motion drawn as dashed lines. (Reprinted by permission from Macmillan Publishers Ltd. *Nat. Mater.*, Nishimura, S. et al., 7, 707, Copyright 2008.) (b) Calculated unblocked capacity versus channel length in LiFePO$_4$ as determined for various defect concentrations. (c) Calculated variation of the Li-vacancy self-diffusion with defect concentration along the different crystallographic directions at 440 K. (b and c: Reprinted with permission from Malik, R. et al., *Nano Lett.*, 10, 4123. Copyright 2013 American Chemical Society.)

After the successful discovery of $LiFePO_4$, a large amount of polyanionic compounds have been reported in the recent years, resulting from research efforts aiming at either increasing the capacity with the use of more than one electron per 3D metal or enhancing the cell voltage making use of the inductive effect.

The possibility of exchanging two electrons per transition metal has concentrated significant interest in transition metal silicates of formula Li_2MSiO_4 (M = Fe, Mn, Co). This family of polyanionic materials offer a rich crystal chemistry with complex polymorphism that results from subtle connectivity variations between Li^+, Si^{4+}, and M^{2+} tetrahedra.[129] Li_2FeSiO_4 has drawn the most attention and three different polymorphs (see Figure 8.10a through c)[131] have been prepared with varying synthesis conditions.[132–135] All three polymorphs exhibit similar electrochemical properties, with slight differences in operating potential owing to variations in the FeO_4 arrangements (orientation, size, and distortion).[134] Practical capacities of 160 mA h g^{-1} are usually obtained, corresponding to the extraction of only one lithium ion per formula unit.[132] Extraction of the second lithium involving the Fe^{3+}/Fe^{4+} couple has been recently reported for the first time at a potential of 4.7 V when cycled at 55°C,[136] as predicted from theoretical calculations[137] (although it has never been achieved at room temperature), with a practical capacity of 204 mA h g^{-1} corresponding to the reversible insertion and extraction of 1.5 Li^+ ions per formula unit. A lowering of the potential from ≈3.0 to ≈2.8 V versus Li^+/Li occurs in all polymorphs after the first cycles due to a structural rearrangement as initially proposed from in situ X-ray diffraction,[132] although the phase transformation kinetics is different for each polymorph.[134] The structure of the phase formed has been recently elucidated when starting from the $P2_1/n$ polymorph and involves inversion of half the SiO_4, FeO_4, and

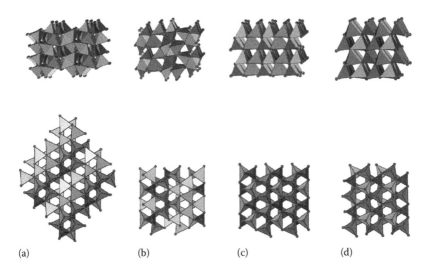

(a)   (b)   (c)   (d)

FIGURE 8.10 Structures of Li_2FeSiO_4 polymorphs showing two orthogonal views. (a) γ_s structure (space group $P2_1/n$), (b) γ_{II} structure ($Pmnb$), (c) β_{II} structure ($Pmn2_1$), and (d) inverse-β_{II} structure ($Pmn2_1$). (Reprinted with permission from Eames, C. et al., *Chem. Mater.*, 24, 2155. Copyright 2012 American Chemical Society.)

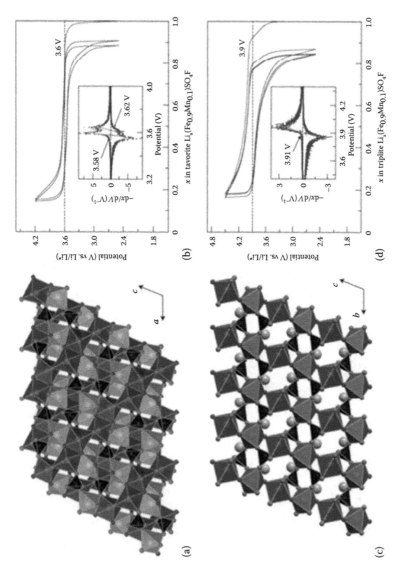

FIGURE 8.11 (a) Crystal structure of tavorite LiFeSO$_4$F, (b) potential–composition curve of tavorite LiFeSO$_4$F, (c) crystal structure of triplite LiFeSO$_4$F, and (d) potential–composition curve of triplite LiFeSO$_4$F. (Adapted by permission from Macmillan Publishers Ltd. *Nat. Mater.*, Barpanda, P. et al., 10, 772, Copyright 2011.)

LiO_4 tetrahedra (see Figure 8.10d).[138] Li_2MnSiO_4 has also been shown to reversibly react with lithium delivering a discharge capacity of 210 mA h g^{-1} at room temperature although it suffers from poor rate capability and drastic capacity fade. The latter are attributed to Jahn–Teller distortion, structural instability and low electronic conductivity.[136] On the other hand, the electrochemical response of Li_2CoSiO_4 is almost negligible.[139]

With the combination of the inductive effect of the PO_4^{3-} group and the high electronegativity of F^-, fluorophosphates have also been recently investigated as cathode materials. $LiFePO_4F$, isostructural with tavorite $LiFePO_4(OH)$, reversibly intercalates lithium at 3 V versus Li^+/Li^0, providing a specific capacity very close to the theoretical one (152 mA h g^{-1}).[140,141] Lithium extraction from $LiFePO_4F$ is not feasible as this would involve oxidation of Fe^{3+} to Fe^{4+} at a too high potential. Other transition metal fluorophosphates with different structures have been explored (M = V, Mn, Ni, Co) with interesting electrochemical properties.[128]

When polyanionic compounds with the PO_4F^{4-} group are stabilized in M^{2+} oxidation state, a different structure is adopted. Li_2FePO_4F can be prepared by ionic exchange from Na_2FePO_4F and crystallizes in a layered structure.[142] This material reversibly intercalates lithium at an average potential of 3.5 V. The crystal structures of Li_2NiPO_4F and Li_2CoPO_4F are 3D.[128] The latter reversibly intercalates lithium delivering a capacity of 60 mA h g^{-1} when the potential limit is fixed at 5.0 V and 110 mA h g^{-1} when increased to 5.5 V,[143] well beyond the stability window of conventional organic electrolytes.

Substitution of the phosphate group PO_4^{3-} by a sulphate group SO_4^{2-} allows achieving higher redox potentials as in $LiFeSO_4F$. This material, with a theoretical capacity of 151 mA h g^{-1}, crystallizes in two different polymorphs. It was first synthesized with a tavorite structure (see Figure 8.11a and b), with lithium extracted at 3.6 V versus Li^+/Li.[144] Good electrochemical performance, even when cycling at high current densities, and excellent capacity retention are obtained without any specific carbon coating and/or downsizing of particles and, despite its strong reactivity with moisture, is considered a promising positive electrode material.[128] Isostructural $LiMSO_4F$ (M = Mn, Ni, Co) have also been prepared although no electrochemical activity was found when cycling up to 5 V, most likely due to the higher potential of the M^{3+}/M^{2+} redox couples.[145] It has been recently discovered that, depending on the reaction conditions used, $LiFeSO_4F$ can also be stabilized in the triplite structure (see Figure 8.11c and d). This polymorph reacts versus Li^+/Li^0 at 3.9 V, which is the highest potential ever reported for the Fe^{3+}/Fe^{2+} couple.[146] Theoretical density functional theory (DFT)+U calculations conclude that the difference in potential between both polymorphs lies in the different electrostatic repulsions induced by the configuration of the fluorine atoms around the transition metal in the two compounds (*trans*- vs. *cis*-configurations in tavorite vs. triplite polymorphs).[147]

8.3 NEGATIVE ELECTRODE MATERIALS

The negative electrode materials present in current commercial batteries are mostly carbonaceous, although $Li_4Ti_5O_{12}$ has also reached practical application. The redox mechanism for such cases is similar to the one exhibited by the positive electrode materials mentioned earlier and rooted in reversible topotactic insertion of lithium ions into the crystal structure, and thus, the degree of lithium uptake (and hence the electrochemical capacity) is limited by the structural changes that the crystal structure is able to withstand.

Alternatively, other redox reaction mechanisms allow achievement of higher capacities (see Figure 8.12). Metals and semimetals that can electrochemically form alloys with lithium have been investigated for long as the capacities derived from alloying reactions can reach extremely high values (e.g., 8365 mA h cm^{-3} and 3590 mA h g^{-1} for silicon compared to 975 mA h cm^{-3} and 372 mA h g^{-1} for graphite). However, the practical utilization of alloy-based electrodes has been hindered by the large volume changes induced by the alloying process itself.[149] These are the source of a huge increase in the volume of the electrode with concomitant loss of adhesion and cohesion. Diverse strategies have been designed to partially circumvent this issue, but although negative electrodes containing a minor amount of alloying materials have been commercialized that yield some improvements in capacity with respect to graphite, the achievement of good cycle life for fully alloy-based electrodes remains a challenge.

A second alternative to reaction mechanisms based on insertion is rooted on the so-termed *conversion reaction*. It is exhibited by binary compounds of general formula M_aX_b, X being a nonmetallic element (X = O, S, N, P, H, F…) that reacts with lithium according to $M_aX_b + (b \cdot n)Li \rightleftharpoons aM + bLi_nX$ (where n is the formal oxidation state of X anions) involving full reduction of the metal ion (and hence large capacities) to yield metallic nanoparticles embedded in a matrix of a lithium binary compound.

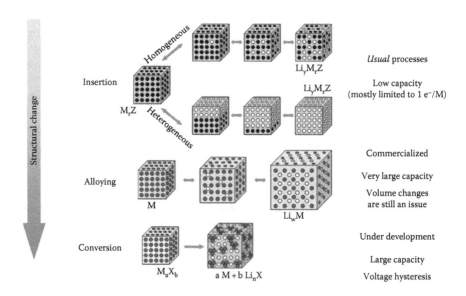

FIGURE 8.12 Schematic representation of different reaction mechanisms observed in electrode materials for lithium batteries. Black circles, voids in the crystal structure; dark gray circles, metal; light gray circles, lithium. (Palacín, M. R., *Chem. Soc. Rev.*, 38, 2565, 2009. Reproduced by permission of The Royal Society of Chemistry.)

This topic has captured a great deal of attention in the recent years, although current fundamental bottlenecks in energy efficiency seem difficult to overcome.

8.3.1 REDOX MECHANISM BASED ON LITHIUM INSERTION

8.3.1.1 Carbonaceous Materials

Carbonaceous materials can be considered as the *standard* negative electrode in the lithium-ion battery technology and are yet, by far, the most commercially used. Many types of carbon materials are available and the intercalation process is strongly influenced by macro- and microstructural features including the specific surface, surface chemistry,[150] morphology, crystallinity, and orientation of the crystallites.[151] Graphite exfoliation related to solvent cointercalation for conventional propylene carbonate (PC)-based electrolytes resulted in the use of hard carbon in the first generation of commercial batteries. Nonetheless, the use of graphite electrodes was soon enabled by the development of ethylene carbonate (EC)-based electrolytes, which form an effective SEI that prevents exfoliation.

Carbonaceous negative electrode materials present a large variety of degrees of graphitization and layer ordering. In addition to the pure hexagonal 2H (ABABAB) graphene layer stacking, graphites may also exhibit some rhombohedral 3R (ABCABC) stacking and are commonly described by giving the relative fractions of 2H, 3R (always less than 40%), and random stacking. Disordered stacking arrangements include both the turbostratic, when graphitic planes are still parallel but shifted or rotated, and the case in which the planes are not parallel, commonly termed unorganized carbon. Highly disordered carbons achieve larger lithium intercalation values, but their higher irreversible capacity has resulted in commercial batteries being mostly based on graphitic carbon. In this case, lithium uptake yields LiC_6 (with 372 mA h g^{-1} and 975 mA h cm^{-3} gravimetric and volumetric capacity) and is concomitant to a transformation to AAAA stacking.

All carbon materials can be lithiated to a certain extent, but the amount of lithium reversibly incorporated in the carbon lattice (the reversible capacity), the faradaic losses during the first cycle (the irreversible capacity) and the potential–composition profile can exhibit some differences.[152] Current commercial carbonaceous negative electrode materials can roughly be classified into three categories[153]: hard carbon and natural and synthetic graphite.

Natural graphite, obtained from different natural ores, exhibits some advantages such as low cost, low and flat potential profile, and high coulombic efficiency but usually exhibits low rate capability derived from its high anisotropy. Synthetic graphite is more expensive because of the high temperature needed for its preparation from soft carbon precursors (>2800°C) yet widely used in different forms and textures: mesocarbon microbeads (MCMB; see Figure 8.13a), mesophase carbon fiber (MCF), vapor-grown carbon fiber (VGCF; see Figure 8.13b), massive artificial graphite (MAG),[154] etc. Hard carbons can deliver high capacity since the random alignment of small-dimensional graphene layers provides significant porosity able to accommodate lithium,[155] yet the rate capability is usually limited and the irreversible capacity is higher than that of graphite while its volumetric capacity is penalized by a lower density. Moreover, its lower operation potential entails a certain risk of lithium

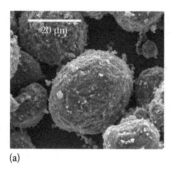 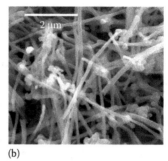

(a)   (b)

FIGURE 8.13   Scanning electron microscopy images depicting a typical microstructure for (a) MCMB and (b) VGCF synthetic graphite electrode materials.

plating at high rates due to larger cell polarization, which could derive into dendrite formation and exothermic shortcut.

The intensive research aimed at developing carbonaceous negative electrode materials with optimized performance in the past has resulted in a deep understanding of the key factors that control carbon's electrochemical performance. As a result of that, current research is mostly technological in character and aimed at specific improvement of existing industrial products.

8.3.1.2   Transition Metal–Based Materials

The capacity values exhibited by carbonaceous materials (significantly higher than those of positive electrode materials) combined to their more affordable prices were the source for a smaller quest in the field of insertion materials for negative electrodes than for positive ones. The number of alternative insertion negative electrodes based on transition metals is limited because of the need of a high potential difference between positive and negative electrodes. As a result of that, the number of suitable compounds is relatively limited and they are mainly based on the Ti^{4+}/Ti^{3+} redox couple.[156] Most of them show insertion potentials between 1.5 and 2 V versus lithium, and although a significant amount of research has been devoted to different TiO_2 polymorphs,[157,158] only $Li_4Ti_5O_{12}$ has reached the commercial stage.

The higher operation potential induces a severe penalty in energy with respect to carbonaceous anodes but does also have some positive attributes: The risk of lithium plating is suppressed and it enables the use of aluminum current collectors for the negative electrode (which would form alloys with lithium at low potential) instead of copper that is heavier and more expensive.

$Li_4Ti_5O_{12}$ exhibits very small changes in the lattice volume upon lithium intercalation/deintercalation[159] and has hence been termed *zero strain material*. This compound is a defect spinel that can be described as $[Li]_{8a}[Li_{1/3}Ti_{5/3}]_{16d}[O_4]_{32e}$ and intercalates lithium at 1.55 V versus Li^+/Li^0 with a theoretical capacity of 175 mA h g^{-1} according to

$$Li_4Ti_5O_{12} + 3Li^+ + 3e^- \rightarrow Li_7Ti_5O_{12}$$

The good performance of this material at high charge/discharge rates has been widely reported. Though nanopowders deliver better power performance,[160] high power rates are achieved even with micron-sized particles of $Li_4Ti_5O_{12}$. Since $Li_4Ti_5O_{12}$ is insulating, intensive research aiming at increasing its electronic conductivity has been carried out. Yet it has recently been shown that excellent high-rate cycling performance can be achieved without even the need of conducting additives, due to the formation of electronically conductive pathways concomitant to reduction of titanium (see Figure 8.14), which has important implications in maximization of the energy density of the electrodes through minimization of the amount of their inactive components.[161,162]

In contrast to the previously reported positive attributes, recent concerns have raised about the presumed stability of titanium-based compounds. Indeed, their well-known catalytic activity may result in undesired reaction with the electrolyte at unexpectedly high potential values.[163–165] Reports on gassing for large cells containing $Li_4Ti_5O_{12}$ negative electrode materials[166] have been rationalized through experiments proving electrolyte solvent reactivity with $Li_4Ti_5O_{12}$ even in the absence of any applied current, which results in the formation of H_2, CO_2, and CO.[167] Fortunately, surface modification of $Li_4Ti_5O_{12}$ through coating with carbon seems to be a simple and very effective strategy to suppress such reactivity, and hence, large-scale application of this compound does not seem to be much compromised.

Aside from the phases mentioned earlier, vanadium compounds have also been investigated in some detail in recent years after reports by Samsung of the possible use of lithium vanadium oxide ($Li_{1.1}V_{0.9}O_2$) as a negative electrode material.[168] It operates at a very low potential (close to 0.1 V vs. Li^+/Li^0) and exhibits significant capacity, which, coupled to its larger density, results in a theoretical volumetric capacity of 1369 mA h cm^{-3} compared to 975 mA h cm^{-3} for graphite. Interestingly,

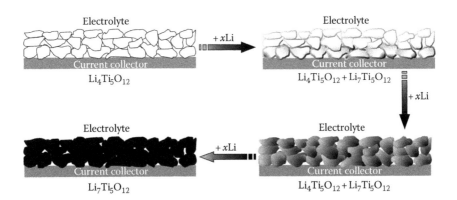

FIGURE 8.14 Scheme for the proposed lithiation mechanism in carbon-free $Li_4Ti_5O_{12}$ electrodes. In the early stage of reduction, the highly conducting phase $Li_{4+x}Ti_5O_{12}$ is formed in the vicinity of the current collector. The early propagation of this phase across all active particle surfaces allows eventual lithiation of the whole electrode. (Kim, C., Norberg, N., Alexer, C., Kostecki, R., and Cabana, J.: *Adv. Funct. Mater.* 2013. 23. 1214. Copyright Wiley-VCH Verlag GmbH & Co. KGaA. Reprinted with permission.)

whereas lithium cannot be intercalated in stoichiometric $LiVO_2$, substituting as little as 3% of the vanadium by lithium renders tetrahedral sites in the alkali metal layers energetically accessible by lithium and promotes a two-phase intercalation process to form Li_2VO_2. While electrolyte reduction upon the first cycle is also observed, its behavior upon cycling does not seem to be compromised. Nonetheless, further reports on performance, especially in full cells against standard positive electrode materials and cost estimations, would be needed in order to ascertain its prospects of practical use.

8.3.2 ALLOYING-TYPE MATERIALS

It has been known for long that alloying reactions of metallic lithium with metallic or semimetallic elements are electrochemically feasible in a lithium cell at room temperature in a nonaqueous-based electrolyte.[169] The huge theoretical capacities expected for the formation of the Li-rich Li_xM binaries are very appealing (see Figure 8.15). All alloys presented in the figure yield volumetric capacities at least twice than that of graphite, and hence, all of them can be considered of future interest. However, if we take into account not only theoretical capacity values but also abundance, price, and toxicity, silicon and tin seem to be the most suitable choices.[170] In contrast to a generic insertion process of Li^+ ions in an open oxide framework ($MO_2 + Li^+ \rightarrow LiMO_2$) that comes with a 33% rise in the number of atoms, a dense metallic particle such as an Al particle has to accommodate one Li per Al ($Li + Al \rightarrow LiAl$: 100% rise in atom count), while Sn atoms in a Sn particle undergo a drastic *dilution* by a factor of 5 ($4.4Li + Sn \rightarrow Li_{4.4}Sn$: atom population × 5.4). As a consequence of this, the unit cell volume increases drastically and this results in electrode disintegration. However, this fact should by no means be considered as an

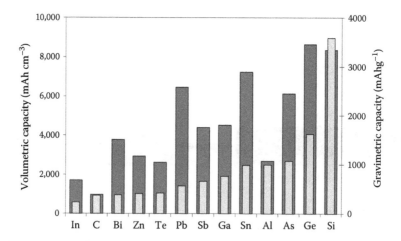

FIGURE 8.15 Specific gravimetric and volumetric capacity and capacity density for selected alloying reactions. Values for graphite are given as a reference. (Palacín, M. R. *Chem. Soc. Rev.*, 38, 2565, 2009. Adapted by permission of The Royal Society of Chemistry.)

absolute hitch for their practical application. Two of the most efficient electrochemical systems (lead/acid and Ni/Cd) involve redox couples entailing large volume expansions (50% for $PbO_2/PbSO_4$, 120% for $Pb/PbSO_4$, and 130% for $Cd/Cd(OH)_2$) and, despite this fact, can successfully sustain prolonged cycling. Thus, electrode design, formulation, and processing are key issues to be addressed. Definite data/experiment is lacking that provides evidence as to whether the pulverization of the particles during lithium alloying is either due to large volume changes or associated with important structural stresses (i.e., grain boundaries and phase transitions) during alloying. So far, it can hence be considered that the loss of electrical contact results from the combined effects of both phenomena. The strategies used to circumvent these issues[171] are summarized in the following:

1. *Amorphous materials*: Even though based on a simple stacking model an amorphous metallic material will always have lower density than the crystallized form, the gain in *free space* is not enough to significantly reduce the extent of the volumetric changes during large Li^+ uptake. However, the intrinsic absence of phase transitions and isotropic expansion seem to have a very positive impact on the cycling performance. This was nicely shown with amorphous binary Si–Sn films, the trick being that materials have to remain amorphous upon cycling.[172]

2. *Porous particles*: The volume change in dense materials might be accommodated by using porous particles in which the available voids could be filled during expansion. A way to prepare such materials would be extracting M from Si_xM alloys mimicking the process applied for the synthesis of Raney's nickel catalyst. Similarly, by using electrodes consisting of nanowires directly deposited onto the current collector, easy strain relaxation allows them to increase in diameter and length without breaking.[173] This has been reported for silicon, with very high capacities (ca. 3500 mA h g^{-1} at C/5 and 2100 mA h g^{-1} at C) that are stable for 20 cycles.[174] In this case, and since every nanowire is connected to the current collector, the need for binders or conducting additives, which add extra weight, is eliminated.

3. *Small particles*: The use of small particles, less prone to break upon stress, is a similar alternative that has been much deeply explored.[175] Ductile elements such as Sn can be cycled as small reacting dots within a matrix that prevents its agglomeration. Together with the need of limiting the side reactions with electrolytes at the surface of small particles (high specific surface of contact), this triggered the research on composite materials such as Si and Sn nanoparticles embedded in a conducting and dense matrix (usually carbonaceous). The relatively good performance of these composites, with capacities of 1500 mA h g^{-1} after 20 cycles for Si/C composites,[176] is not yet fully understood from a fundamental point of view. Raman spectroscopy[177] proves that Si particles are subjected to a compressive force when embedded in a dense carbon matrix, suggesting an important role of the tensile strength around the particles. However, the role of the C coating could be more complex in the sense that it modifies the surface of the Si particles leading to a modified SEI, which could play an important role in

the cyclability. Sn particles embedded in other matrixes, such as amorphous glasses (phosphates, borates, etc.), have also been intensively studied after pioneering work by Fujifilm[178] and yield higher both gravimetric and volumetric capacities than graphite, the problem being instead the high irreversibility of reactions occurring on the first cycle. A concept similar to those of active particles trapped inside a conducting matrix is to have active atoms squeezed in between inactive grains at the grain boundaries.

4. *Thin films*: Efficient interparticle contact can also be maintained through improved contact with the conducting substrate. Indeed, the first lithiation induces an increase in thickness that is constrained in one direction (perpendicular to the film). Nonetheless, curling of the films during reaction may be an issue that suggests the existence of important stress in the expanding material. As already mentioned, the use of amorphous deposits is likely preferred in order to avoid any anisotropic expansion of oriented grains within the films during alloying. The volumetric expansion and the stress are so strong that a relatively thick substrate (i.e., inactive mass) is required, the overall capacity being thus drastically reduced and generally not exceeding that of graphite. The use of thin substrates results in pulverization of the films and complete capacity loss after a few cycles.

5. *Intermetallics*: Multinary metallic materials (MM′) have also been studied, which entail displacement reactions with the formation of a composite containing the displaced metal M′ (Mn, Fe, Co, Ni, Cu, Nb, etc.) that does not alloy with lithium, together with a Li–M alloy/intermetallic compound. This is the case of tin ($M'_x Sn$)- and antimony ($M'_x Sb$)-based compounds that are transformed during the first discharge into metallic M′ nanoparticles and lithium-based nanoalloys $Li_y Sn$ (y = 3.5 or 4.4) or $Li_y Sb$ (y = 3), respectively,[179] either through direct reaction or involving intermediate ternary compounds. The nanocomposite electrode formed must be considered as the real electrode material for subsequent reversible cycling. Moreover, most of the approaches described earlier for electrodes based on pure metals can be applied to multimetallic systems, with similar drastic improvements. For instance, the cycling efficiency of Cu_2Sb/Li cells can be increased through the use of sputtered films or by making Cu_2Sb/C composites[180] that prevent coalescence of the particles and loss of electrical contacts. The same approach was pursued in the Co–Sn–C composite (Nexelion) commercialized by Sony in 2005.

6. *Voltage control (limited reduction)*: On a completely different scenario, Obrovac and Krause[181] demonstrated that a limitation in discharge voltage (0.17 V vs. Li⁺/Li⁰) of a $Si_{crystallized}$/Li cell can be efficient in limiting the amount of reacted Si. The nonreacted Si remains crystallized while the reacted part turns amorphous and remains amorphous during Li uptake and removal (Figure 8.16). Excellent cycling efficiency is then observed (950–1000 mA h g⁻¹) over at least 70 cycles. This approach can be seen as a way to create and maintain a nonactivated substrate that helps in limiting the lost of integrity of the electrode/particles, or as a nonreacted spherical substrate covered by a layer/film of active material.

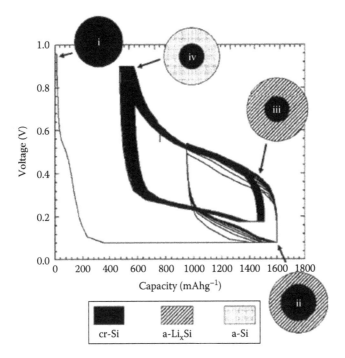

FIGURE 8.16 Voltage versus capacity profile of a cell cycled with a limitation in discharge voltage after 10 conditioning cycles. cr-Si denotes crystalline silicon, a-Li$_x$Si denotes partially lithiated silicon, and a-Si denotes amorphous silicon. (Obrovac, M.N. and Krause, L.J., *J. Electrochem. Soc.*, 154, A103, Copyright 2007. Reproduced with permission from The Electrochemical Society.)

The drawback of all the previously mentioned strategies is the loss of capacity linked to the use of substrates (dense films, limited reaction), voids (porous powders), or matrix (grain boundaries, composite materials, high carbon loads), all inactive or kept unreacted, therefore yielding much lower capacity values than the metal itself.

As an alternative, much attention has been recently devoted to playing with binders and electrode processing. Indeed, PVDF being the most common binder used for carbonaceous negative electrodes, a large amount of the previously mentioned studies deals with its use; moreover, its chain polymeric network was believed to help in preventing particle disconnection. Some attempts were further developed to use elastomeric binders in an attempt to better accommodate the volumetric expansion. Yet the reports of very good performance of micron-sized silicon particles[182] using a very brittle polymer such as carboxymethylcellulose (CMC) came as a surprise and were further rationalized through elucidation of the silicon–binder interactions.[183,184]

Another difference between PVDF and CMC lies in the fact that the former is processed in organic solvents, while the latter requires the use of water. This may have implications with respect to the degree of surface oxidation of silicon particles and traces of water may be difficult to eliminate and result in the formation of HF. However, it is overall considered a positive attribute since it avoids the use of toxic organic solvents during electrode processing. The polar hydrogen bonds between the carboxy groups of CMC and SiO_2 on the particle surface were proposed to re-form if locally broken and hence exhibit a self-healing effect (Figure 8.17).[185,186]

The good results achieved with CMC prompted the study of other binders exhibiting carboxy groups such as polyacrylic acid (PAA) or polysaccharides, which seem to exhibit even better behavior.[187]

Aside from the previously mentioned considerations, the stability of the SEI is much more critical for alloy based materials than for conventional insertion materials since mechanical stability may be more difficult to achieve. Indeed, breaking of the SEI may occur if its endurance limit is exceeded by the amplitude of the interface stresses, and hence new bare surfaces will be exposed to the electrolyte on which SEI needs to be further grown. This will ultimately cause failure of the electrode through electrolyte consumption and loss of electrode porosity with a decrease in the effective surface area of the electrode and rise of electrode impedance/polarization.[188] In this sense, SEI engineering has recently emerged as a key determinant parameter, and efforts in this direction will certainly grant future improvements in the quest for sustained cycling with close to theoretical capacity for alloy-based materials.

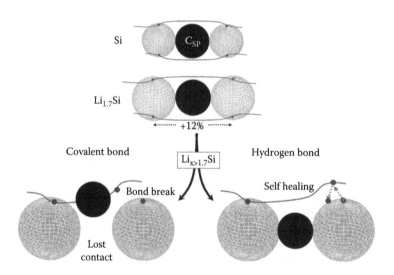

FIGURE 8.17 Schematic model showing the evolution in the CMC-Si bonding as the Li uptake proceeds from top to bottom according to Bridel et al.[186] Up to around 1.7 Li/Si both covalent and hydrogen bonding can sustain the particle volume changes, the overall swelling being buffered by electrode porosity. Beyond 1.7 Li/Si, the maximum CMC stretching ability is reached, and only the hydrogen-type Si–CMC interaction allows preservation of the efficient network through a proposed self-healing phenomenon. Csp denotes carbon black additive. (J.S. Bridel et al., *J. Electrochem. Soc.*, 158, A750, Copyright 2011. Reproduced with permission from The Electrochemical Society.)

8.3.3 CONVERSION-TYPE MATERIALS

The study of materials entailing conversion reactions has boosted in the last decade and the reports dealing with binary M–X compounds (X = O, N, F, S, P, H) have boomed after high reversibility and cycling ability were proved for oxides.[189] As mentioned earlier, these reactions involve full reduction of the transition metal to the metallic state and thus enable full utilization of all the redox potentials of the host metals yielding remarkably high capacity values (Figure 8.18). Though in some cases the reduction is initiated through an intermediate insertion compound with a limited capacity, full reduction invariably brings about the in situ formation of a nanocomposite electrode in which metal nanoparticles are embedded in a matrix of the lithium binary compound (Li_nX). The key to the reversibility of the conversion reaction seems to lie in the large amount of interfacial surface that makes nanoparticles very active toward the decomposition of the lithiated matrix when reverse polarization is applied in the battery, yet full reversibility is difficult to achieve. It is worth emphasizing that in some cases, such as X = P, the redox centers are not exclusively located on the transition metal, but electron transfer occurs also into bands that have a strong anion contribution. This phenomenon is directly correlated with the covalency of the M–X bond and the actual potential at which this occurs has been shown to depend on both the transition metal and the anionic species so that, in principle, the reaction potential can easily be tuned. In practice, it mostly ranges between 0.5 and 1 V versus Li^+/Li^0, the main exception being fluorides, which react at remarkably high potential values (close to 3 V) that would enable their use as positive electrode materials.

Transition metal oxides are, by far, the family of compounds that has deserved the most attention. Co_3O_4 and CoO have been intensively explored and one of the most investigated compounds is Fe_2O_3 since its high theoretical capacity (1007 mA h g^{-1}), coupled to its low toxicity and cost, certainly makes it an attractive electrode material

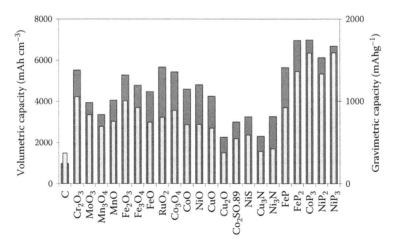

FIGURE 8.18 Specific gravimetric and volumetric capacity and capacity density for selected compounds that exhibit a conversion reaction mechanism with lithium.

candidate. Middle and late transition metal phosphides have also raised considerable interest in view of their high capacities coupled to their smaller polarization. Among these, NiP_2 seems the most attractive phase with five lithium ions per formula unit reversibly uptaken, which leads to reversible capacities of 1000 mA h g^{-1} at an average potential of 0.9 V.[190] The main drawbacks of transition metal phosphides when compared to oxides are capacity fading and environmental issues. Conversion reactions have also been reported for some nitrides and sulphides and even hydrides.[191] Overall, we have witnessed the appearance of a myriad of reports on studies on diverse materials, either pure or composites, exhibiting diverse particle sizes and shapes including nanoarchitectured electrodes and the reader is referred to comprehensive reviews for details on each specific compound.[159,192] Such efforts have unfortunately largely focused on improving performance rather than on understanding fundamental issues. Even if these reactions may seem simple at a first glance, there are still many issues to be understood and the application of such materials in commercial devices is strongly handicapped by several very critical inefficiencies, which are deeply rooted in the reaction mechanisms:[192]

1. *Coulombic efficiency*: Despite the enhanced reactivity shown by the electrochemically formed Li_nX/M nanocomposites, the specific capacity recovered upon the first reoxidation (i.e., the coulombic efficiency) is, in the vast majority of the cases, drastically lower than that generated upon the first reduction (see Figure 8.19). Three causes for this coulombic inefficiency can be envisaged: (1) irreversible electrolyte decomposition, (2) incomplete reconversion due to the presence of inactive or electrically disconnected Li_nX/M regions, or (3) reconversion to other phases than the initial compound. The latter has been, reported in some cases, such as Co_3O_4 in which lower formal oxidation states for the metal (i.e., CoO) are observed upon the first reoxidation. Nonetheless, full understanding of the determining factors behind the first cycle inefficiencies remains to be achieved. The coulombic efficiencies are remarkably improved beyond the first cycle. While this may be explained basing on an increased yield of the conversion reaction after an initial, forming cycle of the electrode into nanoparticles, first cycle losses are observed even when the initial electrode material is already in the form of nanoparticles. Thus, other factors, such as the polymeric layer formed from electrolyte decomposition (see end of this section), may play a role in that respect.

2. *Capacity retention*: The conversion processes entail massive structural reorganization and volumetric changes that can lead to particle isolation and cracking as a result of electrode grinding and to a subsequent fading of the capacity after a few cycles. As in the case of alloy-based electrodes, attempts to address this issue have been made for conversion reaction materials through the engineering of the electrode. A number of strategies have been developed such as forming nanocomposites with a considerable proportion of carbon or increasing the amount of polymeric binder. The former has the advantage that carbon is also active at low potentials at which conversion occurs, thereby partially compensating the decrease in capacity associated with the dilution of the active phase. Other complementary

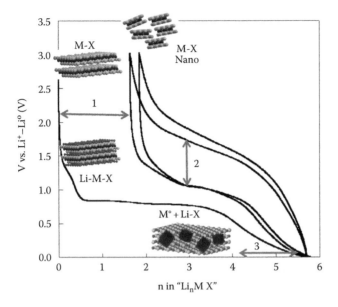

FIGURE 8.19 Typical potential versus composition profile of the first two and half cycles for an electrode containing a material that reacts through a conversion reaction, measured against a Li counterelectrode. The different processes that occur during this complex reaction are indicated at appropriate potentials. The light gray, dark gray, and black balls depict X, Li, and M, respectively. Arrows denote (1) first cycle coulombic inefficiency, (2) potential hysteresis, and (3) extra capacity. (Cabana, J., Monconduit, L., Larcher, D., and Palacín, M. R.: *Adv. Mater.*, 2010. 22. E170. Copyright Wiley-VCH Verlag GmbH & Co. KGaA. Adapted with permission.)

strategies include finding the most suitable polymer binder and performing heat treatments of the composite electrodes prior to testing. In general, all these approaches result in remarkable improvements indicating that a considerable amount of carbon coupled with adequate choice of binder or film pretreatment at moderate temperatures might be a suitable baseline strategy for conversion-based electrodes. Multiple efforts have been directed toward solving the cycle life problem by design at the nanoscale of the M_aX_b particles. However, the benefit of such approach remains to be fully proved. Indeed, particle reorganization and volume changes upon cycling are still likely to modify these nanostructures in the midterm. In addition, comparative studies of the performance of electrodes made of nano- and micronsized particles, respectively, offer mixed results, with the latter sometimes even outperforming the former. Enhanced electrolyte decomposition, which can eventually create a barrier to lithium diffusion and increase particle isolation, and even dissolution have been shown as factors that can handicap electrode performance in nanometric powders.

3. *Potential hysteresis*: The main drawback for practical application of these compounds is the large voltage hysteresis displayed between charge and

discharge that severely limits the energy efficiency of the electrode (Figure 8.19). This effect is more prominent in the first cycle, which seems to be related to the higher crystallinity of the pristine material with respect to the amorphous nature of the binary compound formed after the first reoxidation[193] since differences in free energy, and therefore in equilibrium reaction potential, are expected between crystalline and amorphous materials. The fact that the ratio of surface-to-bulk sites and the crystalline/amorphous character of the composite does not significantly change beyond the first cycle points at other factors significantly contributing to the hysteresis. Dependence of the first charge–second discharge hysteresis on the anionic species has been reported, and since the lowest values are observed for phosphides and hydrides, a correlation between covalency and voltage hysteresis seems to exist. Ascertaining whether this correlation is coincidental and what its exact role on the reaction mechanism is an issue that remains to be addressed. Although the relative kinetic and thermodynamic contributions to the hysteresis are not yet clarified, evidence that polarization is not a pure kinetic problem is provided by the fact that conversion electrodes can simultaneously show large polarization and fast kinetics (i.e., when nanostructured current collectors are used; Figure 8.20).

Indeed, DFT calculations on the FeF_3 system point at a fundamental difference in the reaction path upon conversion and deconversion that could be a more prominent thermodynamic factor.[195] Moreover, the fact that considerable surface/interfacial energy[196] needs to be provided to the system

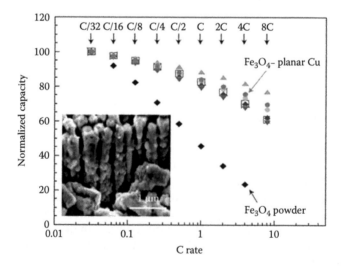

FIGURE 8.20 Rate capability plot for Fe_3O_4 deposits on Cu-nanostructured electrodes, compared to one grown on planar Cu and a plastic electrode film made with commercial Fe_3O_4. (Reprinted by permission from Macmillan Publishers Ltd. *Nat. Mater.*, Taberna, P.L. et al., 5, 567, Copyright 2006.)

during the reaction to create Li_nX/M nanocomposites with a large contact interface between two components is certainly another point to consider.

4. *Extra capacity at low potential*: Analysis of the capacity values reported upon the first reduction generally reveals an additional capacity with respect to theoretical values, its signature being a sloping curve that follows the conversion plateau (Figure 8.19). Two different phenomena, namely, electrolyte decomposition[197] and interfacial storage,[198] have been proposed to account for the charge associated with this step. The nature of the former has been analyzed extensively, using an array of analytical tools. These proved that reduction below 1 V versus Li^+/Li^0 leads to a thin solid layer surrounded by a gel-like polymeric one that can be several nanometers thick, which coat the particles, possibly contributing to the preservation of their integrity. The inner layer was found to be composed mostly of LiF, Li_2CO_3, and some alkylcarbonates, similar to the SEI formed on carbon, while the outer polymeric gel appears to contain polyethylene oxide (PEO) oligomers.[199] Inversion of the polarization to charge the battery up to 3–4 V brings about the consumption of the gel-like polymeric layer formed upon discharge, yet no evidence is available as to what species are formed upon polymeric layer dissolution and whether they stay dissolved in the electrolyte. Apart from the charge consumed for the reduction of the solvent, this polymeric layer was also proposed to enable additional lithium storage on its surface through a two-phase capacitive behavior of the Li_nX/M interface that would allow for the storage of Li^+ ions on the lithium compound side, whereas electrons would be localized on the metallic side, thereby formally leading to a charge separation. Such interfacial storage was supported by theoretical calculations[200] and allowed to rationalize nuclear magnetic resonance (NMR) results on RuO_2 reduction.[201] Nonetheless, recent studies do indicate that this interfacial storage would only account for a small percentage of the experimentally observed capacity, the rest being faradaic in character and fully related to electrolyte decomposition enhanced by the metal nanoparticles generated upon reduction.[202] Thus, the design of a stable SEI layer for these materials will also be crucial if bottlenecks associated with potential hysteresis can be overcome in the future.

8.4 FUTURE PROSPECTS

Although batteries have been available for long, it is only in the last decades that fundamental understanding of the underlying structure–property relationships that govern their performance has been achieved. This was critical for the development of classical lithium-ion batteries based on intercalation reactions in 1991 that quickly conquered the portable electronics market. The technology for such applications (e.g., mobile phones [2–4 W h], laptop computers [30–100 W h]) is now fully mastered and able to evolve and gradually incorporate new materials. Recognizing that the marriage between lithium batteries and portable electronics is a clear success story, we should be aware that the pace in battery evolution is far beyond in other fields of application. Recent upward spikes in the price of oil and gasoline increased awareness of the decreasing energy resources–demand gap, and the need to decrease CO_2 emissions has paved the way to governmental incentives both toward the implementation of electric vehicles (EVs) and hybrid electric vehicles (HEVs) and the use of renewable energies. However, prospects in achieving both electrically powered means of transportation and the deployment of renewable energies (5–10 kW h for cars to a few 100 kW h for buses and MW for grid storage applications) are still plagued with severe limitations in terms of energy density, performance, and cost. The criteria to choose the battery required for a specific application are the amount of energy (in terms of either specific energy [W h kg^{-1}] or energy density [W h dm^{-3}]) and electrical power (expressed in W kg^{-1} or W dm^{-3}) the battery needs to provide to the electrical load (Figure 8.21), its size and mass, reliability, durability, safety, and cost. However, the relative importance of each factor is rather application dependent.

Mature as the lithium-ion technology may seem at first sight, the quest for improved materials has never stopped and new compounds have reached the market in a larger or lesser extent such as $LiNi_{1-y-z}Mn_yCo_zO_2$, $LiFePO_4$ at the positive electrode, and $Li_4Ti_5O_{12}$ at the negative. All commercial Li-ion electrode materials are so far based on intercalation reactions (see Figure 8.22), which are intrinsically limited although the new developments in positive electrode materials such as Li-rich layered compounds hold promise for the achievement of increased energy densities through a substantial gain in capacity. On the other hand, further attainment of increased energy densities through the realization of a 5 V battery must be accompanied with the development of new electrolytes (liquid or solid) able to withstand such potentials without decomposing. Regarding the negative side, even if the current alloy-based electrodes must necessarily be limited in capacity, important gains in performance have been achieved, which holds promise for gradual penetration of this type of electrodes, mostly based in silicon. The advent of conversion reaction electrodes seems much more dubious in the short term, since the bottlenecks in energy efficiency do not seem easy to overcome and their benefits with respect to alloys are less evident. If promises of fast charge for positive electrode materials become a reality, the risk of lithium plating may still hold for materials with low operation potential and a switch to a negative electrode material for higher voltage may be

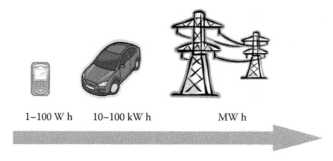

1–100 W h   10–100 kW h   MW h

FIGURE 8.21 Different applications for batteries and corresponding ranges of energy stored.

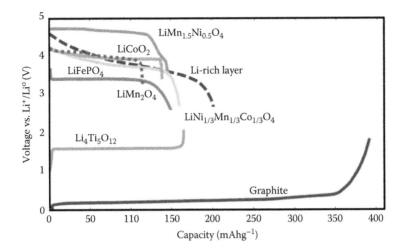

FIGURE 8.22 Voltage profiles of selected intercalation materials in lithium half-cells.

desirable, the only current viable alternatives being titanium-based electrodes, with a hope for the future of organic electrodes. Indeed, this radically different research approach has reemerged in recent years.[203] While electroactivity in organic materials is intrinsically different from that of redox-active inorganic compounds and deals with addition/withdrawal of electrons to/from electron-poor/electron-rich functional groups,[204–206] it is expected that such materials would in general exhibit lower cost and lower environmental footprint than their inorganic counterparts, to the expense of a poorer thermal stability, lower packing density, and solubility issues with conventional electrolytes. The future will tell whether developments arise that could give rise to a paradigm shift toward sustainability of electrode materials.

If a breakthrough in energy density is to be achieved, the use of air or sulfur as positive electrode material in combination with lithium metal negative electrodes appears as the most attractive option, as the theoretical capacities of these materials are extremely appealing. Nonetheless, these systems are still under development at a more fundamental scale and facing severe difficulties as the dissolution of intermediate species in the case of Li–S batteries or the reactivity of conventional electrolytes in the case of Li–air systems.[207]

Overall, the lithium-ion cell that is the *star* in the electronic market falls somewhat short of requirements for the construction of large efficient batteries. Indeed, these intrinsically bring about more stringent requirements, especially in terms of safety that currently translate into a narrower spectrum of possible electrode materials. Building bigger batteries is far from simply entailing up scaling current portable technologies. Within the current state of the art, it looks as though safety is not anymore hindering the lithium-ion technology based on intercalation from entering the transportation quest, with *suitable* electrode materials such as $LiFePO_4$ or $LiMn_2O_4$ at the positive electrode (although with a moderate energy density). With respect to the negative electrode materials, graphite is still a very good choice, since the risk of lithium plating seems to be mastered with affordable cost and good capacities both gravimetric and volumetric. In contrast, entering the field of stationary applications is still a challenge for lithium-ion technology. Indeed, these include storage of energy from distributed renewable (and intermittent!) sources, uninterrupted power supply (UPS), and load leveling to compensate fluctuations in demand (day/night or seasonal), which imply further increase in orders of magnitude in the amount of energy stored. These applications are currently covered by either lower-cost technologies with more limited performance such as lead or nickel based or more costly and performing high-temperature Na/S (especially in Japan). Even if efforts are currently devoted to building large 2 MW Li-ion cells for grid applications, generalized implementation seems somewhat unlikely, due to the concern of lithium availability and price. Indeed, there might be insufficient lithium available to sustain such a generalized field of application in the long term, as this would switch dependency from oil to lithium resources unless we endeavor to develop a very efficient recycling infrastructure. Even if prospects of other battery chemistries are beyond the scope of this chapter, a technology based on sodium, with significant larger abundance and lower cost, and operating at room temperature[208–211] could have good prospects for generalized application in large-scale storage.

Batteries are, however, alive chemical reactors and gaining control of materials properties requires deep fundamental understanding of the processes taking place at all levels. Thus, it is difficult to predict the rate of success for the different alternatives in each range of application. On the other hand, it is clear that with the current dedicated research efforts and revolutions already witnessed in the field of batteries, there is yet hope for important breakthroughs able to fulfill the needs of both existing and newly emerging applications.

ACKNOWLEDGMENTS

The authors acknowledge Jordi Cabana and Dominique Larcher for valuable discussions and Gurpreet Singh and Damien Saurel for critical reading of the manuscript. MRP is grateful to Ministerio de Ciencia e Innovación (Spain) for the financial support through grant MAT2011-24757. MCC wishes to acknowledge the financial support from the Departamento de Desarrollo Económico y Competitividad of the Gobierno Vasco and from the Ministerio de Economia y Competitividad (Spain) through grant ENE2013-44330-R.

REFERENCES

1. Y. Nishi, *Chem. Record*, 2001, **1**, 406.
2. J. B. Goodenough and Y. Kim, *Chem. Mater.*, 2010, **22**, 587.
3. E. Peled, *J. Electrochem. Soc.*, 1979, **126**, 2047.
4. K. Xu and A. von Cresce, *J. Mater. Chem.*, 2011, **21**, 9849.
5. P. Verma, P. Maire and P. Novak, *Electrochim. Acta*, 2010, **55**, 6332.
6. D. Aurbach, *J. Power Sources*, 2000, **89**, 206.
7. K. Edström, T. Gustafsson and J. O. Thomas, *Electrochim. Acta*, 2004, **50**, 397.
8. E. Ligneel, B. Lestriez, A. Hudhomme and D. Guyomard, *J. Eur. Ceram. Soc.*, 2009, **29**, 925.
9. B. Lestriez, *C. R. Chim.*, 2010, **13**, 1341.

10. H. Zheng, J. Li, X. Song, G. Liu and V. S. Battaglia, *Electrochim. Acta*, 2012, **71**, 258.
11. H. Zheng, L. Zhang, G. Liu, X. Song and Vin, *J. Power Sources*, 2012, **217**, 530.
12. C.-J. Bae, C. K. Erdonmez, J. W. Halloran and Y.-M. Chiang, *Adv. Mater.*, 2013, **25**, 1254.
13. X. Huang, *J. Solid State Electr.*, 2011, **15**, 649.
14. C. J. Orendorff, T. N. Lambert, C. A. Chavez, M. Bencomo and K. R. Fenton, *Adv. Energy Mater.*, 2013, **3**, 314.
15. J. M. Tarascon and M. Armand, *Nature*, 2001, **414**, 359.
16. R. Malik, F. Zhou and G. Ceder, *Nat. Mater.*, 2011, **10**, 587–590.
17. M. Thomas, P. Bruce and J. Goodenough, *Solid State Ionics*, 1985, **17**, 13–19.
18. C. Delmas, C. Fouassier and P. Hagenmuller, *Physica B+C*, 1980, **99**, 81.
19. K. Mizushima, P. Jones, P. Wiseman and J. Goodenough, *Mater. Res. Bull.*, 1980, **15**, 783.
20. *Modern Batteries. An Introduction to Electrochem Power Sources*, ed. C. Vincent and B. Scrosati, Butterworth-Heinemann, Burlington, MA, 2nd edn., 1997.
21. D. D. MacNeil and J. R. Dahn, *J. Electrochem. Soc.*, 2001, **148**, A1205.
22. J. N. Reimers and J. R. Dahn, *J. Electrochem. Soc.*, 1992, **139**, 2091.
23. T. Ohzuku and A. Ueda, *J. Electrochem. Soc.*, 1994, **141**, 2972.
24. A. Van der Ven, M. K. Aydinol and G. Ceder, *J. Electrochem. Soc.*, 1998, **145**, 2149.
25. Z. Chen, Z. Lu and J. R. Dahn, *J. Electrochem. Soc.*, 2002, **149**, A1604.
26. G. G. Amatucci, J. M. Tarascon and L. C. Klein, *J. Electrochem. Soc.*, 1996, **143**, 1114.
27. Y. Shao-Horn, S. Levasseur, F. Weill and C. Delmas, *J. Electrochem. Soc.*, 2003, **150**, A366.
28. E. Antolini, L. Giorgi and M. Carewska, *J. Mater. Sci. Lett.*, 1999, **18**, 325.
29. *Lithium Ion Batteries: Fundamentals Performance*, ed. M. Wakihara and O. Yamamoto, Wiley, Weinheim, 2008.
30. J. Cho, Y. Kim and B. Park, *Chem. Mater.*, 2000, **12**, 3788.
31. L. Fu, H. Liu, C. Li, Y. Wu, E. Rahm, R. Holze and H. Wu, *Solid State Sci.*, 2006, **8**, 113.
32. I. D. Scott, Y. S. Jung, A. S. Cavanagh, Y. Yan, A. C. Dillon, S. M. George and S.-H. Lee, *Nano Lett.*, 2011, **11**, 414.
33. T. Ohzuku, A. Ueda and M. Kouguchi, *J. Electrochem. Soc.*, 1995, **142**, 4033.
34. T. Ohzuku and Y. Makimura, *Chem. Lett.*, 2001, **30**, 642.
35. A. Rougier, P. Gravereau and C. Delmas, *J. Electrochem. Soc.*, 1996, **143**, 1168.
36. T. Ohzuku, A. Ueda and M. Nagayama, *J. Electrochem. Soc.*, 1993, **140**, 1862.
37. C. Delmas, J. P. Péres, A. Rougier, A. Demourgues, F. Weill, A. Chadwick, M. Broussely, F. Perton, P. Biensan and P. Willmann, *J. Power Sources*, 1997, **68**, 120.
38. J. Reed and G. Ceder, *Chem. Rev.*, 2004, **104**, 4513.
39. A. Armstrong, N. Dupre, A. Paterson, C. Grey and P. Bruce, *Chem. Mater.*, 2004, **16**, 3106.
40. A. R. Armstrong and P. G. Bruce, *Nature*, 1996, **381**, 499.
41. E. Rossen, C. Jones and J. Dahn, *Solid State Ionics*, 1992, **57**, 311.
42. Z. Lu, D. MacNeil and J. Dahn, *Electrochem. Solid-State Lett.*, 2001, **4**, A200.
43. K. Kang, Y. Meng, J. Breger, C. Grey and G. Ceder, *Science*, 2006, **311**, 977.
44. Y. Hinuma, Y. S. Meng, K. Kang and G. Ceder, *Chem. Mater.*, 2007, **19**, 1790.
45. A. V. der Ven and G. Ceder, *Electrochem. Commun.*, 2004, **6**, 1045.
46. J. Bréger, N. Dupré, P. J. Chupas, P. L. Lee, T. Prfen, J. B. Parise and C. P. Grey, *J. Am. Chem. Soc.*, 2005, **127**, 7529.
47. J. Reed and G. Ceder, *Electrochem. Solid-State Lett.*, 2002, **5**, A145.
48. M. S. Whittingham, *Chem. Rev.*, 2004, **104**, 4271.
49. H. H. Li, N. Yabuuchi, Y. S. Meng, S. Kumar, J. Breger, C. P. Grey and Y. Shao-Horn, *Chem. Mater.*, 2007, **19**, 2551.
50. M. Guilmard, L. Croguennec, D. Denux and C. Delmas, *Chem. Mater.*, 2003, **15**, 4476.
51. M. Guilmard, L. Croguennec and C. Delmas, *Chem. Mater.*, 2003, **15**, 4484.

52. M. Yoshio, H. Noguchi, J. Itoh, M. Okada and T. Mouri, *J. Power Sources*, 2000, **90**, 176.
53. N. Yabuuchi and T. Ohzuku, *J. Power Sources*, 2003, **119–121**, 171.
54. Y. Wang, J. Jiangand and J. Dahn, *Electrochem. Commun.*, 2007, **9**, 2534.
55. N. Yabuuchi, Y. Makimura and T. Ohzuku, *J. Electrochem. Soc.*, 2007, **154**, A314.
56. D. Zeng, J. Cabana, J. Bréger, W.-S. Yoon and C. P. Grey, *Chem. Mater.*, 2007, **19**, 6277.
57. B. L. Ellis, K. T. Lee and L. F. Nazar, *Chem. Mater.*, 2010, **22**, 691.
58. S.-C. Yin, Y.-H. Rho, I. Swainson and L. F. Nazar, *Chem. Mater.*, 2006, **18**, 1901.
59. J. Choi and A. Manthiram, *J. Electrochem. Soc.*, 2005, **152**, A1714.
60. Y.-K. Sun, S.-T. Myung, B.-C. Park, J. Prakash, I. Belharouak and K. Amine, *Nat. Mater.*, 2009, **8**, 320.
61. K. Numata, C. Sakaki and S. Yamanaka, *Chem. Lett.*, 1997, **8**, 725.
62. Z. Lu, D. D. MacNeil and J. R. Dahn, *Electrochem. Solid-State Lett.*, 2001, **4**, A191.
63. J.-S. Kim, C. Johnson and M. Thackeray, *Electrochem. Commun.*, 2002, **4**, 205.
64. H. Yu and H. Zhou, *J. Phys. Chem. Lett.*, 2013, **4**, 1268.
65. M. Sathiya, K. Ramesha, G. Rousse, D. Foix, D. Gonbeau, A. S. Prakash, M. L. Doublet, K. Hemalatha and J.-M. Tarascon, *Chem. Mater.*, 2013, **25**, 1121.
66. M. Thackeray, S.-H. Kang, C. Johnson, J. Vaughey and S. Hackney, *Electrochem. Commun.*, 2006, **8**, 1531.
67. J. Bréger, M. Jiang, N. Dupré, Y. S. Meng, Y. Shao-Horn, G. Ceder and C. P. Grey, *J. Solid State Chem.*, 2005, **178**, 2575.
68. D. Kim, J. Gim, J. Lim, S. Park and J. Kim, *Mater. Res. Bul.*, 2010, **45**, 252.
69. J. Bareño, M. Balasubramanian, S. H. Kang, J. G. Wen, C. H. Lei, S. V. Pol, I. Petrov and D. P. Abraham, *Chem. Mater.*, 2011, **23**, 2039.
70. K. A. Jarvis, Z. Deng, L. F. Allard, A. Manthiram and P. J. Ferreira, *Chem. Mater.*, 2011, **23**, 3614.
71. H. Koga, L. Croguennec, P. Mannessiez, M. Ménétrier, F. Weill, L. Bourgeois, M. Duttine, E. Suard and C. Delmas, *J. Phys. Chem. C*, 2012, **116**, 13497.
72. C. S. Johnson, N. Li, C. Lefief, J. T. Vaughey and M. M. Thackeray, *Chem. Mater.*, 2008, **20**, 6095.
73. T. A. Arunkumar, Y. Wu and A. Manthiram, *Chem. Mater.*, 2007, **19**, 3067.
74. R. Xiao, H. Li and L. Chen, *Chem. Mater.*, 2012, **24**, 4242.
75. A. R. Armstrong, M. Holzapfel, P. Novak, C. S. Johnson, S.-H. Kang, M. M. Thackeray and P. G. Bruce, *J. Am. Chem. Soc.*, 2006, **128**, 8694.
76. Z. Lu and J. R. Dahn, *J. Electrochem. Soc.*, 2002, **149(7)**, A815.
77. N. Tran, L. Croguennec, M. Ménétrier, F. Weill, P. Biensan, C. Jordy and C. Delmas, *Chem. Mater.*, 2008, **20**, 4815.
78. H. Koga, L. Croguennec, M. Ménétrier, K. Douhil, S. Belin, L. Bourgeois, E. Suard, F. Weill and C. Delmas, *J. Electrochem. Soc.*, 2013, **160**, A786.
79. M. Sathiya, G. Rousse, K. Ramesha, C. P. Laisa, H. Vezin, M. T. Sougrati, M.-L. Doublet, D. Foix, D. Gonbeau, W. Walker, A. S. Prakash, M. Ben Hassine, L. Dupont and J.-M. Tarascon, *Nat. Mater.*, 2013, **12**, 827.
80. M. Gu, I. Belharouak, J. Zheng, H. Wu, J. Xiao, A. Genc, K. Amine, S. Thevuthasan, D. R. Baer, J.-G. Zhang, N. D. Browning, J. Liu and C. Wang, *ACS Nano*, 2013, **7**, 760.
81. A. Boulineau, L. Simonin, J.-F. Colin, C. Bourbon and S. Patoux, *Nano Lett.*, 2013, **13**, 3857.
82. C. Johnson, N. Li, J. Vaughey, S. Hackney and M. Thackeray, *Electrochem. Commun.*, 2005, **7**, 528.
83. S.-H. Park, S.-H. Kang, C. Johnson, K. Amine and M. Thackeray, *Electrochem. Commun.*, 2007, **9**, 262.
84. J. Cabana, C. S. Johnson, X.-Q. Yang, K.-Y. Chung, W.-S. Yoon, S.-H. Kang, M. M. Thackeray and C. P. Grey, *J. Mater. Res.*, 2010, **25**, 1601.

Sakaebe, M. Shikano, H. Kojitani, K. Tatsumi and Y. Inaguma, *J. Power* 1, **196**, 6934.

..u A. Manthiram, *Chem. Mater.*, 2003, **15**, 2954.

87. T. Ohzuku, M. Kitagawa and T. Hirai, *J. Electrochem. Soc.*, 1990, **137**, 769.

88. Y. J. Lee, F. Wang, S. Mukerjee, J. McBreen and C. P. Grey, *J. Electrochem. Soc.*, 2000, **147**, 803.

89. H. Bjork, T. Gustafsson, J. O. Thomas, S. Lidin and V. Petricek, *J. Mater. Chem.*, 2003, **13**, 585.

90. J. Goodenough, *J. Solid State Electr.*, 2012, **16**, 2019.

91. M. Thackeray, W. David, P. Bruce and J. Goodenough, *Mater. Res. Bull.*, 1983, **18**, 461.

92. J. Rodriguez-Carvajal, G. Rousse, C. Masquelier and M. Hervieu, *Phys. Rev. Lett.*, 1998, **81**, 4660.

93. D. H. Jang, Y. J. Shin and S. M. Oh, *J. Electrochem. Soc.*, 1996, **143**, 2204.

94. R. Gummow, A. de Kock and M. Thackeray, *Solid State Ionics*, 1994, **69**, 59.

95. *Handbook of Battery Materials*, ed. C. Daniel and J. Besenhard, Wiley, Weinheim, 2011.

96. G. Amatucci, N. Pereira, T. Zheng, I. Plitz and J. Tarascon, *J. Power Sources*, 1999, **81–82**, 39.

97. Y. S. Lee and M. Yoshio, *Electrochem. Solid-State Lett.*, 2001, **4**, A155.

98. A. Manthiram, *J. Phys. Chem. Lett.*, 2011, **2**, 176.

99. S. B. Chikkannanavar, D. M. Bernardi and L. Liu, *J. Power Sources*, 2014, **248**, 91.

100. K. Amine, H. Tukamoto, H. Yasuda and Y. Fujita, *J. Power Sources*, 1997, **68**, 604.

101. Q. Zhong, A. Bonakdarpour, M. Zhang, Y. Gao and J. R. Dahn, *J. Electrochem. Soc.*, 1997, **144**, 205.

102. J. Cabana, M. Casas-Cabanas, F. O. Omenya, N. A. Chernova, D. Zeng, M. S. Whittingham and C. P. Grey, *Chem. Mater.* 2012, **24**, 2952.

103. J. Liua and A. Manthiram, *J. Phys. Chem. C*, 2009, **113**, 15073.

104. A. K. Padhi, K. S. Nanjundaswamy and J. B. Goodenough, *J. Electrochem. Soc.*, 1997, **144**, 1188.

105. N. Ravet, Y. Chouinard, J. Magnan, S. Besner, M. Gauthier and M. Armand, *J. Power Sources*, 2001, **97–98**, 503.

106. H. Huang, S.-C. Yin and L. F. Nazar, *Electrochem. Solid-State Lett.*, 2001, **4**, A170.

107. A. Yamada, S. C. Chung and K. Hinokuma, *J. Electrochem. Soc.*, 2001, **148**, A224.

108. J. M. Tarascon, *Phil. Trans. R. Soc. A*, 2010, **368**, 3227.

109. R. Malik, A. Abdellahi and G. Ceder, *J. Electrochem. Soc.*, 2013, **160**, A3179.

110. C. Delacourt, P. Poizot, J.-M. Tarascon and C. Masquelier, *Nat. Mater.*, 2005, **4**, 254.

111. A. Yamada, H. Koizumi, S.-I. Nishimura, N. Sonoyama, R. Kanno, M. Yonemura, T. Nakamura and Y. Kobayashi, *Nat. Mater.*, 2006, **5**, 357.

112. N. Meethong, H.-Y. S. Huang, W. C. Carter and Y.-M. Chiang, *Electrochem. Solid-State Lett.*, 2007, **10**, A134.

113. P. Gibot, M. Casas-Cabanas, L. Laffont, S. Levasseur, P. Carlach, S. Hamelet, J. M. Tarascon and C. Masquelier, *Nat. Mater.*, 2008, **7**, 741.

114. G. Kobayashi, S.-I. Nishimura, M.-S. Park, R. Kanno, M. Yashima, T. Ida and A. Yamada, *Adv. Funct. Mater.*, 2009, **19**, 395.

115. N. Meethong, Y.-H. Kao, M. Tang, H.-Y. Huang, W. C. Carter and Y.-M. Chiang, *Chem. Mater.*, 2008, **20**, 6189.

116. M. Wagemaker, D. P. Singh, W. J. Borghols, U. Lafont, L. Haverkate, V. K. Peterson and F. M. Mulder, *J. Am. Chem. Soc.*, 2011, **133**, 10222.

117. B. Han, A. V. der Ven, D. Morgan and G. Ceder, *Electrochim. Acta*, 2004, **49**, 4691.

118. J. Jamnik and J. Maier, *Phys. Chem. Chem. Phys.*, 2003, **5**, 5215.

119. K. T. Lee, W. H. Kan and L. F. Nazar, *J. Am. Chem. Soc.*, 2009, **131**, 6044.

120. W. Dreyer, J. Jamnik, C. Guhlke, R. Huth, J. Moskon and M. Gaberscek, *Nat. Mater.*, 2010, **9**, 448.

121. C. Delmas, M. Maccario, L. Croguennec, F. Le Cras and F. Weill, *Nat. Mater.*, 2008, **7**, 665.

122. D. Morgan, A. Van der Ven and G. Ceder, *Electrochem. Solid-State Lett.*, 2004, **7**, A30.

123. M. S. Islam, D. J. Driscoll, C. A. Fisher and P. R. Slater, *Chem. Mater.*, 2005, **17**, 5085.

124. S. Nishimura, G. Kobayashi, K. Ohoyama, R. Kanno, M. Yashima and A. Yamada, *Nat. Mater.*, 2008, **7**, 707.

125. R. Malik, D. Burch, M. Bazant and G. Ceder, *Nano Lett.*, 2010, **10**, 4123.

126. L. Laffont, C. Delacourt, P. Gibot, M. Y. Wu, P. Kooyman, C. Masquelier and J. M. Tarascon, *Chem. Mater.*, 2006, **18**, 5520.

127. A. Yamada, M. Hosoya, S.-C. Chung, Y. Kudo, K. Hinokuma, K.-Y. Liu and Y. Nishi, *J. Power Sources*, 2003, **119–121**, 232.

128. C. Masquelier and L. Croguennec, *Chem. Rev.*, 2013, **113**, 6552.

129. S. Okada, S. Sawa, M. Egashira, J. Ichi Yamaki, M. Tabuchi, H. Kageyama, T. Konishi and A. Yoshino, *J. Power Sources*, 2001, **97–98**, 430.

130. K. Amine, H. Yasuda and M. Yamachi, *Electrochem. Solid-State Lett.*, 2000, **3**, 178.

131. C. Eames, A. R. Armstrong, P. G. Bruce and M. S. Islam, *Chem. Mater.*, 2012, **24**, 2155.

132. A. Nyten, S. Kamali, L. Haggstrom, T. Gustafsson and J. O. Thomas, *J. Mater. Chem.*, 2006, **16**, 2266.

133. S. Nishimura, S. Hayase, R. Kanno, M. Yashima, N. Nakayama and A. Yamada, *J. Am. Chem. Soc.*, 2008, **130**, 13212.

134. C. Sirisopanaporn, C. Masquelier, P. G. Bruce, A. R. Armstrong and R. Dominko, *J. Am. Chem. Soc.*, 2011, **133**, 1263.

135. C. Sirisopanaporn, A. Boulineau, D. Hanzel, R. Dominko, B. Budic, A. R. Armstrong, P. G. Bruce and C. Masquelier, *Inorg. Chem.*, 2010, **49**, 7446.

136. T. Muraliganth, K. R. Stroukf and A. Manthiram, *Chem. Mater.*, 2010, **22**, 5754.

137. A. Saracibar, A. Van der Ven and M. E. Arroyo-de Dompablo, *Chem. Mater.*, 2012, **24**, 495.

138. A. R. Armstrong, N. Kuganathan, M. S. Islam and P. G. Bruce, *J. Am. Chem. Soc.*, 2011, **133**, 13031.

139. C. Lyness, B. Delobel, A. R. Armstrong and P. G. Bruce, *Chem. Commun.*, 46, 2007, 4890.

140. T. N. Ramesh, K. T. Lee, B. L. Ellis and L. F. Nazar, *Electrochem. Solid-State Lett.*, 2010, **13**, A43.

141. N. Recham, J.-N. Chotard, J.-C. Jumas, L. Laffont, M. Armand and J.-M. Tarascon, *Chem. Mater.*, 2010, **22**(46), 1142.

142. B. Ellis, W. Makahnouk, Y. Makimura, K. Toghill and L. Nazar, *Nat. Mater.*, 2007, **6**, 749.

143. E. Dumont-Botto, C. Bourbon, S. Patoux, P. Rozier and M. Dollé, *J. Power Sources*, 2011, **196**, 2274.

144. N. Recham, J.-N. Chotard, L. Dupont, C. Delacourt, W. Walker, M. Armand and J.-M. Tarascon, *Nat. Mater.*, 2010, **9**, 68.

145. P. Barpanda, N. Recham, J.-N. Chotard, K. Djellab, W. Walker, M. Armand and J.-M. Tarascon, *J. Mater. Chem.*, 2010, **20**, 1659.

146. P. Barpanda, M. Ati, B. C. Melot, G. Rousse, J.-N. Chotard, M.-L. Doublet, M. T. Sougrati, S. A. Corr, J.-C. Jumas and J.-M. Tarascon, *Nat. Mater.*, 2011, **10**, 772.

147. M. Ben Yahia, F. Lemoigno, G. Rousse, F. Boucher, J.-M. Tarascon and M.-L. Doublet, *Energy Environ. Sci.*, 2012, **5**, 9584.

148. M. R. Palacín, *Chem. Soc. Rev.*, 2009, **38**, 2565.

149. M. Thackeray, J. Vaughey, C. Johnson, A. Kropf, R. Benedek, L. Fransson and K. Edstrom, *J. Power Sources*, 2003, **113**, 134.

150. D. Aurbach, B. Markovsky, I. Weissman, E. Levi and Y. Ein-Eli, *Electrochim. Acta*, 1999, **45**, 67.

151. M. Endo, C. Kim, K. Nishimura, T. Fujino and K. Miyashita, *Carbon*, 2000, **38**, 183.

152. S. Flandrois and B. Simon, *Carbon*, 1999, **37**, 165.

153. Z. Ogumi and H. Wang, *Lithium-Ion Batteries*, Springer, New York, 2006, p. 49.
154. T. Nishida, *Lithium-Ion Batteries*, Springer, New York, 2009, p. 329.
155. T. Zheng, J. S. Sue and J. R. Dahn, *Chem. Mater.*, 1996, **8**, 389.
156. Z. Yang, D. Choi, S. Kerisit, K. Rosso, D. Wang, J. Zhang, G. Graff and J. Liu, *J. Power Sources*, 2009, **192**, 588.
157. T. Froschl, U. Hormann, P. Kubiak, G. Kucerova, M. Pfanzelt, C. Weiss, R. Behm, N. Husig, U. Kaiser, K. Lfester and M. Wohlfahrt-Mehrens, *Chem. Soc. Rev.*, 2012, **41**, 5313.
158. M. V. Reddy, G. V. S. Rao and B. V. R. Chowdari, *Chem. Rev.*, 2013, **113**, 5364.
159. T. Ohzuku, A. Ueda and N. Yamamoto, *J. Electrochem. Soc.*, 1995, **142**, 1431.
160. A. D. Pasquier, C. Huang and T. Spitler, *J. Power Sources*, 2009, **186**, 208.
161. C. Kim, N. Norberg, C. Alexer, R. Kostecki and J. Cabana, *Adv. Funct. Mater.*, 2013, **23**, 1214.
162. M. Song, A. Benayad, Y. Choi and K. Park., *Chem. Commun.*, 2012, **48**, 516.
163. R. Dominko, E. Baudrin, P. Umek, D. Arcon, M. Gaberscek and J. Jamnik, *Electrochem. Commun.*, 2006, **8**, 673.
164. S. Brutti, V. Gentili, H. Menard, B. Scrosati and P. Bruce, *Adv. Energy Mater.*, 2012, **2**, 322.
165. M. Fehse, F. Fischer, C. Tessier, L. Stievano and L. Monconduit, *J. Power Sources*, 2013, **231**, 23.
166. I. Belharouak, G. Koenig, Jr., T. Tan, H. Yumoto, N. Ota and K. Amine, *J. Electrochem. Soc.*, 2012, **159**, A1165.
167. Y. B. He, B. Li, M. Liu, C. Zhang, W. Lu, C. Yang, J. Li, H. Du, B. Zhang, Q. Yang, J. Kim and F. Kang, *Sci. Rep.*, 2012, **2**, 913.
168. N. Choi, N. Kim, J. Yin and R. Kim, *Mater. Chem. Phys.*, 2009, **116**, 603.
169. A. Dey, *J. Electrochem. Soc.*, 1971, **118**, 1547.
170. C. Park, J. Kim and H. Sohn, *Chem. Soc. Rev.*, 2010, **39**, 3115.
171. D. Larcher, S. Beattie, M. Morcrette, K. Edstrom, J. Jumas and J. Tarascon, *J. Mater. Chem.*, 2007, **17**, 3759.
172. T. Hatchard and J. Dahn, *J. Electrochem. Soc.*, 2004, **151**, A1628.
173. M. Zamfir, H. Nguyen, E. Moyen, Y. Lee and D. Pribat, *J. Mater. Chem. A*, 2013, **1**, 9566.
174. C. Chan, H. Peng, G. Liu, K. McIlwrath, X. F. Zhang, R. Huggins and Y. Cui, *Nat. Nanotechnol.*, 2008, **3**, 31.
175. J. Yang, Y. Takeda, N. Imanishi and O. Yamamoto, *J. Electrochem. Soc.*, 1999, **146**, 4009.
176. S. Ng, J. Wang, D. Wexler, K. Konstantinov, Z. Guo and H. Liu, *Angew. Chem. Int. Edit.*, 2006, **45**, 6896.
177. J. Saint, M. Morcrette, D. Larcher, L. Laffont, S. Beattie, J. Peres, D. Talaga, M. Couzi and J. Tarascon, *Adv. Funct. Mater.*, 2007, **17**, 1765.
178. Y. Idota, T. Kobota, A. Matsufuji, Y. Maekawa and T. Miyasaka, *Science*, 1997, **276**, 1395.
179. C. M. Ionica, P. E. Lippens, J. O. Fourcade and J. C. Jumas, *J. Power Sources*, 2005, **146**, 481.
180. H. Bryngelsson, J. Eskhult, L. Nyholm and K. Edstrom, *Electrochim. Acta*, 2008, **53**, 7226; S.-W. Song, P. R. Reade, E. J. Cairns, J. T. Vaughey, M. M. Thackeray and K. A. Striebel, *J. Electrochem. Soc.*, 2004, **151**(7), A1012–A1019.
181. M. N. Obrovac and L. J. Krause, *J. Electrochem. Soc.*, 2007, **154**, A103.
182. J. Li, R. B. Lewis and J. R. Dahn, *Electrochem. Solid-State. Lett.*, 2007, **10**, A17.
183. B. Lestriez, S. Bahri, I. Su, L. Roue and D. Guyomard, *Electrochem. Commun.*, 2007, **9**, 2801.
184. N. Hochgatterer, M. Schweiger, S. Koller, P. Rainmann, T. Wohrle, C. Wurm and M. Winter, *Electrochem. Solid-State Lett.*, 2008, **11**, A76.

185. J. S. Bridel, T. Azaïs, M. Morcrette, J.-M. Tarascon and D. Larcher, *J. Electrochem. Soc.*, 2011, **158**, A750.
186. J. S. Bridel, T. Azais, M. Morcrette, J. Tarascon and D. Larcher, *Chem. Mater.*, 2010, **22**, 1229.
187. I. Kovalenko, B. Zdyro, A. Magasinski, B. Hertzberg, Z. Milicev, R. Burtovy and G. Yushin, *Science*, 2011, **334**, 75.
188. Y. Oumellal, N. Delpuech, D. Mazouzi, N. Dupre, J. Gaubicher, P. Moreau, P. Soudan, B. Lestriez and D. Guyomard, *J. Mater. Chem.*, 2011, **21**, 6201.
189. P. Poizot, S. Laruelle, S. Grugeon, L. Dupont and J. Tarascon, *Nature*, 2000, **407**, 493.
190. F. Gillot, S. Boyanov, L. Dupont, M. Doublet, M. Morcrette, L. Monconduit and J. M. Tarascon, *Chem. Mater.*, 2005, **17**, 6317.
191. Y. Oumellal, A. Rougier, G. Nazri, J. Tarascon and L. Aymard, *Nat. Mater.*, 2008, **7**, 916.
192. J. Cabana, L. Monconduit, D. Larcher and M. R. Palacín, *Adv. Mater.*, 2010, **22**, E170.
193. O. Delmer, P. Balaya, L. Kienle and J. Maier, *Adv. Mater.*, 2008, **20**, 501.
194. P. L. Taberna, S. Mitra, P. Poizot, P. Simon and J. Tarascon, *Nat. Mater.*, 2006, **5**, 567.
195. R. E. Doe, K. A. Persson, Y. S. Meng and G. Ceder, *Chem. Mater.*, 2008, **20**, 5274.
196. K. Croue, J. Jolivet and D. Larcher, *Electrochem. Solid-State Lett.*, 2012, **15**, F8.
197. S. Laruelle, S. Grugeon, P. Poizot, M. Dolle, L. Dupont and J. M. Tarascon, *J. Electrochem. Soc.*, 2002, **149**, A627.
198. P. Balaya, H. Li, L. Kienle and J. Maier, *Adv. Funct. Mater.*, 2003, **13**, 621.
199. G. Gachot, S. Grugeon, M. Armand, S. Pilard, P. Guenot, J. M. Tarascon and S. Laruelle, *J. Power Sources*, 2008, **178**, 409.
200. Y. F. Zhukovskii, E. A. Kotomin, P. Balaya and J. Maier, *Solid State Sci.*, 2008, **10**, 491.
201. E. Bekaert, P. Balaya, S. Murugavel, J. Maier and M. Menetrier, *Chem. Mater.*, 2009, **21**, 856.
202. A. Ponrouch, P. L. Taberna, P. Simon and M. Palacin., *Electrochim. Acta*, 2012, **61**, 13.
203. P. Poizot and F. Dolhem, *Energy Environ. Sci.*, 2011, **4**, 2003.
204. Z. Song and H. Zhou, *Energy Environ. Sci.*, 2013, **6**, 2280.
205. Y. Liang, Z. Tao and J. Chen, *Adv. Energy Mater.*, 2012, **2**, 742.
206. M. Armand, S. Grugeon, H. Vezin, S. Laruelle, P. Ribiere, P. Poizot and J. M. Tarascon, *Nat. Mater.*, 2009, **8**, 120.
207. P. G. Bruce, S. A. Freunberger, L. J. Hardwick and J.-M. Tarascon, *Nat. Mater.*, 2011, **11**, 19.
208. V. Palomares, P. Serras, I. Villaluenga, K. B. Hueso, J. Carretero-González and T. Rojo, *Energy Environ. Sci.*, 2012, **5**, 5884.
209. V. Palomares, M. Casas-Cabanas, E. Castillo-Martínez, M. H. Han and T. Rojo, *Energy Environ. Sci.*, 2013, **6**, 2312.
210. S. W. Kim, D.-H. Seo, X. Ma, G. Ceder and K. Kang, *Adv. Energy Mater.*, 2012, **2**, 710.
211. B. L. Ellis and L. F. Nazar, *Curr. Opin. Solid State Mater. Sci.*, 2012, **16**, 168.
212. J. B. Goodenough and K. S. Park, *J. Am. Chem. Soc.*, 2013, **135**, 1167.

9 All-Solid-State Rechargeable Batteries

Hirokazu Munakata and Kiyoshi Kanamura

CONTENTS

9.1 INTRODUCTION

The development of energy storage devices with high energy conversion efficiency is essential for effective use of energy. It is ideal that excess energy is stored and then used as needed. At present, lithium-ion batteries are in the mainstream of substantial battery development, which has been started from lead–acid batteries through the development and improvement of new electrode and electrolyte materials. In both gravimetric and volumetric energy densities, lithium-ion batteries are remarkably superior to other rechargeable batteries. Hence, light and portable lithium-ion batteries have been used as power sources in various kinds of electric devices. In recent years, they are used for large-scale applications such as electric vehicles and cogeneration systems. The excellent energy densities of lithium-ion batteries are mainly due to high voltage of more than 3 V. This is achieved by using organic electrolyte solutions, in which cathode materials with high potentials and anode materials with low potentials can be used. Actually, other rechargeable batteries have been developed based on aqueous electrolyte solutions and their voltages are strongly limited by a thermodynamic stability window of water (1.23 V).

In the recent decade, the energy density of lithium-ion batteries has been greatly improved and reached to over 200 W h kg^{-1}, but the rapid progress of electric devices still requires higher energy density. For example, electric vehicles need 500 W h kg^{-1} in order to drive 500 km by one charging, which is comparable to the usability of gasoline cars. According to the estimation based on conventional battery design, this energy density may be accomplished by using 300 mA h g^{-1} class cathode material and 1000 mA h g^{-1} class anode material. To meet these upcoming demands, new electrode materials with higher capacity have been studied. However, on the other hand, we are now faced with safety problems. This is mainly due to flammable organic solvents in electrolyte solutions. One approach is the use of flame retardants. This is somewhat effective but decreases lithium-ion conductivity of the electrolyte solution. In order to solve the issues of safety fundamentally, all-solid-state rechargeable batteries using nonflammable solid electrolytes should be developed. In this chapter, materials for all-solid-state rechargeable batteries are reviewed, and future development of battery technology that is expected by using solid electrolytes is described.

9.2 CLASSIFICATION AND CHARACTERISTICS OF SOLID ELECTROLYTES

Solid electrolytes are classified into two major categories: organic solid electrolytes (Table 9.1)[1-5] and inorganic solid electrolytes (Table 9.2).[6-15] As listed in Table 9.1, the former category includes gel electrolytes impregnated with organic electrolyte

TABLE 9.1
Typical Organic Solid Electrolytes

Composition (Polymer Matrix Electrolyte)	Ionic Conductivity (S cm^{-1})	Temperature (°C)
Gel polymer electrolytes		
Poly(ethylene oxide)$_8$–LiClO$_4$ (EC+PC)	10^{-3}	20
Poly(ethylene oxide)$_8$–LiClO$_4$ (PC)	8×10^{-4}	20
Poly(vinylidene fluoride)–LiN(CF$_3$SO$_2$)$_2$ (EC+PC)	1.5×10^{-3}	20
Poly(ethylene glycol acrylate)–LiClO$_4$ (PC)	10^{-3}	20
Poly(acrylonitrile)–LiClO$_4$ (EC+PC)	10^{-3}	20
Poly(methyl methacrylate)–LiClO$_4$ (PC)	2.3×10^{-3}	25
Intrinsic solid polymer electrolytes		
Linear type		
Poly(ethylene oxide)$_8$–LiClO$_4$	10^{-8}	20
Poly(oxymethylene)–LiClO$_4$	10^{-8}	20
Poly(propylene oxide)$_8$–LiClO$_4$	10^{-8}	20
Poly(dimethyl siloxane)–LiClO$_4$	10^{-4}	20
Branch type		
Poly[(2-methoxy)ethylglycidyl ether]$_8$–LiClO$_4$	10^{-5}	20
Poly(methoxy polyethyleneglycol methacrylate)$_{22}$–LiCF$_3$SO$_3$	3×10^{-5}	20
Polystyrene–poly(ethylene glycol) methyl ether methacrylate-LiClO$_4$	2×10^{-4}	30
Poly[bis((methoxyethoxy) ethoxy)-phosphazene])$_4$–LiN(CF$_3$SO$_2$)$_2$	5×10^{-5}	20

EC, ethylene carbonate; PC, propylene carbonate.

TABLE 9.2
Typical Inorganic Solid Electrolytes

Composition	Ionic Conductivity (S cm⁻¹)	Temperature (°C)
Sulfate type		
$Li_{10}GeP_2S_{12}$ (crystal)	1.2×10^{-2}	R.T.
$Li_{3.25}Ge_{0.25}P_{0.75}S_4$ (crystal: thio-LISICON)	2.2×10^{-3}	25
$0.03Li_3PO_4$–$0.59Li_2S$–$0.38SiS_2$ (glass)	6.9×10^{-4}	R.T.
$57Li_2S$–$38SiS_2$–$5Li_3PO_4$ (glass)	1.0×10^{-3}	R.T.
$70Li_2S$–$30P_2S_5$ (glass-ceramic)	3.2×10^{-3}	R.T.
$Li_7P_3S_{11}$ (glass-ceramic)	1.0×10^{-2}	R.T.
Oxide type		
$Li_{0.34}La_{0.51}TiO_3$ (crystal: perovskite)	2×10^{-5}	R.T.
$Li_{1.3}Al_{0.3}Ti_{1.7}(PO_4)_3$ (crystal-NASICON)	7×10^{-4}	25
$Li_7La_3Zr_2O_{12}$ (crystal: garnet)	3×10^{-4}	25
$Li_6BaLa_2Ta_2O_{12}$ (crystal: garnet)	4.4×10^{-5}	22
$Li_{2.9}PO_{3.3}N_{0.46}$ (LIPON) (glass)	3.3×10^{-6}	25
$50Li_4SiO_4$–$50Li_3BO_3$ (glass)	4×10^{-6}	25
Li_2O–Al_2O_3–SiO_2–P_2O_5–TiO_2 (glass-ceramic)	3×10^{-4}	25
$Li_{1.5}Al_{0.5}Ge_{1.5}(PO_4)_3$–$0.05Li_2O$ (glass-ceramic)	7.25×10^{-4}	R.T.
Nitrate type		
LiN_3	6×10^{-3}	25

R.T., room temperature.

solutions in polymer matrices and intrinsic polymer electrolytes without organic solvents (plasticizer). The lithium-ion batteries using gel electrolytes are generally called lithium-ion polymer batteries and have already been put into practical use. Since the risk of electrolyte leakage is reduced by using gel electrolytes, aluminum laminate films are used for packaging the batteries, and hence, the degree of freedom in battery design such as shape and flexibility is increased compared with conventional lithium-ion batteries using electrolyte solutions. When using intrinsic polymer electrolytes, it is possible to improve the safety of batteries and further extend the cell design. However, the intrinsic polymer electrolytes have not been used in practical batteries yet due to low lithium-ion conductivity. At least 10^{-3} S cm⁻¹ is necessary for practical use. The advantage of using organic solid electrolytes is that it can produce flexible batteries. New forms of battery such as wearable batteries are expected according to the development of new devices. However, organic solid electrolytes are basically flammable and many of them do not have sufficient thermal stability. In addition, self-discharge reactions involving the decomposition of electrolyte solutions proceed in the batteries using gel electrolytes, which is the same as in conventional liquid electrolyte batteries. Therefore, there are still many safety issues to be solved.

On the contrary, inorganic solid electrolytes are basically thermally stable and nonflammable. These characteristics are very important to improve the safety of batteries, particularly in large-scale batteries with relatively high capacity. Another notable feature of inorganic solid electrolytes is that the transport number of lithium ion is almost 1. The transport number is defined as the fraction of total current carried by a particular charge carrier. Thus, lithium ions can be transported more effectively in a solid electrolyte having a larger transport number. Actually, this parameter relates to not only the transfer efficiency of lithium ions but also the life and safety of batteries. It is known that the transport number of lithium ion in organic electrolyte solutions is about 0.4. The current corresponding to the remaining 0.6 relates to the transport of solvent molecules and counteranions. Thus, side reactions such as gas evolution and formation of resistant films on an electrode occur more than a little in addition to charge and discharge reactions. Since those side reactions can be minimized when using inorganic solid electrolytes, it is possible to achieve long life in batteries. From this point of view, the development of all-solid-state batteries has been under investigation for many years. For example, excellent reliability has been demonstrated such as the lithium–iodine battery system used in cardiac pacemakers. Recently, $LiCoO_2$ thin-film battery system using a glass electrolyte called LIPON has been put to practical use, in which 100,000 cycles of charge and discharge are guaranteed.[16] The applications of this type of batteries are expected to continue to grow in the future as power sources for radio frequency (RF) tags and smart cards although the capacity is small. Actually, a clear application of all-solid-state batteries using inorganic solid electrolytes has not still been established at present, but they are expected to contribute greatly to the efficient use of renewable energy due to the low self-discharge property that cannot be achieved by conventional battery systems. The thermal stability of inorganic solid electrolytes also contributes to extend the application of batteries. It is not preferable to use conventional lithium-ion batteries at temperatures above 40°C.[17,18] In particular, it should be avoided to keep the charged batteries at high temperatures. Although safety concerns are generally reduced by a safety circuit in commercially available batteries, the internal resistance is increased by the decomposition of electrolyte solution and associated gas generation as the temperature is high. Also in all-solid-state batteries, some degradation is expected to occur at the electrode/electrolyte interface with increasing temperature, but its amount is extremely small as compared with the batteries using organic electrolyte solutions. A rise in temperature rather improves the cell performance by increasing the lithium-ion conduction of inorganic solid electrolytes. Hence, inorganic solid electrolytes extend the operating temperature range of batteries and then will allow batteries to be used in new applications in which they have been difficult to be used until now, but of course lithium-ion conductivity in the low-temperature region below room temperature has to be improved. It is expected that new concepts and technologies such as directly mounting the battery onto a substrate in a reflow process will emerge in the future.

9.3 DEVELOPMENT OF ALL-SOLID-STATE BATTERIES USING INORGANIC SOLID ELECTROLYTES

All-solid-state batteries using inorganic solid electrolytes are categorized to thin-film and bulk types. The thin-film-type batteries are usually prepared using vapor deposition methods such as sputtering and pulsed laser deposition and have a typical structure in which thin films of electrolyte and electrode are alternately stacked. Oxide-based glasses have been mainly used as the electrolyte due to easiness to handle and chemical stability. Although the ionic conductivity of oxide-based glasses is very low, approximately 10^{-6} S cm^{-1} at room temperature, it is possible to operate the cells by reducing the effective resistance by making their thin layers. It is necessary to increase the amount of electrode active material in order to increase the energy density of the cell. However, the lithium-ion and electron conductivities of electrodes become insufficient when the electrode thickness is increased since the electrodes fabricated by vapor deposition methods are difficult to contain electrolyte and conductive materials. For this reason, there is a limit to the thickness of the electrode in thin-film-type batteries. In order to obtain practical capacity, it is required to make bulk-type batteries using the particles of electrode active materials as in conventional battery systems. In the case of conventional batteries using electrolyte solutions, the electrolyte solution penetrates into the electrode and then the electrochemically active electrode/electrolyte interface for charge and discharge reaction is spontaneously formed. In contrast, it is basically difficult to form a continuous interface between the electrode and electrolyte when using inorganic solid electrolytes having no fluidity (Figure 9.1). This is the main reason for the low output of all-solid-state batteries. If it is possible to sinter the mixture of solid electrolyte and electrode active material, their contact area can be greatly increased. However, many inorganic solid electrolytes require sintering at high temperatures, resulting in degradation of electrode active materials. Sulfide-based glass electrolytes are solid but have flexibility. This feature enables to make the composites with electrode active materials by compression molding as shown in Figure 9.2, in which heat treatment is not always needed, so that the composites including the conductive additives such as carbon can be also prepared. Some sulfide-based glass electrolytes have a lithium-ion conductivity as high as 10^{-3} S cm^{-1} at room temperature. Therefore, they are attracting much attention as electrolytes suitable for bulk-type batteries, and active researches are being conducted.[19] Recently, the performance of all-solid-state batteries using sulfide-based solid electrolytes is improved year by year, via progressive approaches such as preparation of fine particles and surface modification based on nanoionics (Figure 9.3). On the other hand, there is a serious

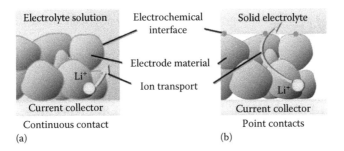

FIGURE 9.1 Effect of electrolyte fluidity on the formation of the electrochemical interface: (a) conventional electrolyte solution and (b) solid electrolyte.

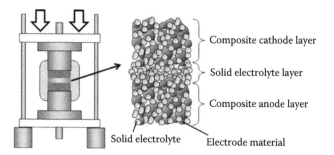

FIGURE 9.2 Preparation of all-solid-state battery composed of solid electrolyte and electrode materials by press molding.

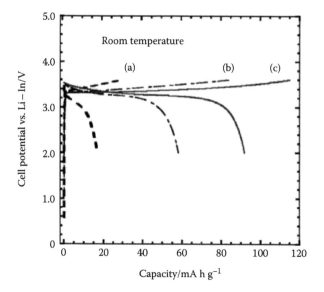

FIGURE 9.3 Charge and discharge performance of all-solid-state rechargeable lithium batteries using $80Li_2S-20P_2S_5$ solid electrolyte, Li–In anode, and (a) pristine $LiCoO_2$, (b) $LiNbO_3$-coated $LiCoO_2$, and (c) $80Li_2S-20P_2S_5$-coated $LiCoO_2$, measured at 130 mA cm^{-2} (Adapted from *J. Power Sources*, 196, Sakuda, A., Hayashi, A., Ohtomo, T., Hama, S., and Tatsumisago, M., 6735–6741, Copyright 2011, with permission from Elsevier.)

problem to generate H_2S gas reacted with water. For practical use, this problem has to be solved to ensure the safety of batteries.

9.4 HIGH-DIMENSIONAL STRUCTURES FOR BATTERIES

The energy and power densities of all-solid-state batteries are still low at present but will be comparable to those of current lithium-ion batteries by progress in research and development. Similarly, the performance of current lithium-ion batteries will be also improved and become less expensive day by day. Therefore, the specific applications that can utilize the advantages of all-solid-state batteries should be clarified for their practical use. Although improvement in battery safety is one of the major advantages provided by using inorganic solid electrolytes, it will be required to consider *all-solid-state* from different viewpoints and then find additional advantages. For example, the high mechanical stability is a major attraction of all-solid-state and enables to prepare various cell structures. So far, the batteries developed to date, including lead–acid batteries as well as lithium-ion batteries, are formed in a 2D manner, in which an electrolyte layer is sandwiched with anode and cathode layers. The kinetics of the charge and discharge reaction, that is, the output of batteries, is dependent on not only the charge transfer rate at the electrode/electrolyte interface but also the lithium-ion diffusion in an electrode layer. Therefore, even if a thicker electrode is prepared in an attempt to obtain higher cell capacity, the electrode is no longer fully functional since the diffusion of the lithium ion becomes difficult as increasing the electrode thickness. This is particularly significant when charge or discharge is performed at large currents, that is, high power densities. Aqueous electrolyte solutions such as potassium hydroxide solution used in nickel–hydrogen batteries and sulfuric acid solution used in lead–acid batteries have high ionic conductivity greater than 0.5 S cm^{-1}. Therefore, ion diffusion inside the electrode does not become a serious problem even in thick electrodes. On the other hand, the ionic conductivity of organic electrolyte solutions used in lithium-ion batteries is about 10^{-2} S cm^{-1}, which is smaller by one order of magnitude or more than that of aqueous electrolyte solutions. For this reason, the practical thickness of an electrode is limited up to ~100 μm in lithium-ion batteries. Hence, the power and energy densities of a battery are largely dependent on the ionic conductivity of the electrolyte. Actually, their balance is taken into account according to their application in practical batteries. For example, the electrode in lithium-ion batteries for portable devices is relatively thick and dense since energy density is required rather than power density in that application. On the other hand, a relatively thin electrode is used in applications that focus on high output such as hybrid and electric cars. When using a solid electrolyte with lower lithium-ion conductivity, this ion transport problem becomes more serious. However, the relationship between power and energy densities described herein is applied just to the batteries formed in the current 2D manner. Therefore, if the battery structure is changed, this relationship will be different. Figure 9.4a shows a schematic illustration of a new battery structure, called *3D battery*.[20,21] It is a structure in which cylindrical cathode and anode are alternately arranged. Both cathode and anode are formed respectively on a current collector in the form of a pinholder. The advantage of this structure is that the amount of electrode active material can be increased by maintaining a constant distance between the cathode and anode cylinders. Namely, even when increasing the capacity of the battery by forming a tall cylindrical electrode, it is possible to maintain the power density of the battery. In this electrode configuration, the ion diffusion between cathode and anode is expanded from current 1D to 2D as shown in Figure 9.4b, resulting in a higher power density. Thus, it is suitable for making a cell using an electrolyte material having low ion conductivity. Although the concept of 3D battery has been known for a long time, its practical development has not been started until recently due to many technical issues. At present, various techniques such as imprint and printing

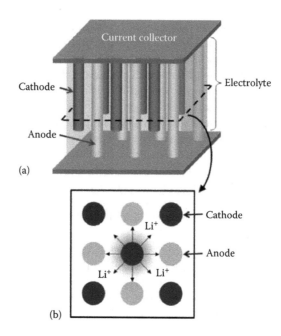

(a)

(b)

FIGURE 9.4 Schematic illustrations of (a) a 3D battery structure and (b) Li^+ diffusion in a 3D battery.

methods have been examined.[22,23] The most difficult point in the fabrication of a 3D battery is how to combine positive and negative electrodes without internal shorting. Since it is difficult to use a separator in 3D batteries with current technology, an electrolyte solution cannot be used basically. For this reason, solid electrolytes are promising candidates. Figure 9.5 shows the preparation process of a 3D battery using

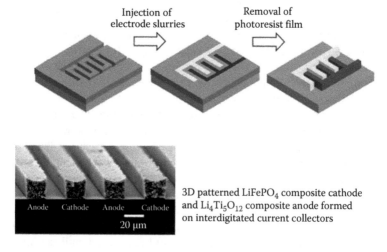

3D patterned $LiFePO_4$ composite cathode and $Li_4Ti_5O_{12}$ composite anode formed on interdigitated current collectors

FIGURE 9.5 Preparation process of 3D batteries by lithographic techniques.

the lithographic technique.[24] When preparing an electrode substrate with patterned current collectors for cathode and anode, a photoresist film is usually removed completely. However, a photoresist can be used as a support wall to prepare a 3D cathode and anode by leaving only the gap space between the patterned current collectors. A 3D cathode and anode can be prepared by injection of electrode slurries for cathode and anode onto corresponding current collectors under the support of the photoresist wall. The prepared 3D electrodes work as a 3D battery after the removal of the photoresist wall and following injection of an electrolyte. There is a great advantage that the electrode slurries for conventional battery production can be applied. If the photoresist film has ionic conductivity, a 3D battery can be obtained without the previously mentioned replacement process.

9.5 FABRICATION OF STRUCTURED INORGANIC SOLID ELECTROLYTES FOR 3D BATTERIES

In the development of 3D batteries, it is necessary to fabricate a 3D matrix from either electrode or electrolyte materials. In each case, a high mechanical strength enough to keep the structure of the 3D battery is required for the 3D matrix. Organic solid electrolytes are soft and flexible, so a 3D matrix should be prepared from electrode materials. On the other hand, inorganic solid electrolytes are rigid. Therefore, their 3D matrices can be fabricated, and then 3D batteries are prepared by filling them with electrode materials. When flexible sulfide-based glass electrolytes are used, it may be possible to fill them into the 3D matrix of the electrode. However, it will be necessary to devise ways to form a continuous electrode/electrolyte interface. Figure 9.6 shows a 3D matrix of $Li_{0.35}La_{0.55}TiO_3$ (LLT), which is one of the oxide-based ceramic electrolytes.[25] A lot of holes are formed alternately on the upper and lower surfaces of the electrolyte membrane. It is possible to obtain a 3D battery structure corresponding to Figure 9.4 by filling cathode and anode materials into the holes on the upper and lower surfaces, respectively. However, as described earlier, it is difficult to achieve good electrode/electrolyte contact by filling electrode materials simply to the holes since both electrode and electrolytes materials are solid without fluidity. Accordingly, new

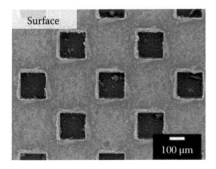 

FIGURE 9.6   3D matrix of $Li_{0.35}La_{0.55}TiO_3$ solid electrolyte for 3D batteries.

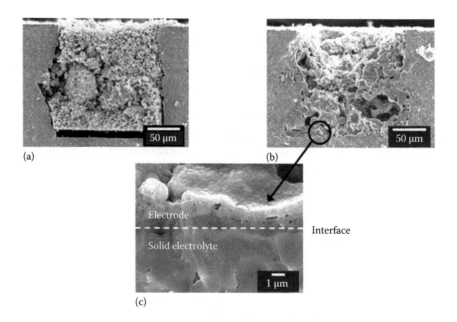

FIGURE 9.7 Electrode/electrolyte interfaces formed by the introduction of (a) only particles of the electrode material and (b) mixture of particles and precursor sol of electrode material. (c) A magnified image of (b).

ideas are required for interface formation. For example, it is possible to form a good interface by applying the sol-gel technique as shown in Figure 9.7. An important point is to use a mixture of precursor sol and particles of electrode material, instead of using only the particles. The precursor sol works as a glue to form a continuous interface between electrode and electrolyte materials. In order to construct an appropriate electrochemical interface, the kind of precursor sol is also considered. Figure 9.8 shows the ac impedance spectra for the $LiMn_2O_4$/LLT composite electrodes prepared using acetate- and nitrate-based precursor sols of $LiMn_2O_4$. A smaller interfacial resistance for nitrate-sol means that it works as a better glue for interface formation. Some kind of precursor sols may form impurities at the interface. In that case, the interfacial resistance is of course increased even when the contact between the electrode and electrolyte is well formed physically. Consequently, all-solid-state batteries of the bulk type that work at room temperature can be developed by applying 3D matrix and sol-gel techniques even when using oxide-based ceramic electrolytes (Figure 9.9). Similarly to current sandwich-type batteries, the amount of electrode materials has to be increased to obtain a higher energy density. However, a more precise design is required in all-solid-state 3D batteries. The detailed design varies depending on the arrangement of the electrodes, but it is basically required to form a cathode and an anode with a thickness of 50 μm or less in order to obtain a practical power density when using current electrode materials without conductive agents. Furthermore, the thickness of the solid electrolyte between a cathode and an anode is preferably as small as possible.

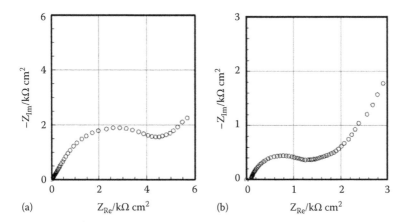

FIGURE 9.8   AC impedance spectra for $LiMn_2O_4$/LLT composite electrodes prepared using (a) acetate- and (b) nitrate-based precursor sols of $LiMn_2O_4$.

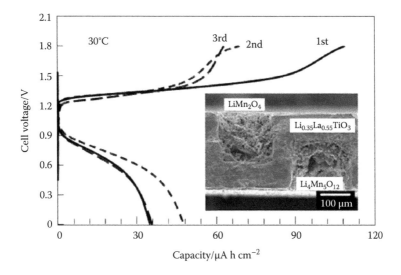

FIGURE 9.9   Charge–discharge curves of all-solid-state battery prepared by introducing $LiMn_2O_4$ cathode and $Li_4Mn_5O_{12}$ anode to 3D patterned $Li_{0.35}La_{0.55}TiO_3$ solid electrolyte, measured at a constant current density of 6.25 mA cm^{-2}.

Ceramic electrolytes having high mechanical strength are suitable for the construction of such 3D structures as compared with other solid electrolytes. Fabrication of structured inorganic solid electrolytes is one of the effective methods used to develop all-solid-state 3D batteries although many factors have to be considered and optimized, such as inexpensive molding techniques and heat treatment conditions for obtaining dense-sintered solid electrolyte 3D bodies.

9.6 BIPOLAR ALL-SOLID-STATE BATTERIES

It is possible to stack all-solid-state batteries as shown in Figure 9.10. This kind of structure is called a bipolar cell, in which single cells are connected in series and the cell components independent from the energy density such as current collectors and package film can be significantly reduced. Therefore, a greatly higher energy density can be achieved than that obtained in series connecting packed single cells. It has been very difficult to form the bipolar structure in current cells with electrolyte solutions having fluidity. In contrast, the formation of a bipolar structure is easy since the electrolytes used for all-solid-state batteries are solid. Hence, it can be said that the bipolar battery is the one utilizing the characteristics of a solid electrolyte effectively. Employing the bipolar structure probably leads to cost reduction in the preparation process as well as in the material. The advantage is found also on the side of battery use. Portable electric devices are typically incorporated with the electronic circuits to increase the voltage of battery. A bipolar battery can output a high voltage, so it is expected to omit such voltage boosting circuits. As demonstrated here, the application and value of a battery will be greatly extended by the formation of a bipolar structure, and an all-solid-state battery can contribute significantly to its practical development.

9.7 SUMMARY

Batteries play important roles in supporting the convenience of various devices and have become indispensable in our life. The improvement of energy and power densities while increasing the safety and reliability is a current issue of batteries, but the extensions of operation temperature and degree of freedom in shape are essentially necessary for future battery development. *All-solid-state* is a key technology to meet those upcoming demands and will lead to the practical use of 3D and bipolar batteries. This movement will also change the manufacturing technologies of battery and various methods for constructing 3D matrices such as imprint and lithography will be introduced. However, it should be noted that the complicated process will lead to the increase in manufacturing cost. The fundamental aspects such as the optimization of

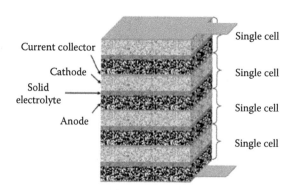

FIGURE 9.10 Schematic illustration of bipolar-type all-solid-state battery.

sintering conditions for solid electrolytes and the formation of appropriate electrode/electrolyte interface are still under investigation, and the convenience and usefulness of batteries will spread extensively through the application of solid electrolytes.

REFERENCES

1. J. Y. Song, Y. Y. Wang and C. C. Wan, *J. Power Sources*, 1999, **77**, 183–197.
2. A. M. Stephan, *Eur. Polym. J.*, 2006, **42**, 21–42.
3. F. B. Dias, L. Plomp and J. B. J. Veldhuis, *J. Power Sources*, 2000, **88**, 169–191.
4. K. Murata, S. Izuchi and Y. Yoshihisa, *Electrochim. Acta*, 2000, **45**, 1501–1508.
5. T. Niitani, M. Shimada, K. Kawamura and K. Kanamura, *J. Power Sources*, 2005, **146**, 386–390.
6. M. Tatsumisago and A. Hayashi, *Kagaku*, 2012, **67**, 19–23.
7. N. Kamaya, K. Homma, Y. Yamakawa, M. Hirayama, R. Kanno, M. Yonemura, T. Kamiyama, Y. Kato, S. Hama, K. Kawamoto and A. Mitsui, *Nat. Mater.*, 2011, **10**, 682–686.
8. R. Kanno and M. Murayama, *J. Electrochem. Soc.*, 2001, **148**, A742–A746.
9. F. Mizuno, A. Hayashi, K. Tadanaga and M. Tatsumisago, *Solid State Ionics*, 2006, **177**, 2721–2725.
10. Y. Inaguma, C. Liquan, M. Itoh and T. Nakamura, *Solid State Commun.*, 1993, **86**, 689–693.
11. H. Aono, E. Sugimoto, Y. Sadaoka, N. Imanaka and G. Adachi, *J. Electrochem. Soc.*, 1989, **136**, 590–591.
12. R. Murugan, V. Thangadurai and W. Weppner, *Angew. Chem. Int. Ed.*, 2007, **46**, 7778–7781.
13. V. Thangadurai and W. Weppner, *J. Power Sources*, 2005, **142**, 339–344.
14. X. Yu, J. B. Bates, G. E. Jellison, Jr. and F. X. Hart, *J. Electrochem. Soc.*, 1997, **144**, 524–532.
15. X. Xu, Z. Wen, X. Wu, X. Yang and Z. Gu, *J. Am. Ceram. Soc.*, 2007, **90**, 2802–2806.
16. THINERGY® Micro-Energy Cells (MECs) from Infinite Power Solutions, Inc. (IPS), http://www.cytech.com/products-ips, accessed on November 21, 2014.
17. M. C. Smart, B. V. Ratnakumar, J. F. Whitacre, L. D. Whitcanack, K. B. Chin, M. D. Rodriguez, D. Zhao, S. G. Greenbaum and S. Surampudi, *J. Electrochem. Soc*, 2005, **152**, A1096–A1104.
18. C. L. Campion, W. Li and B. L. Lucht, *J. Electrochem. Soc*, 2005, **152**, A2327–A2334.
19. A. Sakuda, A. Hayashi, T. Ohtomo, S. Hama and M. Tatsumisago, *J. Power Sources*, 2011, **196**, 6735–6741.
20. J. W. Long, B. Dunn, D. R. Rolison and H. S. White, *Chem. Rev.*, 2004, **104**, 4463–4492.
21. J. F. M. Oudenhoven, L. Baggetto and P. H. L. Notten, *Adv. Energy Mater.*, 2011, **1**, 10–33.
22. D. Golodnitsky, M. Nathan, V. Yufit, E. Strauss, K. Freedman, L. Burstein, A. Gladkich and E. Peled, *Solid State Ionics*, 2006, **177**, 2811–2819.
23. A. Izumi, M. Sanada, K. Furuichi, K. Teraki, T. Matsuda, K. Hiramatsu, H. Munakata and K. Kanamura, *Electrochim. Acta*, 2012, **79**, 218–222.
24. K. Yoshima, H. Munakata and K. Kanamura, *J. Power Sources*, 2012, **208**, 404–408.
25. M. Kotobuki, Y. Suzuki, H. Munakata, K. Kanamura, Y. Sato, K. Yamamoto and T. Yoshida, *J. Electrochem. Soc.*, 2010, **157**, A493–A498.

10 New Trend in Liquid Electrolytes for Electrochemical Energy Devices

Hajime Matsumoto

CONTENTS

10.1 INTRODUCTION

Electrochemical energy devices, such as secondary batteries and electric double layer capacitors, have already been widely used as an energy storage device for various portable electronic devices. Since the development of the lithium-ion battery (LIB), its higher energy storage ability has enabled its long time operation and also its miniaturization and weight reduction, which has provided a greater convenience in our daily life. Not only small devices but also much larger electric vehicles significantly depend on the research and development of LIBs with a high capacity. One of the important features of LIBs compared to various other secondary batteries, such as the lead–acid battery and nickel–metal hydride battery, is the higher output cell voltage (over 4 V). To realize this high cell voltage, organic solvents with an electrochemical stability, such as carbonate solvents, have been used as the electrolyte media instead of water.[1] However, developing new electrolytes has always been necessary to improve the safety issue of the LIB because the typical organic solvents are generally flammable liquids. On the other hand, the research and development of a post-LIB, such as the lithium–air (Li–air) battery, lithium–sulfur (Li–S) battery, and magnesium metal battery, has become gradually active due to the potentially higher theoretical capacity than that of the LIBs. However, these new battery systems require a totally new electrolyte appropriate for each device. For example, the lithium metal anode adopted in the Li–air and Li–S batteries needs a suitable electrolyte, which can suppress dendritic lithium formation during the charge and discharge cycle.[2] Also, the magnesium metal anode has exhibited a good charge and discharge cycle capability only in tetrahydrofuran-containing Grignard reagents; however, the oxidation potential of the reagent is not high enough to combine with a high potential cathode due to the existence of chloride or bromide.[3] These facts suggest that the research and development of a new electrolyte will be quite important not only to improve the safety of the current LIBs but also to realize the next-generation secondary batteries.

In this chapter, the recent trend in electrolytes for electrochemical devices will be reviewed. Especially, the new materials so-called zero solvents, such as ionic liquids (ILs) and ambient temperature alkali metal molten salts, will be the focus because these new electrolytes might have a high potential.

10.2 CONVENTIONAL ELECTROLYTE FOR A LITHIUM-ION BATTERY

The typical components of a lithium secondary battery (LIB) are schematically illustrated in Figure 10.1. The most important feature of the LIBs is their high output voltage around 4 V using a carbon material as the anode and a transition metal oxide as the cathode. The conventional electrolytes for LIBs are basically composed of mixed carbonate solvents, such as dimethyl carbonate and ethyl carbonate, and a lithium salt based on PF_6^-. These materials were selected based on their electrochemical stability toward a cathode and an anode. For example, the oxidation

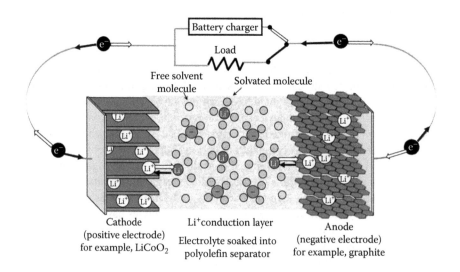

FIGURE 10.1 Schematic illustration of an LIB.

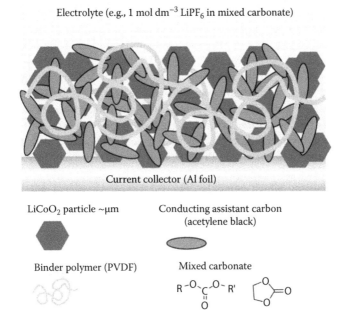

Electrolyte (e.g., 1 mol dm^{-3} LiPF$_6$ in mixed carbonate)

Current collector (Al foil)

LiCoO$_2$ particle ~μm

Conducting assistant carbon (acetylene black)

Binder polymer (PVDF)

Mixed carbonate

FIGURE 10.2 Schematic illustration of a composite electrode.

potential of PF$_6^-$ is over +6.0 V versus Li/Li$^+$, which is more positive than the redox potential of an oxide cathode, such as lithium cobalt oxide (ca. +4.2 V vs. Li/Li$^+$). The carbonate solvents act not only as a liquid media for the Li$^+$ conduction but also as an essential film-forming agent on the anode and cathode, which protects further decomposition of the electrolyte components. The surface film, which is usually called as the solid electrolyte interphase (SEI), also serves as a molecular sieve layer for the required desolvation process of the solvated lithium ion during the intercalation reaction at the electrode surface. It is not an exaggeration to say that only the carbonate electrolyte has been used in practical batteries. Apart from such a function of the carbonate solvent, liquid electrolytes are still important for use in LIBs. As shown in Figure 10.2, the actual electrode for LIBs is composed of submicron-sized particles of the active materials, such as lithium cobalt oxide, and conductive auxiliary particles, such as acetylene black, in order to achieve a high output power density. The liquid electrolyte can easily soak into the nanospace of the composite electrode in order to maintain a good contact between the surface of the active material and the electrolyte.[1]

10.3 NEW ELECTROLYTE SYSTEM BASED ON ZERO SOLVENT

10.3.1 IONIC LIQUIDS

The use of carbonate electrolytes is essential for the current LIBs as stated in the previous section; however, a thermal runaway due to the flammability of the organic solvents might occur. The usefulness of the SEI derived from the carbonates might not be directly applicable to different battery systems such as the magnesium metal battery. Therefore, a new liquid electrolyte system is significantly desired not only to solve the safety issue but to also allow smooth operation of the next-generation batteries. One of the zero solvents, so-called ILs, has been extensively investigated for such a requirement during the past two decades. The ILs are composed only of ionic species without any neutral molecule; therefore, the ILs initially attracted much attention as a less-flammable electrolyte. The typical ionic species in ILs, which are composed of onium cations, such as 1-ethyl-3-methylimidazolium (C_2mim^+) and N-methyl-N-butylpyrrolidinium (C_4pyr^+) and perfluoroanions, such as BF_4^- and *bis*(trifluoromethylsulfonyl)amide (Tf_2N^-), have been extensively investigated as a novel liquid with various unique features, such as less flammability, less volatility, and thermal stability, which cannot be obtained from conventional molecular liquids (Figure 10.3).[4] During the early stage of the research and development of the ILs for LIBs, the use of a conventional composite cathode for organic solvent electrolytes was found to be compatible with the ILs containing a certain amount of lithium salt. This means that the oxidation limiting potential of the ILs composed of perfluoroanions, such as BF_4^- and Tf_2N^-, were significantly positive to operate a 4 V class cathode, such as lithium cobalt oxide ($LiCoO_2$). On the other hand, the reduction limiting potential of the aliphatic quaternary ammonium cation was apparently sufficiently negative to use a lithium metal anode.[5] This means that the charge and discharge of $Li/LiCoO_2$ cell at a 4 V level could be obtained with only the use of the aliphatic quaternary ammonium–based ILs. These early studies suggested the possibility of the ILs as an attractive electrolyte candidate for new LIBs; however, at the same time, various problems were also revealed. One of the major defects of using ILs was the incompatibility of the carbon anode, which has been an essential component to construct LIBs, especially in the 1-ethyl-3-methylimidazolium systems due to the intercalation reaction of the aromatic cation instead of Li^+. Another intrinsic defect of the ILs was their viscous nature due to the strong columbic interaction between the cation and anion. The viscosity of the ILs is at least 10 times higher than that of conventional electrolytes. Therefore, the lithium-ion transport in viscous ILs was much lower than that in conventional organic solvents. However, the problem derived

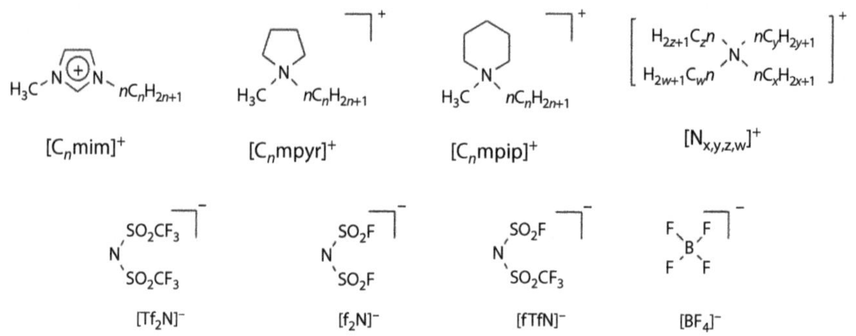

FIGURE 10.3 Typical ionic structure of ILs.

from the high viscosity of ILs seems not to be critical because the viscosity of liquids significantly decreased with the increasing operating temperature. Considering the good thermal stability of the ILs, the liquid media seem well suited for a relatively large scale and stationary storage system.[6]

After these early studies, these problems seemed to be solved to a certain degree with the advent of new anionic species, such as *bis*(fluoromethylsulfonyl)amide (f_2N^-). The ILs composed of this anion and 1-ethyl-3-methylimidazolium (C_2mim^+) had a relatively low viscosity at room temperature (15 mPa s) and also the viscosity increase usually observed in the ILs system with the addition of a certain amount of a lithium salt (ca. 1 mol dm^{-3}) was about twice, which was much lower than that of four times observed in the corresponding ILs composed of Tf_2N^-. However, the most striking feature was that the carbon anode was compatible even in the [C_2mim][f_2N] without any additives to form an SEI on the carbon anode. Furthermore, the charge transfer resistance of various electrodes in the ILs, which is one of the important factors to control the battery performance, was predominantly low among the various C_2mim-ILs composed of various perfluoroanions. Such a good performance must depend on the structure of f_2N^-, which might be suitable to reduce the interaction between Li^+ and f_2N^-. Recent fundamental studies of the [f_2N]$^-$ in terms of theoretical calculations and also in view of solution chemistry imply that a significant difference indeed existed between the [f_2N]$^-$ and the other perfluoroamide, such as [Tf_2N]$^-$, during the interaction between Li^+ and [f_2N]$^-$ and also between [f_2N]$^-$ and a lithium metal and the complex structure around a Li^+. Since the first report on the ILs based on f_2N^- for the LIBs, the research and development of LIBs using ILs has become more and more active along with the verification of the thermal stability of the LIBs using ILs.[7]

10.3.2 INTERMEDIATE TEMPERATURE MOLTEN SALT COMPOSED OF ALKALI METAL CATION

The melting points of the ILs are indeed below room temperature; however, the existence of the onium cation, such as 1-ethyl-3-methylimidazolium, is not an essential ionic species for constructing lithium battery electrolytes. If the room-temperature lithium salt composed of the perfluoroanion could be applied to LIBs, the concentration of the lithium cation would always be constant inside the battery, which means no concentration polarization occurs and also the electrochemical stability of the anode, such as lithium metal, does not need to be considered (Figure 10.4). However, the melting point of such a lithium salt, if it existed, is generally much higher than a few hundred degrees Celsius. Recently, new ILs composed of a hybrid amide anion, such as the (fluorosulfonyl)(trifluoromethylsulfonyl)amide ($fTfN^-$), have been reported. The $fTfN^-$ has the potential ability to lower the melting point of various onium salts, which did not form ILs with Tf_2N^- and f_2N^- due to its structural asymmetry. Furthermore, the melting point of the lithium salt (LifTfN) of 100°C is the lowest temperature among the lithium salts composed of perfluoroanions.[8] Though the viscosity of the alkali metal molten salt is over two orders of magnitude higher than the corresponding ILs, a stable charge and discharge of the conventional cathode and anode for LIBs was observed at 110°C at relatively high current density. One of the

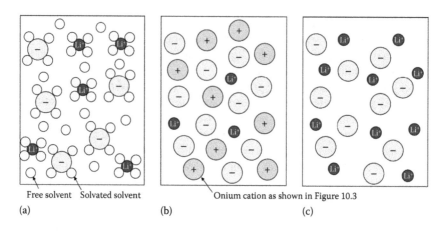

Free solvent Solvated solvent   Onium cation as shown in Figure 10.3

(a)   (b)   (c)

FIGURE 10.4 Schematic illustration of various liquid electrolytes for LIBs composed of low-coordinating anion such as perfluoroanion as shown in Figure 10.3. (a) Organic electrolytes; (b) ILs; (c) lithium molten salt.

reasons for such stable cycling might be due to a very high Li^+ transference number (>0.94). The performance of a cell containing the viscous molten lithium salt was not superior to that with a liquid electrolyte, such as ILs, and also a conventional organic solvent electrolyte; however, various useful information regarding the function of perfluoroanions will be obtained without considering the effect of an organic solvent or onium cations in the ILs. At this time, LifTfN is the only example of a single lithium salt, which can be applied to conventional composite electrodes for LIBs. On the other hand, a sodium-ion battery system has been reported that uses mixed alkali metal salts, such as Naf_2N-Kf_2N.[9] The melting point of the mixed molten salt is lower than that of the corresponding single salt. Also, the viscosity of the mixed molten salt is generally reduced at the same temperature. The mixing technique might be effective to improve the physical properties of the alkali metal salts; however, the concentration polarization of the sodium ion is inevitable. Therefore, low-melting and low-viscosity single molten salts composed of a perfluoroanion will be desired for the ideal zero-solvent electrolyte in energy storage devices.

10.3.3 ORGANIC IONIC PLASTIC CRYSTALS

As stated earlier, preparing new low-coordinating anions and introducing structural asymmetry into the anion have been the key strategy to reduce both melting point and viscosity of zero solvents, which must be the most important points to apply zero solvents to LIB systems. Therefore, onium salts composed of perfluoroanions such as Tf_2N^- and BF_4^- with higher melting point than room temperature usually were not pursued as an electrolyte media. However, recent studies revealed that some of those onium salts exhibited relatively high ionic conductivity even in solid state.[10–12] Such a solid salt has a potential as a unique solid-state conductor with doping a certain amount of lithium salts. The striking features of such a unique salt usually indicate *plastic crystal phase*. The cations or anions exhibit rotatory

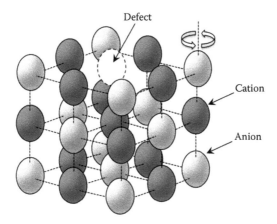

FIGURE 10.5   Schematic illustration of rotator motion of ions in organic ionic plastic crystals.

motion; however, translational motion is prohibited in the plastic crystal phase.[13] An existence of a small amount of lattice defects might support the movement of Li⁺ by hopping mechanism (Figure 10.5). Plastic crystal phase is also seen in organic molecule such as succinonitrile, which is sticky solid at room temperature. The addition of a small amount of alkali metal salts in the succinonitrile also exhibited relatively high conductivity ($>10^{-3}$ S cm⁻¹).[14] However, such organic molecule is usually volatile and flammable like organic solvents. Therefore, organic ionic plastic crystals might also be a much attractive candidate as a less-flammable and nonvolatile solid electrolyte media like ILs and alkali metal molten salts. Further studies on such unique solid electrolyte have been much desired; however, the molecular design for preparing superior organic ionic plastic crystals, which exhibit high ionic conductivity in a wider temperature range, has not yet been well established. One of the good examples of such organic plastic crystals, *N,N*-diethyl-*N*-propyl-*N*-methylammonium trifluoromethyltrifluoroborate, which indicates transparent, self-standing, and flexible film over 100°C (−40°C to 95°C), suggests that the rugby-ball-shaped ion might be preferable.[12] A recent review is recommended to understand the researches on this area.[15]

10.4   ANOTHER RELATED TOPIC—SUPER-CONCENTRATED ELECTROLYTE

As stated in the previous sections, two zero-solvent systems for electrochemical energy devices were briefly reviewed. The preparations of novel perfluoroanions were quite important not only for the electrochemical stability but also for the improvement of the physical properties, such as the transport properties, and thermal properties of the resulting zero solvents. Due to the low coordination ability and high oxidation stability of the perfluoroanions, such as PF_6^- and BF_4^-, they have been actually used as a counter anion of the lithium salts in organic solvents for the LIBs. The concentration of the lithium salts of these used organic electrolytes is ca. 1.0 mol dm^{-3}. At such a high concentration, the number of free solvent molecules per lithium salt is quite low (<2). This suggests that the existence of a few solvent molecules per lithium salt significantly enhances the diffusion ability of the ionic species, which are tightly solvated by three to four solvent molecules. Recently, much denser electrolytes (>3 mol dm^{-3}) have attracted much attention due to their unique properties. An equimolar mixture of LiTf$_2$N (lithium bis(trifluoromethylsulfonyl) amide) and triglyme (triethylene glycol dimethyl ether) has exhibited a specific nature like the zero solvent. This liquid is called a *solvated IL*, which means that no free triglyme molecule is present in the liquids.[16] The thermal stability and also the electrochemical stability of the solvated ILs were much superior to the conventional electrolyte composition, in which free triglyme molecules were present. The solvated ILs indeed contains a flammable organic solvent; however, the liquid is expected to be a unique electrolyte for next-generation batteries such as the Li–S batteries, which do not use the conventional organic solvent electrolyte due to the poor chemical stability of the sulfur.[17] Not only in the next-generation battery but also in conventional LIBs, a highly concentrated electrolyte (>3 mol dm^{-3}) could defy conventional knowledge about the electrolyte for LIBs. For example, unusual organic solvent like dimethoxyethane, in which a carbon anode has never worked, can be successfully applied in the super-concentrated region.[18] It is no exaggeration to say that an organic molecule at such a high concentration no longer acts as a liquid media but as an additive agent to the alkali metal salts.

10.5   SUMMARY

In this chapter, recent trends in liquid electrolytes based on zero solvents for electrochemical energy storage devices, especially LIBs, were briefly reviewed. The low-coordinating perfluoroanion is a key component to apply these zero solvents as a novel electrolyte. Not only the physicochemical properties and electrochemical properties of the zero solvents but also the resulting performance of an LIB is significantly affected by the structure of the perfluoroanions. The bis(fluorosulfonyl) amide anion was the best anion among the reported anions during the past decade.[7] The safety issue derived from the use of flammable organic solvents as an electrolyte media was one of the motivations for the research and development of zero-solvent systems. However, as for the current LIBs, the thermal stability of ILs with active materials, such as a charged cathode based on a transition metal oxide and an anode based on carbon, was not always much better than the zero solvent alone. The research and development of thermally stable composite electrodes will be needed to improve the actual safety of the LIBs. On the other hand, recent studies have revealed that these new electrolytes open the possibility of next-generation batteries, that is, the so-called post-LIBs. The electrode materials for such new devices generally cannot be used in conventional organic solvents due to the lack of a substantial compatibility. The research and development of a new electrolyte for such new devices will be the most important objective because such a new electrolyte must be essential for investigating a cathode and an anode suitable for post–lithium battery systems. The new electrolyte introduced here, such as zero solvents, solvated ILs, and a super-concentrated electrolyte, will be an attractive candidate not only for such demands but also for unknown novel applications.[19]

REFERENCES

1. M. Yoshio, R. J. Brodd and A. Yokozawa, eds., *Lithium-Ion Batteries: Science and Technologies,* Springer, LLC, New York, 2009.
2. B. Scrosati, K. M. Abraham, W. A. Schalkwijk and J. Hassoun, eds., *Lithium Batteries: Advanced Technologies and Applications,* Wiley, New York, 2013.
3. D. Aurbach, Z. Lu, A. Schechter, Y. Gofer, H. Gizbar, R. Turgeman, Y. Cohen, M. Moshkovich and E. Levi, *Nature*, 2000, **407**, 724.
4. B. Kirchner, ed., *Ionic Liquids,* Springer, Berlin, 2009.
5. H. Matsumoto, *Electrochemical Aspects of Ionic Liquids*, ed. H. Ohno, Wiley, Hoboken, NJ, 2nd edn., 2011, ch. 4.
6. H. Sakaebe and H. Matsumoto, *Electrochemical Aspects of Ionic Liquids*, ed. H. Ohno, Wiley, Hoboken, NJ, 2nd edn., 2011, ch. 14.
7. H. Matsumoto, *Electrolytes for Lithium-Ion Batteries*, eds. T. R. Jow, K. Xu, O. Borodin and M. Ue, Springer, New York, 2014, ch. 4.
8. K. Kubota and H. Matsumoto, *J. Phys. Chem. C*, 2013, **117**, 18829.
9. A. Fukunagawa, T. Nohira, Y. Kozawa, R. Hagiwara, S. Sakai, K. Nitta and S. Inazawa, *J. Power Sources*, 2012, **209**, 52.
10. D. R. MacFarlane, J. Huang and M. Forsyth, *Nature*, 1999, **402**, 792.
11. Y. Abu-Lebdeh, P.-J. Alarco and M. Armand, *Angew. Chem. Int. Ed.*, 2003, **42**, 4499.
12. Z. B. Zhou and H. Matsumoto, *Electrochem. Commun.*, 2007, **9**, 1017.
13. H. Ishida, T. Iwachido and R. Ikeda, *Ber. Bunsenges. Phys. Chem.*, 1992, **96**, 1468.
14. P.-J. Alarco, Y. Abu-Lebdeh and M. Armand, *Nat. Mater.*, 2004, **3**, 476.
15. J. M. Pringle, P. C. Howlett, D. R. MacFarlane and M. Forsyth, *J. Mater. Chem.*, 2010, **20**, 2056.
16. K. Ueno, K. Yoshida, M. Tsuchiya, N. Tachikawa, K. Dokko and M. Watanabe, *J. Phys. Chem. B*, 2012, **116**, 11323.
17. N. Tachikawa, K. Yamauchi, E. Takashima, J.-W. Park, K. Dokko and M. Watanabe, *Chem. Commun.*, 2011, **47**, 8157.
18. Y. Yamada, M. Yaegashi, T. Abe and A. Yamada, *Chem. Commun.*, 2013, **49**, 11194.
19. M. Armand, F. Endres, D. R. MacFarlane, H. Ohno and B. Scrosati, *Nat. Mater.*, 2009, **4**, 621.

11 Organic Electrode Active Materials for Rechargeable Batteries

Recent Development and Future Prospects

Masaru Yao and Tetsu Kiyobayashi

CONTENTS

11.1 INTRODUCTION

Rechargeable lithium batteries are widely used as a major power source for daily-use portable electric devices. Typical rechargeable lithium batteries are composed of a metal-oxide-based positive electrode (e.g., $LiCoO_2$) and a graphite-based negative electrode, for which various materials have been proposed to increase their energy densities. Meanwhile, the recent concern about the resource limitations and environmental burden has compelled us to use rare metal–free and less resource-intensive materials in the batteries, especially for the positive electrodes. Many attempts have been made to reduce the use of rare metals by partially replacing them with other abundant elements, for example, partially substituting Fe or Mn for Co. A different approach is to totally forgo any metallic element in the electrode active materials by replacing them with redox active organic compounds that essentially contain no scarce metals. Using these redox active organic materials would alleviate the restrictions posed by the resource scarcity. The preparation processes of organic materials can be less energy intensive as well when compared to inorganic materials because the latter are usually processed at higher temperatures than the former. Furthermore, we have recently revealed that certain organic compounds can work as a positive electrode active material not only in the lithium systems but also in systems in which sodium or magnesium ions are a charge carrier, providing the possibility of developing post-lithium rechargeable batteries (see Section 11.4).

This chapter outlines the characteristics of these redox active organic compounds mostly as *positive* electrode materials in the lithium and other systems by reviewing the literature and the authors' recent studies. Section 11.4.1.4 is an exception in which the properties of an organic salt as a negative electrode are described. Also discussed in this chapter are several techniques used to analyze the redox reaction mechanism of organic materials.

11.2 POLYMERIC MATERIALS

11.2.1 π-CONJUGATED POLYMERS

Since the 1980s, several types of polymers have been investigated during the research on organic materials for rechargeable batteries. The first category includes classical π-conjugated conductive polymers, represented by polyaniline, polypyrrole, and polythiophene (Figure 11.1). These polymers are characterized by their *ion-doping* properties. As the electrode active materials, these polymers are charged and discharged through the anion doping (insertion and deinsertion) accompanied by the redox reactions as shown in Figure 11.2. This mechanism differs from that of the conventional inorganic active materials, such as lithium cobalt oxide ($LiCoO_2$),

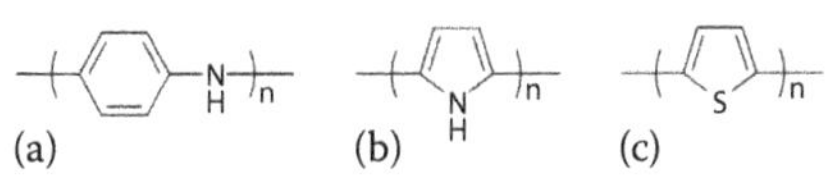

FIGURE 11.1 Representative examples of π-conjugated polymers as positive electrode materials (a: polyaniline, b: polypyrrole, c: polythiophene).

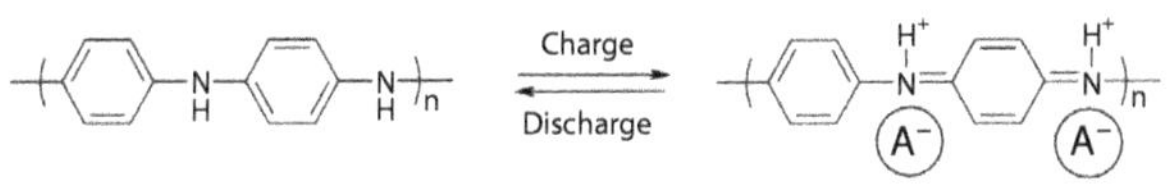

FIGURE 11.2 Possible charge and discharge mechanism of polyaniline.

in which the insertion and deinsertion reaction of the cation occurs, that is, Li^+, during the discharge and charge process. The discharge potential of a representative polyaniline can be as high as ca. 3.7 V versus Li^+/Li under certain conditions, which is comparable to that of the current $LiCoO_2$, 3.8 V versus Li^+/Li.[1] If all the aniline units participate in the redox reaction, the charge and discharge capacity would theoretically reach ca. 300 mA·h·g^{-1}, more than twice the practical capacity of $LiCoO_2$, 140 mA·h·g^{-1}. The experimentally observed capacities of polyanilines are, however, only about 100 mA·h·g^{-1}. The capacity of the polymers in this category is, in principle, dictated by the upper limit of the anion doping level, above which the polymer decomposes or side reactions take place. The most π-conjugated polymers have a similar doping limit, posing a barrier to increasing the capacities. If the mass of the anions that are involved in the reaction is taken into consideration, the available real gravimetric capacity further decreases.

11.2.2 PENDANT-TYPE POLYMERS

The electrode performance of pendant-type polymers that carry redox sites connected to the saturated hydrocarbon main chains has been investigated. The structures are different from the π-conjugated polymers (Section 11.2.1), which have redox sites in their main chains. This category includes a series of radical polymers, carrying 2,2,6,6-tetramethylpiperidine-N-oxyl (TEMPO) as a pendant group, synthesized by Nakahara et al. of the NEC Corp. (Japan) (Figure 11.3),[2,3] and a polymer carrying the ferrocene moiety, an organometallic compound, synthesized by Masuda et al. of Kyoto University (Japan) (Figure 11.4).[4] Although the redox reaction in either polymer entails the anion insertion and deinsertion as in the π-conjugated polymers, the pendant-type polymers do not suffer from a severe limitation of the doping level, that is, their utilization ratio, an indicator of how much of their redox sites are working, can be higher than those of the π-conjugated polymers. For example, the utilization ratio of a radical polymer becomes almost 100%.[3] The discharge potential is relatively high, ca. 3.6 V versus Li^+/Li. Its output power performance (i.e., properties at a high current density) and cycle-life stability are excellent (Figure 11.3). However, despite the high utilization ratio, the discharge

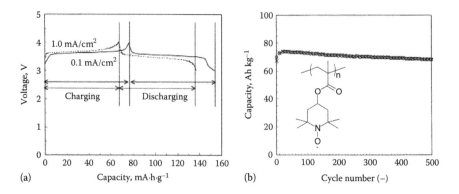

FIGURE 11.3   Charge and discharge performance of a radical polymer (PTMA). (a) Charge and discharge curves of a PTMA/Li cell. (b) Its cycle-life performance. (Adapted from *Chem. Phys. Lett.*, 359, Nakahara, K., Iwasa, S., Satoh, M., Morioka, Y., Suguro, M., and Hasegawa, E., 351, Copyright 2002, with permission from Elsevier.)

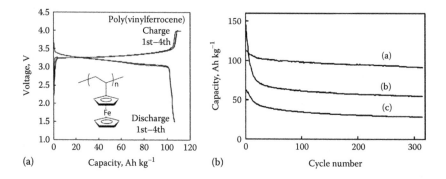

FIGURE 11.4   Charge and discharge performance of a cell using a polymer carrying ferrocene moiety, poly(vinylferrocene). (a) Typical charge and discharge curves of a poly(vinylferrocene/Li cell (current density: 70 mA·g^{-1}). (b) Cycle-life performance (a: poly(vinylferrocene), 200 mA·g^{-1}; b: poly(ethynylferrocene), 250 mA·g^{-1}; c: poly(ferrocene), 290 mA·g^{-1}). (Tamura, K., Akutagawa, N., Satoh, M., Wada, J., and Masuda, T.: *Macromol. Rapid Commun.* 2008. 29. 1944. Copyright Wiley-VCH Verlag GmbH & Co. KGaA. Adapted with permission.)

capacity barely reaches about 100 mA·h·g^{-1}, which will not be easily increased due to the intrinsic high molecular weight of the constituent monomer. The ferrocene polymer also has a high utilization ratio of more than 80% and relatively stable cycle-life performance (Figure 11.4).[4]

The authors also examined a pendant-type polymer, poly(N-vinylcarbazole) (PVK).[5] The positive electrode using PVK as the active material has the initial discharge capacity of about 120 mA·h·g^{-1} at a mean potential of 3.7 V versus Li$^+$/Li, and is stably cycled (Figure 11.5). The possibility of applying this polymer as a redox active *binder* was also examined using a composite electrode with lithium iron phosphate (LiFePO$_4$). The capacity of the LiFePO$_4$ electrode is augmented by replacing

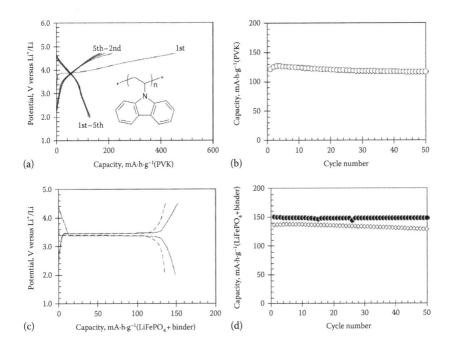

FIGURE 11.5 (a) Charge and discharge curves of the PVK electrode during initial cycles. (b) Cycle-life performance (current density: 20 mA·g⁻¹, potential range: 2.0–4.7 V vs. Li⁺/Li). (c) Charge and discharge curves of the composite electrodes (solid line: LiFePO₄-PVK; broken line: LiFePO₄-PVDF). (d) Cycle-life performances (●: LiFePO₄-PVK; ◇: LiFePO₄-PVDF; current density: 30 mA · g⁻¹(LiFePO₄), potential range: 2.0–4.7 V vs. Li⁺/Li). (Adapted from *J. Power Sources*, 202, Yao, M., Senoh, H., Sakai, T., and Kobayashi, T., 364, Copyright 2012, with permission from Elsevier.)

polyvinylidene fluoride (PVDF), the conventional binder, with PVK, the possible redox active binder. This methodology of using a redox active polymer as a binder will increase the total energy density of the batteries.

11.2.3 DISULFIDE POLYMERS

The third category is a series of organosulfur compounds bearing disulfide bonds, R–S–S–R'. Elemental sulfur, S, itself undergoes a charge and discharge reaction, of which the capacity can theoretically be very high at 1672 mA·h·g⁻¹ if the full redox reaction occurs between S and Li_2S, that is, $S + 2Li^+ + 2e^- \rightleftarrows Li_2S$. In reality, no one has yet succeeded in delivering capacities close to the theoretical value of the elemental sulfur with a satisfactory cycle life. To solve this problem, Visco et al. of the University of California (United States) prepared compounds in which the redox active sulfur is incorporated into polymers as a disulfide bond (Figure 11.6).[6] A lithium ion is inserted into the disulfide bond of these polymers during the reduction or the discharge process if used as a positive electrode. The theoretical capacities would be 300–500 mA·h·g⁻¹, from which the experimentally observed values are still very

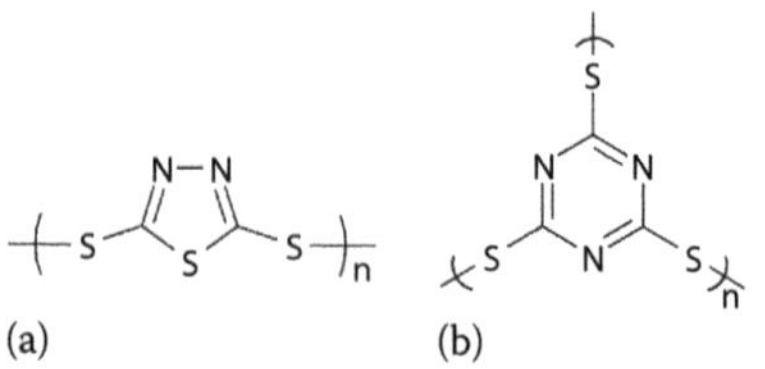

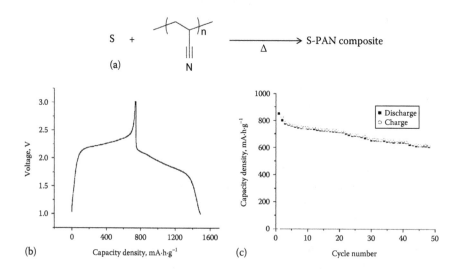

FIGURE 11.6 Representative organosulfur polymers carrying disulfide bonds: (a) dimercapto dithiazole polymer and (b) trithiocyanuric acid polymer.

low, and the discharge potential is not very high, ca. 2 V versus Li^+/Li. The cycle life is also not satisfactory, a reason for which would be that the lithium insertion into the disulfide bonds fragmentizes the polymers, resulting in the dissolution of the redox active sites into the electrolyte solutions during cycling.

11.2.4 SULFUR–POLYMER COMPOSITES

Recently, sulfur–polymer composites have been drawing attention as a superior positive electrode material for rechargeable lithium batteries. What is interesting about this composite is that it is simply prepared by calcinating elemental sulfur with a raw polymer without resorting to the organic synthetic method. A representative example is the composite of polyacrylonitrile (PAN) reported by Wang et al. of the Chinese Academy of Sciences (China) (Figure 11.7).[7] While the discharge potential is comparable to those of the disulfide-type polymers (~2 V vs. Li^+/Li, see Section 11.2.3),

FIGURE 11.7 (a) Preparation scheme of the S-PAN composite. (b) Charge and discharge curve of S-PAN. (c) Cycle performance. A gel-type electrolyte was used in the charge and discharge test. (b and c: Wang, J., Yang, J., Xie, J., and Xu, N.: *Adv. Mater.* 2002. 14. 963. Copyright Wiley-VCH Verlag GmbH & Co. KGaA. Adapted with permission.)

what remarkably distinguishes this sulfur–polyacrylonitrile (S-PAN) composite is its experimentally observed very high capacity, ca. 800 mA·h·g^{-1}. In addition, contrary to many disulfide polymers that suffer from poor cycle stabilities, this composite maintains 600 mA·h·g^{-1} after about 50 cycles just in the first report. What brought these excellent properties is currently unknown because neither the chemical structure nor the redox mechanism is known. Many groups are trying to further extend the cycle life.[8]

11.2.5 Quinone Polymers

Quinone-based materials have been studied as another type of organic active material, since the quinone skeleton undergoes a two-electron redox reaction, which should lead to a high discharge capacity (Figure 11.8).[10–12] One frequently comes across the quinone skeleton in biochemical molecules, some of which play an important role as redox centers in biological electron-transport systems. To anchor the quinone skeleton in the electrode of batteries, polymerization is one approach.[1,9,10] For example, Foos et al. of EIC Laboratories, Inc. (United States) examined the battery performance of a polymer in which benzoquinone monomers are directly bound to each other (Figure 11.9a).[9] The discharge potentials are close to 3 V versus Li$^+$/Li, higher than those of the sulfur-based polymers. The experimentally observed capacity is about one quarter the theoretical upper limit of 506 mA·h·g^{-1}, which is estimated based on the two-electron redox of the benzoquinone moiety. Owen et al.

FIGURE 11.8 Redox reaction of 1,4-benzoquinone.

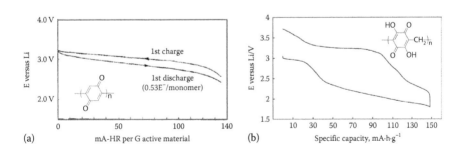

(a)
mA-HR per G active material

(b)
Specific capacity, mA·h·g^{-1}

FIGURE 11.9 Charge and discharge behaviors of benzoquinone polymers. (a) Charge and discharge curves of poly(1,4-benzoquinone) (PQ). (Adapted from Foos, J.S. et al., *J. Electrochem. Soc.*, 133, 836, 1986. With permission of The Electrochemical Society.) (b) Those of poly(2,5-dihydroxy-1,4-benzoquinone-3,6-methylene) (PDBM). (Adapted from *J. Power Sources*, 119–121, Gall, T.L., Reiman, K.H., Grossel, M., and Owen, J.R., 316, Copyright 2003, with permission from Elsevier.)

of the University of Southampton (United Kingdom) synthesized a benzoquinone polymer by the coupling reaction between a benzoquinone derivative and formaldehyde (Figure 11.9b).[10] While its theoretical capacity would be 352 mA·h·g^{-1}, the observed capacity is about 150 mA·h·g^{-1}. Sun et al. of Wuhan University (China) synthesized a coordinated polymer containing the benzoquinone moiety.[11] The observed capacity of 170 mA·h·g^{-1} is also significantly lower than the theoretical value of 350 mA·h·g^{-1}. The desired two-electron redox reaction per quinone unit has not yet been realized in polymers consisting of the simple quinone skeletons. What resulted in these limited utilization ratios is not clear. Amorphous regions that are characteristic of polymers may inhibit the electronic and/or ionic conductions.

In contrast, polymers carrying larger quinone skeletons tend to behave differently. For instance, Häringer et al. of the Paul Scherrer Institute (Switzerland) synthesized a polymer having the naphthoquinone skeleton (Figure 11.10a).[12] Although they evaluated it without a lithium-metal negative electrode in an electrolyte inappropriate for rechargeable batteries, the polymer is able to be discharged at 2.6 V versus Li$^+$/Li with the relatively high capacity of 220–290 mA·h·g^{-1}, which is close to the theoretical one, 313 mA·h·g^{-1}. As another example, the anthraquinone (AQ)-based polymer (Figure 11.10b) was reported by Zhan et al. of Wuhan University (China).[13] Its discharge potential is 2.1 V versus Li$^+$/Li and the initial capacity is 198 mA·h·g^{-1}, achieving 90% of the theoretical one, that is, 225 mA·h·g^{-1}. The cycle performance is also relatively good. Yoshida et al. of Kyoto University (Japan) reported a polymer carrying an orthoquinone-type pyrene derivative (Figure 11.10c).[14] As shown in Figure 11.10d, the observed capacity is about 230 mA·h·g^{-1}, which is also close to the theoretical value of 262 mA·h·g^{-1}, calculated by taking the ratio of the pyrene skeleton in the polymer into consideration. The discharge potential is relatively high, ca. 2.5 V versus Li$^+$/Li, among the quinone derivatives. The output power characteristics and the cycle-life performance are also superior. If the ratio of the pyrene skeleton in the polymer is increased, higher capacity (max 310 mA·h·g^{-1}) could be expected.

Polymers having large macrocyclic quinone structures tend to enjoy high utilization ratios and long cycle stabilities compared to those having a smaller quinone skeleton. Although room for improvement still remains in their discharge potentials, these types of polymers have a possibility to become a high performance organic active material by the proper molecular design.

11.3   LOW MOLECULAR WEIGHT COMPOUNDS

Polymerization is an approach to prevent the redox active organic materials from leaching out of the electrode into the electrolyte solution. However, as described in Section 11.2, extracting their full capacity from polymers is often difficult, a reason for which might be, we suspect, related to the amorphous nature hindering the ionic and/or electronic conductions. Looking at the problem from a different angle, one might as well directly use an organic molecular crystal with a low molecular weight if the dissolution of the compound into the electrolyte is suppressed by properly designing the molecule. This strategy would be, for example, introducing into the molecule some substituents that have a low affinity for the electrolyte solutions, enhancing the intermolecular attraction in the crystal by making use of the van der

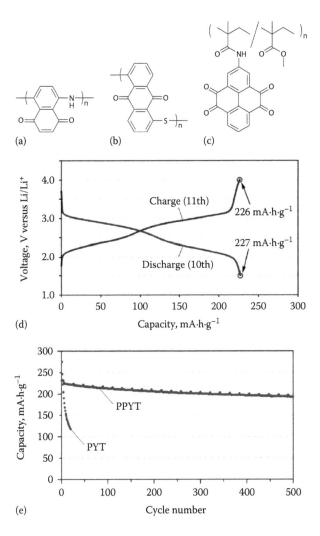

FIGURE 11.10 (a–c) The chemical structures of the representative macrocyclic polymer. (d) Charge and discharge curves of the electrode using polymer-bound pyrene-4,5,9,10-tetraone (PPYT) (Figure 11.10c) at 45°C. (e) Cycle-life performance of the PPYT electrode at 45°C. PYT stands for the monomer pyrene-4,5,9,10-tetraone. (Reprinted with permission from Nokami, T. et al., *J. Am. Chem. Soc.*, 134, 19694. Copyright 2012 American Chemical Society.)

Waals force, π–π interaction, and hydrogen bonding. As shown in the following section, a feature of these low molecular weight compounds is their relatively high utilization ratio as a redox center in the electrode when compared to polymers.

This section explores the electrode performance of some low molecular weight compounds. When dealing with organic compounds, one often encounters situations where techniques familiar in the field of current mainstream inorganic materials are not always applicable to elucidate the charge and discharge reaction mechanism.

Therefore, some methods pertinent to investigating organic electrode materials are also introduced based mainly on the authors' recent experiments.

11.3.1 Quinone Derivatives

11.3.1.1 2,5-Dimethoxy-1,4-Benzoquinone

The authors' trial to build an electrode with the simplest quinone, 1,4-benzoquinone (BQ, theoretical capacity: 496 mA·h·g⁻¹), was unsuccessful because BQ easily sublimates from the electrode. Our first example is thus 2,5-dimethoxy-1,4-benzoquinone (DMBQ, Figure 11.11).[15,16] In the crystalline state, each DMBQ molecule is connected to its neighboring molecules by a π–π interaction as well as weak hydrogen bonding.[15,17] These intermolecular forces stabilize the crystalline state, leading to decreased solubility in the electrolyte solution. The peripheral methoxy groups can act as protective groups for the radical intermediate that is conceived to be formed during the redox process. Figure 11.11a shows the initial discharge curve of the positive electrode prepared using DMBQ. The discharge curve is characterized by two plateau regions of the potential at around 2.8 and 2.5 V versus Li⁺/Li, reflecting the stepwise two-electron reduction reaction of DMBQ. The observed discharge capacity of 312 mA·h·g⁻¹ is more than twice the practical capacity of the conventional LiCoO₂ (ca. 140 mA·h·g⁻¹), and remarkably close to the theoretical value of 319 mA·h·g⁻¹, which assumes the full two-electron transfer of DMBQ. Although the average potential of 2.6 V versus Li⁺/Li of DMBQ is lower than that of LiCoO₂, 3.8 V versus Li⁺/Li, the gravimetric energy density (i.e., capacity × potential) of DMBQ is calculated to be still 1.6 times greater than that of LiCoO₂. As for the cycle stability, no drastic capacity decay was observed during 30–40 cycles (Figure 11.11b).

In order to determine the charge and discharge mechanism of organic active materials, it is important to analyze the crystallographic change upon cycling aside from the electrochemical properties. Some organic crystals undergo structural phase transitions through redox reactions; therefore, the reversibility of these transitions is crucial for the cycle stability of the material. What is the case for the DMBQ electrode? Figure 11.12 shows the comparison of the ex situ XRD profiles of the

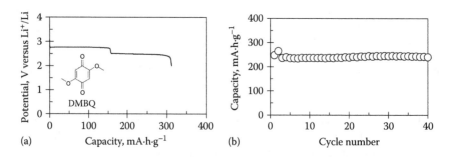

FIGURE 11.11 Battery performance of the DMBQ electrode. (a) Initial discharge curve. (b) Cycle-life performance. (Adapted from *J. Power Sources*, 195, Yao, M., Senoh, H., Yamazaki, S., Siroma, Z., Sakai, T., and Yasuda, K., 8336, Copyright 2010, with permission from Elsevier. The cycle data were updated by the authors.)

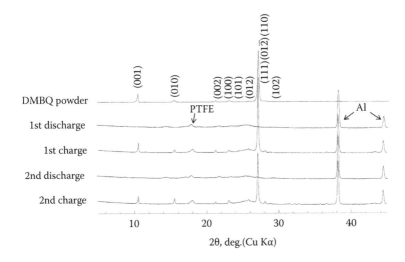

FIGURE 11.12 Ex situ XRD profiles of the DMBQ electrode for several states of charge and discharge. (Adapted from *J. Power Sources*, 195, Yao, M., Senoh, H., Yamazaki, S., Siroma, Z., Sakai, T., and Yasuda, K., 8336, Copyright 2010, with permission from Elsevier. Some lines were added by the authors.)

DMBQ electrode at several states of charge and discharge cycles along with that of the crude DMBQ powder. During the first discharge (i.e., reduction of the DMBQ electrode), the XRD signals become broad. After the subsequent charge (i.e., oxidation), the initial profile is restored. The change in the XRD profile is reversible at least for the initial few cycles of charge and discharge. This observation indicates that the change in the bulk structure of this compound via the charge and discharge process is reversible, which is considered to be important to realize a stable charge and discharge cycling.

In general, organic crystals have a low electrical conductivity, so that most of the molecules situated deep inside a bulk crystal are electrically isolated from the outside. The fact that the utilization ratio of the DMBQ electrode is fairly close to 100% then poses a question: How can almost all the DMBQ molecules in the crystal participate in the charge and discharge process? To obtain a theoretical insight into the electronic conduction mechanism of the DMBQ electrode, the electronic states were calculated based on the density functional theory (DFT). In the crystal, the DMBQ molecules are one-dimensionally stacked (labeled *d-1* direction) by the π–π interaction and horizontally aligned along the *d-2* and *d-3* directions by the hydrogen bonding as shown in Figure 11.13a.[15,17] The densities of electronic states (DOS) calculated along these directions are shown in Figure 11.13. The overlapping of the π-orbitals broadens the energy levels, leading to an electronic band structure along the d-1 direction. On the other hand, the widths of the energy levels along the d-2 and d-3 directions are narrower, indicating less overlapping of the orbitals along these directions. As the present calculation is primitive in that no effect of the energy level relaxation via electron transfer is taken into account, a detailed discussion on the reaction mechanism is premature at present. Nonetheless, one can speculate that the

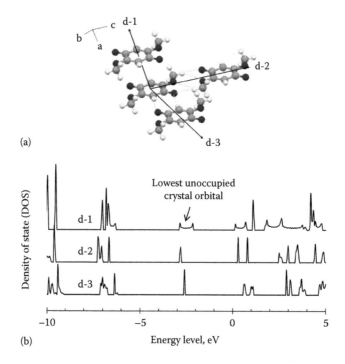

FIGURE 11.13 (a) Crystallographic structure of DMBQ. (Adapted from *J. Power Sources*, 195, Yao, M., Senoh, H., Yamazaki, S., Siroma, Z., Sakai, T., and Yasuda, K., 8336, Copyright 2010, with permission from Elsevier; Bock, H. et al., *Z. Naturforsch. Teil B Chem. Sci.*, 51, 153, 1996.) (b) Calculated density of electronic states (DOS) of the DMBQ crystal. d-1, d-2, and d-3 represent the directions in the DMBQ crystal in Figure 11.13a along which the DOS was calculated (B3LYP/6-31G*, periodic boundary condition).

energy band along the π-stacking direction originating from the lowest unoccupied molecular orbital (LUMO) of DMBQ initiates the reaction by accepting electrons to provide a conduction band-like electronic structure.

While the carrier ion is typically Li^+ in the current rechargeable lithium batteries, some organic positive-electrode active materials in the lithium systems store and release anions, such as PF_6^- or ClO_4^-, instead of Li^+, during the charge and discharge (see, e.g., Sections 11.2.1 and 11.2.2). Hence, identifying the carrier ion is important in order to understand the charge and discharge mechanism of organic materials. Several techniques can be applicable to quantitatively analyze the carrier ions, namely, inductively coupled plasma (ICP), ion chromatography, energy dispersive x-ray spectroscopy (EDX), and nuclear magnetic resonance (NMR). We now introduce our result obtained by ex situ ^7Li-NMR spectroscopy of the DMBQ electrode. Figure 11.14a shows the change in the stoichiometric ratio of Li^+ to DMBQ at given states in the first two cycles, determined by quantitatively measuring the amount of Li^+ in the extracted solutions from the DMBQ electrodes using ^7Li-NMR. The ratio Li^+/DMBQ should alternate between zero and two if two Li^+ ions are stored and released during the discharge and charge, respectively. After the first discharge, the ratio increases

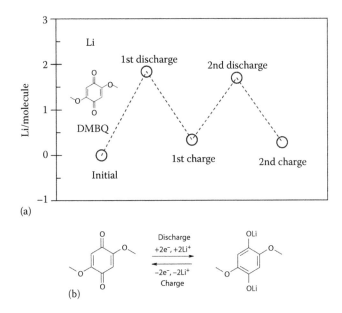

FIGURE 11.14   (a) Stoichiometric ratio of Li^+ to DMBQ in the electrode determined by ex situ ^{7}Li-NMR measurement at each state of charge and discharge. (b) Two-electron redox reaction of DMBQ in a lithium electrolyte system.

to 1.8, that is, ~2. When the discharged electrode is recharged, the ratio returns to the initial value close to zero, which increases again to about two after the second discharge. This change in the Li^+/DMBQ corroborates the two-electron redox reaction of DMBQ accompanied by the two-Li^+ storage and release shown in Figure 11.14b, and also proves that the charge carrier in this system is Li^+.

11.3.1.2 Polycyclic Quinone Derivatives

We also examined the electrode performances of quinone derivatives with larger polycyclic structures, that is, 9,10-anthraquinone (AQ) and 5,7,12,14-pentacenetetrone (PT) (Figure 11.15).[18–20] The highly developed π-systems of these planar polycyclic quinines should induce a strong π–π intermolecular interaction in the crystal, which we consider important to make a good electrode active material as discussed in Section 11.3.1.1. These molecules undergo multielectron redox reactions during the charge and discharge processes. As shown in Figure 11.15a, the initial experimental discharge capacities are both close to their theoretical capacities (AQ: 257 mA·h·g^{-1} and PT: 317 mA·h·g^{-1}) based on the assumption of the full reduction reaction of their carbonyl groups (>C=O). While the discharge capacity of the positive electrode with AQ rapidly decreases during the charge and discharge cycles, the PT electrode maintained about 80% of the initial capacity after 100 cycles (Figure 11.15b). The larger π-system of PT than AQ probably plays a role in more effectively preventing the active material from leaching out of the electrode.

The electronic states of AQ and PT crystals were calculated based on the DFT with the atomic coordinates extracted from a crystallographic database. As shown

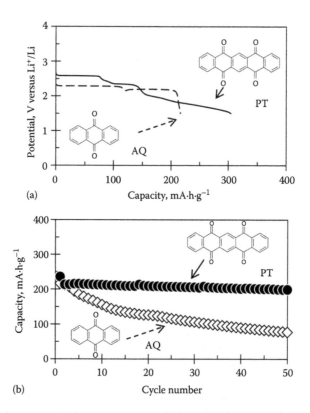

FIGURE 11.15 Battery performances of 9,10-anthraquinone (AQ) and 5,7,12,14-pentacenetetrone (PT). (a) Initial discharge curves in GBL/LiTFSA. (Adapted from Yao, M. et al., *ITE-IBA Lett.*, 4, 52, 2011. With permission from International Technology Exchange Society; Yao, M. et al., *Int. J. Electrochem. Sci.*, 6, 2905, 2011. With permission from ESG.) (b) Cycle-life performances in tetraglyme/LiTFSA. (Adapted from *Mater. Sci. Eng. B*, 177, Yao, M., Yamazaki, S., Senoh, H., Sakai, T., and Kobayashi, T., 483, Copyright 2012, with permission from Elsevier.)

in Figure 11.16a and b, either in the AQ or in the PT crystal, the molecules are stacked one dimensionally by a π–π interaction to form a columnar structure[21,22] as in DMBQ (Figure 11.13a). The DOS calculated for the stacked columns are shown in Figure 11.16c. The energy levels of the many orbitals of AQ and PT expand, forming electronic band structures along the π-stacked directions due to the overlapping of the π-orbitals. These electronic band structures seem to be related to the high utilization ratio of these molecules as in the case of DMBQ in Section 11.3.1.1.

11.3.2 INDIGO ANALOGUES

Since ancient times, human beings have been fascinated by indigo's dark-blue color to use for dyeing and printing; a famous example is blue jeans. This plant-derived dye is a heterocyclic organic molecule, of which several peripherally substituted

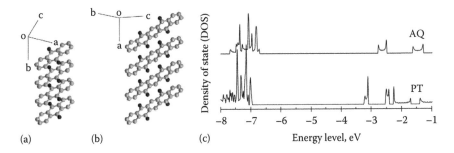

(a)   (b)   (c)   Energy level, eV

FIGURE 11.16   (a and b) Crystal structures of AQ and PT. (From Slouf, M., *J. Mol. Struct.*, 611, 139, 2002; Kafer, D. et al., *Cryst. Growth Des.*, 8, 3053, 2008.) (c) Calculated DOS. (Adapted from *Mater. Sci. Eng. B*, 177, Yao, M., Yamazaki, S., Senoh, H., Sakai, T., and Kobayashi, T., 483, Copyright 2012, with permission from Elsevier.)

derivatives are known. The indigo derivatives accept two electrons per molecule to form the leuco-indigo derivatives. Knowing that the electrochemical potentials of this reaction are in the range of the positive electrode of lithium batteries, we developed the idea of using indigos as an electrode active material.

The electrode made of pristine indigo (without peripheral substitution) is able to be discharged at an average of 2.3 V versus Li$^+$/Li with the initial capacity of 200 mA·h·g^{-1}, remarkably close to the theoretical upper-bound capacity of 204 mA·h·g^{-1}. However, the discharge capacity rapidly decreases upon cycling, presumably due to the high solubility of indigo in the electrolyte solution of the organic solvent. On the other hand, albeit a low capacity (~100 mA·h·g^{-1}), indigo carmine (5,5′-indigodisulfonic acid disodium salt), which is used as a blue dye for food and drugs, enjoys a very long cycle life; no significant capacity decay after more than 1000 cycles (Figure 11.17).[23] Not only is this long cycle life remarkable among the low molecular weight organic compounds but it is also comparable to the cycle life achieved by the current inorganic materials. Unlike most of the low molecular weight compounds, this water-soluble derivative hardly dissolves in ordinary organic solvents due to the polar peripheral sulfonate groups. We consider that a reason for this cycle stability is its low solubility in the electrolyte solution, which should suppress the loss of the active material from the electrode during the charge and discharge process. The result of the indigo derivatives points to the prospect that organic compounds, often regarded as fragile, can be a robust electrode active material if properly designed.

11.3.3   AROMATIC CARBOXYLIC ACIDS ANHYDRIDE

Another example of macrocyclic compounds has been reported by Sun et al. of Wuhan University (China) who investigated the battery performances of 1,4,5,8-naphthalenetetracarboxylic dianhydride (NTCDA) and 3,4,9,10-perylen etetracarboxylic dianhydride (PTCDA) (Figure 11.18).[24] The theoretical capacities of these molecules can reach 400 and 273 mA·h·g^{-1} for NTCDA and PTCDA, respectively, if one could successfully elicit the four-electron transfer reaction from these molecules. The initial capacities are, however, about half the theoretical

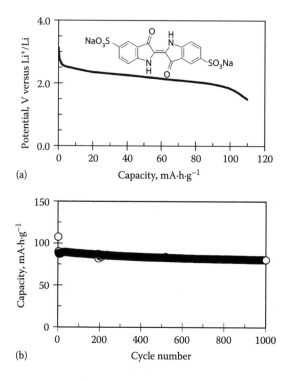

(a)

(b)

FIGURE 11.17 Battery performance of the electrode using indigo carmine. (a) Initial discharge curve. (b) Cycle-life performance. (Adapted from Yao, M. et al., *Chem. Lett.*, 39, 950. Copyright 2010. With permission of The Chemical Society of Japan. Data updated by the authors.)

values at the average potentials of 2.4 V versus Li$^+$/Li. While the lower molecular weight NTCDA suffers from a poor cycle stability due to its dissolution into the electrolyte solution, PTCDA, the higher molecular weight counterpart, better withstands the cycling, that is, the PTCDA electrode maintains 50% of the initial capacity after 90 cycles. This observation again suggests that having a large π-conjugated ring structure can be effective in lengthening the cycle life of the electrode. Sun et al. also prepared composites of PTCDA and sulfur by calcination. An improved cycle stability was reported.

11.3.4 Triquinoxalinylene

Triquinoxalinylene derivatives (Figure 11.19) were synthesized by Sugimoto et al. of Osaka Prefecture University (Japan).[25] The experimental capacity of the unsubstituted derivative for the initial cycles was ca. 400 mA·h·g^{-1}, fairly close to the theoretically expected one of 418 mA·h·g^{-1}, in which full reduction of the aromatic nitrogen moieties is assumed. The average discharge potential is 2 V versus Li$^+$/Li. When subjected to cycling, the charge and discharge curve does not retain its initial shape, implying irreversible side reactions during the redox process. Not until after

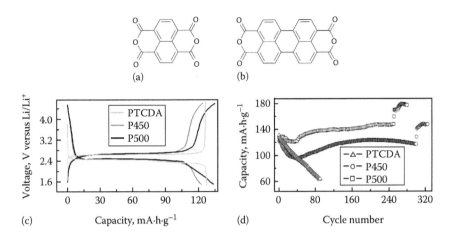

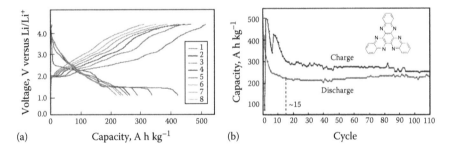

FIGURE 11.18 The chemical structures of (a) 1,4,5,8-naphthalenetetracarboxylic dianhydride (NTCDA) and (b) 3,4,9,10-perylenetetracarboxylic dianhydride (PTCDA). (c) Potential profiles of the initial charge and discharge processes (100 mA·g^{-1}) for PTCDA and its composites with sulfide polymers (P500 and P450). (d) Cycle performances (100 mA·g^{-1}, after 250 and 300 cycles, the current density was decreased to 25 mA·g^{-1} for P500 and P450, respectively). (Han, X., Chang, C., Yuan, L., Sun, T., and Sun, J.: *Adv. Mater.* 2007. 19. 1616. Copyright Wiley-VCH Verlag GmbH & Co. KGaA. Adapted with permission.)

FIGURE 11.19 (a) Charge and discharge curves of the electrode using triquinoxalinylene. (b) Cycle performance. (Adapted from Matsunaga, T. et al., *Chem. Lett.*, 40, 750. Copyright 2011. With permission of The Chemical Society of Japan.)

15 cycles does the monotonous decrease in the capacity cease at a steady state of around 220 mA·h·g^{-1}. The relatively low coulombic efficiency also implies a side reaction in the electrode.

11.3.5 Low Molecular Weight Radicals

Morita et al. of Osaka University (Japan) synthesized a macrocyclic radical, trioxotriangulene (TOT).[26] While the TEMPO-type radical polymers (Section 11.2.2) store and release anions during the redox, it is Li$^+$ that serves as a charge carrier in the TOT. Among the examined derivatives, the brominated one excels in

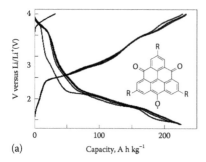
(a)

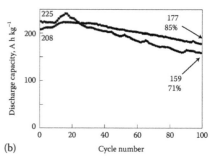
(b)

FIGURE 11.20 (a) Charge and discharge curves of tribromotrioxotriangulene (Br_3TOT). (b) Cycle performance of the discharge processes of Br_3TOT (1.4–4.0 V vs. Li^+/Li). (Adapted by permission from Macmillan Publishers Ltd., *Nat. Mater.*, Morita, Y. et al., 10, 947. Copyright 2011.)

battery performance (Figure 11.20), namely, the electrode has an average potential of 2.3 V versus Li^+/Li with the discharge capacity of about 200 $mA·h·g^{-1}$, and retained 85% of its initial capacity after 100 cycles. This radical molecule firmly stacks on each other in the crystal and only slightly dissolves in the electrolyte solutions, which seems to partially contribute to the good cycle performance.

11.3.6 RHODIZONIC ACID DILITHIUM SALT

The group of Tarascon and Poizot of the University of Picardie Jules Verne (France) examined the battery properties of a rhodizonic acid dilithium salt, a compound that can be synthesized from natural products (Figure 11.21).[27] The redox reaction takes place at 2.2 V versus Li^+/Li with the initial capacity of 580 $mA·h·g^{-1}$, which is quite close to the theoretical value of 589 $mA·h·g^{-1}$, in which a four-electron transfer is assumed. However, exploiting this 4-electron reaction results in a capacity drop to

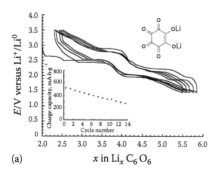
(a)

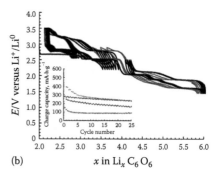
(b)

FIGURE 11.21 (a) Potential profile of the electrode using rhodizonic acid dilithium salt. Inset shows the capacity retention. (b) Potential versus composition profile galvanostatically cycled in several potential windows. Inset shows the capacity retention. (Chen, H., Armand, M., Demailly, G., Dolhem, F., Poizot, P., and Tarascon, J.-M.: *ChemSusChem.* 2008. 1. 348. Copyright Wiley-VCH Verlag GmbH & Co. KGaA. Adapted with permission.)

70% after 10 cycles. The cycle stability is enhanced when the amount of charge is limited to 300 mA·h·g^{-1} by lowering the cutoff potentials in the charge process, in which the valence of the molecule is considered to alternate between -4 and -6. The high ionicity of this molecule presumably contributes to suppressing its dissolution into the electrolyte solution.

11.3.7 RUBEANIC ACID

Rubeanic acid has the potential to work as an active material for the positive electrode of rechargeable lithium batteries as reported by Satoh et al. of Murata Manufacturing Co., Ltd. (Japan) (Figure 11.22).[28] One can expect the discharge capacity of 446 mA·g^{-1} if each thiocarbonyl site receives two electrons. Its low molecular weight helps this acid have a high gravimetric theoretical capacity. The observed initial capacity, ca. 600 mA·h·g^{-1}, exceeds the theoretical one, raising the concern for side reactions. While the redox potential, ca. 2 V versus Li$^+$/Li, is comparable to those of many other sulfur-related materials, the cycle-life performance is better; the electrode maintains 450 mA·h·g^{-1} after 20 cycles, despite the concern for the previously mentioned side reactions.

11.3.8 TETRATHIAFULVALENE ANALOGUES

Inatomi et al. of Panasonic Corp. (Japan) and Misaki et al. of Ehime University (Japan) focused on a series of tetrathiafulvalene (TTF) derivatives, well-known organic electronic conductors.[29–31] Their discharge capacities are around 150 mA·h·g^{-1}. The discharge potentials are relatively high (>3 V vs. Li$^+$/Li) compared to those of the sulfur-based polymers (~2 V vs. Li$^+$/Li, see Section 11.2.3). The difference in the redox potentials between the TTF analogues and the polymers probably stems from the reaction mechanism. That is, it is anions that are exchanged in the TTF analogues whereas the sulfur-based polymers store and release Li$^+$ during the redox reaction. The problem with the TTF monomer's short cycle life due to its high solubility into the electrolyte solution can be circumvented

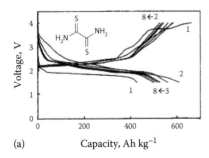

(a)   Capacity, Ah kg^{-1}

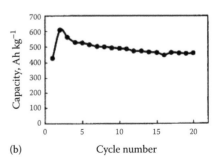

(b)   Cycle number

FIGURE 11.22 (a) Charge and discharge curves of the electrode using rubeanic acid. (b) Cycle performance. (Reproduced from Satoh, M. et al., *The 51st Battery Symposium in Japan*, Nagoya, 2010, Abstract #3G24. With permission of The Electrochemical Society of Japan.)

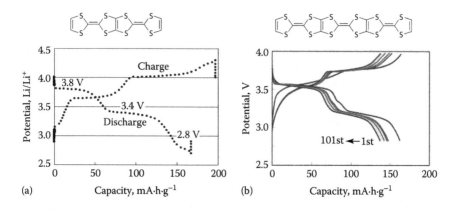

FIGURE 11.23 Charge and discharge curves of Li cells using (a) 2,5-*bis*(1,3-dithiol-2-ylidene)-1,3,4,6-tetrathiapentalene (TTP) and (b) 2,2′-bi[5-(1,3-dihthiol-2-ylidene)-1,3,4,6-tetrathiapentanylidene] (TTPY). (Inatomi, Y., Hojo, N., Yamamoto, T., Watanabe, S., and Misaki, Y.: *ChemPlusChem*. 2012. 77. 973. Copyright Wiley-VCH Verlag GmbH & Co. KGaA. Reproduced with permission.)

by forming its dimer (TTP) and trimer (TTPY) (Figure 11.23). For example, TTPY retains 84% of its initial capacity after 100 cycles.

11.4 BEYOND LITHIUM

The foregoing sections deal with the possibility of applying organic materials to rechargeable lithium batteries. Not only do the transition metals that are used in the current lithium batteries pose a resource problem, but lithium itself, the inevitable charge carrier, can also be viewed as a rare metallic element. Substituting another charge carrier for lithium is thus one of the directions in developing rechargeable post-lithium batteries. Given its enormous abundance on earth, an alternative is sodium, although a drawback is the 0.3 V higher redox potential of Na than Li; cf. the standard redox potentials versus normal hydrogen electrode (NHE), E_{NHE}^{Y}, are −3.03 and −2.71 V for Li and Na, respectively. Another possibility of the post-lithium is to use a multivalent ion as a charge carrier, whereby one can expect an increase in the capacity. Using magnesium, a divalent ion, is an option because magnesium resources in the earth's crust are also more widely distributed than lithium. The 0.6 V higher redox potential of Mg than Li ($E_{NHE}^{\circ}(Mg)/V = -2.4$) may be compensated by the higher theoretical volumetric capacity of metallic Mg (3833 mA·h·cm^{-3}) than Li (2062 mA·h·cm^{-3}) if we succeeded in using a metallic negative electrode.

A challenge in utilizing Na or Mg is to determine such electrode materials that properly work at the desired redox potential with sufficient capacity and cycle life in the sodium or magnesium electrolyte solutions. The variation in these electrode active materials is still limited probably due to the larger ionic radius of Na$^+$ and the higher surface charge density of Mg^{2+} than those of Li$^+$ that impede the movement in the electrodes and/or severely distort the structure of the active materials. This may be where organic compounds come in. We have recently revealed that several organic molecules functioned not only in the lithium systems but also in the sodium and magnesium systems.

11.4.1 SODIUM SYSTEMS

11.4.1.1 Indigo Carmine

Indigo carmine (5,5′-indigodisulfonic acid disodium salt) is introduced in Section 11.3.2 as a very long cycle-life active material for the positive electrode in lithium systems. As shown in Figure 11.24, indigo carmine works as a positive electrode material even in a cell composed of a metallic sodium negative electrode and a sodium-based electrolyte solution.[32] The discharge capacity of the electrode is ~100 mA·h·g⁻¹, nearly equal to that observed in the lithium system. The long cycle life of the indigo carmine electrode observed in the lithium system (Figure 11.17) is also manifested in the sodium system. The average potential is 1.8 V versus Na⁺/ Na, which is 0.4 V lower than that in the lithium system, 2.2 V versus Li⁺/Li. Most of the 0.4 V difference in the redox potential between the Li and Na systems is explained by the 0.3 V difference in the redox potential of the negative electrodes, that is, $E^{\circ}_{NHE}(Li)/V = -3.03$ whereas $E^{\circ}_{NHE}(Na)/V = -2.71$. In other words, in terms of the hypothetical NHE basis, the redox potentials of the positive electrodes made of indigo carmine would be −0.8 and −0.9 V in the Li and Na systems, respectively. The fact that the difference in the redox potentials of indigo carmine in the Li and Na systems is small (0.1 V) implies that the thermodynamic free energy change caused by the cation insertion into this molecular crystal is comparable whichever species, Li or Na, is involved.

The carrier ion inserted in and released from the indigo carmine electrode was confirmed to be the Na⁺ ion using ex situ ion chromatography. Figure 11.25 shows the change in the Na⁺ concentration of the extracted solutions by rinsing the electrodes with a solvent at given states of charge/discharge in the first two cycles. The Na⁺ concentration increases after the first discharge, which returns to the initial level when the discharged electrode is recharged. Similar increases and decreases in the concentration

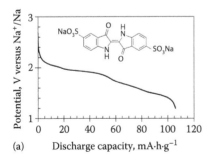
(a)

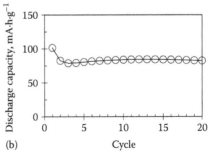
(b)

FIGURE 11.24 (a) Initial discharge curve of the electrode using indigo carmine in a sodium electrolyte solution. (b) Cycle performance of the electrode. (Adapted from Yao, M. et al., *PRiME 2012*, Honolulu, Abstract #1861. Copyright 2012. With permission of The Electrochemical Society.)

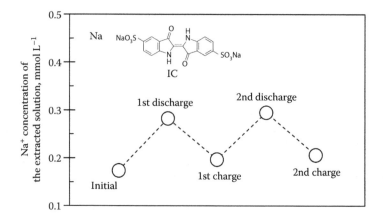

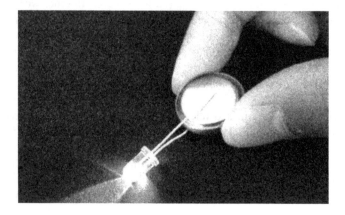

FIGURE 11.25 Concentration change of Na$^+$ in the extracted solutions from the indigo carmine electrode at different states of charge and discharge cycles, determined by ex situ ion chromatography.

FIGURE 11.26 LED lamp powered by an indigo carmine-Na coin cell.

are also observed in the subsequent cycle, proving that the Na$^+$ ion is reversibly inserted in and released from the indigo carmine electrode. Indeed, a more detailed quantitative analysis reveals that the stoichiometric ratio of Na to the indigo carmine molecule oscillates between two and four during the charge and discharge. (Note that an indigo carmine molecule, in the first place, contains two Na ions in its two –SO$_3$Na groups that are, we consider, not involved in the electrochemical reactions.)

Figure 11.26 is an LED lamp powered by a coin-type cell using the indigo carmine–positive electrode and metallic sodium–negative electrode.

11.4.1.2 2,5-Dimethoxy-1,4-Benzoquinone

Section 11.3.1.1 describes DMBQ as a high capacity active material in the lithium system. DMBQ can also serve as a positive electrode material in the sodium system.[33] The initial discharge curve (Figure 11.27) of the DMBQ electrode in a sodium

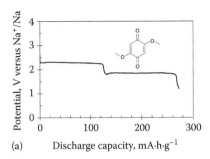

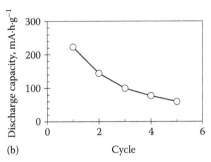

(a) Discharge capacity, mA·h·g⁻¹

(b) Cycle

FIGURE 11.27 (a) Initial discharge curve of the DMBQ electrode in a sodium electrolyte solution. (b) Cycle performance of the DMBQ electrode. (Adapted from Yao, M. et al., *The 54th Battery Symposium in Japan*, Abstract #3D01, 2013. With permission from The Electrochemical Society of Japan.)

electrolyte solution is characterized by two potential plateau regions, reflecting the two-electron redox reaction. The obtained capacity of 270 mA·h·g⁻¹ is less than what is observed in the Li system, 312 mA·h·g⁻¹. Nonetheless, it attains 85% of the theoretical capacity (319 mA·h·g⁻¹) when a two-electron redox of DMBQ is assumed. The average discharge potential is 2.0–2.1 V versus Na⁺/Na, which is 0.5–0.6 V lower than that in the lithium system, 2.6 V versus Li⁺/Li. Although the gap is somewhat greater than in the case of indigo carmine mentioned in the previous section, a significant part of the difference in the redox potential between Na and Li (0.5–0.6 V) reflects the redox potentials of the Na and Li negative electrodes. As for the charge carrier in the DMBQ–Na system, the EDX measurement of the electrodes qualitatively indicates the rise and fall in the amount of Na in the electrode upon cycling. (An example of a more quantitative application of EDX is described in Section 11.4.2.) Unlike in the lithium system (Figure 11.11b), the DMBQ electrode does not tolerate well the charge and discharge cycles; the capacity drops to a quarter of the initial value after five cycles. The reason why DMBQ does not endure the cycling in the sodium system while indigo carmine does is not presently known (Section 11.4.1.1).

11.4.1.3 Rhodizonic Acid Disodium Salt

Okada et al. of Kyusyu University (Japan) are trying to apply the disodium salt of rhodizonic acid, of which the dilithium salt is mentioned in Section 11.3.6.[34] The capacity in the sodium system is about 250 mA·h·g⁻¹ (Figure 11.28a), less than half that observed in the lithium system, that is, 580 mA·h·g⁻¹, but the rationale behind this result is unknown. The multiple potential plateaus in the charge and discharge profile imply that more than one Na ion moves in and out of the salt. The less the electrode is discharged, the longer the cycle life (Figure 11.28b), as observed in the lithium system (Figure 11.21b).

11.4.1.4 Terephthalic Acid Disodium Salt—Negative Electrode

Most of this chapter is dedicated to the description of organic materials for the *positive electrode*. In order to attain a high gravimetric and volumetric capacity of a whole battery, using metallic lithium as the negative electrode would be ideal.

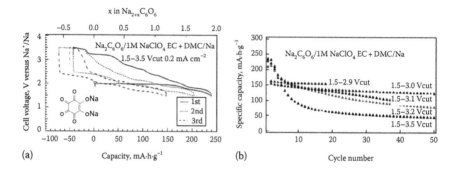

FIGURE 11.28 (a) Charge and discharge profiles of the electrode using rhodizonic acid disodium salt. (b) Cycle stability of the electrode cycled in several potential windows. (Adapted from Chihara, K. et al., *PRiME 2012*, Honolulu, Abstract #1838. Copyright 2012. With permission of The Electrochemical Society.)

However, in current rechargeable lithium batteries, carbon-based materials are often used for the negative electrode due to the technical and safety problems of applying metallic lithium. Two issues we have to confront when using carbons are as follows: (1) using carbons drastically decreases the capacity of the negative electrode both gravimetrically and volumetrically compared to the metallic lithium, and (2) the redox potentials of carbon-based materials are so close to that of the metallic lithium that one has to tackle the harmful *dendrite deposition* during the charging process, as in the case of using metallic lithium. For the rechargeable sodium batteries, the variation in the candidate material is more limited than for the lithium batteries, for example, nongraphitizable carbon (hard carbon) and tin-based alloys. Hence, the R&D of the negative electrode is as important as that of the positive electrode, not least with the fears that metallic sodium is supposedly more chemically active, that is, dangerous, than metallic lithium.

Hu et al. of the Chinese Academy of Sciences (China) investigated the battery properties of the terephthalic acid disodium salt,[35] of which the dilithium salt was reported by the group of Armand and Tarascon from the University of Picardie Jules Verne (France) in Ref. [36]. In the experiment performed by the authors of this chapter, the salt has the discharge potential of 0.5 V versus Na^+/Na with the initial discharge capacity of ca. 170 $mA \cdot h \cdot g^{-1}$ as shown in Figure 11.29. A significant fraction of the capacity obtained in the initial charge (i.e., reduction) process is irreversible, implying side reactions. Although the details are yet to be elucidated, the analogous shape of the charge and discharge profiles in the lithium and sodium systems implies the similarity in the reaction mechanism. These properties well reproduce the result originally reported by Hu et al.[35] Terephthalic acid literally exemplifies the wide *potential* of organic materials.

11.4.2 DMBQ for Mg System

As mentioned in the introductory paragraph of Section 11.4, magnesium can be another candidate for the post-lithium batteries. Few inorganic compounds

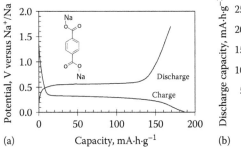
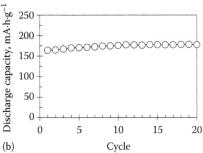

(a)   (b)

FIGURE 11.29 Battery performance of the electrode using terephthalic acid disodium salt in a sodium electrolyte solution. (Originally reported by Zhao, L. et al., *Adv. Energy Mater.*, 2, 962, 2012.) (a) Typical charge and discharge curves and (b) cycle performance. (These figures were made based on the experiments performed by the authors of this chapter.)

electrochemically and reversibly react with Mg^{2+} to function as an electrode active material, presumably because the divalent magnesium ion more strongly interacts with the rigid lattice than monovalent cations, so that it hardly diffuses into the bulk electrode material. Organic compounds may be able to more readily accommodate Mg^{2+} into their flexible lattice, where the molecules interact with each other, generally by weak intermolecular forces. We now describe the result of our recent trial to use 2,5-dimethoxy-1,4-benzoquinone as a positive electrode material in a magnesium system (cf. Section 11.3.1.1 for Li and Section 11.4.1.2 for Na).

Figure 11.30 shows the typical charge and discharge curves of the DMBQ electrode in a magnesium electrolyte solution.[37] The discharge curve consists of two potential plateau regions at around 1.1 and 0.8 V versus Mg_{quasi}, each of which is 1.7 V lower than the plateaus observed in the DMBQ–Li system, 2.8 and 2.5 V versus Li^+/Li (see Figure 11.11a). This unexpectedly high 1.7 V gap should be 0.7 V if the difference in the standard redox potentials of the negative electrode were solely reflected, $E°_{NHE}(Li^+/Li)/V = -3.03$ and $E°_{NHE}(Mg^{2+}/Mg)/V = -2.37$. The inexplicable

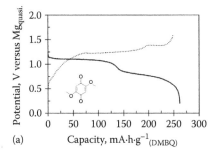
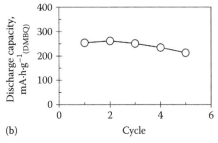

(a)   (b)

FIGURE 11.30 (a) Typical charge and discharge curves of the DMBQ electrode in a magnesium electrolyte solution (0.5 M $Mg(ClO_4)_2$/GBL, solid line: discharge, dashed line: charge). (b) Cycle-life performance of the DMBQ electrode. (Adapted from Sano, H. et al., *Chem. Lett.*, 41, 1594. Copyright 2012. With permission of The Chemical Society of Japan.)

additional 1 V drop can, of course, stem from the difference in the thermodynamic state of the DMBQ positive electrode between the Li and Mg systems. However, we suspect that the negative electrode may also be responsible, that is, the conceived reaction $Mg \rightleftharpoons Mg^{2+} + 2e^-$ is not taking place on the surface of the negative electrode. For this reason, we denote Mg_{quasi} instead of Mg^{2+}/Mg for indicating the potential in this section. Elucidating the enigmatic 1 V gap can be a key to realize the rechargeable magnesium batteries, although a further discussion is beyond the scope of this chapter.

The capacity of 250 mA·h·g^{-1}, observed in the second discharge, is 78% of the theoretical value of 319 mA·h·g^{-1} based on the assumption of a two-electron transfer per DMBQ molecule. This observation implies that a significant fraction of the DMBQ in the electrode undergoes the two-electron redox reaction with divalent Mg^{2+} as with monovalent Li$^+$ (Section 11.3.1.1) and Na$^+$ (Section 11.4.1.2). Investigating the mechanism as to how the monovalent carbonyl groups (>C=O) in DMBQ accommodate divalent Mg^{2+} can be an interesting research subject.

The cyclability is far from any application, at best; the capacity significantly drops in a few cycles. The DMBQ electrode in the Mg system nonetheless retains 85% of the initial capacity after five cycles. What results in the poor cycle stability is not yet clear.

Although not very quantitative, the EDX measurement can be a convenient tool to determine the carrier ions as this technique enables one to directly identify the constituent elements in the solid-state electrode. Figure 11.31 shows the change in the Mg concentration in the DMBQ electrode during the initial two cycles based on the assumption that the entire electrode contains only Mg, C, O, and F. During the first discharge, the Mg concentration increases from the initial negligible value to 3.7 atom%, which should be 3 atom% if one DMBQ molecule in the electrode stores one magnesium ion in the electrode prepared with the formula described in Ref. [37]. The magnesium disappears from the electrode during the subsequent

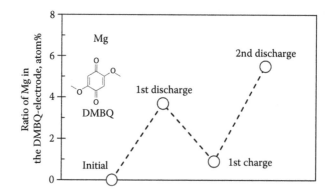

FIGURE 11.31 Atomic content ratio of Mg to Mg + C + O + F in the DMBQ electrode during cycling obtained by EDX measurement. (Adapted from Sano, H. et al., *Chem. Lett.*, 41, 1594. Copyright 2012. With permission of The Chemical Society of Japan.)

charge process but increases again during the next discharge process. Although the observed concentrations in the discharged states significantly deviate from those expected, they are in agreement by rule of thumb. The XRD measurement (not shown) reveals that the DMBQ crystal transforms into an amorphous-like structure when the electrode is reduced during the discharge process. The initial crystalline structure reemerges when the discharged electrode is oxidized again in the subsequent charge process. These results as well as what is described in Sections 11.3.1.1 and 11.4.1.2 suggest that, in whichever system of Li, Na, or Mg, the DMBQ electrode reversibly stores and releases such an amount of cations that balances the charge variation caused by the two-electron transfer of the DMBQ molecule. The redox-active organic crystals have the potential to serve as an active material not only in lithium systems but also in other systems including multivalent ions.

11.5 PROSPECTS AND CHALLENGE

This chapter has examined the possibility of applying redox active organic materials to rechargeable batteries by reviewing the past and very recent reports complemented by the authors' studies. If we substitute organic compounds for the inorganic active materials that are commonly used in the current rechargeable batteries, the benefits we would gain are the following:

Rare metal–free: Most organic materials do not essentially contain any scarce metals. Substituting organic compounds for inorganic materials may thus alleviate the resource problems involved in the current rechargeable lithium batteries. The environmental burden can be further reduced since some organic materials are derived from natural compounds.

High gravimetric capacity: Some organic molecules exceed $LiCoO_2$, a current representative electrode material, in gravimetric capacity, whereby the energy density of batteries can be increased.

Not only Li: The flexible structure of organic compounds seems more versatile in accommodating cations other than Li^+, such as Na^+ or Mg^{2+}, than the rigid framework of inorganic ones. Because lithium can be regarded as a scarce metal, searching for a proper electrode material in the organic compounds may be a shortcut to develop a less resource-sensitive battery, when combined with Na or Mg.

The impact of these benefits would be greater if adopted to large-scale applications, for example, electric vehicles, load-leveling systems, and in-home electricity storage. However, organic materials cannot be considered a silver bullet. The problems we have to confront are the following:

Low volumetric capacity: The *gravimetric* capacity (i.e., in $mA \cdot h \cdot g^{-1}$) of organic materials can be high due to their constituent light elements (C, H, O, N, S, etc.). However, exactly their low density, a flip side of the coin of lightness, results in the low *volumetric* capacity (i.e., in $mA \cdot h \cdot cm^{-3}$). The densities of most organic materials are <2 $g \cdot cm^{-3}$ whereas those of inorganic metal oxides are 3–5 $g \cdot cm^{-3}$, which would translate into a few times lower volumetric capacity of the former even if their gravimetric capacities are equal.

Low discharge potential: The discharge potentials of organic compounds are, by and large, lower than those of the inorganic competitors. Some organic molecules with a high discharge potential often suffer from a low capacity. Because the low discharge potential means a low working voltage and low energy density of the battery, we must discover a high-potential organic material without sacrificing the capacity.

Low cycle stability: The current rechargeable batteries endure hundreds cycles of charge and discharge. Many organic compounds, especially of low molecular weight with high capacity and potential, significantly degrade within relatively short cycles when we try to fully exploit their capabilities. As often said, no matter how we reduced the initial production cost, it is the cycle life that matters at the end of day in terms of the final results. When two products have an equal initial price but the former has half the longevity of the latter, then the real *cost* of the former should be considered twice the latter, or more if taking into account the environmental burden that is usually not internalized in the cognitive framework of *economy*.

Some of these are so essential that solving the problems per se must be intractable, for example, increasing the density of a compound without altering any other properties is impossible. Yet there should be strategies to address some other issues. Adding more redox sites to a molecule through synthetic techniques will increase the capacity to make up for the low volumetric capacity. The redox potential will be raised by introducing electron-withdrawing groups and/or heterocyclic rings. To the notion that organic compounds last less than inorganic ones, indigo carmine (Sections 11.3.2 and 11.4.1) provides counterevidence. The leaching out of the active materials from the electrode into the electrolyte solution is proved to be suppressed by introducing polar ionic groups, having large π-systems, oligomerization of the redox active monomers, etc. Aside from coping with the electrode materials themselves, innovations in the design of the battery systems are also called for in order to explore the potential abilities of organic materials as an energy carrier. For instance, if the dissolution to the liquid electrolyte matters, why not the solid electrolyte? Using a quasi-solid electrolyte based on an ionic liquid and nanosized silica powder is reported to suppress the dissolution of the redox molecules.[38] Furthermore, if the solid electrolyte completely blocks the dissolution, then the electrodes need not be *solid* any more. The authors have proposed a two-compartment cell in which a thin solid electrolyte plate separates the positive and negative *electrode solutions* as in the redox-flow cell.[39] What is demonstrated in this prototype cell is the concept that the redox active organic molecules, saturated and dispersed in the electrode solutions, are progressively subjected to the redox reaction at the interface, which consequently suppresses the cycle degradation. Although not all of these measures may be simultaneously fulfilled, we believe they are not necessarily involved in the trade-off relation as well.

REFERENCES

1. P. Novák, K. Müller, K. S. V. Santhanam and O. Hass, *Chem. Rev.*, 1997, **97**, 207.
2. K. Nakahara, S. Iwasa, M. Satoh, Y. Morioka, M. Suguro and E. Hasegawa, *Chem. Phys. Lett.*, 2002, **359**, 351.
3. K. Nakahara, J. Iriyama, S. Iwasaa, M. Suguro, M. Satoh and E. J. Cairns, *J. Power Sources*, 2007, **165**, 398.
4. K. Tamura, N. Akutagawa, M. Satoh, J. Wada and T. Masuda, *Macromol. Rapid Commun.*, 2008, **29**, 1944.
5. M. Yao, H. Senoh, T. Sakai and T. Kobayashi, *J. Power Sources*, 2012, **202**, 364.
6. M. Liu, S. J. Visco and L. C. De Jonghe, *J. Electrochem. Soc.*, 1991, **138**, 1891.
7. J. Wang, J. Yang, J. Xie and N. Xu, *Adv. Mater.*, 2002, **14**, 963.
8. T. Miyuki, T. Okuyama, T. Kojima, A. Kojima and T. Sakai, *The 52nd Battery Symposium in Japan*, Tokyo, 2011, Abstract #4B12.
9. J. S. Foos, S. M. Erker and L. M. Rembetsy, *J. Electrochem. Soc.*, 1986, **133**, 836.
10. T. L. Gall, K. H. Reiman, M. Grossel and J. R. Owen, *J. Power Sources*, 2003, **119–121**, 316.
11. J. F. Xiang, C. X. Chang, M. Li, S. M. Wu, L. J. Yuan and J. T. Sun, *Cryst. Growth Des.*, 2008, **8**, 280.
12. D. Häringer, P. Novak, O. Haas, B. Piro and M. C. Pham, *J. Electrochem. Soc.*, 1999, **146**, 2393.
13. Z. Song, H. Zhan and Y. Zhou, *Chem. Commun.*, 2009, **4**, 448.
14. T. Nokami, T. Matsuo, Y. Inatomi, N. Hojo, T. Tsukagoshi, H. Yoshizawa, A. Shimizu, H. Kuramoto, K. Komae, H. Tsuyama and J. Yoshida, *J. Am. Chem. Soc.*, 2012, **134**, 19694.
15. M. Yao, H. Senoh, S. Yamazaki, Z. Siroma, T. Sakai and K. Yasuda, *J. Power Sources*, 2010, **195**, 8336.
16. M. Yao, H. Senoh, M. Araki, T. Sakai and K. Yasuda, *ECS Trans.*, 2010, **28**(8), 3.
17. H. Bock, S. Nick, W. Seitz, C. Nather and J. W. Bats, *Z. Naturforsch. Teil B Chem. Sci.*, 1996, **51**, 153.
18. M. Yao, H. Senoh, K. Kuratani, T. Sakai and T. Kobayashi, *ITE-IBA Lett.*, 2011, **4**, 52.
19. M. Yao, H. Senoh, T. Sakai and T. Kobayashi, *Int. J. Electrochem. Sci.*, 2011, **6**, 2905.
20. M. Yao, S. Yamazaki, H. Senoh, T. Sakai and T. Kobayashi, *Mater. Sci. Eng. B*, 2012, **177**, 483.
21. M. Slouf, *J. Mol. Struct.*, 2002, **611**, 139.
22. D. Kafer, M. E. Helou, C. Gemel and G. Witte, *Cryst. Growth Des.*, 2008, **8**, 3053.
23. M. Yao, M. Araki, H. Senoh, S. Yamazaki, T. Sakai and K. Yasuda, *Chem. Lett.*, 2010, **39**, 950.
24. X. Han, C. Chang, L. Yuan, T. Sun and J. Sun, *Adv. Mater.*, 2007, **19**, 1616.
25. T. Matsunaga, T. Kubota, T. Sugimoto and M. Satoh, *Chem. Lett.*, 2011, **40**, 750.
26. Y. Morita, S. Nishida, T. Murata, M. Moriguchi, A. Ueda, M. Satoh, K. Arifuku, K. Sato and T. Takui, *Nat. Mater.*, 2011, **10**, 947.
27. H. Chen, M. Armand, G. Demailly, F. Dolhem, P. Poizot and J.-M. Tarascon, *ChemSusChem*, 2008, **1**, 348.
28. M. Satoh, T. Koizumi, Y. Miura, H. Mokudai and T. Sukigara, *The 51st Battery Symposium in Japan*, Nagoya, 2010, Abstract #3G24.
29. Y. Inatomi, N. Hojo, T. Yamamoto, S. Watanabe and Y. Misaki, *ChemPlusChem*, 2012, **77**, 973.
30. Y. Inatomi, N. Hojo, T. Yamamoto, M. Shimada and S. Watanabe, *213th ECS Meeting*, Phoenix, 2008, Abstract #167.
31. Y. Inatomi, N. Hojo, T. Yamamoto, M. Shimada and S. Watanabe, *214th ECS Meeting*, Honolulu, 2008, Abstract #408.

32. M. Yao, H. Senoh, H. Sano, K. Kuratani, T. Kobayashi and H. Sakaebe, *PRiME 2012*, Honolulu, 2012, Abstract #1861.

33. M. Yao, K. Kuratani, H. Senoh, N. Takeichi and T. Kobayashi, *The 54th Battery Symposium in Japan*, Osaka, 2013, Abstract #3D01.

34. K. Chihara, N. Chujo, A. Kitajou, E. Kobayashi and S. Okada, *PRiME 2012*, Honolulu, 2012, Abstract #1838.

35. L. Zhao, J. Zhao, Y.-S. Hu, H. Li, Z. Zhou, M. Armand and L. Chen, *Adv. Energy Mater.*, 2012, **2**, 962.

36. M. Armand, S. Grugeon, H. Vezin, S. Laruelle, P. Ribière, P. Poizot and J.-M. Tarascon, *Nat. Mater.*, 2009, **8**, 120.

37. H. Sano, H. Senoh, M. Yao, H. Sakaebe and T. Kobayashi, *Chem. Lett.*, 2012, **41**, 1594.

38. Y. Hanyu and I. Honma, *Sci. Rep.*, 2012, **2**, 453.

39. H. Senoh, M. Yao, H. Sakaebe, K. Yasuda and Z. Siroma, *Electrochim. Acta*, 2011, **56**, 10145.

12 Materials for Metal–Air Batteries

Vladimir Neburchilov and Haijiang Wang

CONTENTS

12.1 INTRODUCTION

Electrochemical energy storage technologies play a key role in the movement of the world toward green energy and efficient usage of electricity in micro- and smart-grids. These technologies can compensate for the gaps between demand and supply of electricity and be used in the transportation sector, providing a reliable, low-cost, and clean energy source. Metal–air batteries (MABs) represent the most efficient energy storage technology with high round-trip efficiency, a long life cycle, fast response at peak demand/supply of electricity, and decreased weight due to the use of atmospheric oxygen as one of the main reactants. The most developed MABs with technical potential and market perspectives are zinc–, lithium–, aluminum–, and magnesium–air batteries (ZnAB, LiAB, AlAB, MgAB).[1–3] However, reasonable performance was achieved only for rechargeable ZnABs. MgABs and AlABs are basically used as primary batteries. MABs are now considered the best potential substitutes for currently used rechargeable nickel–cadmium (NiCd), nickel metal hydride (NiMH), and Li-ion batteries for different applications due to some limitations in terms of the safety (Cd for NiCd batteries and Li ion), raw material resources (lithium for Li-ion batteries), cost efficiency, and weight (Li-ion, NiMH batteries). MABs have high energy density compared to other batteries, low cost, possible fuel recycling, long shelf life, no ecological issues, and flat discharge voltage.[1–5]

The main technical characteristics of MABs are outlined in Table 12.1.[1] The comparison of emerging (e.g., LiAB) and established (e.g., ZnAB) MABs with conventional rechargeable batteries such as Li-ion, NiCd, and NiMH shows the advantage of ZnABs and LiABs in terms of their better balance of energy density and specific energy (Figure 12.1).[5] LiABs and MgABs have the maximal theoretical specific energy density, 13,000 and 6,800 W h/kg, respectively (for comparison, gasoline, e.g., has an energy density of 13,000 W h/kg, see Ref. [2]). However, these batteries have some limitations, for example, LiABs have unstable anodes that react with an electrolyte, incomplete discharge due to deposition of discharged products in porous cathodes, low cycle life, insufficient electrical efficiency because of the higher over-potential at the charging mode than that of at the discharging, formation of lithium carbonates or alkyl carbonates, and safety issues.[1,2,6–10] MgABs could be rechargeable with limited cycle life, but AlABs are not rechargeable. The main challenges of

TABLE 12.1
Technical Characteristics of MABs

MAB	Voltage (Theoretical/ Practical), V	Mass Specific Energy Density (Theoretical/Practical), W h/kg	Volumetric Energy Density, W h/L
LiAB	3.4/2.4	13,000/2,200 (primary battery with lithium/water)	100
ZnAB	1.6/1.0–1.2	1,300/200–300[1] (130–180[2])	225–330[1] (130–160[2])
MgAB	3.1/1.2–1.4	6,800/2,700[1] (Mg/O$_2$ cell for undersea use)	
AlAB	2.7/1.1–1.4	8,100/200–250[1] (alkaline)	150–200[1]

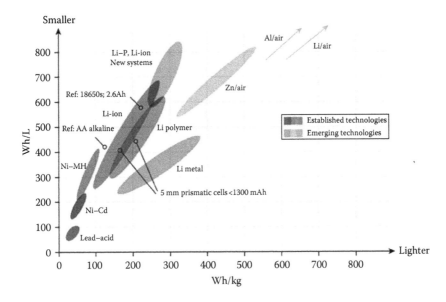

FIGURE 12.1 Comparison of established and emerging energy storage technologies. (Adapted from Zhang, L.-L. et al., *Int. J. Smart Nano Mater.*, 4, 27, 2013. With permission.)

MgABs and AlABs are low durability of their anodes, low irreversibility, and high self-discharge rate.[1–4] ZnABs have less specific energy density than LiABs, MgABs, and AlABs but have higher volumetric energy density. They are mostly practically used for different applications due to their lower cost, longer cycle life, absence of safety issues, and unrestricted outdoor usage.[11–17]

The main components of MABs are air cathode (air-breathing bifunctional electrodes [BEs]), anodes, electrolytes, and separators, which mainly determine battery cycle life and cost efficiency. Therefore, the main focus in this chapter will be on these components, their current status and perspectives. ZnABs also have technical issues related to a low durability of BEs at high oxygen evolution potentials, carbonation of alkaline electrolytes for aqueous batteries, zinc dendrite formation on anodes at the charging mode, nonuniform zinc dissolution, and low zinc solubility in electrolyte.[1–4,11–17] The most successful solutions to these issues were recently found in the rechargeable ZnABs developed by Energy Storage (EOS),[18,19] ReVolt,[20,21] and Fluidic Energy[22] with a long cycle life. EOS's ZnAB *Aurora* with neutral aqueous electrolyte[18] demonstrated stable performance during 5000 cycles without degradation of its BEs.[19] It is very important that usage of the stable and cheap neutral aqueous electrolyte allowed for EOS to avoid the common problems of carbonation of the electrolyte for alkaline MABs. Fluidic Energy developed a flow ZnAB with ionic liquid (IL) electrolyte with an energy density 11 times higher than that of a Li-ion battery at 3 times less cost.[21] An IL, in contrast to an aqueous alkaline electrolyte, does not evaporate at the operation temperature of ZABs, is stable (without alkaline electrolyte carbonation issues), is conductive, and has a wider electrochemical window (water decomposes at the potential over 1.23 V).[22] ZnABs are commercially available for different applications. These batteries consist of inexpensive and available fuel—zinc in contrast to expensive nickel or lithium (Zn—US$2/kg; Ni—US$21.4/kg[23]; and Li—US$8/kg[24,25]). World zinc resources are estimated to be 1.8 gigatons, with 200 megatons economically available in 2008.[26] The development of air cathodes in the 1970s led to significant improvements in ZnABs performance, resulting in low operating costs and cell weight.

12.2 ZINC–AIR BATTERY COMPONENTS

12.2.1 Oxygen Positive Electrodes

12.2.1.1 Air Cathodes

The oxygen reduction reaction (ORR) in aqueous solutions can proceed by two pathways: a direct four-electron pathway (oxygen directly reduces to OH^-) and a peroxide two-electron pathway (with oxygen reduction to HO_2^- at the first stage with the following reduction of HO_2^- to OH^-.

The chemistry of ZnABs is given as follows[1]:

Anode:

$$Zn \rightarrow Zn^{2+} + 2e \qquad (12.1)$$

$$Zn^{2+} + 4OH^- \rightarrow Zn(OH)4^{2-} + 2e^- (E_o = 1.25\,V) \qquad (12.2)$$

$$\text{In solution: } Zn(OH)_4{}^{2-} \rightarrow ZnO + H_2O + 2OH^- \qquad (12.3)$$

$$\text{Cathode: } O_2 + 2H_2O + 4e^- \rightarrow 4OH^- (E_o = 0.4\,V) \qquad (12.4)$$

$$\text{Overall: } 2Zn + O_2 \rightarrow 2ZnO\ (E_o = 1.65\,V) \qquad (12.5)$$

During a discharging mode, zinc oxidizes with the release of electrons and atmospheric oxygen reduces to hydroxyl ions in alkaline electrolyte. During a charging mode, the resulting zinc oxide reduces to metal on the zinc electrode. The design of the typical ZnAB is given in Figure 12.2.

12.2.1.1.1 MnO$_2$-Based ORR Catalysts

Manganese oxide (MnO_x)-based catalysts are characterized by the two-electron ORR and have a sufficient stability in a concentrated alkaline solution and high mass activity for peroxide decomposition, due to simultaneous oxidation and reduction of the surface manganese ions[27,28] (i.e., Mn^{4+}/Mn^{3+} for the mixed manganese-based catalyst[29]). Mao et al.[30] established that the presence of MnO_x, including Mn_2O_3, Mn_3O_4, Mn_5O_8, and $MnOOH$, on Nafion®-modified Au electrodes enhanced the first reduction current peak of O_2 to HO_2^- and decreased the second peak of HO_2^- to OH^- without a potential shift. MnO_x also promoted oxygen disproportionation

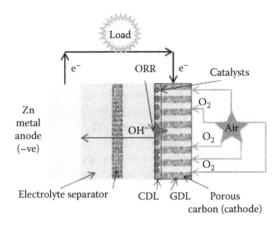

FIGURE 12.2 ZnAB design. (Adapted from Rahman, M.A. et al., *Electrochem. Soc.*, 160, 759, 2013. Copyright 2014, The Electrochemical Society. With permission.)

and resulted in an overall four-electron reduction of O_2 on MnO_x/Nafion-modified Au electrodes.

The majority of the ZnAB air cathode patents analyzed in this review are based on MnO_x.[30–59] In recent years, significant progress has been made in the improvement of the durability and ORR activity of these carbon-supported air cathodes. Yang and Xi[58] identified the ORR pathway for nanoporous amorphous MnO_x as the two-electron process (load 0.85 mg/cm²). In a three-electrode cell, it demonstrated a higher current density ($j > 100$ mA/cm²) with an oxygen backfeed than with nitrogen in 1 M KOH. The catalyst was produced by reacting sodium permanganate with disodium fumarate and the high ORR activity was explained by the high concentration of lattice defects and active sites in the amorphous material $Na_{0.10}MnO_{1.96}\cdot0.7\ H_2O$. The mean oxidation state of manganese was determined to be 3.82.

Due to different morphologies and surface states, the properties of MnO_2 are influenced by the method of fabrication[59], which includes both chemical and electrochemical methods. The chemical methods include the heating of manganese nitrate[30–32] and the reduction and heating of $KMnO_4$.[31–34] Zoltowski et al.[51] disclosed the use of potassium permanganate to catalyze activated carbon, wherein most of the permanganate was reduced to MnO_2 by the carbon. Armstrong et al.[32] disclosed a similar admixture of potassium permanganate and activated carbon, wherein the potassium permanganate was reduced in situ, by either heating or the introduction of hydrogen peroxide, to form MnO_2. Bach et al.[35] disclosed the sintering of potassium permanganate at 250°C–700°C in an oxidizing atmosphere, which produced a mixture of the oxides MnO_2, Mn_2O_3, and Mn_3O_4. Hoge et al.[36,37] used potassium permanganate as the catalyst for a carbon-supported air cathode. Mn(II) isopropoxide has also been used as the catalyst for an air cathode and the design of an air cathode utilizing a MnO_2-based catalyst.[38]

Air Energy Resources Inc. (AER) also used the Mn(II) precursor isopropoxide, but they utilized the more advanced sol–gel method for fabrication. The air cathode composition of 5% $MnO_2 + 75\%$ C (a mixture of 30% Ketjenblack carbon black EC-600JD [EC-600JD] from AKZO with a specific surface area of 1200 m^2/g and 70% AB50 Shawinigan carbon black [AB-50] from Chevron with a BET surface area of 70–90 m^2/g) + 20% polytetrafluoroethylene (PTFE)[39] employs a novel approach for water management in the coating by combining two carbon blacks with low and high BET surface area. This approach leads to an optimum balance of hydrophobic/hydrophilic properties and allows for good oxygen adsorption. Sol was formed by the addition of water to an alcoxide/alcohol solution. Gel was produced by heat treatment and evaporation of the alcohol. The pyrolysis of the gel and active carbon mixture in air at T = 150°C–250°C produces the final composition. AER used the micelle method and combined active carbon and carbon black for additional improvement of the air cathode: 5% MnO_2 (Mn^{3+}/Mn^{4+}) + 70% C (60% active Calgon Carbon PWA + 40% carbon black) + 25% PTFE (Teflon 30B). Design of the gas diffusion layer (GDL) was also improved through the addition of low and high BET carbon blacks (30% EC-600JD + 70% AB-50).

Ndzebet designed an air cathode with a combination of high and low BET surface carbon blacks with the following composition[41]: MnO_2 + C (activated carbon, BET—900 m^2/g + carbon black (Black Pearls 2000 [BP2000]) from Cabot, BET—1500 m^2/g) + PTFE. The particle size distribution of the MnO_2 catalyst was established to be 20–26 μm. A performance test in 30% KOH achieved a potential of 1.15 V at 150 mA/cm^2 (over 15 h).

Koshiba et al.[42] obtained a limiting current of 11 mA/cm^2 (compared to commercial catalysts at 50–100 mA/cm^2) with a ZnAB air cathode composed of 30% MnO_2 + 20% activated carbon + 20% carbon black + 30% PTFE. The MnO_2 was produced by heat treatment of γ-manganese oxyhydroxide at 250°C–450°C with the decomposition products of Mn_5O_8 + β-MnO_2. This patent used the combination of the activated carbon and carbon black (30% MnO_2 + 20% active carbon + 20% carbon black + 30% PTFE) to improve performance. Sun and Wang[44] used a catalyst with 20% MnO_2 + 70% activated carbon + 10% PTFE with a GDL composed of 15% carbon black + 85% PTFE. After 1 h of discharging, a voltage of 1.32 V was measured at a load of 620.

The chemical fabrication methods for MnO_2 require significant time (about 130 h) and are limited to a maximum BET surface area of 400 m^2/g to minimize the risk of spontaneous combustion. Thus, Duracell Inc.[45] used the milling of electrolytic or chemical γ-MnO_2 (EMD or CMD, respectively) for the air cathode (11% γ-MnO_2 + 41% BP2000 + 48% PTFE). The ZnAB with this cathode had a 430 mA h capacity when 1.0 V was used as the cutoff voltage. The other Duracell cathode composition[46] of MnO_2 + PTFE + 2%–20% absorbent (gelling material) was used in the air cathode, giving a limiting current of 27.5 mA at T = 66°C.

It was discovered that another manganese compound, MnOOH, had higher ORR activity than oxides such as Mn_2O_3, Mn_3O_4, and Mn_5O_8 in 0.1 M KOH electrolyte, due to an increased number of active centers.[59] Mn_3O_4 and γ-MnOOH were prepared by chemically oxidizing $MnSO_4$ with H_2O_2. The compounds α-Mn_2O_3 and Mn_5O_8 were made by thermally oxidizing Mn_3O_4 under an O_2-gas atmosphere at 1173 and 703 K, respectively. The best air cathode, 28.5% (9% MnO_x + 1% Ni)/C, has a specific current density of 43.8 A/cm^2, which is about 1.5 times higher than MnO_x/C. The measurements of ORR Tafel slopes and mass activities (MA) were carried out at E = 0 V (NHE) in O_2-saturated 1 M KOH and T = 25°C.[49] The measurements of ORR Tafel slopes and MA were carried out at E = 0 V (NHE). The best air cathode, MnO_x + Ni/C, has a SA of 43.8 mA/cm^2, which is 1.5 times higher than MnO_x/C and a 37–51 nm particle size.

12.2.1.1.2 Ag-Based ORR Catalysts

Silver (Ag) has been demonstrated to be an active component for the ORR in alkaline solutions.[60–72] Wu et al. investigated Ag ORR catalyst supported on carbon nanocapsules (CNC).[62] This catalyst was produced by a simple method that combined precipitation of AgCl and reduction to Ag in a hydrogen atmosphere. The CNC (333 m^2/g, 15–30 nm) surpassed the ORR activity of Vulcan XC-72 carbon black (XC-72) from Cabot. High performance was demonstrated (j = 200 mA/cm^2 at E = 0.8 V) resulting from dense packing that was possible due to the uniform size and the higher conductivity (30% KOH, T = 25°C). Ag/CNC catalyst demonstrated better performance (E = 0.99 V at 200 mA/cm^2) than the commercially available Mn and MnCo catalyzed air cathodes. The galvanostatic discharge for Ag/CNC at j = 200 mA/cm^2 showed a moderate decrease in performance after 80 h. Yang and Zhou[63] also established relatively good stability and insignificant voltage deterioration of Ag on Ni foam during a 120 h galvanostatic discharge at T = 40°C.

Expensive platinum was found to have the highest ORR activity; however, it was insufficiently stable in alkaline electrolytes.[65] The ORR activity and stability of Pd-, Ir-, Co-, Ru-, and Ni-based electrodes (thermal method of fabrication) were investigated as well.

Ag-based air cathode improvement was achieved through doping with tungsten carbide (W_2C), which allowed for a more promising composition for ORR in ZnAB, AgW_2C/C. A comparison of the ORR activity of W_2C/C and Ag/C confirms the synergetic effect of the W_2C additive on Ag. This catalyst was produced by the intermittent microwave heating (IMH) method. $AgNO_3$ was mixed with H_2O_2, IPA, and W_2C/C in the ratio W_2C:Ag = 1:1 before IMH treatment. Considering the porous character of the Ag coating, Ag-coated WC particles are an alternative for AgWC/C worth evaluating.[67]

Gillette Corporation[68] developed a new advanced catalyst consisting of 5% $AgMnO_4$ + MnO_2, based on the high stability of MnO_x in alkaline solution and its high activity for H_2O_2 decomposition. The dual catalyst ($AgMnO_4$) was prepared by reducing silver permanganate. MnO_2 was prepared by reducing $KMnO_4$ with hydrazine or by heating $Mn(NO_3)_2$. Testing of $AgMnO_4$ + (5% or 10%) MnO_2 air cathodes in a three-electrode cell demonstrated performances of 28 and 50 mA/cm^2, respectively, at E = 0.16 V (SHE). The limiting current (51 mA/cm^2) occurred at E = 0.25 V (Hg/HgO) or 1.1 V (vs. Zn). Another useful approach described in the patent involved the utilization of varying Teflon concentrations in the GDL and catalyst layer depending on the purpose of the layer:

- High (30%–70%) PTFE content in the GDL to prevent the cathode from wetting through
- Low (10%–30%) PTFE content in the catalyst layer to promote optimal wetting of the catalyst layer

A catalyst consisting of a combination of Ag and Raney alloy (Ni–Al) was suggested by Goldstein et al.[69] The design of this cathode includes the GDL (C+PTFE, with a carbon loading of 6–10 mg/cm^2), which was pressed onto a Ni foam current collector and the catalyst layer (Ni – Al+Ag+PTFE=5:1 wt) at a loading of 24 mg$_{Me}$/cm^2). Catalyst testing showed a peak current of 10 A at E=0.9 V during 5 h of cell discharge (a peak current of 200 mA/cm^2 at the air cathode).

Zhong[70] used silver oxide (Ag$_2$O) as the ORR active component in the development of catalyst composed of Ag$_2$O+10% LaNiO$_3$ (lanthanum nickelate). Performance degradation of the ZnAB was not observed during 500 h of testing in 32% KOH. Alupower, Inc. suggested a catalyst containing 5% Ag+15% BP2000+10% Daxad (sodium salt of polymerized naphthalene sulfuric acid)+60% Teflon RPM T-30. The Daxad additive was used to increase the Ag adsorption on the carbon black.[71]

Basically, carbon-supported Ag catalysts show high ORR activity as a result of its activity for decomposition of H$_2$O$_2$, which would otherwise accumulate during the two-electron ORR pathway on carbon.[64]

12.2.1.1.3 Mixed Valence CoO$_x$-MnO$_x$ ORR Catalysts

Cobalt oxides were also identified as active catalysts for the ORR in alkaline solutions. Oyshinsky et al.[73] suggested novel multifunctional air cathode compositions with mixed valence components (5% [2.5%–7.5%] MnO$_x$+5% [2.5%–7.5%] CoO$_x$/C and 15% CoO$_x$+5% MnO$_x$/C, 20% CoTMPP+15% CoO$_x$+5% MnO$_x$). CoO$_x$ was utilized because it has Co^{3+} ions located on the octahedral lattice sites and Co^{2+} on the tetrahedral sites. These catalysts were developed with a ratio of high and low valences in the range of 1:2–2:3, where the two components of the catalyst (Co^{3+} and Co^{2+}) were responsible for the two activation steps of the ORR:

- The first is for the two-electron stage of the peroxyl ion formation:

$$O_2 + H_2O + 2e^- = H_2O^- + OH^- \qquad (12.6)$$

- The second is for the two-electron stage of peroxyl ion decomposition

$$H_2O^- + 2e^- + H_2O = 3HO^- \qquad (12.7)$$

As the pores of this catalyst are not through-holes, the peroxide, formed during the two-electron pathway of ORR, can only diffuse into the bulk solution. During this slow diffusion, the peroxide oxidizes the Teflon bonding between the catalyst particles, carbon, and other catalyst components. This decomposition, the main cause of ZAFC degradation, blocked the internal space of the pores, increased the resistance, and reduced the active surface area of the ORR catalyst. The catalyst composed of 2.5% MnO$_x$+7.5% CoO$_x$/C showed the highest ORR activity (120 mA/cm^2 at E=−0.1 V). The fabrication procedure for the 5% CoO$_x$/C catalyst consisted of mixing NH$_4$OH and carbon in an ultrasonic bath before adding an aqueous solution of CoSO$_4$. The final solution was then treated with NaOH, which resulted in the formation of the catalyst deposit. The procedure for the preparation of the catalyst with 20% CoTMPP/C was similar.

12.2.1.1.4 Metal Tetramethoxyphenyl Porphyrin–Based ORR Catalysts (CoTMPP, FeTMPP-Cl/C)

The carbon-supported pyrolyzed macrocycles show good ORR activity in alkali electrolytes and are currently used in mechanically rechargeable MABs.[73–81] One popular nonnoble air cathode is based on cobalt and iron tetramethoxyphenyl porphyrin (CoTMPP and FeTMPP).[73,78–80]

CoTMPP has higher electrochemical stability and promotes the two-electron ORR pathway, while FeTMPP has inferior ORR activity and promotes direct four-electron oxidation. The stability of the metal macrocyclic complex depends on the metal and it decreases accordingly: Co > Fe > Mn.[81] One common approach to combine the advantages of a more stable CoTMPP and a more ORR active FeTMPP is to fabricate a mixture, FeTMPP/CoTMPP.[81–83] The ORR activity increases due to the formation of a face-to-face structure, accelerating the destruction of the O–O bond in the oxygen molecule.[84–87] The effect of heat treatment on increasing the stability of N_4 chelates, discovered by Jahnke et al.[88] in 1976, is currently a very popular method for increasing the stability of Co/FeTMPP-based catalysts.

12.2.1.1.4.1 FeTMMP-Cl-Based ORR Catalysts
Gojkovic et al.[89] established that pyrolyzed FeTMMP-Cl catalyst, heat treated at T < 200°C, has the Fe^{3+}/Fe^{2+} redox peak for CV curves in alkaline (1 M NaOH at 25°C) and acidic solutions. This macrocyclic complex decomposes at T > 700°C. The increase in ORR activity for the catalyst after heat treatment corresponds to the decrease in activation energy for the ORR (RRDE, 0.1 M H_2SO_4). The number of electrons exchanged per oxygen molecule (3.45–4) depends on the potential but not on the pyrolysis temperature. At low pyrolysis temperatures, the ORR on FeTMMP-Cl proceeds by the direct four-electron pathway, resulting in lower polarization.[78] The rate of the ORR on FeTMPP-Cl/BP in alkaline electrolyte is higher than in acidic solution and it is comparable with that of Pt; however, in an acidic solution, Pt is more active than FeTMPP-Cl/BP. The effect of sulfate, perchlorate, and phosphate on ORR activity was not determined.[89]

12.2.1.1.4.2 CoTMPP-Based ORR Catalysts
CoTMPP has higher stability than FeTMPP, which has led to its use for ZnAB air cathodes. Iliev et al.[91] also established that the heat treatment of CoTMPP/C in argon at T=460°C–810°C improves its ORR activity in an alkali electrolyte. The ORR activity after heat treatment at T=750°C–850°C remains high and stable, with transport limitations increasing with time. Heat treatment of CoTMPP increases its long-term stability in both alkali[92,93] and acid[94–96] electrolytes.

The work of Mocci et al.[77] improved understanding of the ORR activity of CoTMPP in alkaline solutions. ORR activity is determined by the simultaneous presence of a metal precursor, active carbon, and a source of nitrogen. The ORR activities of pyrolyzed mixtures of $CoCO_3 + TMPP + C$ and $Co_3O_4 + CoTMPP + C$ (in N_2 at 800°C) were shown to be higher than CoTMPP/C. The findings demonstrate the key role of carbon during pyrolysis, but not as a structural component for CoTMPP/C (pyrolysis in N_2 at 900°C). Mocci et al.[77] developed another interesting modification of CoTMPP—10% $(CoCO_3(Co_3O_4) + TMPP)/C$. The choice of Co_3O_4

was based on its presence in the products of CoTMPP pyrolysis. It was established that $CoCO_3$ was the best Co precursor for achieving maximum ORR activity for 10% $(CaCO_3(Co_3O_4) + TMPP)/C$. The optimal fabrication method for this catalyst consists of pyrolysis of the $CoCO_3 + TMPP$ or $Co_3O_4 + TMPP/C$ mixture at 800°C in inert gas with the addition of carbon black after heat treatment (pyrolysis at 900°C yields catalyst with lower ORR activity). The catalyst with a molar ratio of Co/TMPP = 1:1 and weight ratio of $CoCO_3/C = 5\%$ demonstrated the highest ORR activity. Conversely, increasing the pyrolysis temperature for $Co_3O_4 + TMPP/C$ fabrication decreases the Co_3O_4 surface area, which leads to a drop in ORR activity. It was shown that Co_3O_4 does not significantly improve ORR activity (it is active itself[77]) and the Co- and N-containing molecular moieties do not interact for a synergetic effect. The presence of carbon during pyrolysis more strongly affects ORR activity, as it is responsible for the partial reduction of Co_3O_4, which determines the interaction between Co ions and nitrogen moieties. The performance of 10% $(CaCO_3 (Co_3O_4) + TMPP)/C$ (heat treatment at 200°C) is higher than Co_3O_4 (0.5 and 0.1 mA/cm², respectively) at 0.3 V (SHE), 1 M KOH, and 25°C and it is the same at T > 600°C.

An air cathode consisting of 10% CoTMPP on activated carbon showed a performance of 200 mA/cm² at E = 200 mV(Hg/HgO) in 7 M KOH at room temperature.[91] The catalyst was heat treated at T = 460°C–850°C for 5 h in Ar. A 2500 h durability test of this cathode demonstrated an increase in the initial potential of the electrodes heat treated at 460°C, 610°C, 730°C, and 810°C from 120 mV (Hg/HgO) to 300, 220, 180, and 170 mV. Thus, the best stability for this electrode was achieved with a heat treatment at 730°C–810°C (810°C being optimal).

One method for improving the ORR activity of CoTMPP-based catalysts is through the addition of MnO_2, which is active for H_2O_2 decomposition.[90] This work was conducted in collaboration with Powerzinc Electric Inc. (Shanghai, China). The $MnO_x + CoTMPP/C$ (BP2000) catalyst promotes the two-electron pathway of the ORR with the rate-limiting step being the formation of peroxide in 1 M KOH at T = 25°C (RRDE in half cell). The ORR activity of CoTMPP was improved through modification with MnO_x, one of the most commonly used ORR catalysts for ZnABs.[90] Suspensions of $KMnO_4/C$ and $MnSO_4/KOH$ (pH8) were mixed to produce MnO_x/C (BP2000), as detailed by the following reaction:

$$2MnO_4^- + 3Mn^{2+} + 2H_2O = 5MnO_2 + 4H^+ \qquad (12.8)$$

In this work, carbon black, carbon nanotubes (CNT), and BP2000 were investigated as the support for CoTMPP/C catalysts. BP2000, activated by 30% H_2O_2, supplied the highest peak current for oxygen reduction. H_2O_2 oxidizes the carbon support during ORR but preliminary treatment of the carbon in H_2O_2 increases ORR activity. The tetragonal structure of Mn_2O_3 changes to cubic MnO_x at T = 800°C, with particle sizes of 5–30 nm. At T = 900°C, the cubic MnO_x granules become spherical. This catalyst demonstrated a performance of j = 500 mA/cm² at E = −0.498 V (Hg/HgO) in a half cell with 1 M KOH at T = 25°C. A ZAFC with a $MnO_x + CoTMPP/C$ air cathode (heat treatment at T = 800°C, catalyst mass loading of 14.6 mg/cm²) demonstrated a maximum output current density of 216.3 mA/cm² at a 1 V output potential in 30% KOH.[90]

A similar CoTMPP doping procedure with MnO_x was developed to fabricate 4% CoTMPP + 15% C (BP2000) + 60% Teflon RTM T-30 + MnO_2 (Co_3O_4).[80] Two methods were used to prepare this CoTMPP-based electrode: a wet powder process and a dry powder process. The wet method produced a catalyst with higher performance than the dry method (discharge voltage of 1 V at 500 mA/cm^2, in comparison with 200 mA/cm^2 for the dry method). After soaking the electrode in 7 M KOH at 25°C overnight and applying a duty cycle for 1 h (20 s—200 mA/cm^2, 45 s—50 mA/cm^2, 45 s—0 mA/cm^2), the voltage was E = −0.29 V (Hg/HgO) at j = 0.4 mA/cm^2. This performance was stable during 200 h of testing.

CoTMPP air cathodes demonstrate high stability not only in ZnAB conditions but also in hot alkaline electrolyte. The developed air cathode (10% CoTMPP + 90% carbon [Shawinigan] + Nafion) shows good performance at a current density of 450 mA/cm^2 (an initial voltage of 0.53 V [SHE], 0.77 V after 3 h, and 0.54 V after 134 days).[97,98]

Another interesting approach for improving the ORR activity of CoTMPP-based catalysts was through doping with CoO_x. Catalysts with compositions consisting of 1%–5% (50% [20% CoTMPP/C] + 50% [15% CoO_x + 5% MnO_x])/C and 1%–5% (7.5% CoO_x + 2.5% MnO_x)/C have a performance of E = −0.09 and −0.11 V, respectively, at j = 200 mA/cm^2, see Ref. [73].

12.2.1.1.5 Metal Nitride–Based ORR Catalysts

Miura et al.[99] investigated the nitride-based ORR catalysts Mn_4N, CrN, Fe_2N, Co_3N, and Ni_3N in alkaline electrolyte. All catalyst had a GDL composed of 70% acetylene carbon black (AB-7) from Denka (specific surface area [SSA] 49 m^2/g) + 30% PTFE. The cathode with a 60% Mn_4N/C + 40% Furnace Carbon Black 3000B (FCB 3000B) from Mitsubishi Kasei, catalytic coating, and 15% PTFE had a maximum ORR activity of j = 2400 mA/cm^2 at −125 mV (Hg/HgO) or 0.8 V (RHE). X-ray photoelectron spectroscopy (XPS) analysis showed the presence of a thin oxide on the electrode surface. Mn_4N promotes the direct four-electron ORR mechanism and CO_3N decomposes HO^{2-} into OH^-. This air cathode supplied stable performance during a 50 h test in galvanostatic conditions at j = 300 mA/cm^2 in 9 M NaOH at 80°C.

12.2.1.1.6 Mixed Oxides of Transition Metal–Based ORR Catalysts

Mixed oxides on a spinel, perovskite, or pyrochlore structure are largely used for ZAFC ORR catalysts and their performances are discussed in this section.

12.2.1.1.6.1 Spinel-Type-Based Catalysts $NiCo_2O_4$ (spinel)

Spinels are a group of oxides with the formula AB_2O_4, where A is a divalent metal ion (such as Mg, Fe, Ni, Mn, or Zn) and B is a trivalent metal ion (such as Al, Fe, Cr, or Mn). Some of these oxides display good stability and ORR activity in alkaline solutions. Analysis of their ORR activity and stability makes it possible to select the components required for the creation of composite ORR catalysts with an optimal balance of the most important characteristics.[63,99] The stability and activity of some oxides, including La/La_2O_3, Ti_2O_3/TiO_2, Ni_2O_3/NiO_2, and Co_2O_3/CoO_2 in alkaline solution, are shown in Table 12.2.[100] This stability is given by the Pourbaix diagram,

TABLE 12.2

Properties of Semiconducting Oxides

Oxide	Electrical Conductivity	Corrosion Resistance at pH = 14	Oxygen Reduction	Potential, V versus NHE
La/La$_2$O$_3$	Poor	Good	Poor	−2.069
Ti$_2$O$_3$/TiO$_2$	Poor	Good	Poor	−0.556
V$_2$O$_4$/V$_2$O$_3$	Poor	Poor	Poor	−0.666
Cr$_2$O$_3$/CrO$_2$	Poor	Poor	Poor	1.284
MoO$_2$/MoO$_3$	Poor	Poor	Poor	−1.09
W$_2$O$_3$/WO$_3$	Poor	Poor	Poor	−0.029
Mn$_2$O$_3$/MnO$_2$	Fair	Doubtful	Fair	1.014
Co$_2$O$_3$/CoO$_2$	Poor	Good	Poor	1.44
Lower Co$_2$NiO$_4$/Co$_2$NiO$_4$	Good	Good	Good	1.4
Ni$_2$O$_3$/NiO$_2$	Poor	Good	Poor	1.434

Source: Adapted with permission from Roche, I. et al., *J. Phys. Chem. C*, 111, 1434. Copyright 2007 American Chemical Society.

which provides the equilibrium metal ion concentration for solutions of <10^{-6} ions/L at pH = 14. The properties of the thermally prepared electrodes are given in Table 12.2.[62] NiO has a higher ORR activity than Co$_3$O$_4$, with respective current densities of 2.74 × 10^{-3} and 1.79 × 10^{-3} mA/cm^2. However, in contrast to Co$_3$O$_4$, NiO has four times higher H$_2$O$_2$ current efficiency and promotes the two-electron ORR pathway. Furthermore, unlike the majority of the thermally prepared oxides, Co$_3$O$_4$ is unstable during ORR (Co$_2$O$_3$ has a good stability).[99] It was shown that spinel NiCo$_2$O$_4$-based catalyst combines the desirable properties of NiO$_x$ and CoO$_x$.

Cobalt has two oxides, CoO (cubic structure) and Co$_3$O$_4$ (spinel), while Ni only has NiO (cubic). CoO and NiO have similar lattice parameters, so the formation of a solid solution between them is not problematic. Tseung et al.[101] discovered a correlation between the electrocatalytic activity and the structure of NiCo$_2$O$_4$. High, medium, and low ORR activities correspond to a spinel structure, to a mixed structure of spinel and traces of cubic, and to a structure of spinel with appreciable cubic, respectively. Since Ni does not have a spinel structure, its presence in spinel NiCo$_2$O$_4$ is through its solubility into the Co$_3$O$_4$ matrix, with a temperature-dependent solubility limited at T > 400°C. NiCo$_2$O$_4$ is more active than CoO and NiO in 75% KOH at T = 200°C; however, heat treatment increases the ORR activity at temperatures below 400°C, at which point cubic traces appear. Despite the observed structural changes, a correlation between cathode surface area and catalyst performance was not identified. The maximum corrosion current obtained in electrolyte, in the absence of O$_2$, and at potentials ranging from E = 1000–600 mV, was 20 μA. The spinel phase is metastable, with phase transition from the spinel to cubic structure occurring at T = 450°C. The performance of the ZnABs with a NiCo$_2$O$_4$ + PTFE air cathode is j = 200 mA/cm^2 at a voltage of ~0.77 V in 5 M KOH at room temperature.[101]

$$Mn_xCo_{3-x}O_4 \quad (0 < x < 1) \tag{12.9}$$

$Mn_xCo_{3-x}O_4$ has high electrical conductivity and ORR/oxygen evolution reaction (OER) activity. Rios et al.[102] showed that $Mn_xCo_{3-x}O_4$ has the Mn^{4+}/Mn^{3+} redox couple located in octahedral sites. Changes in the Mn^{4+}/Mn^{3+} content, as a function of x, were correlated to pH of zero charge (pH_z), rest potential ($E_{i=0}$), activation energy of conductance (E_a), and to the electrocatalytical parameters of the ORR and OER. Manganese catalyzed the ORR but inhibited the OER. The correlations between the electrocatalytic activity and the cationic distributions were investigated for the oxygen reactions in 1 M KOH at 25°C. The corrected (real) ORR activity increases with higher Mn content. ORR polarization curves have two Tafel zones with slopes of −60 and −120 mV below and above −50 to −100 mV versus HgO/Hg, respectively. The ORR mechanism occurs via parallel paths of the four-electron and two-electron pathways. The OER activity increases with decreasing x (i.e., 1 < 0.75 < 0.5 < 0.25 < 0). Only one Tafel slope value (b_a) of ~60 mV was measured. The OER mechanism includes fast electrosorption of OH^- and the OH radical, followed by the slow electrodesorption of OH^- into H_2O_2, which is the rate-determining step. In contrast, Co_3O_4 has an ORR-active surface of Co^{3+} cations, but Mn strongly inhibits the oxidation of OH^- ions. The performance of an RDE with a $Mn_xCo_{3-x}O_4$+PTFE coating is j = 100 mA/cm² at 0.2 V.

Stability of the spinel-based catalyst in alkaline solutions was improved through electrodeposition of polypyrrole (PPy), which is electronically conductive. Gautier et al.[103] utilized a multilayered design for $Ni_{0.3}Co_{2.7}O_4$ catalysts: glassy carbon (GC) (support)/first layer of PPy/second layer of $Ni_{0.3}Co_{2.7}O_4$/third (external layer) of PPy. The external layer of PPy protects the spinel-based catalyst against dissolution during operation in the alkaline solution. This catalyst shows an ORR activity of j = −1.4 mA/cm² at −0.5 V (SCE) in an oxygen-saturated 2.5 mM KOH + 0.8 M KCl solution at room temperature. The $Ni_{0.3}Co_{2.7}O_4$ has an SSA of 22 m²/g. The work[4] demonstrated that the maximum peroxide formation on $Ni_xCo_{3-x}O_4$, which takes place on the active sites such as the Co^{3+}/Co^{2+} couples, occurred at x = 0.3. The other spinel, $Cu_{1.4}Mn_{1.6}O_4$/PPy-based composition, showed good ORR activity in acidic solutions as well.[104] The spinel $(CoFe_2O_4)$/PPy with 3–30 nm particle sizes (fabricated by the microemulsion method with the use of sodium dodecyl sulfate [SDS] surfactant) showed a stable performance of j = −1.5 mA/cm² at E = −0.5 V (SHE) over 8 h in an oxygen-saturated 5 mM KOH + 0.5 MK_2SO_4 electrolyte at T = 25°C. The bulk resistance was 4.5 ± 1.7 Ω for the pure polymer and 2.7 ± 0.8 Ω for the spinel/PPy composite.[105]

12.2.1.1.6.2 Perovskite-Type ORR Catalysts Perovskites are promising nonnoble ORR catalysts for ZAFCs. Co- and Mn-containing perovskites were investigated for oxygen reduction.[101,106–113] $LaMnO_3$ and $LaCoO_3$ have high ORR activity but insufficient chemical and electrochemical stability, as XRD analysis showed an additional phase of the lanthanum hydroxide in the composition of their structure. The additional phase indicates the instability of lanthanum. Conversely, the Fe-based perovskite, $La_{0.6}Sr_{0.4}Fe_{0.6}Mn_{0.4}O_3$, had an optimal balance of stability and ORR activity. For example, the Fe-based perovskite exhibited a performance of j = 500 mA/cm² at E = −260 mV (Hg/HgO) during 70 h of testing. The ORR activity and chemical stability of 11 types of carbon, in a 9 M NaOH solution containing hydrogen peroxide, were also investigated. An evaluation was completed after 30 days and it was found that the stability and cathode performance of the investigated carbons correspondingly decreased and increased with increasing SSA. The EC-600JD (SSA 1270 m²/g) and BP2000 (SSA 1475 m²/g) have much lower chemical stability than other popular low surface area carbon blacks such as XC-72 (SSA 254 m²/g) and AB-7 (SSA 49 m²/g).

LaNiO$_3$

Matumoto et al.[114] established the high activity of perovskite lanthanum nickel oxide (LaNiO$_3$). It has higher ORR activity in comparison with Pt at potentials from −150 to + 100 mV (vs. Hg/HgO) in 1 M NaOH. The current densities for LaNiO$_3$ and Pt at −75 mV (vs. Hg/HgO) were 2×10^{-5} and 10^{-5} mA/cm², respectively.

Lamminen et al.[115] tested LaNiO$_3$ electrodes in 7 M KOH and compared their ORR activity with CoTMPP and Pt air cathodes. The electrodes were manufactured by the rolling method, the best of which were tested in long-term tests ranging from 425 to 660 h. The decay in potential during the 660 h run was 0.041 mV/h.

LaMnO$_3$

Hayashi et al.[116] used the reverse micelle (RM) and amorphous malate precursor (AMP) methods for the fabrication of LaMnO$_3$ catalysts. The stability of these electrodes was tested at galvanostatic conditions (j = 300 mA/cm²). The performance of the catalyst LaMnO$_{3+\delta}$ (RM method) was maintained for 140 h (j = 300 mA/cm² at—80 mV (vs. Hg/HgO). The higher oxidation state of LaMnO$_{3.15}$ is more active for the ORR than oxides with stoichiometric ratios. Masayoshi et al.[117] used another modification of the reduction–oxidation precipitation in the RM method for the fabrication of LaMnO$_3$, which produced a catalyst with higher ORR activity than a catalyst produced by the hydrolysis precipitation in reverse micelle (HP-RM) method with the same particle size.

LaCoSrO$_3$

The doping of LaCoO$_3$ with Sr induces reversible behavior in 45% KOH at room temperature with a performance of j = 2 mA/cm² at 500 mV (vs. DHE).[118] The ORR mechanism on La$_{1-x}$Sr$_x$CoO$_3$ and Nd$_{1-x}$Sr$_x$CoO$_3$ at x = 0.5 occurs via two parallel paths[119,120]:

$$\frac{1}{2O_2} + H_2O = 2e^- + 2OH^- \qquad (12.10)$$

$$O \text{ (lattice of oxides)} + H_2O = 2OH^- + 2e^- \qquad (12.11)$$

La$_{0.6}$Ca$_{0.4}$CoO$_3$

Wang et al.[121] modified the widely studied catalyst, La$_{0.6}$Ca$_{0.4}$CoO$_3$,[122–124] by the amorphous citrate method.[125] The (La$_{0.6}$Ca$_{0.4}$CoO$_3$/C + PTFE)/C cathode has a maximum ORR activity of 20 mA at $E = -200$ mV (vs. Hg/HgO) in 7 M KOH at 25°C.

The investigation of $Nd_{0.5}Sr_{0.5}CoO_3$ (SSA 17 m^2/g) and $La_{0.5}Sr_{0.5}CoO_3$ (SSA 20 m^2/g) ORR catalysts established that the rate-limiting step of ORR is dissociative oxygen chemisorption. Unlike $LaNiO_3$, the ORR activity of $Nd_{0.5}Sr_{0.5}CoO_3$ is significantly lower than Pt in 45% KOH at 25°C.[126]

12.2.1.1.6.3 Pyrochlore-Type-Based ORR Catalysts (A2B2O6O′) The cubic pyrochlore structure consists of a B_2O_6 framework with corner-shared BO_6 octahedral.[127]

$$Pb_2M_{2-x}Pb_xO_{7-y} \quad (M = Ru \text{ or } Ir)$$

The oxide cubic pyrochlore $Pb_2Ir_2O_{7-y}$ is a good ORR catalyst in alkaline electrolyte and for the OER in acidic solution.[127] The ORR activities at 25°C and 60°C are similar, most likely due to decreasing electrode hydrophobicity and oxygen solubility with increasing temperature. The performance of this catalyst was $j = 6 \times 10^{-3}$ mA/cm^2 at -100 mV (vs. Hg/HgO), while the Tafel slope of ORR polarization at $j > 6 \times 10^{-3}$ mA/cm^2 is 60 mV/decade. Note that this catalyst does not reduce oxygen in an acidic solution. The iridium pyrochlore oxide–based catalyst has a higher stability in strong alkaline solutions in comparison with ruthenium pyrochlore oxide, while having similar ORR activity in alkaline and in partially acidic (pH > 2) solutions. Pyrochlore ($A_2B_2O_6O′$) has an active surface for OH$^-$ species on the O′ sites (for exchange with an adsorbed $O_2^-{}_{ads}$ during ORR in alkaline solution).[127]

$$Pb_2Ru_2O_{6.5}$$

Lead ruthenate pyrochlore $(Pb_2Ru_2O_{6.5})$[128] is effective in air cathodes because it promotes the four-electron ORR pathway. It can also be used as a self-supported catalyst, which avoids carbon support oxidation. The performance of this catalyst for oxygen reduction in alkaline solution was increased by the use of a hydrogel overlayer, which was a mixture of poly(dimethyldially ammonium) chloride and Nafion.

Finally, there are several other interesting ORR catalysts, which are not directly included in the aforementioned group of catalysts that should be noted. These include ORR catalysts on the basis of Ni, Co, and Fe hydroxides and two Pt-based ORR catalysts.[129] Blanchart and Van Der Poorten[130] obtained a stable voltage of 0.69 V in 6 M KOH at $T = 50$°C for catalyst 5% Pt/C. Henry et al. added Pt to $Ag + MnO_2$, but the performance obtained with $(Ag + Pt + MnO_2 + PTFE)/C$, $j = 30$ mA/cm^2 at 1 V, was not high enough for commercial use.[131]

12.2.2 BIFUNCTIONAL ELECTRODES

Secondary (rechargeable) ZnABs use BEs, which have active components for both the OER and the ORR. The BEs operate in a wide range of potentials, from 0.6–0.7 V (RHE) during the ORR (discharge mode)[1] to over 2.1 V (RHE) during the OER (charge mode). Additional requirements of BEs include higher stability of the

ORR component in comparison with a primary ZnAB and a fast OER on the ORR component. Westinghouse Electric Corp. developed[132-137] promising BEs based on (CuSO$_4$ + NiWO$_4$ + WC + Co)/C.[133,134] The composition was subsequently modified to (3% Ag + 7% FeWO$_4$ + 7% WC + 12% Co + 7% NiS)/C. The carbon in this electrode consisted of a mixture of two different carbon blacks with high and low SSAs, which provided a proper balance of hydrophilic and hydrophobic properties. 22% PTFE was used as the binding agent and the Ag ORR catalyst (load of Ag—2 mg/cm^2) was stabilized with nickel sulfide.[135] The properties of the electrode can be summarized as follows[137]:

- Ag is the ORR catalyst and it is more active than Co.
- Ag, in the presence of cobalt and nickel, improves ORR activity.
- Ag, in the presence of sulfides (NiS), forms Ag$_2$S, which is relatively stable in alkali electrolyte.
- Hydrated cobalt oxide promotes decomposition of the hydroxides produced during the ORR.
- WC and Ni–Fe hydrated oxides (the latter forms during charging) are ORR catalyst with high activity.

Another of Westinghouse's improved BEs was developed by Liu and Demczyk[138] and consisted of (AgCoWO$_4$ + WC + WS$_2$ + NiS + 10%–15% Co)/C + 20% PTFE. The electrode composition does not allow gas pockets to form between the air cathode and electrolyte, due to gas evolution, during discharge. It is by this mechanism that the chemical reaction between electrolyte and air cathode is interrupted and decreases cell performance.

Shepard et al.[139] developed other BEs with complex ORR catalyst compositions similar to the OER catalyst from Westinghouse: ORR catalyst ([0.3%–2%] CoTMPP/C + [1%–4%] Ag + [1%–7%] NiS [or WS$_2$] + [4%–10%] LaNi$_{1-x}$Co$_x$ + [18%–32%] Co$_x$O$_y$) + OER catalyst ([1%–20%] WC + [1%–20%] Co + [1%–7%] FeWO$_4$ [or CoWO$_4$]/C [AB-50]). The fabrication method of this electrode included the following procedures:

- The Co + WC mixture was sintered.
- AB-50 was treated with CoTMPP and Ag.
- The Ag/C was prepared by mixing a carbon black and silver nitrite solution followed by reduction with hydrazine.
- The CoTMPP solution was added to the silverized carbon black (Ag/C) and subsequently sintered at 750°C for 1 h in inert gas.
- KOH was added as a wetting agent and ammonium carbonate was added for pore formation.
- Different concentrations of the ORR and OER catalysts were used in various layers for protection against the formation of gas pockets between the air electrode and electrolyte (2% more OER catalyst was used on the airside and 0.6% more ORR catalyst was used on the electrolyte side).
- The three layer electrodes were dried at 85°C for 120 min and hot pressed at T = 300°C and p = 0.5 ton/in.^{2} for 10 min.

TABLE 12.3

Composition of the Catalyst and GDLs of the Multilayer BE

GDL—a porous layer, composed of 65% carbon (ammonium bicarbonate) + 35% PTFE, that allows gas penetration while preventing liquid penetration

AL—composed of hydrophobic and hydrophilic pores (the latter for ORR) containing 10%–30% catalyst ($MnSO_4 + La_2O_3$ or $MnO_2 + La_2O_3$) + 50%–60% high SA carbon black + 15% PTFE

OEL—requires hydrophilic pores for electrolyte penetration and the OER and contains 45% OER catalyst + 50% high SA carbon black + 5% PTFE. Ammonium bicarbonate was also utilized for pore formation

Source: Adapted from Coetzer, I. and Thackeray, M.M., US Patent, 4,288,506, 1981.

The electrode design, detailed in Table 12.3, consists of an active layer (AL) with two sublayers (0.02 and 0.025 in thick) and a 0.015 in thick wet-proofed GDL layer.[141]

The oxygen evolution layer (OEL) has high-surface-area carbon black and pore former to control electrolyte flooding. Two of the examined catalyst compositions for this electrode were 63.5% Vulcan XC500 (XC-500) from Cabot + 13% $MnSO_4$ + 8.5% La_2O_3 + 15% PTFE and 69% XC-500 + 8% MnO_2 + 8% La_2O_3 + 15% PTFE.[140] If the AL and OEL are combined into a single layer, then the optimal ratio of OER catalyst/ORR catalyst is 40:60 in the bifunctional catalyst. Figure 12.3 shows that the ORR activity of the combined catalyst ($MnSO_4 + La_2O_3$, SSA 12.5 m^2/g) in alkaline electrolyte exceeds the activity of the same catalysts separately. The test was carried out in a half cell with a Ni counter electrode. Although the stability of La_2O_3 was better than that of $MnSO_4$, it was found that the stability of the combined catalyst was higher.[140]

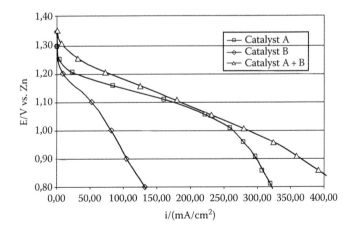

FIGURE 12.3 ORR polarization curves for air electrodes in 6.6 M KOH (A) $MnSO_4$ and (B) La_2O_3, and (A + B) $La_2O_3 + MnSO_4$. (Adapted from Trugve, B., US Patent, 20071666602A1, 2007.)

The replacement of $MnSO_4$ with the same concentration of MnO_2 led to a decrease in the discharge voltage (vs. Zn at 100 mA/cm^2) from 0.98 to 0.88 V. Increasing the percentage of ORR catalyst led to an increase in the discharge voltage and a decrease in the stability during a cycling test, whereas an increase in the percentage of OER catalyst led to the opposite situation. The best ratio of OER catalyst to ORR catalyst was 5%–20%:5%–15%. The replacement of the OER/ORR active La_2O_3 with Ag led to an increase in charge/discharge stability. The cathode 19% Ag + 8% $MnSO_4$ showed higher stability and discharge voltage than pure Ag, while 19% Ag + 8% La_2O_3 has sufficient stability but a low discharge voltage.

Coetzer and Thackeray[141] investigated the use of different carbides for BEs in the alkaline electrolyte of Li-ion batteries. Several carbides, including TaC, WC, W_2C, TiC, Cr_3C_2, Cr_7C_3, $Cr_{23}C_6$, MoNiC, and WCoC, were tested for the evaluation of their stability and ORR activity in alkaline electrolyte. TiC and VC showed the lowest ORR activity, MnC_3 also demonstrated low ORR activity at low temperatures, and both Cr_2O_3 and FeC were unstable in alkaline electrolyte and had low ORR activity. The best ORR activity was obtained with a composition of $TaC + WC + W_2C + TiC$.

Zhimin[142] developed a three-electrode secondary ZnAB, consisting of an ORR cathode with CoTMPP and an OER electrode composed of 30% Ag + 70% $LaNiO_3$. The addition of Ag to $LaNiO_3$ decreased the formation of ZnO, by decreasing the affinity of $LaNiO_3$ for ZnO.

$$La_{1-x}A_xFe_{1-y}Mn_yO_3 \quad (A = Sr \text{ or } Ca)$$

Perovskites are the group of oxides with formula ABO_3, where A is a divalent metal ion (i.e., Ce, Ca, Sr, Ca, Pb, and rare earth metals) and B is a tetrahedral metal (i.e., Ti, Nb, Fe). Shimizu et al.[143] suggested promising bifunctional perovskite-based catalysts, with high ORR and OER activity, on the basis of $La_{1-x}A_xCoO_3$ (A = Sr or Ca)[142–144] and $La_{0.8}Sr_{0.2}BO_3$ (B = Co, Mn, or Fe).[145,146] In these works, it was discovered that the comparative activity of $La_{0.8}Sr_{0.2}BO_3$ (B = Co, Mn, or Fe) catalysts for ORR and OER decreased Co > Mn > Fe and Co > Fe >> Mn, respectively. Unfortunately, even though the Co-based catalysts were shown to have the highest OER and ORR activity, they did not have sufficient stability in the alkaline electrolyte.[145] This was one of the main reasons for the development of the Co-free catalyst with the composition of $La_{1-x}A_xFe_{1-y}Mn_yO_3$. The type and concentration of Sr or Ca at the position A-site in the perovskite structure strongly affects the ORR activity. For example, $La_{0.8}Sr_{0.2}Fe_{1-y}Mn_yO_3$ catalyst has maximum ORR activity at y = 0.6 and maximum OER activity at y = 0.2. The $La_{1-x}Ca_xFe_{0.8}Mn_{0.2}O_3$ catalyst demonstrates maximum ORR (200–300 mA/cm^2 at −300 mV (Hg/HgO) in 7 M KOH at 25°C with air flow) and OER activities and maximum BET surface area at x = 0.4.[145]

OER activity depends on the type of cation on the A-site and is ranked accordingly: Sr > Ca > Ba > La. Increasing Mn concentration to 60% leads to an increase in ORR activity, but the maximum OER activity is at y = 0.2. The ORR reduction on Co-based perovskite catalysts in 7 M KOH proceeds via the two-electron path and it is rate determined by the HO_2^- decomposition reaction: $O_2 + H_2O + 2e = HO_2^- + OH^-$. The rate of HO_2^- decomposition and OER and ORR activity (normalized per unit surface area) increase for the Co-free perovskite catalyst, $La_{1-x}Ca_xFe_{0.8}Mn_{0.2}O_3$. Thus, the HO_2^- decomposition reaction is rate determining for both processes, the ORR and OER. Note that Sr increases the surface areas of the mixed perovskite-based catalysts.[147]

$$La_{0.6}Ca_{0.4}Co_{0.8}B_{0.2}O_3 \quad (B = Mn, Fe, Co, Ni, or Cu)$$

Shimizu et al.[147] developed $La_{0.6}Ca_{0.4}Co_{0.8}B_{0.2}O_3/C$ (B=Mn, Fe, Co, Ni, or Cu) with a maximum bifunctional activity for B=Fe (7 M KOH at 25°C, j=200 mA/cm² for the ORR and 300 mA/cm² for the OER at −150 and +620 V [Hg/HgO], respectively). The Fe-based catalyst also demonstrated a high activity for the decomposition of the HO_2^- intermediate. These electrodes were synthesized by the amorphous citrate precursor (ACP) method. The effect of the B-site metal on the SSA and the HO_2^- decomposition activity in 7 M KOH at 25°C was determined based on the rates of oxygen evolution from the decomposition of H_2O_2 in 7 M KOH at 80°C.[147] The majority of oxygen reduced on the developed electrode proceeds via the two-electron mechanism, but only under working current densities the ORR rate is determined by the HO_2^- decomposition rate. The four-electron mechanism of ORR was only observed at low current densities. The electrode with B=Fe produces the maximum SSA of 28 m²/g and rate of HO_2^- decomposition, while electrodes with B=Mn, Co, Ni, and Cu had SSAs of 18, 18, 17, and 4 m²/g, respectively.

$$La_{0.6}Ca_{0.4}CoO_{3-\delta} \text{ and } La_{0.7}Ca_{0.3}CoO_{3-\delta}$$

Lippert et al.[148] tested $La_{0.6}Ca_{0.4}CoO_{3-\delta}$ and $La_{0.7}Ca_{0.3}CoO_{3-\delta}$ as bifunctional air catalysts for ZnAB. These catalysts, which have an amorphous structure, mixed orientation, or single orientation, were produced by pulsed laser deposition. Catalysts with mixed or single orientations showed higher OER/ORR activity in a ZnAB in comparison to those with an amorphous structure.

12.2.2.1 Core (LaNiO₃)–Corona (Nitrogen-Doped Carbon Nanotube)

Core–corona bifunctional catalyst (CCBC) (Figure 12.4a) developed by Chen et al.[149] was very promising. CCBC has 35.8 wt% lanthanum nickelate ($LaNiO_3$) at core and 64.2 wt% nitrogen-doped carbon nanotubes (NCNTs) at shell. CCBC demonstrated higher performance than a conventional catalyst $LaNiO_3$ for the OER and Pt/C for the ORR. CCBC shows 3 and 13 times greater ORR and OER current, respectively, upon comparison to state-of-the-art Pt/C after the full-range degradation test. CCBC has the stable discharge potential after 75 cycles in comparison with the loss of 20% and 56% for Pt/C and $LaNiO_3$, respectively (Figure 12.4b).

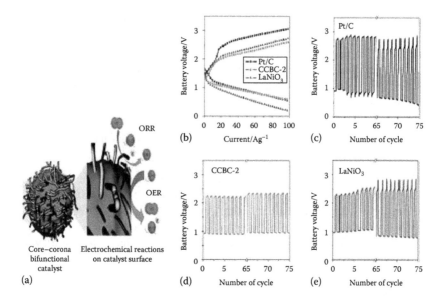

FIGURE 12.4 (a) CCBC design and rechargeable ZnAB performance with air cathodes Pt/C, CCBC (65.2wt% LaNiO$_3$+34.8 wt% NCNT), and LaNiO$_3$. (b) Discharge and charge polarization curves of Pt/C, CCBC-2, and LaNiO$_3$. C–D cycling of (c) Pt/C, (d) CCBC-2, and (e) LaNiO$_3$. (Adapted with permission from Chen, Z. et al., *Nano Lett.*, 124, 1946. Copyright 2012 American Chemical Society.)

The main ORR catalysts, air cathodes, and BEs for ZnABs are given in Table 12.4. A comparison of the ORR catalysts developed for ZnAB shows that electrodes based on Ag, FeTMPP, Mn$_4$N, spinel MnxCo$_{3-x}$O$_4$,[102] manganese nitride Mn$_4$N,[99] and lead ruthenate pyrochlore (Pb$_2$Ru$_2$O$_{6.5}$)[128] promote the four-electron ORR pathway and have the highest ORR activity. However, the catalysts with the best durability in concentrated KOH electrolyte have been based on MnO$_2$,[31–36] LaMnO$_3$,[112] LaCoO$_3$,[113] and LaNiO$_3$[114] and core (LaNO$_3$)–corona (NCNT),[149] all of which demonstrated a higher durability than Pt. Additional improvements in the durability of LaNiO$_3$ have been achieved by coating the catalyst with electronically conductive PPy.[103]

12.2.3 ELECTROLYTES

Aqueous alkaline solution, IL, and composite polymer electrolytes are the main types of electrolytes for ZnABs. Alkaline composite polymer electrolyte (polyethylene oxide [PEO]–polyvinyl alcohol [PVA]–glass fiber mat) has a high ionic conductivity and electrochemical stability (±0.2 V).[150] Among conventional aqueous alkaline electrolytes (KOH, NaOH, LiOH), KOH has the highest ionic conductivity but it reacts with atmospheric CO$_2$ and has the carbonation of the electrolyte.[151] Carbonation of the alkaline electrolyte decreases at its absorption by the mixture

TABLE 12.4

ZnAB Air Cathodes

Assignee	Catalyst Composition	Air-Cathode Design	Performance	Catalyst Fabrication Method	References
MnO₂-based ORR catalyst					
Rayovac Corp. (United States)	MnO_2 + C + PTFE	Standard[a]		$KMnO_4$ reduction	[33]
R. B. Dopp	Mn isopropoxide + activated C + 12% PTFE	Standard[a]			[38]
AER Energy Resources Inc. (United States)	5% MnO_2 + 75% C (mixture of 30% EC-600JD and 70% AB-50) + 20% PTFE	Standard[a]	No data	Sol–gel	[39]
AER Energy Resources Inc. (United States)	5% MnO_2 (Mn^{3+}/Mn^{4+}) + 70% C (60% PWA + 40% carbon black) (PTFE–Teflon 30B) GDL: 30% EC-600JD + 70% AB-50	Standard[a]	No data	Micelle encapsulation with two nonionic surfactants	[40]
Rayovac Corp. (United States)	MnO_2 + C (activated carbon + BP2000) + PTFE Particle size distribution MnO_2 −20 to 26 μm	Standard[a]	Voltage of 1.15 V at 150 mA/cm² is more stable over 15 h than the prior art in 30% KOH	Sol–gel (reduction of $KMnO_4$ by sodium formate at pH7)	[41]
Matsushita Electrical Industrial Co., Ltd.	30% MnO_2 + 20% active carbon + 20% carbon black + 30% PTFE	Standard[a]	No data	Heat treatment of the gamma manganese oxyhydroxide at 250°C–450°C with product: Mn_5O_8 + beta MnO_2 in air 250°C–300°C	[42]

(Continued)

TABLE 12.4 (*Continued*)
ZnAB Air Cathodes

Assignee	Catalyst Composition	Air-Cathode Design	Performance	Catalyst Fabrication Method	References
Gore Enterprise Holdings, Inc. (Newark, DE)	20% MnO_2 + 66% C + 14% PTFE	Standard[a]	Limiting current of 11 mA/cm² (commercial 50–100 mA/cm²)		[43]
F. Sun and F. Wang	Catalyst layer: 20% MnO_2 + 70% active carbon + 10% PTFE GDL: 15% carbon black + 85% PTFE	Standard[a]	Discharging voltage cell 1.32 V after 1 h at impedance for a 620 Ω load and 1000 Hz		[44]
Duracell Inc. (United States)	11% gamma MnO_2 + 41% C (BP2000) + 48% PTFE	Standard[a]	ZnAB output of 430 mA h to 1 V cutoff is 7.5% better than prior art. Deliver the first 400 mA h of capacity at 300–400 mV higher than prior art	Milling of MnO_2 and a carbon black	[45]
Duracell Inc. (United States)	MnO_2 cathode + PTFE + 2%–20% absorbent material such as the gelling material used in the anode.	Standard[a]	Limiting current of 27.5 mA at 66°C		[44]
Duracell Inc. (United States)	MnO_2	Standard[a]	Energy density of ZnAB 0.1–0.3 W h/g MnO_2 at a 1 A discharge to 0.8 V cutoff voltage		[49]

(*Continued*)

TABLE 12.4 (*Continued*)
ZnAB Air Cathodes

Assignee	Catalyst Composition	Air-Cathode Design	Performance	Catalyst Fabrication Method	References
I. Roche et al.	MnO_2	Standard[a]	Discharging voltage after 1 h at 620 and 150 Ω 1000 Hz AC for the stack with 10 cells: 3-layer cathode: 1.31, 1.25, 4.0 V 2-layer cathode: 1.3, 1.18, 7.5 V		[48]
Ag-based ORR catalysts					
C.-Y. Wu et al.	Ag/CNC	Standard[a]	$j = 200$ mA/cm² at $E = 0.8$ V (30% KOH, T = 25°C). Galvanostatic discharge on Ag/CNC at $j = 200$ mA/cm² with a moderate performance decrease after 80 h		[62]
Y. Yang et al.	Ag on Ni foam	Standard[a]	Insignificant voltage deterioration during 120 h galvanostatic discharge at T = 40°C		[63]
H. Meng and P. K. Shen	AgW_2C/C	Standard[a]	Test on RRDE in 1 M KOH, 25°C: $j = 6$ mA/cm² at 200 mV (Hg/HgO). For comparison: $j = 4.5$ for Ag/C, $j = 5.8$ for Pt/C	Microwave intermittent heating method (IMH)	[66]
Gillette Corp. (United States)	$AgMnO_4 + 5\%–10\%$ $MnO_2 +$ $C + PTFE$	GDL: 30%–70% PTFE Catalyst layer: 10%–30% PTFE	Three-electrode cell test $AgMnO_4 + 5\%$ MnO_2—0.16 V (SHE) at 28 mA/cm², $AgMnO_4 + 10\%$ MnO_2 at 50 mA/cm² LSV shows the limiting current, 51 mA/cm², was achieved at −0.25 V versus Hg/HgO or 1.1 V versus Zn	MnO_2 forms by heating $Mn(NO_3)_2$. Dual catalyst $AgMnO_4$ prepared by reducing silver permanganate	[68]

(Continued)

TABLE 12.4 (Continued)
ZnAB Air Cathodes

Assignee	Catalyst Composition	Air-Cathode Design	Performance	Catalyst Fabrication Method	References
Luz Electric Fuel Israel Ltd.	Raney Ag catalyst + PTFE = 5:1 (wt) (24 mg/cm²)	Blocking layer (air side) Carbon + PTFE (6–10 mg/cm²) pressed onto Ni mesh or foam	After 5 h of discharge the cell showed a peak current of 10 A at 0.9 V (200 mA/cm² peak current at the air cathode)		[69]
Zinc–Air Power Corp.	Ag_2O + 10% $LaNiO_3$	Standard[a]	No degradation during 500 h in 32% KOH		[70]
Alupower, Inc. (United States)	5% Ag + 15% BP2000 + 10% Daxad + 60% Teflon RPM T-30	Impregnated layers and fine nickel mesh precoated with an adhesive were passed through an oven		Daxad was an additive to increase Ag adsorption on carbon black	[50]
CoTMPP-based ORR catalysts					
Ovonic Battery Company, Inc. (Rochester, United States)	50% (20% CoTMPP/C) + 50% (15% CoO_x + 5% MnO_x/C) 2.5% MnO_x + 7.5% CoO_x/C	Standard[a]	j = 120 mA/cm² at voltage 0.1 V for 2.5% MnO_x + 7.5% CoO_x/C and, 50% CoTMPP + 50% (15% CoO_x + 5% MnO_x)	Mix NH_4OH + carbon Add Co (Mn) SO_4 Mix 1 and 2 Add NaOH Washing Filtration Dry at 80°C	[73]
Reves, Inc. (United States)	4% CoTMPP + 15% BP2000 + 60% Teflon RTM T-30	Standard[a]	j = 500 mA/cm² (wet coating) and 200 mA/cm² (dry powder process) at 1 V discharge in KOH	Single roll pressing with impregnation: CoTMPP + H_2O + PTFE	[79]

(Continued)

TABLE 12.4 (*Continued*)
ZnAB Air Cathodes

Assignee	Catalyst Composition	Air-Cathode Design	Performance	Catalyst Fabrication Method	References
	MnO$_2$ and/or AgNO$_3$, (Pt, Co$_3$O$_4$)			Sinter 200°C–350°C Laminating porous Teflon sheets at nip pressure 100 lb/in. at T = 250°C–350°C Note: Sintering of the impregnated polymer binder into pores of the foamed support	
Lutz Electric Fuel Cell Israel Ltd.	10% CoTMPP/C + Nafion + FEP + FEP-coated PTFE fibers	Standard[a]	E = −0.29 V (Hg/HgO) at j = 0.4 mA/cm², stable during 200 h test. Electrode was soaked overnight in 7 M KOH at 25°C and duty cycled for 1 h (20 s—200 mA/cm², 45 s—50 mA/cm², 45 s—0 mA/cm²)	Coprecipitation	[80]
Power zinc Electric (China)	CoTMPP + MnO$_x$/C	Standard[a]	j = 500 mA/cm² at E = −0.498 V (Hg/HgO) in 1 M KOH at T = 25°C (half cell) j = 216 mA/m² at E = 1 V in 30% KOH (ZnAB)		[70]

(*Continued*)

TABLE 12.4 (Continued)
ZnAB Air Cathodes

Assignee	Catalyst Composition	Air-Cathode Design	Performance	Catalyst Fabrication Method	References
Nitride-based ORR catalysts					
N. Miura et al.	60% Mn_4N/C + PTFE	Standard[a]	At −125 mV (Hg/HgO) or 0.8 V (RHE) performance was stable at j = 2400 mA/cm² for 5 h, at j = 300 mA/cm² for 50 h in 9 M NaOH, at 80°C	Coprecipitation	[99]
Spinel-based ORR catalyst					
W. J. King et al.	$NiCo_2O_4$ spinel	Standard[a]	0.87 V in 5 M KOH and room temperature	Pyrolysis of the metal salts	[101]
E. Rios et al.	$Mn_xCo_{3-x}O_4$ + PTFE (0 < x < 1) spinel	Standard[a]	RRDE at 2500 rpm j = 100 mA/cm² at E = 0.2 in 1 M KOH at T = 25°C	Pyrolysis of the metal nitrites	[102]
Perovskite-based ORR catalysts					
T. Hyodo et al.	Perovskites $LaMnO_3$ $LaCoO_3$ $LaNiO_3$ $LaCrO_3$ $LaFeO_3$ $La_{0.8}Sr_{0.2}FeO_3$ $La_{0.6}Sr_{0.4}Fe_{0.6}Co_{0.4}O_3$ $La_{0.6}Sr_{0.4}Fe_{0.6}Mn_{0.4}O_3$		Performance, mA/cm² at −160 mV (Hg/HgO) 1266 1006 468 344 273 519 682 922		[112]
M. Hayashi et al.	$LaNiO_3$		j = 300 mA/cm² at E = −80 mV (Hg/HgO) for 140 h in 8 M KOH at 60°C	RM	[116]

(Continued)

TABLE 12.4 (Continued)
ZnAB Air Cathodes

Assignee	Catalyst Composition	Air-Cathode Design	Performance	Catalyst Fabrication Method	References
A. C. Tseung and H. L. Bevan	$LaCoSrO_3$		$j = 2$ mA/cm^2 at 500 mV (DHE) in 45% KOH at room temperature		[118]
Pyrochlore-based ORR catalysts					
A. Gibeney and D. Zuckerbrod	$Pb_2M_{2-x}Pb_xO_7$		$j = 6 \times 10^{-3}$ at −100 mV (Hg/HgO) in 1 M NaOH at 25°C		[136]
Metal hydroxide–based ORR catalyst					
High-Density Energy Inc. (United States)	Ni, Co, Fe hydroxide + carbon black + PTFE	Standard[a]	No data		[130]
Pt-based catalyst					
Gould Inc. (United States)	$Ag + Pt + MnO_2 + C + PTFE$	Standard[a]	No data		[133]
G. Henry et al.	10% Pt/C	Standard[a]	Cell voltage of 0.69 V in 6 M KOH at $T = 50$°C		[131]
BEs					
Westinghouse Electric Corp.	Iron–air fuel cell (similar to ZnAB) with alkaline electrolyte: $CuSO_4$, $NiWO_4$, WC + 20% Co WS_2 + WC or WC + 1%–20% Co WS + C + PTFE WC + Ag + C + PTFE (FEP)	Hydrophillic layer $CuSO_4$, $NiWO_4$, WC + 20% Co or WS_2 + WC Hydrophobic layer—FEP sheet	Test of WC + Ag + C + PTFE + FEP after duty cycle (16 h charge at 25 mA/cm^2 and 8 h discharge at 25 mA/cm^2) at −0.3 V (Hg/HgO). Initial j = 100 mA/cm^2, j = 40 mA/cm^2 after 934 h. Stable performance during first 750 h, voltage drop 50 mV at 25 mA/cm^2		[134]

(Continued)

TABLE 12.4 (Continued)
ZnAB Air Cathodes

Assignee	Catalyst Composition	Air-Cathode Design	Performance	Catalyst Fabrication Method	References
Westinghouse Electric Corp.	30 parts Ag + 30 parts WC (coated with 12% Co) + 32 parts PTFE + 90 parts carbon black	Catalyst impregnated into 45%–95% of collector's pores	Half cell test (Ni counter electrode): no deterioration in performance during 134 charging cycles in KOH		[135]
Westinghouse Electric Corp.	3% (5%–10%) Ag (ORR) + (~7% [10%–15% $FeWO_4$] + 7% [10%–15%] WC + ~12% [10%–15%] Co [OER] + ~54% C) + ~22% PTFE, Ag loading—2 mg/cm²	First layer—GDL: EC-600JD + PTFE; Second layer: current collector impregnated with low SA carbon black + ORR catalyst + PTFE; Third layer: catalyst + low SA carbon black + High SA carbon black + PTFE	No data	Reduction of $AgNO_3$ with hydrazine	[136]
C.-T. Liu and B. Demczyk	[ORR–Ag] + [OER–$CoWO_4$ + WC + WS_2 + NiS + 10%–15% Co] + 20% PTFE	Standard[a]			[138]
AER Energy Resources, Inc.	ORR catalyst ([0.3%–2%] CoTMPP + [4%–10%] $LaNi_{1-x}Cox$ + [1%–4%]Ag + [18%–32%[Co_xO_y]) + OER catalyst ([1%–20%]WC + [1%–20%]Co + [1%–7%] $FeWO_4$ + [1%–7%][NiS) + AB-50 + PTFE	$E_{1\ OER}$ for ORR catalyst > $E_{2\ OER}$ of OER catalyst; ORR catalyst concentration on the electrolyte side > ORR concentration on the air side; OER catalyst concentration on the electrolyte side < OER concentration on the air side			[141]

(Continued)

TABLE 12.4 (*Continued*)
ZnAB Air Cathodes

Assignee	Catalyst Composition	Air-Cathode Design	Performance	Catalyst Fabrication Method	References
ReVolt Technologies AS (Norway)	Catalyst layer: 63.5% XC500 + 15% PTFE + 13% MnSO$_4$ + 8.5% La$_2$O$_3$ 15% PTFE + 69% XC500 + 8% MnO$_2$ + 8% La$_2$O$_3$ 58% XC500 + 15% PTFE + 19% AgNO$_3$ + 8% MnSO$_4$ GDL: 65% C + 35% PTFE	Standard[a]	Current densities (mA/cm^2) at E = 1 V vs. Zn in 6.6 M KOH. (three-electrode cell) for the electrodes: 13% MnSO$_4$ + 8% La$_2$O$_3$ (270), La$_2$O$_3$ (280), MnSO$_4$ (80)	Coprecipitation	[140]
Zinc–Air Power Corp.	OER electrode 30% Ag + 70% LaNiO$_3$	The single functional electrode and third electrode are separated	Charge/discharge capacity is stable for tricell designed ZnAB (40 cycles) in 32% KOH	Coprecipitation	[142]
Y. Shmizu et al.	La$_{1-x}$A$_x$Fe$_{1-y}$Mn$_y$O$_3$ (A = Sr, Ca)		J = 200–300 mA/cm^2 at E = −300 mV (Hg/HgO) in 7 M KOH, 25°C, air flow		[146]
Y. Shimizu et al.	La$_{0.6}$Ca$_{0.4}$Co$_{0.8}$Fe$_{0.2}$O$_3$		j = 200 mA/cm^2 at E = −150 (Hg/HgO) in 1 M NaOH at 25°C		[147]

[a] Standard ZAFC design: Teflon film (air side) + current collector + gas diffusion layer + catalyst layer.

Source: Adapted from *J. Power Sources*, 195, Neburchilov, V., Wang, H., Martin, J., and Qu, W., 1271–1291, Copyright 2010, with permission from Elsevier.

piperazine with 2-(2 aminoethylamino) ethanol (AEEA) and monoethanolamine in a rotating packed bed.[152] Recently, ILs have been used for ZnABs, for example, the polymer gel electrolyte (PGE) IL ethyl-3-methylimidazolium trifluoromethanesulfonate (EMI-TF) + *bis*(trifluoromethanesulfonyl)imide (EMI-TFSI) + polymer matrix (poly(vinylidene fluoride-co-hexafluoropropylene) [PVDF + HFP]) with ionic conductivity 10–3 S/cm, anodic stability at 2.8 V (Zn^{2+}/Zn), thermal stability at T = −50°C + 100°C, and without volatile solvents.[153] The zinc deposition from 10 mol% zinc (dicyanamide) in IL 1-ethyl-3-methylimidazolium dicyanamide [emim][dca] and $Zn(dca)_2$ in [emim][dca] + 3wt% H_2O is smooth without dendrites.[154]

12.2.4 ANODES

Zinc anodes play a crucial role in the advanced performance of ZnABs as their not having uniform dissolution during the discharge of ZnAB leads to its shape change. Moreover, uncontrolled zinc deposition during charge is the reason for a dendrite's growing causing short circuits.[153] Porous[154,155] and foam[156] zinc with high SSA significantly improves the performance of ZnABs. The usage of Zn–Ni and Zn–In alloy anodes increases the cycle life (100 cycles) and decreases the dendrite formation.[156] The typical side reaction, hydrogen evolution on zinc anodes, can be depressed by the usage of zinc alloy with Hg, PbO, and Cd.[157,158]

12.2.5 SEPARATORS

Separators in ZnABs allow the selective transport of the hydroxyl ions from the air cathode to a zinc anode. They have to be porous, chemically stable in electrolyte, and electrically nonconductive, and have a high ionic conductivity. Separators for ZnABs are made predominately from PEO, PVA, polyolefin, and polypropylene.[156,160–163] The PEO and PVA separators demonstrate better performance in ZnABs. Also, membranes such as MCM-41 and polysulfuric-1 can be successfully used in ZnABs.[164,165] Saputra et al.[164] demonstrated the improvement of the ZnAB performance with MSM-41 membrane.

12.3 MAGNESIUM–AIR BATTERY COMPONENTS

MgABs have low cost, high voltage, and long storage life and are easy recycling. Environmentally friendly electrolyte such as seawater can be used for MgAB. The main drawbacks of MgABs are the parasitic hydrogen evolution, thus the necessity of hydrogen inhibitors, and the low durability of Mg anodes. The advantage of MgABs is the presence of all reactants in the electrolyte. Seawater provides high conductivity, which allows the use of several cells in MgABs. MgABs have a threefold higher current capacity and a less aggressive electrolyte, which increases the lifetime of the air cathodes, similar to that of ZnABs (Table 12.5).[166]

The general design of MgAB is given in Figure 12.5. The chemistry of the MgABs is given as follows[3]:

TABLE 12.5

Comparative Characteristics of MgAB and ZnAB

	Battery Type	
Characteristic	MgAB	ZnAB
Specific gravity	1.74	7.13
SHE	−2.363	−0.763
Current capacity, A h/kg	2200	740
Electrolyte	Salt water	KOH
pH electrolyte	6–8	13–14
Anode	Mg alloy(AM60 with Mg > 90%)	Zn alloy (Zn 99.99%)

Source: Adapted from Lei, Z. et al., WO Patent, WO2007/112563, 2007. With permission.

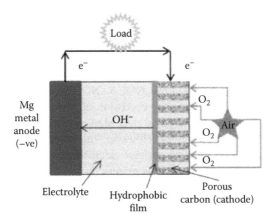

FIGURE 12.5 MgAB design. (Adapted from Rahman, M.A. et al., *Electrochem. Soc.*, 160, A1759, 2013. Copyright 2014, The Electrochemical Society. With permission.)

$$\text{Anode: } 2Mg + 4OH^- \rightarrow 2Mg(OH)_2 + 4e^- \quad (E_o = -2.69\,V) \qquad (12.12)$$

$$\text{Cathode: } O_2 + 2H_2O + 4e \rightarrow 4OH^- \quad (E_o = 0.40\,V) \qquad (12.13)$$

$$\text{Overall: } 2Mg + O_2 + 2H_2O \rightarrow Mg(OH)_2 \quad (E_o = 3.09\,V) \qquad (12.14)$$

$$\text{Overall parasitic reaction: } Mg + H_2O = Mg(OH)_2 + H_2 \qquad (12.15)$$

The parasitic reaction in MgABs by hydrogen evolution has to avoid the use of hydrogen inhibitors. This enables an increase in power efficiency and safety, and decreases battery resistance and gas pressure. One of the first examples of the successful commercialization of MgABs was the MagPower Systems Inc. power generator eMAG™ 250 (12 V/250 W).[166] It generates electricity through the combination of a saltwater electrolyte, air (oxygen), and magnesium.

12.3.1 Air Cathodes

One of the key factors determining the high efficiency of MgABs is the air diffusion cathode. This cathode has low cost, durable current collector, and internal electrical resistance. One of the most effective air cathodes has a catalyst layer comprising CoTMPP + FeTMPP/C and GDL (70 wt% C + 30 wt% PTFE).[166] It has the same performance as the commercial eVionyx air cathode and outperforms the air diffusion cathodes with CoTMPP catalyst from Fuel Cell Technology Inc. in 10 wt% NaCl solution at T = 20°C. Porous metal foam is used to increase the air cathode's performance. This air cathode comprises a hydrophobic GDL facing the atmosphere, a first catalyst embedded layer, a metal foam layer, and a second catalyst embedded layer facing the electrolyte environment.[167]

12.3.2 Electrolytes

MgAB electrolytes (NaCl, KHCO$_3$, NH$_4$NO$_3$, NaNO$_3$, Na$_2$SO$_4$, MgCl$_2$, MbBr$_2$, and Mg(ClO$_4$)$_2$[168,169]) have to minimize the corrosion of Mg anodes, anodic polarization, and coagulation of the discharge product Mg(OH)$_2$. MgABs use basically neutral chloride or seawater electrolytes. However, the corrosion of Mg anodes and the side hydrogen evolution reaction in these electrolytes required the development of a new prospective electrolyte, such as organic boron Mg complex salt (OBMC) and IL-based electrolytes.[170] Mg anodes have a higher reversibility in organometallic Mg salt–based electrolytes in comparison with aqueous electrolytes such as Grignard salts (RMgX, R ¼ alkyl, aryl groups; X ¼ Cl, Br)[171,172] but at the narrow electrochemical window that do not allow to use them in rechargeable MgABs. This problem was solved in the development of organic boron Mg salt–based (BPh$_2$Cl, BPhCl$_2$, B[(CH$_3$)2N], BPh$_3$, BEt$_3$, BBr$_3$, BCl$_3$, and BF$_3$) with the higher reversibility Mg deposition–dissolution process and anodic stability compared to the Grignard salt electrolyte (1.9 vs. 1.5 V).[172] MgAB with modified boron electrolyte (Mg$_2$Cl$_3$ + MgCl + Ph$_2$Mg + tetracoordinated boron anion [Mes$_3$BPh]) with the high ionic conductivity showed good Mg deposition reversibility (ca.100%) and the high anodic potential 3.5 V versus Mg.[173,174] The usage of magnesium perchlorate and the additive 1-butyl-3-methylimidazolium in the boron-type electrolyte leads to less anodic polarization.[175]

The novel nonaqueous gel trihexyl(tetradecyl)phosphonium-based IL-based electrolyte [P6,6,6,14][Cl] was used in MgAB to increase its performance by the formation of the conductive interfacial film organophosphonium–magnesium(hydroxyl) chloride on Mg anodes with a gel-like structure. MgAB with the IL [P6,6,6,14] [Cl] electrolyte and Mg anodes covered by a gel-like structure film can discharge at 0.05 mA/cm^2 and support ORR at i > 0.8 mA/m^2, see Ref. [176].

12.3.3 ANODES

The low durability of magnesium anodes, which causes low coulombic efficiency and irreversible polarization, is one of the main limitations of alkaline MgABs. Anodes' self-discharge leads to the heating and degradation of MgAB.[177] The usage of sea urchin magnesium anodes with a high SSA allowed for a decrease in anodic polarization and the fast coagulation of discharge product $Mg(OH)_2$.[178] Magnesium anodes are irreversible in organic polar electrolytes due to their passivation. The significant improvement of MgAB's anodes was achieved by the use of magnesium alloys Mg–Li–Al–Ce, demonstrating better durability and less self-corrosion than the Mg anode in 3.5% NaCl electrolyte.[179] The nanoparticle-based MoS_2 anode with a graphene structure provides a high discharge capacity of 170 mA h/g and long cycle life of 50 cycles. Further improvement of MgAB performance was achieved using the treatment of Mg alloy anode AZ31 in the IL electrolyte trihexyl(tetradecyl) phosphonium–*bis* 2,4,4-trimethylpentylphosphinate (P6,6,6,14M3PPh) with the formation of the protective interfacial layer (solid electrolyte interface [SEI]) that stabilized the metal–electrolyte interface and prolonged the lifetime of MgAB with an aqueous NaCl electrolyte.[180–182]

12.4 ALUMINUM–AIR BATTERY COMPONENTS

AlABs are cheap and use a nonalkaline electrolyte that does not have carbonation. AlABs are mainly nonrechargeable due to the fact that their anodes are easily oxidized. Recently, a company named Phinergy developed AlABs for electric vehicles (EVs). Their AlAB demonstrated significant driving range per charge using a special cathode and potassium hydroxide electrolyte.[183] The main limitations of AlABs are as follows:

- Low aluminum corrosion stability requires aluminum alloys exclusively.
- Limited shelf life (which could be solved by storing the electrolyte in a tank outside the battery).
- AlABs require reciprocating stirring to avoid gel formation in the electrolyte.

The typical design of AlAB is given in Figure 12.6. The chemistry of the AlABs is given as follows[183]:

$$\text{Anode: } Al + 3OH^- \rightarrow Al(OH)_3 + 3e^- \quad (E_o = -1.66\,V) \qquad (12.16)$$

$$\text{Cathode: } O_2 + 2H_2O + 4e^- \rightarrow 4OH^- \quad (E_o = 0.40\,V) \qquad (12.17)$$

$$\text{Overall: } 4Al + 3O_2 + 6H_2O \rightarrow 4Al(OH)_3 \quad (E_o = 2.71\,V) \qquad (12.18)$$

The typical AlAB consists of the following components (the percentage in the brackets is the cost distribution): cell module (15%), cathodes (47%), anodes (12%), electrolyte (15%), makeup water and reaction product storage (4%), air handling and

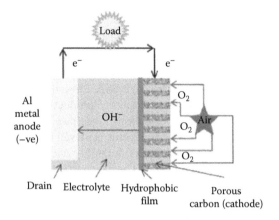

FIGURE 12.6 AlAB design. (Adapted from Rahman, M.A. et al., *Electrochem. Soc.*, 160, A1759, 2013. Copyright 2014, The Electrochemical Society. With permission.)

cooling system (16%), and crystallizer (5%). The projected fuel efficiency in electrical vehicles with AlABs is 15% comparable to vehicles with an internal combustion engine (ICE).[184]

12.4.1 Air Cathodes

Air cathodes in AlABs usually consist of carbon supports with a GDL and cobalt catalyst. Yardney developed several air cathodes with multilayer design for possible use in AlABs[185] and now supplies them for different MABs. One of the main producers of AlABs is ALUPOWER Inc. (subsidiary of Yardney Technical Products Inc.). The company developed a key portable mechanically rechargeable AlAB capable of powering military equipment or recharging secondary NiCd and Li-ion batteries.[186–188] It has the following technical characteristics: 2.85 kW h for 34 h, 24 VDC or 12 VDC output, 5 A peak or 3.5 A continuous rating at 24 VDC, T = −18°C + 40°C, refueling—recycle at depot.[189] The other commercial air cathode consists of Ni mesh as current collector, microporous Teflon layer bonded to the air side of the electrode, the carbon-supported catalyst MnO_x/C, and fibrous carbon. MnO_x was used to reduce oxygen from the environment air.[190]

12.4.2 Electrolytes

The main electrolytes for AlABs are sodium chloride and seawater.[1] Al anodes corrode in alkaline electrolytes. However, AlAB with a gelled 0.6 M KOH electrolyte using a hydroponics gelling agent demonstrates the high capacity and power density of 105 mA h/g and 5.5 mW/cm² respectively.[191]

12.4.3 Anodes

Al anodes easily oxidize, so more electrochemically stable aluminum alloys with tin, gallium, and indium were developed for AlABs.[192–194] Technical-grade aluminum alloys (Al—0.1% In, Al—0.2% Sn, and Al—0.1% In—0.2% Sn) could serve as suitable anodes for AlAB with sodium chloride electrolyte. Al–In alloy demonstrates improved performance compared to Al anodes as In as alloying component reduces anodic polarization, decreases side hydrogen evolution reaction rate, and increases the anode efficiency. Several modified Al anodes, such as Al—0.05 wt% Ga—0.1 wt% Sn,[193] and composite Al anodes with the protective Zn alloy coating and corrosion inhibitor (In, Ca, Pb, Tl)[194] also demonstrate advanced performance in chloride electrolytes.

12.5   LITHIUM–AIR BATTERY COMPONENTS

LiABs have the highest theoretical energy density of 13,000 W h/kg (based only on the lithium metal anode) and 5,200 W h/kg (including both the lithium metal anode and oxygen)[1] cell voltage among all rechargeable batteries (Li ion—200 W h/kg, ZnAB—1084 W h/kg).[195] These LiAB features make this battery the most attractive energy storage technology for different applications.[196–200] However, the practical energy density of LiABs is significantly lower than the required energy density for vehicles of 1700 W h/kg (practical energy density for gasoline).[200] The main challenges of LiABs are as follows:

* Self-discharge of lithium metal due to its corrosion
* Passivation of the carbon air cathode due to the deposition of discharge products L_2O_2/Li_2O (in nonaqueous electrolyte) or LiOH (in aqueous alkaline electrolyte)
* Reduction of the practical discharge capacity and energy density due to deposition of discharge products Li_2O_2/Li_2O on the air cathode while reducing its porosity and oxygen access
* Ineffective discharge at voltages above 2.2 V at the high cathode polarization
* Safety issues (fire and explosion)
* Decomposition of main carbonate- or ether-based nonaqueous electrolytes during the discharge with the formation of nonelectrochemically reversible $LiCO_3$ instead of the reversible and insoluble $L_{i2}O_2$ that limits rechargeability and cycle life
* Low conductivity of discharged products Li_2O and Li_2O_2 increases electronic resistance
* Clogging of pores of carbon-based air cathode during discharge and its passivation
* High cost

LiABs are divided into four main types depending on the electrolyte used: (1) nonaqueous (with aprotic solvent), (2) aqueous, (3) hybrid (with a combination of aqueous and nonaqueous electrolytes), and (4) solid-state electrolyte (Figure 12.7).[8] The practical feasibility of rechargeable LiAB with nonaqueous electrolyte has been proven by Abraham.[195]

PolyPlus Battery Co,[201–204] IBM,[6] and Yardney[205] developed different LiABs. PolyPlus in 2004 developed a pioneering aqueous and hybrid (aqueous + aprotic electrolyte) electrolyte LiAB with a lithium-protected electrode (LPE) consisting

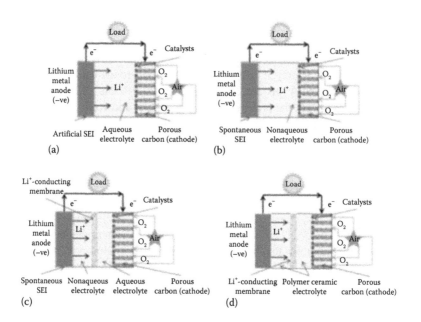

FIGURE 12.7 Design of LiABs. (a) Aqueous electrolytic type, (b) aprotic/nonaqueous electrolytic type, (c) mixed (aprotic/aqueous) type, and (d) solid-state type. (Adapted from Rahman, M.A. et al., *Electrochem. Soc.*, 160, A1759, 2013. Copyright 2014, The Electrochemical Society. With permission.)

of lithium coated by protective solid-state interfacial layer (Cu_3N, $Li_{3-x}PO_{4-y}N_y$ [LIPON]) and the solid electrolyte (glass–ceramic) made by Ohara Inc. (Japan).[206] In spite of the formation of soluble discharge product LiOH in aqueous LiAB, this product precipitates during discharge process at the achievement of its solubility limit and the cathode clogging has to be minimized. The more advanced PolyPlus hybrid LiAB, as compared with the aqueous LiAB, has an additional advantage— the formation of a natural ionic conductive SEI at the point of direct contact of the lithium anode with aprotic electrolyte.[204] Yardney developed a nonaqueous LiAB with electrolyte m of $LiPF_6$ in EC/DEC/DMC with a high specific capacity. This battery has a specific capacity of 2500 mA h/g at j = 0.05 mA/cm² and only 200 mA h/g at j = 2 mA/cm² (see Ref. [207]). Yardney recently developed a hybrid LiAB with increased energy density.[204]

In this section, catalysts for the ORR and OER, anodes, aqueous and nonaqueous electrolytes, and oxygen-selective water barrier membranes (OSWBs) as the main components of LiABs will be reviewed. A great progress have been achieved in the development of air cathodes and BEs for alkaline aqueous ZnABs (Section 12.2).

12.5.1 Nonaqueous LiABs

A nonaqueous LiAB consists of a negative electrode (carbon or lithium alloy), a positive electrode (oxygen electrode), a nonaqueous electrolyte (solution of lithium salt in

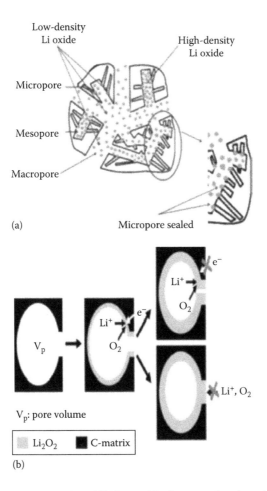

FIGURE 12.8 (a) Accommodation of lithium oxides in pores of various sizes. (b) Schematic illustration of the pore filling during discharge. The growing Li_2O_2 layer leads to cathode passivation by electrical isolation (top right) and pore blocking (bottom right). (Adapted from *J. Power Sources*, 220, Capsony, D., Bini, M., Mustarelli, P., Quartarone, E., and Mustarelli, P., 252–263, Copyright 2012, with permission from Elsevier.)

aprotic solvent), air-dehydration membranes, and a separator (glass fiber or Gelgard separator)[208,209] (Figure 12.8a).

The main advantage of nonaqueous battery over aqueous LiABs is the formation of SEI on the surface of lithium anodes. Main challenges of nonaqueous LiABs are low reversibility of electrochemical reactions, low durability of lithium anode, electrolyte decomposition, as well as electrode volume changes (e.g., at the formation of insoluble Li_2O_2 in pores) leading to pulverization of active electrode materials or mechanical disintegration of an electrode. The main reaction mechanisms of nonaqueous LiABs are presented as follows[208–212]:

Overall:

$$2Li^+ + O_2 + 2e^- \leftrightarrow L_2O_2 \quad (E_o = 2.96\,vs.\,Li/Li^+) \tag{12.19}$$

Anode reaction[213]:

$$Li \leftrightarrow Li^+ + e^- \tag{12.20}$$

Cathode reactions[210,211]:

$$O_2 + 2e^- + 2Li^+ \rightarrow Li_2O_2(E_o = 3.10)\text{-two-electron}$$

$$ORR\ (Li_2O_2\ \text{insoluble, insulator}) \tag{12.21}$$

$$O_2 + 4e^- + 4Li^+ \rightarrow 2Li_2O\ (E_o = 2.9\,V)\text{-four-electron}$$

$$ORR\ (Li_2O\ \text{is insoluble}) \tag{12.22}$$

The detailed mechanism of ORR reaction in nonaqueous Li–air cell was suggested in the fundamental work by Bruce et al.[212]

$$O_2 + e^- \rightarrow O_2^{\cdot-}\ (\text{superoxide}) \tag{12.23}$$

$$Li^+ + O_2^{\cdot-} \rightarrow LiO_2\ (\text{lithium superoxide}) \tag{12.24}$$

$$2LiO_2 = Li_2O + O_2\ (\text{chemical reaction}) \tag{12.25}$$

12.5.1.1 Oxygen Positive Electrodes

Catalysts for nonaqueous LiABs could be divided into two groups. The first group of catalyst was developed before the discovery of a carbonate electrolyte decomposition in 2010[214] due to the reaction of oxygen radical anions, formed during ORR with carbonate-based solvents. The second group of catalysts was developed for Li_2O_2 formation and decomposition for LiAB with relatively stable electrolytes. The ORR catalysts for LiABs and ZnABs are similar, so the main ORR catalysts (Table 12.5) for ZnABs can be used in LiABs as well.

12.5.1.1.1 *ORR Electrocatalysts for LiAB without Approval of Discharge Product Li₂O₂*

LiABs with a carbonate-based electrolyte have discharge products such as lithium carbonate, lithium alkyl, and/or LiOH instead of Li_2O_2.[214,215] However, both lithium carbonate and Li_2O_2 are insoluble and form a nonconductive layer on catalytic sites, so the electrocatalysts for LiAB with a carbonate electrolyte could be used in the case of Li_2O_2 formation in LiAB with more stable nonaqueous electrolytes. The role of catalysts for these LiABs is very crucial as their round-trip efficiency is low due to a significant difference 1 V between the discharge and charge voltage. Catalysts for oxygen (ORR, OER) need to have a significant pore volume to accommodate the discharge products, optimized balance between electronic (>1 S/cm) and ionic conductivity (>10^{-2} S/cm), high electrochemical, chemical, and mechanical stability, and fast diffusion of atmospheric oxygen in the electrode porous matrix.[6–8] The primary groups of ORR electrocatalysts for LiABs with carbonate electrolytes are presented in the following.

12.5.1.1.1.1 Carbon-Based Electrocatalysts Pristine carbon–based oxygen catalysts in nonaqueous LiABs demonstrate ORR/OER catalytic activity in contrast to an aqueous LiAB.[10] Modification of a pristine carbon was implemented mainly in three directions: 1D nanotubes, nanofibers, and nanorods; 2D graphite and graphene; and porous carbon with 3D architecture (Table 12.7). The optimized carbon material for LiAB should have a balance between the SSA, pore size, and pore volume, as micropores of carbon with a high SSA can easily be blocked by the discharge product Li_2O_2 and carbon pores with a large diameter are flooded by electrolyte limiting the oxygen transport in an electrode porous matrix[7,197,216] (Figure 12.8). For example, the carbon Super P (Timcal) with a low SSA of 62 m^2/g and mesoporosity (50 nm) shows the higher capacity of 3000 mA h/g[218–220] than the activated carbon with a high SSA of 2100 m^2/g. Among commercial carbons blacks, the Ketjenblack EC-600JD has the highest pore volume of 2.47 cm^3/g and the specific surface capacity of 2600 mA h/g.[196] Therefore, this comparison shows the key role of pore volume/diameter in the accommodation of the insoluble discharge product Li_2O on the solid–liquid–gas triple-junction region formed by carbon, electrolyte, and oxygen.[221,222] LiABs with graphene and free-standing hierarchical porous carbon-based catalysts demonstrate the highest specific capacity of 8,700–15,000 and 11,060 mA h/g, respectively. A binder-free carbon-based air cathode has a high pore volume as pores are not filled by a binder.[223] LiAB air cathodes with low coating thickness and carbon loading have a higher specific capacity considering carbon weight in a formula for its calculation. The degradation of LiAB capacity with the current density in LiAB was also determined by flooding air-cathode pores and the corresponding dropping of oxygen diffusion.

An oxygen catalyst with a bimodal design (Figure 12.8) provides tunnels for oxygen transport and mesopores on tunnel walls for oxygen diffusion where ORR occurs (on triple joints).[224,225] The graphene-based hierarchical catalyst with surface defects, functional groups, and interconnected pore channels demonstrated the highest discharge specific capacity of 15,000 mA h/g.[225] This high specific capacity can be attributed to the hierarchical structure, fast oxygen diffusion, and preferable formation of isolated Li_2O_2 preventing its agglomeration and the blocking electrode pore (Figure 12.9). The use of graphene nanosheets (GNSs) in LiAB air cathode also provides a higher discharge specific capacity (8700 mA h/g) and cycling stability, and lower overpotential than that of porous carbon–based air cathodes (BP2000 [1900 mA h/g] and Vulcan XC72 [1050 mA h/g]).[226]

Nonaqueous LiAB with functionalized carbon materials by nitrogen and sulfur doping demonstrates the enhanced specific capacity.[227–230] LiAB with the nitrogen-doped Ketjenblack–Calgon activated carbon–based air cathode provides twice

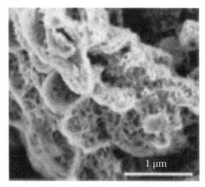

FIGURE 12.9   SEM images of a graphene air electrode at different magnifications. (Adapted with permission from Xiao, J. et al., *Nano Lett.*, 11, 5071. Copyright 2011 American Chemical Society.)

the specific capacity than that with a pristine carbon.[227] The N-doped CNTs also showed the higher specific discharge capacity of 866 mA h/g in comparison with a pristine CNT capacity of 590 mA h/g due to the presence of defective sites on the CNT surface.[228] The use of the sulphur-doped graphene as the cathode in LiAB[230] leads to the formation of Li_2O_2 nanorods ($LiPF_6$–tetraethylene glycol dimethyl ether [TEGDME]) and less cathode degradation, initial discharge capacities of the sulphur-doped and pristine graphene electrodes drops from 4300 and 8700 to 3500 and 220 mA h/g (after the second cycle), respectively.

12.5.1.1.1.2 Transition Metal Oxide Catalysts   Transition metal oxide–based ORR/OER catalysts, such as MnO_2/C[217,231–237] and their composites,[228,238–240] Co_3O_4,[241,242] and 3D $NiCo_2O_4$ nanowire array/carbon cloth (NCONW/CC),[243] are one of the main catalysts for LiABs. Air cathode 10 wt% α-MnO_2 54 wt% C Super P+36 wt% PVDF+on Ni foam demonstrated the discharge capacity 600 mA h/g after 50 cycles at the current rate of 70 mA/g, low working voltage 2.6–2.7 V, and low electrical energy efficiency of 65%, due to blocking the flow of reactants (oxygen and Li+) in the mesopores of the cathode coating during the precipitation of insoluble discharge products. Cheng and Scott obtained the high discharge specific capacity of 4750 mA h/g_{carbon} upon cycling to 4.3 V for LiAB with MnO_2/C-based catalyst (particle size of 50 nm) due to uniform catalyst distribution in a carbon matrix, leading to an enhanced ORR activity and improved electrical connection between the catalyst and current collector.[217]

The LiAB charging potential is compared to the discharge potential showing that the Li_2O_2 formation (reaction 19) is not reversible due to the dominated electrolyte decomposition. Another Mn_3O_4-based oxygen electrode was proposed for LiAB with electrolyte 1-butyl-1-methylpyrrolidinium *bis*(trifluoromethylsulfonyl)imide (BMP-TFSI) (solvent)+lithium *bis*(trifluoromethanesulfonyl)imide (LiTFSI) (conducting salt)+DMSO as superoxide and Li-ion-coordinating solvent additive.[244] LiAB with hierarchical porous Co_3O_4 catalyst with free-standing hollowed flakes deposited on Ni foam (with the size of mesopores controllable by the calcination temperature) has a capacity of 2460 mA h/g, with its dropping to 1000 mA h/g after 35 cycles. The medium capacity of this battery is explained by the bimodal design of its catalyst with macroporous channels surrounded by the flakes providing mesopores that provide two level channels for the O_2 diffusion.[245] The comparison of dense hollow porous (DHP) Co_3O_4, FeO_x, NiO, and MnO_x-based catalyst in LiAB shows superior performance of DHP Co_3O_4 with the long cycle life of 100 cycles and capacity of 2000 mA h/g.[246] The mesoporous $NiCo_2O_4$-based oxygen electrode demonstrates high electrocatalytic activity for both ORR and OER due to its high electronic conductivity and nanoflake structure. LiAB battery with the ether-based electrolyte (1M LiTFSI in TEGDME) and $NiCo_2O_4$ catalyst exhibits a good rate capability and a specific capacity of 1673, 1560, 976, and 790 mA h/g at 0.1, 0.2, 0.3, and 0.5 mA/cm^2, respectively. The $NiCo_2O_4$-based LiAB shows a lower overpotential and higher specific capacity (1560 mA h/g) than that of carbon (Super P)-based LiAB (1000 mA h/g) and Pt/C-based LiAB (450 mA h/g). Also this battery demonstrates a good round-trip efficiency of 67% due to the fast oxygen transport and triple-phase (solid–liquid–gas) junction.[247]

12.5.1.1.1.3 Noble Metal–Based Electrocatalysts Noble metals, such as Pt,[248,249] Au,[249–251] Ag,[252,253] and Pd,[254–256] or their alloy-based catalysts for LiAB do not have a practical application due to their high weight and cost.

The bifunctional PtAu electrode with Pt catalyst for charge process and Au catalyst for discharge process decreases the charge overvoltage and enhances the round-trip efficiency.[248,249] LiABs with carbonate-based electrolyte (1 M $LiPF_6$ in propylene carbonate [PC]) and carbon-supported Pd/C and PdO/C (Norit carbon black) nanocatalysts showed degradation of their initial discharge capacities (after 10 cycles) from 855 and 556 mA h/g_{solids} to 38 and 336 mA/g, respectively, due to loss of catalyst, solvent, and agglomeration of the catalyst. LiAB with a PdO/C oxygen electrode is more stable in the cycling conditions. LiAB with Pd/C catalyst has the discharge at a voltage of 2.65 V and the charge at 3.40 V. In contrast, the LiAB with the PdO/C-based oxygen electrode had a higher discharging voltage (2.75 V) but lower charging voltage (3.15 V).[257] Peng et al. adopted dimethyl sulfoxide as the electrolyte and a porous gold electrode as the cathode and achieved 95% capacity retention after 100 cycles and there were no side reactions observed during cycling; this encourages further studies on nonaqueous Li–O_2 batteries to the maximum extent.[258]

12.5.1.1.1.4 Nonnoble Metal–Based Electrocatalysts The carbon-supported CuFe catalyst (CuFe/C [Ketjenblack EC-600JD]) in the nonaqueous Li–air cell catalyzes two-electron ORR to form Li_2O_2, which reduces discharge polarization, promotes the chemical disproportionation of Li_2O_2 that leads to an apparent four-electron ORR, induces reduction of electrolyte solvents, and increases the open-circuit voltage (OCV) recovery rate of the Li–air cells as a result of the chemical disproportionation of Li_2O_2.[259] LiAB with electrolyte (5 M $LiSO_3CF_3$/tri(ethylene glycol)dimethyl ether [triglyme] electrolyte) and composite oxygen electrode 98 wt% C (Ketjenblack EC-600JD)+2 wt% SiO_2 (fumed silica S5130, Sigma-Aldrich) exhibited the increase in the discharge capacity on 50% and rate capability

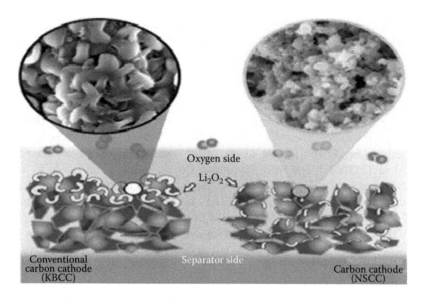

FIGURE 12.10 Structure and morphology of conventional carbon-based (KB) and nanosilica-carbon-based cathode (NSCC) after LiAB discharge. (Adapted with permission from Xia, C. et al., *J. Phys. Chem. C*, 117, 19897. Copyright 2013 American Chemical Society.)

and decrease in the charge overvoltage by 0.2 V. The authors of this study connected this LiAB performance with the formation of nanosize Li_2O_2 deposits during discharge, compared to micrometer-size Li_2O_2 deposits without nanosilica resulting in the pore clogging (Figure 12.10).[260]

12.5.1.1.2 ORR Electrocatalysts for Li_2O_2 Formation and Decomposition

LiABs with a relatively stable ether-based electrolyte demonstrate the formation of Li_2O_2 at the partial electrolyte decomposition with the formation of polyethers/esters detected during the first cycle.[261] The carbon-based ORR catalysts demonstrate different activities for Li_2O_2 formation, for example, the carbon black BP2000 decomposes ether-based electrolyte stronger than Super P and Ketjenblack.[261] Some other side reactions significantly affect LiAB performance, for example, the reaction between discharged Li_2O_2 with PVDF binder leads to PVDF crosslinking[262] or reaction of lithium superoxide with PFDF dehydrofluorinated PVDF with the LiOH formation. During the charge of LiAB, in contrast to discharge, the only reaction of Li_2O_2 decomposition occurs without side reactions. MnO_2 nanowire–based oxygen electrode exhibits the high activity for Li_2O_2 decomposition and the low charge voltage.[263] DME-based LiAB with Co_3O_4 catalyst also demonstrates the low discharge/charge overvoltage.[264] In contrary to the aforementioned results, McCloskey et al. did not observe any changes in the charge voltage for LiAB with Pt-, Au-, and MnO_2-based catalysts due to the formation of insoluble and not mobile intermediates (LiO_2) or product Li_2O_2, while *true* catalysts require mobile reactants for the delivering to active sites and removing of reaction products from active sites.[265]

In spite of significant efforts in the development of stable electrolytes for LiABs, their further improvement is required as LiAB OER/ORR columbic efficiency in relatively stable electrolytes is only about 60%.[266] Different discharged products have different electronic conductivities (Li_2O_2 has orders of magnitude less resistivity than that of Li_2O or Li_2CO_3[267]) that lead to their different decomposition voltages at charge. Morphology, defects/vacancies, and particle size of Li_2O_2 affect LiAB performance as well. The decrease in a particle size Li_2O_2 reduces the OER potential.[266] Defects and vacancies in Li_2O_2 decrease its resistivity, probably due to the formation of a more conductive oxygen-rich surface layer,[268] which leads to low-onset potential of Li_2O_2 decomposition. The main catalysts for oxygen positive electrodes are presented in Table 12.6.

12.5.1.1.3 Oxygen Positive Electrode Design

Oxygen electrode of LiABs consists of a carbon-supported catalyst (e.g., MnO_2, Co_3O_4, $NiCo_2O_4$ on carbon Super P, Ketjenblack, etc.), binder (PVDF, PEO, PTFE, cellulose), and current collectors Al–Ni mesh or foam.[269–271] The optimization of their content, distribution, and interaction plays the key role in the electrode performance. The electrochemical ORR and OER in nonaqueous LiABs related to the formation/decomposition of Li_2O_2 determine the design of the oxygen positive electrodes. This electrode design significantly affects the LiAB performance,[272,273] for example, a significant electrode void volume for the accommodation of Li_2O_2 leads to a high specific capacity.[274] Binder-free design of positive electrode[223,240,275–277] allows for an increase in the utilization of an electrode catalyst and maximization of the pore volume for the accommodation of discharge product Li_2O_2 due to the absence of pore clogging by a binder. The increase in the binder concentration enhances the electrode porosity.[278] The other promising direction for improvement of the LiAB performance is the use of the bimodal design of the catalyst layer structure (Figure 12.11), providing easier oxygen transport through tunnels to pores in walls where Li_2O_2 forms on the surface of triple junctions (electrolyte–catalyst–gas).[225] The minimization of the gas transport limitation decreases the polarization of the oxygen electrode. The optimization of the catalyst distribution in the coating of an air-breathing electrode, lower catalyst concentration on the air side and higher on the electrolyte side, leads to the increase in the current density by 38%.[279] Finally, proper carbon selection for a positive electrode catalyst significantly improves its specific capacity. The degradation of nanostructured carbon–based oxygen electrode occurs when partial utilization of the internal volume of carbon pores occurs due to filling by discharge products Li_2O_2.[223,234] For the optimized carbon, a high pore volume and diameter is preferable to a high SSA, for example, the use of carbon black KB EC-600JD with the high SSA 1325 m^2/g and pore volume 2.47 m^3/g in LiAB positive electrodes leads to the higher specific capacity of 2600 mA h/g compared to the use of other carbon black (KB EC-300J, Vulcan, Super P, etc.) However, only graphene usage in the LiAB positive electrode allows it to get its highest specific capacity of 8,700–15,000 mA h/g.[225,226]

TABLE 12.6
Catalyst Materials for LiABs

Catalyst for Oxygen Electrode	Battery Design/Electrolyte Type	Pore Diameter, nm/Pore Volume, cm³/g	SSA, m²/g	Specific Capacity mA h/g/Cycles/Energy Density	References
Carbon Super P	1 M LiClO₄/PC	50/—	62	1,736	[219]
Carbon Vulcan XC-72	1 M LiClO₄/PC	2/1.74 mL/g	250	762	[219]
Activated carbon	1 M LiClO₄/PC	2/—	2100	414	[219]
CNTs	1 M LiClO₄/PC	10/—	40	583	[219]
Graphite	1 M LiClO₄/PC	—	6	560	[219]
Ball mill graphite	1 M LiClO₄/PC	—	480	1,136	[219]
Mesocellular carbon foam	1 M LiClO₄/PC	30/—	824	2,500	[219]
Carbon Super P	EC/DMC/EMC	—/0.32	62	2,150	[196]
Carbon black KB EC-600JD	EC/DMC/EMC	—/2.47	1325	2,600	[196]
Carbon black KB EC-300JD	EC/DMC/EMC	—/1.98	890	956	[196]
Carbon Denka black	EC/DMC/EMC	—/0.23	60	757	[196]
Carbon Ensaco 250 G	EC/DMC/EMC	—/0.18	62	579	[196]
Carbon Super P	LiPF₆ in PC	—	62	3,400	[217]
Carbon acetylene black	LiPF₆ in PC	—	75	3,900	[217]
Carbon black SX, Norit	LiPF₆ in PC	—	800	4,400	[217]
Carbon black BP2000 (Cabot)	PC/EC-based electrolyte	—	—	1,900 at a current rate of 75 mA/g (Vulcan XC72–1053 mA h/g$_{carbon}$)	[226]
GNS	PC/EC-based electrolyte	—	—	8,700[226] at a current rate of 75 mA/g	[226]
Graphene—functionalized and hierarchical GNSs (conductivity 2–10⁴ S/m) + PTFE	Ether-based electrolyte (LiTFSI in triglyme)	—/0.84	590	15,000 at 0.1 mA/cm² and oxygen pressure 2 atm with a plateau at around 2.7 V	[225]

(Continued)

TABLE 12.6 (*Continued*)
Catalyst Materials for LiABs

Catalyst for Oxygen Electrode	Battery Design/Electrolyte Type	Pore Diameter, nm/Pore Volume, cm³/g	SSA, m²/g	Specific Capacity mA h/g/Cycles/Energy Density	References
80 wt% C (free standing hierarchical porous carbon)+20 wt% PVDF (Kunar 2801) on the Ni mesh	0.21 mL of TEGDME with 1 M LiPF$_6$	—	—	11,060 at a current rate of 0.2 mA/cm^2 (2.0–4.7 V at a current rate of 2 mA/g)	[10]
Composite single-wall carbon nanotube (SWNT)+nanofiber (CNF)	NA	—	—	2,500 (thickness [d] 220 μm), 400 (d=20 μm)	[223]
10 wt% α-MnO$_2$ 54 wt% C Super P (Timcal)+36 wt% copolymer PVDF (Kunar 2801)/Ni foam	LiPF$_6$ in PC	—	—	600 after 50 cycles at a rate of 70 mA/g	[217]
Mesoporous 40 wt% NiCo$_2$O$_4$+50 wt% Super P+10 wt% PTFE	1 M LiTFSI in TEGDME	—	—	1,560 mA h/g at 0.2 mA/cm^2 Round-trip efficiency of 67%	[247]
Co$_3$O$_4$/Ni foam (bimodal catalyst)	NA/1.0 M LiTFSI in TEGDME	—	54.8	2,460 drops to 1,000 after 35 cycles at 0.05 mA/g in O$_2$/35	[245]
80 wt% Co$_3$O$_4$/C (AB):PVDF=1:1 w/w (dense hollow structure DHP, 100 nm particles)	1 M LiPF$_6$ in a mixture solution of EC/DMC (1:1 w/w)	—	—	2,000 mg h/g/100 cycles at 200 mA/g	[246]
83 wt% C (Ketjenblack EC-600JD)+2 wt% SiO$_2$+15 wt% PTFE (carbon loading of 0.6 mg/cm^2)	0.5 M LiSO$_3$CF$_3$/triglyme electrolyte	—/4.5 cm^3/g$_{carbon}$	—	3,800 at 0.24 mA/cm^2 and 9,600 at 0.03 mA/cm^2 (discharge efficiency of 63%, 47%, and 32% at 0.06 mA/cm^2 (100 mA/g), 0.14 mA/cm^2 (220 mA/g), and 0.24 mA/cm^2 (380 mA/g)	[260]
40 wt% Pd/C, 40 wt% PdO/C (carbon black SX, Norit)	1 M LiPF$_6$ in PC electrolyte	—	—	855 mA h/g$_{solids}$ for Pd/C	[257]

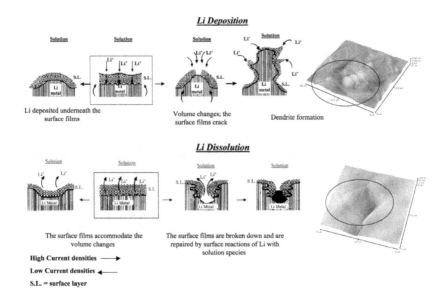

FIGURE 12.11 Morphology and failure mechanisms for lithium anode at the LiAB discharge/charge. (Adapted with permission from Aurbach, D. et al., *Solid State Ionics*, 148, 405. Copyright 2002 American Chemical Society.)

12.5.1.2 Electrolytes

Nonaqueous electrolytes with soluble lithium salts ($LiPF_6$, $LiSO_3CF_3$, $LiBF_4$, $LiAsF_6$)[280–283] for LiABs are divided into five groups depending on the solvent used:

- Organic carbonate-based electrolyte (PC, ethylene carbonate [EC], diethyl carbonate [DEC], and dimethyl carbonate [DMC])
- Ether-based electrolyte (TEGDME, dimethoxyethane, dioxolane, tetrahydrofuran)
- IL
- Polysiloxanes

12.5.1.2.1 Carbonate-Based Electrolyte

The carbonate-based electrolyte (PC, EC, DEC, and DMC) has a high oxidation potential and stability, low viscosity, and high Li-ion diffusion. However, carbonate electrolytes decompose due to the reaction with the lithium anode during discharge. The product of this reaction is mainly Li_2CO_3, which is not electrochemically reversible and insoluble. The reason for the decomposition of carbonate electrolytes is the reaction of Li_2O_2 with carbonates or CO_2 in air or carbonates with oxygen radicals, which were formed in the beginning of ORR. The formation of irreversible discharge products accumulates in the porous air-cathode coating, decreasing the columbic efficiency, rechargeability, and cycle life.[280–282]

12.5.1.2.2 Ether-Based Electrolytes

In contrast to carbonate electrolytes, ether-based electrolytes (tetrahydrofuran, dioxolane, dimethoxyethane, and TEGDME) have higher stability at the high potential >4.5 V versus Li/Li$^+$, better stability at the point of contact with the lithium anode, lower volatility, and greater safety. However, the stability of the ether-based electrolyte can be only during the first five cycles, and then Li$_2$O$_2$ was replaced by Li$_2$CO$_3$.[261] The modified form of ether-based electrolytes, oligoether functionalized silane, enables further enhancement of the stability toward ORR discharge species.[284] The alternative for carbonate- and ether-based electrolytes is a room-temperature ionic liquid (RTIL), which is stable at the point of contact with the lithium anode during the battery operation and does not have any formation irreversible products.[233,285–288] RTIL in aprotic LiAB has a low flammability and vapor pressure, wide potential window, hydrophobic nature, and high thermal stability. Zhang et al. reported a high capacity of 4080 mA h/g of carbon for LiAB with the composite electrolyte RTIL + silica – PVdF-HFP and the air-cathode α-MnO$_2$ catalyst.[233] The higher specific capacity of 5360 mA h/g was achieved in LiAB after a 56-day test with a hydrophobic and conductive IL consisting of 1-alkyl-3-methylimidazolium cation and perfluoroalkylsulfonyl imide anion and carbon-based cathode.[233] Li$^+$-containing RTIL EMI-TFSI provides the high efficiency during recharging of the oxygen electrode (Au electrode), yielding LiO$_2$ and without electrode passivation.[288] However, RTILs for LiABs require further improvements as well to provide simultaneously higher electrochemical stability and low viscosity. Now, pyrrolidinium-based RTIL has a higher durability than imidazolium-based RTIL but less discharge capacity.[289]

12.5.1.2.3 Polysiloxanes

Polysiloxanes are a good alternative for carbonate- and ether-based electrolytes for LiABs and have a high conductivity and ionic transport and low viscosity. The presence of the Si–O group in their composition increases the stability of polysiloxanes at the high energy density.[290]

12.5.1.2.4 Additives

The composition of the electrolyte has a significant effect on LiAB performance. Several additives were developed to increase the ORR rate and the discharge specific capacity:

- Large cations, for example, A$^+$PF$_6^-$ (where A is tetrabutylammonium) in the hexaflurophospate electrolyte increases reversibility of the O$_2$/O^{2-} redox couple. The combination of small Li$^+$, K$^+$ cations, and large TBA salts increases the solubility of discharge products
- Catalysts (quaternary ammonium salts of a heterogeneous catalysis and transfer of solid phase to liquid phase[291,292])
- Promoters of discharge capacity (tetrabutylammonium),[292] as the ammonium cation combines with peroxide anions forming soluble N(Bu$_4$)O$_2$,

decreasing the polarization of Li_2O_2 reduction and cathode discharge capacity, respectively

- Promoters (perfluorinated compounds) of solubility and diffusivity of oxygen in an electrolyte[293] to increase diffusion-limited current of ORR on air-breathing electrode.

12.5.1.3 Anodes

Lithium is the best anode material for LiAB due to its high specific capacity of 3860 mA h/g and low negative potential of −3.04 versus SHE[294] However, lithium anodes have two main technical challenges, those being dendrite formation and low charge/discharge efficiency[295] due to the continuous growth of the SEI and low lithium utilization. According to SEI film theory,[296,297] the reason for the dendrite formation is the preferential lithium growth on sites with higher Li^+ ion conductivity and current density sufficient to break SEI. Cracks in SEI open the lithium surface, which leads to its dissolution (Figure 12.11).[298] The nonuniform distribution of current density leads to significant enhancement of local current density at the edges on Li surface and their heating, increasing explosion risk. SEI consists of the 50A dense inner layer of Li_2O and the porous layer formed by electrolyte reduction products LiF, LiOH, and Li_2CO_3. These problems were partially solved by the use of lithium alloys such as Li–Ga,[299] Li–Al, Li–Na,[300] and Li–Mg with higher durability and less dendrite formation than that of lithium anodes. Additional improvement of lithium anodes was also achieved by using dendrite inhibitors HF[177,301] and CO_2 in electrolyte,[302,303] surfactants poly(ethylene glycol)dimethyl ether (PEGDME)[304] for uniform lithium deposition,[305,306] stabilizers of rechargeability with SnI_2 and AlI_3 in electrolyte,[307] inorganic fillers SiO_2 and Al_2O_3 for the trapping of liquid impurities, and ultrathin plasma polymer coating to stabilize lithium–electrolyte interface.[308] It is interesting to note that organic additives affect the inner layer and inorganic additives influence the outer layer of SEI.[294]

12.5.2 Aqueous LiAB

An aqueous LiAB has a protected lithium electrode (PLE), excluding the direct contact of lithium anode and aqueous electrolyte (lithium salt in water) (Figure 12.8b).[202,309] The cell discharge reactions in aqueous and basic electrolytes are given as follows:

$$\text{Anode: } Li \leftrightarrow Li^+ + e^{-1} \tag{12.26}$$

Cathode (discharge)[195]:

$$4Li + O_2 + 2H_2O = 4LiOH \ (E_o = 3.45\,V \text{ vs. } Li/Li^+) \text{ (basic electrolyte)} \tag{12.27}$$

$$4Li + O_2 + 4H^+ = 4LiOH + 2H_2O + Li \ (E_o = 4.27\,V \text{ vs. } Li/Li^+)$$

$$\text{(acidic electrolyte)} \tag{12.28}$$

Seawater $(pH\ 8.2)^{195}\ 4Li + O_2 + 2H_2O = 4LiOH\ (E_o = 3.79\ V\ vs.\ Li/Li^+)$ (12.29)

$$Overal: 2Li + H_2O + O = 2LiOH\ (E_o = 3.35\ V) \qquad (12.30)$$

Parasitic self-discharge corrosion reaction:

$$2Li + 2H_2O = 2LiOH + H_2 \qquad (12.31)$$

During the discharge process, oxygen reduces at the positive electrode with the formation of hydroxyl ions (OH^-) and Li^+ ions are generated at lithium anode. $LiOH \cdot H_2O$ precipitates at the achievement of its solubility limit (5.25 M at 25°C).[7] During the charge process, O_2 evolves at the positive air-breathing electrode and Li is deposited at the negative electrode. The LiAB with an acidic aqueous electrolyte does not have significant distributions due to significant technical challenges with the development of its durable components. The main difference in the reaction mechanism in aqueous LiAB compared to nonaqueous LiAB is the formation of the soluble LiOH at concentration <5.25 M at 25°C. During the charge process, the oxygen evolves on the positive electrode and lithium deposits on the negative electrode. The Li–air cell with basic and acidic electrolyte has the following theoretical specific energy density: 1300 W h kg/1520 W h/L and 1400 W h kg/1680 W h, respectively.[222] PolyPlus first developed PLE, using water-stable glass-ceramic electrolyte ($Li_{1+x+y}Al_xTi_{2-x}Si_yP_{3-y}O_{12}$ or LTAP, Ohara Inc., Japan), for the protective coating of lithium anodes.[202–204,310] PolyPlus' primary aqueous LiAB demonstrated 0.5 mAh/cm^2 for 230 h at 100% coulombic efficiency.[202] Due to the nonuniform current distribution, uneven Li plating/stripping is possible. PolyPlus included the porous reservoir in its battery design for the cathode to accommodate the discharge product, $LiOH \cdot H_2O$. Recently, several developers, like Toyota,[311] EDF,[312] (100 cycles at 0.1 mA/cm^2 and 40 cycles at 0.2 mA/cm^2), Zhang et al.[309,313] (Figure 12.12b), and Imanishi et al.[314,315] paid attention on PLE concept for aqueous LiABs.

12.5.2.1 Oxygen Positive Electrodes

ORR and OER electrocatalysts are the components of BEs for rechargeable LiABs. ORR during the discharge mode of LiAB is a complex process including several steps, such as oxygen diffusion from air to catalytic sites in porous BE, oxygen absorption on a catalyst, electron transfer from anode to oxygen molecules, breaking oxygen bonds, and diffusion of hydroxyl ions from catalyst surface to electrolyte. OER during the battery charging includes the reverse processes compared to ORR. The ORR and OER catalysts mainly determine LiAB performance, power and energy density (discharge/charge currents 0.1–0.5 A/cm^2 vs. 10 mA/cm^2 for Li-ion batteries), and energy efficiency. The active and durable air cathode for aqueous LiAB is one of the main challenges. Air cathodes based on nonnoble catalysts Cu,[316] titanium nitride (TiN)[317] a low performance. A Cu-based air cathode in O_2-saturated $LiNO_3$ aqueous solution after 12 h continuous discharge demonstrates low ORR due

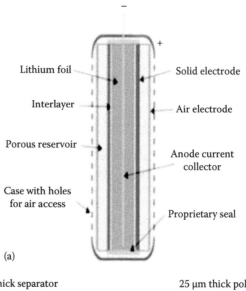

(a)

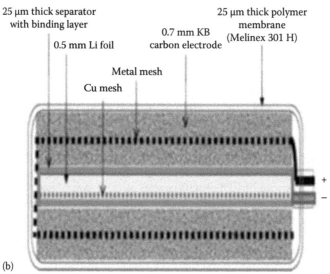

(b)

FIGURE 12.12 Cross-section of aqueous LiABs. (a) PolyPlus Battery Co design. (b) Zhang's group design. (a: Adapted from Visco, M.J. et al., *The 210th Electrochemical Society Meeting*, Cancun, Mexico, 2006, Abstract #0389. Copyright 2014, The Electrochemical Society. With permission; b: Adapted from *J. Power Sources*, 195, Zhang, J., Wang, D., Xu, W., Xiao J., and Williford, R.E., 4332–4337. Copyright 2014, with permission from Elsevier.)

to the copper-corrosion mechanism. High ORR catholic current was observed at an onset potential of 3.80 V versus Li/Li^{+316} The single LiAB with TiN-based air cathode cell exhibited a discharge curve with a voltage plateau of 2.85 V at the current density of 0.5 mA/g in weak acidic solution that demonstrates the necessity of the improvement of its electrochemical performance.[317]

12.5.2.2 Electrolytes

An alkaline or acidic electrolyte could be used for aqueous LiAB.[318–323] An alkaline electrolyte is preferable due to the better stability of lithium anodes. The carbonation of alkaline electrolytes, its common shortcoming, could be avoided by using purified air or oxygen selective membranes. The typical aqueous electrolyte for LiAB is the $LiOH–LiCl–H_2O$[309] with different pH and LiCl concentration.[323] Several aqueous solutions such as $LiNO_3$, LiCl, and LiOH were tested for the selection of potential aqueous electrolyte for LiAB with LPE (lithium coated by the lithium-ion conducting solid electrolyte $Li_{1+x+y}Al_xTi_{2-x}Si_yP_{3-y}O_{12}$ (LATP)) and the protective solid-state interfacial layer (phosphors oxynitrides LiPON). This LiAB (Li–Al/$Li_{3-x}PO_{4-y}N_y$/LATP/aqueous 1 M LiCl/Pt) demonstrates stable OCV of 3.64 V at 25°C and no cell resistance change for 1 week.[318] Zhang et al. used the same aqueous LiCl electrolyte in LiAB (Li/PEO18LiTFSI/LTAP/aqueous LiCl/Pt), which exhibits stable OCV of 3.7 V at 60°C for 2 months.[319]

LiAB with aqueous electrolyte 0.5 M Li_2SO_4 and $LiMn_2O_4$ cathode and lithium-protected electrode with the coating consisting of a gel polymer electrolyte (GPE) and a lithium superionic conductor (LISICON) film with ionic conductivity of Li^+ ions is about 0.1 mS/cm. Due to the *crossover* effect of Li^+ ions in the coating, this aqueous rechargeable lithium battery (ARLB) has the output voltage of 4.0 V, energy density of 446 W h/kg, about 80% higher than that for traditional Li-ion battery, and power efficiency of 95%. The redox reactions in this LiAB are as follows[320]:

$$\text{Anode reaction: } Li^+ + e^- \leftrightarrow Li$$

$$\text{Cathode reaction: } LiMn_2O_4 \leftrightarrow Li_{1-x}Mn_2O_4 + xLi^+ + xe^- \qquad (12.32)$$

$$\text{Total reaction: } LiMn_2O_4 \leftrightarrow Li_{1-x}Mn_2O_4 + xLi$$

LiAB $LiTi_2(PO_4)_3$/Li_2SO_4/$LiFePO_4$ with the similar Li_2SO_4 electrolyte shows high capacity retention over 90% after 1000 cycles for 10 min and 85% after 50 cycles for 8 h.[321]

The similar LiAB with 0.5 M Li_2SO_4 aqueous solution as the electrolyte demonstrates the charge and discharge capacities at a current density of 10 A/g (about 90°C) were 6% and 95% of the total capacity (118 mA h/g) power density 10,000 W k/g. Battery exhibits the high capacity retention of porous $LiMn_2O_4$ is 93%. This improved performance is determined by nanograins, porous morphology, and high crystalline structure air-cathode catalyst and the absence of Mn^{2+} from dissolution.[322] The stability of the Li-ion-conducting glass ceramics, $Li_{1+x+y}Ti_{2-x}Al_xSi_yP_{3-y}O_{12}$ (LTAP), in alkaline solution LiOH and LiOH–LiCl. LTAP conductivity decreased in 0.057 M LiOH aqueous solution at 50°C for 3 weeks and was not changed in the solution LiCl–LiOH–H_2O (pH 8–9). Thus, LiCl saturated aqueous solution can be used as the electrolyte, because the content of OH^- ions in the LiCl saturated aqueous solution does not increase via the cell reaction of $Li + 1/2O_2 + H_2O \rightarrow 2LiOH$, and LTAP is stable under a deep discharge state.[323]

12.5.2.3 Anodes

PolyPlus Battery Co (United States) developed one of the first successful PLE—anode with the solid-state interfacial layer (Cu_3N, LIPON) and glass–ceramic membrane ($Li_{1+x+y} Al_xTi_{2-x}Si_yP_{3-y}O_{12}$ or LATP) (Ohara Inc., Japan).[201,202] The usage of the sublayer in an anode allows to protect upon direct contact with lithium. The chemical stability of LPA is higher in neutral LiCl or $LiNiO_3$ electrolyte[324,325] and low in acidic (HCl, lithium acetate, with phosphate buffer) and alkaline electrolytes (LiOH).[324]

12.5.3 Hybrid Electrolyte LiAB

The hybrid electrolyte LiAB (Figure 12.9)[316,326–337] uses two different electrolytes that are separated by a Li^+ conducting glass–ceramic membranes: LISICON ($Li_{14}Zn(GeO_4)_4$, 10^{-4} S/cm) or sodium (Na) superionic conductor (NASICON) (LTAP, 1.3×10^{-3} S/cm).[7] The anolyte is a lithium-stable nonaqueous electrolyte and the catholyte is the alkaline aqueous electrolyte. The hybrid LiAB does not have issues related to the decomposition of aprotic electrolyte and low cycle life.[333,334] The hybrid LiAB with alkaline aqueous electrolytes, in contrast to LiABs with aprotic electrolytes has the higher ORR rate and soluble reduction products OH^- (instead of insoluble L_2O_2 in aprotic electrolytes)[327–332]. Zhou et al. used two electrolytes, 1 M $LiClO_4$ in EC/DMC and 1 M KOH separated by glass–ceramic LISICON[327] and air-breathing electrode 85 wt% Mn_3O_4/carbon (85 wt%) + 15 wt% PTFE. It achieved 500 h continuous discharge performance and a special capacity of 50,000 mA h/g based on total mass of catalytic electrode (carbon + binder + catalyst). The design of this battery was improved by adding capacitor cathode combining capacitor character and LiAB character, which is now nominated as a *lithium–air capacitor–battery based on a hybrid electrolyte* Li|$LiClO_4$ in EC/DMC|LISICON|4 mL 1 M KOH||Mn_3O_4/C, In this system, Li ions mainly diffuse from the lithium anode to the additional capacitor electrode and the electrons obtained by the air cathode are almost negligible. The capacitor cathode and air cathode play a key role at the high demanded power and energy, respectively.[328] Mathiram et al. developed LiAB Li|1 M $LiPF_6$ in EC/DEC(1:1 v/v)|NASICON membrane 0.15 µm|0.1 M H_3PO_4 + 1 M LiH_2PO_4|40 wt% Pt/C, which demonstrates a discharge capacity of 221 mA h/g at a current density of 0.5 mA/cm with good cycle life (aqueous electrolyte could also be 90 wt% acetic acid + H_2O + Li acetate salt).[337]

12.5.3.1 Oxygen Positive Electrode

Metal-free GNS-based air cathodes showed a high discharge voltage similar to 20 wt% Pt/carbon black in LiAB with hybrid electrolyte ($LiClO_4$/ED/DEC) and LISICON solid-state electrolyte (membrane between the organic and aqueous electrolytes) at a discharging current density of 0.5 mA/cm² for 24 h. Its charge voltage increased from about 3.55 to 3.98 V and the discharge voltage decreased from 3.0 to 2.78 V after the 50th cycle. This advanced performance was explained by sp3 bonding associated with edge, defect sites in GNSs and removal of adsorbed functional groups, and crystallization of the GNS surface into a graphitic structure on heat treatment. GNSs have a high thermal conductivity (~5000 W/mK), high electrical conductivity (103–104 S/m), and high SSA (calculated theoretical value 2630 m^2/g).[332]

12.5.3.2 Electrolytes

Hybrid LiAB combines nonaqueous and aqueous electrolytes described in Sections 12.5.1.2 and 12.5.2. Wang et al.[326] developed the hybrid LiAB (Li|organic electrolyte|LISICON|10 mL 1 M KOH|Mn$_3$O$_4$) demonstrating 500 h continuous discharge performance and the special capacity of 50,000 mA h/g based on total mass of catalytic electrode (carbon + binder + catalyst).

He at al. developed LiAB with hybrid electrolyte Li|organic liquid electrolyte|LiGC|1 M LiClO$_4$|Pt with 85% voltage efficiency at room temperature.[329] This LiAB with electrolyte ≤0.05 M LiOH has a discharge voltage of 3.5 V (at 0.05 mA/cm^2) and a voltage efficiency up to 84%. The addition of LiClO$_4$ into electrolyte slightly lowered the discharge voltage to 3.3 V and significantly decreased the internal resistance and slowed the electrolyte pH increase due to the fast or long-term discharge of the air electrode in the LiAB. It was found that solubility of oxygen is a crucial factor determining the ORR catalyst activity. That explains the decrease of discharge/charge voltage at the increase of LiOH concentration due to the dropping the oxygen solubility with a LiOH concentration (Table 12.7).

Cheng et al. developed the flow hybrid LiAB Li|1.2 M LiPF$_6$ + 0.85 M + and CH$_3$COOLi (1:1 v/v)|carbon nanofoam with high rate capability (5 mA/cm^2) and power density of 7.64 mW/cm^2 at a constant discharge current density of 4 mA/cm^2. The electrolyte storage units and oxygen exchange can be separated to increase the battery efficiency.[335]

12.5.4 Oxygen Selective Membranes

LiAB does not store the main reactant oxygen for ORR on a cathode using ambient air. However, air has impurities, such as H$_2$O and CO$_2$, which are involved in side reactions decreasing battery performance. H$_2$O reacts with a lithium anode, leading

---

TABLE 12.7
Discharge Voltages Depending on Catalyst and Electrolytes

Catalyst	Electrolyte	Current Density Discharge, mA/cm^2	Discharge Voltage, V	References
Pt/C	DI H$_2$O	0.05	3.53	[336]
Pt/C	DI H$_2$O	0.5	3.27	[336]
Pt/C	0.01 M LiOH	0.05	3.4	[336]
Pt/C	0.05 M LiOH	0.05	3.31	[336]
Pt/C	1 M LiCl	0.03	3.2	[314]
CoMn$_2$O$_4$-graphene composite	1 M LiOH	0.025	2.95	[338]
CoMn$_2$O$_4$-graphene composite	1 M LiOH	0.4	2.8	[338]
Mn$_3$O$_4$/C	1 M KOH	0.5	2.8	[327]

to battery failure and safety issues due to the generation of hydrogen. CO_2 is the reason for the decrease in rechargeability due to the formation of the insoluble and electrochemically nondecomposable Li_2CO_3. The modern LiAB uses pure oxygen or dry air using a dehumidifier. However, the usage of an oxygen cylinder decreases the energy density of LiAB. Therefore, the usage of porous or nonporous OSWBs can solve problems related to the negative impact of air impurities on LiAB performance. Practical OSWB membranes should provide (1) oxygen selectivity, (2) sufficient oxygen permeation rate, (3) compatibility with the electrolyte, (4) chemical and mechanical durability, (5) simple integration, and (6) cost efficiency.[294] The development of porous oxygen membranes (microporous PTFE or silicalite zeolite) has not been successful so far because only nonporous membranes can practically be used in LiABs. The main types of nonporous OSWBs are given in Table 12.8.

LiABs using purified air on OSWB membranes such as Melinex 300 H,[341,342] polydimethylsiloxane,[344] and hydrophobic zeolite[345] demonstrate the advanced

TABLE 12.8
Properties of OSWBs

Membrane	Operational RH Limit, %/ Life Time, Day	O_2 Permeance, mol/m² s Pa/O_2/ H_2O Selectivity	Specific Capacity, mA h/g	References
Heat sealable (Melinex 301 H, Wilmington, DuPont Teijin Film)	21/>30	—	—	[309,339,340]
Polyperfluoropropylene oxide coperfluoroformaldehyde (PEPO) (support— polymer Celgard 2500)	—	1.7×10^{-10}/3.9 at 50°C	—	[341]
Polydimethylsiloxane (support—Teflon substrate)	20/16.3	1.62×10^{-6}/3.6	789 mA h/ g_{carbon}	[342]
Hydrophobic zeolite	20/21	—	1022 mA h/ g_{carbon}	[343]
Silicone rubber (polysiloxane and methacrylate– polysiloxane copolymer) (support—air electrode)	—	—	—	
Silicon rubber (Semicosil 964)	42.9/—	—	570 (151 mA h/g for LiAB without a membrane)	
Teflon-coated fiberglass cloth (TCFC)	Sufficient O_2 diffusivity at 0.2 mA/cm²	—	—	[344,345]

performance with reasonable membrane lifetime (about 30 days). However, further improvement of the technical and cost efficiency of OSWB membranes is required.

12.6 SUMMARY

The significant progress in the development of highly efficient MABs was done last decade. MABs with higher energy density, low cost, absence of safety problems, fuel recycling, plentiful fuel resources, less weight, and inexpensive production make them the ideal substitute for the rechargeable NiCd, Li-ion, and NiMH batteries. Optimistic high specific energy and power density, cycle life, specific capacity, and cost efficiency were achieved. Companies such as EOS (United States) and PolyPlus Battery Co. (United States) achieved precommercialization stage for aqueous ZnAB and aqueous Li-ion batteries, respectively.

1. The most commercialized MABs are primary and secondary ZnABs. ZnAB components, such as zinc alloy anodes, BEs, separator, and electrolytes, in comparison with LiAB are more stable toward moisture, durable, and cost effective. Their market potential is principally determined by the durability of their electrodes (air cathodes, BEs, anodes). Several Zn–air technologies are at the precommercialization stage with companies (EOS, ReVolt) preparing to launch them on the market after intensive development of electrolytes and advanced air cathodes/BEs. EOS ZnAB *Aurora* achieved the precommercialization stage and demonstrated the stable performance after >5000 cycle durability test. The most promising ZnAB components are as follows:
   a. Air cathodes
      i. The composition and architecture of ZnAB cathodes and their ORR catalysts (MnO_x, CoO_x-MnO_x, Ag, CoTMPP/FeTMPP, metal nitrides, spinel, perovskites, and pyrochlore) significantly affect battery performance.
      ii. MnO_x-based catalyst is characterized by the two-electron ORR pathway, a high activity for peroxide decomposition, but an insufficient durability in concentrated alkaline solution. Rayovac Corp. (United States) developed a stable MnO_2-based cathode with a voltage 1.15 V at 150 mA/cm^2 (15 h durability test in 30% KOH).
      iii. Ag-based catalysts (Ag:W_2C = 1:1) have higher ORR activity (direct four-electron ORR pathway) and could be used in the BE, as Ag is active for ORR and WC is active for both OER and ORR in alkaline solution.
      iv. CoTMPP and FeTMPP correspondingly promote the two- and four-electron ORR pathways and have high and low stability in alkali, respectively. The composite catalyst CoTMPP–FeTMPP/C combines the advantages of both compounds. Powerzinc Electric Inc. (China) developed CoTMPP+MnO_2 catalyst with a high performance of j = 216 mA/cm^2 at 1.0 V in 30% KOH and T = 25°C.

     v. Spinel-based catalysts, such as $NiCo_2O_4$, $Mn_xCo_{3-x}O_4$ ($0 < x < 1$), and $CoFe_2O_4$, have higher ORR activity than the separate CoO_x, MnO_x, and NiO-based catalysts. Air-cathode $NiCo_2O_4$+PTFE in 5M KOH shows performance $j = 200$ mA/cm^2 at -0.87 V.

     vi. Perovskite-based ORR catalysts, such as $LaMnO_3$ and core ($LaNiO_3$)–corona (NCNT), have high ORR activity and sufficient stability in the alkaline electrolyte. The air cathode with core ($LaNiO_3$)—corona (NCNT) can be used in a BE, demonstrating the stability of the discharge potential during 75 cycles.

  b. Zinc alloys (Zn–Ni, Zn–In, Zn–Hg, Zn–PbO) anodes have higher durability and depressed-side hydrogen evolution reaction.

  c. PEO separators provide optimized transport of hydroxyl ions from air cathodes to Zn anodes.

  d. IL-based electrolytes demonstrate stability in a wide range of temperatures, the absence of evaporation and carbonation issues, and smooth deposition of Zn. EOS company developed the neutral aqueous chloride electrolyte of rechargeable ZnABs without carbonation.

2. Rechargeable aqueous and nonaqueous LiABs have an excellent potential to be a future energy storage system. However, the absence of cost-effective, durable electrodes, oxygen selective membranes, stable electrolytes, cheap and highly conductive glass-ceramic membranes for lithium protected anodes (in aqueous LiAB) requires the significant further development and improvement of LIABs to achieve their pre-commercialization stage.

REFERENCES

1. D. Linden and T. B. Reddy, *Handbook of Batteries*, 3rd edn., McGraw-Hill, New York, 2001, p. 1200.
2. A. J. Bard, M. Stratmann and P. R. Unwin, *Encyclopedia of Electrochemistry*, Wiley-VCH, New York, 2007, p. 7720.
3. M. A. Rahman, X. Wang and C. Wen, *J. Electrochem. Soc.*, 2013, **160**, A1759–A1771.
4. F. Beck and P. Ruetschi, *Electrochim. Acta*, 2000, **45**, 2467–2482.
5. L.-L. Zhang, Z.-L. Wang, D. Xu, X.-B. Zhang and L.-M. Wang, *Int. J. Smart Nano Mater.*, 2013, **4**, 27–46.
6. G. Girishkumar, B. McCloskey, A. C. Luntz, S. Swanson and W. Wilcke, *J. Phys. Chem. Lett.*, 2010, **1**, 2193–2203.
7. J. Christensen, P. Albertus, R. S. Sanchez-Carrera, T. Lohmann, B. Kozinsky, R. Liedtke, J. Ahmed and A. Kojic, *J. Electrochem. Soc.*, 2012, **159**, R1–R30.
8. J. S. Lee, S. T. Kim, R. Cao, N. S. Choi, M. Liu, K. T. Lee and J. Cho, *Adv. Energy Mater.*, 2011, **1**, 34–50.
9. J. Wang, Y. Li and X. Sun, *Nano Energy*, 2013, **2**, 443–467.
10. Z. Wang, D. Xu, X.-B. Zhang and X.-B. Zhang, *Chem. Soc. Rev.*, 2014, **43**, 7746–7786.
11. Y. Li, M. Gong, Y. Liang, J. Feng, J.-E. Kim, H. Wang, G. Hong, B. Zhang and H. Dai, *Nat. Commun.*, 2013, **4**, 1805.
12. C. Chakkaravarthy, A. K. Abdul Waheed and H. V. K. Udupa, *J. Power Sources*, 1981, **6**, 203–228.
13. V. Neburchilov, H. Wang, J. Martin and W. Qu, *J. Power Sources*, 2010, **195**, 1271–1291.
14. R. Cao, J.-S. Lee, M. Liu and J. Cho, *Adv. Energy Mater.*, 2012, **2**, 816–829.

15. L. Jorissen, *J. Power Sources*, 2006, **155**, 23–32.
16. H. Abbasi, S. Salehi, R. Ghorbani, F. Tarabi and M. Amidpour, *Iranica J. Energy Environ.*, 2013, **4**, 110–115.
17. F. Cheng and J. Chen, *Chem. Soc. Rev.*, 2012, **41**, 2172–2192.
18. EOS Energy Storage Co., corporative website, http://www.eosenergystorage, (accessed September 6, 2013).
19. S. Amendola, M. Binder and M. B. Phillip, *US Pat.*, US20120040254A1, 2012.
20. ReVolt Technology Co., corporative website, http://www.revolttechnology.com, (accessed September 6, 2013).
21. T. Burchardt, J. P. McDougall, R. F. Ngamga, H. Studiger and W. G. Less, *US Pat.*, US20100330437A1, 2010.
22. Tyler Hamilton, Working paper, Fluidic energy, *MIT Technology Review*, http://www.technologyreview.com, (accessed September 5, 2013).
23. London Metal Exchange website, www.lme.com, (accessed September 2, 2013).
24. J. Garthwaite, What the looming lithium squeeze means for electric car batteries, GIAGOM, 2009, http://gigaom.com/cleantech/what-the-looming-lithium-squeeze-means-for-electric-car-batteries, (accessed September 2, 2013).
25. H. Kempf, EV World, 2008, http://www.evworld.com/article.cfm, (accessed October 6, 2013).
26. Z A. C. Tolcin, Zinc statistics and information 2009, http://minerals.er.usgs.gov/minerals/pubs/commodity/zinc/mcs-2009-zinc.pdf, (accessed on October 2, 2013).
27. B. Kanungo, K. M. Parida and B. R. Sant, *Electrochim. Acta*, 1981, **26**, 1157–1167.
28. S. Liompart, L. T. Yu, J. C. Mas, A. Mendiboure, and R. Vignaud, *J. Electrochem. Soc.*, 1990, **137**, 371–377.
29. J. P. Brenet, *J. Power Sources*, 1979, **4**, 183–190.
30. L. Mao, T. Sotomura, K. Nakatsu, K. Nobuharu, D. Zhang and T. Ohsaka, *J. Electrochem. Soc.*, 2002, **149**, A504–A507.
31. G. A. Deborski, *US Pat.*, 4,256,545, 1978.
32. W. A. Armstrong, *US Pat.*, 3,948,684, 1976.
33. J. L. Passaniti and R. B. Dopp, *US Pat.*, 5,378,562, 1995.
34. J. L. Passaniti and R. B. Dopp, *US Pat.*, 5,308,711, 1994.
35. S. Bach, N. Baffier, M. Henry, and J. Livage, *French Pat.*, 2,659,075, 1991.
36. W. H. Hoge, *US Pat.*, 4,906,535, 1988.
37. W. H. Hoge, *US Pat.*, 5,032,473, 1991.
38. R. B. Dopp, *US Pat.*, 5,656,395, 1997.
39. N. Golovin, *US Pat.*, 6,444,609, 2002.
40. N. Golovin, *US Pat.*, 6,428,931, 2002.
41. R. Ndzebet, *US Pat.*, 6,780,347, 2002.
42. N. Koshiba, H. Hayakawa, K. Momose and A. Ohta, *US Pat.*, 4,595,643, 1984.
43. R. L. Sassa, D. Zuckerbrod, W. Buerger and J. E. Bacino, *US Pat.*, 6,921,606, 2005.
44. F. Sun and F. Wang, *US Pat.*, 6,248,476, 2001.
45. L. Borbelly and F. Wang, *US Pat.*, 6,248,476, 1990.
46. J. J. McEvoy, *US Pat.*, 4,585,710, 1986.
47. G. S. Kelsey, P. Chalilpoyil, P. D. Trainer, A. Kaplan, G. Cintra, V. H. Vu and J. D. Sillesky, *US Pat.*, 6,207,322, 1998.
48. I. Roche, E. Chainet, M. Chatanetet and J. Vondrak, *J. Phys. Chem. C*, 2007, **111**, 1434–1443.
49. I. Roche, E. Chainet, M. Chatanetet and J. Vondrak, *J. Appl. Electrochem.*, 2008, **38**, 1195–1201.
50. J. Vondrak, J. B. Klapste, J. Velicka, M. Sedlarikova, J. Reiter, I. Roche, E. Chainet, J. F. Fauvarque and M. Chatenet, *J. New Mater. Electrochem. Syst.*, 2005, **8**, 209–212.
51. P. Zoltowski, D. M. Drazic and L. Vorkapic, *J. Appl. Electrochem.*, 1973, **3**, 271–283.

52. P. Bezdička, T. Grygar, B. Klápště and J. Vondrak, *Electrochim. Acta*, 1999, **45**, 913–920.

53. B. Klápště, J. Vondrák and J. Velická, *Electrochim. Acta*, 2002, **47**, 2365–2369.

54. J. Vondrák, B. Klápště, J. Velická, M. Sedlarikova and R. Cerny, *J. Electrochem. Solids*, 2003, **8**, 44–47.

55. J. Vondrák, B. Klápště, J. Velická, M. Sedlarikova, V. Novak and J. Reiter, *J. New Mater. Electrochem. Syst.*, 2005, **8**, 1–4.

56. J. Vondrack, M. Sedlarykova and V. Novak, *J. New Mater. Electrochem. Syst.*, 1998, **1**, 25–30.

57. E. A. Ticianelli, F. H. B. Lima and F. M. Calegaro, *209th ECS Spring Meeting*, Report 301, Denver, CO, May 7–12, 2006.

58. J. Yang and J. J. Xi, *Electrochem. Commun.*, 2003, **5**, 306–311.

59. K. Matsuki and H. Kamada, *Electrochim. Acta*, 1986, **31**, 13–18.

60. D. Sepa, M. Vojnovic and A. Damjanovic, *Electrochim. Acta*, 1970, **15**, 1355–1366.

61. F. H. B. Lima, C. D. Sanches and E. A. Ticianelli, *J. Electrochem. Soc.*, 2005, **152**, A1466–A1473.

62. C.-Y. Wu, P.-W. Wu and Y.-M. Lin, *J. Electrochem. Soc.*, 2007, **154**, B1059–B1062.

63. Y. Yang and Y. Zhou, *J. Electroanal. Chem.*, 1995, **397**, 271–278.

64. K. Kinoshita, *Carbon*, John Willey & Sons Inc., New York, 1988, p. 533.

65. C.-C. Chang and T.-C. Wen, *Mater. Chem. Phys.*, 1997, **47**, 203–210.

66. H. Meng and P. K. Shen, *Electrochem. Commun.*, 2006, **8**, 588–594.

67. K. Hohne and K. Mund, *US Pat.*, 3,94,0510, 1973.

68. E. Curelop, S. Lu, S. McDevitt and J. Sunstrom, *US Pat.*, 6,632,557, 1999.

69. J. Goldstein, N. Naimer, E. Khasin and A. Brokman, *US Pat.*, 5,190,833, 1993.

70. Z. Zhong, *US Pat.*, 6,383,675, 2002.

71. R. Bhaskara, *US Pat.*, 5,053,375, 1990.

72. E. Ndzebet, *US Pat.*, 2003146414, 2001.

73. S. R. Ovshinsky, C. Fierro, B. Reichman, W. Mays, J. Strebe, M. A. Fetcenko, A. Zallen and T. Hicks, *US Pat.*, 7,097,933, 2003.

74. M. Maya, C. Oreccia, M. Strano, P. Tosco and M. Vanni, *Electrochim. Acta*, 2000, **46**, 423–432.

75. C. Mocci and S. Trassatti, *J. Mol. Catal. A Chem.*, 2003, **204**, 713–720.

76. X.-Y. Xie, Z.-F. Ma, and X.-X. Ma, *J. Electrochem. Soc.*, 2007, **154**, B733–B738.

77. C. Mocci, A. C. Tavares, S. Trassati, P. Tosco, M. Manzoli and F. Boccuzzi, *Energy and Electrochem: Processes for a Cleaner Environ*, ed. E. W. Boomam, C. M. Doyle, C. Comnellis and J. Winnick, The Electrochemical Society, Pennington, NJ, 2001, V. 2001–2023, p. 363.

78. S. Lj. Gojkovic, S. Gupta and R. F. Savinell, *J. Electrochem. Soc.*, 1998, **145**, 3493–3499.

79. W. Yao and T. Tsai, *US Pat.*, 6,368,751, 2002.

80. N. Naimer, E. E. Khasin, J. R. Goldstein and J. Sassen, *US Pat.*, 5,242,765, 1993.

81. R. Jiang, L. Xu, S. Dong and S. Dong, *Fenxi Huaxue*, 1985, **13**, 270–275.

82. L. Zhang, J. Zhang, D. P. Wilkinson and H. Wang, *J. Power Sources*, 2006, **156**, 171–182.

83. D. Chu and R. Jiang, *Solid State Ionics*, 2002, **148**, 591–599.

84. F. C. Anson, *Anal. Chem.*, 1980, **52**, 1192–1198.

85. J. P. Collman, P. Denisevich, Y. Konai, M. Marrocco, C. Koval and F. C. Anson, *J. Am. Chem. Soc.*, 1980, **102**, 6027–6036.

86. M. Lefèvre, J. P. Dodelet and P. Bertrand, *J. Phys. Chem. B*, 2000, **104**, 11238–11247.

87. C. N. Shi and F. C. Anson, *Inorg. Chem.*, 1990, **29**, 4298–4305.

88. H. Jahnke, M. Schonborn and G. Zimmermann, *Top. Curr. Chem.*, 1976, **61**, 133–181.

89. S. Lj. Gojkovic, S. Gupta and R. F. Savinell, *J. Electroanal. Chem.*, 1999, **462**, 63–72.

90. A. Li, H. Wang, W. Qu, Q. Ren, V. M. Schmidt and L. Huang, *217th ECS Meeting*, Vancouver, April 25–30, 2010, Abstract #760.

91. I. Iliev, S. Gamburzev and A. Kaisheva, *J. Power Sources*, 1986, **17**, 345–352.

92. S. Gambuzev, A. Kaisheva and I. Iliev, *29th ISE Meeting*, Budapest, August 28 to September 2, 1978, Ext., Abstract #106, Part II.
93. R. J. van Veen and C. Visser, *Electrochim. Acta*, 1979, **24**, 921–928.
94. V. S. Bagotsky, M. R. Tarasevich, K. A. Radyushkina, O. A. Levina and S. I. Andrusyova, *J. Power Sources*, 1978, **2**, 233–240.
95. K. Wiesener and A. Fuhrmann, *Z. Phys. Chem. (Leipzig)*, 1980, **261**, 411.
96. Y. Kiros, *Int. J. Electrochem. Sci.*, 2007, **2**, 285–300.
97. F. Solomon, Y. Genodman and J. Irizarry, *US Pat.*, 4,877,694, 1989.
98. J. H. Zagal, F. Bedioui and J. P. Dodelet, *N₄-Macrocyclic Metal Complexes*, Springer, New York, 2006, p. 801.
99. N. Miura, H. Horiuchi, Y. Shimitzu and N. Yamazoe, *Nippon Kagaku Kaishi*, 1987, **4**, 617–621.
100. A. C. Tseung and S. Jasem, *Electrochim. Acta*, 1977, **22**, 31–34.
101. W. J. King, A. C. Tseung, J. R. Gancedo, V. de la Garza Guadarrama, H. Nguyen and P. Chartier, *Electrochim. Acta*, 1994, **19**, 485–491.
102. E. Rios, J. L. Gautier, G. Poillerat and P. Chartier, *Electrochim. Acta*, 1998, **44**, 1491–1497.
103. J. L. Gautier, J. F. Marco, M. Gracia, J. R. Gancedo, V. de la Garza Guadarrama, H. Nguyen-Cong and P. Chartier, *Electrochim. Acta*, 2002, **48**, 119–125.
104. H. Nguyen, V. de la Garza Guadarrama, J. L. Gautier and P. Cartier, *Electrocihim. Acta*, 2003, **48**, 2389–2395.
105. R. N. Singh, B. Lal, M. Malviya and T. Seiyama, *Electrochim. Acta*, 2004, **49**, 4605–4612.
106. N. Miura, Y. Shimizu, N. Yamazoe and T. Seiyama, *Nippon Kagaku Kaishi*, 1985, **4**, 644–650.
107. N. Miura, Y. Shimizu, N. Yamazoe and N. Yamazoe, *Nippon Kagaku Kaishi*, 1986, 751–755.
108. Y. Shimizu, K. Uemura, H. Matsuda, N. Miura and N. Yamazoe, *J. Electrochem. Soc.*, 1990, **137**, 3430–3433.
109. Y. Teraoka, H. Kakebayashi, I. Moriguchi and S. Kagawa, *Chem. Lett.*, 1991, **88**, 673–676.
110. T. Hyodo, N. Miura and N. Yamazoe, *Mater. Res. Soc. Sump. Proc.*, 1995, **393**, 79–84.
111. T. Hyodo, M. Hayashi, N. Yamazoe and N. Yamazoe, *J. Electrochem. Soc.*, 1996, **143**, L266–L267.
112. T. Hyodo, M. Hayashi and N. Yamazoe, *J. Appl. Electrochem.*, 1997, **27**, 745–746.
113. T. Hyodo, Y. Shimizu, N. Miura and N. Yamazoe, *Denki Kagaku*, 1994, **62**, 158–164.
114. Y. Matumoto, H. Yoneyama and H. Tamura, *Chem. Lett.*, 1975, 661–662.
115. J. Lamminen, J. Kivisaari, M. J. Lampinen, M. Viitanen and J. Vuorisalo, *J. Electrochem. Soc.*, 1991, **138**, 905–908.
116. M. Hayashi, H. Uemura, K. Shimanoe, N. Miura and N. Yamazoe, *J. Electrochem. Solid-State Lett.*, 1998, **1**, 268–270.
117. Y. Masayoshi, S. Kengo, Y. Teraoka, N. Yamazoe, *Catalysis Today*, 2007, **126**, 313–319, 235.
118. A. C. Tseung and H. L. Bevan, *Electroanal. Chem. Interf. Electrochem.*, 1973, **45**, 429–438.
119. T. Kudo, H, Obayshi and M. Yoshide, *J. Electrochem. Soc.*, 1977, **124**, 321–325.
120. T. Kudo, H. Obayshi and T. Gejo, *J. Electrochem. Soc.*, 1975, **122**, 159–163.
121. X. Wang, P. J. Sebastian, M. A. Smit, H. Yang and S. A. Camboa, 2003, **124**, 278–284.
122. J. Ponce, J.-L. Rehspringer, G. Poillerat and J. L. Gautier, *Electrochim. Acta*, 2001, **46**, 3373–3380.
123. R. N. Singh and B. Lal, *Int. J. Hydrogen Energy*, 2002, **27**, 45–55.

124. S. K. Tiwari, P. Chartier and R. N. Singh, *J. Chem. Soc. Faraday Trans.*, 1995, **91**, 1871–1875.
125. H. Arai, S. Muller and O. Haas, *J. Electrochem. Soc.*, 2000, **147**, 3584–3591.
126. K. L. K. Yeung and A. C. Tseung, *J. Electrohem. Soc.*, 1978, **125**, 878–882.
127. J. B. Goodenough, R. Manoharan and M. Paranthaman, *J. Am. Chem. Soc.*, 1990, **112**, 2076–2082.
128. J. Prakash, D. Tryk and E. Yeager, *J. Power Sources*, 1990, **29**, 413–422.
129. Y.-K. Shun and C.-L. Lou, *US Pat.*, 6,127061, 2000.
130. A. P. O. Blanchart and C. J. E. Van Der Poorten, *US Pat.*, 4,696,872, 1987.
131. G. Henry, *US Pat.*, 4,333,993, 1982.
132. B. G. Demczyk and C. T. Liu, *J. Electrochem. Soc.*, 1982, **129**, 1159–1164.
133. E. S. Buzzet, *US Pat.*, 3,977,901, 1976.
134. J. Chottiner, *US Pat.*, 4,152,489, 1979.
135. C.-T. Liu and J. F. Jackovitz, *US Pat.*, 5,318,862, 1994.
136. A. Gibeney and D. Zuckerbrod, *Power Sources*, ed. J. Thomson, Academic, New York, 1983, vol. 9, p. 143.
137. D. Tryk, W. Alfred and E. Yeager, *First Report for the Period October 9, 1980 to April 1, 1983, Prepared by Western Reserve University*, Subcontract 1377901 for Lawrence Livermore National Laboratory, Livermore, CA, July 15, 1983.
138. C.-T. Liu and B. Demczyk, *US Pat.*, 4,444,852, 1984.
139. V. R. Shepard, Y. G. Smalley and R. D. Bentz, *US Pat.*, 5306579, 1994.
140. B. Trugve, *US Pat.*, 20071666602A1, 2007.
141. J. Coetzer and M. M. Thackeray, *US Pat.*, 4,288,506, 1981.
142. Z. Zhimin, *US Pat.*, 6,383,675, 2002.
143. Y. Shimizu, H. Matsude, A. Nemoto, N. Miura and N. Yamazoe, *Prog. Batteries Batt. Mater.*, 1993, **12**, 108.
144. A. N. Jain, S. K. Tiwari, P. Chartier and R. N. Singh, *Faraday Trans.*, 1995, **91**, 1871–1875.
145. T. Hyodo, T. Shmizu, N. Miura and N. Yamazoe, *Denki Kagaku*, 1994, **62**, 158–163.
146. Y. Shimizu, A. Nemoto, T. Hyodo, N. Miura and N. Yamazoe, *Denki Kagaku*, 1993, **61**, 1458–1460.
147. Y. Shimizu, H. Matsuda, N. Miura and N. Yamazoe, *Chem. Lett.*, 1992, **21**, 1033–1036.
148. T. Lippert, M. J. Montenegro, A. Wokaun, A. Weidenkaff, S. Muller, P. R. Willmott and A. Wokaun, *Prog. Solid State Chem.*, 2007, **35**, 221–231.
149. Z. Chen, A. Yu, D. Higgins, H. Li and H. J. Wang, *Nano Lett.*, 2012, **124**, 1946–1952.
150. E. Deis, F. Holzer and O. Haas, *Electrochem. Acta.*, 2012, **47**, 3995–4010.
151. A. A. Mohamad, *J. Power Sources*, 2006, **159**, 752–757.
152. H. H. Chen and C. S. Tan, *J. Power Sources*, 2006, **162**, 1431–1436.
153. J. J. Xu, H. Ye and J. Huang, *Electrochem. Commun.*, 2005, **7**, 1309–1317.
154. Q. H. Tian, L. Z. Cheng, J. X. Liu and X. Y. Guo, *Adv. Mater. Res.*, 2012, **416**, 35–40.
155. N. Shaigan, W. Qu and T. Takeda, *ECS Trans.*, 2010, **28**, 35–44.
156. J.-F. Drillet, M. Adam, S. Barg, A. Herter, D. Koch, V. M. Schmidt and M. Wilhelml, *ECS Trans.*, 2010, **28**, 13–17.
157. S. W. Lee, K. Sathiyanaraanan, S. W. Eom and M. S. Yun, *J. Power Sources*, 2006, **160**, 1436–1439.
158. X. G. Zhang, *J. Power Sources*, 2006, **163**, 591–597.
159. P. N. Ross, *US Pat.*, 4,842,963, 1989.
160. A. Lewandowski, K. Skorupska and J. Malinska, *Solid State Ionics*, 2000, **133**, 265–271.
161. J. Fauvarquem, S. Guinot and N. Bouzir, *Electrochim. Acta*, 1995, **40**, 2449–2455.
162. A. Appleby and M. Jacquier, *J. Power Sources*, 1976, **1**, 17–34.
163. L. Dewi, K. Oyaizu, H. Nishide and T. Eishun, *J. Power Sources*, 2003, **115**, 149–152.
164. H. Saputra, R. Othman and A. G. E. Satjipto, *J. Mater. Sci.*, 2011, **367**, 152–157.

165. MagPower Systems Inc., http://www.magpowersystems.com, (accessed September 16, 2013).
166. Z. Lei, H. Liu, J. Zhang, D. Ghosh, J. Jung, Y. Chung and B. Downing, *WO Pat.*, WO2007/112563, 2007.
167. S. Venkatesan, H. Wang, B. Aladjov, S. Dhar and S. R. Ovshinsky, *US Pat.*, 6,835,489 B2, 2002.
168. W. Li, C. Li, C. Zhou, H. Ma and J. Chen, *Angew. Chem. Int. Ed.*, 2006, **118**, 6155–6158.
169. S. Sathyanarayana and N. Munichandraiah, *J. Appl. Electrochem.*, 1981, **11**, 33–39.
170. C. Liebenow, *J. Appl. Electrochem.*, 1997, **27**, 221–225.
171. C. Liebenow, Z. Yang and P. Lobitz, *Electrochem. Commun.*, 2000, **2**, 641–645.
172. Y.-S. Guo, J. Yang, J. L. Wang and J. Wang, *Electrochem. Commun.*, 2010, **12**, 1671–1673.
173. D. Aurbach, H. Gizbar, A. Schechter, O. Chusid, H. E. Gottlieb, Y. Gofer and I. Goldberg, *J. Electrochem. Soc.*, 2002, **149**, A115–A121.
174. Y.-S. Guo, F. Zhang and J. Yang, *Energy Environ. Sci.*, 2012, **5**, 9100–9106.
175. Z. Jiang, R. Sirotina and N. Iltchev, *US Pat.*, 8,211,578, 2012.
176. T. Khoo, A. Somers, A. A. J. Torriero, D. R. MacFarlane, P. C. Howlett and M. Forsyth, *Electrochim. Acta*, 2013, **87**, 701–708.
177. T. Zhang, N. Imanishi, N. Sammes, D. R. MacFarlane, P.C. Howlett and M. Forsyth, *J. Electrochem. Soc.*, 2010, **157**, A214–A218.
178. G. B. Appetecchi and S. Passerini, *Electrochim. Acta*, 2000, **45**, 2139.
179. Y. Ma, N. Li, D. Li, M. Zhang and X. Huang, *J. Power Sources*, 2011, **196**, 2346–2350.
180. W. Li, C. Li, C. Zhou, H. Ma and J. Chen, *Angew. Chem. Int. Ed.*, 2006, **45**, 6009–6012.
181. P. C. Howlett, W. Neil, T. Khoo, J. Sun, M. Forsyth1 and D. R. MacFarlane, *Isr. J. Chem.*, 2008, **48**, 313–318.
182. T. Khoo, P. C. Howlett, M. Tsagouria, D. R. MacFarlane and M. Forsyth, *Electrochim. Acta*, 2011, 58, 583–588.
183. M. Tamez and J. H. Yu, *J. Chem. Educ.*, 2007, **84**, 1936A–1941A.
184. S. Yang and H. Knicle, *J. Power Sources*, 2002, **112**, 162–173.
185. W. H. Hoge, *US Pat.*, 4,885,217, 1987.
186. R. Bhaskara, *US Pat.*, 5,225,291, 1990.
187. R. P. Hamlen and P. F. Connolly, *US Pat.*, 4,626,482, 1985.
188. R. P. Hamlen, T. Zoltner, W. Kobasz, M. V. Rose and Alupower Inc., *US Pat.*, 4,745,529, 1986.
189. Yardney Technical Products Inc., http://www.yardney.com, (accessed September 16, 2013).
190. L. W. Niedrach and H. Alford, *J. Electrochem. Soc.*, 1965, **112**, 117–124.
191. A. A. Mohamad, *Corros. Sci.*, 2008, **50**, 3475–3479.
192. M. Nestoridia, D. Pletchera and R. J. K. Wood, *J. Power Sources*, 2008, **178**, 445–455.
193. I. Smoljko, S. Gudić, N. Kuzmanić and M. Kliški, *J. Appl. Electrochem.*, 2012, **42**, 969–977.
194. T. A. Yager and E. Kruglick, *US Pat.*, 20,120,251,897, 2012.
195. K. M. Abraham and Z. Jiang, *J. Electrochem. Soc.*, 1996, **143**, 6–12.
196. M.-K. Song, S. Park, F. M. Alamgir, J. Cho, and M. Liu, *Mater. Sci and Eng. R*, 2011, **72**, 203–252.
197. D. Capsony, M. Bini, P. Mustarelli, E. Quartarone and P. Mustarelli, *J. Power Sources*, 2012, **220**, 252–263.
198. A. Kraystberg and Y. Ein-Eli, *J. Power Sources*, 2011, **196**, 8860–8893.
199. P. Richard and Z. Xiangwu, *J. Power Sources*, 2011, **196**, 4436–4444.
200. B. Ritcher, D. Goldston, and G. Crabtree, American Physical Society website: http://www.aps.org/energyefficiencyreport/report/aps-energyreport.pdf, Energy Future: Think Efficiency, 1, 2008 (accessed October 2, 2013)

201. S. J. Visco, E. Nimon and B. Katz, *The 12th International Meeting on Lithium Batteries*, Nara, June 27 to July 2, 2004, Abstract #53.
202. S. J. Visco, E. Nimon and B. Katz, *The 210th Electrochemical Society Meeting*, Cancun, 2006, Abstract #0389.
203. S. J. Visco, Y. S. Nimon and B. D. Katz, *US Pat.*, 7,282,302, B2, 2007.
204. S. Visco, L. De Jonghe, E. Nimon, A. Petrov and K. Pridatko, *WO Pat.*, WO/2010/005,686, 2010.
205. A. Dobley, C. Morein, R. Roark and M. Abraham, *42nd Power Source Conference*, Philadelphia, PA, June 12–15, 2006.
206. X. L. Ji, K. T. Lee and L. F. Nazar, *Nat. Mater.*, 2009, **8**, 500.
207. A. Dobley, C. Morein, T. Dillon and C. Morein, *44th Power Sources Conference*, Las Vegas, NV, June 14–17, 2010, pp. 143–145.
208. J. A. Zhang, W. Xu, X. H. Li and W. Liu, *J. Electrochem. Soc.*, 2010, **57**, A940–A946.
209. J. Kumar and B. Kumar, *J. Power Sources*, 2009, **194**, 1113–1119.
210. C. O. Laoire, S. Mukerjee, M. A. Hendrickson, E. J. Plichta and M. A. Hendrickson, *J. Phys. Chem. C*, 2009, **113**, 20127–20134.
211. C. O. Laoire, S. Mukerjee and K. M. Abraham, *J. Phys. Chem. C*, 2010, **114**, 9178–9186.
212. Z. Peng, S. A. Freunberger, P. G. Bruce, Y. Chen, V. Giordani, F. Bardé, P. Novák, D. Graham, J.-M. Tarascon and P. G. Bruce, *Angew. Chem. Int. Ed.*, 2011, **50**, 6351–6355.
213. M. Winter and R. J. Brodd, *Chem. Rev.*, 2004, **104**, 4245–4269.
214. F. Mizuno, *Symposium on Energy Storage Beyond Lithium Ion III: Materials Perspectives*, Oak Ridge National Laboratory, Oak Ridge, TN, 2010.
215. A. Debart, J. Bao and P. G. Bruce, *J. Power Sources*, 2007, **174**, 1177–1182.
216. A. Debart, M. Holzapfel and P. Bruce, *J. Am. Chem. Soc.*, 2006, **128**, 1390–1393.
217. H. Cheng and K. Scott, *J. Power Sources*, 2010, **195**, 1370–1376.
218. C. Tran, X. Q. Yang and D. Y. Qu, *J. Power Sources*, 2010, **195**, 2057–2063.
219. H. Yang, P. He and Y. Y. Xia, *Electrochem. Commun.*, 2009, **11**, 1127–1130.
220. K. F. Blurton and A. F. Sammlls, *J. Power Sources*, 1979, **4**, 263–279.
221. J. S. Lee, S. T. Kim, R. Cao, N.-S. Choi, M. Liu, K. T. Lee and J. Cho, *Adv. Energy Mater.*, 2011, **1**, 34–38.
222. J. P. Zheng, R. Y. Liang, M. Hendickson and E. J. Plichta, *J. Electrochem. Soc.*, 2008, **155**, A432–A437.
223. G. Q. Zhang, J. P. Zheng, R. Liang, C. Zhang, B. Wang, M. Hendrickson and E. J. Plichta, *J. Electrochem. Soc.*, 2011, **157**, A953–A956.
224. R. E. Williford and J. G. Zhang, *J. Power Sources*, 2009, **194**, 1164–1170.
225. J. Xiao, D. H. Mei, X. Li, W. Xu, D. Wang, G. L. Graff, W. D. Bennett, Z. Nie, L. V. Saraf, I. A. Aksay, J. Liu and J.-G. Zhang, *Nano Lett.*, 2011, **11**, 5071–5078.
226. Y. L. Li, J. J. Wang, D. Geng, R. Li and X. Sun, *Chem. Commun.*, 2011, **47**, 9438–9440.
227. P. Kichambare, J. Kumar, B. Kumar and S. Rodrigues, *J. Power Sources*, 2011, **196**, 3310–3316.
228. Y. L. Li, J. J. Wang, J. Liu, D. Geng, J. Yang, R. Li and X. Sun, *Electrochem. Commun.*, 2011, **13**, 668–672.
229. Y. L. Li, J. J. Wang, X. Li, D. Geng, M. N. Banis, R. Li and X. L. Sun, *Electrochem. Commun.*, 2012, **18**, 12–15.
230. Y. L. Li, J. J. Wang, D. Geng, M. N. Banis, Y. Tang, D. Wang, R. Li, T.-K. Sham and X. Sun, *J. Mater. Chem.*, 2012, **22**, 20170–20174.
231. T. Ogasawara, A. Debart, P. G. Bruce, P. Novák and P. G. Bruce, *J. Am. Chem. Soc.*, 2006, **128**, 1390–1393.
232. V. M. B. Crisostomo, J. K. Ngala, S. L. Suib, A. Dobley, C. Morein, C. H. Chen and X. Shen, *Chem. Mater.*, 2007, **19**, 1832–1839.

233. D. Zhang, R. S. Li, T. Huang, S. Alia, A. Dobley, C. Morein, C.-H. Chen, X. Shen and S. L. Suib, *J. Power Sources*, 2010, **195**, 1202–1206.
234. J. Read, *J. Electrochem. Soc.*, 2002, **149**, A1190–A1195.
235. D. Zhang, Z. H. Fu, A. S. Yu et al., *Electrochem. Soc.*, 2010, **157**, A862–A865.
236. L. Wang, X. Zhao, Y. H. Lu, M. W. Xu, D. W. Zhang, R. S. Ruoff, K. J. Stevenson and J. B. Goodenough, *J. Electrochem. Soc.*, 2011, **158**, A1879–A1882.
237. E. M. Benbow, S. P. Kelly, J. W. Reutenauer and S. L. Suib, *J. Phys. Chem. C*, 2011, **115**, 22009–22017.
238. J. K. Ngala, S. Alia, A. Dobley, V. M. B. Crisostomo and S. L. Suib, *Chem. Mater.*, 2007, **19**, 229–234.
239. J. Xiao, W. Xu, D. Y. Wang and J. G. Zhang, *J. Electrochem. Soc.*, 2010, **157**, A294–A297.
240. L. Trahey, C. S. Johnson, J. T. Vaughey, S. H. Kang, L. J. Hardwick, S. A. Freunberger, P. G. Bruce and M. M. Thackeray, *Electrochem. Solid State Lett.*, 2011, **14**, A64–A66.
241. Y. M. Cui, Z. Y. Wen and Y. Liu, *Energy Environ. Sci.*, 2011, **4**, 4727–4734.
242. J. Ming, Y. Wu and Y.-K. Sun, *Nanoscale Res. Lett.*, 2012, **7**, 1–4.
243. W.-M. Liu, T.-T. Gao and Y. Yang, *Phys. Chem. Chem. Phys.*, 2013, **15**, 15806–15810.
244. M.-M. Augustin and O. Yezerska, J. Derendorf, M. Knipper, D. Fenske, T. Plaggenborg and J. Parisi, *ECS Trans.*, 2013, **4**, 1–10.
245. G. Zhao, Z. Xu and K. Sun, *J. Mater. Chem. A*, 2013, **41**, 12862–12867.
246. J. Ming, Y. Wu, Y. J. B. Park, J. K. Lee, F. Zhao and K. Sun, *Nanoscale*, 2013, **5**, 10390–10396.
247. L. Zhang, S. Zhang, K. Zhang, G. Xu, X. He, S. Dong, Z. Liu, C. Huang, L. Gu and G. Cui, *Chem. Commun.*, 2013, **49**, 3540–3543.
248. Y. C. Lu, H. A. Gasteiger, E. Crumlin, R. McGuire Jr. and Y. Shao-Horn, *J. Electrochem. Soc.*, 2011, **157**, A1016–A1025.
249. Y. C. Lu, H. A. Gasteiger, M. C. Parent, V. Chiloyan and Y. Shao-Horn, *Electrochem. Solid State Lett.*, 2010, **13**, A69–A72.
250. Y. C. Lu, D. G. Kwabi, K. P. C. Yao and Y. Shao-Horn, *Energy Environ. Sci.*, 2011, **4**, 2999–3007.
251. Y. C. Lu, Z. C. Xu, H. A. Gasteiger, S. Chen, K. Hamad-Schifferli and Y. Shao-Horn, *J. Am. Chem. Soc.*, 2010, **132**, 12170–12171.
252. S. Lee, S. L. Zhu, C. C. Milleville, C. Y. Lee, P. W. Chen, K. J. Takeuchi, E. S. Takeuchi and A. C. Marschilok, *Electrochem. Solid State Lett.*, 2010, **13**, A162–A164.
253. A. C. Marschilok, S. L. Zhu, C. C. Milleville, S. H. Lee, E. S. Takeuchi and K. J. Takeuch, *J. Electrochem. Soc.*, 2011, **158**, A223–A226.
254. A. K. Thapa, K. Saimen and T. Ishihara, *Electrochem. Solid State Lett.*, 2010, **13**, A165–A167.
255. D. Zhu, L. Zhang, M. Song, X. F. Wang and Y. G. Chen, *Chem. Commun.*, 2013, **49**, 9573–9575.
256. A. K. Thapa and T. Ishihara, *J. Power Sources*, 2011, **196**, 7016–7020.
257. H. Cheng and K. Scott, *Appl. Cat. B Environ.*, 2011, **108**, 140–151.
258. Z. Q. Peng, S. A. Freunberger, Y. H. Cheng and P. G. Bruce, *Science*, 2012, **337**, 563–566.
259. X. M. Ren, S. S. Zhang, D. T. Tran and J. Read, *J. Mater. Chem.*, 2011, **21**, 10118–10125.
260. C. Xia, M. Wlystaletzko, C. Xia, M. Waletzko and K. Peppler, *J. Phys. Chem. C*, 2013, **117**, 19897–19904.
261. S. A. Freunberger, Y. H. Chen, N. E. Drewett, L. J. Hardwick, F. Bardé and G. Bruce, *Angew. Chem. Int. Ed.*, 2011, **50**, 8609–8613.
262. W. Xu, V. V. Viswanathan, D. Y. Wang, S. A. Towne, J. Xiao, Z. M. Nie, D. H. Hu and J. G. Zhang, *J. Power Sources*, 2011, **196**, 3894–3899.
263. V. V. Giordani, S. A. Freunberger, P. G. Bruce, J. M. Tarascon and D. Larcher, *Electrochem. Solid State Lett.*, 2010, **13**, A180–A188.

264. Y. Cui, Z. Wen, S. Sun, Y. Lu and J. Jin, *Solid State Ionics*, 2012, **225**, 598–603.
265. B. D. McCloskey, R. Scheffler, A. Speidel, D. S. Bethune, R. M. Shelby and A. C. Luntz, *J. Am. Chem. Soc.*, 2011, **133**, 18038–18041.
266. B. D. McCloskey, D. S. Bethune, R. M. Shelby, G. Girishkumar and A. C. Luntz, *J. Phys. Chem. Lett.*, 2011, **2**, 1161–1166.
267. P. Albertus, G. Girishkumar, B. McCloskey, R. S. Sanchez-Carrera, B. Kozinsky, J. Christensen and A. C. Luntz, *J. Electrochem. Soc.*, 2011, **158**, A848–A851.
268. D. Radin, J. F. Rodriguez, F. Tian and D. J. Siegel, *J. Am. Chem. Soc.*, 2012, **134**, 1093–1103.
269. A. Debart, A. J. Paterson, J. Bao and P. G. Bruce, *Angew. Chem. Int. Ed.*, 2008, **120**, 4597–4600.
270. C.H. San, C.W. Hong *J. Electroch. Soc.* 2012, **159**, K116–K121.
271. D. Beattie, D. M. Manolescu and S. L. Blair, *J. Electrochem. Soc.*, 2009, **156**, A44–A47.
272. J. Xiao, D. H. Wang, W. Xu, D. Y. Wang, R. E. Williford, J. Liu and J. G. Zhang, *J. Electrochem. Soc.*, 2010, **157**, A487–A492.
273. L. Jin, L. P. Xu and C. Morein, *Adv. Funct. Mater.*, 2010, **20**, 3373–3382.
274. R. R. Mitchell, B. M. Gallant, C. V. Thompson and Y. Shao-Horn, *Energy Environ. Sci.*, 2011, **4**, 2952–2958.
275. Z. L. Wang, D. Xu, J. J. Xu, L. L. Zhang and X. B. Zhang, *Adv. Fun. Mater.*, 2012, **22**, 3699–3705.
276. Y. Yang, Y. Q. Sun, Y. S. Li and Z. W. Fu, *J. Electrochem. Soc.*, 2011, **158**, B1211–B1216.
277. Y. Wang and H. Zhou, *Energy Environ. Sci.*, 2011, **4**, 1704–1707.
278. S. R. Younesi, S. Urbonaite and K. Edstrom, *J. Power Sources*, 2011, **196**, 9835–9838.
279. P. Andrei, J. P. Zheng and E. J. Plichta, *J. Electrochem. Soc.*, 2010, **157**, A1287–A1295.
280. Z. Hong, M. Wei, T. Lan and G. Cao, *Nano Energy*, 2012, **1**, 466–471.
281. K. Xu, *Chem. Rev.*, 2004, **104**, 4303–4417.
282. L. Ji, H. Zheng, A. Ismach, Y. Zhang, Z. Tana, S. Xunb, E. Lina, V. Battagliab, V. Srinivasanb and Y. Zhanga, *Nano Energy*, 2012, **1**, 164–171.
283. X.-H. Yang, P. He and Y.-Y. Xia, *Electrochem. Commun.*, 2003, **150**, A1351–A1356.
284. Z. Zhang, J. Lu, R. S. Assary, R. S. Assary, P. Du, H.-H. Wang, Y.-K. Sun, Y. Qin, K. C. Lau, J. Greeley, H. Iddir, L. A. Curtiss and K. Amine, *J. Phys. Chem. C*, 2011, **115**, 25535–25542.
285 T. Kuboki, T. Okuyama, T. Ohsaki and N. Takami, *J. Power Sources*, 2005, **146**, 766–769.
286. J. S. Wilkes and M. J. Zaworotko, *J. Chem. Soc. Chem. Commun.*, 1992, **13**, 965–967.
287. J. Fuller, R. T. Carlin and R. A. Osteryoung, *J. Electrochem. Soc.*, 1997, **144**, 3881–3886.
288. C. J. Allen, S. Mukerjee and K. M. Abraham, *J. Phys. Chem. Lett.*, 2011, **2**, 2420–2424.
289. H. W. Zimmermann, *Zeitschrift für Physikalische Chemie*, 2011, **225**, 1–13.
290. R. S. Assary, L. A. Curtiss, P. C. Redfern, Z. Zhang and K. Amine, *J. Phys. Chem. C*, 2011, **115**, 12216–12223.
291. F. Mizuno, S. Nakanishi, Y. Kotani, S. Yokoishi and H. Iba, *Electrochemistry*, 2010, **78**, 403–405.
292. S. S. Zhang, D. Foster and J. Read, *Electrochim. Acta*, 2011, **56**, 1283–1287.
293. S. S. Sandhu, J. P. Fellner and G. W. Brutchen, *J. Power Sources*, 2007, **164**, 365–371.
294. Y. Shao, F. Ding, J. Liu, W. X. Park, J.-G. Zhang, Y. Wang and J. Liu, *Adv. Funct. Mater.*, 2013, **23**, 987–1004.
295. D. Aurbach, I. Weissman and A. Zaban, *Electrochim. Acta*, 1994, **39**, 51–71.
296. D. Aurbach, B. Markovsky and M. D. Levi, *J. Power Sources*, 1999, **81**, 95–111.
297. D. Aurbach, A. Zaban and Y. Gofer, *J. Power Sources*, 1995, **54**, 76–84.
298. D. Aurbach, E. Zinigrad, Y. Cohen and H. Teller, *Solid State Ionics*, 2002, **148**, 405–416.
299. R. D. Deshpande, J. C. Li, Y. T. Cheng and M. W. Verbrugge, *J. Electrochem. Soc.*, 2011, **158**, A845–A849.

300. J. K. Stark, Y. Ding and P. A. Kohl, *J. Electrochem. Soc.*, 2011, **158**, A1100–A1105.
301. S. Shiraishi, K. Kanamura and Z. Takehara, *J. Electrochem. Soc.*, 1999, **146**, 1633–1639.
302. T. Osaka, T. Momma, Y. Matsumoto, Y. Uchida, *J. Power Sources*, 1997, **68**, 497–500.
303. T. Osaka, T. Momma, T. Tajima and Y. Matsumoto, *J. Electrochem. Soc.*, 1995, **142**, 1057–1060.
304. R. S. Thompson, D. J. Schroeder, C. M. Lopez, S. Neuhold and J. T. Vaughey, *Electrochem. Commun.*, 2011, **13**, 1369–1372.
305. S. Yoon, J. Lee, S. O. Kim and H.-J. Sohn, *Electrochim. Acta*, 2008, **53**, 2501–2506.
306. M. Ishikawa, M. Morita and Y. Matsuda, *J. Power Sources*, 1997, **68**, 501–505.
307. Y. Matsuda, M. Ishikawa, S. Yoshitake, M. Ishikawa, S. Yoshitake and M. Morita, *J. Power Sources*, 1995, **54**, 301–505.
308. Z. Takehara, Z. Ogumi, Y. Uchimoto, K. Yasude and H. Yoshiba, *J. Power Sources*, 1993, **44**, 377–383.
309. J. Zhang, D. Wang, W. Xu, J. Xiao and R. E. Williford, *J. Power Sources*, 2010, **195**, 4332–4337.
310. S. J. Visco, Y. Nimon, L. De Jonghe, B. Katz and A. Petrov, *US Pat.*, 2007/0037058, 2007.
311. K. Suto, S. Nakanishi, H. Iba and K. Nishio, *15th International Meeting on Lithium Batteries*, Montreal, Canada, 2010.
312. P. Stevens, G. Toussaint and M. Mallouki, *ECS Trans.*, 2010, **28**, 1–12.
313. J. Xiao, J. Hu, D. Wang, D. Hu, W. Xu, G. L. Graff, Z. Nie, J. Liu, J.-G. Zhang, *J. Power Sources*, 2011, **196**, 5674–5678.
314. S. Hasegawa, N. Imanishi and O. Yamamoto, *J. Power Sources*, 2009, **189**, 371–377.
315. T. Zhang, N. Imanishi and S. Nasegawa, *J. Electrochem. Soc.*, 2008, **155**, A965–A969.
316. Y. G. Wang and H. S. Zhou, *Chem. Commun.*, 2010, **46**, 6305–6307.
317. Y. Shimonishia, T. Zhanga, P. Johnson, *J. Power Sources*, 2010, **195**, 6187–6191.
318. S. Hasegawa, N. Imanishi, T. Zhang, J. Xie, A. Hirano, Y. Takeda and O. Yamamoto, *J. Power Sources*, 2009, **189**, 371–377.
319. T. Zhang, N. Imanishi and N. Sammes, *J. Electrochem. Soc.*, 2010, **157**, A214.
320. X. Wang, Y. Hou, R. Holze, A. Hirano, J. Xie, Y. Takeda, O. Yamamoto and N. Sammes, *Sci. Reports*, 2013, **3**, 1–5.
321. J.-Y. Luo, W.-J. Cui and Y.-Y. Xia, *Nat. Chem.*, 2010, **2**, 760–765.
322. Q. Qu, L. Fu, Y. Wu, J. Maier, L. Li, S. Tian, Z. Li and Y. Wu, *Energy Environ. Sci.*, 2011, **4**, 3985–3989.
323. Y. Shimonishia, T. Zhanga, N. Sammes, D. J. Lee, A. Hirano, Y. Takeda, O. Yamamoto and N. Sammes, *J. Power Sources*, 2011, **196**, 5128–5132.
324. D. Aurbach, Y. Talyosef, B. Markovsky, N.-S. Choi1, M. Liu, K. T. Lee and J. Cho, *Electrochim. Acta*, 2004, **50**, 247–254.
325. Y. Shimonishi, T. Zhang, P. Johnson, K. Saitob, Y. Nemotob and M. Arakawab, *J. Power Sources*, 2010, **195**, 6187–6191.
326. S. J. Visco, B. D. Katz, Y. S. Nimon and L. D. Dejonghe, *US Pat.*, 7, 282, 295, 2007.
327. Y. G. Wang and H. S. Zhou, *J. Power Sources*, 2010, **195**, 358–361.
328. Y. G. Wang and H. S. Zhou, *Energy Environ. Sci.*, 2011, **4**, 4994–4999.
329. H. S. Zhou, Y. G. Wang and H. Q. Li, *ChemSusChem*, 2010, **3**, 1009–1019.
330. P. He, Y. G. Wang and H. S. Zhou, *Electrochem. Commun.*, 2010, **12**, 1686–1689.
331. P. He, Y. G. Wang and H. S. Zhou, *J. Power Sources*, 2011, **196**, 5611–5616.
332. E. J. Yoo and H. S. Zhou, *ACS Nano*, 2011, **5**, 3020–3026.
333. D. H. Kim, S. I. Son and H. K. Ryu, *US Pat.*, 2013/0011750, 2013.
334. S. Uesaka, *International WO Pat.*, 2011/052440, 2011.
335. X. J. Chen, A. Shellikeri and Q. Wu, *J. Electrochem. Soc.*, 2013, **160**, A1619–A1623.
336. H. He, W. Niu and Y. Kim, *Electrochim. Acta*, 2012, **67**, 87–31.
337. L. Li, X. Zhao and A. Manthiram, *Electrochem. Commun.*, 2012, **14**, 78–81.

338. L. Wang, X. Zhao, Y. H. Lu, M. W. Xu, D. W. Zhang, R. S. Ruoff, K. J. Stevenson and J. B. Goodenough, *J. Electrochem. Soc.*, 2011, **158**, A1379–A1382.
339. J. G. Zhang, J. Xiao, W. Xu, D. Wang, R. E. Williford and J. Liu, *US Pat.*, US2011/0059355 A1, 2011.
340. J. G. Zhang, J. Xiao, W. Xu, D. Wang and R. E. Williford, *US Pat.*, US 2011/0059364 A, 2011.
341. T. A. Reynolds, D. J. Brose and M. N. Golovin, *US Pat.*, 5,985,475, 1999.
342. J. Zhang, W. Xu and W. Liu, *J. Power Sources*, 2010, 195, 7438–7444.
343. O. Crowther, B. Meyer, M. Morgan and M. Salomon, *J. Power Sources*, 2011, **196**, 1498–1502.
344. O. Crowther, D. Keeny, D. Moureau, B. Meyera, M. Salomona and M. Hendrickson, *J. Power Sources*, 2011, **202**, 347–351.
345. O. Crowther, D. Chua, W. Eppley, B. Meyer, M. Salomon, A. Driedger and M. Morgan, *US Pat.*, US 2011/0177401 A, 2011.

13 Photocatalysts for Hydrogen Production

Jiefang Zhu

CONTENTS

13.1 INTRODUCTION

In the twenty-first century, mankind is facing two serious challenges of energy and environment. The former determines whether we can survive, while the latter is closely related to the quality of our survival. On one hand, the energy consumption has increased about 50% along with the ascending population and gross domestic product (GDP) during the past 20 years. The primary source for the energy consumption is fossil fuels (including petroleum, coal, and natural gas). On the other hand,

CO_2, SO_2, and NO_2 emission from consuming fossil fuels leads to environmental problems like global warming, acid rain, acid fog, and hazy weather. Therefore, in order to realize the sustainable development of human beings, the formidable task is to identify an alternative energy system, which is clean and renewable.

Hydrogen is a storable, efficient, clean, and environmentally friendly energy carrier. It reacts with oxygen giving off heat, with water as the only product. In this way, the chemical energy stored in hydrogen can be released *via* being burned in a combustion engine, or more efficiently *via* being oxidized in a fuel cell. Although hydrogen is abundant on earth, molecular hydrogen is not the stable form in nature. It normally bonds with carbon, oxygen, nitrogen, or sulfur, which means that extracting it from hydrogen sources needs enough driving force (i.e., energy) to break these chemical bonds. Therefore, hydrogen economy needs to deal at least with five issues, namely, energy supply, hydrogen sustainable source, storage, transportation, and cost. Figure 13.1 schematically illustrates hydrogen production pathways using different energy supplies and hydrogen sources.[1] Currently, about 95% of hydrogen originates from fossil fuels, mainly through steam reforming and gasification, which requires large amounts of energy and releases a vast amount of CO_2, exacerbating nonrenewable source consumption and environmental pollution. Because of the high cost of the raw materials and energy consumption,

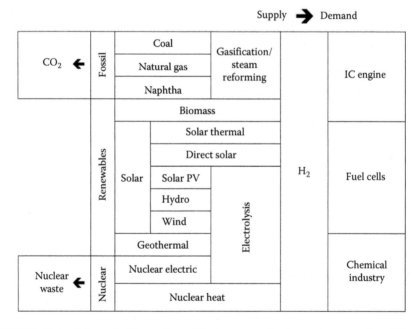

FIGURE 13.1 Schematic illustration of hydrogen production pathways using different energy supplies and hydrogen sources. The main application areas of hydrogen and unwanted emissions from selected processes are indicated on the right- and left-hand side, respectively. (With kind permission from Springer Science+Business Media: Zhu, J., Chakarov, D., and Zäch, M., in: Zang, L. (ed.), *Energy Efficiency and Renewable Energy through Nanotechnology*, Springer, London, 2011, ch. 13, pp. 441–486.)

most hydrogen is just used as a chemical raw material (e.g., for making ammonia, removing sulfur from gasoline, or converting heavy hydrocarbons into gasoline and diesel). In order to make hydrogen economy a viable option for our energy future, hydrogen must be produced from a sustainable carbon-neutral (or carbon-free) source. Favorable feedstocks include H_2O, H_2S, and biomass, which are the hydrogen sources dealt with in this chapter. The first two options are carbon free. The last one can be considered as a carbon-neutral source, since even if the biomass is consumed for hydrogen production with CO_2 emission, CO_2 can be converted back to biomass by photosynthesis within a short timescale (see Figure 13.2). This means that when CO_2 emission from biomass consumption does not exceed nature's capacity to convert CO_2 to biomass, hydrogen production from biomass would not contribute to global warming.[2] Thus, biomass also belongs to a renewable hydrogen source. Ninety-five percent of hydrogen is produced from fossil fuels, and the remaining five percent is presently produced by electrolysis and pyrolysis of water. Although water is a perfectly clean and renewable hydrogen source, electrolysis and pyrolysis of water involve electricity and heat generation, respectively, and their efficiency and cost are not satisfactory at all.

Photocatalytic hydrogen production is an environmentally friendly technique, which produces hydrogen by using light (preferably sunlight) as the only energy input to directly split hydrogen-containing compounds (HCCs). The ultimate goal of this technique is to use sunlight for water splitting. Since solar energy is clean, sustainable, and enormous, achieving this goal will eventually solve our energy shortage. However, sustained hydrogen production from water by sunlight has yet to be achieved. Hydrogen production by ultraviolet (UV) light as a pioneer work can provide fundamental knowledge, maintaining its significance in the renewable energy generation. Hydrogen production can also be applied to other sustainable hydrogen sources than water. Hydrogen production by decomposition of H_2S (carbon-free) pollutant and by reformation of biomass (carbon-neutral) derivatives is an alternative to water splitting. The photoactive materials, which are heroes involved in all the photocatalytic hydrogen production processes mentioned earlier, are called photocatalysts and will be the protagonists in this chapter.

In this chapter, we will briefly introduce the working mechanism and historical development of photocatalytic hydrogen production and the construction of and the requirements for semiconductor photocatalysts in hydrogen production. An overview of various semiconductor photocatalysts will be presented, and their

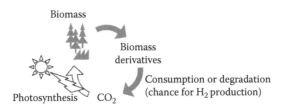

FIGURE 13.2 A carbon-neutral cycle for biomass. (Shimura, K. and Yoshida, H., *Energy Environ. Sci.*, 4, 2467, 2011. Adapted by permission of The Royal Society of Chemistry.)

merits and shortcomings will be analyzed. Furthermore, different photocatalytic hydrogen systems will be described. Finally, we conclude by outlining possible challenges that need to be addressed on the way.

13.2 MECHANISM OF PHOTOCATALYTIC HYDROGEN PRODUCTION

Figure 13.3 illustrates the mechanism of photocatalytic hydrogen production: a semiconductor can absorb a photon with energy equal to or higher than its bandgap (E_g), which is called intrinsic absorption. This absorption causes an electron to jump from the valence band (VB) to the conduction band (CB), leaving a hole (h^+) in the VB and introducing an excess electron (e^-) in the CB. The pair of photogenerated h^+ and e^- can recombine, emitting the absorbed energy as heat and/or light, which leads to low quantum efficiency. The e^-–h^+ pair can also move to the semiconductor's surface to drive oxidation reaction (by h^+) and reduction reaction (by e^-), both of which are called photocatalytic reactions. In photocatalytic hydrogen production from the imaging substance HD (representing a molecular with hydrogen and electron donor), e^- reduces the electron acceptor, H^+, to produce hydrogen, while h^+ oxidizes the electron donor (D^-).

13.3 DEVELOPMENT OF PHOTOCATALYTIC HYDROGEN PRODUCTION

After the initial work of Fujishima and Honda, who found that water splitting could be performed by using a TiO_2 photoanode in 1972,[3] photocatalytic hydrogen production started to attract intense attention. In the twenty-first century, it had limited progress, due to the slow development of photocatalysts. In the last decade, Ti, Nb, Ta, and Ca oxides, oxynitrides, and oxysulfides have shown inspiring achievements in photocatalytic hydrogen production. Meanwhile, energy and environment have become hot topics worldwide. Photocatalytic hydrogen production has come back again as a hot spot of research interest. The rise of nanoscience and nanotechnology is infusing new life into this field.[4] The development of photocatalytic hydrogen

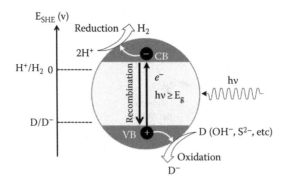

FIGURE 13.3 Illustration of hydrogen production process on photocatalyst.

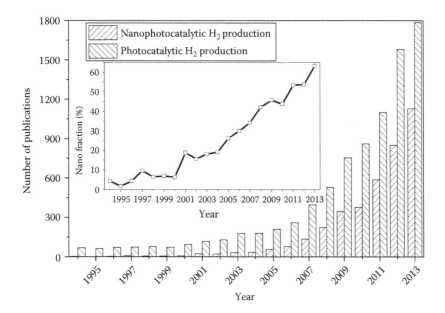

FIGURE 13.4 The number of publications on (nano)photocatalytic H_2 production sorted by year. The inset shows the fraction of publications on photocatalytic H_2 production, which deal with nanoaspects. (Data were collected from the *Web of Science*, and entries until March 9, 2014, have been considered.)

production and the contribution of nanoscience and nanotechnology to it can be illustrated by the publication record shown in Figure 13.4.

13.4 PHOTOCATALYSTS FOR HYDROGEN PRODUCTION

13.4.1 TYPICAL ELEMENTAL MAKEUP OF PHOTOCATALYSTS

Most photocatalysts are composed of metal oxides, metal sulfides, metal nitrides, oxynitrides, oxysulfides, or composites thereof. In these photocatalysts, metal cations show the highest oxidation state with d^0 or d^{10} electron configuration, while nonmetal anions (O, S, and N) are in their lowest states (Figure 13.5). The bottom of the CB of a semiconductor photocatalyst is formed by d and sp orbitals of the metal cations, while the top of the VB in metal oxides consists of O 2p orbitals, which lies around +3 V (vs. normal hydrogen electrode [NHE]). The tops of the VBs in metal sulfides and metal nitrides are contributed by S 3p and N 2p orbitals, respectively, which are both located at more negative potential than that of O 2p orbitals in metal oxides. Alkali, alkaline earth, and some transition metal (Y, La, and Gd) ions can participate in the construction of a crystal structure of semiconductor photocatalysts, but they do not make any contribution to the band structure. Nowadays, more and more ordinary or unexpected elements show up in the makeup of photocatalysts (e.g., C in graphene and g-C_3N_4), and thereby, this makeup element group is getting expanded along with the development of novel photocatalysts.

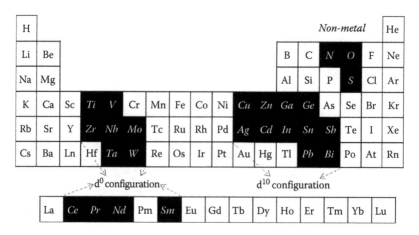

FIGURE 13.5 Element construction of photocatalysts for hydrogen production.

13.4.2 REQUIREMENTS FOR PHOTOCATALYSTS IN HYDROGEN PRODUCTION

The structure and electronic properties of photocatalysts determine their applicability and performance in hydrogen production. As one requirement, photocatalysts, from the thermodynamic point of view, should have a suitable bandgap and band edge positions to run the necessary reactions (including hydrogen production). According to the thermodynamic requirement for photocatalytic hydrogen production from HD, the potential of the CB bottom of the semiconductor photocatalyst should be higher (i.e., more negative) than that of H^+/H_2 ($E_{H^+/H_2} = 0$ V vs. NHE at pH 0), while the potential of the VB top should be lower (i.e., more positive) than that of D/D$^-$ (e.g., $E_{O_2/H_2O} = 1.23$ V vs. NHE at pH 0 for water splitting). Apart from this, if we consider thermodynamic losses and all kinds of resistances at involved steps, overpotential is needed to ensure a certain kinetic requirement. For this reason, although water splitting has its standard potential of -1.23 eV, water splitting photocatalysts need a bandgap larger than 2 eV. Figure 13.6 lists the band edge positions for commonly used semiconductor photocatalysts, in comparison with the reduction potential of H^+/H_2 and the oxidation potential of O_2/H_2O.[5] Another requirement is that the bandgap of the semiconductor photocatalyst should match the light source, that is, the bandgap should be narrow enough to be excited by the incident photons. For example, anatase TiO_2 has its bandgap of 3.2 eV, corresponding to its band edge absorption at 387 nm (λ [nm] = 1240/E_g [eV]), which means it can only absorb photons with wavelengths shorter than 387 nm. Therefore, anatase TiO_2 is not a suitable photocatalyst for use with solar irradiation, since only 4% of the solar irradiation spectrum can be absorbed by TiO_2, as shown in Figure 13.7a. For water splitting, even if all the light up to 400, 600, and 800 nm were utilized (assuming 100% quantum efficiency), the solar conversion efficiency would be only 2%, 16%, and 32%, respectively,[6] as shown in Figure 13.7b. In order to realize water splitting using sunlight, the ideal photocatalyst need have a bandgap smaller than 3.1 eV to utilize visible light ($\lambda \geq 400$ nm), but larger than 2.0 eV in order to include the necessary overpotentials.

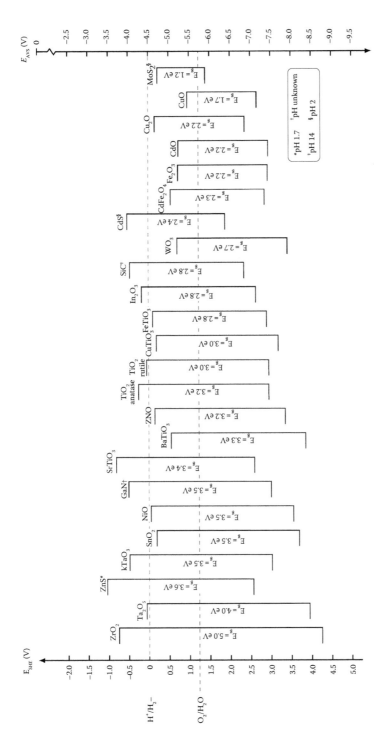

FIGURE 13.6 VB and CB edge positions of various semiconductors with respect to the standard hydrogen electrode (SHE) scale and the vacuum reference energy scale E_{AVS}. All values were reported being tested at pH 1, unless otherwise noted. (With kind permission from Springer Science+Business Media: J. Zhu, in: R. Meyers (ed.), *Encyclopedia of Sustainability Science and Technology*, Springer, New York, 2012, ch. 855, pp. 7881–7901.)

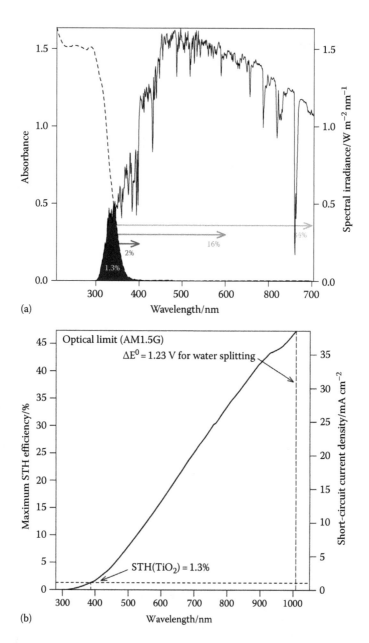

FIGURE 13.7   (a) TiO$_2$ absorption spectrum, solar spectrum, and maximum solar light conversion efficiencies up to 200, 400, and 800 nm for water splitting with 100% quantum efficiency. (b) Dependence of STH efficiency and solar photocurrent density of photoelectrodes on their bandgap absorption edges under AM 1.5 G irradiation.

Apart from the bandgap and band position requirements for photocatalysts used in a specific hydrogen production reaction and under a certain irradiation, the ideal photocatalyst should be highly photocatalytically active, but chemically stable; highly absorptive, but photocorrosion resistant; highly charge carrier transportable; nontoxic; abundant; and cheap. Unfortunately, no photocatalyst has been found to meet all these requirements. We need always keep the balance between these requirements.

13.4.3 Titanium Dioxide (TiO_2)

In the early stage of photocatalysis research,[3] UV-responsive TiO_2 attracted lots of interest, due to its abundance, stability, and nontoxicity. Since the early 1970s, TiO_2 photocatalysis has been attracting extensive and in-depth research interests, covering aspects of morphology,[7–15] size,[16] crystal structure,[16–22] modification,[23–28] computation modeling, and theoretical calculation.[29–32] TiO_2 has four main crystal phases: anatase (tetragonal), rutile (tetragonal), brookite (orthorhombic), and TiO_2 (B) (monoclinic). Among them, rutile is the most stable phase existent in nature, while anatase is the most photocatalytically active phase. People believe that the mixture of different TiO_2 phases can give a higher photocatalytic activity than its component phase of TiO_2, due to the separation of photogenerated electron–hole pairs between different phases. For example, Degussa (Evonik) P25 TiO_2 is composed of about 80% anatase and 20% rutile and is considered as a reference with high photocatalytic activity. Zhang et al. controlled the TiO_2 phase composition via the calcination temperature and thereby achieved the highest photocatalytic activity for hydrogen production by TiO_2 of anatase and rutile,[19] as shown in Figure 13.8. Zhu et al. synthesized an unusual bicrystalline TiO_2 of anatase and TiO_2 (B), which showed a higher photocatalytic activity than P25.[21,22]

Improving photon efficiency and utilizing visible light are the two main issues for TiO_2 photocatalysis in hydrogen production. Sensitization of TiO_2 for visible-light absorption has been a research hot spot since the concept of photocatalysis was established. At the beginning of this century, nonmetal (including N^{33} and C^{34}) doping was considered to have a better effect on TiO_2 sensitization than traditional metal doping, due to its uplifted VB composed of 2p orbitals of O and the nonmetal dopant. Thereafter, codoping TiO_2 with different nonmetals, or transition metal and nonmetal, has been tried. Luo et al. codoped TiO_2 with Br and Cl. Br^- and Cl^- codoping caused the absorption edge of TiO_2 to shift to a lower energy. The photocatalytic activity of doped TiO_2 with mixed anatase/rutile phases exceeded that of commercial TiO_2 photocatalyst, P25 for water splitting into H_2 and O_2.[35] Chen et al. enhanced TiO_2 solar absorption by introducing a disorder in the surface layers of TiO_2 through hydrogenation.[36] The disorder-engineered TiO_2 nanocrystals exhibited substantial solar-driven photocatalytic activity in the production of hydrogen with the use of a sacrificial reagent. Huang et al. recently developed several methods to prepare a series of black TiO_2 with its absorption shifted to visible or even infrared (IR) wavelengths.[37–41] Different from the black TiO_2 nanocrystals prepared by the high H_2-pressure process,[36] these TiO_2 nanocrystals have a core–shell structure with a crystalline TiO_2 core and an amorphous TiO_2 shell, the disorder of which led to the black color of TiO_2. The shell contained oxygen vacancies or nonmetal dopants

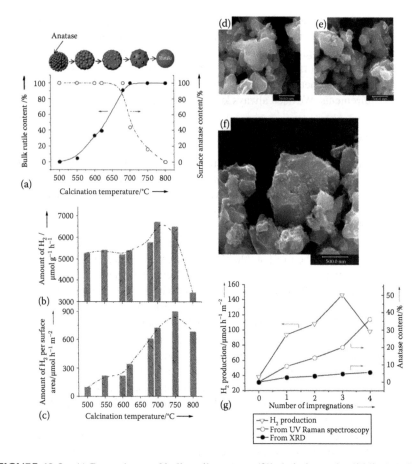

FIGURE 13.8 (a) Dependence of bulk rutile content (filled circles and solid line) and surface anatase content (open circles and broken line) on the calcination temperature. (b) TiO_2 samples calcined at different temperatures and their corresponding overall and (c) surface-specific photocatalytic activity. Scanning electron micrographs of (d) $TiO_2(R)$, (e) $TiO_2(A)/TiO_2(R)$-1, and (f) $TiO_2(A)/TiO_2(R)$-4. (g) The photocatalytic activities for evolved H_2 per surface area of $TiO_2(A)/TiO_2(R)$-n samples with increasing amount of the anatase phase on the surface of rutile TiO_2. The rate of evolved H_2 per surface area of $TiO_2(R)$ before depositing anatase TiO_2 is added for comparison (data point at 0 impregnations). The anatase contents estimated from X-ray diffraction (XRD) and UV Raman spectroscopy for $TiO_2(A)/TiO_2(R)$-n samples are also displayed. (Zhang, J., Xu, Q., Feng, Z., Li, M., and Li, C.: *Angew. Chem. Int. Ed.* 2008. 47. 1766–1769. Copyright Wiley-VCH Verlag GmbH & Co. KGaA. Adapted with permission.)

(H, N, S, or I). This structure can result in 80% absorption of the solar spectrum by the black TiO_2. A wide absorption range in the solar spectrum, chemical and physical stability, proper concentration, low recombination, and fast transportation of charge carriers gave the black TiO_2 samples high efficient utilization of solar light. The photocatalytic H_2 generation of the N-doped black titania is 15.0 mmol h^{-1} g^{-1} under 100 mW cm^{-2} of full sunlight and 200 μmol h^{-1} g^{-1} under 90 mW cm^{-2} of

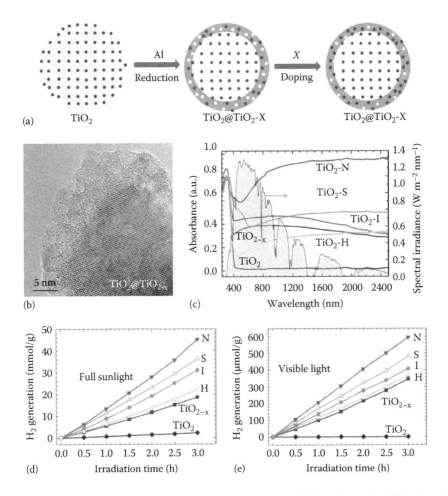

FIGURE 13.9 (a) Schematic core–shell structures of $TiO_2@TiO_{2-x}$ (denoted as TiO_{2-x}) and $TiO_2@TiO_2-X$ (denoted as TiO_2-X) with the Ti^{4+}, oxygen vacancies, and X sites in green, white, and orange, respectively. (b) High resolution transmission electron microscopy (HRTEM) image of black TiO_{2-x} nanocrystal. (c) Diffuse reflectance spectra of TiO_2-X (X=H, N, S, or I) and solar spectral irradiance (right). Evaluation of the photocatalytic activities of TiO_2-X (X=H, N, S, or I). (a) Full-sunlight- and (b) visible-light-driven photocatalysis for H_2 generation. (Lin, T. et al., *Energy Environ. Sci.*, 7, 967, 2014. Adapted by permission of The Royal Society of Chemistry.)

visible-light irradiation, making this sample as one of the best visible-light-responsive photocatalysts,[39] as shown in Figure 13.9. They reported an innovative two-step method to prepare a core–shell nanostructured S-doped rutile TiO_2 (R'-TiO_2-S). This modified black rutile TiO_2 sample exhibits remarkably enhanced absorption in visible and near-IR regions and efficient charge separation and transport. As a result, the unique sulfide surface (TiO_{2-x}:S) boosts the photocatalytic water splitting with a steady solar hydrogen production rate of 0.258 mmol h^{-1} g^{-1}. Black titania is also an

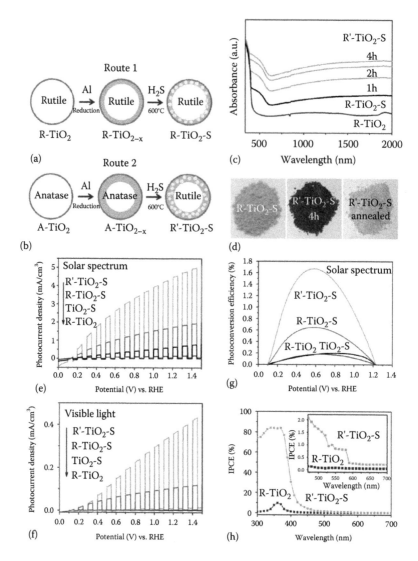

FIGURE 13.10 (a and b) Two schematic synthetic routes of rutile TiO$_2$ with sulfide surface. The shaded area is the disordered layer, while the white region is the crystalline core. (c) Diffuse reflectance spectrum of R-TiO$_2$, R-TiO$_2$-S, and R'-TiO$_2$-S with different sulfidation time. (d) Photographs of as-prepared R-TiO$_2$-S, R'-TiO$_2$-S-4h, and R'-TiO$_2$-S annealed at 800°C in Ar atmosphere. Photoelectrochemical properties of R'-TiO$_2$-S, R-TiO$_2$-S, TiO$_2$-S, and R-TiO$_2$ electrodes: chopped J–V curves under (e) simulated solar light illumination and (f) visible-light illumination using a three-electrode setup (TiO$_2$ working, Pt counter, Ag/AgCl reference electrode, scan rate of 10 mV/s) in 1 M NaOH electrolyte (pH 13.6). Photoelectrochemical properties of R'-TiO$_2$-S, R-TiO$_2$-S, TiO$_2$-S, and R-TiO$_2$ electrodes: (g) photoconversion efficiency as a function of applied potential and (h) incident photon to current efficiency (IPCE) spectra in the region of 300–700 nm at 0.65 V$_{RHE}$. Inset: IPCE spectra in the region of 420–700 nm. (Adapted with permission from Yang, C. et al., *J. Am. Chem. Soc.*, 135, 17831. Copyright 2013 American Chemical Society.)

excellent photoelectrochemical electrode exhibiting a high solar-to-hydrogen (STH) conversion efficiency of 1.67%, as shown in Figure 13.10. The sulfide surface shell is proved to be an effective strategy for enhancing solar light absorption and photoelectric conversion.[41] They also reported a mass production approach to synthesize black titania by aluminum reduction. The obtained sample possesses a unique crystalline core–amorphous shell structure ($TiO_2@TiO_{2-x}$). Black titania absorbs 65% of the total solar energy by improving visible and IR absorption, superior to pristine TiO_2 (5%). The unique core–shell structure ($TiO_2@TiO_{2-x}$) and high absorption boost the photocatalytic water splitting. Black titania is also an excellent photoelectrochemical electrode exhibiting a high STH efficiency (1.7%), as shown in Figure 13.11. The Al-reduced amorphous shell is shown to be an excellent candidate to absorb more solar light and perform more efficient photocatalysis.[40]

There are also many other binary metal oxides that have been used in photocatalytic hydrogen production, such as ZnO, Fe_2O_3, WO_3, SnO_2, CeO_2, NiO, CdO, PdO, V_2O_5, MoO_3, Cr_2O_3, In_2O_3, Ga_2O_3, Cu_2O, CuO, Bi_2O_3, ZrO_2, and Ta_2O_5. Although all of them have inherent shortcomings, some of them are good coupling partners to be used in conjunction with another semiconductor.

13.4.4 PEROVSKITES

After the discovery of the TiO_2 photocatalyst, perovskites (e.g., $SrTiO_3$, $CaTiO_3$, $BaTiO_3$) have been attracting broad attention in photocatalytic hydrogen production. Several hundred oxides possess the perovskite structure with the general formula ABX_3. $SrTiO_3$ has a cubic unit cell, and the framework structure contains corner-sharing TiO_6 octahedra with a Sr cation occupying 12-coordinate sites. It contains Ti at the cube corners (coordinates 0, 0, 0), Sr at the body center (½, ½, ½), and O at the edge centers (½, 0, 0; 0, ½, 0; 0, 0, ½). Different from photoelectrochemical hydrogen production, the sustained photogeneration of hydrogen on $SrTiO_3$ single-crystal surfaces without platinum coating or platinum counter electrode was first reported by Wagner and Somorjai.[42] Hydrogen yields far exceed the monolayer amounts (~10^{15} molecules cm^{-2}) expected from a surface stoichiometric reaction. The reaction takes place on illumination of the crystal with bandgap irradiation (hv ≥ 3.2 eV) in aqueous alkaline solution. The rate of hydrogen evolution increased with increasing hydroxide concentration in the solution. This result shows that strongly reductive (hydrogen production) as well as oxidative (oxygen evolution) reactions can be carried out and sustained on illuminated oxide semiconductor particles, without using a photoelectrochemical cell. This work starts up $SrTiO_3$ application in hydrogen production. Recently, semiconductor–semiconductor coupling and metal doping[43,44] were adopted to expand the visible-light response of $SrTiO_3$ for hydrogen production. $BaTiO_3$ has a tetragonal structure, due to the twisting of TiO_6 octahedra. Both $SrTiO_3$ and $BaTiO_3$ have a bandgap of 3.3 eV. Like $SrTiO_3$, doping $BaTiO_3$ with metal cations will extend its visible-light response for hydrogen production. Due to the similarity of $SrTiO_3$ and $BaTiO_3$, they are sometimes compared with regard to their preparation, modification, and performance in photocatalytic hydrogen production.[45–47] Similar to $SrTiO_3$, $CaTiO_3$ with a bandgap of 3.5 eV was also found to decompose water into hydrogen and oxygen under irradiation of UV light in aqueous

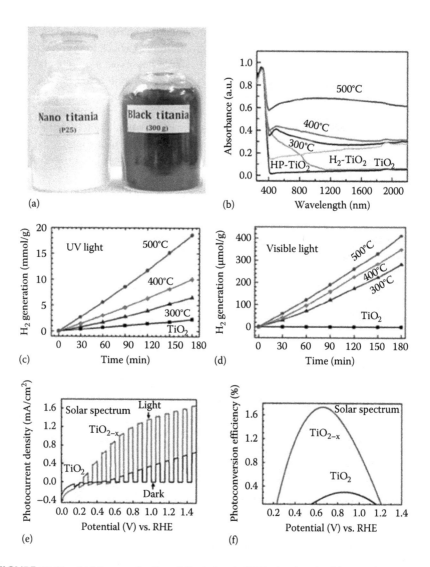

FIGURE 13.11 (a) Mass production of black titania (TiO_{2-x}) using the Al-reduction method. (b) Absorption spectra of TiO_{2-x} samples reduced at different temperatures (300°C, 400°C, and 500°C), the high-pressure hydrogenated black titania,[36] H_2-reduced titania (H_2–TiO_{2-x}), and pristine titania (TiO_2). H_2 generation of black TiO_{2-x} under (c) UV-light and (d) visible-light irradiation. Photoelectrochemical properties of pristine TiO_2 and Al-reduced black TiO_{2-x} electrodes: (e) chopped J–V curves under simulated solar light illumination using a three-electrode setup (TiO_2 working, Pt counter, Ag/AgCl reference electrode, scan rate of 10 mV s^{-1}) in a 1 M NaOH electrolyte (pH 13.6), (f) photoconversion efficiency as a function of applied potential.   (*Continued*)

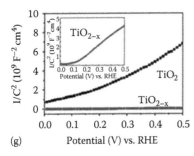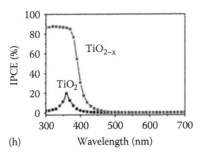

(g)   Potential (V) vs. RHE   (h)   Wavelength (nm)

FIGURE 13.11 (Continued) (g) Mott–Schottky plots collected at a frequency of 5 kHz in the dark, and (h) IPCE spectra in the region of 300–700 nm at 0.65 V_{RHE}. (Wang, Z. et al., *Energy Environ. Sci.*, 6, 3007, 2013. Adapted by permission of The Royal Society of Chemistry.)

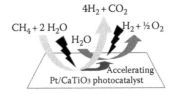

FIGURE 13.12 Illustration of the synergistic effect from photocatalytic steam reforming of methane and photocatalytic water decomposition simultaneously. (Shimura, K. and Yoshida, H., *Energy Environ. Sci.*, 3, 615, 2010. Adapted by permission of The Royal Society of Chemistry.)

KOH solution.[48] In this work, the author also showed that the direct measurement of the photocurrent spectrum corresponding to the efficiency of water decomposition is an effective method to survey photocatalytic activities of materials. $Pt/CaTiO_3$ can have a high photocatalytic activity for water splitting under UV-light irradiation. $Pt/CaTiO_3$ photocatalysts exhibited a higher production rate of hydrogen in a flowing mixture of water vapor and methane than in a flow of water vapor only (Figure 13.12), indicating photocatalytic steam reforming of methane and photocatalytic water decomposition can promote each other.[49] This is important to the design of photocatalytic applications to pursue a synergistic effect.

13.4.5 TITANATES

Alkali-metal titanates having a chemical formula of $M_2Ti_nO_{2n+1}$ (M = Na, K, Rb, and n = 2, 3, 4, 6) were employed as semiconductor oxide supports for the photoassisted decomposition of water under Xe lamp. In their association with oxidized Ru, the photocatalytic activity increased with increasing n (M = K). It was suggested that the photocatalytic activity was associated with the framework of TiO_6 octahedra forming the tunnel structures.[50] After that, $M_2Ti_6O_{13}$ (M = Na, K, or Rb) with tunnel structures and $Cs_2Ti_6O_{13}$ with a layer structure were employed to decompose water.[51] The titanates with the tunnel structures showed high photocatalytic activities and photoexcited radical formation. The size effect of the alkaline metal atoms has a significant effect on the extent of distortion of TiO_6 octahedra, which form the rectangular tunnel structures. The rectangular tunnel structures of the hexatitanates have three kinds of TiO_6 octahedra: the position of a Ti ion deviates from the center of gravity of six oxygen atoms surrounding the Ti ion, thus producing a dipole moment in each TiO_6. The total dipole moments of three TiO_6 octahedra in $M_2Ti_6O_{13}$ (M=Na, K, or Rb) increase in the order of Na>K>Rb. The order is the same as that of the photocatalytic activity and the efficiency for the photoexcited charge formation. This agreement suggests that the internal local fields due to the dipole moments play an important role in photoexcitation.

13.4.6 TANTALATES

The 5d orbital of Ta can construct a highly negative CB of alkali tantalates $ATaO_3$ (A=Li, Na, or K) and alkaline-earth tantalates ATa_2O_6 (A=Mg, Ca, Sr, or Ba) with corner-sharing TaO_6 octahedra, making them suitable for hydrogen production by water splitting.[52–54] Among $ATaO_3$ (A=Li, Na, or K), $NiO/NaTaO_3$ showed the highest photocatalytic activity, while among different crystal phases of ATa_2O_6 (A=Mg, Ca, Sr, or Ba), orthorhombic $BaTa_2O_6$ gave the best performance, which can be further improved by adding $Ba(OH)_2$ and/or loading Ni. Doping $NaTaO_3$ with lanthanides

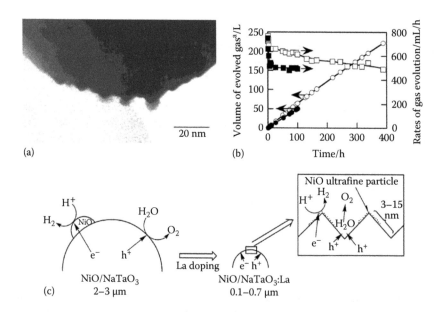

FIGURE 13.13   (a) TEM image of NiO (0.5 wt%)/NaTaO₃:La(1.5%). (b) Photocatalytic water splitting over NiO (0.2 wt%)/NaTaO₃:La-(2%) in pure water (closed marks) and 1 mmol L^{-1} of an aqueous NaOH solution (open marks). (a) The amounts of evolved gas were determined by volumetric measurement. Catalyst, 1.0 g; reactant solution, 390 mL; inner irradiation cell made of quartz: 400 W high-pressure mercury lamp. (c) Mechanism of highly efficient photocatalytic water splitting over NiO/NaTaO₃:La photocatalysts. (Adapted with permission from Kato, H. et al., *J. Am. Chem. Soc.*, 125, 3082. Copyright 2003 American Chemical Society.)

(La, Pr, Nd, Sm, Gd, or Dy) can improve its photocatalytic activity in water splitting by decreasing the $NaTaO_3$ particle size, increasing its surface area, and restraining catalyst deactivation. Under UV-light irradiation, 1 mol% La-doped $NiO/NaTaO_3$ showed a high photocatalytic activity for water splitting into H_2 (5.9 mmol h^{-1}) and O_2 (2.9 mmol h^{-1}).[55] The maximum apparent quantum yield of the NiO (2 wt%)/$NaTaO_3$:La (2 mol%) photocatalyst was 56% at 270 nm. The factors affecting the highly efficient photocatalytic water splitting were examined by using various characterization techniques. Electron microscope observations revealed that the particle sizes of $NaTaO_3$:La crystals (0.1–0.7 μm) were smaller than that of the nondoped $NaTaO_3$ crystal (2–3 μm) and that an ordered surface nanostructure with many steps was created by lanthanum doping. The small particle size with a high crystallinity was advantageous to increase the probability of the reaction of photogenerated electrons and holes with water molecules. Transmission electron microscopy (TEM) observations and extended x-ray absorption fine structure analyses indicated that NiO cocatalysts were loaded on the edge of the nanostep structure of $NaTaO_3$:La photocatalysts as ultrafine particles. The H_2 evolution proceeded on the ultrafine NiO particles loaded on the edge, while the O_2 evolution occurred at the groove of the nanostep structure. Thus, the reaction sites for H_2 evolution and O_2 evolution were separated from each other over the ordered nanostep structure. The small particle size and the ordered surface nanostep structure of the $NiO/NaTaO_3$:La photocatalyst powder contributed to the highly efficient water splitting into H_2 and O_2,[56] as shown in Figure 13.13.

13.4.7 NIOBATES

Nb_2O_3 does not show any photocatalytic activity for hydrogen production, but many niobates do. $NaNbO_3$ was prepared by three methods, namely, solid-state reaction (SSR), hydrothermal (HT), and polymerized complex (PC) methods, and the relationships between the photocatalytic activity and the particle size and morphology was investigated. The photocatalytic activity was evaluated by H_2 evolution from an aqueous methanol solution and pure water in the presence of the Pt (0.5 wt%)/$NaNbO_3$ and RuO_2 (1.25 wt%)/$NaNbO_3$, respectively. It is found that the sample prepared by PC with smallest particles exhibits the highest photocatalytic activity in both reactions.[57] Layered $K_4Nb_6O_{17}$ has two kinds of interlayers (I and II) that alternate. Water is reduced to hydrogen by photogenerated electrons from one interlayer, in which nanosized cocatalysts (Ni or Au) are selectively introduced by ion-exchange or intercalation reactions, while water oxidation to oxygen by photogenerated holes occurs at the other interlayer. In this way, hydrogen and oxygen evolutions are separated by niobate sheets.[58] Photocatalytic properties of $Cs_2Nb_4O_{11}$ consisting of NbO_6 octahedra and NbO_4 tetrahedra were investigated. The bandgap of $Cs_2Nb_4O_{11}$ was 3.7 eV. Pretreated NiO-loaded $Cs_2Nb_4O_{11}$ showed a high activity for water splitting under UV-light irradiation.[59]

13.4.8 (OXY)NITRIDES

There are some transition metal (oxy)nitrides with d^0-electronic configurations that have the potential to split water by absorbing light with wavelengths up to ∼600 nm.

However, overall water splitting has not yet been achieved by these 600 nm class photocatalysts. It is very likely that defects that are inevitably formed during preparation decrease their photocatalytic performance.[60] Tantalum(V) nitride (Ta_3N_5) has an orthorhombic anosovite (Ti_3O_5) structure, which generates both H_2 and O_2 from H_2O in the presence of suitable sacrificial agents by absorbing light with wavelengths up to $\sim$600 nm (bandgap $\sim$2.1 eV).[61] Since its H_2-production activity is much lower than its O_2-generating one, it is important to enhance the former. This can be improved by high-pressure high-temperature treatments and flux-assisted nitridation and posttreatments. $LaMO_2N$ (M = Ti, Zr), $LaMON_2$ (M = Nb, Ta), and $AEMO_2N$ (AE = alkaline-earth metal, Ca, Sr, Ba; M = Nb, Ta) are perovskite oxynitrides.[62] Most of them can absorb light with wavelengths up to $\sim$600 nm or even longer, and moreover, the absorption edge (i.e., bandgap) can be tuned in a wide range from UV to near-IR region by the formation of solid solutions with other perovskite oxides and oxynitrides. In addition to chemical and thermal stabilities, which are both superior to nitrides, this nature makes them very attractive in visible-light-responsive water splitting.[63]

13.4.9 METAL SULFIDES

Most of metal sulfide photocatalysts consist of metal cations with d^{10} configuration, as shown in Figure 13.14. The CB of metal sulfides is formed by d and sp orbitals of metal cations, while the VB is composed of S 3p orbitals, which are more negative than O 2p orbitals. This band structure endows metal sulfides with enough negative CBs for water reduction and narrow bandgaps for visible-light absorption. Sulfide

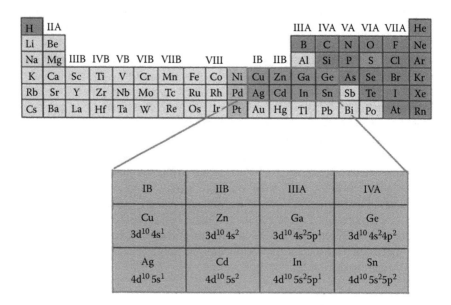

FIGURE 13.14 Elements used for constructing most of the sulfide photocatalysts for hydrogen production. (Zhang, K. and Guo, L.J., *Catal. Sci. Technol.*, 3, 1672, 2013. Adapted by permission of The Royal Society of Chemistry.)

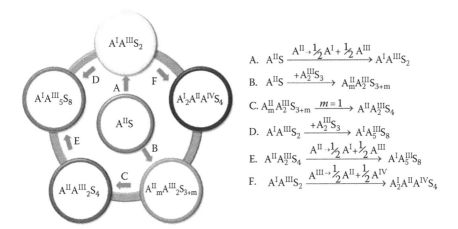

A. $A^{II}S \xrightarrow{A^{II} \to \frac{1}{2}A^{I} + \frac{1}{2}A^{III}} A^{I}A^{III}S_2$

B. $A^{II}S \xrightarrow{+A_2^{III}S_3} A_m^{II}A_2^{III}S_{3+m}$

C. $A_m^{II}A_2^{III}S_{3+m} \xrightarrow{m=1} A^{II}A_2^{III}S_4$

D. $A^{I}A^{III}S_2 \xrightarrow{+A_2^{III}S_3} A^{I}A_5^{III}S_8$

E. $A^{II}A_2^{III}S_4 \xrightarrow{A^{II} \to \frac{1}{2}A^{I} + \frac{1}{2}A^{III}} A^{I}A_5^{III}S_8$

F. $A^{I}A^{III}S_2 \xrightarrow{A^{III} \to \frac{1}{2}A^{II} + \frac{1}{2}A^{IV}} A_2^{I}A^{II}A^{IV}S_4$

FIGURE 13.15 Genealogy of sulfide semiconductor photocatalysts and the corresponding relationship between semiconductors with different formulations. (Zhang, K. and Guo, L.J., *Catal. Sci. Technol.*, 3, 1672, 2013. Adapted by permission of The Royal Society of Chemistry.)

photocatalysts can be divided into several groups, such as IIB–VIA, IIB–IIIA–VIA, IA–IIIA–VIA, and IB–IIB–IVA–VIA sulfide photocatalysts, according to their elemental compositions and material formulations. All these sulfide semiconductors can be considered as derivatives of the zinc-blende structure, which are obtained by substituting IIB atoms with other metallic elements (Figure 13.15), reducing the symmetries of the corresponding crystal structures. Thus tetragonal crystal systems, the symmetries of which are lower than cubic crystal systems, are gradually presented in sulfides such as chalcopyrite and stannite structures.[64] Most sulfide photocatalysts suffer from their instability due to photocorrosion, since they have a more negative VB constructed by S ions, which is easily oxidized by photogenerated holes. Normally, sacrificial reagents are needed during photocatalytic hydrogen production to consume photogenerated holes. However, this is not an economical strategy. Coupling metal sulfides with other semiconductors can enhance the stability of the former, but satisfactory photocatalytic efficiency has not yet been achieved by the composite photocatalysts.

13.4.10 GRAPHENE

The CB bottom of reduced graphene oxide (RGO), which is mainly formed by the antibonding π^* orbital, has a higher energy level (-0.52 eV vs. NHE, pH 0) than that needed for the H_2 production, while the VB top of RGO, which is mainly composed of O 2p orbital, varies with the reduction degree. The bandgap of RGO decreases with increasing reduction degree, suggesting that RGO with a suitable bandgap for water splitting might be obtained by tuning its reduction level,[65] as shown in Figure 13.16.

A graphite oxide (GO) semiconductor photocatalyst with an apparent bandgap of 2.4–4.3 eV was, for the first time, reported steadily catalyzing H_2 generation from a 20 vol% aqueous methanol solution and pure water under UV- or visible-light irradiation, even without any cocatalyst,[66] as shown in Figure 13.17. The authors also

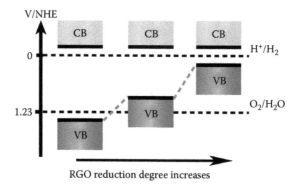

FIGURE 13.16 Energy-level diagrams of RGO with different reduction degrees in comparison with the potentials for water reduction and oxidation. (Xie, G., Zhang, K., Guo, B., Liu, Q., Fang, L., and Gong, J.R.: *Adv. Mater.* 2013. 25. 3820–3839. Copyright Wiley-VCH Verlag GmbH & Co. KGaA. Adapted with permission.)

suggested that GO can have different electronic properties by varying the oxidation level. The conduction and VB edge levels of GO from appropriate oxidation are suitable for both the reduction and the oxidation of water. The CB edge shows little variation with the oxidation level, and the VB edge governs the bandgap of GO, which increases with the oxygen content. The photocatalytic activity of GO specimens with various oxygenated levels was measured in methanol and $AgNO_3$ solutions for evolution of H_2 and O_2, respectively. The H_2 evolution was strong and stable over time. When $NaIO_3$ was used as a sacrificial reagent, strong O_2 evolution was observed over the GO specimens. The authors demonstrate that chemical modification can easily modify the electronic properties of GO for water splitting.[67] Matsumoto et al. also reported hydrogen evolution from an aqueous suspension of GO nanosheets or a GO photoelectrode under UV irradiation.[68]

13.4.11 GRAPHITIC CARBON NITRIDE

Graphitic carbon nitride ($g\text{-}C_3N_4$) has attracted intense attention since the pioneering work by Wang et al. on visible-light water splitting over $g\text{-}C_3N_4$.[69] $g\text{-}C_3N_4$ is composed of abundant elements and is thermally and chemically stable. The proposed structure of $g\text{-}C_3N_4$ is a 2D framework of tri-*s*-triazine connected via tertiary amines. It is identified as a visible-light-responsive polymeric semiconductor with a bandgap of ~2.7 eV, corresponding to an optical absorption till ~460 nm, as well as band edges suitable for both water reduction and oxidation. The photocatalytic hydrogen production performance of $g\text{-}C_3N_4$ can be further improved by morphology modulation, nanostructure design, dye sensitization, doping, and heterojunction construction.[70]

13.4.12 Z-SCHEME PHOTOCATALYST SYSTEMS

Inspired by the two-step photoexcitation (Z-scheme) mechanism of natural photosynthesis in green plants, the water splitting reaction is broken up into two

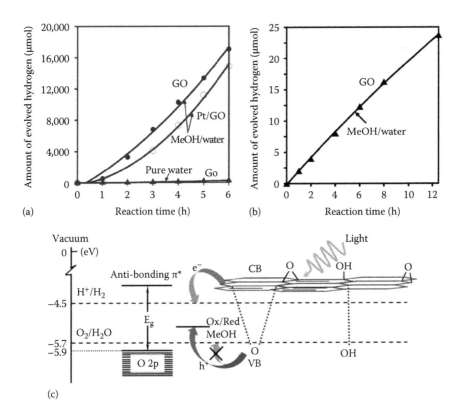

FIGURE 13.17 (a) Time course of H_2 evolution from a 20 vol% aqueous methanol solution (MeOH/water) or pure water with suspended photocatalysts (GO or Pt/GO) (0.5 g) under mercury-lamp irradiation. (b) Time course of H_2 evolution from the methanol solution with 0.5 g of suspended GO under visible-light irradiation. (c) Schematic energy-level diagram of GO relative to the levels for H_2 and O_2 generation from water. During photocatalytic reaction the CB of GO with a high overpotential relative to the level for H_2 generation exhibits fast electron injection from the excited GO into the solution phase, whereas the holes in the VB of GO do not interact with the water molecules for O_2 generation, but are exhausted by the hole scavenger (methanol) instead. The top of the valence energy level was obtained using the density functional theory for GO with 12.5% O coverage. (Yeh, T.-F., Syu, J.-M., Cheng, C., Chang, T.-H., and Teng, H.: *Adv. Funct. Mater.* 2010. 20. 2255–2262. Copyright Wiley-VCH Verlag GmbH & Co. KGaA. Adapted with permission.)

stages: one for H_2 evolution and the other for O_2 evolution, which cooperate with each other by using an electron mediator (a shuttle redox couple, Red/Ox) in the solution. This Z-scheme system requires the lower energy to drive each photocatalysis process, extending the working wavelengths ($\approx$660 nm for H_2 evolution and $\approx$600 nm for O_2 evolution), compared to that ($\approx$460 nm) in conventional water splitting systems based on one-step photoexcitation in a single semiconductor material, as shown in Figure 13.18. The development in this field has been reviewed recently.[71,72]

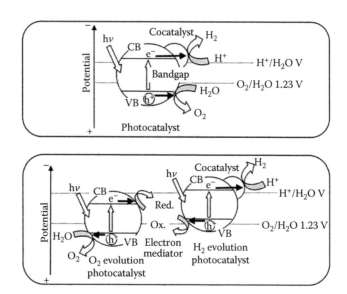

FIGURE 13.18 Single- and two-particulate photocatalyst systems for water splitting into H_2 and O_2. (Adapted from Kudo, A., *MRS Bull.*, 36, 32–38. Copyright 2011, Materials Research Society. With permission.)

13.4.13 POLYOXOMETALATES

Polyoxometalates (POMs), complex early transition metal–oxygen clusters, have peculiar optoelectronic properties and high reduction potential, playing a role as excellent electron pools.[73] They can work as molecular photocatalysts or cooperate with another semiconductor photocatalyst. The majority of processes involved in the photochemistry of POMs are oxidative. However, hydrogen production can be performed during the photooxidation of organic substrates in water under anaerobic conditions, since the photoreduced POMs can simultaneously reduce protons from the organic substrates to molecular hydrogen.[74–76] The solar-assisted conversion of HCCs into hydrogen was performed using POM-based photocatalysts, such as isopolytungstates (IPTs) and silicotungstic acid (STA). Upon exposure to solar photons, IPT aqueous solutions containing various HCCs (e.g., alcohols, alkanes, organic acids, and sugars) produce hydrogen gas and corresponding oxygenated compounds. The presence of small amounts of colloidal platinum increases the rate of hydrogen evolution by one order of magnitude. The tests demonstrated a steady-state production of hydrogen gas for several days. IPT immobilized on granules of anion exchange resins with quaternary ammonium active groups show good photocatalytic activity for hydrogen production from water–alcohol solutions exposed to near-UV or solar radiation.[77] Illumination of polytungstate (PT) solutions containing organic compounds results in the formation of photoreduced tungsten species and the evolution of hydrogen. Prolonged irradiation of this solution leads to the formation of hydrogen and acetaldehyde in a stoichiometric ratio. The addition of colloidal platinum to the system results in the increase in rate of hydrogen evolution. A mechanism of the

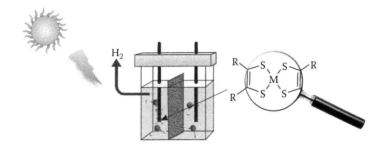

FIGURE 13.19 Schematic diagram of hydrogen production by using a dithiolene complex photoelectrode in a photoelectrocatalytic cell. (Adapted from *Coord. Chem. Rev.*, 256, Zarkadoulas, A., Koutsouri, E., and Mitsopoulou, C.A., 2424–2434, Copyright 2012, with permission from Elsevier.)

process might involve the formation of two-electron reduced PT and its subsequent reoxidation yielding hydrogen in the presence of colloidal platinum.[78] Photocatalytic H_2 formation from aqueous glycerin was achieved over giant POM complexes Na_2 $_1K_4[\{P_2W_{15}Ti_3O_{57.5}(OH)_3\}4Cl]\cdot104H_2O$ and $Na_{18}H_{15}[\{P_2W_{15}Ti_3O_{59}(OH)_3\}4\{\mu_3\text{-}Ti(OH)_3\}4Cl]\cdot105H_2O$. This is the first example of the photocatalytic formation of H_2 from glycerin and is also the first example of a photochemical application of giant POM complexes. The reaction mechanism can be explained by redox reactions between the complexes, water, and glycerin.[79]

13.4.14 DITHIOLENE COMPLEXES

Dithiolene complexes have recently attracted researchers' interest for the photosplitting of water (Figure 13.19). This is mainly based on their reversible redox behavior that makes them of potential interest for multielectron chemistry in hydrogen and oxygen production. Moreover, these complexes have intense absorption bands in the visible region and are quite cheap (depending on the transition metal used) and soluble in a variety of solvents. Their electrochemical and photochemical properties can be finely tuned by the appropriate design of the ligands that are resistant to hydrogenation. Although dithiolene complexes are not good homogeneous photocatalysts, their monoanion turned out to be a good alternative to platinum catalyst for H^+ reduction, especially if composed of abundant, inexpensive metals like Co or Fe. The disadvantages of the photocatalytic activity of the dithiolene complexes in homogeneous solutions can be overcome if they are attached to semiconductors such as TiO_2 or if they are used as an electrode in a photoelectrocatalytic cell. In these circumstances, dithiolene complexes show high activity in the reduction of protons in aqueous solutions.[80]

13.5 PHOTOCATALYTIC HYDROGEN SYSTEMS

With the development of new photocatalysts, different photocatalytic hydrogen production systems have been designed to adapt to the photocatalysts used. For example, sulfide photocatalysts are easily photocorroded. But their photostability during photocatalytic hydrogen production can be significantly improved in S^{2-}-containing solutions, since S^{2-} from the solution can sacrifice to replace photooxidable sulfide in the photocatalyst lattice. Generally speaking, photocatalytic hydrogen production can be performed in two systems: (1) in pure water, which is the ultimate goal, but is difficult, and (2) in a system with various sacrificial reductants (i.e., electron donors), including biomass, H_2S and S^{2-}, that efficiently accelerate the consumption of photo-generated holes thereby enhancing the hydrogen production.

13.5.1   DIRECT WATER SPLITTING

Currently, most photocatalysts for direct water splitting are UV-light-responsive metal oxides, which are composed of O and metal with d^0 (Ti^{4+}, Zr^{4+}, Nb^{5+}, Ta^{5+}, W^{6+}) or d^{10} (Ga^{3+}, In^{3+}, Ge^{4+}, Sn^{4+}, Sb^{5+}, Zn^{2+}) cation configuration. Recently, some visible-light-responsive photocatalysts were also developed. Special interest focuses on the solid solution photocatalysts (including oxynitrides), since their band structure and optical properties can be tuned by varying their composition during the preparation. There are many challenges and several issues to pay attention to in direct water splitting. First of all, both H_2 and O_2 should form with a stoichiometric amount, 2:1 (molar ratio). In addition, amounts of H_2 and O_2 evolved should increase with irradiation time.[58] Furthermore, an efficient photocatalyst for direct water splitting can be also very active in catalyzing the back reaction to form water (downhill reaction). In order to avoid this back reaction, we need to spatially separate H_2 and O_2 evolution, or design special photocatalysts inactive for back reaction.[81]

13.5.2   HYDROGEN PRODUCTION FROM REFORMING BIOMASS

As we mentioned in the beginning, biomass can be considered a carbon-neutral and renewable hydrogen resource. Since it is difficult to obtain a perfect direct water splitting system with satisfactory rate, biomass can be added in water as sacrificial organic reductants (electron donors, e.g., methanol, ethanol, glycerol, saccharides, and methane) to accelerate hydrogen production. During this process, these biomass-derived oxygenated compounds are eventually oxidized to CO_2. When the experimental conditions are deliberately controlled, the oxidation can be selective rather than complete, and the added value of the carbon-containing by-products (intermediates) can be obtained.[2,81–83] Currently, research in biomass reforming photocatalyst mainly focuses on TiO_2, and visible-light-responsive photocatalysts should be exploited, in order to utilize solar energy.

13.5.3   HYDROGEN PRODUCTION FROM H_2S

Photocatalytic hydrogen production from H_2S decomposition benefits both environmental protection and energy exploitation.[84] Photocatalytic H_2S decomposition can be performed by indirect and direct methods. Currently, most of the research focuses on indirect photocatalytic H_2S decomposition. H_2S can easily be absorbed in alkaline solution to form S^{2-}, which can be oxidized during hydrogen production from water reduction. The main steps can be described as follows:

$$\text{Absorption of } H_2S: \quad H_2S + 2OH^- \rightarrow S^{2-} + 2H_2O \quad (13.1)$$

$$\text{Oxidation of } S^{2-}: \quad 2S^{2-} + 2h^+ + xH_2O \rightarrow S_2^{2-}\left(\text{or } SO_x^{2-}\right) \quad (13.2)$$

$$\text{Reduction of proton}: \quad 2H^+ + 2e^- \rightarrow H_2 \uparrow \quad (13.3)$$

The photocatalysts used in this system are mainly metal sulfides, since they are more resistant to photocorrosion in the solution with S^{2-}, S_2^{2-}, and SO_x^{2-} electron donors. The challenge of this technique lies in the posttreatment of the reaction solution, since it contains multiple components (i.e., the products from S^{2-} oxidation, S_2^{2-}, $S_2O_3^{2-}$, SO_3^{2-}, $S_2O_6^{2-}$, and SO_4^{2-}), which are difficult to separate, purify, and recycle. Direct photocatalytic H_2S decomposition leads to S recycling during hydrogen production:

$$H_2S \rightarrow H_2 \uparrow + S \downarrow \quad (13.4)$$

Li et al. reported that H_2S can be decomposed stoichiometrically into hydrogen and sulfur by CdS-based photocatalysts under visible-light irradiation using ethanolamine as a H_2S solvent and reaction media. The hydroxyls of the reaction media were found to be crucial for the hydrogen production. The potential of H_2S splitting in ethanolamine was lowered and the photogenerated electrons could be fully used to reduce protons for hydrogen production. Important factors (solution, temperature, H_2S concentration, and cocatalyst) affecting hydrogen production were also investigated,[85] as shown in Figure 13.20.

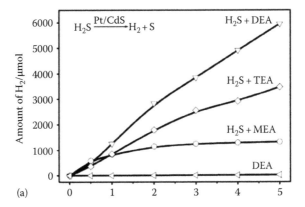

FIGURE 13.20 (a) Photocatalytic H_2 production under visible-light irradiation over Pt/CdS (0.20 wt% Pt) in different solutions. DEA, diethanolamine; TEA, triethanolamine; MEA, monoethanolamine. Reaction conditions: volume of solution, 100 mL; concentration of H_2S, 0.30 M; amount of catalyst, 0.025 g; reaction temperature, 30°C; light source, 300 W Xe lamp with a cutoff filter ($\lambda > 420$ nm). *(Continued)*

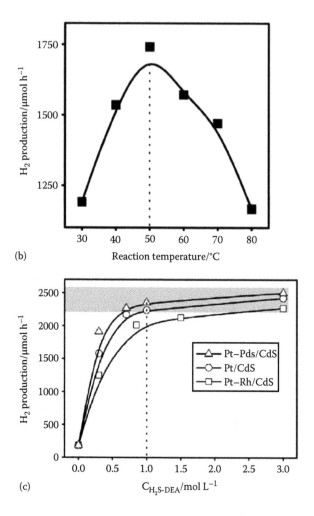

(b)

(c)

FIGURE 13.20 (Continued) (b) Dependence of H_2 production rate upon reaction temperature under visible-light irradiation. Reaction conditions: reaction solution, 100 mL DEA; concentration of H_2S, 0.30 M; catalyst, 0.025 g Pt (0.2 wt%)/CdS; light source, 300 W Xe lamp with a cutoff filter ($\lambda > 420$ nm). (c) Dependence of H_2 production rate under visible-light irradiation upon H_2S concentration on Pt (0.2 wt%)/CdS, Pt–PdS (0.2–0.2 wt%)/CdS, and Pt–Rh (0.2–0.2 wt%)/CdS. Reaction conditions: reaction solution, 100 mL DEA; amount of catalyst, 0.025 g; reaction temperature, 30°C; light source, 300 W Xe lamp with a cutoff filter ($\lambda > 420$ nm). (Adapted from *J. Catal.*, 260, Ma, G., Yan, H., Shi, J., Zong, X., Lei, Z., and Li, C., 134–140, Copyright 2008, with permission from Elsevier.)

13.6 CONCLUSION

Energy exploitation should meet the requirements from both economic development and environmental protection. New, clean, and renewable energy resources are not only the complements to current energy resources but also the basis of our future energy infrastructure. Hydrogen might be an ideal energy carrier, if it is produced in a sustainable way. Photocatalytic hydrogen production can make the dream come true. Photocatalytic hydrogen production is highly challenging and strategically significant. The key to boost this technology lies in the development of photocatalysts. Although there has been some progress in the R&D of photocatalytic materials in the last decade, the efficiency of photocatalytic hydrogen production still needs major improvement, especially under solar irradiation. In the near future, research on photocatalysts for hydrogen production should consider the following issues: (1) emphasizes fundamental research, which includes the mechanism of the basic steps involved (e.g., charge carrier generation, separation, transportation, and reaction) and the relationship between the structure and performance of photocatalysts (this knowledge is important for the design of efficient photocatalysts; currently, the R&D of photocatalytic materials is mainly carried out in a trial and error fashion through vast experiments, since we lack enough fundamental knowledge), (2) strengthens multidisciplinary cross-linking and interacts with other related research fields (the study of photosynthesis, photovoltaic cell, and dye-sensitized solar cell can also aid in the design of visible-light-responsive photocatalysts), and (3) by virtue of the latest achievements in material science, prepares stable, efficient, and visible-light-responsive photocatalysts (by using multicomponent doping and intercalation, the synergetic effect from sensitization, photoelectric conversion, and catalysis can be maximized in the prepared photocatalysts with high efficiency and visible-light response).

REFERENCES

1. J. Zhu, D. Chakarov and M. Zäch, *Energy Efficiency and Renewable Energy through Nanotechnology*, ed. L. Zang, Springer, London, 2011, ch. 13, pp. 441–486.
2. K. Shimura and H. Yoshida, *Energy Environ. Sci.*, 2011, **4**, 2467–2481.
3. A. Fujishima and K. Honda, *Nature*, 1972, **238**, 37–38.
4. J. F. Zhu and M. Zach, *Curr. Opin. Colloid Interface Sci.*, 2009, **14**, 260–269.
5. J. Zhu, *Encyclopedia of Sustainability Science and Technology*, ed. R. Meyers, Springer, New York, 2012, ch. 855, pp. 7881–7901.
6. R. Abe, *J. Photochem. Photobiol. C*, 2010, **11**, 179–209.
7. H. Yu, J. Yu, B. Cheng and S. Liu, *Nanotechnology*, 2007, **18**, 065604.
8. S. H. Lim, N. Phonthammachai, S. S. Pramana and T. J. White, *Langmuir*, 2008, **24**, 6226–6231.
9. C. Song, W. Yu, B. Zhao, H. Zhang, C. Tang, K. Sun, X. Wu, L. Dong and Y. Chen, *Catal. Commun.*, 2009, **10**, 650–654.
10. Y. Lu, M. Hoffmann, R. S. Yelamanchili, A. Terrenoire, M. Schrinner, M. Drechsler, M. W. Möller, J. Breu and M. Ballauff, *Macromol. Chem. Phys.*, 2009, **210**, 377–386.
11. X. W. Lou and L. A. Archer, *Adv. Mater.*, 2008, **20**, 1853–1858.
12. Y. Kondo, H. Yoshikawa, K. Awaga, M. Murayama, T. Mori, K. Sunada, S. Bandow and S. Iijima, *Langmuir*, 2007, **24**, 547–550.
13. Z. Liu, D. D. Sun, P. Guo and J. O. Leckie, *Chem. Eur. J.*, 2007, **13**, 1851–1855.
14. I. Paramasivam, A. Avhale, A. Inayat, A. Bösmann, P. Schmuki and W. Schwieger, *Nanotechnology*, 2009, **20**, 225607.
15. J. Yu and M. Zhou, *Nanotechnology*, 2008, **19**, 045606.
16. J. Zhang, Q. Xu, M. Li, Z. Feng and C. Li, *J. Phys. Chem. C*, 2009, **113**, 1698–1704.
17. J. Zhang, M. Li, Z. Feng, J. Chen and C. Li, *J. Phys. Chem. B*, 2005, **110**, 927–935.

18. J. Shi, J. Chen, Z. Feng, T. Chen, Y. Lian, X. Wang and C. Li, *J. Phys. Chem. C*, 2006, **111**, 693–699.

19. J. Zhang, Q. Xu, Z. Feng, M. Li and C. Li, *Angew. Chem. Int. Ed.*, 2008, **47**, 1766–1769.

20. G. Li, N. M. Dimitrijevic, L. Chen, J. M. Nichols, T. Rajh and K. A. Gray, *J. Am. Chem. Soc.*, 2008, **130**, 5402–5403.

21. J. F. Zhu, J. L. Zhang, F. Chen and M. Anpo, *Mater. Lett.*, 2005, **59**, 3378–3381.

22. J. F. Zhu, J. L. Zhang, F. Chen, K. Iino and M. Anpo, *Top. Catal.*, 2005, **35**, 261–268.

23. J. K. Zhou, L. Lv, J. Yu, H. L. Li, P.-Z. Guo, H. Sun and X. S. Zhao, *J. Phys. Chem. C*, 2008, **112**, 5316–5321.

24. C.-S. Kuo, Y.-H. Tseng, C.-H. Huang and Y.-Y. Li, *J. Mol. Catal. A: Chem.*, 2007, **270**, 93–100.

25. Z. Wang, F. Zhang, Y. Yang, B. Xue, J. Cui and N. Guan, *Chem. Mater.*, 2007, **19**, 3286–3293.

26. Y. Wang, C. Feng, Z. Jin, J. Zhang, J. Yang and S. Zhang, *J. Mol. Catal. A: Chem.*, 2006, **260**, 1–3.

27. G. Liu, Y. Zhao, C. Sun, F. Li, G. Q. Lu and H.-M. Cheng, *Angew. Chem. Int. Ed.*, 2008, **47**, 4516–4520.

28. F. L. Zhang, Y. H. Zheng, Y. N. Cao, C. Q. Chen, Y. Y. Zhan, X. Y. Lin, Q. Zheng, K. M. Wei and J. F. Zhu, *J. Mater. Chem.*, 2009, **19**, 2771–2777.

29. M. Aizawa, Y. Morikawa, Y. Namai, H. Morikawa and Y. Iwasawa, *J. Phys. Chem. B*, 2005, **109**, 18831–18838.

30. Y. Morikawa, I. Takahashi, M. Aizawa, Y. Namai, T. Sasaki and Y. Iwasawa, *J. Phys. Chem. B*, 2004, **108**, 14446–14451.

31. H. G. Yang, C. H. Sun, S. Z. Qiao, J. Zou, G. Liu, S. C. Smith, H. M. Cheng and G. Q. Lu, *Nature*, 2008, **453**, U638–U634.

32. P. J. D. Lindan and C. J. Zhang, *Phys. Rev. B*, 2005, **72**, 7.

33. R. Asahi, T. Morikawa, T. Ohwaki, K. Aoki and Y. Taga, *Science*, 2001, **293**, 269–271.

34. S. U. M. Khan, M. Al-Shahry and W. B. Ingler, *Science*, 2002, **297**, 2243–2245.

35. H. M. Luo, T. Takata, Y. G. Lee, J. F. Zhao, K. Domen and Y. S. Yan, *Chem. Mater.*, 2004, **16**, 846–849.

36. X. Chen, L. Liu, P. Y. Yu and S. S. Mao, *Science*, 2011, **331**, 746–750.

37. H. Cui, W. Zhao, C. Yang, H. Yin, T. Lin, Y. Shan, Y. Xie, H. Gu and F. Huang, *J. Mater. Chem. A*, 2014, **2**, 8612–8616.

38. Z. Wang, C. Yang, T. Lin, H. Yin, P. Chen, D. Wan, F. Xu, F. Huang, J. Lin, X. Xie and M. Jiang, *Adv. Funct. Mater.*, 2013, **23**, 5444–5450.

39. T. Lin, C. Yang, Z. Wang, H. Yin, X. Lu, F. Huang, J. Lin, X. Xie and M. Jiang, *Energy Environ. Sci.*, 2014, **7**, 967–972.

40. Z. Wang, C. Yang, T. Lin, H. Yin, P. Chen, D. Wan, F. Xu, F. Huang, J. Lin, X. Xie and M. Jiang, *Energy Environ. Sci.*, 2013, **6**, 3007–3014.

41. C. Yang, Z. Wang, T. Lin, H. Yin, X. Lü, D. Wan, T. Xu, C. Zheng, J. Lin, F. Huang, X. Xie and M. Jiang, *J. Am. Chem. Soc.*, 2013, **135**, 17831–17838.

42. F. T. Wagner and G. A. Somorjai, *Nature*, 1980, **285**, 559–560.

43. Y. Qin, G. Wang and Y. Wang, *Catal. Commun.*, 2007, **8**, 926–930.

44. T. Takata and K. Domen, *J. Phys. Chem. C*, 2009, **113**, 19386–19388.

45. Y. Liu, T. H. Ji, R. Z. Zhu and J. Y. Sun, *Spectrosc. Spec. Anal.*, 2010, **30**, 3290–3294.

46. J. P. Zou, L. Z. Zhang, S. L. Luo, L. H. Leng, X. B. Luo, M. J. Zhang, Y. Luo and G. C. Guo, *Int. J. Hydrogen Energy*, 2012, **37**, 17068–17077.

47. L. Li, X. Liu, Y. L. Zhang, P. A. Salvador and G. S. Rohrer, *Int. J. Hydrogen Energy*, 2013, **38**, 6948–6959.

48. H. Mizoguchi, K. Ueda, M. Orita, S.-C. Moon, K. Kajihara, M. Hirano and H. Hosono, *Mater. Res. Bull.*, 2002, **37**, 2401–2406.

49. K. Shimura and H. Yoshida, *Energy Environ. Sci.*, 2010, **3**, 615–617.
50. Y. Inoue, T. Kubokawa and K. Sato, *J. Phys. Chem.*, 1991, **95**, 4059–4063.
51. S. Ogura, M. Kohno, K. Sato and Y. Inoue, *Appl. Surf. Sci.*, 1997, **121–122**, 521–524.
52. H. Kato and A. Kudo, *Chem. Phys. Lett.*, 1998, **295**, 487–492.
53. H. Kato and A. Kudo, *Catal. Today*, 2003, **78**, 561–569.
54. H. Kato and A. Kudo, *J. Phys. Chem. B*, 2001, **105**, 4285–4292.
55. A. Kudo and H. Kato, *Chem. Phys. Lett.*, 2000, **331**, 373–377.
56. H. Kato, K. Asakura and A. Kudo, *J. Am. Chem. Soc.*, 2003, **125**, 3082–3089.
57. G. Li, T. Kako, D. Wang, Z. Zou and J. Ye, *J. Phys. Chem. Solids*, 2008, **69**, 2487–2491.
58. A. Kudo and Y. Miseki, *Chem. Soc. Rev.*, 2009, **38**, 253–278.
59. Y. Miseki, H. Kato and A. Kudo, *Chem. Lett.*, 2005, **34**, 54–55.
60. Y. Moriya, T. Takata and K. Domen, *Coord. Chem. Rev.*, 2013, **257**, 1957–1969.
61. G. Hitoki, A. Ishikawa, T. Takata, J. N. Kondo, M. Hara and K. Domen, *Chem. Lett.*, 2002, **31**, 736–737.
62. S. G. Ebbinghaus, H.-P. Abicht, R. Dronskowski, T. Müller, A. Reller and A. Weidenkaff, *Prog. Solid State Chem.*, 2009, **37**, 173–205.
63. K. Maeda and K. Domen, *MRS Bull.*, 2011, **36**, 25–31.
64. K. Zhang and L. J. Guo, *Catal. Sci. Technol.*, 2013, **3**, 1672–1690.
65. G. Xie, K. Zhang, B. Guo, Q. Liu, L. Fang and J. R. Gong, *Adv. Mater.*, 2013, **25**, 3820–3839.
66. T.-F. Yeh, J.-M. Syu, C. Cheng, T.-H. Chang and H. Teng, *Adv. Funct. Mater.*, 2010, **20**, 2255–2262.
67. T.-F. Yeh, F.-F. Chan, C.-T. Hsieh and H. Teng, *J. Phys. Chem. C*, 2011, **115**, 22587–22597.
68. Y. Matsumoto, M. Koinuma, S. Ida, S. Hayami, T. Taniguchi, K. Hatakeyama, H. Tateishi, Y. Watanabe and S. Amano, *J. Phys. Chem. C*, 2011, **115**, 19280–19286.
69. W. Xinchen, K. Maeda, A. Thomas, K. Takanabe, X. Gang, J. M. Carlsson, K. Domen and M. Antonietti, *Nat. Mater.*, 2009, **8**, 76–80.
70. S. Cao and J. Yu, *J. Phys. Chem. Lett.*, 2014, **5**, 2101–2107.
71. A. Kudo, *MRS Bull.*, 2011, **36**, 32–38.
72. R. Abe, *Bull. Chem. Soc. Jpn.*, 2011, **84**, 1000–1030.
73. R. Sivakumar, J. Thomas and M. Yoon, *J. Photochem. Photobiol. C*, 2012, **13**, 277–298.
74. A. Ioannidis and E. Papaconstantinou, *Inorg. Chem.*, 1985, **24**, 439–441.
75. P. Argitis and E. Papaconstantinou, *J. Photochem.*, 1985, **30**, 445–451.
76. C. L. Hill, *Comprehensive Coordination Chemistry II*, ed. J. A. McCleverty and T. J. Meyer, Pergamon, Oxford, 2003, pp. 679–759.
77. N. Muradov and A. T-Raissi, *J. Sol. Energy Eng.*, 2006, **128**, 326–330.
78. M. I. Rusmatov, N. Z. Muradov, A. D. Guseinova and Y. V. Bazhutin, *Int. J. Hydrogen Energy*, 1988, **13**, 533–538.
79. H. Hori, K. Koike, Y. Sakai, H. Murakami, K. Hayashi and K. Nomiya, *Energy Fuels*, 2005, **19**, 2209–2213.
80. A. Zarkadoulas, E. Koutsouri and C. A. Mitsopoulou, *Coord. Chem. Rev.*, 2012, **256**, 2424–2434.
81. M. Bowker, *Green Chem.*, 2011, **13**, 2235–2246.
82. M. D. Melo and L. A. Silva, *J. Braz. Chem. Soc.*, 2011, **22**, 1399–1406.
83. M. Cargnello, A. Gasparotto, V. Gombac, T. Montini, D. Barreca and P. Fornasiero, *Eur. J. Inorg. Chem.*, 2011, **2011**, 4309–4323.
84. V. Preethi and S. Kanmni, *Mater. Sci. Semicond. Process.*, 2013, **16**, 561–575.
85. G. Ma, H. Yan, J. Shi, X. Zong, Z. Lei and C. Li, *J. Catal.*, 2008, **260**, 134–140.

14 Photocatalytic CO$_2$ Reduction

Ana Primo, Ştefan Neaţu, and Hermenegildo García

CONTENTS

14.1 INTRODUCTION

Photocatalysis aims at converting light into chemical energy.[1–5] This typically can be achieved by using a solid material (*photocatalyst*) that is able to absorb photons of a wavelength range to produce reactive chemical species on its surface. Most of the photocatalysts are semiconductor solid materials that are characterized by the existence of a valence band fully occupied by electrons and an empty conduction band, there being a bandgap energy between these occupied and empty bands (Scheme 14.1).[5]

Electronic transitions from the valence to the conduction bands can be promoted by light absorption provided that the energy of the excitation photons is higher than the bandgap energy of the semiconductor. This photon absorption leads to a short-lived (in the submicrosecond timescale or shorter) state of *local* charge separation in which electrons populate the conduction band in close proximity to positive electron holes created in the valence band of the material. For instance, in the case of TiO$_2$, this charge separation can be understood as one electron initially localized on an oxygen atom where the photon is absorbed being transferred to the Ti^{4+} ion. This short-lived state of charge separation has various possible pathways to decay

421

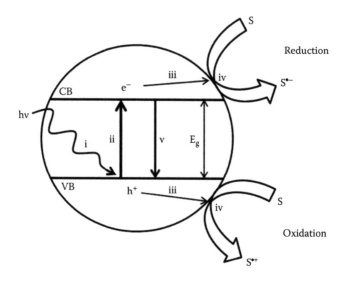

SCHEME 14.1   Elementary steps of a photocatalytic process. For the meaning of the Roman numerals, see text. E_g stands for energy gap.

including prompt recombination (*back electron transfer* in the geminate pair) but also random walk migration of the charges out of the site in which the photon was originally absorbed.

In this way, the general mechanism of a photocatalytic process has several elementary steps including (i) light absorption, (ii) the promotion of one electron from the valence to the conduction band, (iii) charge migration from the site of charge separation to the external surface, and (iv) reaction at the external surface of electrons (e^-) and holes (h^+). Electron/hole recombination, either at the site where the photon has generated charge separation (v) or later upon migration of the charges, is an unwanted energy waste process since the overall conversion of the energy of the photon is into heat. Scheme 14.1 summarizes the possible elementary steps occurring after light absorption. Charge recombination is the main deactivation process and it should be minimized as much as possible.

In order to maintain the charge balance in the process, the consumption rate of e^- and h^+ must be equal during operation of the photocatalytic reaction, and this means that the overall reaction rate is controlled by the slowest of the two processes, either electron or hole capture. In many photocatalytic reactions in the absence of h^+ quenchers, h^+ is the species controlling the overall reaction rate of the process. Besides the primary photocatalytic process, the increase in the overall efficiency of a photocatalytic system may require the presence of cocatalysts located at the external surface in intimate contact with the semiconductor.[6-8] The role of cocatalysts is to store and manage the transfer of e^- and h^+ much faster than in their absence, avoiding the accumulation of charges in the semiconductor. Typically, noble metal nanoparticles (NPs) such as Pt or Au act as cocatalysts for e^-, and RuO_2, CoO_2, and other metal oxides are adequate h^+ cocatalysts. The importance of cocatalyst is being increasingly recognized in the field, and notable increments in the photocatalytic activity of one order of magnitude or more can be achieved with its presence. In this way, the complete design of a photocatalyst should contain the photoactive semiconductor and two cocatalysts to handle the reduction and oxidation semireactions by handling e$^-$ and h$^+$.

14.2 APPLICATIONS OF PHOTOCATALYSIS TO ENVIRONMENTAL REMEDIATION AND SOLAR FUELS PRODUCTION

When the photocatalytic reaction is carried out at the ambient atmosphere or in water suspension, the universal electron trapping agent is atmospheric oxygen due to its lower reduction potential compared to the other species present in the media.[9] The h$^+$ quencher must correspond to the species that is oxidized easier and this species can be frequently water. What is interesting is that either by oxygen reduction or by water oxidation, the reactive species that are formed can be coincident and are generally described as reactive oxygen species (*ROS*). ROS include hydroxyl radicals (•OH), hydroperoxyl radicals (HOO•), hydrogen peroxide (H$_2$O$_2$), and superoxide (O$_2$•$^-$), among other possible ROS (Scheme 14.2). ROS can be generated from molecular oxygen by reduction with e$^-$ from 0 to -1 oxidation state or by oxidation of water from -2 to -1 state by reaction with h$^+$. In both cases, most of the ROS have one or two oxygen atoms in the -1 oxidation state that is highly reactive.

Hydroxyl radical is one of the most aggressive species and can react virtually with any organic compound and many inorganic species.[10,11] The three general reaction pathways available to hydroxyl radicals are summarized in Scheme 14.3 and include electron abstraction to form hydroxide anion, hydrogen abstraction to form water, and electrophilic attack to unsaturated multiple bonds. Considering the high oxidation potential of •OH radical ($E_{ox} = +2.80$ V), many electron-rich compounds can undergo electron transfer leading to the generation of the corresponding radical cation that subsequently can react further with oxygen, which are the first steps of oxidative degradation. In addition, the strength of the O–H bond ($E_{O-H} = 463$ kJ mol^{-1})

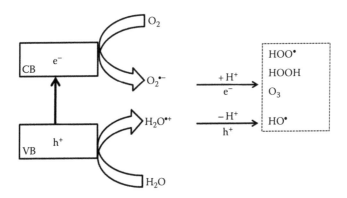

SCHEME 14.2   General mechanism for the generation of ROS from molecular oxygen and H$_2$O.

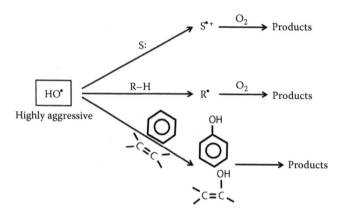

SCHEME 14.3   Reaction pathways triggered by hydroxyl radicals.

is responsible for the ability of this oxygen radical to abstract one H atom from most C–H bonds in organic compounds (E_{C-H}=413 kJ mol^{-1}) leading to the formation of C-centered radicals that can react with molecular oxygen in a reaction chain mechanism leading to peroxyl radicals and triggering also the oxidative degradation of organic pollutants (Scheme 14.4).

According to the previous comments on the oxidative degradation of organic compounds by ROS, photocatalysis is one of the most powerful techniques to degrade organic pollutants in the atmosphere, in the aqueous phase, or even in solid media. For this reason, photocatalysis was initially applied to waste water treatment, air cleaning, and environmental remediation in general.[1–4,9]

However, since the seminal work of Fujishima and Honda showing the possibility of photoelectrocatalytic generation of hydrogen from water[12] and later the reduction of CO_2,[13] there has been an increasing interest in applying photocatalysis for the production of what is called *solar fuels*.[1,14–20] The direct use of Sun energy in transportation would be problematic not only for the daily cycle with days and nights

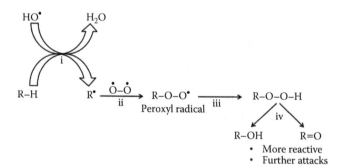

SCHEME 14.4   Autoxidation of organic compounds mediated by C-centered radicals initiated by hydroxyl radicals: (i) generation of C-centered radical; (ii) trapping of R• by molecular oxygen; (iii) hydrogen abstraction; and (iv) formation of oxidized products.

and the dependence on weather conditions but also because the low power density of the solar light would make it necessary to harvest light in a large surface to collect enough photons. Therefore, it seems more viable to accumulate the energy of the Sun in electrical devices or, more simply, in a chemical compound that can release energy with high power on demand. Thus, one viable possibility to replace fossil fuels would be based on the generation from sunlight of chemicals that can be used in transportation (*solar fuels*). Among them, the most studied by far is hydrogen generation by water reduction or water splitting. But as it will be commented in later text, other possible solar fuels are gaining increasing interest, particularly those derived from CO$_2$ reduction such as methanol, methane, and CO.[17,21,22]

14.3 PROPERTIES OF A PHOTOCATALYST

An ideal photocatalyst should have all the desirable properties of a solid catalyst including large surface area, high porosity, large density of active sites, and stability, but it has additional requirements derived from the interaction of light with the solid. Specifically, light absorption is a prerequisite in order to even consider the possibility of photocatalysis. Most of the photocatalysts that have been studied, such as TiO$_2$ and ZnO, exhibit a strong absorption in the UV region below 400 nm. Accordingly, these solids can be excited only with UV light and require artificial light to exhibit high photocatalytic activity.

It happens that frequently, and particularly for solar fuels production as in the present case, the excitation light source must be exclusively natural sunlight. Sunlight reaching the Earth surface contains only a small percentage of UV light depending on latitude, height, weather conditions, etc., that can be considered only about 4% as average. About 48% of the sunlight falls into the visible region and is not absorbed by white solids, as they are most common photocatalysts such as TiO$_2$ and ZnO. The white color indicates the lack of light absorption of any visible wavelength and the reflection of all the visible photons giving the white appearance. For this reason, the search of materials that can act as solar light photocatalysts, including the modification of photocatalysts that are only active under UV irradiation, has been a topic of continued interest and is closely related to develop colored photocatalysts with narrow bandgap energy falling in the visible range. Excitation with visible light photons requires that the bandgap of the semiconductor is about 1.5 eV that is much smaller than many of the photocatalysts that are commonly used.

14.4 TITANIUM DIOXIDE AS PHOTOCATALYST

Titanium dioxide with a bandgap for the most active anatase phase of about 3.4 eV has been by far the most studied photocatalyst.[23] Extensive studies have led to the conclusion that crystallinity and crystal phase strongly influence the efficiency of the photocatalytic process, the most active samples of this metal oxide corresponding to the anatase crystal phase, while the photocatalytic activity decreases considerably for the amorphous materials and even for the rutile form.[23–26] It has also been found that the combination of anatase with a minor percentage (below 20%) of rutile renders the most active titania photocatalyst.[25,26] This combination of anatase contaminated by some rutile is found in the commercial Evonik TiO_2 sample P25 obtained by flame hydrolysis of $TiCl_4$ that undergoes hydrolysis at high temperatures with the H_2O molecules formed in situ by combustion of H_2 and O_2 (Equation 14.1). The process renders in a very short period of time large quantities of highly reproducible titania samples that are constituted mostly by anatase with about 20% of rutile and renders one of the most efficient photocatalysts.

$$TiCl_4 + 2H_2 + O_2 \rightarrow TiO_2 + 4HCl \qquad (14.1)$$

There are other commercially available titania samples that exhibit very different photocatalytic activities, generally lower, among them and with respect to Evonik P25. This wide range in photocatalytic activity may derive from variations in the presence of impurities (Fe and other metals can dope titania), crystallinity and the presence of defects, crystal phase, particle size distribution, surface area and porosity, particle morphology, etc. All these parameters are known to affect considerably the photocatalytic activity by providing charge recombination centers, sites for electron trap, increased light absorption in the visible range, stronger substrate adsorption on the surface, etc. For this reason, the photocatalytic activity is notably influenced by the preparation procedure, the nature of the precursors, and the final properties of the titania.

Considering the advantages of titania in terms of photocatalytic activity with UV light, availability, stability, and lack of toxicity, there has been an intense research aimed at overcoming its main limitation, that is, the lack of visible light photoresponse. The goal has been to expand the activity of TiO_2 toward the visible in such a way that the efficiency for solar light irradiation increases by implementing response into the visible zone.[27]

The main strategy for this has been titania doping by replacing a low percentage of Ti or O atoms in the lattice by other heteroatoms.[28–34] The initial studies were focused on the influence of metal doping on the photocatalytic activity. The rationale behind doping is shown in Scheme 14.5 and consists in reducing the bandgap by introducing additional empty levels (below the conduction band) by doping with metal ions or additional full levels (above the valence band) by doping with nonmetallic elements. The list of the most studied metallic and nonmetallic dopants includes Fe, Cr, Cu, Ag, Pt, and Pd as well as N, C, and S.

One of the main problems regarding doping of titania is how to reproduce the results of photocatalytic activity and, not infrequently, contrasting reports in which the photoactivity after doping was found to be higher or lower than that of the material before doping was observed. There are basically two different synthetic methods, either doping during the sol–gel synthesis of titania or by postsynthesis modification of crystalline titania. In the first methodology, there is a final thermal treatment to increase the crystallinity of the material to enhance the photocatalytic activity. As commented earlier, amorphous titania generally obtained in the sol–gel synthesis at moderate temperature exhibits poor photocatalytic activity and it is necessary to calcine at adequate temperatures above 300°C to increase the crystallinity of the solid obtained by sol–gel and, as consequence, to increase its photocatalytic activity. It is during this thermal annealing where doping heteroatoms can migrate and become

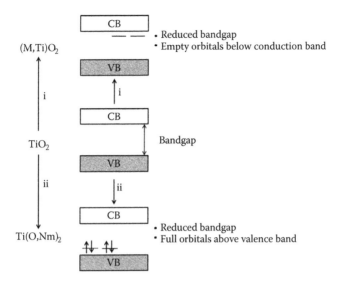

SCHEME 14.5 Two complementary strategies based on TiO$_2$ doping by metals (i: introducing empty levels below the conduction band) or by nonmetallic elements (ii: providing electrons with energy above the valence band). Both possibilities lead to an effective narrowing of the bandgap and shift photon absorption for the doped material to the visible region.

partially or totally expelled out of the lattice as a consequence of the crystallinity increase. If the thermal treatment is carried out under milder conditions to minimize heteroatom rearrangement and generation of extraframework dopant atoms, then the presence of defects in the lattice can still be sufficient to decrease substantially the efficiency of the photocatalytic reaction.

The second methodology for doping consisting in postsynthetic modification of preformed titania has as the main drawback the robustness and stability of TiO$_2$ phases that make necessary harsh conditions to graft the heteroatom. One typical procedure is to expose titania in a fixed-bed reactor placed in an oven at the required temperature to a gas flow of the dopant precursor. This procedure frequently leads to heteroatom doping on a shallow depth of the particle surface, with dopant elements weakly bound to the TiO$_2$ lattice and can be removed or rearranged in a large extent during the operation of the photocatalyst.

In fact, one of the main problems of doping is the process called photocorrosion that consists in the modification of the material during operation as consequence of the continuous charge separation/recombination process, mostly involving the loosely bound heteroatom due to the intragap energy levels that it provides. Thus, the weakest component of the lattice corresponding to the doping heteroatom is submitted to continuous oxidation/reduction stress. This photocorrosion leads to the leaching and solubilization of part of the lattice and, not surprisingly, generally involving the doping heteroatom. Thus, any doped TiO$_2$ has to be carefully surveyed not only for the initial activity of the material with respect to undoped titania but also importantly for its stability after long-term operation. It may happen that the activity of the fresh material decays significantly during the course of the photocatalysis.

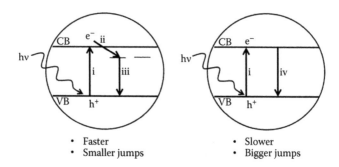

- Faster
- Smaller jumps

- Slower
- Bigger jumps

SCHEME 14.6   Possible negative effect of doping by accelerating e⁻/h⁺ charge recombination: (i) charge separation; (ii) electron migration to intra bandgap levels; (iii) fast charge recombination; and (iv) slow charge recombination.

On the other hand, it has been also claimed that if the loading percent of doping increases beyond an optimal value, then a detrimental effect of doping could appear due to the enhancement of the unfavorable e⁻/h⁺ recombination taking place in these centers. This negative doping effect is due to the position of the levels in the intra-bandgap energy allowing e⁻ or h⁺ to use these levels as intermediate stations in their recombination from the conduction to the valence bands, making this recombination faster (Scheme 14.6). This also makes the process of doping more difficult, since low doping values may not produce the desired photocatalytic enhancement, while high doping percentage could be detrimental for the photocatalytic efficiency. It should be reminded that as commented earlier the exact doping percentage may be difficult to control depending on the doping process.

14.5   PLASMONIC PHOTOCATALYSIS

An alternative to doping that has emerged more recently to expand the visible light photoresponse consists in depositing Au NPs on the surface of TiO_2.[35,36] Au, Ag, and Cu NPs exhibit an absorption band in the visible as a consequence of the collective oscillation of electrons in the particle. This visible band is called surface plasmon band and is responsible for the color of these metal NPs, pink in the case of Au, yellow for Ag, and red for Cu. These metal NPs can be formed by reduction of the corresponding metal salts dissolved in water using citric or ascorbic acids or by using metal hydrides and can be obtained with fairly homogeneous distribution in sizes below 10 nm. The exact position of the surface plasmon band depends on several factors including the dielectric constant of the support and medium, the size and morphology of NPs, and the charge density of the NPs. For instance, metal nanorods exhibit a surface plasmon band with two different maxima corresponding to the oscillation of electrons in the direction of the longest axis of the rod or perpendicular to it, the latter appearing at shorter wavelengths than the former.

Excitation of the surface plasmon band produces large electric fields near the metal NP that can have various effects,[37] as for instance, a notable enhancement of the Raman spectra of those molecules on the surface or spatially close to these

NPs, being in this way possible to record the Raman spectrum of a single molecule. Excitation of the metal NP surface plasmon band can also produce a local instantaneous increase in the temperature on the NP surface estimated in about 500°C that can produce even the melting of the NP. This sudden temperature increase decays in the microsecond timescale. Another phenomenon promoted by surface plasmon band excitation is the ejection of electrons from the metal NP to the medium, and these electrons can be injected into the semiconductor conduction band when the metal NP is supported on the surface of the semiconductor.

Au is a noble metal that is among the most stable as a metal and difficult to oxidize. In addition, Au NPs can be easily grafted on the surface of TiO$_2$ following several procedures and, in particular, the one reported by Haruta and termed as deposition/precipitation.[38] This protocol leads reproducibly to Au NPs of a size distribution around 5 nm that are strongly bound to the surface of TiO$_2$. The resulting material has been used as catalyst to promote selective reduction of nitro groups and low temperature CO oxidation, among other reactions.[39] Thus, the combination of reliable preparation protocols together with high (photo)stability makes Au/TiO$_2$ an interesting material for application in the field of photocatalysis. It should be mentioned that the weight percentage of Au NPs to achieve the highest activity can be very low, sometimes between 0.25 and 0.5 wt.% and, therefore, the cost of the precious metal is not unaffordable.

It happens that Au/TiO$_2$ is among the most efficient photocatalysts for hydrogen generation from water, exhibiting visible light activity.[40] It has been proposed that by excitation with visible light at the surface plasmon band of Au NPs, hot electrons ejected from Au NPs can populate the conduction band of the semiconductor leading to efficient charge separation upon excitation with visible light. Scheme 14.7 summarizes the proposed mechanism to rationalize the visible light photocatalytic activity of Au/TiO$_2$.

Besides visible light activity introduced by the presence of Au NPs grafted on the TiO$_2$ surface, other effect explaining the high photocatalytic activity of

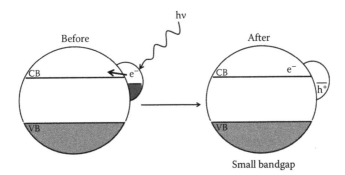

SCHEME 14.7 Mechanistic proposal for the visible light photoresponse of Au/TiO$_2$ arising from photoexcitation of the metal surface plasmon band on the surface of TiO$_2$ and injection of hot electrons from Au NP to the conduction band of TiO$_2$. Note that in this case, the oxidation potential of h$^+$ located on Au NPs should be much lower than h$^+$ on TiO$_2$ valence band.

Au/TiO$_2$ is the role of Au NPs as catalytic centers for efficient e$^-$ and h$^+$ transfer to the substrate. As commented earlier, current design of photocatalysts includes not only photoactive semiconductors but also their modification supporting on their surface NPs that can act as oxidation or reduction centers. These NPs should increase the rate of charge transfer to the final substrate. In this regard, it has been known for a long time that the presence of Pt in the system increases considerably (up to a factor of 100 times) the rate of hydrogen generation exhibited by TiO$_2$.[41] This catalytic effect of Pt was believed to arise from the role of these Pt NPs to store electrons from the conduction band of TiO$_2$, acting as electron buffer, and the subsequent highly efficient generation of H$_2$ gas by Pt NPs. It should be commented that Au NPs can also act as cocatalyst center playing the same role as Pt NPs storing in the NPs many e$^-$ or h$^+$ and favoring the transfer of this charge to the substrate, although for this role Pt has been found to be a more efficient photocatalyst.[42] A major difference between Pt NPs and Au NPs, besides the higher catalytic activity of Pt for hydrogenation in general, is that Pt NPs do not exhibit any surface plasmon band and, therefore, Pt NPs cannot be used to produce visible light photoresponse in TiO$_2$.

A third role of Au NPs besides photosensitization of TiO$_2$ upon light absorption at the surface plasmon band and cocatalyst center for H$_2$ generation is to increase the efficiency of charge separation by disfavoring e$^-$/h$^+$ recombination. Since Au NPs and TiO$_2$ are independent microdomains that are separated by a surface (see Scheme 14.7), once the electron crosses the interface as in the elementary process of electron injection from Au NPs to the conduction band of TiO$_2$, recombination becomes very difficult. It is believed that the combination of these three effects is what renders Au/TiO$_2$ as one of the most efficient photocatalysts under visible light irradiation.

14.6   SPECIFIC FEATURES OF CO$_2$ REDUCTION

Among the possible solar fuels that can be prepared photocatalytically, the one that has attracted larger attention has been hydrogen generation from water. From a simplistic mechanism, this reaction only requires the transfer of one electron, generated photocatalytically on the semiconductor surface, to a H$^+$ with the formation of a hydrogen atom. Two hydrogen atoms will bind to form a dihydrogen molecule basically without activation energy barrier. Hydrogen has been identified as the optimal transportation fuel since this element is widely available. It has a large energy power and its massive use will not influence the environment since its oxidation in fuel cells will lead to water as the only by-product.[43,44] However, there are still several technology gaps, including hydrogen storage and efficient fuel cells that make the use of hydrogen far from application (Figure 14.1). The problems associated with the use of hydrogen as fuel have motivated the interest in developing also the photocatalytic production of other solar fuels, particularly those derived from CO$_2$, such as methanol, CO, and methane.

There are, however, remarkable differences in the photocatalytic CO$_2$ reduction compared to hydrogen generation from water that need to be commented in this section. Figure 14.2 summarizes the differences between H$_2$ generation from water and CO$_2$ reduction.

Advantages	Knowledge Gaps
• High energy density • H$_2$O as only by-product • No environmental drawback	• Efficient generation from primary energy resources • Room temperature storage • Durable fuel cells

FIGURE 14.1   Advantages and current gaps in hydrogen technology for transportation.

Advantages of H$_2$ Generation	Knowledge Gaps in Photocatalytic CO$_2$ Reduction
• Possibility of overall water splitting • No by-products • H$_2$ on the gas phase	• Need of a reagent providing hydrogen • Complex product distribution • Gaseous and liquid products

FIGURE 14.2   Differences between hydrogen generation from water and photocatalytic CO$_2$ reduction.

The first one is that while water can undergo simultaneous reduction to hydrogen and oxidation to oxygen and, in this way, the overall water splitting into hydrogen and oxygen is possible, the formation of methanol, methane, and other compounds derived from CO$_2$ requires the addition of a sacrificial hydrogen donor compound that in the overall photocatalytic cycle should act as a hole scavenger. In this way, an ideal wanted CO$_2$ reduction process will be the one indicated in Equation 14.2 in which H$_2$O is oxidized by h$^+$ to oxygen. However, the problem in this case is how to reduce CO$_2$ selectively in the presence of water.

$$CO_2 + 2H_2O \rightarrow CH_3OH + \frac{3}{2}O_2 \quad \Delta G^0 = 702.27 \text{ kJ/mol} \quad (14.2)$$

According to thermodynamic potentials, water reduction to hydrogen should take place at lower potentials than CO$_2$ reduction and, therefore, besides providing conduction band electrons with high reduction potential, the photocatalyst has to exhibit active sites with kinetic selectivity toward CO$_2$ with respect to water. Preferential or simultaneous H$_2$O reduction taking place along CO$_2$ reduction is particularly problematic when the photocatalytic CO$_2$ reaction is performed in aqueous media. Under these conditions, hydrogen evolution can be the prevalent photocatalytic process taking place. Although the knowledge is still limited and there is at the moment no conceptual rationalization, it appears that the presence of Pt NPs on TiO$_2$ favors the generation of H$_2$ from H$_2$O, while CuOx as cocatalysts is more selective for CO$_2$ reduction leading mainly to CH$_4$. This change in the photocatalytic process depending on the cocatalyst present on TiO$_2$ is a clear example of how the activity and efficiency of a semiconductor can be controlled and improved by the presence of additional centers that direct the product formation.

With regard to the architecture of the active sites for CO_2 reduction, one requirement that derives from thermodynamics is that the reduction potential of elementary steps decreases significantly as the transfer of the electron is accompanied by simultaneous transfer of a proton as indicated in Equations 14.3 through 14.8. Since the reduction of CO_2 to methanol requires the transfer of six electrons and protons, thermodynamics predicts that the reduction potential should decrease continuously, the reaction being less endoergonic, as the reaction being considered involves the simultaneous transfer of several protons and electrons. It should be noted that Equations 14.3 through 14.8 require that protons and electrons do not react among them to form hydrogen, but both should be added to CO_2, suggesting that the ideal architecture of the active sites for photoreduction should have acid sites near the center where the conduction band electrons are transferred to CO_2 without interference between the electron and proton flow. In a simplistic way, Equations 14.3 through 14.8 can be rationalized considering that the combination of one or more H^+ and e^- is the driving force that renders CO_2 reduction less unfavorable.

$$CO_2 + e^- \rightarrow CO_2^- \qquad\qquad E^0_{redox} = -1.90 \text{ V} \qquad\qquad (14.3)$$

$$CO_2 + 2H^+ + 2e^- \rightarrow HCOOH \qquad E^0_{redox} = -0.61 \text{ V} \qquad\qquad (14.4)$$

$$CO_2 + 2H^+ + 2e^- \rightarrow CO + H_2O \qquad E^0_{redox} = -0.53 \text{ V} \qquad\qquad (14.5)$$

$$CO_2 + 4H^+ + 4e^- \rightarrow HCHO + H_2O \qquad E^0_{redox} = -0.48 \text{ V} \qquad\qquad (14.6)$$

$$CO_2 + 6H^+ + 6e^- \rightarrow CH_3OH + H_2O \qquad E^0_{redox} = -0.38 \text{ V} \qquad\qquad (14.7)$$

$$CO_2 + 8H^+ + 8e^- \rightarrow CH_4 + 2H_2O \qquad E^0_{redox} = -0.24 \text{ V} \qquad\qquad (14.8)$$

Another notable difference between hydrogen generation from water and photocatalytic CO_2 reduction is that in the latter case, a complicated mixture of products can be obtained. The possible products include not only oxalic and formic acids, formaldehyde, and methanol but also elemental carbon, CO, CH_4, and C_2–C_4 and saturated and unsaturated hydrocarbons including ethylene, ethane, and propane. Of these products, the most valuable one would be methanol, which is liquid at room temperature and contains the highest energy density among oxygenated derivatives. In general, liquid products at room temperature are easier to handle than gases. On the other hand, considering the current vast consumption of natural gas and that methane technology is in use, its formation from CO_2 reduction will also be of considerable interest. However, it is still unclear how to gain selectivity on the product formation during the photocatalytic reduction of CO_2. This lack of knowledge derives from the poor understanding of the reaction mechanism of the photocatalytic CO$_2$ reduction and the paucity of theoretical studies providing rationale to the product distribution.

One problem related to the possible complex product distribution that could be formed in the photocatalytic CO$_2$ reduction is to provide full analytical data not only of the gas phase but also of the liquid phase and the solid photocatalyst. Careful mass balances should be performed to assess that the percentage of C-containing products that have not been analyzed is negligible. Product analysis can be specially complex for those reactions carried out in water. Very frequently, basic pH values are employed to increase CO$_2$ solubility in water, but under these conditions, carboxylic acids such as formic and oxalic acids that could be formed will be present as carboxylate salts and should be analyzed by ion chromatography. On the other hand, formaldehyde and methanol are highly water soluble and must be analyzed and quantified by liquid chromatography with refraction index detectors. Also products can be in the gas phase and they should be determined and quantified by gas chromatography using standards and adequate calibration protocols. At least two types of detectors (and possibly carrier gases) should be needed to detect H$_2$, CH$_4$, and also higher hydrocarbons and alcohols. Finally, if the amount of products is low compared to the mass of the photocatalyst, it is a good practice to characterize possible products that could remain adsorbed on the solid surface. Among the possible products that could be strongly adsorbed on the photocatalyst, the most common ones are elemental carbon deposits and carboxylic acids that bind strongly to the TiO$_2$ surface.

Not uncommonly, CO$_2$ conversions are in the range of a few μmol g^{-1} h^{-1}. In these cases with low product formation, a good practice is to present data of blank controls submitting the photocatalyst to illumination in the absence of CO$_2$. It could happen that organic compounds deposited on the photocatalyst surface are the origin of this small amount of products observed in the gas phase rather than CO$_2$, which is much less reactive than the impurities.

14.7 HIGHLY ACTIVE TITANIA PHOTOCATALYSTS WITH UV LIGHT

A general strategy that has been used to prepare some highly efficient photocatalysts is based on isolation and entrapment of photoactive species in an inert matrix that provides large surface area and an absorption capability near the photocatalytic center. In addition, the solid matrix can also have special properties in terms of polarity and dielectric constant that can enhance the efficiency of some elementary steps and, particularly, charge separation and migration of the carriers that are two key steps in photocatalysis.

As highly porous and large surface area matrix, zeolites and related periodic mesoporous aluminosilicates have been often used.[45] In both cases, the rigid lattice of these solids constituted by SiO$_4$ or AlO$_4^-$ tetrahedra sharing the corners define cages and cavities that can be accessed from the exterior of the particle, allowing intracrystalline diffusion of substrates and products reaching the photoactive sites. In this context, small clusters of TiO$_2$ or even independent Ti atoms have been encapsulated inside zeolites or grafted on the lattice of these materials.[46,47] By reducing the particle sizes down to the nanometric scale, dramatic variations in the photocatalytic behavior have been observed.[45] In this way, it has been shown that the photocatalytic activity per Ti atom for CO_2 reduction is the highest when Ti is included in the zeolite pores.[46] This is not surprising considering that by preparing small clusters of TiO_2 almost virtually all Ti atoms become accessible and can interact with CO_2 and H_2O. In contrast, in conventional NPs there are thousands of Ti atoms forming part of the framework of the particle that will not be at the external surface of the material in contact with substrates.

The main problem of this strategy of active site entrapment is, however, that by reducing the particle size of TiO_2, a point is reached in which *quantum confinement* effects start to operate, and although this can be beneficial from the point of view of the lifetime of charge separation, the onset of the absorption becomes significantly shifted toward shorter wavelengths. This blue shift is a reflection of the increase in the bandgap of small TiO_2 clusters with respect to conventional TiO_2 NPs. On one hand, this increase in the bandgap potential is associated with higher reduction potential of electrons that is a positive parameter for CO_2 reduction, but on the other hand, this increase in the bandgap makes photoexcitation of the material by sunlight not possible and CO_2 photoreduction by TiO_2 clusters can only be achieved with artificial UV light. Scheme 14.8 summarizes the advantages and disadvantages of entrapped small TiO_2 clusters compared to conventional TiO_2 NPs.

Furthermore, as commented earlier, one of the main limitations of the current state of the art in photocatalytic CO_2 reduction is the lack of tools to control the product distribution in the photoreduction. In this regard, it has been shown that isolated Ti atoms grafted in mesoporous aluminosilicates form selectively methanol, while the presence of Ti aggregates in the material changes the selectivity of the photocatalyst toward the formation of CH_4.[48] According to this observation, it should be possible to prepare different photocatalysts based on Ti atoms embedded on porous silicates exhibiting the optimal selectivity toward the target product distribution, particularly methanol.

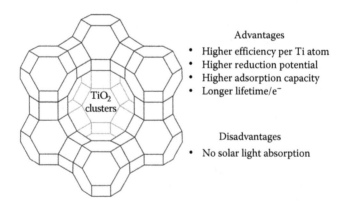

SCHEME 14.8 Model of TiO_2 clusters encapsulated inside the cavities of zeolite Y and list of advantages and disadvantages of the strategy based on the use of this type of photocatalysts.

14.8   SOLAR LIGHT TITANIA PHOTOCATALYSTS FOR CO$_2$ REDUCTION

One of the TiO$_2$-based photocatalysts that has been reported to exhibit among the highest photocatalytic activity for CO$_2$ reduction by H$_2$O in the gas phase upon exposure to the natural sunlight has been an array of N-doped TiO$_2$ that has been modified by Cu and Pt (Scheme 14.9).[49] Since this material constitutes a paradigmatic example of a titania photocatalyst designed for optimal performance under solar light irradiation, it is worth to comment it in some detail to provide understanding on the current directions to enhance titania activity.

Thin titanium metal foils (<1 mm) can be anodized in the presence of fluoride salts in highly viscous media such as ethylene glycol under controlled electrochemical conditions of maximum voltage and current to form nanotube arrays of titania.[50,51] In the reported photocatalytic material, the depth of the TiO$_2$ array is about 130 μm, with a pore diameter of 95 nm and a wall thickness of 20 nm, approximately. This porosity ensures a large surface area for interaction with CO$_2$ and H$_2$O as well as an easy accessibility of substrate to most of the titania.

N doping was achieved during electrochemical oxidation of the titanium metal foil using NH$_4$F as electrolyte, but also part of the N doping is due to dissolved atmospheric N$_2$ present during the anodization of Ti foil as some controls using HF as electrolyte have shown. The photocatalytic activity of these arrays tends to be poor due to the low crystallinity of the as-prepared material. However, annealing at temperatures of 460°C or 600°C makes the material to become mostly in the anatase crystal phase, increasing considerably the photocatalytic efficiency of the array. This annealing has to be performed in the absence of oxygen since otherwise most of the doping by N will be lost.

Structuring of titania as nanotube arrays is important for the overall efficiency since it is known that the morphology of titania can play an important role in its photocatalytic activity. Besides doping and chemical composition, the photocatalytic activity of TiO$_2$ also depends to a large extent on the morphology and size of the particles and their spatial structuring.[52] Studies by transient absorption spectroscopy detecting the charge separation state have determined that samples with long aspect

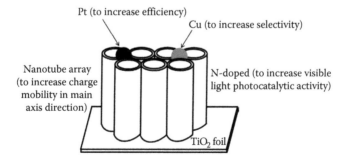

SCHEME 14.9   Simplistic representation of the N-doped TiO$_2$ nanotube arrays modified with Pt and Cu.

ratio exhibit enhanced charge mobility along the longest direction of the particle.[53] In this way, by structuring titania as nanotubes, mobility of the charges should take place preferentially along the nanotube main direction and the mean free path of electrons increases in this direction as compared to spherical titania particles, where no preferential charge migration should take place.

On the other hand, N doping increases the photocatalytic activity of the nanotube array to the visible light as a consequence of the shift of the onset of the absorption band toward longer wavelengths (Scheme 14.5).

The nanotube array photocatalytic system also comprises Pt or Cu metals that have been deposited on top of the surface by dc sputtering and results in the formation of islands of these metals at the pore openings of the arrays, without blocking the access to the nanotubes. The role of these metals is to act as cocatalysts and favor gas evolution, particularly of the reduction semireaction. The nature of this metal exerts a notable influence on the product distribution arising from the photocatalytic CO_2 reduction. Although the way in which the nature of the metal influences the outcome of the photocatalytic reduction is not yet understood, it seems that as a general rule, Pt favors hydrogen evolution with respect to CO_2 reduction in comparison with other metals. We have mentioned that evolution of H_2, even in larger quantities than the gaseous products derived from CO_2 reduction, should be expected for a mixture of CO_2 and H_2O. Cu NPs deposited on TiO_2, on the other hand, decrease the formation of H_2 gas, frequently without penalizing CO_2 conversion. The product distribution for a TiO_2-based catalyst containing either Pt or Cu NPs, rather than a combination of the two, has some remarkable differences.

In general, it can be said that Cu NPs on the photocatalyst favor the selectivity toward methane. This Cu NP selectivity to methane contrasts with electrocatalytic data for CO_2 reduction at Cu anodes that have shown a high selectivity toward methanol formation. A nanotube titania array that was covered only by independent Cu and Pt NPs exhibited the highest performance with a productivity of about 160 µL g^{-1} h^{-1} that is about 20 times higher than that reported for titania NPs upon UV-light irradiation. These comparisons among the activity of different photocatalyst should be, however, taken carefully since differences in the spectrum and power of the excitation source and even geometry of the photoreactor should influence the apparent activity of the sample.

In this way, further improvements in this system with more strict periodic modulation and multiwall configuration of the nanotubes arrays as well as increased CO_2 pressure inside the photoreactor allow to go further in the productivity of the photocatalyst. Under these conditions, values as high as 160 µmol g^{-1} h^{-1} can be achieved.[54]

Concerning the actual numbers of productivity values, it should be commented that different factors, but particularly light intensity, reactor design, and operation conditions, can determine that the same material renders different productivity efficiency. In fact, one of the main problems in photocatalysis, where there is a need of light absorption in opaque powders that interact with gaseous substrates but also can form liquid products, is a convenient engineering of the process including adequate reactor design in which the photocatalyst should be in a shallow bed highly exposed to the photons. Similarly, implementation of continuous flow operation with sufficiently long contact times inside the reactor would be an achievement in the area of photocatalytic solar fuel production. Thus, productivity values have only a relative importance because they are strongly associated with the irradiation procedure and photoreactor system. Caution has to be taken when comparing the activity of different materials just by considering productivity numbers without paying attention to other conditions of the photocatalytic process.

As already commented, besides doping, other alternative to increase the solar light photocatalytic activity of titania is to use metal NPs with surface plasmon band. This strategy using a Au/TiO$_2$ photocatalyst in which Au NPs are supported on anatase has rendered one of the most active photocatalysts for hydrogen generation from water, and a similar photocatalytic activity has been described for CO$_2$ reduction using Pt–Cu/TiO$_2$.[55]

14.9 OTHER PHOTOCATALYSTS FOR CO$_2$ REDUCTION BESIDES TIO$_2$

TiO$_2$ as photocatalyst has several virtues including availability, robustness, and a high activity under UV irradiation. However, besides the lack of visible light photoresponse, an additional problem of TiO$_2$ is the relatively low reduction potential of conduction band electrons in comparison with the high reduction potential needed for the primary electron injection to CO$_2$ (Equation 14.3). For this reason, there is an obvious interest in exploring the photocatalytic activity of other semiconductors for CO$_2$ reduction.

The list of semiconductors that have been evaluated for photocatalytic CO$_2$ reduction includes ZnO, probably the second most used photocatalyst, and other transition metal oxides and sulfides, such as CdS in the form of quantum dots. Other different semiconductors including BiVO$_4$, Bi$_2$WO$_6$, InTaO$_4$, Zn$_{1.7}$GeN$_{1.8}$O, ZnAl$_2$O$_4$, and ZnGaNO have been tested as photocatalysts for CO$_2$ reduction by H$_2$O.[56] Although an exhaustive review on the activity of these materials is out of the scope of this chapter, the reader should be aware that besides the composition, also the spatial structuring of the material in the form of rods, tubes, ribbons, and layers should play an important role in determining the final photocatalytic activity, similarly as it has been described for the case of TiO$_2$ nanotubes. This influence is not only due to changes in the surface area available but also due to preferential directions for charge migration, the absence of interparticle boundaries that limit charge mobility, or even the exposure of preferential crystallographic planes that could favor the location of electrons or holes in contrast with adsorbates. Also as in the case of titania, further modification of the material by depositing cocatalysts is a general strategy to enhance their photoactivity. In this section, we will describe the results achieved with two semiconductors alternative to TiO$_2$ that are among those that have been exhibited a promising activity.

Layered double hydroxides (LDHs) are materials having a brucite-like structure that is formed by the stacking of sheets constituted by octahedra in which oxygen is at the corners and a divalent and a tri-/tetravalent cation at the centers (Scheme 14.10). The presence of trivalent (more rarely tetravalent) cation creates an excess of net positive charge in the layer that needs the presence of a counter anion in the intergallery space to ensure the electroneutrality of the solid. When the composition of the LDH

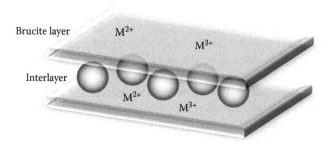

Brucite layer   M²⁺

Interlayer

M²⁺

M³⁺

M²⁺

M³⁺

SCHEME 14.10   Simplistic representation of an LDH with a brucite-like structure.

includes Ti, the material can be viewed as a layered titanate with a relatively high proportion of some divalent dopant, and therefore, it is expected that Ti-containing LDHs should exhibit photocatalytic activity. The interest in LDHs derives from their easy synthesis that can be considered frequently as a precipitation in basic aqueous media of a solution containing the mixture of soluble salts such as nitrates of the di- and polyvalent cations. In addition, the intergallery anion can be submitted to ion exchange to introduce the desired one. Actually, LDHs have a spontaneous tendency to adsorb CO_2 from the ambient and CO_3^{2-} is one of the preferred counter anions present in this type of layered materials. Thus, LDHs can be in principle considered as promising photocatalysts particularly for CO_2 reduction.

LDHs consisting of Zn^{2+} and/or Cu^{2+} as divalent ions and Al^{3+} or Ga^{3+} have been synthesized and tested as photocatalysts to convert gaseous CO_2 to CH_3OH or CO by UV/Vis irradiation using H_2 as reducing agent.[57,58] The use of H_2 increases significantly the efficiency of the photocatalytic CO_2 reduction with respect to water since the enthalpy of the reaction is less unfavorable and also kinetically the reaction mechanism is simpler. $[Zn_3Al(OH)_8]_2^+$ in the CO_3^{2-} form exhibits notable activity for CO_2 photoreduction, forming CO as the major product at a rate of 620 nmol g^{-1} h^{-1}.[59] In contrast, CH_3OH was observed as the major product using $[Zn_{1.5}Cu_{1.5}Ga(OH)_8]_2^+CO_3^{2-} \cdot H_2O$ as photocatalyst with a formation rate of 170 nmol g^{-1} h^{-1}.[59] The presence of Cu in the photocatalyst generally tends to favor in these materials the formation of CH_3OH over CH_4 or CO. In contrast to the product distribution found in these LDHs, the reference catalyst Cu–ZnO with similar chemical composition gives rise to 30 nmol g^{-1} h^{-1} of CO under the same conditions. Similarly, Ga_2O_3 affords CO in 47 nmol g^{-1} h^{-1} using the same setup. Thus, the way in which the photocatalytic efficiency and product distribution are modified by the composition and the structure of the photocatalyst cannot at the present be rationalized, and in situ studies and computational calculations are necessary to gain some understanding in this important point, as we will comment in Section 14.10.

Studies on the tuning of the photocatalytic properties of LDHs have been extended by introducing Cu ions in the composition of the layers as well as in the interlayer space.[60] Thus, the photocatalytic reduction of CO_2 to CH_3OH can be carried out using Zn–Ga or Zn–Cu–Ga hydroxide layers in the presence of H_2. In this case, the presence of structural water in the interlamellar spaces is detrimental for the photocatalysis and the activity can be increased by partial desorption of this water. Using a CO$_2$-to-H$_2$ mol ratio of 1:10, $[Zn_3Ga(OH)_8]_2^+(CO_3)^{2-}$•H$_2$O resulted in the generation of CO as major product with a productivity rate of 80 nmol g^{-1} h^{-1} under UV/Vis light irradiation.[60] In contrast, analogous material containing Cu in its composition, namely, $[Zn_{1.5}Cu_{1.5}Ga(OH)_8]_2^+$ CO$_3^{2-}$•H$_2$O, yielded CH$_3$OH as major product with higher efficiency at a rate of 170 nmol g^{-1} h^{-1}.[60] Partial removal of hydration water molecules in the material by heating at 150°C further increased the rate of CH$_3$OH formation by a factor of 1.8 to reach a value of 310 nmol g^{-1} h^{-1}.[60] Apparently, the availability of interlayer sites bound to Cu is crucial to control the selectivity in CO$_2$ photoreduction toward the formation of methanol. Similar enhancement of CH$_3$OH selectivity by the presence of Cu was observed for other layered materials such as $[Zn_3Ga(OH)_8]_2^+[Cu(OH)_4]^{2-}$•H$_2$O and $[Zn_{1.5}Cu_{1.5}Ga(OH)_8]_2^+$ $[Cu(OH)_4]^{2-}$•H$_2$O as photocatalysts. Overall, the earlier results serve to illustrate the versatility of LDH structure for tuning the composition of the two metals in the layer and in the interlayer space to optimize the performance of the solid with respect to the product distribution and photocatalytic activity. In addition, the basicity of the material contributes favorably to adsorb on its surface a large concentration of gaseous CO$_2$.

Recently, it has been reported that LDHs containing surface basic sites have a high water tolerance and show improved activity for the photocatalytic conversion of CO$_2$ into CO compared to a simple metal hydroxide in water.[61] Since the presence of water can lead to preferential photocatalytic hydrogen generation, it is believed that the presence of basic sites can favor the reaction of CO$_2$ by strong adsorption on the photocatalyst. Among the various LDHs tested for the photoreduction of CO$_2$, Mg–In LDH (Mg/In = 3) exhibited a yield of CO and O$_2$ of 3.21 and 17.0 μmol, respectively, which is much higher than that of the independent Mg(OH)$_2$ and In(OH)$_3$ hydroxides that afford negligible amounts of CO and O$_2$.[61]

14.10 MECHANISM INSIGHTS BY IN SITU SPECTROSCOPY

The current state of the art in the photocatalytic CO$_2$ reduction is needed for the understanding of the reaction mechanism. This knowledge can be obtained by theoretical calculations combined with in situ measurements. In this context, and considering that CO$_2$ adsorption on the surface should be the first elementary step in the photocatalytic reaction, it is of interest to determine what are the species formed in this adsorption that subsequently would evolve toward the photoproducts. This is particularly relevant considering that CO$_2$ is the acid dehydrated form of carbonic acid and that there are two stable species, namely, bicarbonate and carbonate, that are the basic forms of this acid. In this context, the influence of pretreatments in titania nanotubes having or not deposited platinum NPs has shown that different CO$_2$ adsorbates can be formed on the surface depending on the conditions of the thermal annealing to which the nanotubes were submitted.[62] In this way, heating under oxygen or air increases the electron density of surface oxygen that becomes stronger Lewis bases. In addition, this treatment also increases the electropositivity of Ti^{4+} ions that are converted into strong Lewis acids (Scheme 14.11). As a result of this treatment, CO$_2$ adsorption on the surface of oxidized titania nanotubes leads to the

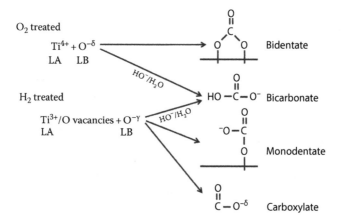

SCHEME 14.11 Species detected spectroscopically upon CO_2 adsorption on titania nanotubes depending on the pretreatment of the solid.

formation of bidentate carbonate and bicarbonate in the case that water or hydroxyl groups are available (Scheme 14.11).[62]

In contrast, reduction of the surface of titania nanotubes by hydrogen treatment at high temperature forms Ti^{3+} ions associated with O vacancies in the lattice, exhibiting very weak acidity due to the lower Coulombic charge.[62] On the other hand, oxygen is still strong Lewis base sites.[62] The result of this hydrogen reduction is that besides the detection of monodentate carbonate and bicarbonate if water is available, new species in which CO_2 has higher electron density, denoted as carboxylate, are present. Scheme 14.12 summarizes the different species formed from CO_2 adsorption depending on the surface treatment. Considering that titania nanotubes conveniently modified is one of the most efficient photocatalyst for CO_2 reduction, this

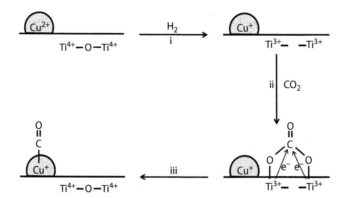

SCHEME 14.12 Mechanistic proposal to rationalize the detection of CO formed spontaneously upon CO_2 adsorption on Cu(I) supported on TiO_2: (i) reductive modification of the surface; (ii) CO_2 adsorption; and (iii) generation of CO bonded to Cu(I) sites.

type of studies by in situ FTIR spectroscopy shows the complexity and the variety of species that can be formed on the surface, most probably, with different photocatalytic reactivities. For instance, in principle, carbonates and bicarbonates should be more difficult to reduce than CO_2 in the gas phase due to the fact that attack by hydroxide forming (bi)carbonate stabilizes considerably CO_2.

Besides adsorption that is the first event in the photocatalytic process, it is of considerable interest to detect some other reaction intermediates on the surface of the photocatalyst. In this context and connected with the known influence that the presence of Cu has on the product distribution, it has been possible to detect the formation from CO_2 of CO bonded to Cu^+ sites (Scheme 14.12). Thus, in situ diffuse reflectance IR spectroscopy on hydrogen-treated Cu(I) supported on TiO_2 has been able to detect the presence of CO that is generated on the active surface spontaneously in some extent even in the dark.[63] Scheme 14.12 summarizes the proposed reaction mechanism that requires the presence of Ti^{3+} associated with oxygen vacancies as well as the presence of Cu^+ ions on the surface. According to this in situ study, the key point of photocatalysis would be to generate Ti^{3+} by photoinduced charge separation. If instead of photocatalytically, Ti^{3+} ions are formed by an alternative reductive treatment, similar steps as in photocatalysis could take place. The difference is that in some cases Ti^{3+} ions are acting as a real reactant and is consumed stoichiometrically rather than acting as a catalytic center with several turnovers. In spite of some isolated in situ spectroscopic studies, it is clear that considering the complexity of the reaction mechanism many more species should be formed and, therefore, it is still necessary to complement photocatalytic data with additional in situ spectroscopy measurements aimed at the detection of reaction intermediates and their transformation by reduction or oxidation with conduction band electrons or valence band holes.

14.11 CONCLUSIONS AND FUTURE PROSPECTS

As it can be deduced from the previous sections, the present situation is far from having identified an optimal photocatalyst for CO_2 reduction that can have some possibility to be applied commercially. It appears clear that further work trying to improve the solar light photocatalytic activity of TiO_2 and modification with cocatalysts will continue in the future. However, it is very likely that the basic concepts for TiO_2 modification have been already laid out and that the major advances could be the preparation in a reliable way of a TiO_2 material that encompasses all the features needed for optimal photocatalytic activity upon solar light irradiation.

In contrast to the case of TiO_2, much more effort should be devoted to explore the photocatalytic activity of other (either known or novel) types of photocatalysts. Besides metal oxides, sulfides, and chalcogenides, combinations of double metal oxides and layered materials are worth to be studied. One line of research that is currently emerging that holds considerable promise is the use of carbonaceous materials as photocatalysts. The main advantage of these carbonaceous materials is sustainability and optimization of natural resources since they do not contain metals and can be obtained from renewable biomass. Among this type of photocatalysts, fullerenes, carbon nanotubes, and graphenes are those that are currently under the spotlight and can serve to develop an array of semiconductors paralleling silicon-based materials, but being far more accessible and without requiring the amount of energy characteristic of silicon metallurgy.

At the moment, the study of the photocatalytic CO_2 reduction has been based on the screening of a large number of materials, without having any predictive capability in what concerns the photocatalytic activity or the product distribution. To overcome this limitation, theoretical calculations and in situ studies are needed to gain insight into the reasons that determine the selectivity of the process and the key steps that control product formation. Considering that from the point of view of solar fuels the most valuable product would be methanol, efforts should be made to understand the way in which its formation can be favored, probably by suitable cocatalysts that can collect electrons from the semiconductor and can act as buffers of electrons and protons.

Since the advantages of solar fuel production from CO_2 are considerable, it can be envisioned that this research line will continue to be developed until finally a commercial process based on a highly efficient photocatalyst is implemented at industrial scale.

REFERENCES

1. D. W. Bahnemann, *Res. Chem. Intermediat.*, 2000, **26**, 207–220.
2. M. A. Fox and M. T. Dulay, *Chem. Rev.*, 1993, **93**, 341–357.
3. A. Fujishima, X. Zhang and D. A. Tryk, *Surf. Sci. Rep.*, 2008, **63**, 515–582.
4. J. M. Herrmann, *Catal. Today*, 1999, **53**, 115–129.
5. A. Mills and S. LeHunte, *J. Photochem. Photobiol. A*, 1997, **108**, 1–35.
6. A. Kudo, *Catal. Surv. Asia*, 2003, **7**, 31–38.
7. A. Kudo and M. Sekizawa, *Chem. Commun.*, 2000, 1371–1372.
8. W. Zhao, C. C. Chen, X. Z. Li, J. C. Zhao, H. Hidaka and N. Serpone, *J. Phys. Chem. B*, 2002, **106**, 5022–5028.
9. M. R. Hoffmann, S. T. Martin, W. Y. Choi and D. W. Bahnemann, *Chem. Rev.*, 1995, **95**, 69–96.
10. M. Anbar, D. Meyerste and P. Neta, *J. Phys. Chem.*, 1966, **70**, 2660.
11. J. De Laat and H. Gallard, *Environ. Sci. Technol.*, 1999, **33**, 2726–2732.
12. A. Fujishima and K. Honda, *Nature*, 1972, **238**, 37–38.
13. T. Inoue, A. Fujishima, S. Konishi and K. Honda, *Nature*, 1979, **277**, 637–638.
14. A. J. Esswein and D. G. Nocera, *Chem. Rev.*, 2007, **107**, 4022–4047.
15. Y. Amao, *ChemCatChem*, 2011, **3**, 458–474.
16. J. R. Bolton, *Science*, 1978, **202**, 705–711.
17. G. Centi and S. Perathoner, *ChemSusChem*, 2010, **3**, 195–208.
18. D. Gust, T. A. Moore and A. L. Moore, *Acc. Chem. Res.*, 2009, **42**, 1890–1898.
19. L. Hammarström and S. Hammes-Schiffer, *Acc. Chem. Res.*, 2009, **42**, 1859–1860.
20. N. Serpone, D. Lawless and R. Terzian, *Solar Energy*, 1992, **49**, 221–234.
21. A. J. Morris, G. J. Meyer and E. Fujita, *Acc. Chem. Res.*, 2009, **42**, 1983–1994.
22. S. C. Roy, O. K. Varghese, M. Paulose and C. A. Grimes, *ACS Nano*, 2010, **4**, 1259–1278.
23. A. L. Linsebigler, G. Q. Lu and J. T. Yates, *Chem. Rev.*, 1995, **95**, 735–758.
24. R. R. Bacsa and J. Kiwi, *Appl. Catal. B*, 1998, **16**, 19–29.
25. D. C. Hurum, A. G. Agrios, K. A. Gray, T. Rajh and M. C. Thurnauer, *J. Phys. Chem. B*, 2003, **107**, 4545–4549.
26. T. Ohno, K. Tokieda, S. Higashida and M. Matsumura, *Appl. Catal. A*, 2003, **244**, 383–391.

27. S. Rehman, R. Ullah, A. M. Butt and N. D. Gohar, *J. Hazard. Mater.*, 2009, **170**, 560–569.
28. C. Burda, Y. B. Lou, X. B. Chen, A. C. S. Samia, J. Stout and J. L. Gole, *Nano Lett.*, 2003, **3**, 1049–1051.
29. J. M. Herrmann, *Acs Symp. Ser.*, 1986, **298**, 200–211.
30. R. Janisch, P. Gopal and N. A. Spaldin, *J. Phys.—Condens. Matter*, 2005, **17**, R657–R689.
31. K. Koci, K. Mateju, L. Obalova, S. Krejcikova, Z. Lacny, D. Placha, L. Capek, A. Hospodkova and O. Solcova, *Appl. Catal. B*, 2010, **96**, 239–244.
32. S. Sakthivel, M. Janczarek and H. Kisch, *J. Phys. Chem. B*, 2004, **108**, 19384–19387.
33. S. Sato, R. Nakamura and S. Abe, *Appl. Catal. A*, 2005, **284**, 131–137.
34. J. F. Zhu, W. Zheng, H. E. Bin, J. L. Zhang and M. Anpo, *J. Mol. Catal. A*, 2004, **216**, 35–43.
35. S. Sakthivel, M. V. Shankar, M. Palanichamy, B. Arabindoo, D. W. Bahnemann and V. Murugesan, *Water Res.*, 2004, **38**, 3001–3008.
36. A. Primo, A. Corma and H. García, *Phys. Chem. Chem. Phys.*, 2011, **13**, 886–910.
37. J. C. Scaiano, J. C. Netto-Ferreira, E. Alarcon, P. Billone, C. J. B. Alejo, C. O. L. Crites, M. Decan, C. Fasciani, M. Gonzalez-Bejar, G. Hallett-Tapley, M. Grenier, K. L. McGilvray, N. L. Pacioni, A. Pardoe, L. Rene-Boisneuf, R. Schwartz-Narbonne, M. J. Silvero, K. G. Stamplecoskie and T. L. Wee, *Pure Appl. Chem.*, 2011, **83**, 913–930.
38. M. Haruta, S. Tsubota, T. Kobayashi, H. Kageyama, M. J. Genet and B. Delmon, *J. Catal.*, 1993, **144**, 175–192.
39. A. Corma and H. García, *Chem. Soc. Rev.*, 2008, **37**, 2096–2126.
40. C. G. Silva, R. Juarez, T. Marino, R. Molinari and H. García, *J. Am. Chem. Soc.*, 2011, **133**, 595–602.
41. B. Ohtani, K. Iwai, S. Nishimoto and S. Sato, *J. Phys. Chem. B*, 1997, **101**, 3349–3359.
42. G. R. Bamwenda, S. Tsubota, T. Nakamura and M. Haruta, *J. Photochem. Photobiol. A*, 1995, **89**, 177–189.
43. G. W. Crabtree, M. S. Dresselhaus and M. V. Buchanan, *Phys. Today*, 2004, **57**, 39–44.
44. S. Dunn, *Int. J. Hydrogen Energy*, 2002, **27**, 235–264.
45. G. Cosa, M. S. Galletero, L. Fernandez, F. Marquez, H. García and J. C. Scaiano, *New J. Chem.*, 2002, **26**, 1448–1455.
46. M. Anpo, H. Yamashita, Y. Ichihashi, Y. Fujii and M. Honda, *J. Phys. Chem. B*, 1997, **101**, 2632–2636.
47. K. Ikeue, H. Yamashita, M. Anpo and T. Takewaki, *J. Phys. Chem. B*, 2001, **105**, 8350–8355.
48. K. Mori, H. Yamashita and M. Anpo, *RSC Adv.*, 2012, **2**, 3165–3172.
49. O. K. Varghese, M. Paulose, T. J. LaTempa and C. A. Grimes, *Nano Lett.*, 2009, **9**, 731–737.
50. D. Gong, C. A. Grimes, O. K. Varghese, W. C. Hu, R. S. Singh, Z. Chen and E. C. Dickey, *J. Mater. Res.*, 2001, **16**, 3331–3334.
51. J. M. Macak, H. Tsuchiya and P. Schmuki, *Angew. Chem. Int. Ed.*, 2005, **44**, 2100–2102.
52. C. Aprile, A. Corma and H. García, *Phys. Chem. Chem. Phys.*, 2008, **10**, 769–783.
53. T. Tachikawa, S. Tojo, M. Fujitsuka, T. Sekino and T. Majima, *J. Phys. Chem. B*, 2006, **110**, 14055–14059.
54. X. Zhang, F. Han, B. Shi, S. Farsinezhad, G. P. Dechaine and K. Shankar, *Angew. Chem. Int. Ed.*, 2012, **51**, 12732–12735.
55. Q. Zhai, S. Xie, W. Fan, Q. Zhang, Y. Wang, W. Deng and Y. Wang, *Angew. Chem. Int. Ed.*, 2013, **52**, 5776–5779.
56. S. Navalon, A. Dhakshinamoorthy, M. Alvaro and H. García, *ChemSusChem*, 2013, **6**, 562–577.
57. F. Cavani, F. Trifiro and A. Vaccari, *Catal. Today*, 1991, **11**, 173–301.

58. S. P. Newman and W. Jones, *New J. Chem.*, 1998, **22**, 105–115.

59. N. Ahmed, Y. Shibata, T. Taniguchi and Y. Izumi, *J. Catal.*, 2011, **279**, 123–135.

60. N. Ahmed, M. Morikawa and Y. Izumi, *Catal. Today*, 2012, **185**, 263–269.

61. K. Teramura, S. Iguchi, Y. Mizuno, T. Shishido and T. Tanaka, *Angew. Chem. Int. Ed.*, 2012, **51**, 8008–8011.

62. K. Bhattacharyya, A. Danon, B. K. Vijayan, K. A. Gray, P. C. Stair and E. Weitz, *J. Phys. Chem. C*, 2013, **117**, 12661–12678.

63. L. Liu, C. Zhao and Y. Li, *J. Phys. Chem. C*, 2012, **116**, 7904–7912.

15 Materials for Reversible High-Capacity Hydrogen Storage

Min Zhu, Liuzhang Ouyang, and Hui Wang

CONTENTS

15.1 INTRODUCTION

The continuous consumption of fossil energy and the resultant server pollution have driven great effort to develop clean and sustainable energy. Hydrogen, as an energy carrier, plays a very important role in the future energy system for its high energy density, being 142 kJ/g, and clean burning by-product, being water.[1] Hydrogen energy can be used in stationary power stations for distributed energy supplier and mobile power supplier for on-board applications. To make use of hydrogen energy, safe and efficient hydrogen storage technology is required with suitable charging/discharging rate under appropriate temperature and pressure conditions, normally near-ambient temperature and pressure, in addition to the environmental and economic considerations. For future on-board applications, the U.S. Department of Energy (DOE) has set the targets as given in Table 15.1.[2]

Over the last two decades, great efforts have been made on the development of high-capacity hydrogen storage in gas, liquid, and solid state. Solid-state hydrogen storage, which stores hydrogen by either chemical or physical sorption in solid materials, exhibits great advantages in volumetric and gravimetric hydrogen storage density as well as safety. However, the trade-off between storage density and sorption rate becomes a great challenge for practical application of hydrogen storage materials in the power supplier systems, in particular for on-board fuel cell application. For this reason, many new materials have been developed for reversible and irreversible hydrogen storage.

For irreversible hydrogen storage materials, such as $NaBH_4$, NH_3BH_3, and $Mg(BH_4)_2$,[3,4] the hydrogen release is generally realized by hydrolysis or thermolysis. The hydrogen capacity, hydrogen release rate, and working temperature can satisfy the application requirements. However, complicated and high-cost processes are required to regenerate the hydrides from the dehydrogenation product, which hinders the application of irreversible hydrogen storage materials. Much effort has been made on developing efficient and economic regeneration methods.[5]

TABLE 15.1
Partial DOE Targets for On-Board Hydrogen Storage Systems

Storage Parameters	Units	2015	Ultimate
System gravimetric capacity	kWh/kg	1.8	2.5
	kg H_2/kg system	0.055	0.075
System volumetric capacity	kWh/L	1.3	2.3
	kg H_2/L system	0.04	0.07
Min/max delivery temperature	°C	−40/85	−40/95 to 105
Cycle life (1/4 tank to full)	Cycles	1500	1500
Min/max delivery pressure for fuel cell	Bar (abs)	5/12	3/12
Min/max delivery pressure for internal combustion engine	Bar (abs)	35/100	35/100
Onboard reversible system efficiency	%	90	90
System fill rate	kg H_2/min	1.5	2

More research interest is focused on the development of high-capacity reversible hydrogen storage systems. In order to achieve high hydrogen storage capacity, hydrides mainly constituted of light elements, such as Li, C, N, B, Ca, K, Mg, Al, and Si, in addition to considering the hydrogen affinity, environmental concern, and their abundance in earth. Until now, Mg-based alloys,[6] complex hydrides,[7] and metal amides (or imides)[8] are most attractive candidates because their capacities exceed the DOE target. However, owing to the nature of H bonding with those elements, being either too strong or too weak, the hydrogenation/dehydrogenation thermodynamics and kinetics of those materials are generally unfavorable, and the working temperature and pressure could not satisfy the application requirements. Besides, some side reactions may take place accompanying the dehydrogenation and result in the release of harmful impurity gas. The side reaction also deteriorates the reversibility. The aforementioned problems hinder practical applications of high-capacity reversible hydrogen storage materials.

In the following sections, we will review the recent research progress of three types of important high-capacity reversible hydrogen storage materials: Mg-based hydrogen storage alloys, complex hydrides, and metal imides. Emphasis will be laid on the improvement of their properties through manipulation and modification of structure by alloying, nanostructuring, catalyzing, and compositing methods.

15.2 MG-BASED HYDROGEN STORAGE ALLOYS

Mg is an alkali metallic element with a proportion of about 2.2% of the earth's crust. It is light with a density of 1.738 g/cm^3, low cost, and environment friendly. When Mg reacts with H_2, it directly forms MgH_2 with hydrogen content of 7.6% in mass. Thus, it is an excellent candidate for practical hydrogen storage. The MgH_2 has two polymorphic structures,[9] the β-MgH_2 of rutile structure (space group $P4_2/mnm$) exists at normal temperature and hydrogen pressure conditions, while a metastable γ-MgH_2 with PbO_2 structure (space group $Pbcn$) can be formed under higher temperature and hydrogen pressure, or some milling conditions.[10,11]

The MgH_2 has a formation enthalpy and entropy of about 75 kJ/mol H and 135 J/mol H·K, respectively.[12,13] This large formation enthalpy is due to the strong bonding between Mg and H. According to the van't Hoff equation $\ln P_e = \Delta H/R - \Delta S/RT$ (where P_e is the equilibrium pressure, T the Kevin temperature, R the universal gas constant, ΔH the reaction enthalpy, and ΔS the reaction entropy), the thermodynamic equilibrium temperature for hydrogen desorption of MgH_2 is 289°C under 0.1 MPa H_2 pressure. Therefore, reducing the desorption enthalpy is the basic choice for lowering the decomposition temperature of MgH_2. This can be achieved by tuning the energy state of Mg and/or MgH_2 through alloying,[14–16] nanostructuring,[17,18] changing reaction path,[19] and exerting stress constraint.[20,21]

The rigorous hydrogen desorption condition due to thermodynamic characteristics becomes even more slashing by involving the poor desorption kinetics of MgH_2. As shown in Figure 15.1, the release of hydrogen includes several steps of Mg–H debonding, H diffusion, surface penetration of H atoms, and recombination of H to H_2 on the surface of alloys.[22] There are several factors that significantly block the dehydrogenation kinetics of MgH_2. The first is high activation energy required for

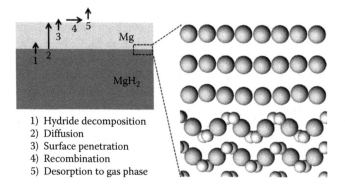

FIGURE 15.1   Reaction steps for hydrogen desorption in the MgH_2–Mg interface (left) and the $MgH_2(101)$–$Mg(100)$ interface structure (right).

the debonding of Mg–H bond. The second is surface oxidation of magnesium and/or formation of magnesium hydroxide, which prevents hydrogen dissociation out of the hydride matrix.[23] Once the disadvantageous layer is removed, the sluggish hydrogen diffusion in magnesium hydride is generally considered to be a controlling factor of dehydrogenation because the diffusion of H in MgH_2 is very slow, and the diffusion coefficient of H in magnesium hydride at 300°C–100°C lies between 10^{-18} and 10^{-24} m^2/s.[24,25] From the energy viewpoint, the activation energy for the desorption of bulk magnesium hydride ranges from 160 to 200 kJ/mol,[26,27] which is too high for practical applications. Therefore, the kinetic improvement by reducing activation energy and enhancing H diffusion has been attempted by adding catalyst,[28–33] nanostructuring,[17,34–36] and forming nanocomposite.[37–39]

15.2.1   Mg-Based Composite

Mg-based composite hydrides have been developed by combining Mg with other hydrogen storage alloys that have excellent thermodynamic and kinetic properties. For example, $LaNi_5$-type alloy,[40] which could easily uptake and release hydrogen at room temperature, was shown to greatly improve the hydrogen storage properties of Mg by preparing Mg–$LaNi_5$ composite. As reported by Liang et al.,[41] the mechanical-milled MgH_2–$LaNi_5$ or Mg–$LaNi_5$ was fully hydrogenated into a composite of MgH_2–LaH_3–Mg_2NiH_4, which shows much better hydrogen desorption kinetics than the binary Mg–La and Mg–Ni alloys due to a synergetic effect in the composite. In fact, Zhu et al. found that the hydrogenation of Mg–$LaNi_5$ nanocomposite followed an autocatalytic mechanism instead of nucleation/growth mechanism for a single-phase alloy.[42]

Similarly, the composites of Mg–Mg_2Ni, Mg–TiFe, Mg–$LaMg_2Ni$, Mg–$Ti_{0.4}Cr_{0.15}Mn_{0.15}V_{0.3}$, and Mg–ZrNiCr were produced with improved hydrogen storage properties. $LaMg_2Ni$ was reported as a compound with superior hydriding/dehydriding kinetics.[43] Xiao et al. found that the hydriding capacity of Mg + x wt.% $LaMg_2Ni$ ($x = 5$, 10, 20, 30) composites are all above 4.1 wt.% at 120°C and above 4.3 wt.% at 180°C within 6000 s.[37] Moreover, the addition of $LaMg_2Ni$ also improved the dehydriding performances of the composites. Yu et al. investigated the effect of

$Ti_{0.4}Cr_{0.15}Mn_{0.15}V_{0.3}$ (termed BCC due to the body-centered cubic structure) alloy on the hydrogen storage properties of MgH_2 and found that the hydrogenated BCC alloy showed superior catalytic properties compared to the quenched and ingot samples.[38] The hydrogenated BCC alloy is much easier to crush into small particles and embed in the MgH_2 aggregates. The BCC alloy not only increases the hydrogen diffusivity in bulk Mg but also promotes the dissociation and recombination of hydrogen molecule. The improvement should be attributed to the synergistic effect among the multiphase nanocrystalline structure, which makes hydrogen easily absorbed/desorbed on the abundant phase boundaries.

Au tried to reveal the hydrogen absorption/desorption mechanism of nanostructured $Mg–LaNi_5$ composite prepared by mechanical alloying.[44] The composite material Mg-20 wt.% $LaNi_5$ absorbed and desorbed 1.9 and 1.7 wt.% hydrogen at 25°C, respectively. It is believed that the dehydriding temperature is dominantly controlled by the thermodynamic configuration of magnesium hydride. The reduction in hydriding and dehydriding temperatures from 200°C to 25°C in the composite materials indicates that the thermal stability of Mg–Ni nanocrystalline/amorphous composite hydrides can be reduced by doping with additives such as La and $LaNi_5$. It could be attributed to the formation of less-stable Mg–La or Mg–Ni–La intermetallic compounds during the mechanical alloying process.

Mg can react with rare earth metals (denoted as RE hereafter) to form large amount of intermetallic compounds, including $Mg_{12}RE$, $Mg_{17}RE_2$, Mg_3RE, and MgRE. Ouyang et al. reported the hydrogen storage properties of Mg_3RE and $Mg_3RENi_{0.1}$ (RE=La, Pr, and mischmetals) alloys.[45–48] The Mg_3RE and $Mg_3RENi_{0.1}$ alloys disproportionated to a composite hydride of MgH_2 and REH_3, which exhibited rapid hydriding kinetics even at room temperature. It is found that the hydriding of Mg_3La and $Mg_3LaNi_{0.1}$ alloys at room temperature is a 1D, diffusion-controlled nucleation and growth process. Comparatively, the hydriding of Mg_3Pr alloy at room temperature is likely to be a 3D interface reaction process, while that of the $Mg_3PrNi_{0.1}$ alloy is an autocatalytic reaction. Besides, these alloys can easily become amorphous by melt spinning and then form nanocomposites by crystallization treatment. By controlling either the disproportionating or crystallization condition, the grain size of composite hydrides can be reduced to about 10 nm. This will be discussed later in combination with nanostructuring.

As described earlier, Mg-based composites, which are formed by adding other hydrogen storage materials of excellent sorption kinetics, show superior hydrogen absorption and desorption performances compared with pure MgH_2. This improvement is attributed to several effects: (1) hydride formation/decomposition is greatly enhanced through the pumping effect of phases, which reacts with hydrogen more easily; (2) H diffusion is enhanced in the nanocomposites; and (3) synergetic effect between different phases including stress constraint effect and boundary chemical mixing effect.

15.2.2 CATALYZED MG-BASED HYDRIDES

Adding catalysts is an efficient way to accelerate the chemical reaction process. A large variety of metals, oxides, halides, and carbon materials have been doped as catalysts to enhance the hydriding and dehydriding kinetic performances of MgH_2, and the hydrogen sorption temperature and rate were significantly improved.

Transition metals (TMs) were first used as catalytic additives by mechanical milling. Liang et al. reported that 3d elements Ti, V, Mn, Fe, and Ni showed different catalytic effects on the reaction kinetics of Mg–H system.[26] The composites containing Ti exhibited the most rapid absorption rate, followed in order by Mg–V, Mg–Fe, Mg–Mn, and Mg–Ni, while most fast desorption rate was obtained for MgH_2–V, followed by MgH_2–Ti, MgH_2–Fe, MgH_2–Ni, and MgH_2–Mn. The activation energy for MgH_2 desorption was reduced drastically; however, the formation enthalpy and entropy of MgH_2 were not altered. Similar results were also reported by Bobet et al. in the reactive mechanical alloyed Mg–M (M = Co, Ni, and Fe) systems under H_2 atmosphere.[28] Reactive alloying of Mg-based materials showed further improved hydrogen storage properties due to the strongly reduced partial size under H_2 atmosphere. A systematic calculation via ab initio density functional theory were made to explain the effect of those TMs.[49] The calculation results show that Ni is the best possible choice.

Metal oxides are another group suitable for catalyzing the hydrogenation/dehydrogenation of MgH_2. Oelerich et al. prepared nanocrystalline MgH_2/Me_xO_y and Mg_2NiH_4/Me_xO_y powders via high-energy ball milling ($Me_xO_y = Sc_2O_3$, TiO_2, V_2O_5, Cr_2O_3, Mn_2O_3, Fe_3O_4, CuO, Al_2O_3, SiO_2).[29] They found that some of the selected oxides lead to an enormous catalytic acceleration of hydrogen sorption compared with the pure nanocrystalline counterpart. In absorption, the catalytic effect of TiO_2, V_2O_5, Cr_2O_3, Mn_2O_3, Fe_3O_4, and CuO is comparable. Concerning desorption, the composite material containing Fe_3O_4 shows the fastest kinetics followed by V_2O_5, Mn_2O_3, Cr_2O_3, and TiO_2. In later study, they discovered an excellent catalyst Nb_2O_5, which is superior in absorption as well as desorption.[30] Thermodynamic properties of the oxide catalysts and experimental desorption rates are listed in Table 15.2. Based on the experimental data, a catalytic mechanism is proposed for TM compounds on Mg hydrogen sorption reaction. The catalytic activity was suggested to be influenced by four distinct physico-thermodynamic properties of the TM compound: a high number of structural defects, a low stability of the compound, a high valence state of the TM ion within the compound, and a high affinity of the transition metal ion to hydrogen.[50]

Halides are another group of catalysts that has a great effect on hydrogen sorption of Mg-based materials. Xie et al. milled MgH_2 nanoparticles with 5 wt.% TiF_3 as a doping precursor in a hydrogen atmosphere.[51] The sample desorbed 4.5 wt.% hydrogen in 6 min under an initial hydrogen pressure of ~0.001 bar at 573 K and absorbed 4.2 wt.% hydrogen in 1 min under ~20 bar hydrogen at room temperature. Here, the TiF_3 helps to dissociate hydrogen molecule to improve the absorption rate. It was found that TiF_3 showed superior catalytic effect over $TiCl_3$ in improving the hydrogen sorption kinetics of MgH_2.[52] The MgH_2 + 4 mol.% TiF_3 sample presented fast absorption kinetics even at 100°C. However, the MgH_2 + 4 mol.% $TiCl_3$ sample exhibited strong temperature dependence in the absorption kinetics, which substantially degrades with decreasing operation temperature. The incorporated fluorine differs significantly from its analog chlorine in terms of bonding state, which is responsible for the improvement of hydrogenation/dehydrogenation behaviors.

TABLE 15.2
Thermodynamic Properties of the Oxide Catalysts and Experimental Desorption Rates

Compound	TM Hydride Stability (kJ/mol H)	Compound Stability (kJ/mol O)	TM Valence	Desorption Rate
Nb_2O_5	−43.5	−379	5	10.2
V_2O_5	−30.5	−310	5	6
Ta_2O_5	−32.6	−409	5	3.6
NbO_2	−43.5	−398	4	3.5
Mn_2O_3	−7.6	−319	3	2.5
NbO	−43.5	−405	2	2
TiO_2	−72	−472	4	1.9
Cr_2O_3	−8	−380	3	1.9
Sc_2O_3	−67	−636	3	1
Al_2O_3	−3.8	−558	3	0.7
CuO		−156	2	0.5
SiO_2		−455	4	0.2
Re_2O_7		−180	7	0
MgO		−635		

Source: Barkhordarian, G. et al., *J. Phys. Chem. B*, 110, 11020, 2006.

Cui and coworkers coated a Ti-based multivalence catalyst, including Ti, TiH_2, $TiCl_3$, and TiO_2, on the surface of ball-milled Mg powders.[33] It was believed that the easier electron transfer among these Ti multivalences played a key role in enhancing the hydrogen recombination for the formation of a hydrogen molecule. Temperature-programmed desorption (TPD) and isothermal dehydrogenation analysis demonstrated that the MgH_2 coated with Ti-based multivalence catalyst had excellent dehydrogenation properties, which could start to release H_2 at about 175°C and release 5 wt.% H_2 within 15 min at 250°C. Moreover, a new mechanism, as shown in Figure 15.2, was proposed that multiple valence Ti sites acted as the intermediate for electron transfers between Mg^{2+} and H^-, which made the recombination of H_2 on surface of Ti compound much easier.

It was found that lithium- or potassium-doped carbon nanotubes (CNTs) can absorb hydrogen at moderate (200°C–400°C) and room temperatures, respectively, under ambient pressure,[53] suggesting carbon-based materials could be beneficial dopants for Mg-based materials. Yao et al. reported that the synergistic effect of metallic couple and CNTs on Mg resulted in an ultrafast kinetics of hydrogenation.[31] TMs significantly enhance the hydrogen kinetics while CNTs remarkably increase the capacity. V is much more effective than Fe or Ti in enhancing the absorption kinetics. However, the combination of two metals proves to be more efficient than a single-metal catalyst. The MgH_2–VTi–CNT system showed ultrafast hydrogenation kinetics and a very large capacity, which is very promising for hydrogen storage applications. A plausible hydrogenation mechanism has also been proposed by the

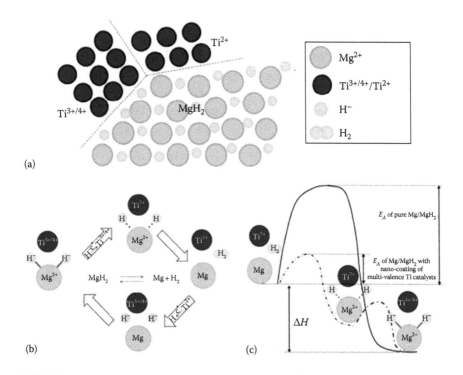

FIGURE 15.2 The schematic diagram of catalytic mechanism in the dehydrogenation of ball-milled MgH_2 doped with $TiCl_3$. (a) Showing the interface between high valence and low valence Ti; (b) showing the electron transfer in the dehydrogenation process; (c) showing the reduced energy needed in the catalytic process. (Cui, J., Wang, H., Liu, J., Ouyang, L., Zhang, Q., Sun, D., Yao, X., and Zhu, M., *J. Mater. Chem. A*, 1, 5603–5601, 2013. Reproduced by permission of The Royal Society of Chemistry.)

authors, which can explain the important phenomenon of coupled catalytic effects of transitional metals V and Ti, with CNT.

Lototskyy et al. made a time-resolved study to uncover the kinetics and mechanism of Mg–H interactions during high-energy reactive ball milling in hydrogen in the presence of various types of carbon, including graphite, activated carbon (AC), multiwalled carbon nanotubes (MWCNT), expandable graphite (EG), and thermally expanded (TEG) graphite.[54] It was found that introducing carbon significantly changes the hydrogenation behavior, which is strongly dependent on the nature and amount of carbon additive. For the materials containing 1 wt.% AC or TEG, and 5 wt.% MWCNT, the hydrogenation becomes superior to that for the individual magnesium. Carbon acts as a carrier of the *activated* hydrogen by a mechanism of spillover. For Mg–G, the hydrogenation starts from an incubation period and proceeds slower. An increase in the content of EG and TEG above 1 wt.% results in the deterioration of the hydrogenation kinetics. The effect of carbon additives has roots in their destruction during the ball milling to form graphene layers encapsulating the MgH_2 nanoparticles and preventing the grain growth, resulting in an increase of absorption/desorption cycle stability and a decrease of the MgH_2 crystallite size in the rehydrogenated Mg–C hybrid materials (40–125 nm) as compared to Mg alone (180 nm).

In summary, a great deal of effort has been done on the catalyzing of MgH_2 and great progress on hydrogen storage property has been achieved, especially in the kinetic aspect. However, the thermodynamics of absorption/desorption of MgH_2 could hardly be improved by catalyzing. The main drawback with respect to applications is the high reaction enthalpy, which requires high operating temperatures during desorption.

15.2.3 NANOSTRUCTURED MG-BASED HYDRIDES

Nanostructuring, including the reduction of either the grain size or geometric size of materials to nanometer range, is one of the most efficient strategies to improve the hydrogen storage properties of Mg-based materials. The improvement should be attributed to the enhanced hydrogen diffusion, energy-state change in nanostructured materials, and nanosize effect.

Ball milling is the simple method often used to fabricate nanocrystalline Mg-based materials for hydrogen storage together with the doping of catalyst. Zaluski et al. produced nanocrystalline Mg and Mg_2Ni with grain sizes of about 20–30 nm by this method.[14,23,55] Nanocrystalline Mg_2Ni readily absorbs hydrogen at low temperatures. The nanocrystalline hydrides show increased slope of the van't Hoff plots due to the higher binding energy of hydrogen in the nanocrystalline alloys. The shortened hydrogen diffusion paths together with strain and disorder also contribute to the enhancement on hydrogen absorption and desorption kinetics.

Nanostructured Mg-based materials can also be prepared by thin-film technology. The $MmNi_{3.5}(CoAlMn)_{1.5}/Mg$ (Mm denotes La-rich mischmetal) multilayer hydrogen-storage thin films have been prepared by direct-current and radio-frequency magnetron sputtering.[34,35,56,57] The dehydrogenation temperature is substantially lowered in the multilayer film, which is attributed to the unique nanocomposite structure. The MmM_5 layers are composed of two regions: an amorphous region with a thickness of ~4 nm at the bottom of the layers and a randomly oriented nanocrystallite region on the top of the amorphous region, and the Mg layers consist of typical columnar crystallite with the [0001] direction nearly parallel to the growth direction. This unique structure can lead to hydrogen-pumping effect and enhanced hydrogen diffusion.

The high interfacial energy (or surface energy) of nanostructured magnesium hydrides is the main factor that destabilizes their kinetics and thermodynamics. Both experiment and theoretical calculation results proved that the hydrogenation/dehydrogenation reaction temperature decreases due to the extra interfacial free energy stored in the boundary.[56,58,59] Ouyang et al. prepared an $Mg_{2.9}Ni$ film with preferential orientated nanocrystalline structure by magnetron sputtering.[58] The film contains Mg and Mg_2Ni grains with their average diameters of 40 and 20 nm, respectively. With a simplified model, the extra interfacial free energy can be calculated in accordance with their phase size. Assuming the thickness of the interfaces to be 1 nm and the shapes of Mg and Mg_2Ni grains to be sphere with their average diameters of 40 and 20 nm, respectively, the calculated extra interfacial free energy is about 5 kJ/mol. By simply involving the stored extra free energy estimated previously, the formation enthalpies of Mg_2NiH_4 and MgH_2 hydrides decreased to −59.5 and −69.5 kJ/mol, respectively. A more accurate calculation was made by regarding energy difference between nanocrystalline Mg and MgH_2.[60] According to this calculation, the enthalpy of dehydriding reaction and the dehydriding temperature also reduced by high-energy ball milling, which resulted from structural deformation and the associated volume change.

Besides nanocrystalline Mg-based alloys, Mg materials with nanosize in one, two, or three dimensions are also fabricated, and their hydrogen storage properties have been extensively studied. Theoretical calculation based on first-principles DFT revealed that the dehydrogenation enthalpy of MgH_2 can be apparently changed only if the particle size of Mg was reduced to 3–5 nm.[61] Li et al. prepared Mg nanowires with the diameters of 30–50, 80–100, and 150–170 nm via a vapor-transport method.[17] They found that thinner Mg–MgH_2 nanowires had a much lower desorption energy than that of thicker nanowires or bulk Mg–MgH_2. They further calculated the relationship between the dehydrogenation enthalpy and the size of MgH_2.[62] According to this, the dehydrogenation enthalpy for the MgH_2 nanowires of φ 1.24, 0.85, and 0.68 nm were calculated to be 61.86, 34.54, and −20.64 kJ/mol H_2, and the hydrogen storage is possible at room temperature with MgH_2 nanowire of φ 0.85 nm. This clearly indicates that nanostructuring can lead to kinetic and thermodynamic destabilization of MgH_2.

Because of the high interfacial energy, nanostructured Mg-based materials have very strong tendency of structure coarsening in hydrogenation/dehydrogenation cycles. In addition, Mg-based alloys are easy to oxidize especially when they are in nanoscale. In order to overcome these drawbacks, the confinement or encapsulation of Mg has been attempted. Jeon et al. prepared an air-stable composite material that consists of metallic Mg nanocrystals (NCs) in a gas-barrier polymer matrix that enables both the storage of a high density of hydrogen (up to 6 wt.% of Mg, 4 wt.% for the composite) and rapid kinetics (loading in <30 min at 200°C).[63] Moreover, nanostructuring of the Mg provided rapid storage kinetics without using expensive heavy-metal catalysts.

It is an often used strategy to confine Mg–MgH_2 clusters within nanoporous material scaffolds, such as aerogel scaffold,[64] carbon scaffold, SBA15, and CMK3.[18,65,66] Konarova et al. synthesized MgH_2 particles by wet impregnation within the pores of the mesoporous materials SBA15 and CMK3.[66] Thermal desorption behavior of MgH_2–CMK3 compounds were studied by varying the MgH_2 loading amount in CMK3. The onset and maximum hydrogen desorption temperatures were significantly influenced by the MgH_2 loading amount. MgH_2–CMK3 compounds with 20 wt.% loading released hydrogen at a temperature of 253°C. Its corresponding decomposition enthalpy is 52.4 ± 2.2 kJ/mol, implying that the thermodynamics of the nano-MgH_2 was greatly changed in comparison with those of coarse-crystalline MgH_2. Similar results are reported by Zhao et al. MgH_2 nanoparticles with a size of ~3 nm were formed by direct hydrogenation of Bu_2Mg inside the pores of a carbon scaffold.[18] The activation energy for the dehydrogenation was lowered by 52 kJ/mol compared to the bulk material, and a significantly reduced reaction enthalpy of

63.8 ± 0.5 kJ/mol and entropy (117.2 ± 0.8 J/mol) was found for the nanoconfined system. It is clear from the above results that both thermodynamic and kinetic properties are altered in the nanoconfined Mg-based systems. Besides, the stability problem of the nanostructure is well solved by the nanoconfinement. However, it should be pointed out that the reaction entropy is also reduced and counteracts the effect of the reaction enthalpy reduction on the reduction of dehydrogenation reaction temperature for the nanoconfined MgH_2.

The aforementioned synthesis of nanostructure is quite complicated. Therefore, to form nanocomposite structure through a more simple way is of great interest. Under condition of rapid solidification, Mg–RE–Ni alloys could be easily amorphousized, which makes them very suitable precursors to fabricate nanocomposites with superior hydrogen storage properties. Spassov et al. reported that some nanocrystalline and/or amorphous Mg–RE–Ni alloys showed much superior hydrogen absorption/desorption properties than the corresponding conventional crystalline alloys.[67–69] Wu et al. studied the effect of solidification rate on the microstructures and hydrogen storage properties of melt-spun Mg–Ni–Mm (Mm = La, Ce-rich mischmetal) alloy and found that the hydrogenation properties showed close connection with the microstructures.[70–74] Lin et al. reported that the melt-spun $Mg_3LaNi_{0.1}$ alloy showed much better hydrogen storage properties compared with those of the as-cast alloys, including both faster kinetics and higher capacity yet thermodynamic improvement.[45] The improvement on absorption kinetics for those melt-spun Mg-based materials can be attributed to the nanosized particles of REH_x and Mg_2Ni embedded in Mg matrix after crystallization, while the enhancement of desorption kinetics is due to the nanosized particles of REH_x and Mg_2NiH_4 embedded in MgH_2 matrix upon hydrogenation.

It should be noted that the nanocomposite can also be in situ formed in some designed systems by controlling the hydriding reaction of the compounds in Mg–RE–Ni system. For example, $Mg_3RENi_{0.1}$ compound can disproportionate into a nanocomposite consisted of MgH_2 and REH_3 as shown in Figure 15.3, and the hydrogen absorption/desorption kinetics is greatly improved.[46,47] Besides, the nanostructure of this kind of nanocomposite is quite stable, and this may be due to the pinning effect of stable REH_3 hydride to the growth of the nanocomposite.

In this kind of nanocomposite, the contents of RE and Ni are very important on the hydrogen storage capacity and hydrogen absorption/desorption kinetics. On one hand, RE and Ni are advantageous elements for the glass-forming ability (GFA) of Mg-based alloys and the hydrogen absorption/desorption kinetics of MgH_2. On the other hand, Ni and RE are much heavier than Mg, which reduces the hydrogen storage capacity for Mg-based alloys. Thus, Ni and RE contents should be designed in a reasonable range by considering the requirement of hydrogen storage capacity and GFA to make use of in situ–formed nanocomposite from amorphous precursor. As reported by Lin et al., Ce was more advantageous for the GFA of Mg-rich Mg–Ce–Ni system than Ni, and the composition region to form amorphous in this system was obtained. According to this, the lowest Ce content is ~5 at.% to obtain the fully amorphous alloy, and the amorphous alloy with the highest Mg content,[74] $Mg_{90}Ce_5Ni_5$, was obtained by melt spinning. With the amorphous alloy as precursor, nanostructured multiphase composite was prepared by crystallizing it in the hydrogenation

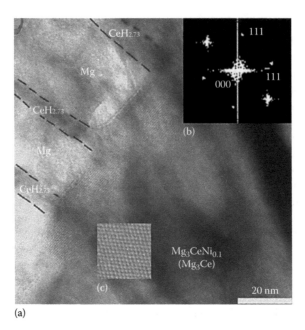

FIGURE 15.3 Transmission electron microscopy (TEM) image of partially hydrogenated $Mg_3CeNi_{0.1}$ alloy, showing (a) laminated Mg and $CeH_{2.73}$ composite, (b) electron diffraction pattern, and (c) high-resolution image of $CeH_{2.73}$.

process. The composite with reversible hydrogen storage capacity of 5.3 wt.% shows much faster kinetics and lower MgH_2 desorption activation energy (109 kJ/mol) than those of induction-melt $Mg_{90}Ce_5Ni_5$ alloy (124 kJ/mol) and bulk MgH_2 (160 kJ/mol). This is because both in situ–formed nanosized Mg_2Ni and $CeH_{2.73}$ act as effective catalysts and significantly improve the hydrogen storage properties of MgH_2.

Similar to the amorphous phase, long-period stacking ordered (LPSO) structure is another phase that could induce nanocrystalline hydrides for hydrogen storage in the Mg-based alloys. Zhang et al. prepared a melt-spun Mg–Y–Ni alloy with an LPSO phase,[75] and the nanostructured YH_2 (or YH_3) and Mg_2Ni induced from the LPSO phase significantly improve the hydrogen storage properties of MgH_2. Liu et al. revealed that different elements are concentrated in the different positions of the LPSO structure.[76] This results in a localized formation of nanocomposite.

Based on the current research progress, nanostructuring is the most efficient strategy to enhance the hydrogen storage properties of Mg-based materials. This is not only due to the nanosize effect but also due to the catalytic and synergetic effects generated from effective doping of efficient catalysts/additives into the nanostructured Mg-based materials. It can be pointed out that nanosize below 10 nm is very beneficial for hydrogen storage properties of MgH_2 from both kinetic and thermodynamic views. However, the nanosized MgH_2 could grow significantly after hydrogenation/dehydrogenation cycles at high temperatures (e.g., ~250°C). This problem can be overcome by nanoconfinement and pining the boundaries.

15.2.4 REACTION COMPOSITE

From a thermodynamic point of view, one of the most effective methods to alter the thermodynamics of hydrogenation/dehydrogenation reactions for MgH_2 is introducing different reaction pathways. Mg–Cu and Mg–Ni systems are first studied by Reilly and Wiswall. The intermetallic compound Mg_2Cu reacts with H_2 at ~300°C to form MgH_2 and $MgCu_2$ by the following reaction (15.1), with $\Delta H°_{298}$ and $\Delta S°_{298}$ being -72.9 ± 4.2 kJ/mol H_2 and 132.3 ± 2.9 J/mol, respectively:[15]

$$2Mg_2Cu + 3H_2 \rightarrow 3MgH_2 + MgCu_2 \tag{15.1}$$

However, the reaction (15.1) is regarded as irreversible because $MgCu_2$ could not absorb hydrogen at moderate conditions, which makes the reaction impossible for practical use. Later, reversible reaction between other Mg-based compounds and hydrogen are explored, among which Mg–Ni system is a typical one suitable for practical applications.

In the Mg–Ni system, Mg and Ni can form two stable compounds: Mg_2Ni and $MgNi_2$. Reilly et al., for the first time, reported that $MgNi_2$ did not absorb hydrogen at temperature up to 350°C and pressure up to 400 psi, while Mg_2Ni could uptake hydrogen at 325°C and 300 psi to generate $Mg_2NiH_{0.3}$ and Mg_2NiH_4 through the following reaction (15.2), with $\Delta H°_{298}$ and $\Delta S°_{298}$ being -64.5 ± 4.2 kJ/mol H_2 and 122.2 ± 6.3 J/mol, respectively:[16]

$$Mg_2Ni + 2H_2 \rightarrow Mg_2NiH_4 \tag{15.2}$$

In this system, the enthalpy and entropy of the Mg_2Ni-H are reduced significantly from -74.5 kJ/mol H_2 and 135 J/mol for the Mg–H, respectively. However, hydrogen storage capacity is also decreased to ~3.6 wt.% from 7.6 wt.% of MgH_2.

Vajo et al. proposed an interesting strategy to alter the thermodynamics of Mg by alloying it with Si.[77] The dehydrogenation reaction enthalpy of MgH_2–Si system is reduced to 36.4 kJ/mol H_2 because of the formation of intermediate compound Mg_2B. However, the MgH_2/Si system was not readily reversible. Hydrogenation of Mg_2Si appears to be kinetically limited because of poor mobility of Si atoms at low temperature.

Another pathway to alter the hydriding/dehydriding thermodynamics of Mg-based alloys is to form a solid solution system. Mg–Cd and Mg–In are the two typical examples. A fully reversible transformation in the Mg–In–H system with reduced hydrogen sorption reaction enthalpy was reported by Zhong et al. The Mg(In) solid solution absorbs hydrogen to form MgH_2 and a disordered Mg(In) compound, as shown in the following reaction:[19]

$$Mg(In) + H_2 = MgH_2 + \beta \tag{15.3}$$

The $Mg_{0.95}In_{0.05}$ solid solution shows reduced hydrogen sorption reaction enthalpy of -68.1 ± 0.2 kJ/mol, which is much lower than that of pure Mg, and a high hydrogen storage capacity of about 5 wt.% is still maintained.

Similarly as the Mg–In system, the $Mg(Cd)_x$ solid solution may decompose upon hydrogenation according to the following equation[78]

$$MgCd_x + H_2 \rightarrow MgH_2 + xCd \qquad (15.4)$$

Since there is a significant hysteresis at lower temperatures for the Mg–5Cd alloy, the calculation of the thermodynamic parameters (reaction enthalpy and entropy) is not accurate. A rough calculation indicates that the enthalpy of hydrogen desorption is reduced to 71 kJ/mol for Mg–5Cd–V–C from 75 kJ/mol for pure magnesium hydride.

In summary, it is clear from the aforementioned reactions that the destabilization of thermodynamics for Mg–H reaction could be obtained through changing the reaction paths with different reaction products. However, the storage capacity loss is unavoidable. Therefore, it is the key issue to find the strategy to realize reversible reaction and keep the hydrogen storage capacity at the same time under mild conditions.

15.3 COMPLEX HYDRIDES

Complex hydride for reversible hydrogen storage was first introduced by Bogdanović and Schwickardi in 1997.[79] Their pioneering studies realized fast dehydriding and feasible rehydrogenation of $NaAlH_4$ under moderate conditions by doping Ti-based compounds. The high reversible hydrogen storage capacity and moderate working temperature (<200°C) stimulate extraordinary research interest to enable $NaAlH_4$ to be a solid hydrogen storage medium for onboard fuel cell. Afterward, lithium borohydride was successively discovered for reversible hydrogen storage by Züttel in 2005.[80] Up to now, intensive research on complex hydrides over more than a decade has undoubtedly broadened the scope of high-capacity hydrogen storage materials and greatly promoted their potential of various hydrogen storage applications.

The term *complex hydride* is normally used to describe a family of ionic hydrogen–containing compounds, which are composed of groups I and II metal cations and *complex* anions (e.g., $[AlH_4]^-$, $[NH_2]^-$, and $[BH_4]^-$ named as alanate, amide, and borohydride, respectively). Unlike the interstitial hydrogen in metal hydrides, the hydrogen atoms of complex hydrides are covalently bonded to the central atoms in the complex group. These complex hydrides (see Table 15.3; amides will be described in the next section and will not be included here) possess high gravimetric and volumetric hydrogen densities, and are regarded as promising hydrogen storage media for stationary and vehicular power systems. Among these alanates and borohydrides, commercially available $NaAlH_4$ and $LiBH_4$ are the most popular and promising complex hydrides for reversible hydrogen storage because of their relatively better comprehensive hydrogen storage performances.

$NaAlH_4$ releases 5.5 wt.% hydrogen by a two-step decomposition mechanism, as shown in Equations 15.5 and 15.6. The first step gives 3.7 wt.% hydrogen, and the second one 1.8 wt.%. The decomposition of NaH must proceed at a temperature more than 400°C that is too high for practical application. Additionally, Claudy et al. showed that there exists an α-$Na_3AlH_6 \rightarrow \beta$-$Na_3AlH_6$ phase transition in the heating.[81]

TABLE 15.3

Crystal Structure and Hydrogen Storage Capacities of Typical Complex Hydrides

Formula	Crystal Structure	Gravimetric Capacity (wt.%)	Reversible Capacity (wt.%)
$LiAlH_4$	$P2_{1/c}$	10.5	4
$NaAlH_4$	$I4_{1/a}$	7.5	5.6
$KAlH_4$	$Pbmn$	5.7	3.5
$Mg(AlH_4)_2$	$P3m1$	9.3	6.9
$Ca(AlH_4)_2$	$Pbca$	7.7	—
$LiBH_4$	$Pnma$ (LT)	18.3	11.3
	$P6_3mc$ (HT)		
$NaBH_4$	$Fm\bar{3}m$	10.4	5
KBH_4	$Fm\bar{3}m$	7.5	—
$Mg(BH_4)_2$	$P6_1$ (LT)	14.8	6.1
	$Fddd$ (HT)		
$Ca(BH_4)_2$	$Fddd$ (LT)	11.4	6.6
	$I\bar{4}2d$ (HT)		
$Al(BH_4)_3$	$C2/c$ (LT)	17	—
	$Pna2_1$ (HT)		

$$NaAlH_4 \rightarrow 1/3\ Na_3AlH_6 + 2/3\ Al + H_2 \qquad (15.5)$$

$$1/3\ Na_3AlH_6 \rightarrow NaH + Al + 1/2\ H_2 \qquad (15.6)$$

The decomposition of $NaAlH_4$ into Na_3AlH_6 has an enthalpy of 37 kJ/mol H_2, while the endothermic heat of the second step is 47 kJ/mol H_2.[82] According to the van't Hoff equation, under 0.1 MPa hydrogen pressure, the desorption temperature of the first step in the decomposition reaction of $NaAlH_4$ is 30°C, and 110°C for the second one. Therefore, the second decomposition is the main thermodynamic barrier and needs to be improved.[83]

The decomposition of $LiBH_4$ into LiH releases 13.8 wt.% hydrogen. Several distinct peaks in the nonisothermal differential scanning calorimetry (DSC) and/or thermogravimetry (TG) heating curves of $LiBH_4$ reflect its multistep decomposition process: at about 100°C, the orthorhombic structure of $LiBH_4$ transforms to hexagonal structure; the melting of $LiBH_4$ at 280°C is accompanied with a small amount of hydrogen release; and the majority of hydrogen is released at a temperature range of 400°C–600°C.[84] The dehydrogenation mechanism investigation demonstrates the formation of $Li_2B_{12}H_{12}$ as an intermediate compound, which is a reaction product between the diborane formed in the decomposition of $LiBH_4$ and the remaining $LiBH_4$.[85] In addition, the dehydrogenation isotherms of $LiBH_4$ indicate an enthalpy ΔH of 74 kJ/mol H_2 and entropy ΔS of 115 J/K mol H_2.

In general, the thermal decomposition of complex hydride involves a multistep process, in which an intermediate hydride is generally formed.[86] The stepwise manner implies unfavorable thermodynamic and kinetic properties for the decomposition of complex hydrides. In addition, the nature of covalent bond in the complex group determines that hydrogen desorption from complex hydride is highly endothermic, and the high decomposition temperature is necessary. In comparison with the dehydrogenation, the rehydrogenation of complex hydrides proceeds under more harsh temperature and pressure conditions. For example, the rehydrogenation of $LiBH_4$ needs the recombination of the end products LiH and B, which involves the breakage of rigid boron lattice, the interdiffusion of Li and B atoms, and the formation of intermediate $Li_2B_{12}H_{12}$. Therefore, the rehydrogenation of $LiBH_4$ requires 873 K temperature, 35 MPa hydrogen pressure, and reaction time for 12 h.[80]

Several effective strategies have been employed to improve the dehydrogenation and rehydrogenation properties of complex hydrides. For the alanates, the most successful approach is doping catalytic additives, while for the borohydrides, the most effective method is to construct a reactive composite system of borohydrides and additives (e.g., metal hydride, oxides, fluorides, elements), which undergoes different dehydrogenation routes with destabilized thermodynamics. In addition, nanoconfinement strategy is also successfully adopted to synthesize nanoscale complex hydrides by loading them into the nanopores of supporting materials, which would avoid the phase segregation in the dehydrogenation/rehydrogenation process and thus improve the hydrogen sorption kinetics and cyclic stability. Lastly, composition modification via ion substitution is used to tailor the thermodynamic properties of complex hydrides. This chapter will focus on the research progress of complex hydrides over the past decades with the emphasis on thermodynamic destabilization and kinetic improvement by the four strategies discussed earlier.

15.3.1 COMPOSITION MODIFICATION OF COMPLEX HYDRIDES

Complex hydrides consisting of a different cation and the same complex anion possess different dehydrogenation properties. For the borohydrides, Nakamori et al.[87] reported that the decomposition temperature of the single-metal borohydrides was empirically related to the Pauling electronegativity of the cation. As shown in Figure 15.4, the higher electronegativity of the cation is, the lower decomposition temperature of complex hydrides will be. Besides, it was expected that borohydrides with electronegativity >1.5 of the cation are thermodynamically unstable. This experimental result is in well agreement with the prediction by first-principles calculations in the literature.[88] Theoretically, of over 700 structures investigated by density functional theory, about 20 multication borohydrides are predicted to form potentially stable compounds with favorable decomposition energies.[88] The (Li/ Na/K)(Al/Mn/Fe)$(BH_4)_4$, (Li/Na)Zn$(BH_4)_3$, and (Na/K)(Ni/Co)$(BH_4)_3$ compounds are found to be the most promising ones.

Motivated by this finding, ion substitution has been proposed to design mixed cation borohydrides with tailored thermodynamic stability, in which the cations have different electronegativity. Mechanochemical synthesis (ball milling) has been applied to synthesize bimetallic borohydrides, such as LiZr$(BH_4)_5$, LiK$(BH_4)_2$, and KSc$(BH_4)_4$.[89–93] Most of them exhibit moderate thermodynamic stabilities between the single-metal borohydrides as predicated. Some typical bimetallic borohydrides

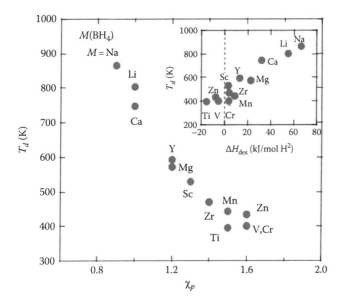

FIGURE 15.4   Decomposition temperature (T_d) as a function of Pauling electronegativity χ_p. The inset shows the correlation between T_d and estimated enthalpy ΔH_{des} for desorption reaction. (Adapted from Nakamori, Y. et al., *J. Phys. Chem. C*, 69, 2292, 2008. Copyright 2008 Pergamon. With permission.)

and their hydrogen storage properties are shown in Table 15.4. It is notable that the thermodynamic stability and the theoretic hydrogen content of the borohydrides can be tuned by appropriate combination of cations with different electronegativity and mole ratio.

With respect to alanates for hydrogen storage, many alkali and alkali earth alanates show different thermodynamic stabilities and decomposition behaviors. For example, the dehydrogenation of $LiAlH_4$ into hexahydride is exothermic, indicating that this is a nonspontaneous process. $KAlH_4$ decomposes into K_3AlH_6 at a much higher temperature (300°C) than the analogous process for $NaAlH_4$. For the $Mg(AlH_4)_2$, the initial dehydrogenation product consists of MgH_2 and Al, without the involvement of $[AlH_6]^-$ as an intermediate phase. Therefore, it is possible to design mixed alanates, containing more than one alkali and/or alkaline earth metals. Table 15.4 also lists several types of mixed alanates, as-synthesized by mechanical milling, with different crystal structures and dehydrogenation temperatures.

Since hydrogen and fluorine atoms are highly similar in size and valence, partial substitution of hydrogen by fluorine is possible to form the so-called hydridofluoride with altered thermodynamic properties.[99] The first-principles calculation was carried out to predict the performance of hydridofluoride. The computed cell parameters of $LiBH_4$ were in good agreement with experimental data reported by Soulie et al.[100] The calculated decomposition enthalpy of $Li_8B_8H_{32-x}F_x$ is dependent on the F-substituted level in $LiBH_4$, decreasing with the increase of F doping content. When the x value of F is 0.25, the $LiBH_{3.75}F_{0.25}$ contains 11.3 wt.% theoretical hydrogen;

TABLE 15.4

Hydrogen Storage Properties of Mixed Cation Complex Hydrides

Hydrides	Ideal Hydrogen Content (Mass%)	Decomposition Temperature (°C)	Practical Capacity (wt.%)	References
LiSc(BH$_4$)$_4$	14.5	220	4.38–6.4[b]	[89]
LiZn$_2$(BH$_4$)$_5$[a]	9.5	127	—	[89]
Li$_4$Al$_3$(BH$_4$)$_{13}$[a]	17.2	70	—	[91]
NaSc(BH$_4$)$_4$	12.6	200	0.97[b]	[94]
NaZn$_2$(BH$_4$)$_5$[a]	8.8	95	—	[89]
NaZn(BH$_4$)$_3$[a]	9.0	103	—	[89]
KSc(BH$_4$)$_4$	11.1	210	4.4[c]	[95]
Na$_2$LiAlH$_6$	7.5	255	3.4[c]	[96]
K$_2$NaAlH$_6$	2.5	300	2.0[c]	[97]
K$_2$LiAlH$_6$	5.1	300	2.3[c]	[97]
LiMg(AlH$_4$)$_3$	9.6	130	4.9[c]	[98]
LiMgAlH$_6$	9.3	190	2.4[c]	[98]

[a] Represents the release of diborane in the decomposition process.
[b] Indicates the value referred to the mixture of borohydrides/alanates with side products.
[c] Indicates the value referred to the pure borohydrides/alanates.

the decomposition enthalpy can be reduced to 36.5 kJ/mol H$_2$ for LiBH$_{3.75}$F$_{0.25}$ by taking the zero-point energy into consideration. This results states that the dehydrogenation temperature would be 100°C at 1 bar equilibrium pressure.

Although the composition modification by ionic substitution successfully decreases the thermodynamic stability of complex hydrides, it is difficult to regenerate the mixed cation complex hydride through the rehydrogenation reaction, and therefore, this destabilizing effect is a one-time event. This shortcoming is hardly overcome because the structural rearrangement of complex structure from the metal elements, intermediate compound, and/or elementary hydride requires extremely strict reaction conditions.

15.3.2 Destabilized Complex Hydride Composites

Complex hydrides would exothermically react with additives and show destabilized thermodynamics because of the formation of energetically favorable intermediate compounds during dehydrogenation. Therefore, constructing reactive hydride composites is widely used to improve the thermodynamic properties of complex hydrides, especially of borohydrides. Similar with the reaction system of MgH$_2$–Si, the dehydrogenation enthalpy of LiBH$_4$ could be reduced due to the formation of metal borides, which is more stable than the elements B. Vajo et al. first explored the hydrogen storage properties of LiBH$_4$–Si and LiBH$_4$–MgH$_2$ systems by applying this strategy.[101] After that, to obtain a reversible LiBH$_4$–M system with superior hydrogen storage properties, various additives have been tentatively introduced into LiBH$_4$, such as elements, metal hydrides, metal halides, and metal oxides. The dehydrogenation mechanism and reversibility of those composites were investigated theoretically and experimentally.

To predict the thermodynamic destabilizing effect of additives and resultant decomposition products, Siegel's group vetted more than 20 candidate reactions based on destabilized $LiBH_4$ and $Ca(BH_4)_2$ borohydrides (listed in Table 15.5) using the first-principles calculation.[102] The theoretical results reveal several reactions having both favorable thermodynamics and relatively high hydrogen densities. Besides, the authors stated that the chemical intuition alone was not sufficient to identify the validity of reaction pathways and proposed three thermodynamic guidelines for predicting the dehydrogenation reactions.

TABLE 15.5
Hydrogen Densities and Calculated Thermodynamic Quantities for Candidate Hydrogen Storage Reactions

No.	Decomposition Reaction	wt.% H_2	$\Delta_r H$ (kJ/mol H_2)	T, $P_{H2}=1$ bar (°C)
1	$2LiBH_4 \rightarrow 2LiH + 2B + 3H_2$	13.8	67	322
2*	$4LiBH_4 + 2AlH_3 \rightarrow 2AlB_2 + 4LiH + 9H_2$	12.4	39.6	83
3	$2LiBH_4 + Al \rightarrow AlB_2 + 2LiH + 3H_2$	8.6	57.9	277
4*	$4LiBH_4 + MgH_2 \rightarrow MgB_4 + 4LiH + 7H_2$	12.4	51.8	206
5	$2LiBH_4 + MgH_2 \rightarrow MgB_2 + 2LiH + 4H_2$	11.6	50.4	186
6	$2LiBH_4 + TiH_2 \rightarrow TiB_2 + 2LiH + 4H_2$	8.6	4.5	—
7	$2LiBH_4 + TiH_2 \rightarrow VB_2 + 2LiH + 4H_2$	8.4	7.2	−238
8	$2LiBH_4 + ScH_2 \rightarrow ScB_2 + 2LiH + 4H_2$	8.9	32.6	26
9*	$2LiBH_4 + CrH_2 \rightarrow CrB_2 + 2LiH + 4H_2$	8.3	16.4	−135
10	$6LiBH_4 + CaH_2 \rightarrow CaB_6 + 6LiH + 10H_2$	11.7	45.4	146
11*	$2LiBH_4 + Mg \rightarrow MgB_2 + 2LiH + 3H_2$	8.9	46.4	170
12*	$2LiBH_4 + 2Fe \rightarrow FeB + 2LiH + 3H_2$	3.9	12.8	−163
13	$2LiBH_4 + 4Fe \rightarrow Fe_2B + 2LiH + 3H_2$	2.3	1.2	—
14	$2LiBH_4 + Cr \rightarrow CrB_2 + 2LiH + 3H_2$	6.3	31.7	25
15	$3Ca(BH_4)_2 \rightarrow 2CaH_2 + CaB_6 + 10H_2$	9.6	41.4	88
16*	$Ca(BH_4)_2 + MgH_2 \rightarrow CaH_2 + MgB_2 + 4H_2$	8.4	47	135
17*	$2Ca(BH_4)_2 + MgH_2 \rightarrow 2CaH_2 + MgB_4 + 7H_2$	8.5	47.9	147
18*	$2Ca(BH_4)_2 + 2AlH_3 \rightarrow 2CaH_2 + 2AlB_2 + 9H_2$	9.1	36.6	39
19	$Ca(BH_4)_2 + ScH_2 \rightarrow CaH_2 + ScB_2 + 4H_2$	6.9	29.2	−20
20	$Ca(BH_4)_2 + TiH_2 \rightarrow CaH_2 + TiB_2 + 4H_2$	6.7	1.1	—
21	$Ca(BH_4)_2 + VH_2 \rightarrow CaH_2 + VB_2 + 4H_2$	6.6	3.8	—
22*	$Ca(BH_4)_2 + CrH_2 \rightarrow CaH_2 + CrB_2 + 4H_2$	6.5	13.1	−180
23*	$Ca(BH_4)_2 + Mg \rightarrow CaH_2 + MgB_2 + 3H_2$	6.4	41.9	111
24*	$Ca(BH_4)_2 + Al \rightarrow CaH_2 + AlB_2 + 3H_2$	6.3	53.4	200

Source: Siegel, D.J. et al., *Phys. Rev. B*, 76, 2007
Note: Reactions denoted with a * will not proceed as written.

Guideline 1: The enthalpy of the proposed destabilizing reaction must be less than the decomposition enthalpies of the individual reactants.

Guideline 2: If the proposed reaction involves a reactant that can absorb hydrogen, the formation enthalpy of the corresponding hydride cannot be greater in magnitude than the enthalpy of the destabilized reaction.

Guideline 3: In general, it is not possible to tune the thermodynamics of destabilized reactions by adjusting the molar fractions of the reactants.

There is only one stoichiometry corresponding to a single-step reaction with the lowest possible enthalpy; all other stoichiometries will release H_2 in multistep reactions, where the initial reaction is given by the lowest reaction enthalpy.

Experimentally, the destabilized $LiBH_4$ composites with several types of additives have been explored up to now. Au et al. preliminarily investigate the metal (Mg–Al–Ca–In)-modified $LiBH_4$ composites prepared by mechanical milling method[103] and found that only Mg and Al exhibited a positive effect on facilitating the hydrogen release from $LiBH_4$. For the $LiBH_4$ 0.2 Mg composite, the on-set dehydrogenation temperature was reduced to as low as 60°C, which is far lower than the temperature (about 280°C) required for the pristine $LiBH_4$. As for the $LiBH_4$ 0.2 Al composite, the dehydrogenation process was significantly accelerated with temperature range of 300°C–400°C. Subsequently, Yang et al. reported the dehydrogenation performance of $LiBH_4$ doped with the metals (Mg–Al–Ti–V–Cr–Sc).[104] It was found that Mg and Al followed the thermodynamically preferred reaction resulting in the formation of MgB_2 and AlB_2, respectively, while the other metals kinetically promoted the decomposition of $LiBH_4$ in a modest manner. Furthermore, Remhof et al. reported the dehydrogenation/rehydrogenation cycling properties of $LiBH_4$–Al composite by hydrogenating the milled LiH–AlB_2 mixture[105–107] and found that the destabilization reaction of this composite was reversible but with deteriorated hydrogen capacity. The in-depth examination reveals that AlB_2 does not completely reproduce in the dehydrogenation process.

Therefore, the lack of high chemical reactivity of B and the decomposition of AlB_2 caused the limited regeneration of $LiBH_4$. For these metal-modified $LiBH_4$ composites mentioned earlier, the majority of hydrogen release still occurs at a temperature over 400°C.

Metal hydrides are advantageous over other reactive additives in terms of hydrogen storage capacity; therefore, the $LiBH_4$–MH_x composites attracted more research attention. The destabilizing reactions and hydrogen storage properties of $LiBH_4$ compositing with several metal hydrides are summarized in Table 15.6. Vajo et al.[101] firstly demonstrated that the mechanically milled mixtures of $LiBH_4$–MgH_2 in a molar ratio of 2:1 dehydrogenated into the LiH and MgB_2, resulting in the reaction enthalpy reduction by 25 kJ/mol H_2 compared with that of pure $LiBH_4$. More importantly, this composite shows superior reversible hydrogen capacity of ca. 8 wt.%, which is 70% of the theoretical value. According to the van't Hoff plot as shown in Figure 15.5, the destabilized $LiBH_4$–MgH_2 system has higher equilibrium pressure than the individual $LiBH_4$ and MgH_2, and especially the equilibrium pressure of $LiBH_4$ is increased by 10 times. Therefore, below 350°C, both $LiBH_4$ and MgH_2 are destabilized by the formation of MgB_2 during dehydrogenation.

TABLE 15.6
Hydrogen Storage Properties of Reversible $LiBH_4 + MH_x$ Composites

Reaction	Hydrogen Capacity (wt.%)			Temp. (°C) [Pressure (MPa)]		$\Delta_r H$ (kJ/mol H_2)	References
	Ideal	Dehy.	Second Dehy.	Dehy.	Rehy.		
$2LiBH_4 + MgH_2 \leftrightarrow 2LiH + MgB_2 + 4H_2$	11.6	8ᵃ	9	280–450	230–350 [88]	42ᶜ	[101]
$6LiBH_4 + CaH_2 \leftrightarrow 6LiH + CaB_6 + 10H_2$	11.7	9.3ᵃ / 7.7ᵇ	8.9ᵃ	227–427	400 [88]	65.5ᵈ	[108,109]
$6LiBH_4 + SrH_2 \leftrightarrow SrB_2 + 6LiH + 10H_2$	9.1	8.7ᵇ	6.0ᵇ	480	450 [86]	48ᵈ	[110]
$4LiBH_4 + YH_3 \leftrightarrow 4LiH + YB_4 + 7.5H_2$	8.5	2ᵃ / 7.2ᵇ	5	350	350 [87]	51ᶜ	[111,112]
$6LiBH_4 + LaH_2 \leftrightarrow 6LiH + LaB_6 + 10H_2$	7.4	5.1ᵇ	4.6ᵇ	260–400	400 [84]	70ᵈ	[113]
$6LiBH_4 + CeH_2 \leftrightarrow 6LiH + CeB_6 + 10H_2$	7.4	6.1ᵃ / 4.6ᵇ	6.0ᵃ	260–400	350 [88]	58ᶜ / 44.1ᵈ	[108,113,114]

ᵃ Desorption at static vacuum.
ᵇ Desorption under hydrogen back pressure.
ᶜ Experimental result.
ᵈ Theoretical calculation.

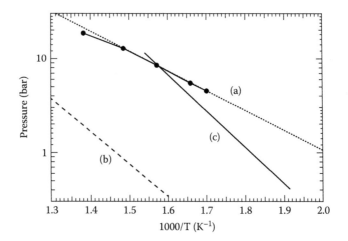

FIGURE 15.5 van't Hoff plots showing the elevated equilibrium pressure for the 2LiBH$_4$. (a) MgH$_2$ system, (b) relative to those of LiBH$_4$, and (c) MgH$_2$. (Reprinted with permission from Vajo, J.J., Skeith, S.L., and Mertens, F., *J. Phys. Chem. B*, 109, 3719–3722. Copyright 2005 American Chemical Society.)

As also shown in Table 15.6, adding other metal hydrides, such as CaH$_2$, SrH$_2$, LaH$_2$, and CeH$_2$, into LiBH$_4$ leads to different dehydrogenated products and hydrogen storage properties. Maximum theoretical hydrogen capacity up to 11.7 wt.% and practical dehydrogenation capacity of ca. 9.3 wt.% are acquired in the LiBH$_4$–CaH$_2$ system. Such high capacity is due to the formation of CaB$_6$ other than CaB$_2$ and thus larger molar ratio for LiBH$_4$ in the dehydrogenation reaction.

It was found that the hydrogen back pressure played a key role on the dehydrogenation route of LiBH$_4$–MH$_x$ composites and the end product. Taking the LiBH$_4$–YH$_3$ system, for example, under vacuum condition, the formation of amorphous B and/or Li$_2$B$_{12}$H$_{12}$ from the decomposition of LiBH$_4$ may exist at the interfaces of liquid LiBH$_4$ and YH$_3$ and block their direct contact, so that the destabilization reaction is disturbed, and metal boride could not be formed. Whereas, the hydrogen back pressure can suppress the direct decomposition of LiBH$_4$ into amorphous B and/or Li$_2$B$_{12}$H$_{12}$. Additionally, the hydrogen back pressure may improve the wetting interaction between liquid LiBH$_4$ and YH$_3$, and suppress the phase separation. Consequently, the destabilizing reaction between LiBH$_4$ and YH$_3$ was promoted, and YB$_4$ was preferably formed under hydrogen atmosphere. It was further found that the increase of hydrogen back pressure remarkably enhanced the overall dehydrogenation reaction rate of LiBH$_4$–YH$_3$ composite, although the dehydrogenation amount decreases. Similar influence was observed as well when argon back pressure was exerted in other destabilized systems, such as LiBH$_4$–MgH$_2$ and LiBH$_4$–CaH$_2$ composites. For example, Bösenberg et al.[115] demonstrated that a suitable combination of temperature and back pressure was necessary to promote the formation of MgB$_2$ in the LiBH$_4$–MgH$_2$ composite.

To further improve the kinetics of destabilizing reaction at lower temperature, adding nanostructured metal hydrides via in situ formation was attempted. It was shown that $LiBH_4$ reacted with in situ–formed GdH_2 at quite low temperature 275°C to facilitate the hydrogen release, and the GdB_4 compound was formed under either vacuum or 0.5 MPa hydrogen back pressure. Sun et al. reported that about 4.0 wt.% hydrogen could be reversibly released at 400°C in the $LiBH_4$–Mg_3La composite doped with $TiCl_3$,[116] and attributed it to the synergistic effect of Mg_3La and $TiCl_3$. This synergistic effect was explained by that the in situ–formed MgH_2 nanoparticles upon hydrogenation of Mg_3La alloy promoted the dehydrogenation of $LiBH_4$, and the addition of $TiCl_3$ together with the nanosized LaH_3 enhanced the dehydrogenation of MgH_2 and amplified the destabilizing effect of MgH_2 on $LiBH_4$. Similar synergistic thermodynamic and kinetic improvement was reported in the $LiBH_4$–La_2Mg_{17}.[117]

From a thermodynamic standpoint, ScH_2 and TiH_2 are predicted to be suitable destabilizers according to the first-principles calculations as shown in Table 15.5. The reaction enthalpy can be reduced to 32.6 and 4.5 kJ/mol H_2 by doping with ScH_2 and TiH_2, respectively, implying a lower dehydrogenation temperature. However, the experimental results demonstrated that the destabilizing reaction did not take place, and the dehydrogenation temperature of $LiBH_4$ was not effectively decreased. This result is due to the high thermal stability of ScH_2 and TiH_2 as indicated by large heat of formation, which remained stable during dehydrogenation.[104,118]

Besides the elements and metal hydrides, other additives such as metal oxides, halides, and fluorides were also selected to be reducing agents to destabilize $LiBH_4$; some typical destabilized systems are shown in Table 15.7. Züttel et al. first found

TABLE 15.7
Hydrogen Storage Properties of $LiBH_4$–Oxide/Halide/Fluoride Composites

Additives		Hydrogen (wt.%)		Condition: Temp.(°C)		Toxic
Type	Amount (wt.%)	First Dehy.	Rehy.	First Dehy.	Rehy.	By-Product
SiO_2	10–25	9–10[a]	—	150–600	—	—
TiO_2	25–80	4–9[a]	3.5–8.3[a]	100–600	600	—
ZrO_2	25	8–9[a]	—	175–600	—	—
V_2O_3	25	8–9[a]	8[a]	175–600	600	—
SnO_2	25	8–9[a]	—	175–600	—	—
Nb_2O_5	50–80	4–6[a]	—	100–600	—	—
Fe_2O_3	50–66.7	5.7–9[a]	—	100–600	—	—
V_2O_5	50–66.7	5.7–9[a]	—	100–600	—	—
$TiCl_3$	10–88	2.8–9.2[a]	3.4[a]	100–600	500	B_2H_6
$CoCl_2$	5–100	10.5–18.3[b]	—	230–600	—	B_2H_6
TiF_3	10–50	6.4–14[a]	0.2–4.0[a]	100–500	350–500	B_2H_6
ZnF_2	10–50	3.7–7[a]	1–4[a]	120–500	500	B_2H_6
Mixture of $MgCl_2/TiCl_3$	30	5[a]	4.5[a]	50–600	600	—

Source: Zuttel, A. et al., *J. Power Sources*, 118, 1, 2003.
[a] Indicates the value referred to the mixture of $LiBH_4$ with by-product.
[b] Indicates the value referred to the pristine $LiBH_4$.

that SiO_2 promoted the hydrogen liberation from $LiBH_4$ at temperature as low as 200°C;[84] SiO_2 actually served as reactant to result in the thermally stable compound Li_4SiO_4. Yu et al. investigated the dehydrogenation properties of TiO_2-doped $LiBH_4$ with various proportions.[119] For the $LiBH_4 + TiO_2$ (mass ratio 1:4) composite, the on-set temperature for dehydrogenation is 150°C and the peak temperature reaches about 290°C. The result demonstrates that $LiBH_4$ reacts with TiO_2 to form $LiTiO_2$ and liberate hydrogen. Yu et al. also systematically investigated the effect of various oxides on the dehydrogenation of $LiBH_4$, and the effectiveness follows the sequence order of $Fe_2O_3 > V_2O_5 > Nb_2O_5 > TiO_2 > SiO_2$.[120] For the ball-milled $LiBH_4 - 2Fe_2O_3$ composite, the dehydrogenation starts from 100°C and the majority of hydrogen (6 wt.%) can be liberated when heating up to 200°C. Generally, the interaction between $LiBH_4$ and oxides (MO_x) followed a redox reaction:

$$LiBH_4 + MO_x \rightarrow LiMO_x + B + 2H_2 \qquad (15.7)$$

Similar reducing reaction is also found in the halide-/fluoride- modified $LiBH_4$ composites. This destabilizing reaction is highly exothermic, and the dehydrogenated products are hard to be rehydrogenated back to $LiBH_4$.

The rehydrogenation of destabilized system containing lithium hydride and metal borides normally requires harsh temperature (>400°C) and pressure (>10 MPa) to promote the decomposition of borides and the recombination of complex anion. To improve the rehydrogenation properties, nanosized additives and catalyst doping were attempted by mechanical milling. Lim et al. investigated the reversibility of ball-milled $6LiBH_4 - CaH_2$ with NdF_5 as catalyst.[121] The composite exhibited much better rehydrogenation properties in terms of rehydrogenation rate and amount than the undoped one. About 9 wt.% hydrogen was absorbed at 450°C under 3 bar H_2 pressure for 12 h. With 15 wt.% NdF_5 dopant, the $6LiBH_4 - CaH_2$ composite showed a superior cyclic capacity of about 6 wt.% up to 9 cycles. In addition, Zhang et al. introduced TiF_3 as a dopant into the $LiBH_4 - SiO_2$ composite to avoid the formation of Li_4SiO_4 and greatly enhanced the reversibility of this composite.[122] The $LiBH_4 - SiO_2 - TiF_3$ system could absorb more than 4.0 wt.% hydrogen at 500°C under 4.5 MPa hydrogen pressure, but different rehydrogenation products were obtained.

It is stated that the rehydrided composite hydrides, such as $LiBH_4 - LaH_2$, $LiBH_4 - CeH_2$, and $LiBH_4 - GdH_2$, normally decompose at higher temperature and the kinetics become deteriorative in the cyclic dehydrogenation.[112,113,123] For instance, the dehydrogenation temperature of $LiBH_4 - LaH_2$ composite was shifted from 350°C to 390°C after the second cycle.[113] With respect to $LiBH_4 - GdH_2$, the dehydrogenation presented poor kinetics and only about 2 wt.% hydrogen was desorbed in the second cycle.[123] Cai et al. found that the reversibility of $LiBH_4 - NdH_2$ composite could only be realized when the size of NdH_2 is less than ca. 10 nm, as shown in Figure 15.6.[124] As the NdH_2 grains grew up and became stable after the first dehydrogenation treatment, the thermodynamic destabilization effect disappeared and the rehydrogenation capacity decreased. These experimental results mean that the thermodynamic and kinetic improvements are diminished due to structure coarsening in the cycling.

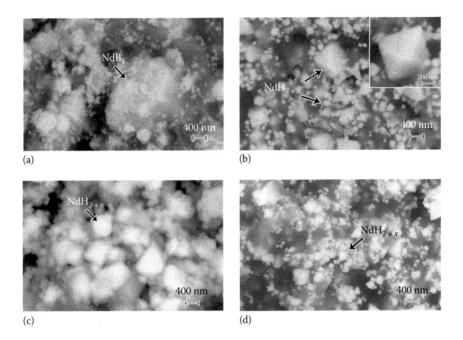

FIGURE 15.6 Back-scattered SEM images of LiBH$_4$–NdH$_{2+x}$. (a) first dehydrogenation, (b) first rehydrogenation, (c) second dehydrogenation, and (d) second rehydrogenation. (Reprinted with permission from Cai, W., Wang, H., Sun, D., and Zhu, M., *J. Phys. Chem. C*, 117, 9566–9572. Copyright 2013 American Chemical Society.)

In summary, doping reactive additive could significantly improve the dehydrogenation of thermodynamic and kinetic properties of LiBH$_4$, but the rehydrogenation of composite hydrides still suffers at rather high temperature and pressure, which can be further improved by nanostructure controlling and catalyst doping.

15.3.3 Catalyzed Alanates

The dehrdrogenation/rehydrogenation reaction of complex hydrides generally involved multisteps of physisorption and chemisorption of hydrogen, dissociation and recombination, interface reaction, and diffusion of metal and hydrogen atoms, all of which cause high activation energy barrier that is responsible for poor kinetics and high dehydrogenation/rehydrogenation temperature. Bogdanović et al. first found that the TiCl$_3$-catalyzed NaAlH$_4$ had more favorable reaction kinetics and reversible storage capacity than the uncatalyzed one.[79] It has been generally recognized that doping catalyst is an indispensable and the most effective method to improve the hydrogen storage properties of NaAlH$_4$, especially for the rehydrogenation. The rehydrogenation back to NaAlH$_4$ is in solid state and is severely limited because of slow interdiffusion rate between the NaH and Al, especially when small Al domains are segregated to large particles of micrometer size at elevated temperature. Therefore, tremendous efforts have been devoted to develop effective catalysts and doping methods. The concentration of dopant is also optimized to reduce the capacity loss. In addition, the catalytic mechanism and cyclic stability of catalyzed $NaAlH_4$ in association with the structural change of catalytic species are widely investigated.

15.3.3.1 Hydrogen Storage Properties of Catalyzed $NaAlH_4$

$TiCl_3$ and $Ti(OBu^n)_4$ in ether solution were first used to promote the dehydrogenation/rehydrogenation of $NaAlH_4$.[79] In comparison to the solution doping, the mechanical doping is more easily realized by milling the additives with $NaAlH_4$ or NaH–Al mixture. This doping method readily achieves fine distribution of catalysts and close contact between the hydride and catalyst and also promotes the conversion of catalytic precursor to active species. Bogdanović et al. prepared the $ScCl_3$- and $CeCl_3$-doped $NaAlH_4$ using the one-step direct synthesis method,[125,126] namely simultaneously milling the NaH–Al powder with dopants under H_2 pressure of 55–85 bar. The obtained Ce-doped $NaAlH_4$ showed outstanding kinetics and superior cyclic properties. Kang et al. prepared Ti-doped $NaAlH_4$ by the two-step milling method, in which metallic Ti was first milled with NaH and then milled again after adding Al powder.[127] This method overcomes the problem that the ductile Al may impede a highly homogeneous distribution of Ti-containing phase during the milling process and improved the active property of Ti-containing species during the $NaAlH_4$ preparation.

Ti oxides, hydrides, chloride, and fluoride are the most used dopants with their high activity on catalyzing dehydrogenation/rehydrogenation of $NaAlH_4$. These titanium-containing compounds actually served as precursor that would react with $NaAlH_4$ in the milling or dehydrogenation process and result in NaCl, Ti, and Al phases in the dehydrogenated product according to the following reaction:[128]

$$TiCl_3 + 3NaAlH_4 \rightarrow Ti + 3Al + 3NaCl + 6H_2 \qquad (15.8)$$

The metallic Ti may further react with NaH and convert to Ti hydride according to Equation 15.5 or react with Al to form active Ti–Al cluster.[83]

$$Ti + 2NaH \rightarrow TiH_2 + 2Na \qquad (15.9)$$

Obviously, these side reactions further reduce the storage capacity. In this aspect, zero-valent Ti dopant is superior to dopants containing titanium in positive oxidation states. To avoid the formation of inactive by-products and the loss of active hydrides, alanates could be doped using TiH_2 or through indirect doping by premilling $TiCl_2$ with LiH. However, this method was proven to be unfavorable with respect to the kinetics; the improved rates were achieved only after 10 cycles.[129] This result suggests that the reactive decomposition of Ti-containing precursors in the doping process serves as an important function.

The direct use of nanosized dopants instead of coarse-grained dopants enables one to achieve a high distribution of active species and decrease the doping amount. Bogdanović et al. remarkably improved the hydrogen storage properties of $NaAlH_4$ by doping Ti nanoparticles.[130] After only two cycles, the dehydrogenated product could absorb a maximum of 4.6 wt.% hydrogen in 15 min under 102°C and 133 bar; the absorption rate is 40 times faster than that of $Ti(OBu^n)_4$-doped $NaAlH_4$ by dry milling. However, the rehydrogenation rate experienced a drastic decrease after 16 cycles and only 90% of maximum capacity was obtained in 45 min. The degenerative property is ascribed to the coarsening of Ti nanoparticles and the diminishment of catalytic effect. In contrast to Ti nanoparticles, the $NaAlH_4$ doped with 6 nm sized TiN exhibited better cyclic properties, and TiN nanocrystals showed high structural stability in the hydriding/dehydriding cycles.[131]

In addition to Ti-containing catalyst, other TM compounds were also explored to improve the hydrogen storage properties of $NaAlH_4$. Anton et al. found that the catalytic effect of V, Ti, Zr, Ce, and Nb halides and fluorides[132] followed a sequence order of $Ti^{4+} > Ti^{3+} > Ti^{2+} > Zr^{4+} > V^{3+} > V^{2+} > V^{4+} > Nb^{4+}$. The catalytic effect was thought to be mainly correlated with the ionic radius of cations, and the highest catalytic activity was acquired as the ionic radius is 0.076 nm, between those of Al^{3+} and Na^+. The catalytic role of rare-earth compounds (e.g., $LaCl_3$, $CeCl_3$, $PrCl_3$) on the dehydrogenation was first investigated by Bogdanović[133] and Zhu et al.[134] Bogdanović found that the dehydrogenation/rehydrogenation properties of $CeCl_3$-doped $NaAlH_4$ were even better than that of the $TiCl_3$-doped one, especially on the rehydrogenation rate under low pressure. Zhu et al. found that there exists a small amount of La_3Al_{11} phase in the hydrogenated product of $NaAlH_4$–10 mol% $LaCl_3$, indicating that $LaCl_3$ also acted as a catalyst precursor and reacted with $NaAlH_4$, and thus the La–Al phase are possibly real catalyst.[134] To verify this hypothesis, Chen et al. directly introduced $CeAl_2$ into $NaAlH_4$.[135,136] As shown in Figure 15.7, almost the same dehydrogenation rate and rehydrogenation rate were obtained for $CeAl_2$-doped and $CeCl_3$-doped

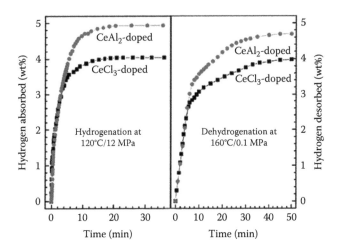

FIGURE 15.7 Comparison on the dehydrogenation and rehydrogenation properties between $CeCl_3$-doped $NaAlH_4$ and $CeAl_2$-doped $NaAlH_4$. (Reprinted with permission from Fan, X., Xiao, X., Chen, L., Li, S., Ge, H., and Wang, Q., *J. Phys. Chem. C*, 115, 2537–2543. Copyright 2011 American Chemical Society.)

NaAlH$_4$, but CeAl$_2$-doped NaAlH$_4$ could reload 4.9 wt.% hydrogen under 12 MPa and 393 K in 20 min, much higher than 3.9 wt.% for CeCl$_3$-doped NaAlH$_4$. Besides, a reversible capacity of more than 4.0 wt.% hydrogen could be achieved at a low hydrogen pressure of 4.0 MPa. With prolonged cycling, the CeAl$_2$ transformed to a more stable species of CeAl$_4$.

Majzoub and Gross[137] and Anton[132] reported that the catalytic activity of TiF$_3$ was similar to that of TiCl$_3$, and ignored the catalytic function of anion. However, Wang et al.[138] found that TiF$_3$ was superior to TiCl$_3$ as a dopant precursor in preparation of catalytically enhanced NaAlH$_4$ system, and proposed the functional anion concept[139] by elucidating the effect of fluorine anion on NaAlH$_4$ according to theoretical calculation results. This concept challenges the traditional viewpoint that the anion of dopant precursors contributes nothing but to generate inactive side products. The calculation result indicates that the fluoride doping results in a substitution of H by F anion in the hydride lattice and accordingly a favorable thermodynamics adjustment. The decomposition enthalpy is reduced from 59.1 kJ/mol of pure Na$_{12}$Al$_4$H$_{24}$ down to 56.0 kJ/mol after substituting one H by F in Na$_{12}$Al$_4$H$_{23}$F$_1$.

15.3.3.2   Catalytic Mechanism of Ti-Doped NaAlH$_4$

Mechanism studies have been carried out on the Ti-doped NaAlH$_4$ to reveal the chemical state of Ti and its action on the dehydrogenation/rehydrogenation reaction of NaAlH$_4$. It has been generally accepted that the Ti dopants was readily reduced by reacting with NaAlH$_4$ under H$_2$ atmosphere, and the real active species were deduced to be TiH$_x$, or Ti–Al alloys, or Al(Ti) solid solution, or Ti$_{15}$Al$_{85}$ phase, or TiAl$_3$ phase according to phase structure analysis by XRD/EDS/NMR/TEM/XANE characteristic methods reported in different works. The complexity on the existing form of Ti causes great trouble to correctly understand the catalytic mechanism of Ti-catalyzed NaAlH$_4$. So far, different viewpoints regarding the catalytic role of Ti species were proposed: (1) reducing the formation enthalpy of H vacancy and H interstices; (2) enhancing the H diffusion by forming Ti defect inside NaAlH$_4$ or on NaAlH$_4$ surface; (3) weakening the Al–H bonding in AlH$_4$ complex; (4) promoting the interfacial reaction or the nucleation of Al; (5) lattice substitution mechanism; and (6) AlH$_3$ vacancy–mediated mechanism.

Sun et al. found that the lattice parameters of NaAlH$_4$ changed with the concentration of dopants because of the incorporation of Ti with a charge between 2$^+$ and 4$^+$ on sodium sides.[140] It was believed that the lattice substitution in association with unit cell volume destabilized the NaAlH$_4$ and facilitated its decomposition. Some ab initio calculation results also confirm that Ti prefers to substitute Na of NaAlH$_4$ surface and attracts neighboring H atom, and thus help to break the Al–H bond. It is also reported that Ti would replace Al atoms according to density function theory calculation. Whatever happens, the energy barrier for the dehydrogenation of substituted NaAlH$_4$ becomes less according to the calculation results. However, the dissolution of Ti in NaAlH$_4$ is rarely found by other researchers. In addition, the lattice substitution mechanism could not explain the role of Ti on the rehydrogenation of NaAlH$_4$.

The AlH$_3$ vacancy–mediated mechanism was first reported by Gunaydin et al.[141] They determined the activation energy for Al mass transport via AlH$_3$ vacancies to be 85 kJ/mol·H$_2$, which is in excellent agreement with experimentally measured values in Ti-catalyzed NaAlH$_4$. This activation energy is also significantly less than the value of 112 kJ/mol·H$_2$ for an alternative decomposition mechanism via NaH vacancies. The calculation results also suggest that bulk diffusion of Al species is the rate-limiting step in the dehydrogenation of Ti-doped NaAlH$_4$, whereas much higher activation energy for undoped NaAlH$_4$ is controlled by other processes, such as breaking up of [AlH]^4 complex, formation/dissociation of H$_2$, and/or nucleation of the product phases.

Regarding the AlH$_3$ mechanism in the dehydriding and rehydriding of NaAlH$_4$, based on the calculation results by using density function theory method, Zhu et al. further pointed out that there was an intermediate state of metastable AlH$_3$ molecule during the dehydriding of NaAlH$_4$; the hydrogen release comes from the decomposition of AlH$_3$ instead of direct decomposition of the matrix.[142] This mechanism involves the following critical steps:

1. At first, AlH$_3$–H separation occurs in AlH$_4$ groups at surface.
2. Then, the AlH$_3$ escapes from the lattice point of surface and it is probably adsorbed at the matrix surface, that is the AlH$_3$ vacancy V^0_{AlH3} is formed at the surface.
3. Next step is the migration and decomposition of AlH$_3$, followed by the growth of Al cluster and the formation of NaAlH$_4$–Al interface.
4. V^0_{AlH3} at the surface or NaAlH$_4$–Al interface migrates into bulk (or AlH$_3$ move from bulk to surface/interface).

Steps 1–4 are reiterated under certain thermodynamic conditions, and thus NaAlH$_4$ is decomposed.

Rehydriding of NaAlH$_4$ is more difficult to take place than the dehydriding under normal experimental condition because of the intrinsic thermodynamic metastability of AlH$_3$ molecules. It was reported that the rehydriding of NaAlH$_4$ will be significantly enhanced if AlH$_3$ molecules could be stabilized.[143] Therefore, the rehydriding of NaAlH$_4$ could be improved by increasing the stability of intermediate AlH$_3$. The suggested AlH$_3$ mechanism is also in line with the catalyzing effect of Ti on NaAlH$_4$ because it was suggested that the TM cluster or TM–Al cluster, located on the NaAlH$_4$ surface, would reduce the formation enthalpy, increase the stability of absorbed AlH$_3$ in rehydriding, and decrease the migration barrier of AlH$_3$ and/or V^0_{AlH3}. This surface catalysis mechanism could also well explain the long-range migration of Al because of the fine distribution of catalytic species in the hydride matrix; even the grain size of Al may increase up to 100–300 nm.

15.3.4 NANOCONFINEMENT OF COMPLEX HYDRIDES

Gutowska et al. successfully infiltrated ammonia borane NH$_3$BH$_3$ into highly porous silica (SBA-15, 7.5 nm in diameter) by solvent-mediated infiltration (SMI) method.[144] This strategy, so-called nanoconfinement, is effective to enhance the dehydrogenation properties of ammonia borane as well as suppress the borazine release.

The enhanced dehydrogenation performance should be related to the physical confinement, support effects, shorter diffusion distances, and particle size effects. This pioneering report encourages subsequent intensive explorations on the nanoconfinement of complex hydrides, such as $NaAlH_4$ and $LiBH_4$. To be suitable nanoscaffolds, these materials should possess high surface area and large total porosity in order to allow large loading ratio of complex hydrides. In addition, good chemical inertness and structure stability are required to avoid chemical reaction with highly reducing complex hydrides and the collapse of porous structure during hydrogenation/dehydrogenation. So far, carbon materials including carbon aerogels (CAs), AC, and ordered mesoporous carbon (MC), metal organic frameworks materials, and mesoporous silica have been explored to be confining media of complex hydrides. Table 15.8 shows the hydrogen storage properties of nanoconfined $LiBH_4$ and $NaAlH_4$ by different loading processes and supporting materials.

Effective infiltration techniques have been developed involving melt infiltration (MI) and SMI.[159] The main point of MI is that the decomposition temperature of selected porous materials should be higher than that of hydrides. For SMI, the prerequisite is a solvent capable of dissolving complex hydrides, such as THF (C_4H_8O) and MTBE ($C_5H_{12}O$). A typical procedure of SMI is illustrated in Figure 15.8. The porous material is submerged in the solution and the pores are infiltrated by the solution. Finally, complex hydrides solidify to form amorphous and/or crystalline solid nanoparticles in the pores upon removal of the solvent by evaporation. Several cycles of infiltration and evaporation may be necessary to increase the loading weight of complex hydrides.

By MI under argon atmosphere at 280°C–300°C, $LiBH_4$ was confined in resorcinol–formaldehyde carbon aerogels (RF-CAs)[147] with an average pore size of 13 nm and a total pore volume of 0.8 cm^3/g. Up to 90% volume of RF-CAs was filled with $LiBH_4$ and the weight loadings are 25–30 wt.%. With the confinement of CA, the dehydrogenation rate of $LiBH_4$ increased by as much as 50 times, and the on-set dehydrogenation temperature was reduced by 75°C. In addition, the hydrogen capacity retention over three sorption cycles was increased from 28% for bulk $LiBH_4$ to 60% confined $LiBH_4$. Confinement of $LiBH_4$ was also realized by using highly ordered porous carbon with an average pore size of 2 nm as scaffold.[151] Confinement in such small pores resulted in the disappearance of structural phase transition and melting transition of $LiBH_4$, as well as the significant decrease of desorption temperature from 460°C to 220°C. Meanwhile, the release of gaseous diborane was eliminated in this nanoconfined $LiBH_4$. In the interest of the dependence of decomposition behavior on the pore size, $LiBH_4$ confined by a series of carbon materials with pore sizes ranging from 2 to 15 nm was systematically investigated by Liu et al.[152] Both the structural phase transition and melting of $LiBH_4$ shifted to a lower temperature range with the decreasing pore size for scaffold and finally vanished below a pore size around 4 nm. The release of poisonous diborane was gradually reduced with the decreasing pore size of carbon materials, and a superior reversibility was achieved under 60 bar hydrogen pressure at 250°C. This result agrees well with the finding by solid-state nuclear magnetic resonance (NMR)[160] that the diffusional mobility of Li^+ and $[BH_4]^-$ tetrahedral was significantly enhanced by the nanoconfinement effect.

TABLE 15.8

Hydrogen Storage Properties of Confined LiBH$_4$ and NaAlH$_4$ into Different Scaffolds

Hydride	Scaffold	Infiltration Method	Load Weight: (wt.%)	Hydrogen Content[a] (wt.%)		Temperatue: (°C) (Pressure: [MPa H$_2$])		References
				First Dehy.	Second Dehy.	Dehy.	Rehy.	
LiBH$_4$	Silica gel	Ball milling	50	12.0	—	260	—	[145]
LiBH$_4$	AC	SMI	28.4	11.3	6.6	300	300 [5]	[146]
LiBH$_4$	CAs	MI	25–50	~5.0	3.5	380	400 [10]	[147]
LiBH$_4$	MC	SMI	33	3.4	—	300	—	[148]
LiBH$_4$	Micro-/macroporous carbon	SMI	30	4.0	0	300	300 [10]	[149]
LiBH$_4$	Nanoporous carbon	MI	30	14	5.8	350	320 [4]	[150]
LiBH$_4$	Highly ordered nanoporous carbon/CA	MI	10–20	11	4	250	350 [6]	[151,152]
LiBH$_4$	Ordered mesoporous silica	SMI	30	8.5	0	105	450 [7]	[153]
NaAlH$_4$	Ordered mesoporous silica	SMI	20	3.0	2.0	180	150 [5.5]	154]
NaAlH$_4$	AC	MI	48.6	4.0	4.0	150	160 [10]	[155]
NaAlH$_4$	High-surface-area graphite	MI	20	6.3	3.3	180	150 [5.5]	[156]
NaAlH$_4$	MC	MI	26	5.0	4.8	180	150 [7]	[157]
NaAlH$_4$	MOF–74(Mg)	MI	21	4	4	200	160 [10.5]	[158]

[a] Hydrogen content is calculated with respect to LiBH$_4$.

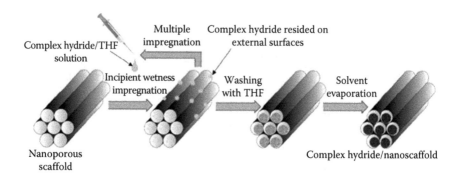

FIGURE 15.8 SMI procedure for complex hydrides loading into nanoporous scaffold. (Reprinted with permission from Fan Zheng, S.Y., Fang, F., Zhou, G.Y., Chen, G.R., Ouyang, L.Z., Zhu, M., and Sun, D.L., *Chem. Mater.*, 20, 3954–3958. Copyright 2008 American Chemical Society.)

Li et al. synthesized a space-confined NaAlH$_4$ system exclusively embedded in ordered MC by melting impregnation plus dehydrogenation/rehydrogenation treatment.[157] Significant improvement in dehydrogenation kinetics and cycling stability was realized. The activation energy was reduced to 46 ± 5 kJ/mol, and the enhanced cycling stability was also achieved by the high-capacity retention of >80% after 15 cycles. These remarkable improvements were attributed to the synergetic effects of both nanoconfinement and chemical catalysis caused by MC, where the nanoconfinement played the most important role. According to the scanning electron microscopy (SEM) analysis, the phase distribution during the dehydrogenation/rehydrogenation cycles is schematically illustrated in Figure 15.9. In the case of pristine NaAlH$_4$, the decomposition products (NaH–Al) agglomerate into micrometer level in size, resulting in the difficulty for the rehydrogenation into NaAlH$_4$. However, for the NaAlH$_4$–MC system, the MC pore serves as a *nanoreactor*, where both NaAlH$_4$ particles and dehydriding products are physically restricted at a nanoscale level. This facilitates the mass transportation of hydrides and metal by shortening the diffusion distance and thus leads to superior kinetics and enhanced cycling stability.

In summary, although great improvements have been made on the dehydrogenation/rehydrogenation temperature and rate, and reversibility of complex hydrides, none of the current complex hydrides can satisfy the requirements of on-board hydrogen storage for fuel cell vehicles, and continued efforts are needed to further overcome the dehydrogenation/rehydrogenation thermodynamic and kinetic limitations of the complex hydrides. As for the thermodynamics, the dehydrogenation of complex hydrides is required to be endothermic with reaction enthalpy of 20–50 kJ/mol H$_2$, which enables the reversible dehydrogenation and rehydrogenation under practical pressure and temperature.[161] However, most of complex hydrides, exclusive of NaAlH$_4$, possess a desorption enthalpy above 50 kJ/mol H$_2$. By designing composite hydrides of complex hydride with reaction additive, the proper dehydriding reaction enthalpy could be attained but the destabilizing reaction has to suffer from more server kinetics due to slow atomic interdiffusion. Further, some alanates and

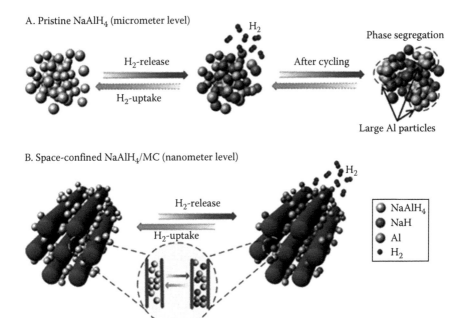

FIGURE 15.9 Illustration of phase distribution in pristine (a) NaAlH₄ and (b) NaAlH₄/MC upon cycling. (Adapted from Li, Y. et al., *Acta Mater.*, 59, 1829, 2011. Copyright 2011 Pergamon. With permission.)

borohydrides, such as $LiAlH_4$, $Zn(BH_4)_2$, $Al(BH_4)_3$, and $Mn(BH_4)_3$, dehydrogenate either endothermically or exothermically.[162,163] Regarding kinetic limitation, owning to the intrinsic stepwise manner and slow atomic diffusion rate in solid state for the dehydriding/rehydriding reaction of complex hydrides, sluggish kinetics is regarded as a much bigger challenge than the high thermodynamic stability. High reaction temperature has to be required to overcome activation energy barriers of multistep kinetic processes and promote the atomic diffusion. Catalyst doping and nanoscale hydride preparation have been proven to be effective in decreasing the reaction temperature and rehydrogenation pressure. Future research emphasis should be laid on the design and synthesis of novel catalyst and nanoscale complex hydride, aiming to achieve fast reaction kinetics under moderate conditions and practical application of complex hydrides.

15.4 METAL AMIDES AND IMIDES

15.4.1 GENERAL PRINCIPLE OF METAL AMIDE AND IMIDE SYSTEMS FOR HYDROGEN STORAGE

Metal amides are a class of coordination compounds composed of a metal center with amide ligands of the form NR_2 (R refer to H or organic groups). Amide ligands have two electron pairs available for bonding. In principle, they can be terminal or bridging. Metal imides are a class of coordination compounds through a substitute of R in metal amides by a metal. In the field of hydrogen storage research, metal amide and imide systems have attracted extensive attention since Chen et al. reported that lithium nitride (Li_3N) could reversibly absorb/desorb ~10.4 wt.% of hydrogen in 2002.[164] This Li–N–H system is different from metal hydrides or nonmetals physisorption materials of high specific surface area. In fact, metal–N–H system is composed of both metallic and nonmetallic elements, which interconvertibly transformed between different chemical phases by the forming and breaking of nonmetal (N) hydrogen bonds.[165] According to Chen's report,[164] the reversible hydrogen storage in Li_3N follows a two-step reaction:

$$Li_3N + 2H_2 \leftrightarrow Li_2NH + LiH + H_2 \leftrightarrow LiNH_2 + 2LiH \qquad (15.10)$$

The standard enthalpy change of the first step in the reaction (15.1) was calculated to be –116 kJ/mol H_2 from the values of standard formation enthalpy of Li_3N, Li_2NH, and LiH. The large reaction enthalpy change leads to a very high temperature (over 430°C) requirement for complete recovery of Li_3N from the hydrogenated state. On the other hand, the second step reaction has a much smaller enthalpy change (–44.5 kJ/mol H_2) and a large amount of hydrogen storage capacity of 6.5 wt.%.[166] Therefore, much more attention had been paid to the second step in the reaction (15.1) as one of the suitable hydrogen storage materials for on-board use.

The significant difference between $LiNH_2$–LiH_2 and the previously reported hydrogen storage materials is that $LiNH_2$ contains H^+ and the LiH contains H^-. In order to lower the operating temperatures and maintain its reversibility and capacity, numerous efforts have been devoted to destabilizing the Li–N–H system through replacing Li with other alkali or alkaline earth metals. Thus, a series of new metal–N–H systems were synthesized, such as Mg–N–H,[167,168] Ca–N–H,[169,170] Li–Mg–N–H,[171,172] Li–Ca–N–H,[173–175] Li–Al–N–H,[176] and Mg–Ca–N–H.[177] Metal–N–H systems can be divided into binary system, ternary system, and multinary system with respect to the number of metal elements in them to be one, two, and three, respectively. In these systems, NH_3 would be released from amide self-decomposition, which is a parallel competing reaction with the H_2 release in the amide/hydride desorption process. NH_3 formation will damage the noble metal catalysts in a fuel cell. Therefore, in order to apply the metal–N–H materials in commercial utilization, besides good hydrogenation/dehydrogenation performance, ammonia release should also be avoided or minimized.

15.4.2 BINARY SYSTEMS

15.4.2.1 Li–N–H System

Li_3N was reported to absorb 10.4 wt.% H_2 with a starting temperature at around 100°C and absorb hydrogen quickly at a temperature range of 170°C–210°C. A total amount of about 9.3 wt.% was achieved after maintaining the sample at 255°C for half an hour. Upon hydrogenation, Li_3N first transforms into $Li_2NH + LiH$ and then

$LiNH_2 + 2LiH$. Li_3N releases large quantities of heat (116 kJ/mol H_2) in the first step of hydrogen absorption. This fast reaction leads to the agglomeration or sintering of reactant and product, which results in the deactivation of reabsorption. To avoid this problem, Hu and Ruckenstein[178] tried to improve the kinetics of Li–N–H system by the partial oxidation of Li_3N. This oxidation is highly effective for hydrogen storage, through which a reversible capacity of 5 wt.% H_2 as well as excellent stability can be reached in only 3 min at a relatively low temperature of 180°C. The desorption process also has two steps: 6.3 wt.% of hydrogen releases at temperatures below 200°C for the first step and the remaining 3 wt.% hydrogen could be desorbed only when the temperature is above 320°C. The reversible hydrogenation/dehydrogenation reaction is shown in Equation 15.10. Chen et al.[164] and Kojima and Kawai[179] had revealed that the standard enthalpy change of reaction (15.11) from the van't Hoff plot was ~66 kJ/mol H_2 through carefully measuring the PCI curves for the Li–N–H system at different temperatures, which is higher than the theoretically calculated value of 45 kJ/mol H_2.[164] Because the enthalpy change of reaction (15.12) is even much higher than the reaction (15.2), the researchers focus mainly on the storage properties of Li_2NH.

$$Li_2NH + H_2 \leftrightarrow LiNH_2 + LiH \qquad (15.11)$$

$$Li_3N + 2H_2 \leftrightarrow LiNH_2 + 2LiH \qquad (15.12)$$

According to the experimentally measured entropy change (~120 J/mol K), ~277°C is required to raise the plateau pressure up to 1.0 bar for the reaction (15.12).[179]

15.4.2.2 Mg–N–H System

Like the system of $LiNH_2$ and LiH, the reaction of $Mg(NH_2)_2$ and MgH_2 leads to a Mg–N–H hydrogen storage material system. Nakamori et al.,[180] Leng et al.,[181] Hu et al.,[168] and Xie et al.[182] revealed that the Mg–N–H system had a quite low onset temperature of hydrogen desorption than that of the Li–N–H system. The corresponding reaction can be written as follows and 7.4 wt.% of hydrogen could be released:

$$Mg(NH_2)_2 + 2MgH_2 \rightarrow Mg_3N_2 + 4H_2 \qquad (15.13)$$

Unlike the desorption reaction step of $LiNH_2$–LiH, the investigation of thermal behavior by Nakamori et al.[180] showed that $Mg(NH_2)_2$ was difficult to react directly with MgH_2, and only self-decomposition of $Mg(NH_2)_2$ was observed together with the release of NH_3 in the heating process. However, Hu et al.[168] demonstrated that the hydrogen desorption from $Mg(NH_2)_2$ and MgH_2 can occur under a mechanochemical reaction condition. It was found that hydrogen could be released from $Mg(NH_2)_2$–$2MgH_2$ shortly after ball milling at room temperature and the evolvement accelerated when extending the milled time; 75% hydrogen was released in 20 h and the remaining hydrogen was released when the time extends to 72 h. In addition, the concentration of NH_3 was below 2 ppm during the whole process. The phase structure of different ball-milled samples was investigated in order to reveal the reaction mechanism of $Mg(NH_2)_2$–MgH_2 system. In the reaction of $Mg(NH_2)_2$ and MgH_2, the MgNH was first formed according to the following reaction:

$$Mg(NH_2)_2 + 2MgH_2 \rightarrow 2MgNH + MgH_2 + 2H_2 \tag{15.14}$$

And further reaction between MgNH and MgH_2 results in the formation of Mg_3N_2, as shown in the following:

$$2MgNH + MgH_2 \rightarrow Mg_3N_2 + 2H_2 \tag{15.15}$$

According to the thermodynamic analysis, the dehydrogenation reaction enthalpy of reaction (15.13) is only 14 kJ/mol (3.5 kJ/mol H_2). Hence, Chen and coworkers postulated that although the reaction was reversible in principle,[169] the rehydrogenation of Mg_3N_2 is difficult at moderate condition.

15.4.2.3 Ca–N–H System

The Ca–N–H system possesses more complicated properties. Ca_3N_2 absorb ~2.36 wt.% hydrogen above 300°C to form CaNH and CaH_2 according to the following reaction:

$$Ca_3N_2 + 2H_2 \rightarrow 2CaNH + CaH_2 \tag{15.16}$$

However, the reaction was not completely reversible even at 600°C and only approximately half of the absorbed hydrogen could be released. In addition, CaNH reacts with CaH_2 to form Ca_2NH, but not Ca_3N_2, as described in the following reaction:

$$2CaNH + 2CaH_2 \leftrightarrow 2Ca_2NH + 2H_2 \tag{15.17}$$

Xiong et al. demonstrated that Ca_2NH could reversibly store ~1.9 wt.% hydrogen at 550°C, but at least 500°C is required to raise the plateau pressure up to 1.0 bar.[169]

15.4.3 Ternary Systems

15.4.3.1 Li–Mg–N–H System

Binary systems have unfavorable thermodynamic properties for practical application although they keep a relatively high hydrogen capacity. A lot of attempts have been made to improve their thermodynamic properties by adjusting composition. Orimo et al.[183–185] improved the Li–N–H system by replacing Li_3N with Mg_3N_2 based on the replacement of Li atom by Mg. A mixture of Li_3N–20 mol% Mg_3N_2 had a reversible hydrogen capacity of 9.1 wt.% according to the following reaction:

$$3Mg(NH_2)_2 + 12LiH \leftrightarrow Mg_3N_2 + 4Li_3N + 12H_2 \tag{15.18}$$

In a different route, Xiong et al.,[173] Luo et al.,[171] and Leng et al.[186] substituted LiH by MgH_2 or replaced $LiNH_2$ with $Mg(NH_2)_2$ to destabilize the Li–N–H system. The dehydrogenation product of either MgH_2–$2LiNH_2$ or $Mg(NH_2)_2$–$2LiH$ is

$Li_2Mg(NH)_2$. $Li_2Mg(NH)_2$ can absorb hydrogen to turn back to $Mg(NH_2)_2$ and $2LiH$. The dehydrogenation/rehydrogenation reaction is shown as follows:

$$2LiNH_2 + MgH_2 \rightarrow Mg(NH_2)_2 + 2LiH \leftrightarrow MgLi_2(NH)_2 + 2H_2 \qquad (15.19)$$

Compared with binary Li–N–H system, ternary Li–Mg–N–H system has higher hydrogen desorption plateau pressure, larger reversible capacity, and much lower hydrogenation/dehydrogenation temperature. In fact, $Li_2Mg(NH)_2$ started to absorb hydrogen at about 90°C, and the absorption accelerated at temperatures above 110°C. When the temperature reached 200°C, approximately 5.0 wt.% hydrogen was absorbed. The rehydrogenated $Mg(NH_2)_2$–$2LiH$ sample could desorb hydrogen at temperatures below to 100°C, and more than 4.5 wt.% of hydrogen was released as the temperature reached 200°C.[187]

The hydrogen desorption enthalpy of $Mg(NH_2)_2$–$2LiH$ was calculated by fitting the pressure–composition–temperature (PCT) curves with the van't Hoff plot. About two-thirds of the hydrogen is released at a relatively higher pressure of 46 bar and the other one-third is desorbed at a lower pressure around 5 bar. The calculated desorption heat for the higher desorption plateau is about −38.9 kJ/mol H_2, which means that the temperature required to release the stored hydrogen is about 90°C at an equilibrium pressure of 1.0 bar. This temperature is suitable for proton exchange membrane (PEM) fuel cell application.[187]

However, the relatively high activation energy of 102 kJ/mol results in a high practical operating temperature (>90°C), which sets a kinetic barrier for on-board applications. This kinetic barrier could be analyzed by the reaction mechanism shown in Figure 15.10. As the activation energy required for the reconstruction of reactant and dissociation of chemical bonds is high, enhancing the attraction of $H^{\delta+}$ and $H^{\delta-}$ or weakening the Li–H and N–H bonds would decrease the activation energy.[187]

15.4.3.2 Other Ternary Systems

The successful exploitation of Li–Mg–N–H had inspirited researchers to continue to explore other ternary systems. Li–Ca–N–H,[173,175] Li–Na–N–H,[166] Na–Mg–N–H,[188] Na–Ca–N–H,[189] and Mg–Ca–N–H[190] have been reported since then. The Li–Ca–N–H system is composed of $Ca(NH_2)_2 + LiH$ or $CaH_2 + LiNH_2$, which is described by the following reactions:

$$CaNH_2 + 2LiH \leftrightarrow CaNH + Li_2NH + 2H_2 \qquad (15.20)$$

$$CaH_2 + 2LiNH_2 \rightarrow CaNH + Li_2NH + 2H_2 \leftrightarrow CaNH_2 + 2LiH \qquad (15.21)$$

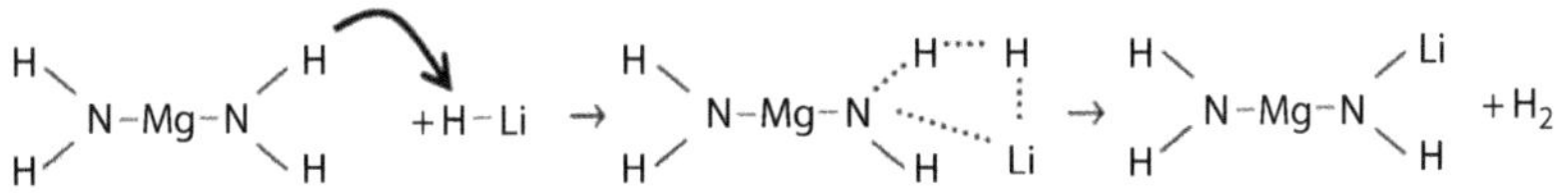

FIGURE 15.10 Reaction mechanism of $Mg(NH_2)_2 + 2LiH$. (Reprinted from *J. Alloys Compd.*, 398, Xiong, Z., Hu, J., Wu, G., Chen, P., Luo, W., Gross, K., and Wang, J., 235–239. Copyright 2005, with permission Elsevier.)

Although Li–Ca–N–H system has a lower theoretical and practically observed gravimetric capacity of 4.3 wt.% than that of both Li–Mg–N–H and Li–N–H, its hydrogen desorption and absorption occurred at a lower temperature.[173]

Similarly, Na–Mg–N–H[188] and Na–Ca–N–H[189] systems had lower-onset absorbing/desorbing temperature than those of Li–Mg–N–H system, but its low hydrogen capacity and large reaction entropy make them not within the consideration of on-board utilizations.

15.4.4 MULTINARY SYSTEMS

Except for LiH and MgH_2, the capacity of other alkali metal hydrides and alkaline-earth metal hydrides are all lower than 5 wt.%, far below those of metal–Al–H and metal–B–H. Hence, the latter can be combined with metal–N–H to form new systems with higher capacity, such as $LiAlH_4$–$LiNH_2$[176,191,192] and $Mg(NH_2)_2$–$LiAlH_4$.[193] However, the chemical process of a multicomponent system is much more complicated than those of the binary and ternary systems. For example, the reaction between $NaNH_2$ and $LiAlH_4$ is a multistep exothermic process involving continuous phase changes (Table 15.9). More than 5 wt.% hydrogen could be desorbed by ball milling at ambient temperatures within a short period of time.[192]

15.4.5 IMPROVING STRATEGIES OF M–N–H SYSTEMS

With respect to the metal–N–H systems mentioned earlier, their hydrogen properties should be improved to meet with the practical utilization. Although the thermodynamically predicated operating temperature is low, the practical dehydrogenation temperature is higher because of the high kinetic barrier. According to the thermodynamics of reaction, for example, the $Mg(NH_2)_2$–2LiH system only requires a temperature of ~90°C at an equilibrium pressure of 1.0 bar, but acceptable hydrogen desorption rate is required to operate at 200°C in practical cases. Besides, NH_3 emission is also of great concern for its toxic effect to PEM fuel cell. Hence, numerous efforts, including modifying material compositions, doping catalyst, and reducing particle or grain size, are the effective methods to improve thermodynamics and kinetic properties for hydrogen storage.

15.4.5.1 Modification by Adjusting Composition

Composition change is an effective approach to tune the thermodynamics of hydrogen storage materials. Liu et al.[194] found that the partial substitution of Mg or Li with Na in $Mg(NH_2)_2$–2LiH system decreased the dehydrogenation temperature by approximately 10°C. The change of the initial phase ratio of $LiNH_2$–MgH_2 mixture was able to improve the hydrogen storage capacity of this system. Xiong et al.[195] investigated the hydrogen storage properties of molar ratio of LiH–$Mg(NH_2)_2$ with 1/1, 2/1, and 3/1. Remarkable differences were observed in the temperature dependence of hydrogen absorption/desorption in these three samples. Of these samples, the peak areas of H_2 are in an order of 1LiH–$Mg(NH_2)_2$ < 2LiH–$Mg(NH_2)_2$ < 3LiH–$Mg(NH_2)_2$. In addition, the desorption of ammonia was largely depressed when increasing the hydride/amide molar ratio. No ammonia can be detected in the

TABLE 15.9
Content of Solid Residue at Different Ball Milling Intervals

Time (min)	Pressure (psi)	Phase							
0	0	$NaNH_2$	$LiAlH_4$						
1	4	$NaNH_2$	$LiAlH_4$	$Li_3Na(NH_2)_4$	$NaAlH_4$				
2	14	$NaNH_2$	$LiAlH_4$	$Li_3Na(NH_2)_4$	$NaAlH_4$				
3	43	$NaNH_2$		$Li_3Na(NH_2)_4$	$NaAlH_4$				
5	63			$Li_3Na(NH_2)_4$	$NaAlH_4$	$LiNa_2AlH_6$	Al	$LiNH_2$		Unknown
7	97			$Li_3Na(NH_2)_4$		$LiNa_2AlH_6$	Al	$LiNH_2$	NaH	Unknown
10	167			$Li_3Na(NH_2)_4$		$LiNa_2AlH_6$	Al	$LiNH_2$	NaH	Unknown
30	178			$Li_3Na(NH_2)_4$		$LiNa_2AlH_6$	Al	$LiNH_2$	NaH	Unknown

Source: Xiong, Z.T. et al., *Catal. Today*, 120, 287, 2007.

3LiH–Mg(NH$_2$)$_2$ sample. It was revealed that the higher content of LiH inhibited the desorption of ammonia but pushed the hydrogen desorption to a higher temperature region. The desorption of 1LiH–Mg(NH$_2$)$_2$ sample is described in the following reaction.

$$LiH + Mg(NH_2)_2 \rightarrow 1/2\ Li_2Mg_2N_3H_3 + H_2 + 1/2\ NH_3 \qquad (15.22)$$

Leng et al.[186] investigated the hydrogen storage properties of a ball-milled mixture of 8LiH and 3Mg(NH$_2$)$_2$. Approximately 7 wt.% of hydrogen desorbed during the heating process up to 450°C, with an onset temperature of 140°C and peak temperature of 190°C. Moreover, almost no ammonia emission was observed in the whole process, according to the reaction shown in Equation 15.23. Besides, the reversibility of the hydrogen absorption/desorption reactions were confirmed to be completed.

$$8LiH + 3Mg(NH_2)_2 \rightarrow Mg_3N_2 + 4Li_2NH + 8H_2 \qquad (15.23)$$

It is worth noting that the hydrogen storage capacity increased to ~9.1% as the molar ratio of LiH–Mg(NH$_2$)$_2$ increased to 4:1,[196] as expressed by the reaction shown in Equation 15.18.

Although the desorption capacity reduces as the molar ratio of LiH–Mg(NH$_2$)$_2$ increases at a relatively low operating temperature of 250°C, a same plateau pressure was observed for all the samples. The results indicated that they possessed a similar thermodynamic nature.[197,198] According to the hydrogen capacity, operating temperature, and ammonia generation, the optimal composition of LiH to Mg(NH$_2$)$_2$ determined through the investigation on the xLiH–Mg(NH$_2$)$_2$ mixtures (x = 1.5, 1.8, 2.0, 2.2, 2.5, 2.7) (Table 15.10) is 2:1. Deviation from this composition leads to either severe ammonia emission or inefficiency of hydrogen storage. The formation of the

---

TABLE 15.10

NH$_3$ Emission and Gravimetric Hydrogen Release of the Investigated Compositions

x Value	NH$_3$ Concentration in Ball Milling Step (ppm)	NH$_3$ Concentration in TPD Test (ppm)	H Atoms Released in Ball-Milling Step[a]	H Atoms Released in Volumetric Method	H Atoms Released Total	Total H Release (wt.%)
1.5			0.48	2.52	3.00	4.4
1.8	1866	6684	0.42	3.10	3.52	5.0
2.0	317	516	0.43	3.24	3.67	5.1
2.2	370	406	0.29	3.39	3.68	5.0
2.5	294	239	0.30	3.34	3.64	4.8
2.7	217	275	0.31	3.41	3.72	4.8

Source: Hu, J. and Fichtner, M., *Chem. Mater.*, 21, 3485, 2009.

[a] Calculated from the pressure increase in the milling vessel via the gas equation.

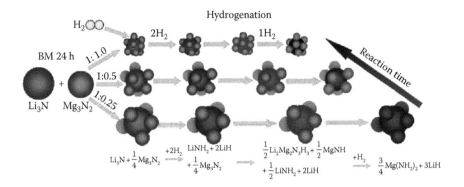

FIGURE 15.11   Schematic hydrogenation reaction processes of the $Li_3N-xMg_3N_2$ composites. (Reprinted with permission from Li, B., Liu, Y., Zhang, Y., Gao, M., and Pan, H., *J. Phys. Chem. C*, 116, 13551–13558, Copyright 2012 American Chemical Society.)

ternary imide $Li_2Mg(NH)_2$ was found in both dynamic and quasi-equilibrium dehydrogenation, which is responsible for the good cycle stability of this system.[199]

Liu et al. investigated the hydrogen storage properties of the $Li_3N-xMg_3N_2$ composites with different stoichiometries ($x=0$, 0.25, 0.5, 1.0) by ball milling and revealed the reaction pathway between the Li_3N and Mg_3N_2 (Figure 15.11). It was confirmed that the $Li_3N-0.25Mg_3N_2$ composite exhibited optimal hydrogen storage performances, which can store reversibly ~8.4 wt.% hydrogen with an onset temperature of 125°C for dehydrogenation.[200]

15.4.5.2   Doping Additive

Catalyst doping is an effective approach to improve the hydrogenation/dehydrogenation kinetics of metal–N–H systems. The effective additives can be divided into the following three main categories:

The first one is the TMs and their compounds. Lohstroh and Fichtner reported that the addition of 2 mol% $TiCl_3$ to the $2LiNH_2-MgH_2$ mixture caused the hydrogen release at lower temperature.[201] However, the catalytic effect disappeared after only two hydrogenation cycles. TM vanadium and vanadium-based compounds (VCl_3) were able to lower the activation energy of the decomposition of $Mg(NH_2)_2$ by weakening the Mg–N bond of $Mg(NH_2)_2$.[202] Tang et al.[203] reported the uniformly distributed metallic Co created by ball milling $LiNH_2-LiBH_4$ together with $CoCl_2$ additive contributes to a high catalytic efficiency by activating the N–H bond, resulting in the enhanced properties of $LiNH_2-LiBH_4$ system. In addition, many chemicals, such as NaH, $NaNH_2$, V_2O_5, TiF_3, TiN, TaN, Li_3N, and $Ti_3Cr_3V_4$ hydrides and graphite-supported Ru nanoparticles, were also added as catalysts for improving the kinetics of the $Mg(NH_2)_2-2LiH$ system.[204–207] But the operating temperatures for hydrogen absorption/desorption are still far from the practical requirement.

The second one is complex hydrides. Recently, Hu et al. reported that the addition of $LiBH_4$ effectively improved the hydrogen absorption/desorption performances of the $Mg(NH_2)_2-2LiH$ mixture. The onset as well as the peak temperatures of hydrogen desorption shifted to lower temperatures and a three-times increase in the hydrogenation/dehydrogenation rates and NH_3 evolution was reduced with the addition of $LiBH_4$. The reductions of both reaction enthalpies and activation energies result in the improvement of hydrogen storage properties by thermodynamic and kinetic analyses.[208,209] Similar improvement on the hydrogen storage properties was also obtained by adding $NaBH_4$,[210] but a different catalytic mechanism was reported, namely that the $NaBH_4$ improved the dehydrogenation kinetics by facilitating the formation of Mg vacancies, which weakens the N–H bonds and promotes the diffusion of atoms and/or ions. The addition of $Ca(BH_4)_2$ and $Mg(BH_4)_2$ showed better hydrogen storage properties than the addition of $LiBH_4$, owing to the fact that a metathesis reaction between $Mg(BH_4)_2$ or $Ca(BH_4)_2$ and LiH readily occurred to form $LiBH_4$ and MgH_2 or CaH_2. The newly formed MgH_2 or CaH_2 reacted with $Mg(NH_2)_2$ to generate MgNH or $MgCa(NH)_2$, which plays the seeding effect on the subsequent dehydrogenation process.[211,212]

The third one is potassium-based additive (KH, KOH, KF, KNH_2, K_2CO_3, K_3PO_4), which has a quite unique effect on the hydrogenation/dehydrogenation of amides. A dramatic improvement was achieved by adding 0.1KH with the hydrogen desorption peak shifting downward about 50°C–132°C in the hydrogenation/dehydrogenation performances of $Mg(NH_2)_2$–2LiH and NH_3 is hardly detectable in the temperature range of 75°C–200°C. In particular, reversible hydrogen release and uptake can be carried out at a temperature as low as 107°C.[213] The result indicated that the kinetic barrier in the ball-milled $Mg(NH_2)_2$–2LiH system is mainly from the interface reaction at the early stage of the dehydrogenation. KH first reacted with $Mg(NH_2)_2$ to form $K_2Mg(NH_2)_4$ through a two-step reaction at non-isothermal conditions,[214] which weakens the N–H bonds and metathesizes rapidly with LiH to regenerate KH. The circular transformation between three K species $[KH \leftrightarrow K_2Mg(NH_2)_4 \leftrightarrow KLi_3(NH_2)_4]$ developed a more energy-favorable pathway for dehydrogenation and thus results in the kinetic enhancement (Figure 15.12).[215] Moreover, the hydrogen storage properties of $Mg(NH_2)_2$–2LiH can be further improved by adding KOH.[216] The $Mg(NH_2)_2$–2LiH–0.07KOH sample can reversibly store 4.9 wt.% hydrogen with an onset and peak dehydrogenation temperature of 75°C and 120°C, respectively, which are the lowest in the current $Mg(NH_2)_2$–2LiH system studied. Moreover, the cycling stability of hydrogenation/dehydrogenation is remarkably improved by KOH doping, only average 0.002 wt.% capacity degradation per cycle within 30 cycles. Structural investigations reveal that doped KOH can react with $Mg(NH_2)_2$ and LiH to convert to MgO, KH, and $Li_2K(NH_2)_3$, which work together to provide the synergistic effects on improving the thermodynamics and kinetics of the $Mg(NH_2)_2$–2LiH.

Besides these three kinds of additives, other additives also improve the kinetic of metal–N–H systems. The peak temperature for hydrogen desorption of the $Mg(NH_2)_2$–2LiH mixture was lowered by 36°C through adding 0.5NaOH.[217] The cocatalytic effects of NaH, $LiNH_2$, and MgO, which formed from NaOH reacts with $Mg(NH_2)_2$ and LiH during ball milling, result in a significant improvement in the hydrogenation/dehydrogenation kinetics of the $Mg(NH_2)_2$–2LiH system. Sudik et al.[218] demonstrated that seeding the component with the dehydrogenated product, namely, $Li_2MgN_2H_2$, is effective to decrease the hydrogen dehydrogenation temperature by 40°C because

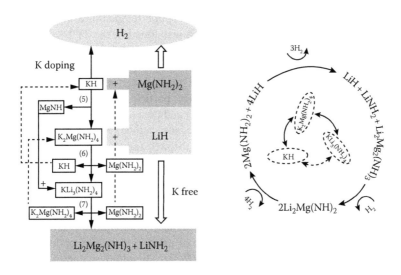

FIGURE 15.12 Representations of proposed pathway of K in the first dehydrogenation step of the $Mg(NH_2)_2$–2LiH system. (Wang, J., Chen, P., Pan, H., Xiong, Z., Gao, M., Wu, G., Liang, C., Li, C., Li, B. and Wang, J.: *Chemsuchem*. 2013. 31. 1–10. Copyright Wiley-VCH Verlag GmbH & Co. Adapted with permission.)

the activation energy reduced from 88.0 to 76.2 kJ/mol for the unseeded and seeded forms. In addition, the liberation of ammonia was significantly reduced. The effect of cycling reveals that 75% capacity can be maintained after 13 charge/discharge cycles. Isothermal desorption kinetics show rapid hydrogen desorption within minutes at 220°C for the seeded sample whereas the unseeded sample requires 2 h. Nevertheless, the seeded $Li_2MgN_2H_2$ diminished after one hydrogenation/dehydrogenation cycle and the peak hydrogen desorption shifted 10°C toward higher temperature. This is caused by the agglomeration and consumption (converting to $Mg(NH_2)_2$ and LiH after hydrogenation) of $Li_2MgN_2H_2$ in cycling. Aguey–Zinsou[219] demonstrated that the reaction kinetics of $LiNH_2$ with LiH were greatly enhanced by BN compound, which can facilitate the diffusion of Li^+ and/or H^- across the interface of $LiNH_2$ and LiH. Hydrogen can be fully desorbed from the mixture ($LiNH_2$+LiH+BN) in less than 7 h, while the pristine ($LiNH_2$+LiH) only desorbs less than half of hydrogen in the same time. Chen et al.[220] used different carbon materials such as single-walled carbon nanotubes (SWNTs), MWNTs, graphite, and AC as additives to improve the hydrogen storage properties of $Mg(NH_2)_2$–2LiH mixture. It was found that about 90% of the hydrogen capacity was released within 20 min with the addition of SWNTs at 200°C, compared to 60 min for the pristine $Mg(NH_2)_2$–LiH material.

15.4.5.3 Reducing the Particle or Grain Sizes

The hydrogenation/dehydrogenation kinetics of the metal–N–H systems was also enhanced by reducing the particle or grain sizes. Isobe et al.[221] investigated the catalytic effect of various Ti additives on the hydrogen desorption performance for Li–N–H system. The result showed that Ti^{nano}, $TiCl_3$, and TiO_2^{nano} had a superior catalytic effect on lowering the hydrogen desorption temperature; furthermore, no ammonia emission was detected, while such was found in the Ti^{micro} and TiO_2^{micro} catalyzed Li–N–H system. They concluded that the particle size of additives plays an important role in improving the kinetics of hydrogen desorption. Liu et al.[222] reported that the $Li_2MgN_2H_2$ sample showed reduced starting hydrogen absorption/desorption temperature through tailoring the particle size from 800 nm (hand milling) to 300–400 nm (ball milling for 3 h) and then to 100–200 nm (ball milling for 36 h), as shown in Figure 15.13. This is because activation energy was decreased and the specific surface area was enlarged through reducing particle size by ball milling. However, the particle size is gradually augmented during the hydrogenation/dehydrogenation cycles. Xie et al.[223] prepared $Mg(NH_2)_2$ with three different particle sizes of 100, 500, and 2000 nm by reactions between Mg, H_2, and NH_3. The ammonia desorption temperature from the decomposition of $Mg(NH_2)_2$ decreases with the particle size. After mixing with LiH, the hydrogen desorption kinetics increases obviously with decreasing the particle size of amide for the desorption activation energy decreases with decreasing the particle size. In addition, due to the reduced diffusion distance and larger specific surface area, the absorption kinetics increases as well with decreasing the particle size. Leng et al.[181] prepared Li_2NH hollow nanospheres by plasma metal reaction based on the Kirkendall effect. Compared to the Li_2NH micrometer particles, the special nanostructure significantly improved

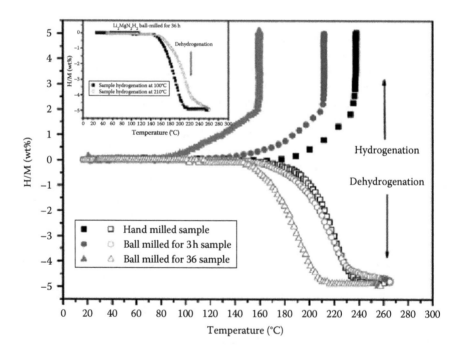

FIGURE 15.13   Hydrogenation/dehydrogenation curves of $Li_2MgN_2H_2$ samples with different milling treatments. (Reprinted with permission from Liu, Y., Zhong, K., Luo, K., Gao, M., Pan, H., and Wang, Q., *J. Amer. Chem. Soc.* 131, 1862–1870. Copyright 2009 American Chemical Society.)

hydrogen storage kinetics, reflecting on decreasing absorption temperature markedly and enhancing the absorption rate dramatically. It can be contributed to the shorter diffusion distance and larger specific surface area in the special hollow nanosphere. Moreover, the hollow nanosphere structure could limit the hydrogenated product of $LiNH_2$ and LiH in the nanometer range to decrease the absorption/desorption temperature dramatically. In addition, Wang et al.[224] found that the addition of triphenyl phosphate could markedly prevent the crystallization of $Mg(NH_2)_2$ and the aggregation of dehydrogenated products to improve the hydrogenation/dehydrogenation kinetic of the $Mg(NH_2)_2-2LiH$ system.

15.5 SUMMARY

Great progress has been achieved to develop high-capacity hydrogen storage materials with enhanced hydriding/dehydriding thermodynamics and kinetic properties. Although none of the hydrogen storage materials could fully meet the harsh requirements of large-scale practical applications, Mg-based hydrogen storage alloys, complex hydrides, and metal imides have respective advantages on the reversibility, capacity, or working temperature and pressure, which are determined by both thermodynamic and kinetic properties. Persistent efforts are needed to further tune the thermodynamic and kinetic properties of current reversible hydrogen storage materials by composition and structure modification via innovative strategies, which should be suitable for massive application.

REFERENCES

1. L. Schlapbach and A. Züttel, *Nature*, 2001, **414**, 353–358.
2. U.S. Department of Energy, Targets for onboard hydrogen storage systems for light-duty vehicles, http://energy.gov/sites/prod/files/2014/03/f11/targets_onboard_hydro_storage_explanation.pdf, September 2009.
3. W. Grochala and P. P. Edwards, *Chem. Rev.*, 2004, **104**, 1283–1315.
4. H. W. Li, Y. Yan, S. Orimo, A. Züttel and C. M. Jensen, *Energies*, 2011, **4**, 185–214.
5. E. Fakioğlu, Y. Yürüm and T. Nejat Veziroğlu, *Int. J. Hydrogen Energy*, 2004, **29**, 1371–1376.
6. I. P. Jain, C. Lal and A. Jain, *Int. J. Hydrogen Energy*, 2010, **35**, 5133–5144.
7. U. Eberle, M. Felderhoff and F. Schüth, *Angew. Chem. Int. Ed.*, 2009, **48**, 6608–6630.
8. P. Chen, Z. T. Xiong, J. Z. Luo, J. Y. Lin, K. L. Tan, *J. Phys. Chem. B*, 2003, **107**, 10967.
9. P. Vajeeston, P. Ravindran, B. C. Hauback, H. Fjellvåg, A. Kjekshus, S. Furuseth and M. Hanfland, *Phys. Rev. B*, 2006, **73**, 224102.
10. P. Vajeeston, P. Ravindran, A. Kjekshus and H. Fjellvåg, *Phys. Rev. Lett.*, 2002, **89**, 175506.
11. R. A. Varin, T. Czujko and Z. Wronski, *Nanotechnology*, 2006, **17**, 3856.
12. J. F. Stampfer, C. E. Holley and J. F. Suttle, *J. Am. Chem. Soc.*, 1960, **82**, 3504–3508.
13. B. Bogdanovic, K. Bohmhammel, B. Christ, A. Reiser, K. Schlichte, R. Vehlen and U. Wolf, *J. Alloys Compd.*, 1999, **282**, 84–92.
14. L. Zaluski, A. Zaluska and J. O. Strom-Olsen, *J. Alloys Compd.*, 1995, **217**, 245–249.
15. J. J. Reilly and R. H. Wiswall, *Inorg. Chem.*, 1967, **6**, 2220–2223.
16. J. J. Reilly and R. H. Wiswall, *Inorg. Chem.*, 1968, **7**, 2254–2256.

17. W. Y. Li, C. S. Li, H. Ma and J. Chen, *J. Am. Chem. Soc.*, 2007, **129**, 6710–6711.
18. Z. Zhao-Karger, J. Hu, A. Roth, D. Wang, C. Kubel, W. Lohstroh and M. Fichtner, *Chem. Commun.*, 2010, **46**, 8353–8355.
19. H. C. Zhong, H. Wang, J. W. Liu, D. L. Sun and M. Zhu, *Scr. Mater.*, 2011, **65**, 285–287.
20. A. Baldi, M. Gonzalez-Silveira, V. Palmisano, B. Dam and R. Griessen, *Phys. Rev. Lett.*, 2009, **102**, 26102.
21. C. J. Chung, S.-C. Lee, J. R. Groves, E. N. Brower, R. Sinclair and B. M. Clemens, *Phys. Rev. Lett.*, 2012, **108**, 106102.
22. M. Martin, C. Gommel, C. Borkhart and E. Fromm, *J. Alloys Compd.*, 1996, **238**, 193–201.
23. A. Zaluska, L. Zaluski and J. O. Strom-Olsen, *J. Alloys Compd.*, 1999, **288**, 217–225.
24. X. Yao, Z. Zhu, H. Cheng and G. Lu, *J. Mater. Res.*, 2008, **23**, 336–340.
25. P. Spatz, H. A. Aebischer, A. Krozer and L. Schlapbach, *Z. Phys. Chem.*, 1993, **181**, 393–397.
26. G. Liang, J. Huot, S. Boily, A. Van Neste and R. Schulz, *J. Alloys Compd.*, 1999, **292**, 247–252.
27. J. F. Fernández and C. R. Sánchez, *J. Alloys Compd.*, 2003, **356–357**, 348–352.
28. J.-L. Bobet, E. Akiba and B. Darriet, *Int. J. Hydrogen Energy*, 2001, **26**, 493–501.
29. W. Oelerich, T. Klassen and R. Bormann, *J. Alloys Compd.*, 2001, **315**, 237–242.
30. G. Barkhordarian, T. Klassen and R. Bormann, *Scr. Mater.*, 2003, **49**, 213–217.
31. X. Yao, C. Wu, A. Du, J. Zou, Z. Zhu, P. Wang, H. Cheng, S. Smith and G. Lu, *J. Am. Chem. Soc.*, 2007, **129**, 15650–15654.
32. Y. Jia, L. Cheng, N. Pan, J. Zou, G. Lu and X. Yao, *Adv. Energy Mater.*, 2011, **1**, 387–393.
33. J. Cui, H. Wang, J. Liu, L. Ouyang, Q. Zhang, D. Sun, X. Yao and M. Zhu, *J. Mater. Chem. A*, 2013, **1**, 5603–5611.
34. L. Z. Ouyang, H. Wang, M. Zhu, J. Zou and C. Y. Chung, *J. Alloys Compd.*, 2005, **404–406**, 485–489.
35. L. Z. Ouyang, H. Wang, C. Y. Chung, J. H. Ahn and M. Zhu, *J. Alloys Compd.*, 2006, **422**, 58–61.
36. S. Y. Ye, S. L. I. Chan, L. Z. Ouyang and M. Zhu, *J. Alloys Compd.*, 2010, **504**, 493–497.
37. X. Xiao, G. Liu, S. Peng, K. Yu, S. Li, C. Chen and L. Chen, *Int. J. Hydrogen Energy*, 2010, **35**, 2786–2790.
38. X. B. Yu, Z. X. Yang, H. K. Liu, D. M. Grant and G. S. Walker, *Int. J. Hydrogen Energy*, 2010, **35**, 6338–6344.
39. J. Yang, M. Ciureanu and R. Roberge, *Mater. Lett.*, 2000, **43**, 234–239.
40. K. H. Buschow and H. H. Van Mal, *J. Less Common Met.*, 1972, **29**, 203–210.
41. G. Liang, J. Huot, S. Boily, A. Van Neste and R. Schulz, *J. Alloys Compd.*, 2000, **297**, 261–265.
42. M. Zhu, Y. Gao, X. Z. Che, Y. Q. Yang and C. Y. Chung, *J. Alloys Compd.*, 2002, **330–332**, 708–713.
43. L. Z. Ouyang, L. Yao, H. W. Dong, L. Q. Li and M. Zhu, *J. Alloys Compd.*, 2009, **485**, 507–509.
44. M. Au, *Mater. Sci. Eng. B*, 2005, **117**, 37–44.
45. H. J. Lin, L. Z. Ouyang, H. Wang, J. W. Liu and M. Zhu, *Int. J. Hydrogen Energy*, 2012, **37**, 1145–1150.
46. L. Z. Ouyang, F. X. Qin and M. Zhu, *Scr. Mater.*, 2006, **55**, 1075–1078.
47. L. Z. Ouyang, X. S. Yang, H. W. Dong and M. Zhu, *Scr. Mater.*, 2009, **61**, 339–342.
48. L. Z. Ouyang, H. W. Dong and M. Zhu, *J. Alloys Compd.*, 2006, **446–447**, 124–128.
49. M. Pozzo and D. Alfè, *Int. J. Hydrogen Energy*, 2009, **34**, 1922–1930.
50. G. Barkhordarian, T. Klassen and R. Bormann, *J. Phys. Chem. B*, 2006, **110**, 11020–11024.

51. L. Xie, Y. Liu, Y. T. Wang, J. Zheng and X. G. Li, *Acta Mater.*, 2007, **55**, 4585–4591.
52. L. P. Ma, X. D. Kang, H. B. Dai, Y. Liang, Z. Z. Fang, P. J. Wang, P. Wang and H. M. Cheng, *Acta Mater.*, 2009, **57**, 2250–2258.
53. P. Chen, X. Wu, J. Lin and K. L. Tan, *Science*, 1999, **285**, 91–93.
54. M. Lototskyy, J. M. Sibanyoni, R. V. Denys, M. Williams, B. G. Pollet and V. A. Yartys, *Carbon*, 2013, **57**, 146–160.
55. A. Zaluska, L. Zaluski and J. O. Ström-Olsen, *J. Alloys Compd.*, 1999, **289**, 197–206.
56. M. Zhu, H. Wang, L. Z. Ouyang and M. Q. Zeng, *Int. J. Hydrogen Energy*, 2006, **31**, 251–257.
57. H. Wang, L. Z. Ouyang, M. Q. Zeng and M. Zhu, *J. Alloys Compd.*, 2004, **375**, 313–317.
58. L. Z. Ouyang, S. Y. Ye, H. W. Dong and M. Zhu, *Appl. Phys. Lett.*, 2007, **90**, 21917–21919.
59. V. Bérubé, G. Radtke, M. Dresselhaus and G. Chen, *Int. J. Energy Res.*, 2007, **31**, 637–663.
60. V. Berube, G. Chen and M. S. Dresselhaus, *Int. J. Hydrogen Energy*, 2008, **33**, 4122–4131.
61. L. Li, B. Peng, W. Ji and J. Chen, *J. Phys. Chem. C*, 2009, **113**, 3007–3013.
62. B. Peng, L. Li, W. Ji, F. Cheng and J. Chen, *J. Alloys Compd.*, 2009, **484**, 308–313.
63. K.-J. Jeon, H. R. Moon, A. M. Ruminski, B. Jiang, C. Kisielowski, R. Bardhan and J. J. Urban, *Nat. Mater.*, 2011, **10**, 286–290.
64. T. K. Nielsen, K. Manickam, M. Hirscher, F. Besenbacher and T. R. Jensen, *ACS Nano*, 2009, **3**, 3521–3528.
65. H. Wang, S. F. Zhang, J. W. Liu, L. Z. Ouyang and M. Zhu, *Mater. Chem. Phys.*, 2012, **136**, 146–150.
66. M. Konarova, A. Tanksale, J. Norberto Beltramini and G. Q. Lu, *Nano Energy*, 2013, **2**, 98–104.
67. T. Spassov and U. Köster, *J. Alloys Compd.*, 1998, **279**, 279–286.
68. T. Spassov and U. Köster, *J. Alloys Compd.*, 1999, **287**, 243–250.
69. T. Spassov, L. Lyubenova, U. Köster and M. D. Baró, *Mater. Sci. Eng. A*, 2004, **375–377**, 794–799.
70. Y. Wu, J. K. Solberg and V. A. Yartys, *J. Alloys Compd.*, 2007, **446–447**, 178–182.
71. Y. Wu, M. V. Lototskyy, J. K. Solberg and V. A. Yartys, *Int. J. Hydrogen Energy*, 2012, **37**, 1495–1508.
72. H. Wu, *Chemphyschem*, 2008, **9**, 2157–2162.
73. L.Z. Ouyang, H. Wang, M. Zhu, unpblished results.
74. H. J. Lin, L. Z. Ouyang, H. Wang, D. Q. Zhao, W. H. Wang, D. L. Sun and M. Zhu, *Int. J. Hydrogen Energy*, 2012, **37**, 14329–14335.
75. Q. A. Zhang, D. D. Liu, Q. Q. Wang, F. Fang, D. L. Sun, L. Z. Ouyang and M. Zhu, *Scr. Mater.*, 2011, **65**, 233–236.
76. J. W. Liu, C. C. Zou, H. Wang, L. Z. Ouyang and M. Zhu, *Int. J. Hydrogen Energy*, 2013, **38**, 10438–10445.
77. J. J. Vajo, F. Mertens, C. C. Ahn, R. C. Bowman and B. Fultz, *J. Phys. Chem. B*, 2004, **108**, 13977–13983.
78. G. Liang and R. Schulz, *J. Mater. Sci.*, 2004, **39**, 1557–1562.
79. B. Bogdanovic and M. Schwickardi, *J. Alloys Compd.*, 1997, **253**, 1–9.
80. S. Orimo, Y. Nakamori, G. Kitahara, K. Miwa, N. Ohba, S. Towata and A. Züttel, *J. Alloys Compd.*, 2005, **404**, 427–430.
81. J.-P. Bastide, B. Bonnetot, J.-M. Létoffé and P. Claudy, *Mater. Res. Bull.*, 1981, **16**, 91–96.
82. B. Bogdanović, R. A. Brand, A. Marjanović, M. Schwickardi and J. Tölle, *J. Alloys Compd.*, 2000, **302**, 36–58.
83. K. J. Gross, G. J. Thomas and C. M. Jensen, *J. Alloys Compd.*, 2002, **330–332**, 683–690.
84. A. Zuttel, P. Wenger, S. Rentsch, P. Sudan, P. Mauron and C. Emmenegger, *J. Power Sources*, 2003, **118**, 1–7.

85. O. Friedrichs, A. Remhof, S. J. Hwang and A. Züttel, *Chem. Mater.*, 2010, **22**, 3265–3268.
86. M. P. Pitt, M. Paskevicius, D. H. Brown, D. A. Sheppard, C. E. Buckley, *J. Am. Chem. Soc.*, 2013, **135**, 6930.
87. Y. Nakamori, H. W. Li, M. Matsuo, K. Miwa, S. Towata and S. Orimo, *J. Phys. Chem. Solids*, 2008, **69**, 2292–2296.
88. J. S. Hummelshoj, D. D. Landis, J. Voss, T. Jiang, A. Tekin, N. Bork, M. Dulak, J. J. Mortensen, L. Adamska, J. Andersin, J. D. Baran, G. D. Barmparis, F. Bell, A. L. Bezanilla, J. Bjork, M. E. Bjorketun, F. Bleken, F. Buchter, M. Burkle and P. D. Burton, *J. Chem. Phys.*, 2009, **131**, 014101–014109.
89. D. Ravnsbæk, Y. Filinchuk, Y. Cerenius, H. J. Jakobsen, F. Besenbacher, J. Skibsted and T. R. Jensen, *Angew. Chem. Int. Ed.*, 2009, **48**, 6659–6663.
90. D. B. Ravnsbæk, L. H. Sørensen, Y. Filinchuk, D. Reed, D. Book, H. J. Jakobsen, F. Besenbacher, J. Skibsted and T. R. Jensen, *Eur. J. Inorg. Chem.*, 2010, **11**, 1608–1612.
91. I. Lindemann, R. Domènech Ferrer, L. Dunsch, Y. Filinchuk, R. Černý, H. Hagemann, V. D'Anna, L. M. Lawson Daku, L. Schultz and O. Gutfleisch, *Chem. Eur. J.*, 2010, **16**, 8707–8712.
92. H. W. Li, S. Orimo, Y. Nakamori, K. Miwa, N. Ohba, S. Towata and A. Züttel, *J. Alloys Compd.*, 2007, **446–447**, 315–318.
93. H. Hagemann, M. Longhini, J. W. Kaminski, T. A. Wesolowski, R. Cerny, N. Penin, M. H. Sorby, B. C. Hauback, G. Severa and C. M. Jensen, *J. Phys. Chem. A*, 2008, **112**, 7551–7555.
94. R. Ccaronernyacute, G. Severa, D. B. Ravnsbaeligk, Y. Filinchuk, V. D'Anna, H. Hagemann, D. Haase, C. M. Jensen and T. R. Jensen, *J. Phys. Chem. C*, 2010, **114**, 1357–1364.
95. R. Černý, D. B. Ravnsbæk, G. Severa, Y. Filinchuk, V. D'Anna, H. Hagemann, D. Haase, J. Skibsted, C. M. Jensen and T. R. Jensen, *J. Phys. Chem. C*, 2010, **114**, 19540–19549.
96. F. Wang, Y. Liu, M. Gao, K. Luo, H. Pan and Q. Wang, *J. Phys. Chem. C*, 2009, **113**, 7978–7984.
97. J. Graetz, Y. Lee, J. J. Reilly, S. Park and T. Vogt, *Phys. Rev. B*, 2005, **71**, 184115.
98. H. Grove, H. W. Brinks, R. H. Heyn, F. J. Wu, S. M. Opalka, X. Tang, B. L. Laube and B. C. Hauback, *J. Alloys Compd.*, 2008, **455**, 249–254.
99. L. Yin, P. Wang, Z. Fang and H. Cheng, *Chem. Phys. Lett.*, 2008, **450**, 318–321.
100. J. P. Soulie, G. Renaudin, R. Cerny and K. Yvon, *J. Alloys Compd.*, 2002, **346**, 200–205.
101. J. J. Vajo, S. L. Skeith and F. Mertens, *J. Phys. Chem. B*, 2005, **109**, 3719–3722.
102. D. J. Siegel, C. Wolverton and V. Ozolins, *Phys. Rev. B*, 2007, **76**.
103. M. Au, A. Jurgensen and K. Zeigler, *J. Phys. Chem. B*, 2006, **110**, 26482–26487.
104. J. Yang, A. Sudik and C. Wolverton, *J. Phys. Chem. C*, 2007, **111**, 19134–19140.
105. A. Remhof, O. Friedrichs, F. Buchter, P. Mauron, J. W. Kim, K. H. Oh, A. Buchsteiner, D. Wallacher and A. Züttel, *J. Alloys Compd.*, 2009, **484**, 654–659.
106. J. W. Kim, O. Friedrichs, J.-P. Ahn, D. H. Kim, S. C. Kim, A. Remhof, H.-S. Chung, J. Lee, J.-H. Shim, Y. W. Cho, A. Züttel and K. H. Oh, *Scr. Mater.*, 2009, **60**, 1089–1092.
107. O. Friedrichs, J. W. Kim, A. Remhof, F. Buchter, A. Borgschulte, D. Wallacher, Y. W. Cho, M. Fichtner, K. H. Oh and A. Züttel, *Phys. Chem. Chem. Phys.*, 2009, **11**, 1515–1520.
108. S.-A. Jin, Y.-S. Lee, J.-H. Shim and Y. W. Cho, *J. Phys. Chem. C*, 2008, **112**, 9520–9524.
109. J.-H. Lim, J.-H. Shim, Y.-S. Lee, J.-Y. Suh, Y. W. Cho and J. Lee, *Int. J. Hydrogen Energy*, 2010, **35**, 6578–6582.
110. D. M. Liu, W. J. Huang, T. Z. Si and Q. A. Zhang, *J. Alloys Compd.*, 2013, **551**, 8–11.
111. J.-H. Shim, Y.-S. Lee, J.-Y. Suh, W. Cho, S. S. Han and Y. W. Cho, *J. Alloys Compd.*, 2012, **510**, L9–L12.

112. J.-H. Shim, J.-H. Lim, S.-u. Rather, Y.-S. Lee, D. Reed, Y. Kim, D. Book and Y. W. Cho, *J. Phys. Chem. Lett.*, 2009, **1**, 59–63.
113. F. C. Gennari, *Int. J. Hydrogen Energy*, 2011, **36**, 15231–15238.
114. P. Mauron, M. Bielmann, A. Remhof, A. Züttel, J. H. Shim and Y. W. Cho, *J. Phys. Chem. C*, 2010, **114**, 16801–16805.
115. U. Bösenberg, D. B. Ravnsbaek, H. Hagemann, V. D'Anna, C. B. Minella, C. Pistidda, W. van Beek, T. R. Jensen, R. Bormann and M. Dornheim, *J. Phys. Chem. C*, 2010, **114**, 15212–15217.
116. T. Sun, H. Wang, Q. Zhang, D. Sun, X. Yao and M. Zhu, *J. Mater. Chem.*, 2011, **21**, 9179–9184.
117. Y. Zhou, Y. Liu, W. Wu, Y. Zhang, M. Gao and H. Pan, *J. Phys. Chem. C*, 2011, **116**, 1588–1595.
118. J. Purewal, S.-J. Hwang, R. C. Bowman, Jr., E. Rönnebro, B. Fultz and C. Ahn, *J. Phys. Chem. C*, 2008, **112**, 8481–8485.
119. X. B. Yu, D. M. Grant and G. S. Walker, *J. Phys. Chem. C*, 2008, **112**, 11059–11062.
120. X. B. Yu, D. M. Grant and G. S. Walker, *J. Phys. Chem. C*, 2009, **113**, 17945–17949.
121. J.-H. Lim, J.-H. Shim, Y.-S. Lee, Y. W. Cho and J. Lee, *Scr. Mater.*, 2008, **59**, 1251–1254.
122. Y. Zhang, W. S. Zhang, M. Q. Fan, S. S. Liu, H. L. Chu, Y. H. Zhang, X. Y. Gao and L. X. Sun, *J. Phys. Chem. C*, 2008, **112**, 4005–4010.
123. F. C. Gennari, L. F. Albanesi, J. A. Puszkiel and P. A. Larochette, *Int. J. Hydrogen Energy*, 2011, **36**, 563–570.
124. W. Cai, H. Wang, D. Sun and M. Zhu, *J. Phys. Chem. C*, 2013, **117**, 9566–9572.
125. B. Bogdanovic, M. Felderhoff, A. Pommerin, F. Schuth, N. Spielkamp and A. Stark, *J. Alloys Compd.*, 2009, **471**, 383–386.
126. J. M. Bellosta von Colbe, M. Felderhoff, B. Bogdanovic, F. Schuth and C. Weidenthaler, *Chem. Commun.*, 2005, **37**, 4732–4734.
127. X. D. Kang, P. Wang and H. M. Cheng, *J. Phys. Chem. C*, 2007, **111**, 4879–4884.
128. F. Schuth, B. Bogdanovic and M. Felderhoff, *Chem. Commun.*, 2004, **20**, 2249–2258.
129. K. J. Gross, E. H. Majzoub and S. W. Spangler, *J. Alloys Compd.*, 2003, **356–357**, 423–428.
130. B. Bogdanovic, M. Felderhoff, S. Kaskel, A. Pommerin, K. Schlichte and F. Schuth, *Adv. Mater.*, 2003, **15**, 1012.
131. J. W. Kim, J. H. Shim, S. C. Kim, A. Remhof, A. Borgschulte, O. Friedrichs, R. Gremaud, F. Pendolino, A. Züttel, Y. W. Cho and K. H. Oh, *J. Power Sources*, 2009, **192**, 582–587.
132. D. L. Anton, *J. Alloys Compd.*, 2003, **356**, 400–404.
133. B. Bogdanović, M. Felderhoff, A. Pommerin, F. Schüth and N. Spielkamp, *Adv. Mater.*, 2006, **18**, 1198–1201.
134. T. Sun, B. Zhou, H. Wang and M. Zhu, *J. Alloys Compd.*, 2009, **467**, 413–416.
135. X. Fan, X. Xiao, L. Chen, L. Han, S. Li, H. Ge and Q. Wang, *J. Phys. Chem. C*, 2011, **115**, 22680–22687.
136. X. Fan, X. Xiao, L. Chen, S. Li, H. Ge and Q. Wang, *J. Phys. Chem. C*, 2011, **115**, 2537–2543.
137. E. H. Majzoub and K. J. Gross, *J. Alloys Compd.*, 2003, **356–357**, 363–367.
138. P. Wang, X. D. Kang and H. M. Cheng, *ChemPhysChem*, 2005, **6**, 2488–2491.
139. L. C. Yin, P. Wang, X. D. Kang, C. H. Sun and H. M. Cheng, *Phys. Chem. Chem. Phys.*, 2007, **9**, 1499–1502.
140. D. Sun, T. Kiyobayashi, H. T. Takeshita, N. Kuriyama and C. M. Jensen, *J. Alloys Compd.*, 2002, **337**, L8–L11.
141. H. Gunaydin, K. N. Houk and V. Ozoliņš, *Proc. Natl. Acad. Sci.*, 2008, **105**, 3673–3677.
142. C. K. Huang, Y. J. Zhao, H. Wang, J. Guo and M. Zhu, *Int. J. Hydrogen Energy*, 2011, **36**, 9767–9771.

143. G. Sandrock, J. Reilly, J. Graetz, W. M. Zhou, J. Johnson and J. Wegrzyn, *J. Alloys Compd.*, 2006, **421**, 185–189.
144. A. Gutowska, L. Li, Y. Shin, C. M. Wang, X. S. Li, J. C. Linehan, R. S. Smith, B. D. Kay, B. Schmid, W. Shaw, M. Gutowski and T. Autrey, *Angew. Chem. Int. Ed.*, 2005, **44**, 3578–3582.
145. J. Kostka, W. Lohstroh, M. Fichtner and H. Hahn, *J. Phys. Chem. C*, 2007, **111**, 14026–14029.
146. Z. Z. Fang, P. Wang, T. E. Rufford, X. D. Kang, G. Q. Lu and H. M. Cheng, *Acta Mater.*, 2008, **56**, 6257–6263.
147. A. F. Gross, J. J. Vajo, S. L. Van Atta and G. L. Olson, *J. Phys. Chem. C*, 2008, **112**, 5651–5657.
148. S. Cahen, J. B. Eymery, R. Janot and J. M. Tarascon, *J. Power Sources*, 2009, **189**, 902–908.
149. N. Brun, R. Janot, C. Sanchez, H. Deleuze, C. Gervais, M. Morcrette and R. Backov, *Energy Environ. Sci.*, 2010, **3**, 824–830.
150. P. Ngene, M. van Zwienen and P. E. de Jongh, *Chem. Commun.*, 2010, **46**, 8201–8203.
151. X. F. Liu, D. Peaslee, C. Z. Jost and E. H. Majzoub, *J. Phys. Chem. C*, 2010, **114**, 14036–14041.
152. X. F. Liu, D. Peaslee, C. Z. Jost, T. F. Baumann and E. H. Majzoub, *Chem. Mater.*, 2011, **23**, 1331–1336.
153. T. Sun, J. Liu, Y. Jia, H. Wang, D. Sun, M. Zhu and X. Yao, *Int. J. Hydrogen Energy*, 2012, **37**, 18920–18926.
154. S. Y. Zheng, F. Fang, G. Y. Zhou, G. R. Chen, L. Z. Ouyang, M. Zhu and D. L. Sun, *Chem. Mater.*, 2008, **20**, 3954–3958.
155. R. D. Stephens, A. F. Gross, S. L. V. Atta, J. J. Vajo and F. E. Pinkerton, *Nanotechnology*, 2009, **20**, 204018.
156. P. Adelhelm, J. Gao, M. H. W. Verkuijlen, C. Rongeat, M. Herrich, P. J. M. van Bentum, O. Gutfleisch, A. P. M. Kentgens, K. P. de Jong and P. E. de Jongh, *Chem. Mater.*, 2010, **22**, 2233–2238.
157. Y. Li, G. Zhou, F. Fang, X. Yu, Q. Zhang, L. Z. Ouyang, M. Zhu and D. Sun, *Acta Mater.*, 2011, **59**, 1829–1838.
158. V. Stavila, R. K. Bhakta, T. M. Alam, E. H. Majzoub and M. D. Allendorf, *ACS Nano*, 2012, **6**, 9807–9817.
159. L. H. Rude, T. K. Nielsen, D. B. Ravnsbæk, U. Bösenberg, M. B. Ley, B. Richter, L. M. Arnbjerg, M. Dornheim, Y. Filinchuk, F. Besenbacher and T. R. Jensen, *Phys. Status Solidi A*, 2011, **208**, 1754–1773.
160. M. H. W. Verkuijlen, P. Ngene, D. W. de Kort, C. Barré, A. Nale, E. R. H. van Eck, P. J. M. van Bentum, P. E. de Jongh and A. P. M. Kentgens, *J. Phys. Chem. C*, 2012, **116**, 22169–22178.
161. S. Satyapal, J. Petrovic, C. Read, G. Thomas and G. Ordaz, *Catal. Today*, 2007, **120**, 246–256.
162. B. Bogdanović, M. Felderhoff and G. Streukens, *J. Serb. Chem. Soc.*, 2009, **74**, 183–196.
163. Y., Nakamori, H. W. Li, K. Kikuchi, M. Aoki, K. Miwa, S. Towata and S. Orimo, *J. Alloys Compd.*, 2007, **446**, 296–300.
164. P. Chen, Z. Xiong, J. Luo, J. Lin and K. L. Tan, *Nature*, 2002, **420**, 302–304.
165. D. H. Gregory, *J. Mater. Chem.*, 2008, **18**, 2321–2330.
166. T. Ichikawa, N. Hanada, S. Isobe, H. Leng and H. Fujii, *J. Phys. Chem. B*, 2004, **108**, 7887–7892.
167. H. Leng, T. Ichikawa, S. Isobe, S. Hino, N. Hanada and H. Fujii, *J. Alloys Compd.*, 2005, **404**, 443–447.
168. J. Hu, G. Wu, Y. Liu, Z. Xiong and P. Chen, *J. Phys. Chem. B*, 2006, **110**, 14688–14692.
169. Z. Xiong, P. Chen, G. Wu, J. Lin and K. L. Tan, *J. Mater. Chem.*, 2003, **13**, 1676–1680.

170. S. Hino, T. Ichikawa, H. Leng and H. Fujii, *J. Alloys Compd.*, 2005, **398**, 62–66.
171. W. Luo, *J. Alloys Compd.*, 2004, **381**, 284–287.
172. H. Leng, T. Ichikawa and H. Fujii, *J. Phys. Chem. B*, 2006, **110**, 12964–12968.
173. Z. Xiong, G. Wu, J. Hu and P. Chen, *Adv. Mater.*, 2004, **16**, 1522–1525.
174. G. Wu, Z. Xiong, T. Liu, Y. Liu, J. Hu, P. Chen, Y. Feng and A. T. S. Wee, *Inorg. Chem.*, 2007, **46**, 517–521.
175. K. Tokoyoda, S. Hino, T. Ichikawa, K. Okamoto and H. Fujii, *J. Alloys Compd.*, 2007, **439**, 337–341.
176. Z. Xiong, G. Wu, J. Hu and P. Chen, *J. Power Sources*, 2006, **159**, 167–170.
177. Y. F. Liu, J. J. Hu, Z. T. Xiong, G. T. Wu, P. Chen, K. Murata and K. Sakata, *J. Alloys Compd.*, 2007, **432**, 298–302.
178. Y. H. Hu and E. Ruckenstein, *Ind. Eng. Chem. Res.*, 2004, **43**, 2464–2467.
179. Y. Kojima and Y. Kawai, *J. Alloys Compd.*, 2005, **395**, 236–239.
180. Y. Nakamori, G. Kitahara and S. Orimo, *J. Power Sources*, 2004, **138**, 309–312.
181. H. Y. Leng, T. Ichikawa, S. Isobe, S. Hino, N. Hanada and H. Fujii, *J. Alloys Compd.*, 2005, **404–406**, 443–447.
182. L. Xie, Y. Li, R. Yang, Y. Liu and X. Li, *Appl. Phys. Lett.*, 2008, **92**, 231910–231913.
183. S. Orimo, Y. Nakamori, G. Kitahara, K. Miwa, N. Ohba, T. Noritake and S. Towata, *Appl. Phys. A*, 2004, **79**, 1765–1767.
184. Y. Nakamori, G. Kitahara, K. Miwa, N. Ohba, T. Noritake, S. Towata, S. Orimo, *J. Alloys Compd.*, 2005, **404**, 396.
185. Y. Nakamori and S.-i. Orimo, *J. Alloys Compd.*, 2004, **370**, 271–275.
186. H. Y. Leng, T. Ichikawa, S. Hino, N. Hanada, S. Isobe and H. Fujii, *J. Phys. Chem. B*, 2004, **108**, 8763–8765.
187. Z. Xiong, J. Hu, G. Wu, P. Chen, W. Luo, K. Gross and J. Wang, *J. Alloys Compd.*, 2005, **398**, 235–239.
188. Z. Xiong, J. Hu, G. Wu and P. Chen, *J. Alloys Compd.*, 2005, **395**, 209–212.
189. Z. Xiong, G. Wu, J. Hu and P. Chen, *J. Alloys Compd.*, 2007, **441**, 152–156.
190. J. Hu, Z. Xiong, G. Wu, P. Chen, K. Murata and K. Sakata, *J. Power Sources*, 2006, **159**, 116–119.
191. Z. Xiong, G. Wu, J. Hu, Y. Liu, P. Chen, W. Luo and J. Wang, *Adv. Funct. Mater.*, 2007, **17**, 1137–1142.
192. Z. T. Xiong, J. J. Hu, G. T. Wu, Y. F. Liu and P. Chen, *Catal. Today*, 2007, **120**, 287–291.
193. Y. Liu, J. Hu, G. Wu, Z. Xiong and P. Chen, *J. Phys. Chem. C*, 2007, **111**, 19161–19164.
194. Y. Liu, J. Hu, Z. Xiong, G. Wu and P. Chen, *J. Mater. Res.*, 2007, **22**, 1339–1345.
195. Z. Xiong, G. Wu, J. Hu, P. Chen, W. Luo and J. Wang, *J. Alloys Compd.*, 2006, **417**, 190–194.
196. Y. Nakamori, G. Kitahara, K. Miwa, S. Towata and S. Orimo, *Appl. Phys. A Mater. Sci. Process.*, 2005, **80**, 1–3.
197. T. Ichikawa, K. Tokoyoda, H. Leng and H. Fujii, *J. Alloys Compd.*, 2005, **400**, 245–248.
198. M. Aoki, T. Noritake, Y. Nakamori, S. Towata and S. Orimo, *J. Alloys Compd.*, 2007, **446–447**, 328–331.
199. J. Hu and M. Fichtner, *Chem. Mater.*, 2009, **21**, 3485–3490.
200. B. Li, Y. Liu, Y. Zhang, M. Gao and H. Pan, *J. Phys. Chem. C*, 2012, **116**, 13551–13558.
201. W. Lohstroh and M. Fichtner, *J. Alloys Compd.*, 2007, **446–447**, 332–335.
202. R. R. Shahi, T. P. Yadav, M. A. Shaz and O. N. Srivastva, *Int. J. Hydrogen Energy*, 2010, **35**, 238–346.
203. W. S. Tang, G. Wu, T. Liu, A. T. S. Wee, C. K. Yong, Z. Xiong, A. T. S. Hor and P. Chen, *Dalton Trans.*, 2008, 2395–2399.
204. Y. F. Liu, J. J. Hu, Z. T. Xiong, G. T. Wu and P. Chen, *J. Mater. Res.*, 2007, **22**, 1339–1345.

205. Q. Wang, Y. Chen, G. Niu, C. Wu and M. Tao, *Ind. Eng. Chem. Res.*, 2009, **48**, 5250–5254.
206. L. P. Ma, P. Wang, H. B. Dai and H. M. Cheng, *J. Alloys Compd.*, 2009, **468**, L21–L24.
207. L. P. Ma, H. B. Dai, Y. Liang, X. D. Kang, Z. Z. Fang, P. J. Wang, P. Wang and H. M. Cheng, *J. Phys. Chem. C*, 2008, **112**, 18280–18285.
208. J. Hu, Y. Liu, G. Wu, Z. Xiong, Y. S. Chua and P. Chen, *Chem. Mater.*, 2008, **20**, 4398–4402.
209. J. Hu, E. Weidner, M. Hoelzel and M. Fichtner, *Dalton Trans.*, 2010, **39**, 9100–9107.
210. C. Liang, Y. F. Liu, Y. Jiang, Z. J. Wei, M. X. Gao and H. G. Pan, *Phys. Chem. Chem. Phys.*, 2011, **13**, 314–321.
211. H. Pan, S. Shi, Y. Liu, B. Li, Y. Yang and M. Gao, *Dalton Trans.*, 2013, **42**, 3802–3811.
212. B. Li, Y. Liu, J. Gu, Y. Gu, M. Gao and H. Pan, *Int. J. Hydrogen Energy*, 2013, **38**, 5030–5038.
213. J. H. Wang, T. Liu, G. T. Wu, W. Li, Y. F. Liu, C. M. Araújo, R. H. Scheicher, A. Blomqvist, R. Ahuja, Z. T. Xiong, P. Yang, M. X. Gao, H. G. Pan and P. Chen, *Angew. Chem. Int. Ed.*, 2009, **48**, 5828–5832.
214. J. Wang, G. Wu, Y. S. Chua, J. Guo, Z. Xiong, Y. Zhang, M. Guo, H. Pan and P. Chen, *Chemsuschem*, 2011, **4**, 1622–1628.
215. J. Wang, P. Chen, H. Pan, Z. Xiong, M. Gao, G. Wu, C. Liang, C. Li, B. Li and J. Wang, *Chemsuschem*, 2013, **31**, 1–10.
216. C. Liang, Y. Liu, M, Gao and H. Pan, *J. Mater. Chem. A*, 2013, **1**, 5031–5036.
217. C. Liang, Y. Liu, Z. Wei, Y. Jiang, F. Wu, M. Gao and H. Pan, *Int. J. Hydrogen Energy*, 2011, **36**, 2137–2144.
218. A. Sudik, J. Yang, D. Halliday and C. Wolverton, *J. Phys. Chem. C*, 2007, **111**, 6568–6573.
219. K.-F. Aguey-Zinsou, J. Yao and Z. X. Guo, *J. Phys. Chem. B*, 2007, **111**, 12531–12536.
220. Y. Chen, P. Wang, C. Liu and H. M Cheng, *Int. J. Hydrogen Energy*, 2007, **32**, 1262–1268.
221. S. Isobe, T. Ichikawa, Y. Kojima and H. Fujii, *J. Alloys Compd.*, 2007, **446**, 360–362.
222. Y. Liu, K. Zhong, K. Luo, M. Gao, H. Pan and Q. Wang, *J. Am. Chem. Soc.*, 2009, **131**, 1862–1870.
223. L. Xie, Y. Liu, G. Li and X. Li, *J. Phys. Chem. C*, 2009, **113**, 14523–14527.
224. J. Wang, J. Hu, Y. Liu, Z. Xiong, G. Wu, H. Pan and P. Chen, *J. Mater. Chem.*, 2009, **19**, 2141–2146.

16 Ammonia-Based Hydrogen Storage Materials

Yoshitsugu Kojima, Hiroki Miyaoka, and Takayuki Ichikawa

CONTENTS

16.1 INTRODUCTION

To solve the problem concerning the exhaustion of fossil fuels and global warming, sustainable society using renewable energy is expected. Renewable energy is environment-friendly and is derived from solar heat, wind power, geothermal, and hydropower, although their power densities are low and fluctuate over time. Therefore, renewable energy can be converted into secondary energy sources like electricity and hydrogen. Electricity and hydrogen are stored in batteries and high-pressure cylinders, respectively. Hydrogen is a secondary energy that stores and transports renewable energy because of its high energy density.

TABLE 16.1

Target of the Hydrogen Storage System for Vehicles in Each Organization

Property	Units	NEDO	DOE
Gravimetric density (system)	mass% H_2	7.5	7.5
Volumetric density (system)	g H_2/L	7.1	7.0
H_2 desorption temperature	K	<373	233–358
Cycle lifetime	Cycles	>1000	1500
Target year	Year	2030	Ultimate

As a result, a fuel cell vehicle has a cruising range of 500–700 km using a 70 MPa high-pressure cylinder and goes on sale in 2015 as scheduled. The high-pressure cylinder results in a gravimetric density of 5.0–6.0 mass% and a volumetric density of 30 kg/m^3 (3.9 kg H_2/100 L inner volume).[1–3] The densities are not sufficient for the propagation of the fuel cell vehicle (FCV).

Hydrogen storage materials can safely store the higher-density hydrogen compared to the gaseous hydrogen storage systems.[1,4–7] Therefore, the systems using hydrogen storage materials are considered suitable for not only on-board application but also stationary uses and as energy carriers.[1,4–7] Recently, various kinds of materials have been studied, where the following properties are required:

1. High gravimetric and volumetric density of hydrogen
2. Suitable thermodynamic properties, which are the reversible hydrogen sorption/desorption under moderate temperature and pressure
3. Rapid kinetics (refueling in 3 min)
4. Easy handling
5. Abundant resources and low cost

With respect to the target of hydrogen storage materials for vehicles, the New Energy and Industrial Technology Development Organization (NEDO) in Japan[3] and the Department of Energy (DOE) in the United States[8] propose that the materials with properties as shown in Table 16.1 should be discovered or developed.

16.2 AMMONIA (NH₃)

16.2.1 PROPERTIES OF NH₃

Figure 16.1 shows gravimetric and volumetric H_2 densities of hydrides. Ammonia is a hydride with the formula NH_3. Gaseous covalent hydrides such as NH_3 store hydrogen with high gravimetric density. Furthermore, NH_3 has a high volumetric hydrogen density of 107.3 kg H_2/m^3 because NH_3 is easily liquefied by compression under about 1.0 MPa at room temperature. The global production of ammonia for 2006 was about 150 million tons. The Haber–Bosch process to produce ammonia from nitrogen and hydrogen was developed by Fritz Haber and Carl Bosch in 1909 as shown in reaction (16.1). This conversion is typically conducted at 20–40 MPa

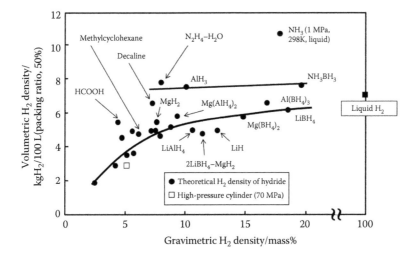

FIGURE 16.1 Volumetric H_2 density vs. gravimetric H_2 density. The packing ratio is assumed to be 50% for the solid state materials.

and 773 K.[9,10] The most popular catalyst is the iron-based catalyst (Fe-K_2O [CaO]-Al_2O_3). The modified Haber–Bosch processes are also developed by Luigi Casale, George Claude, and Giacomo Fauser.[9,10] Part of the industry uses ruthenium rather than an iron-based catalyst (Kellogg Advanced Ammonia Process [KAAP]).

$$N_2 + 3H_2 \rightarrow 2NH_3 \qquad (16.1)$$

As liquefied temperature of NH_3 is 240 K at a pressure of 0.1 MPa, the liquid must be stored under a slightly high pressure or at a slightly low temperature. The standard enthalpy change of decomposition to desorb hydrogen is 30.6 kJ/mol H_2, which is a close value to that of a conventional hydrogen absorbing alloy. However, a high temperature over 673 K and suitable catalysts are required to decompose NH_3 into H_2 and N_2 gases in the current technology.[11,12]

The combustion of ammonia to produce nitrogen and water is exothermic and expressed by Equation 16.2.

$$4NH_3 + 7O_2 \rightarrow 4N_2 + 6H_2O \qquad (16.2)$$

The standard enthalpy change of such combustion is 11.2 (lower heating value [LHV]) to 13.6 kJ/L (higher heating value [HHV]) and 1.3–1.4 times of liquid hydrogen. Ammonia neutralizes the nitrogen oxide (NO_x) pollutants emitted by diesel engines. This technology is called selective catalytic reduction (SCR). Hodgson reported the research about the internal combustion engine using ammonia in 1974.[13] So the problem that a large amount of NO_x may be included in the exhausted gas was resolved. Exhausted NO_x emission from an ammonia engine is below 20% of the emission from a gasoline engine.[13,14] Ammonia is thought to be a nonflammable gas. But it still meets the definition of a material that is toxic by inhalation. According to American Conference of Governmental Industrial Hygienists (ACGIH), an 8 h exposure limit is 25 ppm by volume, and ammonia should be used under control.

Although NH_3 is recognized as an attractive hydrogen carrier in future energy systems because of a high gravimetric and volumetric hydrogen density as shown in Figure 16.1,[15–19] the development of various technologies such as efficient H_2 generation and safe storage is required to utilize NH_3.

16.2.2 Electrolysis of NH_3

Hydrogen gas has been generated by the electrolysis of ammonia–water solution at room temperature. The ammonia concentration in saturated ammonia–water solution is 34.2 mass% at 298 K, and the capacity is limited to 6.1 mass% for hydrogen. The value goes down to 34% of its original hydrogen capacity (17.8 mass%). Then we focus on the direct electrolysis of liquid ammonia. The standard ammonia electrolysis voltage is calculated here according to reaction (16.3).[20] The voltage difference between the anode electrode $\left(E^{N_2} \right)$ and cathode electrode $\left(E^{H_2} \right)$ is described by Nernst's equation (Equation 16.3),

$$E^{N_2} - E^{H_2} = \frac{-\Delta G^0}{3F} + \frac{RT \ln \left(p_{N_2}^{1/2} p_{H_2}^{3/2} \right)}{3F} \tag{16.3}$$

where
   F is the Faraday constant
   R is gas constant

Standard Gibbs free energy change (ΔG^0) of reaction (16.3) is −10.984 kJ/mol NH_3 from the enthalpy of formation, ΔH_f (liquid NH_3) = −67.2 kJ/mol, and the entropies, S^0 (liquid NH_3) = 103.3 J/mol, S^0 (H_2) = 130.7 J/mol, and S^0 (N_2) = 191.6 J/mol. The hydrogen pressure (p_{H_2}) and nitrogen pressure (p_{N_2}) should be equivalent to the equilibrium ammonia vapor pressure (p_{NH_3}) of liquid ammonia, because gas is released while working for the solution pressure of ammonia in the closed liquid ammonia system. The p_{NH_3} at 298.15 K is 9.9 atm. So the second term of 0.039 V is almost equal to the first term of 0.038 V. It indicates that the pressure change largely affects the ammonia electrolysis voltage. Then the standard ammonia electrolysis voltage is determined to be 0.077 V at 1 MPa and 298 K. The electrolysis voltage is only 6% of the water electrolysis (1.23 V).

It was found that hydrogen is desorbed by the electrolysis of NH_3 using KNH_2 as a supporting electrolyte.[20,21] The scheme of the electrolysis of liquid ammonia is shown in Figure 16.2. It was clarified that the concentration of the NH_2^- ion in liquid NH_3 strongly affects the voltage efficiency of ammonia electrolysis. The voltage efficiency increases with the concentration. The current efficiency of electrolysis for liquid ammonia with 1 M KNH_2, which is defined as the ratio of released hydrogen based on the pressure increase to theoretically released hydrogen based on the electric charge, is estimated to be 85%. The average mole ratio of H_2 to N_2 was estimated

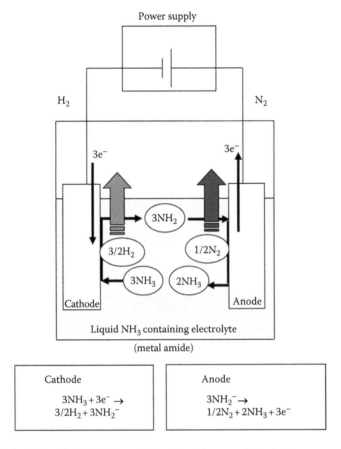

FIGURE 16.2 The principle of electrolysis of liquid ammonia for hydrogen generation.

to be 3.0 by using the area of gas chromatography peaks. This ratio corresponds to the stoichiometry of ammonia decomposition according to the reaction. The experimental electrolysis voltage of more than 1 V is still high for practical use.

The introduction of platinized platinum electrodes proves that it is possible to reduce the cell voltage by improving the surface area of the electrodes.[21] However, to decrease the overpotential of ammonia electrolysis, the effective electrode should be developed with the investigation of the electrode reaction mechanism in the future.

16.2.3 MECHANOCHEMICAL REACTION OF NH₃

Perovskites are well known for their catalytic efficiency in a number of chemical reactions. Therefore, we focused on exploring the catalytic effect of $ATiO_3$ (A = Sr, Ba) to decompose ammonia at room temperature.[22] Commercially available $SrTiO_3$ (STO) and $BaTiO_3$ (BTO) powders of micron size were mechanically milled at 0.6 MPa NH_3 atmosphere for 10 h. In order to confirm the decomposition of NH_3 during milling, we first examined the as-milled gas atmosphere in the milling

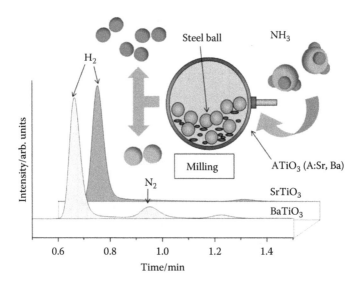

FIGURE 16.3 H$_2$ and N$_2$ peaks in the gas chromatogram (GC) of SrTiO$_3$ and BaTiO$_3$.

vessel by using gas chromatography. Hydrogen gas in the vessel for these as-milled perovskites was detected as shown in Figure 16.3. The perovskite materials SrTiO$_3$ and BaTiO$_3$ show a catalytic effect for the decomposition of ammonia into hydrogen and nitrogen during mechanical milling at room temperature, in which a part of the hydrogen and nitrogen atoms are absorbed in the perovskites in a stable manner without any phase changes.

16.3 NH$_3$–HYDRIDE SYSTEM

In order to decompose NH$_3$ into hydrogen H$_2$ and nitrogen N$_2$, more than 673 K is required, resulting in the conclusion that a practical application of NH$_3$ is limited. To realize a usable H$_2$ storage system by NH$_3$, H$_2$ storage/generation at ambient temperature is necessary. On the basis of the background, the NH$_3$ and MH (M = Li, Na, and K) systems were designed.[17,23] The reaction is expressed by Equation 16.4.

$$NH_3(g) + MH(s) \leftrightarrow MNH_2(s) + H_2(g) \tag{16.4}$$

Theoretically, H$_2$ of 8.1 [H$_2$/NH$_3$ + LiH], 4.9 [H$_2$/NH$_3$ + NaH], and 3.5 mass% [H$_2$/NH$_3$ + KH] can be reversibly stored by this reaction (Table 16.2). The systems desorbed hydrogen with the formation of metal amides even at room temperature by exothermic reactions (ΔH = Li, −43; Na, −21; K, −25 kJ/mol H$_2$).[17,23]

In the hydrogen generating reactions, the reaction kinetics was better in the order of the atomic number on the periodic table, Li < Na < K. The ball milling enhanced the reaction yield for all the systems, where the ball-milled MH (LiH, NaH, KH) materials are named MH* (LiH*, NaH*, KH*) in Table 16.2. Meanwhile, Miyaoka et al. have focused on the system as a technique to chemically compress the

TABLE 16.2
Reactivity of NH₃–MH Systems

Hydride	LiH	LiH*	NaH	NaH*	KH	KH*
H₂ density (mass%)	8.1		4.9		3.5	
H₂ yield (24 h) (%)	12	53	22	60	90	98

* Activated specimen.

hydrogen.[24] In fact, by using hydrogen generation with an exothermic reaction, high-pressure hydrogen of more than 25 MPa can be produced without any heat sources.

After hydrogen generation, alkali-metal amides were recycled back to NH_3 and the corresponding alkali-metal hydrides at 373–573 K under 0.5 MPa H_2 flow for 4 h.[17,23] The reactivity of MNH_2 with H_2 was better following the atomic number of M on the periodic table, Li < Na < K, as shown in Table 16.3. In order to obtain more than 70% of the reaction yield, $LiNH_2$, $NaNH_2$, and KNH_2 required 573, 473, and 373 K, respectively. From the previously mentioned results, the NH_3–KH system exhibits the best reactivity for both hydrogen desorption and absorption reactions among the systems. However, the hydrogen capacity in the NH_3–KH system is less than the other systems.

The reaction enthalpy of H_2 storing is positive (endothermic), which is quite different from the conventional hydrogen storage materials (Figure 16.4). By controlling the partial pressure of NH_3 generated from the recycle process, the regeneration at lower temperature can be realized.

Various kinds of metal chlorides were examined as a potential catalyst to improve the kinetics of the NH_3–LiH system.[25] The reaction yields corresponding to the LiH system with each additive at room temperature are shown in Figure 16.5. The reaction was performed under 0.45 MPa of NH_3 for 1 h with a 1:1 molar ratio of LiH:NH_3. The reaction yield of raw LiH was only 6%. By mechanical milling, the reaction yield was increased and reached about 30%. For the LiH with various additives, it was confirmed that $NiCl_2$, NaCl, TiO_2, and boron nitride (BN) revealed almost no effect, where the increase of the reaction yield compared with the raw LiH was caused by the milling effect. On the other hand, other chloride additives clearly

TABLE 16.3
Reaction Yield of the Reaction MNH₂ with H₂ for 4 h

Temperature (K)	LiNH₂ (%)	NaNH₂ (%)	KNH₂ (%)
323	0	3.6	19
373	0	14	79
473	4.2	94	91
573	71	—	~100

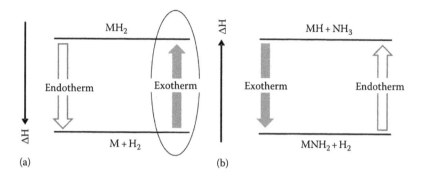

FIGURE 16.4 Reversible hydrogen storage system by endothermic reaction. (a) Conventional hydrogen storage materials and (b) reversible hydrogen storage system with endothermic reaction.

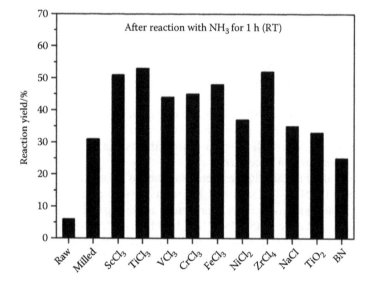

FIGURE 16.5 Reaction yield of the raw, milled, and each additive dispersing LiH for the reactions with NH$_3$ at room temperature.

showed a catalytic effect for the reaction and enhanced the reaction yields, which were more than 15% higher than that of the milled LiH. The best catalyst was TiCl$_3$ among the chlorides, and then the reaction of more than 50% proceeded by only a 1 h reaction time. Figure 16.6 shows the reaction yield of TiCl$_3$-doped LiH as a function of the reaction time, where those of the raw and milled LiH were also evaluated as the comparison. As shown in Figure 16.6, the raw LiH showed about 10% of the reaction yield after 24 h. The reaction became faster by milling; however, the reaction yield was still about 50%. In the case of TiCl$_3$ dispersed in LiH, the reaction

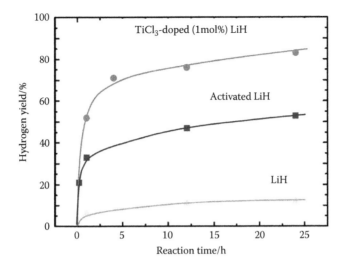

FIGURE 16.6 Reaction yield of the raw, milled, and $TiCl_3$-doped LiH.

kinetics was drastically improved. As a result, the reaction yield after 1 h was 53% and reached up to more than 80% in 24 h (H_2 desorption capacity, 6.1 mass%). The reaction kinetics of titanium chloride $TiCl_3$ dispersed in LiH was about eight times faster than raw LiH, suggesting that $TiCl_3$ possessed an excellent catalytic effect. Further improvement of the kinetics for the NH_3–LiH system containing a higher amount of hydrogen would be a future prospect.

The regeneration reaction was carried out under 0.5 MPa of H_2 flow condition at various temperatures. Figure 16.7 shows the reaction yield of each sample for 1 h at 573 K.[24] The reaction yield obtained for the raw LiH, 30%, was increased to 65% by only milling. Among the additives, $TiCl_3$, VCl_3, $NiCl_2$, and $ZrCl_4$ led to a higher reaction yield, which were more than 80%, than the other samples although the difference was not so large. The activation energy E_a of raw $LiNH_2$ was 77 kJ/mol, and it was clearly reduced to be 56 kJ/mol by the milling effect. In the case of $TiCl_3$ dispersed in $LiNH_2$, the E_a was almost the same as that of the milled $LiNH_2$, indicating that the reduction of E_a was mainly caused by the milling effect. On the other hand, the frequency factor of the $TiCl_3$ dispersed in $LiNH_2$ was larger than that of the milled $LiNH_2$. From the previously mentioned results, it is concluded that the improvement of the kinetics is mainly caused by the physical effects in contrast to the hydrogen desorption process; in other words, the small crystallites and/or particles are formed by milling with the additive.

It was found that when a little amount of KH (5 mol%) was added in the NH_3–LiH hydrogen storage system, the hydrogen desorption kinetics of this system at 373 K was drastically improved by the KH *pseudocatalytic* effect as shown in Figure 16.8. The mechanism of the improvement of hydrogen desorption kinetics for the KH-added NH_3–LiH system is proposed as reactions 16.5 and 16.6.[26]

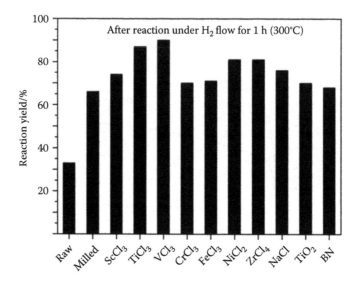

FIGURE 16.7 Reaction yield of the raw, milled, and each additive dispersing $LiNH_2$ for the reactions with H_2 at 573 K.

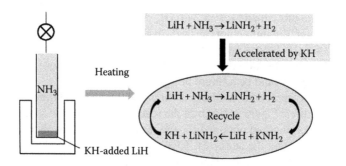

FIGURE 16.8 Pseudocatalytic effect of KH for the NH_3–LiH system.

$$NH_3 + KH \rightarrow KNH_2 + H_2 \qquad (16.5)$$

$$KNH_2 + LiH \rightarrow KH + LiNH_2 \qquad (16.6)$$

In the KH-added NH_3–LiH system, KH can immediately react with NH_3 to form KNH_2 and release H_2. Then at 373 K, the generated KNH_2 reacts with LiH to form $LiNH_2$ by the solid–solid reaction, resulting in the recovery of KH. These two reactions are regarded as cyclic reaction, which would continue until LiH is exhausted.

Mechanical ball milling improves the kinetics of the reaction as mentioned earlier. In fact, the decrease in the crystallite size of LiH by ball milling was confirmed by synchrotron radiation x-ray diffraction (XRD) studies.[18] From a Williamson–Hall plot, the crystallite sizes of commercial and the activated LiH were evaluated to be 70.4 and 27.8 nm, respectively. Because nitrogen atoms need to diffuse in the LiH crystal to be transformed into the $LiNH_2$ crystal, the crystalline size could be an important factor to control the reaction rate. The strain of LiH was increased from 0.20% to 0.34% by ball milling. The results of x-ray photoelectron spectroscopy showed that the mechanical milling was effective to reduce a hydroxide phase from the surface of LiH.[13,14] The surface of LiH is lightly covered with LiOH and a fresh surface of LiH appears due to cracking of the particles by ball milling. Consequently, the fresh surface, the smaller crystallites, and the larger strain (defect) of LiH may accelerate the reactivity of LiH with NH_3.

Ab initio molecular dynamics simulations have been performed to clarify the microscopic mechanism of hydrogen desorption for a system of one Li_2H_2 cluster and one NH_3 molecule.[27] The dynamical reaction process for the H_2 molecule formation from two H atoms, one of which is from the LiH cluster and the other is from the NH_3, and each H atom of H_2 is not equivalent both in geometry and in charge during the reaction (Figure 16.9). The height of the potential barrier becomes lower

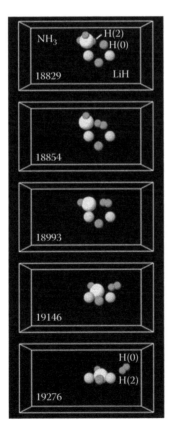

FIGURE 16.9 Snapshots during H_2 desorption at 1000 K. Numbers at the lower left are time steps.

in the order of Li, Na, and K. The H atom in M_2H_2 takes a larger negative value in the order of Li, Na, and K.[28]

In order to clarify the mechanisms of its hydrogenation and dehydrogenation, we have studied the dynamical properties of atoms in the temperature region between 3.4 and 673 K by Raman scattering, using a single-crystalline $LiNH_2$.[29] Figure 16.10 shows the representative Raman scattering spectra from 3.4 to 300 K below 700 cm^{-1}. Below 100 K, new peaks appeared and were depicted by triangles in the figure, indicating that the rotation of the $-NH_2$ group stopped below 100 K. In addition, the vibration energy of Li increased with increasing temperature. This temperature dependence concludes that Li vibration is highly anharmonic with large amplitude.

The gravimetric and volumetric H_2 densities of the NH_3–LiH system are calculated as follows. Under an operating pressure of 70 MPa at 298 K, the volume of a tank (100 L) for gaseous hydrogen allows a storage density of 3.9 kg H_2. Assuming that the density of LiH is 0.78 g/cm^3, when the tank (internal volume 62 L) is filled with liquid NH_3 (density 0.61 g/cm^3) of 38 kg and the other tank (internal volume 46 L, filling fraction 50%) is filled with LiH of 18 kg (LiH/NH_3 = 1), the total gravimetric and volumetric H_2 densities are calculated as 8.1 mass% and 4.2 kg/100 L,

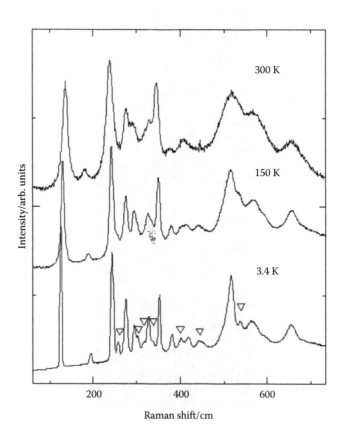

FIGURE 16.10 Raman scattering spectra from 3.4 to 300 K below 700 cm^{-1}.

respectively. The volumetric H_2 density provides 1.8 and 1.1 times of compressed hydrogen at 35 and 70 MPa, respectively. Therefore, the NH_3–LiH system can improve the H_2 densities of a high-pressure hydrogen tank of 70 MPa.

16.4 AMMONIA BORANE (NH_3BH_3)

16.4.1 PROPERTIES

Ammonia borane is the soluble white solid hydride with formula NH_3BH_3 and has emerged as an attractive solid state hydrogen storage material because of the high hydrogen capacity (19.6 mass%).[30] The structure of the solid indicates a close association of the NH and BH centers. The closest H–H distance is 2.02 Å, in which the corresponding interaction is called a dihydrogen bond.[31] In the ammonia borane molecule, the hydrogen atoms on nitrogen have a partially positive charge, denoted as $H^{\delta+}$, and the hydrogen atoms on boron have a partially negative charge, denoted as $H^{\delta-}$ (Figure 16.11). In other words, the hydrogen bonded to nitrogen is a proton and that bonded to boron is protide (hydride). NH_3BH_3 is a crystalline solid at room temperature, whereas the isoelectronic CH_3CH_3 boils at 193 K. Ammonia borane and ammonium borohydride, which is a precursor of NH_3BH_3, were first synthesized by Shore and Parry of the University of Michigan in 1955.[30] This ammonium borohydride (NH_4BH_4) is stable at 223 K but decomposes at 253 K liberating hydrogen (Equation 16.7).[32]

$$NH_4BH_4(s) \rightarrow NH_3BH_3(s) + H_2(g) \tag{16.7}$$

The decomposition of NH_3BH_3 is expressed by Equations 16.8 through 16.10.[33]

$$nNH_3BH_3 \, (s) \rightarrow -(NH_2BH_2)_n- \, (s) + H_2 \, (g) \tag{16.8}$$

$$-(NH_2BH_2)_n- \, (s) \rightarrow -(NHBH)_n- \, (s) + H_2 \, (g) \tag{16.9}$$

$$-(NHBH)_n- \, (s) \rightarrow BN \, (s) + H_2 \, (g) \tag{16.10}$$

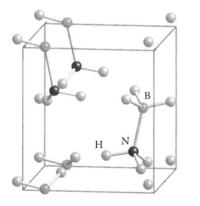
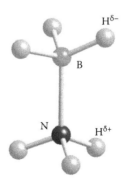

FIGURE 16.11 Structure of ammonia borane.

The stability of NH_3BH_3 can be attributed by the strong proton–protide hydrogen bonding between the $H^{\delta+}$ in $-NH_3$ and the $H^{\delta-}$ in $-BH_3$. Above its melting point around 387 K, NH_3BH_3 decomposes to the polymeric aminoborane $-(NH_2BH_2)_n-$ with the release of hydrogen. After that, the $-(NH_2BH_2)_n-$ releases the hydrogen forming the polymeric iminoborane $-(NHBH)_n-$, while the samples expand by foaming due to polymerization during dehydrogenation. Temperature in excess of 773 K is needed to release the hydrogen from $-(NHBH)_n-$ to form BN. However, not only hydrogen but also aminoborane (NH_2BH_2), borazine $(BHNH)_3$, diborane (B_2H_6), and NH_3 are considerably desorbed during the decomposition process of NH_3BH_3. In fact, we observed a large weight loss of NH_3BH_3 by decomposition at 473 K (Aldrich; purity of NH_3BH_3, 97%; heating rate, 5 K/min; weight loss, 86%). The gas desorption behaviors of NH_3BH_3 were examined by thermal desorption mass spectroscopy (MS) measurements (Aldrich; purity, 97%; heating rate, 5 K/min) and shown in Figure 16.12.[34] Figure 16.13 shows in situ optical microscope observation of NH_3BH_3 at 393 K. H_2 was desorbed by the breakdown of the bubbles above the melting point, which may include hydrogen, borazine, diborane, and ammonia. Furthermore, the hydrogen desorption reactions are exothermic $(-21.7 \text{ kJ/mol } H_2)$,[35] indicating that it is extremely difficult to regenerate NH_3BH_3 from the by-products by applying high H_2 pressure.

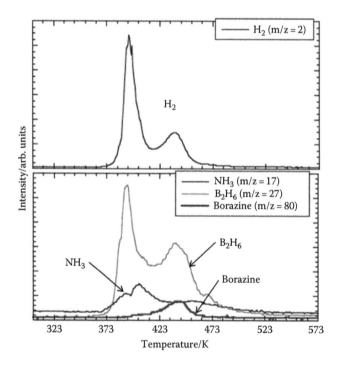

FIGURE 16.12   MS of NH_3BH_3.

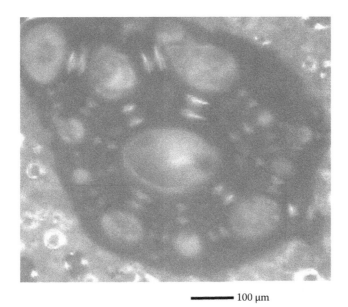

100 μm

FIGURE 16.13 In situ optical microscope observation of NH_3BH_3 at 393 K.

16.4.2 RECYCLE OF THE SPENT FUEL

As a technique of NH_3BH_3 regeneration, a method for the conversion of polyborazylene, which is an intermediate product (spent fuel BNH_x) during dehydrogenation, back to NH_3BH_3 by a single step using hydrazine/ammonia was reported.[36,37] Hydrogenation of hexagonal BN (hBN), which is a final product of NH_3BH_3, was carried out by the following method.[38] Hydrogen-containing hBNs were synthesized by ball milling using ZrO_2 balls under hydrogen (H_2) to remove the effect of Fe contamination, in which the product is named $BN^{nano}H_x$. The annealed hBNs of 300 mg and 20 ZrO_2 balls 8 mm in diameter were put into a milling vessel made of Cr steel with an inner volume of 30 cm^3. And then the hBN was milled under a H_2 atmosphere at room temperature for 80 h. Figure 16.14 shows the results of thermogravimetry (TG) and MS measurements. $BN^{nano}H_x$ desorbed only H_2, which is indicated by mass number 2, in the wide temperature range from 373 to 1173 K without other gases. Due to the desorption of H_2 by heating to 1173 K, about 1.0 mass% weight loss was revealed, as shown in the TG profile of Figure 16.14.

In Figure 16.15, the Fourier transform infrared (FT-IR) spectra of $BN^{nano}H_x$ synthesized by 80 h milling and the host hBN annealed at 473 K are shown. The spectrum of hBN showed characteristic peaks at 3020, 2800, 2530, 2330, and 780 cm^{-1}, as shown by arrows. It seems likely that the broad peak at 3420 cm^{-1} is assigned to the O–H stretching mode, which might indicate the existence of a small amount of remaining water after annealing. Among them, the peaks at 3020, 2800, and 2330 cm^{-1} completely disappeared after milling under a H_2 atmosphere, indicating that these peaks would originate in the host hBN. However, the peak at 780 cm^{-1}

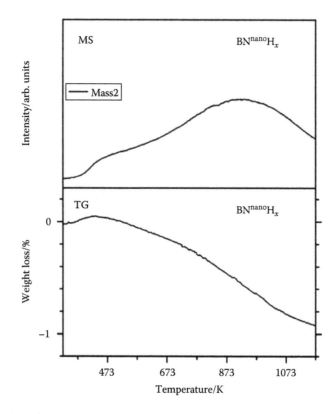

FIGURE 16.14 Thermal gas desorption (MS) profiles of $BN^{nano}XH_x$ (upper) synthesized by the milling for 80 h and TG profiles (lower) of $BN^{nano}H_x$.

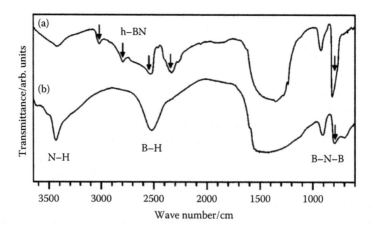

FIGURE 16.15 FT-IR spectra of (a) the annealed hBN and (b) 80 h milled $BN^{nano}H_x$.

remained after milling. This peak was assigned to the absorption of out-of-plane B–N–B on the B_3N_3 ring structure as reported before,[39,40] indicating that the hexatomic ring structure composed of B and N is preserved although the hBN structure is broken by ball milling. On the other hand, two new peaks were observed at 3400 and 2500 cm^{-1}. Even though the peak at 3400 cm^{-1} was located at almost the same position as the peak of the O–H stretching mode, the peak shape was clearly sharper and larger than that shown in the spectrum of the host hBN. Thus, the two peaks originate in the hydrogenated states of $BN^{nano}H_x$, which was previously examined.[39,40] In that paper, the peaks corresponding to the N–H and B–H stretching mode had been observed at 3440 and 2520 cm^{-1}, respectively, where these peak positions were quite consistent with the results in this work. Therefore, the peaks around 3400 and 2500 cm^{-1} are assigned to the N–H and B–H bonding, respectively. In addition, the IR spectrum of this product was similar to that of poly(aminoborane) ($[NH_2BH_2]_n$) formed by heat treatment of ammonia borane (NH_3BH_3).[39] These results indicate that hydrogen atoms are chemisorbed as the B–H and N–H groups at the edges and defects in the ball-milled hBN. By using the reaction of $BN^{nano}H_x$ with hydrazine in the ammonia solution, a small amount of NH_3BH_3 has been generated.[41]

16.5 AMMONIA BORANE–HYDRIDE SYSTEM

16.5.1 ALKALI-METAL AMIDOBORANE

The practical application of ammonia borane is greatly retarded by a higher dehydrogenation temperature based on the slow kinetics at 383–393 K and the release of trace quantities of borazine, diborane, and ammonia with foaming as mentioned earlier.

In order to solve these problems, the composites are synthesized by carrying out a mechanical crushing and mixing of NH_3BH_3 and a metal hydride (MH).[42] Novel alkali-metal amidoboranes (MABs), such as $LiNH_2BH_3$ (LiAB) and $NaNH_2BH_3$ (NaAB), are recently synthesized by the milling method as shown in Figure 16.16.[43] Here, the mechanochemical reactions are expressed by Equations 16.11 and 16.12.

$$LiH + NH_3BH_3 \rightarrow LiNH_2BH_3 + H_2 \qquad (16.11)$$

$$NaH + NH_3BH_3 \rightarrow LiNH_2BH_3 + H_2 \qquad (16.12)$$

Differential scanning calorimetry (DSC) measurements indicated that the heat of hydrogen desorption from $LiNH_2BH_3$ and $NaNH_2BH_3$ is about –3 to –5 kJ/mol H_2,[43] which is significantly less exothermic than that from the pristine NH_3BH_3.

Assuming that the final thermal decomposed product of metal amidoborane is hydride and –NB–, the theoretical H_2 capacities of the novel MABs, $LiNH_2BH_3$, $NaNH_2BH_3$, and KNH_2BH_3, are 10.9, 7.5, and 5.8 mass%, respectively. Figures 16.17 through 16.19 show the thermal decomposition behaviors of LiAB, NaAB, and KAB by analyzing thermogravimetry-mass spectrometry (TG-MS) at temperatures up to 473 K. The weight losses of LiAB, NaAB, and KAB were about 12.0, 7.0, and 5.8 mass%, respectively, while the samples expanded the volume during

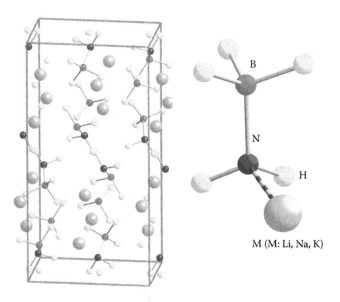

FIGURE 16.16   Structure of lithium amidoborane.

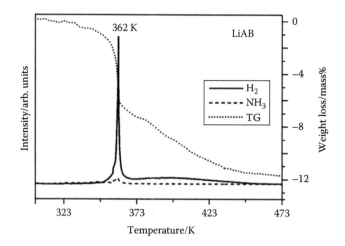

FIGURE 16.17   TG and MS of LiNH$_2$BH$_3$.

dehydrogenation. Hydrogen purity still suffers from the concurrent release of ammonia, which is extremely detrimental for fuel cell operation even at trace level (NH$_3$ would poison the noble Pt catalyst), as shown in Figures 16.17 and 16.18. The weight loss of NaAB was similar to the theoretical value. Meanwhile, the relatively larger weight loss of LiAB indicated the presence of ammonia, which was hard to disregard. KNH$_2$BH$_3$ was synthesized by the chemical route.[44] Typical TG-MS analyses as shown in Figure 16.19 indicated that only hydrogen was released. The suppression of ammonia emission may be due to the fact that KH has a higher reactivity with NH$_3$.

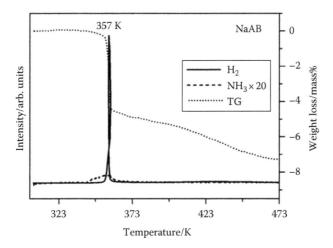

FIGURE 16.18   TG and MS of NaNH$_2$BH$_3$.

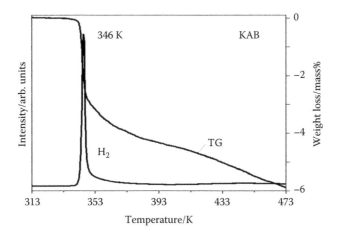

FIGURE 16.19   TG and MS of KNH$_2$BH$_3$.

16.5.2 THERMAL DECOMPOSITION PATHWAY OF SODIUM AMIDOBORANE (NaNH$_2$BH$_3$)

XRD on the products of sodium amidoborane (NaAB) indicated the formation of amorphous phases, which prevented direct determination of their structure. Nuclear magnetic resonance (NMR) spectroscopy is a useful technique to characterize such multicomponent amorphous materials. ^{11}B NMR has been effectively utilized in some of the previous works,[43] and these studies only focused on a given temperature, making it difficult to fully characterize the decomposition pathway of the MABs. Thus, we have reinvestigated in detail the thermal decomposition of NaAB

(NaNH$_2$BH$_3$) by using ^{11}B and ^{23}Na MAS/3QMAS NMR techniques, in combination with a discussion of thermal gas desorption properties.[45]

^{11}B MAS NMR spectra of AB and the as-prepared NaAB are compared in Figure 16.20. AB and NaAB had a peak at −25.4 and −22.9 ppm, respectively, suggesting that both samples had a single boron site. The small difference in chemical shift indicates the small difference in the chemical environment on the boron atoms in NH$_3$BH$_3$ and NaNH$_2$BH$_3$. By increasing the temperature to 352 K, an additional broad peak was observed at 25 ppm. This broad peak grew and finally became dominant above 473 K. This peak can be assigned to threefold boron ([BN$_3$] or [BN$_2$H] species) as in the hBN or polyborazylene. Moreover, the result of ^{11}B cross-polarization magic angle spinning (CPMAS) measurement at 423 K indicates that the boron atom at the 25 ppm peak is close to hydrogen (no figure), suggesting the presence of [BN$_2$H] species as in polyborazylene. ^{11}B 3-quantum magic angle spinning (3QMAS) spectrum of the decomposition product at 473 K showed that the BNH$_3$ peak at −23 ppm in Figure 16.20 became broader above 357 K. This indicates that the BNH$_3$ species are

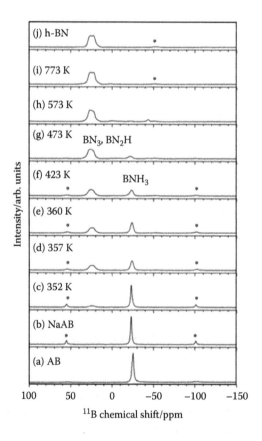

FIGURE 16.20 ^{11}B MAS NMR spectra of (a) as-received AB and NaAB at (b) room temperature, (c) 352 K, (d) 357 K, (e) 360 K, (f) 423 K, (g) 473 K, (h) 573 K, and (i) 773 K and (j) hBN milled under Ar for 1 h (reference). Spinning sidebands are indicated by asterisks.

in the amorphous state above the onset of H_2 desorption. By increasing the temperature to 473 K, the BNH_3 peak became weaker and finally disappeared above 573 K. We note that in addition to the threefold boron species, small amounts of $[BN_3H + BN_2H_2]$ (−10 to 0 ppm) and $[BH_4]$ (−44 ppm) units were also observed above 473 K.

Figure 16.21 shows [23]Na MAS spectra of NaAB at different temperatures on the decomposition process. At room temperature, complex signals having two peak tops were observed. This lineshape seems to result from a large quadrupolar interaction on a single Na site in NaAB. At the onset of the dehydrogenation temperature (352 K), a sharp peak appeared at 17.6 ppm, which increased in intensity with increasing temperature up to 573 K. This peak was assigned to NaH, indicating that NaH was formed from just above the onset of H_2 desorption. On the other hand, the quadrupolar lineshape of NaAB broadened above 357 K, which was again consistent with the amorphization of NaAB. This broad peak shifted slightly to a higher frequency and decreased in intensity up to 573 K. At 573 K, another peak was observed at ~1140 ppm (not shown), which was attributable to the Knight shift of the Na metal formed by the decomposition of NaH. Then most of the [23]Na signal disappeared at 773 K due to the evaporation of the Na metal (no figure).

From the earlier results, the thermal decomposition pathway of NaAB is discussed, assuming that the specimen weight loss is attributable to hydrogen released during the thermal decomposition of NaAB. The first decomposition stage includes

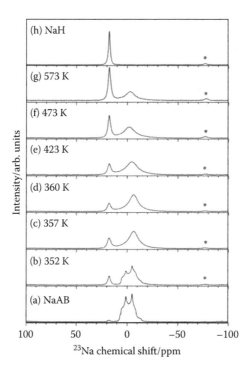

FIGURE 16.21 ^{23}Na MAS NMR spectra of NaAB at (a) room temperature, (b) 352 K, (c) 357 K, (d) 360 K, (e) 423 K, (f) 473 K, and (g) 573 K and (h) NaH as a reference.

steep hydrogen release (333–388 K). Based on the ^{11}B and ^{23}Na MAS/3QMAS NMR results, it is suggested that $NaNH_2BH_3$ molecules in NaAB start to polymerize by the intermolecular reaction between Na^+ or H^+ in a $NaNH_2BH_3$ molecular unit ($NaNH_2$) and H^- in the adjacent molecule (BH_3) at the beginning of dehydrogenation. The decomposition process is described in Equations 16.13 and 16.14.

$$[NaNH_2BH_3]_n \rightarrow NaNH_2[BHNH]_{n-1}BH_3 + (n-1)NaH + (n-1)H_2, \quad (16.13)$$

$$[NaNH_2BH_3]_n \rightarrow NaNH_2[BHNNa]_{n-1}BH_3 + 2(n-1)H_2. \quad (16.14)$$

These possible reactions provide the threefold boron ($[BN_2H]$) with consumption of the terminal $[BNH_3]$ unit by polymerization as shown in Figure 16.22. As a possible model for this polymerization, a chain-like N–B=N–B bond can be created (Figure 16.22a), which is similar to polyiminoborane (PIB), along with the formation of NaH (in reaction 16.13) and H_2 gas release. This is similar to that proposed by ab initio molecular orbital calculation. As another model, the BN benzene ring structure as in hBN or polyborazylene can also be considered (Figure 16.22b). We speculate that the drastic hydrogen release at the first decomposition stage results from the *cooperative* polymerization between the $NaNH_2BH_3$ molecular units. At 388 K, the weight loss reaches 5.3 mass% (Figure 16.18), which is corresponding to 2.8 equiv. H of NaAB.

The fact that NaH is formed above 352 K as one of the decomposition products of NaAB is quite interesting because NaAB is prepared by ball milling NaH and AB at room temperature. Thus, the NaH including state is thermodynamically stable above 352 K, resulting in NaH being isolated again during the decomposition process. Considering the amount of NaH estimated from Figure 16.21, the decomposition pathway up to 388 K can be written as Equation 16.15.

$$NaNH_2BH_3 \rightarrow Na_{0.7}NBH_{1.9} + 0.3NaH + 1.4H_2 \quad (16.15)$$

Here, the hypothetical $Na_{0.7}NBH_{1.9}$ is amorphous, which has partially polymerized structures as proposed earlier (Figure 16.22).

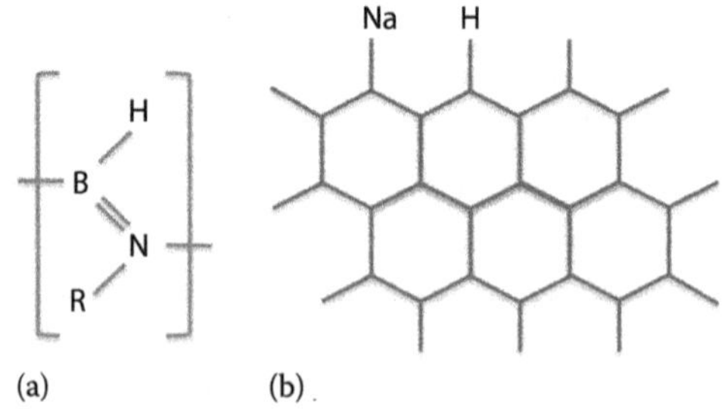

(a)   (b)

FIGURE 16.22 (a) Proposed polymerization of $NaNH_2BH_3$ molecules (R denotes Na or H) and (b) a possible BN benzene ring formation.

The second decomposition stage can be identified from 388 to 473 K with a relatively mild hydrogen release. The weight loss reached 7.5 mass% at 473 K. This corresponds to 3.9 equiv. H of NaAB, implying that the main decomposition product is a tentative material *NaNBH* or NaH/BN mixture. Again, we should stress the presence of NaH and another amorphous material based on ^{23}Na MAS NMR spectra. The presence of NaH is also confirmed by XRD, being consistent with the previous report. These two phases have almost equal amounts of Na atoms at 473 K. Therefore, we conclude that the decomposition pathway up to 473 K would be expressed by reaction (16.16).

$$NaNH_2BH_3 \rightarrow Na_{0.5}NBH_{0.5} + 0.5NaH + 2.0H_2 \qquad (16.16)$$

This suggests that $Na_{0.7}NBH_{1.9}$ changes to $Na_{0.5}NBH_{0.5}$ by releasing additional NaH and H_2 between 388 and 473 K. The hypothetical material $Na_{0.5}NBH_{0.5}$ is also amorphous and may be considered as a mixture of hBN and NaNBH-like structures, both of which contain threefold boron. Its structural framework would be composed of BN benzene rings as in hBN or $[N-B=N-B]_n$ chains similar to PIB, $(NHBH)_n$. The unknown broad peak at −2 ppm on the ^{23}Na MAS NMR spectra at 473 K (Figure 16.21) corresponds to the Na environments in $Na_{0.5}NBH_{0.5}$. Finally, this phase would further decompose into hBN and Na metal (and H_2 gas) above 773 K.

It was also found that the diammoniate of diborane (DADB) and polyaminoborane (PAB) were not formed during the decomposition of NaAB, which are both key compounds on the pyrolysis of ammonia borane.

16.5.3 ALKALI-METAL AMIDOBORANE–HYDRIDE COMPOSITE MATERIALS

MH (M = Li, Na)–MABs (M = Li, Na) composites were prepared by ball milling NH_3BH_3 and the corresponding alkali-metal hydrides with various molar ratios under a H_2 atmosphere for 1 h using a planetary mill at 200 rpm.[46] It was found that the dehydrogenation temperature and the released gas composition significantly depended on the initial molar ratios of NH_3BH_3 and NaH.

Figure 16.23 shows the relation between the H_2 desorption peak temperature and the MH/NH$_3$BH$_3$ ratio. The dehydrogenation peak temperature decreases with the increase in the NaH content. For the composite materials of 3NaH/NH$_3$BH$_3$, the dehydrogenation temperature was reduced to 345 K, which was much lower than the dehydrogenation temperature of NaAB (NaH/NH$_3$BH$_3$) (357 K). The excess of either NaH or LiH can effectively alleviate the problem of volume expansion (foaming) during dehydrogenation (Figure 16.24), and the composite systems having excess nNaH (n: 2, 3) suppressed ammonia gas emission. The H_2 desorption capacities of nNaH/NH$_3$BH$_3$ (n: 2, 3) were 4–5 mass% at 473 K.

Surprisingly, for the sample of a complex composite material such as NaH/0.5LiH/NH$_3$BH$_3$, it started to release hydrogen from a temperature as low as 313 K and reached the first dehydrogenation peak at 337 K (Figure 16.25),[45] which was much lower than the onset dehydrogenation temperatures of LiAB (362 K) and NaAB (357 K). TG-MS measurements indicated the sample was dehydrogenated step by step, and more than 3 mass% of hydrogen can be released until 353 K. When the

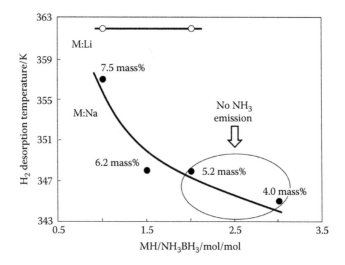

FIGURE 16.23 H_2 desorption temperature vs. MH/NH_3BH_3.

(a)   Foaming   (b)   No foaming

FIGURE 16.24   The comparison of the two samples after heating: (a) $LiNH_2BH_3$, $NaNH_2BH_3$ (expansion) and (b) $NaH/NH_3BH_3 = 2$–3, $NaH/0.5LiH/AB$ (no expansion).

temperature increased further, about 6.2 mass% of hydrogen can be desorbed at 473 K (Figure 16.25). No measurable borazine and diborane were detected throughout the decomposition process. It should be noted that no measurable ammonia could be detected throughout the decomposition process of the $NaH/0.5LiH/NH_3BH_3$ sample, indicating that this specific sample holds satisfactory performance in suppressing the volatile by-products. This is of significant importance for its application in fuel cells. Moreover, the dehydrogenation products of the $NaH/0.5LiH/NH_3BH_3$ sample did not expand and maintained its initial volume, which was completely different from the phenomenon observed during the dehydrogenation process of AB, $LiNH_2BH_3$, and $NaNH_2BH_3$. This provides an ideal way for effective utilization owing to the ease of apparatus design. It is found that the XRD patterns of the as-prepared sample are quite different from that of starting materials. We can further reasonably assign

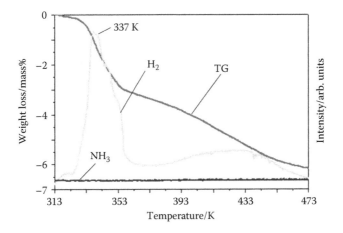

FIGURE 16.25 TG and corresponding MS of 0.5LiH + NaH + NH₃BH₃.

these peaks to the $NaNH_2BH_3$ and/or $LiNH_2BH_3$ phase on the basis of reported data and our NMR results. However, we cannot exclusively assign the peak to one phase due to the similarity of their XRD profile. The NaH phase can be clearly identified for the dehydrogenation sample, which means that the NaH additive remains its phase stability during the heating process.

^{11}B MAS NMR spectra of the samples are presented in Figure 16.26. Two signals can be identified to tricoordinated B^{III} atoms (0–50 ppm) and tetracoordinated B^{IV} atoms (−50 to 0 ppm). The ^{11}B spectrum in the as-prepared sample revealed that $-BH_3$ with a chemical shift at −23.2 ppm was the dominant species. The subtle downfield shift of the sharp peak compared with that in ammonia borane (−22.8 ppm) suggests that the alkali-metal amido group $-NH_2(M)$ may form a stronger donor complex with borane. The ^{11}B spectrum of the dehydrogenated sample showed a boron species at ~27.0 ppm, which was consistent with a trigonal planar N–BH–N environment. This suggests that the final product is a borazine-like or polyborazine-like compound. Noticeably, another peak centered at −43.5 ppm was also detected in the products, indicating the formation of a small amount of $-BH_4$.

16.5.4 Ammonia Borane–Hydride Composite Materials

MgH_2/NH_3BH_3 and CaH_2/NH_3BH_3 samples were prepared by ball milling alkaline-earth MHs with ammonia borane in a molar ratio of 1:1 under Ar atmosphere (0.1 MPa) for 10 h.[47] The gas desorption behaviors of all the products were examined by TG-MS. Figure 16.27 shows the thermal decomposition behaviors of the NH_3BH_3 composites with MgH_2. Pristine NH_3BH_3 releases about 6.3 mass% of H_2 in the temperature range of 343–393 K. Compared to pristine NH_3BH_3, mechanically milling NH_3BH_3 with alkaline-earth MH can significantly improve the dehydrogenation properties, such as lower dehydrogenation temperature and enhanced dehydrogenation quantity. MgH_2/NH_3BH_3 began to release hydrogen at as low as 323 K and most of the hydrogen was released at 351 K. CaH_2/NH_3BH_3 mixtures started to desorb hydrogen at 318 K with

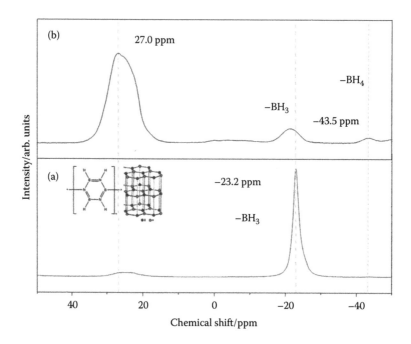

FIGURE 16.26 ^{11}B NMR spectra of NaH/0.5LiH/NH$_3$BH$_3$ sample: (a) NaH/0.5LiH/ NH$_3$BH$_3$ and (b) dehydrogenated NaH/0.5LiH/NH$_3$BH$_3$ after heating.

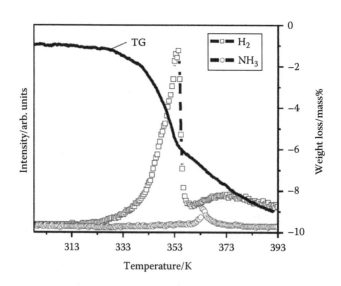

FIGURE 16.27 TG and MS profiles for MgH$_2$/NH$_3$BH$_3$ mixtures.

vigorous hydrogen release at 345 and 361 K. It should be noted that only trace ammonia was detected throughout the decomposition process of the CaH_2/NH_3BH_3 sample compared with the MgH_2/NH_3BH_3 mixture, indicating that this specific sample would hold satisfactory performance in suppressing the volatile by-products. The powder XRD pattern indicated that alkaline-earth MH and NH_3BH_3 mixtures were typical of a multicomponent composed of NH_3BH_3 and corresponding MH. The result seems to show that MH does not react with the NH_3BH_3 during the milling process, which is different from the reported $LiNH_2BH_3$ and $NaNH_2BH_3$ compounds. The FT-IR spectra measured for the ball-milled sample (MgH_2/NH_3BH_3) consists of several strong absorption bands in the NH bending (1380 and 1600 cm^{-1}), BH bending (1060 and 1170 cm^{-1}), BH stretching region (2335 cm^{-1}), and bands in the NH stretching region (3270 cm^{-1}) as shown in Figure 16.28. The dramatic decay of intensity of these bands in the NH stretching and NH bending region was seen upon thermal decomposition of MgH_2/NH_3BH_3 mixtures.

Besides this excellent MH_2/NH_3BH_3 sample, other nonstoichiometric alkaline-earth metal amidoboranes also exhibit superior dehydrogenation performance compared with pristine NH_3BH_3. For instance, ball-milled $MgH_2/2NH_3BH_3$, $MgH_2/3NH_3BH_3$, and $MgH_2/9NH_3BH_3$ samples show vigorous hydrogen release at about 361, 365, and 381 K, respectively. And ball-milled $CaH_2/2NH_3BH_3$, $CaH_2/3NH_3BH_3$, and $CaH_2/9NH_3BH_3$ samples show vigorous hydrogen release at about 355, 358, and 360 K, respectively. It should be pointed out the beginning of decomposition temperatures of nonstoichiometric calcium amidoboranes are lower than 328 K, which is much lower than the onset dehydrogenation temperatures of LiAB (362 K) and NaAB (357 K).

The NH_3BH_3 and NaH composite was prepared at low temperature. The resultant composite holds a hydrogen storage capacity of 11.0 mass% by increasing 3.6 mass%

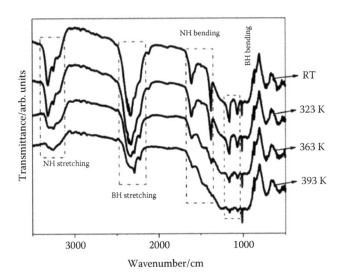

FIGURE 16.28 FT-IR spectra of MgH_2/NH_3BH_3 mixtures heated at different temperatures.

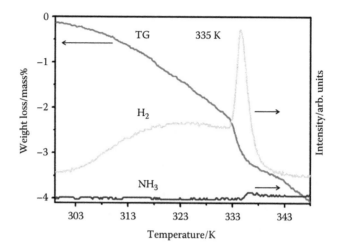

FIGURE 16.29   TG and corresponding MS of NaH–NH$_3$BH$_3$ composite material.

of H$_2$ below 343 K (peak temperature, 335 K), which is much higher than that of NaAB (7.5 mass%) (Figure 16.29). Interestingly, almost all the hydrogen is released below 473 K, and foaming with hydrogen desorption is suppressed.[48]

16.6   CONCLUSION

Hydrogen in NH$_3$ is desorbed by the electrolysis and mechanochemical reaction of NH$_3$ at room temperature. The NH$_3$–MH and NH$_3$BH$_3$–MH systems possess greater properties than that of each component and keep high hydrogen capacity. The NH$_3$–MH and NH$_3$BH$_3$–MH systems can release hydrogen below 373 K. Although the properties of ammonia-based hydrogen storage materials are improving as mentioned in this chapter, further developments are required for practical use in the future.

ACKNOWLEDGMENTS

This work was partially supported by the project *Advanced Fundamental Research Project on Hydrogen Storage* of NEDO, Japan.

REFERENCES

1. Y. Kojima, *Mater. Sci. Forum*, 2010, **654–656**, 2935.
2. A. Züttel, *Mitig. Adapt. Strat. Glob. Change*, 2007, **12**, 343.
3. New Energy and Industrial Technology Development Organization, http://www.nedo.go.jp/news/other/FF_00059.html. Accessed May 1, 2013.
4. Y. Kojima, T. Ichikawa and H. Fujii, *Encyclopedia of Electrochemical Power Sources*, ed. J. Garche, C. Dyner, P. Moseley, Z. Ogumi, D. Rand and B. Scrosati, Elsevier, Amsterdam, 2009, vol. 3, pp. 473–483.

5. Y. Kojima, H. Miyaoka and T. Ichikawa, *New and Future Developments in Catalysis: Batteries, Hydrogen Storage and Fuel Cells*, ed. S. L. Suib, Elsevier, Amsterdam, 2013, pp. 99–136.
6. P. Chen and M. Zhu, *Mater. Today*, 2008, **11**, 36.
7. S. Orimo, Y. Nakamori, J. R. Eliseo, A. Zuttel and C. M. Jensen, *Chem. Rev.*, 2007, **107**, 4111.
8. U.S. Department of Energy, http://energy.gov/sites/prod/files/2014/03/f11/targets_onboard_hydro_storage_explanation.pdf. Accessed November 15, 2014.
9. Y. Kojima, M. Tsubota and T. Ichikawa, *J. Hydrogen Energy Syst. Soc. Jpn.*, 2008, **33**, 20.
10. I. Makino, *Survey Reports on the Systemization of Technologies*, Center of the History of Japanese Industrial Technology, Tokyo, 2008, vol. 12.
11. T. V. Choudhary, C. Sivadinarayana and D. W. Goodman, *Catal. Lett.*, 2001, **72**, 197.
12. S. F. Yin, B. Q. Xu, X. P. Zhou and C. T. Au, *Appl. Catal. A*, 2004, **277**, 1.
13. J. W. Hodgson, *ASME J. Mech. Eng.*, 1974, **96**, 22.
14. T. Ota, *Kagaku Keizai*, 1981, **7**, 2.
15. Y. Kojima and T. Ichikawa, *J. Fuel Cell Technol.*, 2012, **12**, 64.
16. Y. Kojima and T. Ichikawa, *J. Hydrogen Energy Syst. Soc. Jpn.*, 2011, **36**, 34.
17. Y. Kojima, K. Tange, S. Hino, S. Isobe, M. Tsubota, K. Nakamura, M. Nakatake, H. Miyaoka, H. Yamamoto and T. Ichikawa, *J. Mater. Res.*, 2009, **24**, 2185.
18. NH_3 Fuel Association, http://nh3fuelassociation.org/. Accessed November 15, 2014.
19. Y. Kojima and T. Ichikawa, *Ceramics*, 2011, **46**, 187.
20. N. Hanada, S. Hino, T. Ichikawa, H. Suzuki, K. Takai and Y. Kojima, *Chem. Commun.*, 2010, **46**, 7775.
21. B. X. Dong, T. Ichikawa, N. Hanada, S. Hino and Y. Kojima, *J. Alloys Compd.*, 2011, **509**(Suppl. 2), S891.
22. B. Paik, M. Tsubota, T. Ichikawa and Y. Kojima, *Chem. Commun.*, 2010, **46**, 3982.
23. H. Yamamoto, H. Miyaoka, S. Hino, H. Nakanishi, T. Ichikawa and Y. Kojima, *Int. J. Hydrogen Energy*, 2009, **34**, 9760.
24. H. Miyaoka, H. Fujii, H. Yamamoto, S. Hino, H. Nakanishi, T. Ichikawa and Y. Kojima, *Int. J. Hydrogen Energy*, 2012, **37**, 16025.
25. H. Miyaoka, T. Ichikawa, S. Hino and Y. Kojima, *Int. J. Hydrogen Energy*, 2011, **36**, 8217.
26. Y.-L. Teng, T. Ichikawa, H. Miyaoka and Y. Kojima, *Chem. Commun.*, 2011, **47**, 12227.
27. A. Yamane, F. Shimojo, K. Hoshino, T. Ichikawa and Y. Kojima, *J. Mol. Struct.*, 2010, **944**, 137.
28. A. Yamane, F. Shimojo, K. Hoshino, T. Ichikawa and Y. Kojima, *J. Chem. Phys.*, 2011, **134**, 124515.
29. A. Michigoe, T. Hasegawa, N. Ogita, T. Ichikawa, Y. Kojima, S. Isobe and M. Udagawa, *J. Phys. Soc. Jpn.*, 2012, **81**, 094603.
30. S. G. Shore and R. W. Parry, *J. Am. Chem. Soc.*, 1955, **77**, 6084.
31. W. T. Klooster, T. F. Koetzle, P. E. M. Siegbahn, T. B. Richardson and R. H. Crabtree, *J. Am. Chem. Soc.*, 1999, **121**, 6337.
32. B. D. James and M. G. H. Wallbridge, *Progress in Inorganic Chemistry*, ed. S. J. Lippard, John Wiley & Sons, Inc., Hoboken, NJ, 1970, vol. 11, p. 99.
33. H. W. Langmi and G. S. McGrady, *Coord. Chem. Rev.*, 2007, **251**, 925.
34. K. Shimoda, K. Doi, T. Nakagawa, Y. Zhang, H. Miyaoka, T. Ichikawa, M. Tansho, T. Shimizu, A. K. Burrell and Y. Kojima, *J. Phys. Chem. C*, 2012, **116**, 5957.
35. G. Wolf, J. Baumann, F. Baitalow and F. P. Hoffmann, *Thermochim. Acta*, 2000, **343**, 19.
36. K. Ott, *2010 Annual Merit Review Proceedings, Hydrogen Storage*, U.S. Department of Energy, 2010.

37. A. D. Sutton, A. K. Burrell, D. A. Dixon, E. B. Garner, J. C. Gordon, T. Nakagawa, K. C. Ott, J. P. Robinson and M. Vasiliu, *Science*, 2011, **331**, 1426.
38. H. Miyaoka, T. Ichikawa, H. Fujii and Y. Kojima, *J. Phys. Chem. C*, 2010, **114**, 8668.
39. Y. Kojima, Y. Kawai and N. Ohba, *J. Power Sources*, 2006, **159**, 81.
40. D. P. Kim, K. T. Moon, J. G. Kho, J. Economy, C. Gervais and F. Babonneau, *Polym. Adv. Technol.*, 1999, **10**, 702.
41. New Energy and Industrial Technology Development Organization, Advanced fundamental research on hydrogen storage materials, http://hydro-star.kek.jp/lastmeeting/program.html. Accessed November 15, 2014.
42. Y. Kawai, Y. Kojima and T. Haga, Composite hydrogen storing material and hydrogen generating and storing apparatus, *Jpn. Pat.*, 2007-070203, 2005.
43. Z. Xiong, C. Yong, G. Wu, P. Chen, W. Shaw, A. Karkamkar, T. Autrey, M. O. Jones, S. R. Johnson, P. P. Edwards and W. F. David, *Nat. Mater.*, 2008, **7**, 138.
44. H. V. K. Diyabalanage, T. Nakagawa, R. P. Shrestha, T. A. Semelsberger, B. L. Davis, B. L. Scott, A. K. Burrell, W. I. F. David, K. R. Ryan, M. O. Jones and P. P. Edwards, *J. Am. Chem. Soc.*, 2010, **132**, 11836.
45. K. Shimoda, Y. Zhang, T. Ichikawa, H. Miyaoka and Y. Kojima, *J. Mater. Chem.*, 2011, **21**, 2609.
46. Y. Zhang, K. Shimoda, T. Ichikawa and Y. Kojima, *J. Phys. Chem. C*, 2010, **114**, 14662.
47. Y. Zhang, K. Shimoda, H. Miyaoka, T. Ichikawa and Y. Kojima, *Int. J. Hydrogen Energy*, 2010, **35**, 12405.
48. Y. Zhang, K. Shimoda, L. Zeng, T. Ichikawa and Y. Kojima. *Second Asian Symposium on Hydrogen Storage Materials (AHSAM 2012)*, Jeju Seogwipo KAL Hotel, Korea, 2012.

17 Progress in Cathode Catalysts for PEFC

Hideo Daimon

CONTENTS

17.1 INTRODUCTION

From the viewpoints of mitigation in global warming and construction of sustainable society, clean and highly efficient electric energy sources without CO_2 emission are strongly demanded. Polymer electrolyte fuel cell (PEFC) is a clean energy device in which hydrogen and oxygen are reacted on the catalysts and the chemical energy is directly converted into electricity. Since the PEFC can generate electricity at an ambient condition with high efficiency, 1 kW level of cogeneration system for home use (ENE-FARM) has been commercialized from 2009, and the fuel cell vehicle (FCV) will be on the market from 2015 in Japan.[1]

A typical design of the PEFC is illustrated in Figure 17.1. A proton conductive membrane is placed in the center of the cell and anode and cathode electrode layers are formed on both sides of the membrane. Two chemical reactions, that is, hydrogen

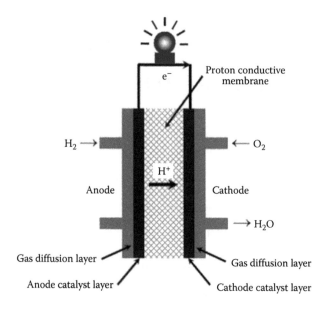

FIGURE 17.1 A typical design of PEFC.

oxidation reaction (HOR) (17.1) and oxygen reduction reaction (ORR) (17.2), occur on the anode and the cathode, respectively. Overall reaction in the PEFC (17.3) forms only water, which is why PEFC is recognized as a clean energy device.

$$2H_2 = 4H^+ + 4e^- \tag{17.1}$$

$$O_2 + 4e^- = 2O^{2-} \tag{17.2}$$

$$2H_2 + O_2 = 2H_2O \tag{17.3}$$

However, at the ambient temperature, sluggish kinetics in the PEFC requires catalysts to accelerate the chemical reactions in the both electrodes. Carbon-supported Pt nanoparticles (NPs) have been the universal choice of catalysts for chemical reactions. Unfortunately, the Pt is a very expensive precious metal (market price of Pt: 48.2 USD/g on February 18, 2014), which accounts for about 50% of the total cost of a fuel cell stack.[2] Of the two chemical reactions, the ORR in the cathode has a higher overpotential than the HOR in the anode, resulting in a much higher Pt usage in the cathode. Therefore, for the cost reduction of the PEFC, it is crucial to decrease the Pt usage in the cathode.

In order to decrease the Pt usage in the cathode, there are two major strategies: one is the enhancement of the ORR activity and the other is improvement of utilization efficiency of the Pt catalyst. In this chapter, Pt–M alloyed catalyst (M: 3d transition metals such as Fe, Co, Ni, and Cu), core/shell structured catalyst, and shape-controlled catalyst realizing the strategies are highlighted. The mechanisms on the enhancement of the ORR activity and the improvement of the utilization efficiency in each catalyst system are explained. Finally, nonprecious metal catalysts for an ultimate cost reduction of the PEFC are briefly reviewed.

17.2 PT–M ALLOYED CATALYSTS

17.2.1 PT–M THIN FILM AND BULK CATALYSTS

Watanabe and coworkers fabricated Pt–M, (M: Fe, Co, and Ni) alloyed thin-film catalysts by a vacuum sputtering technique and demonstrated enhanced ORR activity of the catalysts.[3–5] Figure 17.2 shows linear sweep voltammograms (LSVs) for the ORR of the Pt, Pt–Fe, Pt–Co, and Pt–Ni thin-film catalysts measured in O_2-saturated 0.1 M $HClO_4$ at room temperature with rotation speed of 1500 rpm. The onset potentials for the ORR in the alloyed thin-film Pt–M catalysts shifted toward higher potentials relative to that of the Pt thin-film catalyst, clearly indicating enhanced ORR activity in the Pt–M alloyed catalysts.[5] Compositional analysis using x-ray photoelectron spectroscopy (XPS) revealed that the surface Fe, Co, and Ni atoms were leached out and Pt skin layers with a few monolayer thickness were formed on the Pt–M alloyed under layers after the electrochemical measurements. The XPS analysis also revealed that 4d and 4f binding energies of the Pt skin layers exhibit positive shifts from those of the pure Pt, implying that electronic structures in the Pt skin layers were modified. They claimed that the modified electronic structures enhanced the ORR activity of the Pt skin layers.

The enhancement of the ORR activity in the Pt–M alloyed catalysts was more systematically examined by Stamenkovic et al. using surface-sensitive analytical techniques.[6–11] They prepared polycrystalline Pt and $Pt_{75}M_{25}$ (in atomic%) catalysts by

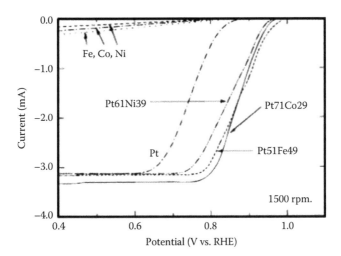

FIGURE 17.2 Polarization curves on rotating disk electrodes of $Pt_{69}Ni_{39}$, $Pt_{71}Co_{29}$, and $Pt_{51}Fe_{49}$ ($\omega = 1500$ rpm) in 0.1 M $HClO_4$ saturated with pure O_2 at room temperature in comparison with that of pure Pt, Ni, Co, and Fe. (Adapted from Toda, T. et al., *J. Electrochem. Soc.*, 146, 3750. Copyright 1999, Electrochemical Society. With permission.)

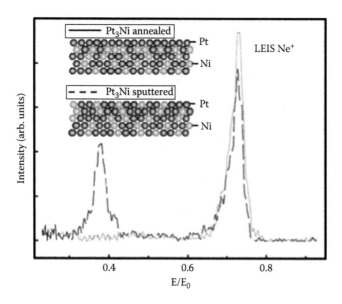

FIGURE 17.3 LEIS spectra of annealed and mildly Ar⁺ inos sputtered Pt₃Ni catalysts. Also shown is a schematic picture of the *Pt skin* electrode and the sputtered Pt₃Ni electrodes. (Adapted with permission from Stamenkovic, V. et al., *J. Phys. Chem. B*, 106, 11970. Copyright 2002 American Chemical Society.)

conventional metallurgy and the catalysts were annealed at 1000 K in ultrahigh vacuum (HUV) or mildly sputtered with Ar⁺ ions, and they correlated the electrochemical properties of the two catalysts with their surface microstructures. Low-energy ion scattering (LEIS) spectra of the annealed and the Ar⁺ ions sputtered $Pt_{75}Co_{25}$ catalysts are depicted in Figure 17.3 with their surface microstructural sketches.[6] The LEIS analysis revealed that the Co atoms exist on the surface of the Ar⁺ ions sputtered catalyst maintaining the same composition as the bulk one, while the surface of the annealed catalyst is covered with a Pt monolayer (Pt skin). According to the low-energy electron diffraction (LEED) study of Gauthier et al., it has been clarified that $Pt_{80}Co_{20}(111)$ bulk alloy annealed at 1350 K in UHV exhibits highly structured compositional oscillations in the first three atomic layers, that is, the outermost layer of the annealed surface is pure Pt, Pt is depleted in the second layer, and Pt is a little bit enriched in the third layer (composition of the fourth Pt layer is almost equivalent with the bulk one),[12] which supports the LEIS analytical result of the annealed polycrystalline $Pt_{75}Co_{25}$ catalyst. In addition, they demonstrated that the Pt skin is thermodynamically stable not only in the bulk alloy but also in $Pt_{75}Co_{25}$ and $Pt_{75}Ni_{25}$ NPs using Monte Carlo (MC) simulation (Figure 17.4[9,11]).

They further demonstrated a fundamental relationship between experimentally determined surface electronic structure (*d* band center) and ORR activity of the polycrystalline $Pt_{75}M_{25}$ alloyed surfaces (M: Ti, V, Fe, Co, and Ni).[8,9,11] As described earlier, the Pt skin layer is created by the annealing in UHV, while the M atoms exist on the surface sputtered with Ar⁺ ions. In the annealed catalysts, the 3d M

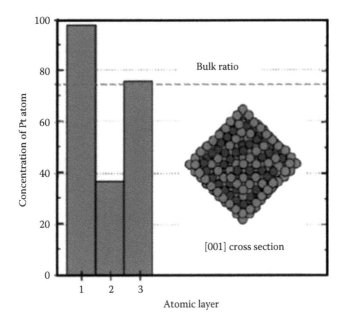

FIGURE 17.4 Segregation profile of $Pt_{75}Ni_{25}$ NP with [111] facet obtained from MC simulations. The cross-section model is related to the stable octahedral particle; dark gray spheres are Ni atoms and gray spheres are Pt atoms. (Adapted from *Electrochim. Acta*, 53, Fowler, B., Lucas, C.A., Omer, A., Wang, G., Stamenkovic, V.R., and Markovic, N.M., 6076–6080, Copyright 2008, with permission from Elsevier.)

atoms are stable even after immersion into an electrochemical environment, that is, 0.1 M $HClO_4$, because the 3d M atoms are protected by the Pt skin layer. However, in the catalysts sputtered with Ar^+ ions, the 3d M atoms dissolve into the acidic circumstance, forming Pt skeleton layer on the surface as illustrated in Figure 17.5.[7] It should be noted that the annealed catalysts exhibit highly structured compositional oscillations in the first three atomic layers (Figure 17.4), while subsurface of the Pt skeleton layer has an equivalent composition to the bulk one (Figure 17.5). Although

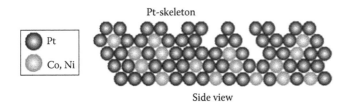

FIGURE 17.5 PtCo surface after exposure to 0.1 M $HClO_4$: PtCo surface contains only Pt atoms in the topmost atomic layer after exposure to the electrolyte. Surface Co atoms are being dissolved forming the Pt skeleton surface. (Adapted with permission from Stamenkovic, V.R. et al., *J. Am. Chem. Soc.*, 128, 8813. Copyright 2006 American Chemical Society.)

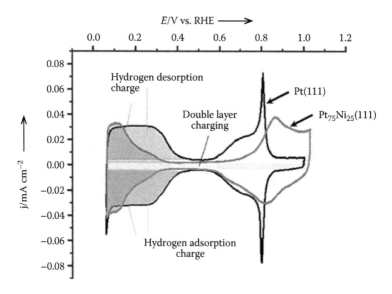

FIGURE 17.6 CVs of Pt(111) and Pt$_{75}$Ni$_{25}$(111) skin surfaces by using a rotating disk electrode in 0.1 M HClO$_4$. (Vliet, D.F., Wang, C., Li, D., Paulikas, A.P., Greeley, J., Rankin, R.B., Strmcnik, D., Tripkovic, D., Markovic, N.M., and Stamenkovic, V.R.: *Angew. Chem. Int. Ed.* 2012. 51. 3139–3142. Copyright Wiley-VCH Verlag GmbH & Co. KGaA. Adapted with permission.)

the polycrystalline Pt, the Pt skin, and the Pt skeleton are composed of pure Pt atoms, their ORR activities increase in the order of Pt ploy < Pt skeleton < Pt skin.[9] This is attributed to the different electronic structures (*d* band center) in the Pt layers caused by their different microstructures in the surface region described earlier. Figure 17.6 shows cyclic voltammograms (CVs) of the Pt surfaces measured in Ar-saturated 0.1 M HClO$_4$ at 298 K, revealing that there is a clear positive shift in OH formation (E > 0.7 V) on the Pt$_{75}$Ni$_{25}$(111) skin surface relative to the pure Pt(111) surface.[9,13]

Oxygen chemisorption energy (ΔE_O) is a good descriptor of the ORR activity of a given surface because the energy correlates with average energy of the d states on the surface atoms (*d* band center). The variation in the oxygen metal bond depends on the coupling strength between oxygen 2p states and metal d states and this coupling forms bonding and antibonding states. The filling of the antibonding state depends on the position of the d states relative to the Fermi level (*d* band center). An upward shift of the *d* band center results in an upward shift of the antibonding states, leading to less filling and thus to a stronger bond. Therefore, this effect gives the relationship as shown in Figure 17.7.[8]

Next, they measured the *d* band centers for the Pt$_{75}$M$_{25}$ alloyed catalysts using synchrotron-based high-resolution photoemission spectroscopy and correlated ORR activity with the *d* band center.[7–9] The ORR activity vs. the *d* band center position exhibits a classical volcano-shaped dependence summarized in Figure 17.8, which well agrees with ORR activity predicted with density function theory (DFT) calculations.[8,9]

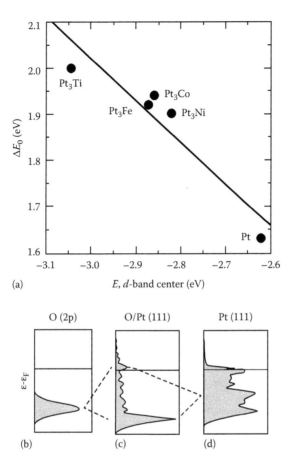

(a)

(b)   (c)   (d)

FIGURE 17.7 (a) The correlation between the d band center and the oxygen adsorption energy. The d band center was calculated (DFT) as the average for the Pt atoms in the two top layers. (b) sp broadened 2p orbital for O(g), (c) projected p density of states of oxygen atoms on Pt(111), and (d) projected d density of states of Pt(111). (Stamenkovic, V., Mun, B.S., Mayrhofer, K.J.J., Ross, P.N., Markovic, N.M., Rossmeisl, J., Greeley, J., and Nørskov, J.K.: *Angew. Chem. Int. Ed.* 2006. 45. 2897–2901. Copyright Wiley-VCH Verlag GmbH & Co. KGaA. Adapted with permission.)

In order to create better catalysts than the pure Pt for the ORR, the overall consequence of this trend is that the catalysts should counterbalance the two opposing effects, that is, relatively strong adsorption energy of O_2 and relatively low coverage by oxygenated species on the topmost Pt surface. Therefore, for the Pt surfaces that bind oxygen too strongly, for example, as in the case of the polycrystalline Pt, the d band center is too close to the Fermi level and the rate of the ORR is limited by the availability of OH and anion-free Pt sites. On the other hand, when the d band center is too far from the Fermi level, as in the case of $Pt_{75}V_{25}$ and $Pt_{75}Ti_{25}$, the surface is less covered by OH and anions, but the adsorption energy of O_2 is too low to enable a high turnover rate for the ORR. Therefore, the resulting volcano dependence can

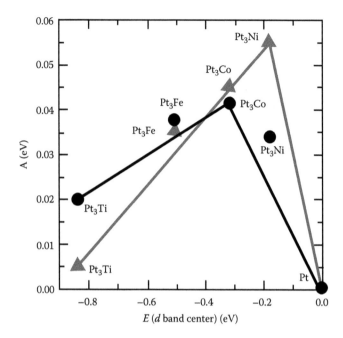

FIGURE 17.8   Activity vs. the experimentally measured d band center relative to platinum. The activity predicted from DFT simulation is shown in gray triangles and the measured activity is shown in black circles. (Stamenkovic, V., Mun, B.S., Mayrhofer, K.J.J., Ross, P.N., Markovic, N.M., Rossmeisl, J., Greeley, J., and Nørskov, J.K.: *Angew. Chem. Int. Ed.* 2006. 45. 2897–2901. Copyright Wiley-VCH Verlag GmbH & Co. KGaA. Adapted with permission.)

be rationalized by the application of the Sabatier principle, that is, for catalysts that bind oxygen too strongly, the rate is limited by the rate of removing surface oxides and anions, while for catalysts that bind oxygen too weakly, the rate is limited by the rate of electron and proton transfer to adsorbed O_2.

They also demonstrated that the Pt skin surfaces are the most active, followed by the Pt skeleton surface and the least active surface is the polycrystalline Pt. Therefore, the most active Pt skin surfaces have the most pronounced shift of the d band center induced by enrichment of the alloying component in the subsurface atomic layer relative to the bulk. In contrast, the Pt skeleton (leached out) surfaces do not have an enriched subsurface layer with 3d M atoms, and thus the level of electronic modification of the topmost Pt layer is not that pronounced as that for the Pt skin surface, which drives slower kinetics for the ORR. Finally, they showed that the Pt skin surface on $Pt_{75}Ni_{25}(111)$ single crystal is 10-fold more active for the ORR than the corresponding Pt(111) single-crystal surface, and 90-fold more active than the current state-of-the-art carbon-supported Pt catalyst (Pt/C).[10] Recently, Chorkendorff and his coworkers reported that the Pt skin with Cu-enriched subsurface shows fivefold improvement in the ORR activity over the Pt(111) single-crystal surface at 0.9 V vs. RHE.[14] It is important to recognize that all these enhancements in the ORR activity are attributed to the modified electronic structure of the Pt skin layers with the underneath subsurface enriched with the 3d transition metals.

17.2.2 PT–M NANOPARTICLE CATALYSTS

A way to decrease the Pt usage in the PEFC cathode with the enhancement of the ORR activity would be to create NP catalysts with electronic properties that mimic the extended Pt skin surfaces. With currently available techniques for the synthesis of nanosized material and for the characterization, it is feasible to tune the electronic properties of the bimetallic NP in the same way as for the Pt–M extended surfaces described in the previous section. Conventional syntheses for the NP catalysts are impregnation and coprecipitation methods where Pt and M precursors are impregnated or coprecipitated on a carbon support followed by reduction with H_2 at an elevated temperature. These two methods are simple and easy for synthesizing the Pt–M NP catalysts, but it is difficult to control size and distribution of the NP. On the contrary, wet chemical syntheses using organic stabilizers, for example, oleylamine (OAm), oleic acid (OA),[15–17] polyvinylpyrrolidone (PVP),[18,19] and polyvinyl alcohol (PVA),[20] and reducing agents, for example, alcohol, amine, and $NaBH_4$, are well established for controlling the size and achieving monodispersity of the NP. In this section, the Pt–M NP catalysts fabricated with wet chemical synthesis are reviewed.

Wang et al. synthesized monodisperse $Pt_{75}Co_{25}$ NP catalysts with size controlled from 3 to 9 nm (Figure 17.9) through an organic solvothermal route and evaluated their ORR activities.[21] Organic stabilizers (OAm) chemisorbed on the NP surface were removed by heat treatment in an oxygen atmosphere at 458 K before the electrochemical measurements. The crystalline sizes of the NP catalysts calculated from the Scherrer equation were close to those from the transmission electron microscopy (TEM) observations, implying the single-crystal nature of the individual NP.

Despite a lack of consensus, it is generally accepted that the mechanism of the Pt size effect is fulfilled through enhanced adsorption of oxygenated species (O^-, OH^-, etc.) in smaller particles due to the decrease of average coordination number and consequently more pronounced oxophilic behavior.[22,23] Oxygenated species adsorbed on the low-coordinated Pt surface sites (such as steps, edges, and kinks) inhibit the ORR.[24] The particle size effect is also reflected in the case of the $Pt_{75}Co_{25}$ NP catalysts with sizes from 3 to 9 nm as summarized CVs in Figure 17.10.[21] The oxidation peak (~0.9 V) in the anodic scan and the reduction peak (~0.8 V) in the cathodic scan exhibit negative potential shifts with decrease of the catalyst size from 9 to 3 nm. These shifts indicate that the smaller $Pt_{75}Co_{25}$ NP catalysts are oxidized at a lower potential, which corresponds to an enhanced adsorption of the oxygenated species and thus decreased ORR activity. The ORR specific activity evaluated at 0.9 V vs. RHE showed an ascending trend as the NP size increases (1.6 mA/cm² with 3 nm, 3.4 mA/cm² with 9 nm), and their specific surface areas showed reverse trend. These two opposite trends lead to a volcano-shaped behavior in size-dependent mass activity as shown in Figure 17.11, and the maximum ORR mass activity of 1420 mA/mg was achieved with 4.5 nm $Pt_{75}Co_{25}$ NP catalyst. The enhancement of the ORR activity is ascribed to the modification of the Pt surface electronic structure

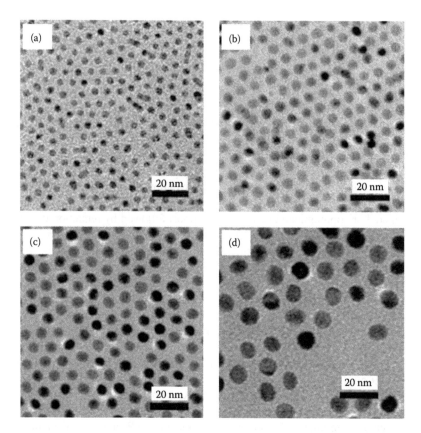

FIGURE 17.9 TEM images of as-synthesized (a) 3 nm, (b) 4.5 nm, (c) 6 nm, and (d) 9 nm $Pt_{75}Co_{25}$ NPs. (Adapted with permission from Wang C., et al., *J. Phys. Chem. C*, 113, 19365. Copyright 2009 American Chemical Society.)

by alloying with 3d transition metal of Co and the improvement factor is in line with that observed in the extended surfaces.[9–11] It should be noted, however, that all of the surface Co would be immediately dissolved under a low pH environment in the PEFC, resulting in the skeleton-type surface morphology with low-coordinated Pt topmost atoms. Therefore, the level of the ORR activity enhancement is likely to depend on the Co concentration in the subsurface layers and the extent of the surface Pt coordination.[7,9,10]

Wang et al. conducted postannealing treatments using the 4.5 nm size $Pt_{75}Co_{25}$ NP catalyst in a reductive atmosphere (Ar + 5% H_2). XRD analysis for as-prepared and various annealed catalysts revealed that all the patterns show characteristic peaks of the disordered fcc phase $Pt_{75}Co_{25}$ and there are no diffraction peak shifts under annealing, confirming the homogeneous alloy composition in the NP catalysts. The size enlargement was insignificant for annealing up to 773 K, while the average particle size increased to 6 nm for 773 K and 13 nm for 873 K annealing, which was consistent with the TEM observations.[25] Changes of the specific surface area and the ORR specific activity in the catalysts with the annealing temperature

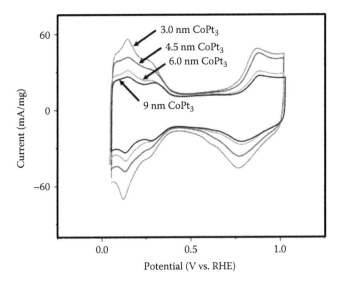

FIGURE 17.10 CVs of $Pt_{75}Co_{25}$ NPs of different sizes (3.0–9.0 nm). (Adapted with permission from Wang, C. et al., *J. Phys. Chem. C*, 113, 19365. Copyright 2009 American Chemical Society.)

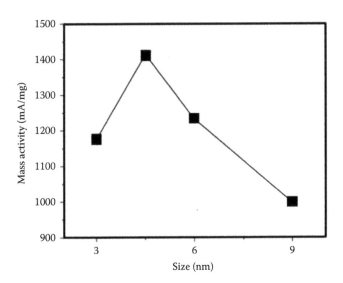

FIGURE 17.11 Mass activities of $Pt_{75}Co_{25}/C$ catalysts. (Adapted with permission from Wang, C. et al., *J. Phys. Chem. C*, 113, 19365. Copyright 2009 American Chemical Society.)

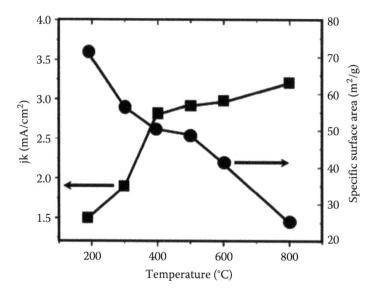

FIGURE 17.12 The plots of specific activity and specific surface area vs. annealing temperature. (Adapted with permission from Wang, C. et al., *ACS Catal.*, 2, 891. Copyright 2012 American Chemical Society.)

are summarized in Figure 17.12.[25,26] The change in surface area is not exactly corresponding to the size change observed by XRD and TEM, where it was shown that the particle size is not significantly increased after annealing at 773 K. However, the specific surface area decreased with annealing temperature lower than 573 K. The divergence could be due to reduction of the surface roughness in the moderately annealed catalysts (<773 K) compared to the nonannealed one. The reduction in the surface roughness not only decreases the surface area but also improves the ORR activity, since it diminishes the surface defects, relaxes the low-coordinated surface Pt atoms, and lowers the oxophilicity. Furthermore, at the moderate annealing temperatures (673–773 K), the compositional profile of the NP catalysts in the near surface region is changed due to the Pt segregation,[9–11] which contributes for the maximum mass activity without significant increase of the particle size (Figure 17.13[25,26]). They performed an MC simulation and claimed that the annealing temperature–dependent ORR activity can be interpreted in a sequential stage mechanism described earlier. It should be emphasized that the ORR mass activities of the $Pt_{75}Co_{25}$ NP catalysts annealed at the moderate temperatures (1420 mA/mg at 673–773 K) are much higher than those reported in the literature for the alloyed PtCo catalysts synthesized with the conventional impregnation or coprecipitation methods, and also higher than the values for a commercially available carbon-supported Pt–Co catalyst.[22,27]

In the Pt–M alloyed NP catalysts, dissolution of the 3d transition metals should be concerned when they are used in a low pH environment of the PEFC. The dissolution of the 3d transition metals significantly depends on the initial composition of the alloyed NP catalysts, which in turn determines the ORR activity of the catalysts.

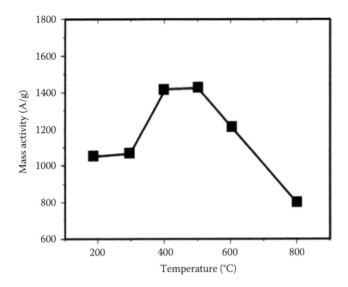

FIGURE 17.13 The plot of mass activity vs. annealing temperature. (Adapted with permission from Wang, C. et al., *ACS Catal.*, 2, 891. Copyright 2012 American Chemical Society.)

The Pt_xNi_{1-x} alloyed catalysts 5 nm in size were synthesized with an organic solvent approach and their ORR activities and Ni composition after the electrochemical measurements are summarized in Figure 17.14.[28] The Pt_xNi_{1-x} alloyed catalysts with intermediate Pt–Ni ratio, that is, $Pt_{50}Ni_{50}$ and $Pt_{33}Ni_{67}$, exhibited higher ORR specific activity than the others, while the surface area was almost the same for all the four catalysts. As a result of the NPs composition variation, a volcano curve was obtained for the ORR mass activity dependence on the alloy composition of the catalysts, showing that the $Pt_{50}Ni_{50}$ exhibited the highest ORR mass activity. Furthermore, the Ni dissolution also possesses a volcano-like dependence on the initial alloy composition, which coincides with the trend in mass activity.

The loss of Ni is not out of expectation since the surface Ni will be dissolved when the Pt_xNi_{1-x} alloyed NP catalysts are exposed to the acidic electrolyte and potential cycling. The formed nanostructures after the Ni dissolution would be more intrinsic for understanding the observed dependence of the ORR activity on the alloyed catalysts composition. Initially, the as-synthesized Pt_xNi_{1-x} alloyed NP catalysts had homogeneous distributions of alloy elements. After the electrochemical measurements, a Ni-depleted Pt–Ni structure with a Pt-rich surface layer was observed in all the catalysts with different alloy compositions, that is, Ni was not detected in the surface region compared to the bulk of the NPs. However, the thickness of the Pt-enriched layer was found to be dependent on the initial particle composition, where the Pt thickness increased with the Ni content. Energy dispersive x-ray spectroscopy (EDX) line profile analysis revealed that the Pt-enriched layer with a thickness of >1 nm (more than five Pt atomic layers) was formed for the $Pt_{25}Ni_{75}$ NP catalyst while ~0.5 nm (two to three Pt atomic layers) Pt-enriched layer was formed for the $Pt_{50}Ni_{50}$ NP catalyst (Figure 17.15[28]). Similar observations were also obtained

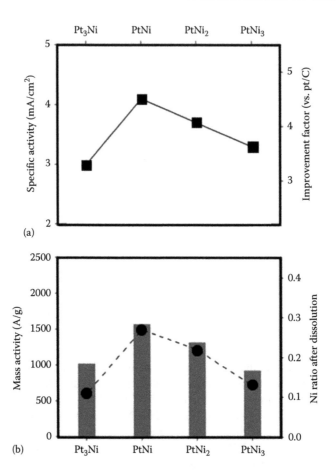

FIGURE 17.14 (a) Summary of specific activities for ORR at 0.9 V and improvement factors vs. Pt/C. (b) Mass activity and the ratio of Ni left after electrochemical characterizations for different alloy compositions. (Wang, C., Chi, M., Wang, G., Vliet, D., Li, D., More, K., Wang, H.H., Schlueter, J.A., Markovic, N.M., and Stamenkovic, V.R.: *Adv. Funct. Mater.* 2011. 21. 147–152. Copyright Wiley-VCH Verlag GmbH & Co. KGaA. Adapted with permission.)

in the reports on the dealloyed Pt–Cu bimetallic catalyst.[29–31] The difference of the thickness in the Pt skeleton structure could be one of the key factors governing the catalytic performance dependence on the alloy composition.[28]

On the basis of these findings in the Pt–M extended surfaces and the NP catalysts, Chao and Stamenkovic et al. synthesized carbon-supported $Pt_{50}Ni_{50}$ alloyed NP catalyst 5 nm in size ($Pt_{50}Ni_{50}$/C) through the solvothermal synthetic route and leached out the surface Ni atoms with an exposure to 0.1 M $HClO_4$ to form the Pt skeleton structure, and finally, the acid treated catalyst was annealed at 673 K to transform the Pt skeleton structure into the Pt skin one.[32] The electrochemical properties for the $Pt_{50}Ni_{50}$/C and the Pt/C catalysts (6 nm in size) are summarized in Figure 17.16.

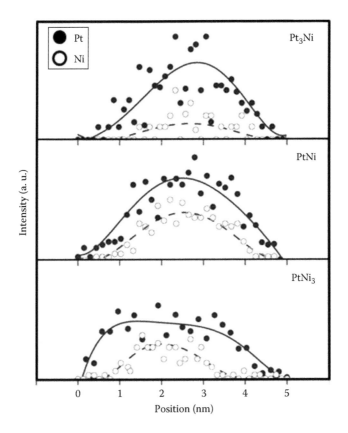

FIGURE 17.15 Composition line profiles crossing typical Pt_xNi_{1-x} NPs after the electrochemical characterization obtained by combinational HAADF-STEM and EDX analyses. (Wang, C., Chi, M., Wang, G., Vliet, D., Li, D., More, K., Wang, H.H., Schlueter, J.A., Markovic, N.M., and Stamenkovic, V.R.: *Adv. Funct. Mater.* 2011. 21. 147–152. Copyright Wiley-VCH Verlag GmbH & Co. KGaA. Adapted with permission.)

The H_{upd} areas decreased in the order of skeleton $Pt_{50}Ni_{50}/C$ (acid treated) < skin $Pt_{50}Ni_{50}/C$ (acid treated/annealed) with respect to the Pt/C catalyst. In addition, onset potentials for the oxides formation positively shifted in the same order, which coincides with the trends observed in the thin-film model catalysts.[9–11,32] The positive shifts are representative of less oxophilic catalyst surfaces due to the formation of the multilayered Pt skin structure and further corresponding to remarkable enhancement in the ORR activity as evidenced by the polarization curves and the Tafel plots in Figure 17.16. They further demonstrated superior durability of the skin structured catalyst under 4000 potential cycles between 0.6 and 1.1 V vs. RHE in 0.1 M $HClO_4$ at 333 K (Figure 17.17). At 0.95 V vs. RHE, the $Pt_{50}Ni_{50}/C$ catalyst with the multilayered Pt skin surface exhibited improvement factors in the ORR mass activity of more than one order of magnitude with respect to the Pt/C catalyst even after the potential cycling durability test.[32]

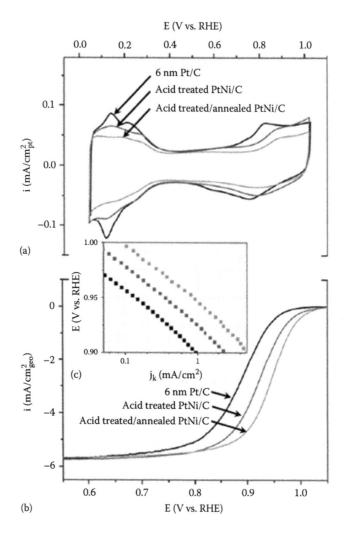

FIGURE 17.16 (a) CVs, (b) polarization curves, and (c) Tafel plots with the specific activity (j_k, kinetic current density) as a function of electrode potential, in comparison with the commercial Pt/C catalyst. (Adapted with permission from Wang, C. et al., *J. Am. Chem. Soc.*, 133, 14396. Copyright 2011 American Chemical Society.)

17.2.3 Summary of Pt–M Alloyed Catalysts

In this section, the Pt–M alloyed thin-film/bulk and NP catalysts were reviewed. The studies of the thin-film/bulk Pt–M model catalysts demonstrated an importance of the Pt skin surfaces for enhancing the ORR activity. The d states of the Pt topmost surface (*d* band centers) in the catalysts are crucial for the overall ORR performance, which are strongly influenced by the component and composition of the 3d transition metals in the subsurface. In order to create better catalysts than the Pt for the ORR, the catalysts should counterbalance the opposing two effects, that is, relatively

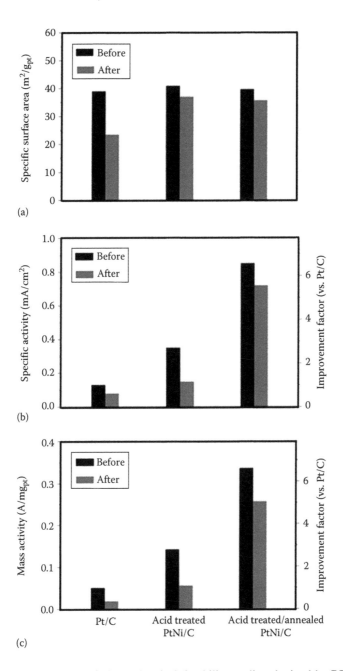

FIGURE 17.17 Summary of electrochemical durability studies obtained by RDE before and after 4000 potential cycles between 0.6 and 1.1 V for the Pt/C and PtNi/C catalysts in 0.1 M HClO$_4$ at 0.95 V and 60°C. (a) Specific surface area, (b) specific activity, and (c) mass activity. Activity improvement factors vs. Pt/C before and after cycling were also shown for specific and mass activities in (b) and (c). (Adapted with permission from Wang, C. et al., *J. Am. Chem. Soc.*, 133, 14396. Copyright 2011 American Chemical Society.)

strong adsorption energy of O_2 and relatively low coverage by the oxygenated species. The d band center of the polycrystalline Pt is too close to the Fermi level and the d band center is too far from the Fermi level in the case of $Pt_{75}V_{25}$ and $Pt_{75}Ti_{25}$ catalysts, resulting in a well-known volcano-shaped ORR activity dependence. The studies showed that the Pt skin surfaces on the $Pt_{75}Co_{25}$ and $Pt_{75}Ni_{25}$ model catalysts moderately counterbalance the two opposing effects and enhance the ORR activity with respect to the polycrystalline Pt catalyst. On the basis of these findings in the systematic studies on the model catalysts, carbon-supported Pt–Ni NP catalysts were designed and synthesized with tuning their sizes and compositions. It was demonstrated that acid leached and annealed $Pt_{50}Ni_{50}$ NP catalyst 5 nm in size with multilayered Pt skin surface shows superior ORR activity and durability to those of the current state-of-the-art Pt/C catalyst.

17.3 CORE/SHELL STRUCTURED CATALYSTS

17.3.1 BENEFITS OF CORE/SHELL STRUCTURED CATALYSTS

The most important feature of the core/shell structured catalysts is a high utilization efficiency of the Pt as illustrated in Figure 17.18. Since the chemical reactions occur on the surface of the catalyst, only the topmost surface of the catalyst contributes to the reactions. In the core/shell structured catalysts, Pt monolayer (Pt_{ML}) is formed on the surface of the core metal NP different from the Pt; thus, the utilization efficiency of the Pt basically becomes 100% regardless of the core NP diameter. In addition, Adzic and coworkers have demonstrated that the ORR activity of the Pt_{ML} is enhanced with an appropriate choice of the core metal.[33–38]

    Changes in the ORR activity of the Pt_{ML} formed on various kinds of precious metal single crystals are depicted in Figure 17.19, indicating that the ORR activity of the Pt_{ML}/Pd(111) overcomes that of the Pt(111).[34,35] Generally, the changes of the ORR activity in the Pt_{ML} can be explained by a strain effect (geometric effect) induced in the Pt_{ML} with difference of the lattice constants between the Pt_{ML} and

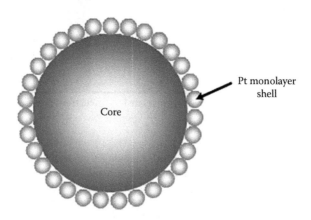

FIGURE 17.18   Schematic illustration of core/shell structured catalyst.

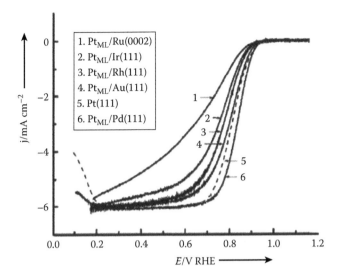

FIGURE 17.19 Polarization curves for O_2 reduction on platinum monolayers (Pt_{ML}) on Ru(0001), Ir(111), Rh(111), Au(111), and Pd(111) in 0.1 M $HClO_4$ solution on disk electrode. (1) Pt_{ML}/Ru(002), (2) Pt_{ML}/Ir(111), (3) Pt_{ML}/Rh(111), (4) Pt_{ML}/Au(111), (5) Pt(111), (6) Pt_{ML}/Pd(111). (Zhang, J., Vukmirovic, M.B., Xu, Y., Mavrikakis, M., and Adzic, R.R.: *Angew. Chem. Int. Ed.* 2005. 44. 2132–2135. Copyright Wiley-VCH Verlag GmbH & Co. KGaA. Adapted with permission.)

the underlayer. It has been reported that the induced strain directly influences the d band center of the Pt_{ML}, that is, compressive and tensile strains lead to downshift and upshift of the d band center, respectively.[34,35,39–41] As described previously, the oxygen chemisorption energy (ΔE_O) is a good descriptor of the ORR activity of a given surface because the energy well correlates with the d band center and results in a nearly linear correlation as shown in Figure 17.20.[34,35] The Pt_{ML}/Au(111) lies at the high d band center end of the plot, thus binds with atomic oxygen more strongly than the Pt(111), whereas the Pt_{ML}/Ru(0001), Pt_{ML}/Rh(111), and Pt_{ML}/Ir(111) lie at the other end, leading to less strong binding than the Pt(111), which gives a volcano-shaped ORR activity dependence on the d band center. It should be noted that the highest ORR activity can be achieved with the Pt_{ML}/Pd(111); therefore, the two opposing effects, that is, relatively strong adsorption energy of oxygen and relatively low coverage by the oxygenated species, are most appropriately balanced in the Pt_{ML}/Pd(111).

17.3.2 CORE/SHELL STRUCTURED NP CATALYSTS

Based on the benefits in the core/shell structured catalysts explained earlier, Pd core/Pt shell structured NP catalyst was synthesized by Adzic and coworkers.[33,35–38] They used Cu underpotential deposition (Cu-UPD) technique to form Cu monolayer shell on the surface of the Pd NP core loaded on a carbon support (Pd/C). After the Cu monolayer shell formation, a Pt^{2+} precursor (K_2PtCl_4) was added

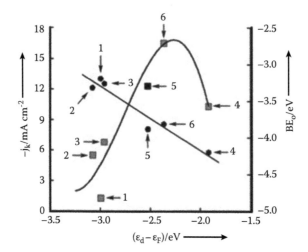

FIGURE 17.20 Kinetic currents (j_k; square symbols) at 0.8 V for O_2 reduction on the platinum monolayers supported on different single-crystal surfaces in a 0.1 M $HClO_4$ solution and calculated binding energies of atomic oxygen (BE_O; filled circles) as functions of calculated d band center ($\varepsilon_d-\varepsilon_F$; relative to the Fermi level) of the respective clean platinum monolayers. Labels: 1 $Pt_{ML}/Ru(0001)$, 2 $Pt_{ML}/Ir(111)$, 3 $Pt_{ML}/Rh(111)$, 4 $Pt_{ML}/Au(111)$, 5 Pt(111), and 6 $Pt_{ML}/Pd(111)$. (Zhang, J., Vukmirovic, M.B., Xu, Y., Mavrikakis, M., and Adzic, R.R.: *Angew. Chem. Int. Ed.* 2005. 44. 2132–2135. Copyright Wiley-VCH Verlag GmbH & Co. KGaA. Adapted with permission.)

to replace the Cu monolayer shell with the Pt one, yielding carbon-supported Pd core/Pt shell structured NP catalyst (Pt/Pd/C). The core/shell structure was verified with the high-angle annular dark-field scanning transmission electron microscopy (HAADF-STEM) chemical mapping analysis as shown in Figure 17.21.[36] The ORR mass activity of the Pt/Pd/C catalyst is 570 A/g-Pt at 0.9 V vs. RHE, which is approximately three times higher than that of a commercial Pt/C catalyst 4 nm in size (200 A/g-Pt) and is also greater than the target of the U.S. Department of Energy in 2015 (440 A/g-Pt).[42]

Due to a lower redox potential of the Pd relative to the Pt (Pd: 0.92 V vs. standard hydrogen electrode [SHE], Pt: 1.19 V vs. SHE), oxidative dissolution of the Pd core is concerned in the acidic circumstance of the PEFC cathode. They fabricated a single cell using the Pt/Pd/C cathode catalyst and found the Pd core dissolution, forming a Pd band in the membrane after 100,000 potential cycles from 0.7 to 0.9 V vs. RHE.[37] In order to suppress the Pd core dissolution, they added a small amount of Au (10 atomic%) into the Pd NP core. Figure 17.22 shows change of the ORR mass activity in a single cell using the $Pt/Pd_{90}Au_{10}/C$ cathode catalyst as a function of the number of potential cycles from 0.7 to 0.9 V vs. RHE. For the $Pt/Pd_{90}Au_{10}/C$ cathode catalyst, the ORR mass activity decreased about 30% after 200,000 cycles, whereas the ORR mass activity of a commercialized Pt/C cathode catalyst showed a terminal loss below 50,000 cycles, implying an improved durability of the $Pt_{ML}/Pd_{90}Au_{10}/C$ cathode catalyst.[37]

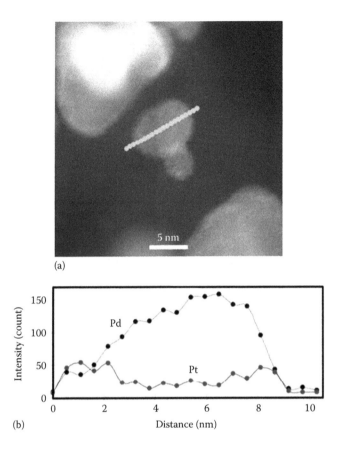

(a)

(b)

FIGURE 17.21 Line profile analysis of a Pt_{ML}/Pd/C NP (a) by scanning EDS (b). (Adapted from *Electrochim. Acta*, 55, Sasaki, K., Wang, J.X., Naohara, H., Marinkovic, N., More, K., Inada, H., and Adzic, R.R., 2645–2652, Copyright 2010, with permission from Elsevier.)

17.3.3 CORE/SHELL STRUCTURED NP CATALYSTS EXCEPT FOR PD CORE

In the previous section, it was demonstrated that the Pd is the most appropriate choice of the core metal for the Pt_{ML} from the viewpoint of the strain effect. In the single-crystal experiments, it was shown that the d band center shift of the Pt_{ML}/Au(111) is too high to enhance the ORR activity (too high tensile strain results in strong binding with oxygenated species). Recently, however, it has been reported that the Au NP core/Pt shell structured catalysts exhibit two to three times higher ORR specific activities with respect to a commercialized Pt/C catalyst.[43–45] Shao et al. claimed that the Au–Au interatomic distance decreases as the Au particle size decreases. Consequently, the surface Au atoms show compressive strain with decrease in the particle size. They calculated strain in the Pt_{ML} formed on the Au NP 3 nm in size and the oxygen binding energy. They found that the energy on the Pt_{ML}/Au$_{3\ nm}$ is equivalent to that on the Pt_{ML}/Pd(1111), which could be an origin of the enhanced ORR specific activity for the Pt_{ML}/Au$_{3\ nm}$ catalyst.[44]

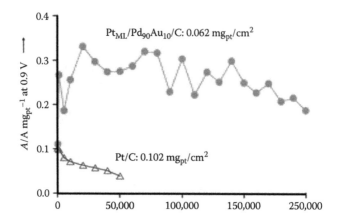

FIGURE 17.22 Pt mass activity for the ORR as a function of number of potential cycles during fuel cell testing of the $Pt_{ML}/Pd_{90}Au_{10}/C$ and Pt/C catalysts containing 0.062 and 0.102 mg-Pt/cm^2, respectively. Limits of the potential cycle were 0.6 and 1.0 V and sweep rate was 50 mV/s. (Sasaki, K., Naohara, H., Cai, Y., Choi, Y.M., Liu, P., Vukmirovic, M.B., Wang, J.X., and Adzic, R.R.: *Angew. Chem. Int. Ed.* 2010. 49. 8602–8607. Copyright Wiley-VCH Verlag GmbH & Co. KGaA. Adapted with permission.)

Inaba and Daimon synthesized Au NP core 3 nm in size with a thiol capping agent and conducted accelerated durability test (ADT) of the carbon-supported $Au_{3\,nm}$ core/Pt shell structured catalyst ($Pt/Au_{3\,nm}/C$) using a potential cycling from 0.6 to 1.0 V vs. RHE in Ar-saturated 0.1 M $HClO_4$ at 333 K.[45] The $Pt/Au_{3\,nm}/C$ catalyst showed an inferior durability to that of a commercialized Pt/C one. X-ray fluorescence compositional analysis and TEM observation revealed that there are little changes in the Au and Pt compositions and in the morphology of the catalyst after the durability test. Therefore, they concluded that the inferior durability was caused from dissolution of the Pt shell into the Au NP core. Recent calculations predict that mixing of the Au and the Pt atoms becomes thermodynamically stable in the case that the Au–Pt particle size is smaller than 6 nm, even though there is a large miscibility gap in the Au–Pt bimetallic bulk system,[46,47] which supports their conclusion.

17.3.4 Summary of Core/Shell Structured Catalysts

The important benefits in the core/shell structured catalysts are the high utilization efficiency of the expensive Pt and, simultaneously, the ability-enhancing ORR activity of the Pt_{ML} shell with a moderate strain effect induced from the core metals. However, usage of the core metals with low redox potentials results in the oxidative dissolution of the cores in the acidic circumstance of the PEFC cathode. As is described in the Au NP core/Pt shell structured catalysts, their ORR activity trends are quite different from those found in the catalysts using single-crystal underlayers. Since it is generally observed that lattice parameters of the metals decrease with decrease of their particle sizes, particularly in a single nanometer region, careful analyses and understandings are required to predict the ORR activity and the durability of the nanosized core/shell structured catalysts.

17.4 SHAPE-CONTROLLED CATALYSTS

17.4.1 BASIC STUDY ON PT SINGLE-CRYSTAL CATALYSTS

The ORR is initiated via adsorption of oxygen molecule on the surface Pt atoms of the catalysts and the oxygen is successively reduced toward water as depicted in Figure 17.23. Since the adsorption starts from hybridization of O_{2p} and Pt_{5d} orbitals, the hybridization energy, that is, adsorption energy, is strongly influenced with surface structure of the Pt. Markovic et al. have extensively studied the ORR of Pt single crystal using rotating hanging meniscus electrode (RHME) and rotating ring disk electrode (RRDE) techniques. They demonstrated that the ORR activity evaluated in 0.1 M $HClO_4$ increases in the sequence $(100) < (111) < (110),$[48] and the activity measured in 50 mM H_2SO_4 increases in the order $(111) < (100) < (110).$[49] Their RHME data showed that the ORR mainly involves four-electron reduction to water as the main product. Kadiri et al., employing the RHME technique, found that the ORR is structure sensitive in solutions containing strongly adsorbing anions such as bisulfate, phosphate, and Cl⁻, indicating that the structure sensitivity arises from structure sensitive adsorption of the anions that impede the reaction and that the ORR activity of the Pt single crystals in the solutions containing strongly adsorbing anions increases in the order $Pt(111) < Pt(100) < Pt(110).$[50,51]

In H_2SO_4 electrolyte, the simplest and the most logical explanation for the higher overpotential for the ORR on the Pt(111) surface shown in Figure 17.24[49] is that its

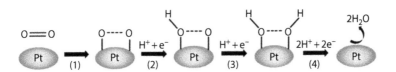

FIGURE 17.23 Elemental reactions in ORR on Pt catalyst.

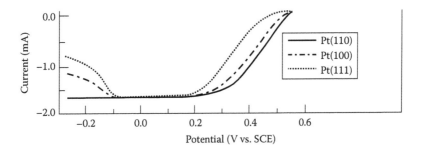

FIGURE 17.24 Polarization curves for ORR on Pt(hkl) measured in 0.05 M H_2SO_4, 50 mV/s, 1600 rpm. (Adapted with permission from Markovic, N.M. et al., *J. Phys. Chem.*, 99, 3411. Copyright 1995 American Chemical Society.)

activity is reduced by the relatively strong adsorption of tetrahedrally bonded bisulfate anions on the (111) sites observed in Fourier transform infrared spectroscopy (FTIR)[52] and radiotracer experiments,[53] both of which report that the onset of bisulfate adsorption at ca. 0.15 V vs. saturated calomel electrode (SCE), which is the same potential as the onset of kinetic currents, can be observed on Pt(111). The absolute magnitude of the bisulfate anion coverage derived from the radiotracer experiments on Pt(111) in 1 mM H_2SO_4 is ca. 0.2 monolayer, which is two to three times higher than on Pt(100). Furthermore, from the FTIR experiments, it has emerged that the bisulfate adsorption on Pt(111) is significantly stronger than on Pt(100)[54] and that the bonding of the bisulfate anion to Pt(111) surface is through all three equivalent oxygen atoms, in contrast to the bisulfate adsorption on Pt(100) and Pt(110) where only one or two oxygen atoms are bonded to the surface, resulting in a reduced adsorption strength. These differences of the bisulfate adsorption in terms of coverage and bonding between Pt(111) and the other two low-index faces are consistent with the significantly lower ORR activity of Pt(111) compared to either Pt(100) or Pt(110) (Figure 17.24). The most active surface for the ORR is Pt(110) plane, as was also found for O_2 reduction in 0.1 M $HClO_4$.[48] The difference in the ORR activity between Pt(110) and Pt(100) in H_2SO_4 electrolyte may derive either from subtle differences in bisulfate adsorption or from other structure-sensitive surface processes, such as the adsorption of oxygen-containing species. The fact that the ORR activity of all three low-index Pt planes is significantly higher in the $HClO_4$ electrolyte, however, suggests that the major differences in the H_2SO_4 electrolyte stem from the structure-sensitive adsorption of the bisulfate anions. It should be pointed out that although the bisulfate adsorption onto the Pt(hkl) surfaces inhibits the reduction of molecular O_2, probably by blocking the initial adsorption of O_2, it does not affect the pathway of the reaction since no H_2O_2 is detected on the ring electrode for any of the surfaces in the kinetically controlled potential region.[49]

17.4.2   PT SIZE EFFECT ON ORR

In the previous section, it was described that the ORR activity of the Pt single-crystal model catalysts in acidic media is strongly affected with structure-sensitive adsorption of the anions in the electrolytes, especially in H_2SO_4 (bisulfate). However, in the real catalysts, electrochemical surface area (ECSA) is an important factor for promoting the ORR. Therefore, nanosized catalysts have been extensively explored and prepared via rational synthetic schemes.[55–57]

However, Gasteiger et al. have shown that ORR specific activity of the Pt catalysts in 0.1 M $HClO_4$ electrolyte declines with increase of the ECSA, that is, with decrease of the Pt particle size as depicted in Figure 17.25a.[22] They demonstrated that more than one order of magnitude variation in the ORR specific activity is obtained in the Pt catalysts having different ECSA, which confirmed that the Pt particle size effect on the ORR in the $HClO_4$ electrolyte is quite apparent. Therefore, ORR mass activity of the Pt catalysts goes through a maximum between ECSA of 70 m²/g-Pt and ECSA of 120 m²/g-Pt as shown in Figure 17.25b.[22]

If the hypothesis of an increasing adsorption strength of OH_{ads} species together with their negative impact on the ORR with decrease of the Pt particle size were

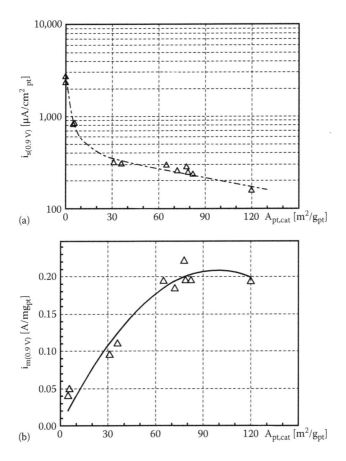

FIGURE 17.25 (a) ORR specific activity and (b) ORR mass activity of polycrystalline Pt (shown at 0 m²/g-Pt), Pt black (shown at 5 m²/g-Pt) and Pt/C catalysts at 0.9 V and 60°C determined by RDE measurements in O_2 saturated 0.1 M $HClO_4$. (Adapted from *Appl. Catal. B: Environ.*, 56, Gasteiger, H.A., Kocha, S.S., Sompalli, B., and Wagner, F.T., 9–35, Copyright 2005, with permission from Elsevier.)

true, the one order of magnitude difference in the ORR specific activity between the polycrystalline Pt catalyst (ECSA: ~0 m²/g-Pt) and the Pt/C catalyst (ECSA: ~80 m²/g-Pt) would show a shift of oxide adsorption and desorption potentials in their CVs. Figure 17.26 demonstrates the CVs of the Pt catalysts with different size, indicating that the oxide adsorption and desorption waves of the Pt/C catalyst occur ca. 80 mV negatively compared to those of the polycrystalline Pt catalyst, while the Pt black catalyst (ECSA: 5 m²/g-Pt) falls in between.[22] Although this may not be a definitive proof for the role of site blocking OH_{ads} species in the ORR,[58–60] it certainly substantiates the hypothesis.

Since Pt atoms positioned at corners and edges have lower coordination numbers compared with those that exist in bulk state, it is believed that the low-coordinated Pt atoms are unstable and stabilized by chemisorption with oxygen-containing species.

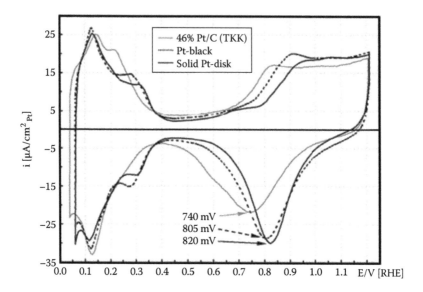

FIGURE 17.26 CVs of carbon-supported Pt, Pt black and polycrystalline Pt in terms of Pt surface area normalized current densities (mA/cm²). Data were recorded in 0.1 M $HClO_4$ at 25°C and 20 mV/s. (Adapted from *Appl. Catal. B: Environ.*, 56, Gasteiger, H.A., Kocha, S.S., Sompalli, B., and Wagner, F.T., 9–35, Copyright 2005, with permission from Elsevier.)

It is easy to imagine that the number of the lower-coordinated Pt atoms increases with the decrease of the particle size. Shao et al. investigated the particle size effect on the ORR of Pt catalysts in the range of 1–5 nm in 0.1 M $HClO_4$ electrolyte.[23] The particle size was carefully controlled via layer-by-layer growth of Pt shell using the Cu-UPD/Pt replacement technique for a Pt/C seed catalyst 1.3 nm in size.[33] Figure 17.27a displays ORR polarization curves of the Pt/C catalysts with one, two, three, and four layers of Pt shell measured in O_2-saturated 0.1 M $HClO_4$ solution using RRDE technique. It can be clearly seen that the half-wave potentials of the Pt/C catalysts with one, two, three, and four layers of Pt shell positively shift compared with the seed catalyst, indicating enhanced ORR activity with increase in the particle size of the Pt catalyst. Figure 17.27b shows ORR mass and specific activities measured at 0.93 V vs. RHE as a function of particle size. The ORR mass activity increases with increase in the particle size and gradually decreases with the maximum activity observed at ca. 2.2 nm by an enhancement of twofold compared with the 1.3 nm seed Pt/C catalyst. The ORR specific activity sharply increases from 1.3 to 2.2 nm by approximately four times. This trend is very similar to the change of the Pt oxide reduction peak potential as a function of the particle size (Figure 17.27c).

It is well known that the surface reactivity of the catalysts correlates well with its catalytic activity.[61–64] Shao et al. investigated the reasons for this structural sensitivity of the ORR focusing on the particle size in the range from 0.8 to 3.0 nm using DFT calculations for cuboctahedrally structured Pt model particles.[23] The dispersion, that is, the ratio between the number of surface atoms and overall number of atoms in the cuboctahedral particles, is plotted in Figure 17.28a. Decreasing the

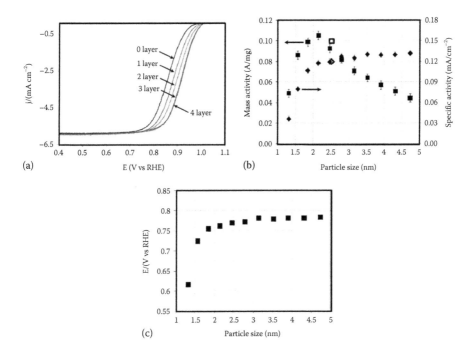

FIGURE 17.27 (a) Polarization curves of Pt/C with different numbers of extra Pt layers in O_2-saturated 0.1 M $HClO_4$. (b) Size dependence of specific activity (black diamond) and mass activity (black square) of Pt /C for ORR at 0.93 V. Specific activity (open black diamond) and mass activity (open black square) of state-of-the-art Pt/C catalyst (TKK, TEC10E50E, 46.7 wt.%) with average particle size of 2.5 nm are also included for comparison. (c) Peak potentials of Pt oxides reduction for Pt/C with different numbers of extra Pt layers obtained from the CVs in N_2-saturated 0.1 M $HClO_4$. (Adapted with permission from Shao, M. et al., *Nano Lett.*, 11, 3714. Copyright 2011 American Chemical Society.)

particle sizes results in improved dispersion and an increase in the number of sites on the surface of the catalyst. The plot also shows the distribution of these surface sites per {111} and {100} facets and the edges of particles. The percentage of predominant (111) sites and also (100) sites decreases with the decrease in the particle size. A sharp drop in the number of these sites is seen below 3 nm, while the opposite trend is seen for the edge sites.

The surface reactivity of these different sites in the particles was examined by calculating oxygen-binding energies on the cuboctahedral Pt particles. Figure 17.28b shows a geometry of the Pt particle 2.6 nm in size. The {100} and {111} facets are labeled with the coordination numbers for the atoms comprising these facets and edges between them. The coordination numbers are 9, 8, 7, and 6 for (111), (100) edge and vertex atoms, respectively. Figure 17.28c shows the relationship between the oxygen binding energy for possible adsorption sites on the particles with sizes between 0.8 and 3 nm. It is evident that, at an identical site, the binding energy decreases with the increase of the particle size up to 2.1 nm. It is also demonstrated that, at the same particle size, the adsorption sites with lower coordination

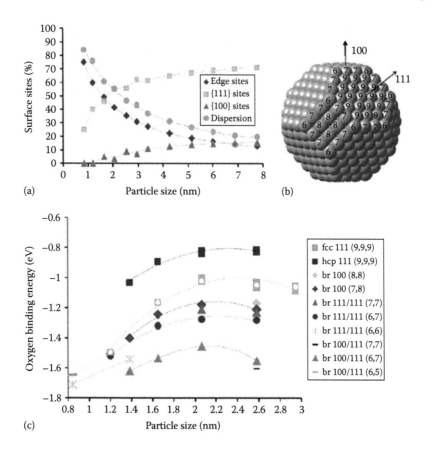

FIGURE 17.28 (a) Size dependence of dispersion and surface percentage of atoms on {111} and {100} facets and on the edges between the facets. (b) Geometry of 2.6 nm Pt NP showing the {111} and {100} facets and the coordination numbers of atoms comprising these facets and edges. (c) Calculated oxygen binding energy as a function of particle size for various adsorption sites. Sites include hollow (face-centered cubic [fcc] and hexagonal close packed) on {111} facets, bridge site on {100} facets, and bridge sites on edges between different facets. Besides site notation, coordination numbers for atoms making sites are given in brackets. (Adapted with permission from Shao, M. et al., *Nano Lett.*, 11, 3714. Copyright 2011 American Chemical Society.)

number of the Pt atoms result in increased binding energy. More importantly, a higher increase in the binding energy can be seen when the coordination number at the edge sites is lowered.

These changes toward stronger binding in Pt–O with decrease of the coordination number can be understood through changes in electronic structures in the surface Pt atoms, in particular, energy of d electrons of the surface Pt atoms.[39–41,61–64] The lower coordination number in the Pt atoms narrows the *d* band and lowers *d* band filling,[63] which consequently causes a stronger Pt–O bond at these sites. The stronger Pt–O bond is reflected with lower potentials of Pt oxides reduction peaks for the smaller

Pt particles having lower coordination numbers (Figure 17.27c), which is coincident with the results of Gasteiger's group (Figure 17.25).

Because of the strong reactivity of the edge sites, these sites are blocked by adsorbed intermediates during the ORR. As the particle size decreases, the oxygen binding to the surface Pt atoms is expected to become weaker due to larger compressive strains.[44] It is interesting to note in Figure 17.28c that all the sites including the {111} facets for the Pt particles below 2.1 nm have a very strong interaction with oxygen. The stronger oxygen binding is due to the increase in percentage of the low-coordinated Pt atoms (edges and corners). Therefore, these two opposing effects lead to a maximum ORR mass activity around 2.2 nm and a sharp drop of the ORR specific activity observed in Figure 17.27b.

As explained earlier, the Pt size effect on the ORR has been recognized experimentally and theoretically. For smaller-size Pt particles, increase of percentage in the edge and corner sites is likely the main reason for the lower ORR specific activity owing to strong oxygen binding energies on these sites. Even though the oxygen bonding energy of the {111} and {100} facet sites increases with the size reduction, resulting in lower ORR specific activity, it can be emphasized that these facet sites in the Pt particles play an important role for improving the ORR activity. These research results could be a guideline for understanding the general roles of the different surface sites in the Pt NPs toward ORR and designing the shape-controlled catalysts.

17.4.3 SHAPE-CONTROLLED NP CATALYSTS

The shapes of Pt and its alloy NPs can be controlled by both thermodynamic and kinetic factors, which are influenced by both the intrinsic structural properties of the Pt and the reaction systems such as solvents, capping agents, and reducing agents. Metallic NPs form facets to minimize surface energy and total excess free energy. The Pt NPs, which have the face-centered cubic (fcc) symmetry, are usually surrounded by three low-index planes, that is, {100}, {110}, and {111} surfaces. Among these three, the {111} planes have the lowest surface energy while the {110} planes have the highest.[56,65] Other reaction conditions including concentration, time, and temperature are also critical. Pt precursors can be chosen from hexachloroplatinic acid (H_2PtCl_6), potassium hexachloroplatinate (K_2PtCl_6), potassium tetrachloroplatinate (K_2PtCl_4), and platinum acetylacetonate ($Pt(acac)_2$), depending on the choice of solvents (either water or organic liquids), reductants, surfactants, and other additives.[66] Reducing agents including borohydride, hydrazine, hydrogen, citrate, and ascorbic acid can be used in the aqueous systems, whereas polyols, diols, and amines are commonly used for the organic ones. Besides these, many other chemicals including both organic and inorganic molecules as well as ions can be used as the capping agents to passivate or activate specific surfaces and influence growth habits of the Pt and its alloy NPs. In principle, the Pt NPs with predetermined shapes having certain low-indexed planes can be synthesized through judicious selections of these reaction parameters.

While the *simple* shapes including tetrahedron, cube, octahedron, and their truncated forms are the morphologies that can be predicted for a face-centered cubic crystal, introduction of defects can effectively break down the crystal symmetry,

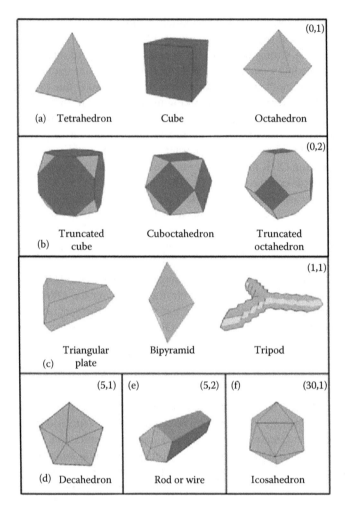

FIGURE 17.29 Selective possible shapes of Pt NPs (a and b) without defects and bounded by (a) one group and (b) two groups of facets; and (c–f) with different numbers of defects. The notation (m, n) represents the number of defects m and different facets n in the crystals. (Adapted from *Nano Today*, 4, Peng, Z. and Yang, H., 143–164, Copyright 2009, with permission from Elsevier.)

resulting in formation of shapes with reduced symmetry such as nanorods, plates, planar tripods, and multipods.[67] Figure 17.29 shows illustration of a few possible shapes of the Pt NPs with {111} and {100} planes exposed. The notation (m, n) represents the number of defects m and different facets n in the Pt NPs.[56] For those shapes without defects, the promotion or inhibition of selective growing surfaces should be a key for those shapes obtained. Design and control of the type and number of defects are important for controlling the shapes with reduced symmetry.

In the practical synthesis, organic capping agents and inorganic ions play pivotal roles for controlling the shape of the Pt NPs. Long carbon chains of organic capping agents are hydrophobic and have a steric hindrance effect to prevent direct contacts among relatively high energy surfaces of the Pt, thus stabilize the Pt NPs. The decrease in total excess free energy due to the adsorption of capping agents effectively prevents the Pt NPs from further growth and Ostwald ripening. When capping agents adsorb selectively onto a given Pt surface, morphology of the Pt NPs can therefore be controlled.[65,68,69] Since Pt NPs have different electronic structures and atomic arrangements for various facets, it is expected that a given capping agent should differently absorb onto these surfaces. The preferred adsorption onto one set of surfaces over the others should result in different growth rates along various given crystallographic directions. The solute atoms would more likely attach to those less protected Pt surfaces, leading to an anisotropic growth. One key criterion in selecting a proper capping agent for the shape control is a proper interaction between the guest molecules and various Pt facets, which should be balanced but selective in the adsorption and desorption processes.

El-Sayed and coworkers first successfully prepared Pt NPs with cube-like and tetrahedron-like shapes by using acrylic acid and polyacrylate as capping agents for a water-based synthetic system, although both size and shape were not uniform in their early work.[70] Even today, synthesis for uniform and high-yield tetrahedral Pt NPs is still challenging.[71] For the organic synthetic systems, OA and OAm have been routinely used to control the shape of the Pt NPs.[72–74]

Inorganic ions and other small molecules, which have been overlooked for quite sometimes on their functions in the shape control, play important roles in the design of the shape-controlled NPs of various metals including Pt. These inorganic species show preferred adsorption to the specific facets of the Pt similar to the role of the organic molecules, which promotes or inhibits further growth along given directions. For example, Ag species have been found to influence the shape of the Pt NPs in a big way. The Ag is thought to adsorb preferentially on (100) over (111) surfaces and alter the growth rates along these directions. As can be seen in Figure 17.30, cubic-, cuboctahedra-, and octahedra-shaped Pt NPs have been prepared by adding different amounts of Ag ions in the reaction mixtures.[75] Cu ions have also been found to strongly interact with the Pt surfaces and used to synthesize Pt nanocubes (Figure 17.31).[76]

The high-index facets of Pt have been shown to have significantly higher ORR activities than their low-index counterparts in acidic solutions,[77,78] which can be attributed to the favorable adsorption of O_2 molecules onto the stepped surfaces.[79] Xia and coworkers have demonstrated the synthesis of Pt concave nanocubes enclosed by high-index facets including {510}, {720}, and {830} by using a simple route based on reduction in an aqueous solution.[78] The key to their shape-controlled synthesis is the use of a Pt pyrophosphato complex as the precursor and control of the reaction rate with a syringe pump, facilitating selective overgrowth of seeds from corners and edges with Br⁻ anion serving as a capping agent to block the {100} facets. The ORR activities of the Pt concave cubes, cuboctahedra, and cubes were measured with RDE technique in O_2-saturated 0.1 M $HClO_4$ at room temperature. As illustrated in Figure 17.32, the ORR-specific activity of the Pt concave cubes is two to three times higher than that of the Pt cuboctahedra and cubes, which demonstrates that the Pt concave cubes exhibited greatly enhanced ORR activity because of its high-index facets.[78]

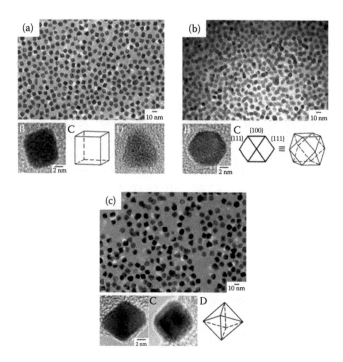

FIGURE 17.30 (a) TEM image of the Pt cubes, (b) Pt cuboctahedra, and (c) Pt octahedra. (Adapted with permission from Song, H. et al., *J. Phys. Chem. B*, 109, 188. Copyright 2005 American Chemical Society.)

It is worth pointing out that for the low-index facets of Pt, the ORR activity increases in the order of (100) and (111) when a nonabsorbing electrolyte such as $HClO_4$ is used.[48] Since Pt cuboctahedra have a mixture of {100} and {111} facets, they are supposed to have a higher ORR activity than Pt cubes enclosed only by {100} facet. In comparison with a commercialized Pt/C catalyst (Pt: 3.2 nm), the Pt concave cubes exhibited 3.6 times higher ORR specific activity than that of the Pt/C catalyst. However, ORR mass activity, that is, kinetic current per unit Pt mass, of the Pt concave cubes was lower than that of the Pt/C catalyst, presumably owing to smaller ECSA associated with their relatively larger particle sizes ca. 20 nm.[78]

17.4.4 Summary of Shape-Controlled Catalysts

In this section, it was demonstrated that the shape of the Pt NPs can be controlled by both thermodynamic and kinetic factors. Pt cuboctahedra have {100} and {111} facets and they showed a higher ORR specific activity than those of Pt cubes enclosed only by {100} facet and a commercialized Pt/C catalyst. Interestingly, Pt concave nanocubes with high-index facets of {510}, {720}, and {830} exhibited a higher ORR specific activity than that of the Pt cuboctahedra. However, it should be noted that ECSAs of the shape-controlled Pt catalysts are smaller than that of the Pt/C catalyst due to their larger particle sizes, resulting in their inferior ORR mass activities to that of the Pt/C

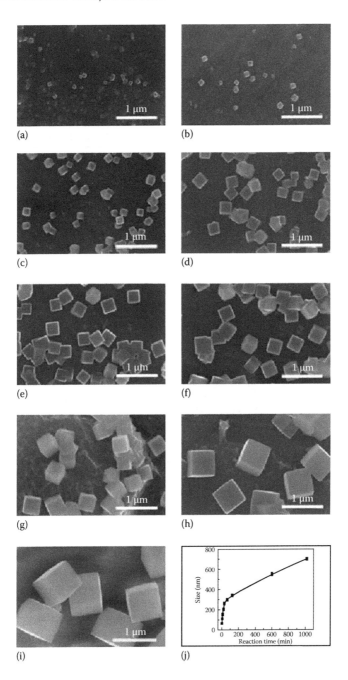

FIGURE 17.31 SEM images of Pt nanocubes formed on Cu foils in an aqueous solution of K_2PtCl_4 at different reaction times: (a) 30 s, (b) 2 min, (c) 5 min, (d) 15 min, (e) 30 min, (f) 60 min, (g) 2 h, (h) 10 h, (i) 17 h. (j) A typical plot of the cube size vs. the reaction time. (Adapted with permission from Qu, L. et al., *J. Am. Chem. Soc.*, 128, 5523. Copyright 2006 American Chemical Society.)

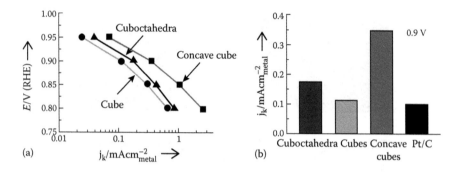

(a)

(b)

FIGURE 17.32 Comparison of ORR specific activity of Pt concave cubes (■), cubes (●), and cuboctahedra (▲) measured at 0.9 V vs. RHE. (Yu, T., Kim, D.Y., Zhang H., and Xia, Y.: *Angew. Chem. Int. Ed.* 2011. 50. 2773–2777. Copyright Wiley-VCH Verlag GmbH & Co. KGaA. Adapted with permission.)

(a)   (b)

FIGURE 17.33 Change in morphology of cubic Pt NPs with ADT. (a) and (b) are SEM images of the cubic Pt NPs before and after ADT, respectively. The ADT was conducted with potential cycling 0.2–1.1 V vs. RHE in N_2-saturated 1.5 M H_2SO_4 at 308 K.

catalyst. As described previously, ORR specific activity of the Pt catalysts declines with the decrease of the Pt particle size owing to the increase of the low-coordinated Pt atoms on the surface, that is, edges and corners. Therefore, careful tuning should be taken when the shape-controlled catalysts with smaller size are synthesized.

Additionally, durability of the shape-controlled Pt catalysts is concerned. Figure 17.33 shows change in morphology of cubic Pt NPs 7 nm in size with a potential cycling durability test (0.2–1.1 V vs. RHE in N_2 saturated 1.5 M H_2SO_4 at 308 K, 1000 cycles). The cubic shape completely disappeared after the durability test, indicating insufficient durability of the shape-controlled Pt catalyst. Therefore, for the shape-controlled Pt catalysts, much effort should be taken to improve their durability.

17.5 NONPRECIOUS METAL CATALYSTS

According to the cost analysis, the fuel cell stack is responsible for more than 50% of the total cost of the PEFC.[2] Although a state-of-the-art PEFC stack uses several highly priced components, the Pt catalyst is the most expensive constituent, which accounts for more than half of the stack cost. Therefore, PEFC is in need of efficient, durable, and inexpensive alternatives to the Pt catalyst. Ideally, the Pt catalysts should be replaced at both electrodes in the PEFC, but substitution at the cathode with nonprecious metal catalysts would have greater impact because the slow kinetics in the ORR at the cathode requires much more Pt than the faster HOR at the anode. As a consequence, development of the nonprecious metal catalysts with high ORR activity has become a major focus on the cost reduction of the PEFC.

In the current researches on the Pt replacement, there are two major candidates for the nonprecious metal catalysts. One is a composite catalyst composed of nitrogen (N), carbon (C), and transition metals M such as Co and Fe (N–C–M catalysts), which is synthesized with pyrolysis of the corresponding precursors.[80–92] The other candidate is partially oxidized metal (Zr, Ta, Ti, Hf, and Nb) carbonitride catalysts.[93–101] In this final section, these two nonprecious metal catalysts are briefly reviewed.

17.5.1 N–C–M Nonprecious Metal Catalysts

Since nitrogen-derived nonprecious metal catalysts (N–C–M catalysts) started with the early works by Jasinski[80] and Yeager,[81,82] numerous attempts have been made to develop the N–C–M catalysts by pyrolyzing the precursors containing transition metals (Fe or Co), a nitrogen source, and a carbon source. Although active sites for the ORR in the N–C–M catalysts have been an ongoing debate, there is no doubt that the activity strongly depends on the type of nitrogen, transition metal precursors, heat treatment temperature, morphology of carbon support, and synthetic conditions.

As to the active sites for the ORR in the N–C–M catalysts, one model is based on the sites where transition metal (Fe or Co) and nitrogen are coordinated. Dodelet and coworkers synthesized the N–C–M catalysts with pyrolyzing iron precursors (Fe acetate or Fe porphyrin) adsorbed on a synthetic carbon made from the pyrolysis of perylene tetracarboxylic dianhydride in a $H_2/NH_3/Ar$ atmosphere at a temperature range 400°C–1000°C. They claimed that Fe-N_x coordinated species play a dominant role for the ORR active sites. They demonstrated that current density of a cathode made with their N–C–M catalyst was equal to that of the Pt cathode with loading of 0.4 mg/cm^2 at cell voltage $\geq$0.9 V.[83,86]

However, it has been pointed out that transition metal nitrogen–coordinated species are no longer stable in the catalysts pyrolyzed at high temperature ($\geq$800°C).[82,85] Bashyam and Zelenay developed polypyrrole–C–Co nonprecious metal catalyst synthesized without pyrolysis and the catalyst showed high ORR activity and respectable durability for a nonprecious metal catalyst.[84] However, its ORR activity remained relatively low; thus, they shifted toward high temperature synthesis using Fe, Co, and heteroatom polymer precursors (polypyrrole and polyaniline). They demonstrated that the polyaniline-derived N–C–M catalyst exhibits high ORR activity with unique durability for a heat-treated nonprecious metal catalyst (Figures 17.34 and 17.35[88,91]).

For the N–C–M nonprecious metal catalysts, it is presumed that the nitrogen content is crucial for high ORR activity and that the transition metals are essential for the growth of the active carbon network–containing nitrogen. Therefore, the

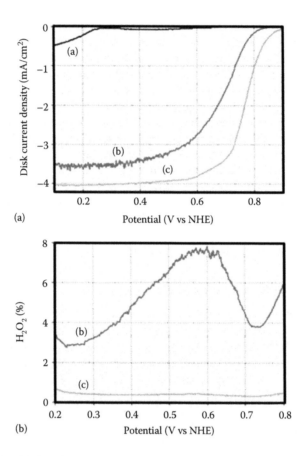

FIGURE 17.34 ORR activity and H_2O_2 yield obtained with PANI–FeCo–C catalyst at different catalyst synthesis stages (a) before pyrolysis; (b) after pyrolysis but before chemical treatment; (c) after chemical treatment. (Adapted from Wu, G. et al., *ECS Trans.*, 16, 159. Copyright 2008, Electrochemical Society. With permission.)

pyrolysis protocol could be quite important in order to synthesize highly active and durable N–C–M catalysts. It has been reported that Fe species are important up to 600°C for the growth of the carbon network, but Fe NPs eliminate the active sites over 700°C.[89] Currently, Nabae et al. have demonstrated a multistep pyrolysis of Fe phthalocyanine and phenolic resin to synthesize highly active and durable N–C–M catalysts.[92] The Fe NPs were removed with acid washing (conc. HCl) after the first pyrolysis at 600°C and the acid washing was repeated every time after higher heat treatments. They conducted durability test of a fuel cell using their nonprecious metal catalyst by introducing air into the cathode and keeping the current density at 200 mA/cm². Although the cell voltage gradually decreased, the performance loss was less than 20% after 500 h generation, indicating that the N–C–Fe catalyst synthesized via the multistep pyrolysis combined with the acid washing has extremely good durability as a nonprecious metal catalyst.[92]

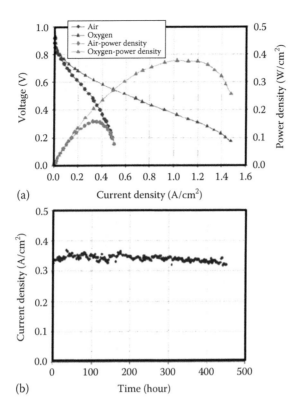

(a)

(b)

FIGURE 17.35 Fuel cell testing of a PANI–FeCo–C cathode catalyst at a loading of 4 mg/cm²; cathode backpressure of 30 psig and 80°C. (a) Polarization curves; (b) life test at 0.40 V using H₂/Air. (Adapted from Wu, G. et al., *ECS Trans.*, 16, 159. Copyright 2008, Electrochemical Society. With permission.)

17.5.2 PARTIALLY OXIDIZED METAL CARBONITRIDE CATALYSTS

Two operating conditions of the PEFC cathode, that is, a low pH of ~1 or less and high potentials of 0.6–0.9 V vs. SHE, would dissolve most metals and this limits the choice of alternative catalysts to Pt. Groups 4 and 5 metal oxide compounds have been regarded as promising candidates for the PEFC cathode catalysts because they are insoluble in acidic media and have a catalytic activity for ORR. The metal oxide compounds (Metal: Zr, Ta, Ti, Hf, and Nb) have been developed by Ota and coworkers[93–101] and partially oxidized metal carbonitrides exhibited the highest ORR activity among the catalysts they developed, that is, the onset potential of the ORR is more than 0.95 V vs. RHE (Figure 17.36).[98]

Ta or Zr oxide–based catalysts were synthesized from transition metal carbonitrides, such as Ta₂CN or Zr₂CN, and they were partially oxidized under very low oxygen pressure.[99,100] The ORR active sites of these catalysts are considered as oxygen vacancy defect sites.[101] Interestingly, such oxide-based catalysts show the ORR activity only when they were synthesized from carbonitrides. Naturally introduced oxygen vacancy

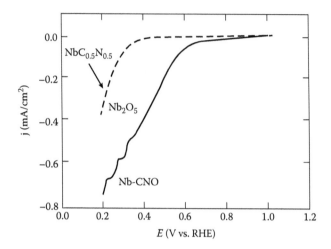

FIGURE 17.36 ORR polarization curves of Nb-based compound catalysts measured in 0.1 M H_2SO_4 at 30°C with a scan rate of 5 mV/s. (Adapted from *J. Power Sources*, 196, Ota, K., Ohgi, Y., Nam, K.D., Matsuzawa, K., Mitsushima, S., and Ishihara, A., 5256–5263, Copyright 2011, with permission from Elsevier.)

sites or oxygen vacancies formed through reduction atmosphere never exhibit the ORR activity.[100] It is considered that carbon or nitrogen involved in the transition metal carbonitrides and/or slow oxidation process could play a definitive role for an emergence of the ORR activity in these catalysts. It was indicated that carbon is deposited on the metal oxide surface during the partial oxidation process of the carbonitrides. The deposited carbon is considered to produce local reductive atmosphere near the oxide surface, playing a role to create the oxygen vacancy sites, and to promote the ORR with providing locally electric conductive paths on the insulative metal oxide surface.

Although the metal oxide–derived nonprecious metal catalysts have strong resistance to the acidic media and exhibited respectable ORR activity, the ORR current is still insufficient due to the low electric conductivity of the oxides. The key technology in enhancing the ORR current is the fine-tuning of the deposited carbon on the oxide surface, which could not only act as electric paths but also increase the active sites for the ORR.

REFERENCES

1. Joint Statement at http://www.meti.go.jp/press/20110113003/20110113003-2.pdf. Accessed on January 2011.
2. Y.-J. Wang, D. P. Wilkinson and J. Zhang, *Chem. Rev.*, 2011, **111**, 7625–7651.
3. T. Toda, H. Igarashi and M. Watanabe, *J. Electrochem. Soc.*, 1998, **145**, 4185–4188.
4. T. Toda, H. Igarashi and M. Watanabe, *J. Electroanal. Chem.*, 1999, **460**, 258–262.
5. T. Toda, H. Igarashi, H. Uchida and M. Watanabe, *J. Electrochem. Soc.*, 1999, **146**, 3750–3756.
6. V. Stamenkovic, T. J. Schmidt, P. N. Ross and N. M. Markovic, *J. Phys. Chem. B*, 2002, **106**, 11970–11979.

7. V. R. Stamenkovic, B. S. Mun, K. J. J. Mayrhofer, P. N. Ross and N. M. Markovic, *J. Am. Chem. Soc.*, 2006, **128**, 8813–8819.
8. V. Stamenkovic, B. S. Mun, K. J. J. Mayrhofer, P. N. Ross, N. M. Markovic, J. Rossmeisl, J. Greeley and J. K. Nørskov, *Angew. Chem. Int. Ed.*, 2006, **45**, 2897–2901.
9. V. R. Stamenkovic, B. S. Mun, M. Arenz, K. J. J. Mayrhofer, C. A. Lucas, G. Wang, P. N. Ross and N. M. Markovic, *Nat. Mater.*, 2007, **6**, 241–247.
10. V. R. Stamenkovic, B. Fowler, B. S. Mun, G. Wang, P. N. Ross, C. A. Lucas and N. M. Markovic, *Science*, 2007, **315**, 493–497.
11. B. Fowler, C. A. Lucas, A. Omer, G. Wang, V. R. Stamenkovic and N. M. Markovic, *Electrochim. Acta*, 2008, **53**, 6076–6080.
12. Y. Gauthier, R. B. Savois and J. M. Bugnard, *Surf. Sci.*, 1992, **276**, 1–11.
13. D. F. Vliet, C. Wang, D. Li, A. P. Paulikas, J. Greeley, R. B. Rankin, D. Strmcnik, D. Tripkovic, N. M. Markovic and V. R. Stamenkovic, *Angew. Chem. Int. Ed.*, 2012, **51**, 3139–3142.
14. S. A. Bondarenko, I. E. L. Stephens, L. Bech and I. Chorkendorff, *Electrochim. Acta*, 2012, **82**, 517–523.
15. S. Sun, C. B. Murray, D. Weller, L. Folks and A. Moser, *Science*, 2000, **287**, 1989–1992.
16. R. Loukrakpam, J. Luo, T. He, Y. Chen, Z. Xu, P. N. Njoki, B. N. Wanjala, B. Fang, D. Mott, J. Yin, J. Klar, B. Powell and C. J. Zhong, *J. Phys. Chem. C*, 2011, **115**, 1682–1694.
17. S. Rudi, X. Tuaev and P. Strasser, *Electrocatalysis*, 2012, **3**, 265–273.
18. Y. Wang and N. Toshima, *J. Phys. Chem. B*, 1997, **101**, 5301–5306.
19. Y. Hu, P. Wu, H. Zhang and C. Cai, *Electrochim. Acta*, 2012, **85**, 314–321.
20. R. Lin, C. Cao, T. Zhao, Z. Huang, B. Li, A. Wieckowski and J. Ma, *J. Power Sources*, 2013, **23**, 190–198.
21. C. Wang, D. Vliet, K. C. Chang, H. You, D. Strmcnik, J. A. Schlueter, N. M. Markovic and V. R. Stamenkovic, *J. Phys. Chem. C*, 2009, **113**, 19365–19368.
22. H. A. Gasteiger, S. S. Kocha, B. Sompalli and F. T. Wagner, *Appl. Catal. B: Environ.*, 2005, **56**, 9–35.
23. M. Shao, A. Peles and K. Shoemaker, *Nano Lett.*, 2011, **11**, 3714–3719.
24. J. J. Mayrhofer, B. B. Blizanac, M. Arenz, V. R. Stamenkovic, P. N. Ross and N. M. Markovic, *J. Phys. Chem. B*, 2005, **109**, 14433–14440.
25. C. Wang, G. Wang, D. Vliet, K. C. Chang, N. M. Markovic and V. R. Stamenkovic, *Phys. Chem. Chem. Phys.*, 2010, **12**, 6933–6939.
26. C. Wang, N. M. Markovic and V. R. Stamenkovic, *ACS Catal.*, 2012, **2**, 891–898.
27. S. Chen, P. J. Ferreira, W. C. Sheng, N. Yabuuchi, L. F. Allard and Y. S. Horn, *J. Am. Chem. Soc.*, 2008, **130**, 13818–13819.
28. C. Wang, M. Chi, G. Wang, D. Vliet, D. Li, K. More, H. H. Wang, J. A. Schlueter, N. M. Markovic and V. R. Stamenkovic, *Adv. Funct. Mater.*, 2011, **21**, 147–152.
29. S. Koh and P. Strasser, *J. Am. Chem. Soc.*, 2007, **129**, 12624–12625.
30. P. Mani, R. Srivastava and P. Strasser, *J. Phys. Chem. C*, 2008, **112**, 2770–2778.
31. P. Strasser, S. Koh, T. Anniev, J. Greeley, K. More, C. Yu, Z. Liu, S. Kaya, D. Nordlund, H. Ogasawara, M. F. Toney and A. Nilsson, *Nat. Chem.*, 2010, **2**, 454–460.
32. C. Wang, M. Chi, D. Li, D. Strmcnik, D. Vliet, G. Wang, V. Komanicky, K. C. Chang, A. P. Paulikas, D. Tripkovic, J. Pearson, K. L. More, N. M. Markovic and V. R. Stamenkovic, *J. Am. Chem. Soc.*, 2011, **133**, 14396–14403.
33. J. Zhang, Y. Mo, M. B. Vukmirovic, R. Klie, K. Sasaki and R. R. Adzic, *J. Phys. Chem. B*, 2004, **108**, 10955–10964.
34. J. Zhang, M. B. Vukmirovic, Y. Xu, M. Mavrikakis and R. R. Adzic, *Angew. Chem. Int. Ed.*, 2005, **44**, 2132–2135.
35. A. U. Nilekar, Y. Xu, J. Zhang, M. B. Vukmirovic, K. Sasaki, R. R. Adzic and M. Mavrikakis, *Top Catal.*, 2007, **46**, 276–284.

36. K. Sasaki, J. X. Wang, H. Naohara, N. Marinkovic, K. More, H. Inada and R. R. Adzic, *Electrochim. Acta*, 2010, **55**, 2645–2652.
37. K. Sasaki, H. Naohara, Y. Cai, Y. M. Choi, P. Liu, M. B. Vukmirovic, J. X. Wang and R. R. Adzic, *Angew. Chem. Int. Ed.*, 2010, **49**, 8602–8607.
38. K. Sasaki1, H. Naohara, Y. M. Choi1, Y. Cai1, W. F. Chen, P. Liu and R. R. Adzic, *Nat. Commun.*, 2012, **3**, 1–9.
39. A. Ruban, B. Hammer, P. Stoltze, H. L. Skriver and J. K. Nørskov, *J. Mol. Catal. A: Chem.*, 1997, **115**, 421–429.
40. M. Mavrikakis, B. Hammer and J. K. Nørskov, *Phys. Rev. Lett.*, 1998, **81**, 2819–2822.
41. B. Hammer and J. K. Nørskov, *Adv. Catal.*, 2000, **45**, 71–129.
42. DOE Target at http://www1.eere.energy.gov/hydrogenandfuelcells/mypp/pdfs/fuelcells. pdf. Accessed on June 2013.
43. M. Inaba, H. Ito, H. Tuji, T. Wada, M. Banno, H. Yamada, M. Saito and A. Tasaka, *ECS Trans.*, 2010, **33**, 231–238.
44. M. Shao, A. Peles, K. Shoemaker, M. Gummalla, P. N. Njoki, J. Luo and C. J. Zhong, *J. Phys. Chem. Lett.*, 2011, **2**, 67–72.
45. M. Inaba and H. Daimon, *ECS Trans.*, 2012, **50**, 65–73.
46. S. Xiao, W. Hua, W. Luo, Y. Wu, X. Li and H. Deng, *Eur. Phys. J. B*, 2006, **54**, 479–484.
47. B. N. Wanjala, J. Luo, B. Fang, D. Mott and C. J. Zhong, *J. Mater. Chem.*, 2011, **21**, 4012–4020.
48. N. M. Markovic, R. R. Adzic, B. D. Cahan and E. B. Yeager, *J. Electroanal. Chem.*, 1994, **337**, 249–253.
49. N. M. Markovic, H. A. Gasteiger and P. N. Ross, *J. Phys. Chem.*, 1995, **99**, 3411–3145.
50. F. El. Kadiri, R. Faure and R. Durand, *J. Electroanal. Chem.*, 1991, **301**, 177–188.
51. N. M. Markovic, T. J. Schmidt, V. Stamenkovic and P. N. Ross, *Fuel Cells*, 2001, **1**, 105–116.
52. P. W. Faguy, N. M. Markovic, R. R. Adzic, C. A. Fierro and E. B. Yeager, *J. Electroanal. Chem.*, 1990, **289**, 245–262.
53. A. Wieckowski, P. Zelenay and K. Varga, *J. Chim. Phys.*, 1991, **88**, 1247–1270.
54. P. W. Faguy, N. M. Markovic and P. N. Ross, *J. Electrochem. Soc.*, 1993, **140**, 1638–1643.
55. R. Ferrando, J. Jellinek and R. L. Johnston, *Chem. Rev.*, 2008, **108**, 845–910.
56. Z. Peng and H. Yang, *Nano Today*, 2009, **4**, 143–164.
57. Y. Bing, H. Liu, L. Zhang, D. Ghosh and J. Zhang, *Chem. Soc. Rev.*, 2010, **39**, 2184–2202.
58. N. M. Markovic, H. A. Gasteiger and P. N. Ross, *J. Electrochem. Soc.*, 1997, **144**, 1591–1597.
59. N. M. Markovic and P. N. Ross, *Surf. Sci. Rep.*, 2002, **45**, 117–229.
60. N. M. Markovic, H. A. Gasteiger and P. N. Ross, *J. Phys. Chem.*, 1996, **100**, 6715–6721.
61. J. Greeley, J. K. Nørskov and M. Mavrikakis, *Annu. Rev. Phys. Chem.*, 2002, **53**, 319–348.
62. B. Hammer and J. K. Nørskov, *Surf. Sci.*, 1995, **343**, 211–220.
63. Y. Xu, A. V. Ruban and M. Mavrikakis, *J. Am. Chem. Soc.*, 2004, **126**, 4717–4725.
64. J. R. Kitchin, J. K. Nørskov, M. A. Barteau, J. G. Chen, *J. Chem. Phys.*, 2004, **120**, 10240–10246.
65. A. R. Tao, S. Habas and P. D. Yang, *Small*, 2008, **4**, 310–325.
66. *Metallic Nanomaterials*, ed. Z. M. Peng, S. C. Yang, H. Yang and C. Kumar, Wiley-VCH Verlag, Weinheim, 2009, vol. 1.
67. S. Maksimuk, X. Teng and H. Yang, *J. Phys. Chem. C*, 2007, **111**, 14312–14319.
68. Y. Yin and A. P. Alivisatos, *Nature*, 2005, **437**, 664–670.
69. X. Peng, L. Manna, W. Yang, J. Wickham, E. Scher, A. Kadavanich and A. P. Alivisatos, *Nature*, 2000, **404**, 59–61.
70. T. S. Ahmadi, Z. L. Wang, T. C. Green, A. Henglein and M. A. El-Sayed, *Science*, 1996, **272**, 1924–1926.

71. M. Inaba, M. Ando, A. Hatanaka, A. Nomoto, K. Matsuzawa, A. Tasaka, T. Kinumoto, Y. Iriyama and Z. Ogumi, *Electrochim. Acta*, 2006, **52**, 1632–1638.
72. J. Ren and R. D. Tilley, *Small*, 2007, **3**, 1508–1512.
73. C. Wang, H. Daimon, Y. Lee, J. Kim and S. Sun, *J. Am. Chem. Soc.*, 2007, **129**, 6974–6975.
74. J. T. Ren and R. D. Tilley, *J. Am. Chem. Soc.*, 2007, **129**, 3287–3291.
75. H. Song, F. Kim, S. Connor, G. A. Somorjai and P. Yang, *J. Phys. Chem. B*, 2005, **109**, 188–193.
76. L. Qu, L. Dai and E. Osawa, *J. Am. Chem. Soc.*, 2006, **128**, 5523–5532.
77. A. Kuzume, E. Herrero and J. M. Feliu, *J. Electroanal. Chem.*, 2007, **599**, 333–343.
78. T. Yu, D. Y. Kim, H. Zhang and Y. Xia, *Angew. Chem. Int. Ed.*, 2011, **50**, 2773–2777.
79. N. M. Markovic, R. R. Adzic, B. D. Cahan and E. B. Yeager, *J. Electroanal. Chem.*, 1997, **377**, 249–259.
80. R. Jasinski, *Nature*, 1964, **201**, 1212–1213.
81. E. Yeager, *Electrochim. Acta*, 1984, **29**, 1527–1537.
82. S. Gupta, D. Tryk, I. Bae, W. Aldred and E. Yeager, *J. Appl. Electrochem.*, 1989, **19**, 19–27.
83. M. Lefevre, J. P. Dodelet and P. Bertrand, *J. Phys. Chem. B*, 2002, **106**, 8705–8713.
84. R. Bashyam and P. Zelenay, *Nature*, 2006, **443**, 63–66.
85. V. Nallathambi, J. W. Lee, S. P. Kumaraguru, G. Wu and B. N. Popov, *J. Power Sources*, 2008, **183**, 34–42.
86. M. Lefevre, E. Proietti, F. Jaouen and J. P. Dodelet, *Science*, 2009, **324**, 71–74.
87. F. Jaouen, J. Herranz, M. Lefevre, J. P. Dodelet et al., *Appl. Mater. Interfaces*, 2009, **1**, 1623–1639.
88. G. Wu, Z. Chen, K. Artyushkova, F. H. Garzon and P. Zelenay, *ECS Trans.*, 2008, **16**, 159–170.
89. Y. Nabae, M. Malon, S. M. Lyth, S. Moriya, K. Matsubayashi, N. Islam, S. Kuroki, M. Kakimoto, J. Ozaki and S. Miyata, *ECS Trans.*, 2009, **25**, 463–467.
90. F. Jaouen, E. Proietti, M. Lefevre, R. Chenitz, J. P. Dodelet, G. Wu, H. T. Chung, C. M. Johnston and P. Zelenay, *Energy Environ. Sci.*, 2011, **4**, 114–130.
91. G. Wu, K. L. More, C. M. Johnston and P. Zelenay, *Science*, 2011, **332**, 443–447.
92. Y. Nabae, M. Sonoda, C. Yamauchi, Y. Hosaka, A. Isoda and T. Aoki, *Catal. Sci. Technol.*, 2014, **4**, 1400–1406.
93. Y. Liu, A. Ishihara, S. Mitsuhima, N. Kamiya and K. Ota, *Electrochem. Solid-State Lett.*, 2005, **8**, A400–A402.
94. S. Doi, A. Ishihara, S. Mitsushima, N. Kamiya and K. Ota, *J. Electrochem. Soc.*, 2007, **154**, B362–B369.
95. J. H. Kim, A. Ishihara, S. Mitsushima, N. Kamiya and K. Ota, *Electrochim. Acta*, 2007, **52**, 2492–2497.
96. A. Ishihara, S. Doi, S. Mitsushima and K. Ota, *Electrochim. Acta*, 2008, **53**, 5442–5450.
97. Y. Ohgi, A. Ishihara, Y. Shibata, S. Mitsushima and K. Ota, *Chem. Lett.*, 2008, **37**, 608–609.
98. K. Ota, Y. Ohgi, K. D. Nam, K. Matsuzawa, S. Mitsushima and A. Ishihara, *J. Power Sources*, 2011, **196**, 5256–5263.
99. Y. Ohgi, A. Ishihara, K. Matsuzawa, S. Mitsushima, M. Matsumoto, H. Imai and K. Ota, *Electrochim. Acta*, 2012, **68**, 192–197.
100. Y. Ohgi, A. Ishihara, K. Matsuzawa, S. Mitsushima, M. Matsumoto, H. Imai and K. Ota, *J. Electrochem. Soc.*, 2013, **160**, F162–F167.
101. H. Imai, M. Matsumoto, T. Miyazaki, S. Fujieda, A. Ishihara, M. Tamura and K. Ota, *Appl. Phys. Lett.*, 2010, **96**, 191905.

18 Fundamentals and Materials Aspects of Direct Liquid Fuel Cells

Yogeshwar Sahai and Jia Ma

CONTENTS

18.1 INTRODUCTION

Fuel cells constitute an attractive class of renewable and sustainable energy sources alternative to conventional energy sources, such as petroleum that has finite reserves. They are intrinsically energy efficient, nonpolluting, silent, and reliable. In some embodiments, a fuel cell is a low-temperature device that provides electricity instantly upon demand and exhibits a long operating life. Fuel cells combine the advantages of both combustion engines and batteries, at the same time eliminating the major drawbacks of both. Similar to a battery, a fuel cell is an electrochemical energy device that converts chemical energy into electricity, and akin to a heat engine, fuel cell supplies electricity as long as fuel and oxidant are supplied to it.

Among the various types of fuel cells developed so far (as seen in Table 18.1), polymer electrolyte fuel cells (PEFCs) have the advantage of high power densities at

569

TABLE 18.1
Fuel Cell Types and Their Features

Fuel Cell Name	Electrolyte	Electrode Reaction	Working Temperature (°C)	Efficiency (%)	Cell Output
Alkaline	Potassium hydroxide aqueous solution	Anode reaction: $H_2 + 2OH^- \rightarrow 2H_2O + 2e^-$ Cathode reaction: $1/2O_2 + H_2O + 2e^- \rightarrow 2OH^-$	150–200	70	300 W to 5 kW
Molten carbonate	Molten salts, like sodium or magnesium carbonate	Anode reaction: $H_2 + CO_3^{2-} \rightarrow H_2O + CO_2 + 2e^-$ $CO + CO_3^{2-} \rightarrow 2CO_2 + 2e^-$ Cathode reaction: $1/2O_2 + CO_2 + 2e^- \rightarrow CO_3^{2-}$	650	60–80	2–100 MW
Phosphoric acid	Molten phosphoric acid	Anode reaction: $H_2 \rightarrow 2H^+ + 2e^-$ Cathode reaction: $1/2O_2 + 2H^+ + 2e^- \rightarrow H_2O$	150–200	40–80	200 kW to 11 MW
Solid oxide	Ceramic compounds of metal oxides, such as YSZ	Anode reaction: $H_2 + O^{2-} \rightarrow H_2O + 2e^-$ $CO + O^{2-} \rightarrow CO_2 + 2e^-$ $CH_4 + 4O^{2-} \rightarrow 2H_2O + CO_2 + 2e^-$ Cathode reaction: $1/2O_2 + 2e^- \rightarrow O^{2-}$	1000	60	100 kW
Polymer electrolyte membrane	Proton exchange membrane	Anode reaction: $H_2 \rightarrow 2H^+ + 2e^-$ Cathode reaction: $1/2O_2 + 2H^+ + 2e^- \rightarrow H_2O$	50–120 (Nafion) 125–220 (PBI)	40–50	100 W to 500 kW

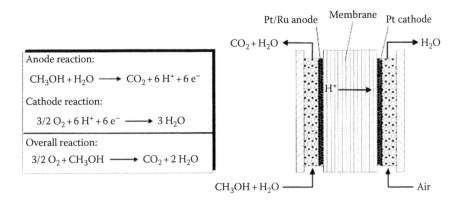

Anode reaction:

$$CH_3OH + H_2O \longrightarrow CO_2 + 6\,H^+ + 6\,e^-$$

Cathode reaction:

$$3/2\,O_2 + 6\,H^+ + 6\,e^- \longrightarrow 3\,H_2O$$

Overall reaction:

$$3/2\,O_2 + CH_3OH \longrightarrow CO_2 + 2\,H_2O$$

FIGURE 18.1 Schematic principle of a direct methanol fuel cell single cell (DMFC). (Adapted from *J. Power Sources*, 84, Baldauf, M. and Preidel, W., 161, Copyright 1999, with permission from Elsevier.)

relatively low operating temperatures and therefore are considered promising power sources for portable and residential applications. Research and development on PEFCs using hydrogen as the fuel have progressed enormously, but their successful commercialization is restricted because of poisoning of the platinum anode by carbon monoxide while using a reformer in conjunction with the PEFC and the safety and storage efficiency of the flammable hydrogen gas.

In order to overcome these difficulties, liquid fuel is used instead of PEFCs. The direct use of liquid fuel in a PEFC simplifies engineering issues, thereby driving down system complexity and hence cost. Figure 18.1 shows the reactions and operation principles of direct liquid fuel cells (DLFCs) using methanol as the fuel.[1]

This chapter reviews the DLFCs reported in the literature. For each fuel cell, a discussion of the fundamentals, the fuel, and the by-products is presented. Finally, the materials and performance aspect of the DLFC system are examined.

18.2 BRIEF SURVEY OF REACTIONS AND ENVIRONMENTAL CONCERNS OF FUELS

Table 18.2[2,3] summarizes the anode, cathode, and overall reactions for each DLFCs with oxygen as the oxidant, as well as the thermodynamic comparison of the DLFCs. The theoretical energy conversion efficiencies of all the DLFCs exceed 90%, which is larger than that of the PEFC fed with H_2 (83%). The theoretical specific energy is proportional to both the number of electrons involved (in oxidation and reduction reactions) and the overall cell voltage (electromotive force) and is inversely proportional to the fuel molecular weight.

Although fuel cells are considered environmentally friendly, the chemicals used as liquid fuels are not completely safe and the fuel cells are not emission-free. As listed in Table 18.1, except for hydrazine and borohydride, all the carbon-containing liquid fuels ideally produce CO_2 which, although a greenhouse gas, is environmentally acceptable. Regarding hydrazine, the ideal products are N_2 and water.

TABLE 18.2
Liquid Fuels for Direct Liquid Fuel Cells, Their Electrochemical Reactions, and Their Thermodynamic Features at 25°C and 1 atm

Fuel	Electrode Reactions	$-\Delta G°$ (kJ mol^{-1})	$-\Delta H°$ (kJ mol^{-1})	Standard Theoretical Potential (V)	Specific Energy (W h kg^{-1})	Pure Compound Capacity (A h kg^{-1})	Theoretical Energy Conversion Efficiency (%)
Borohydride	Anode: $BH_4^- + 8OH^- \rightarrow 8BO_2^- + H_2O + 8e^-$	1267.5	1392.2	-1.24	9295	5668	91
	Cathode: $2O_2 + 4H_2O + 8e^- \rightarrow 8OH^-$			0.40			
	Overall: $BH_4^- + 2O_2 \rightarrow BO_2^- + 2H_2O$			1.64			
Dimethoxymethane	Anode: $(CH_3O)_2CH_2 + 4H_2O \rightarrow 3CO_2 + 16H^+ + 16e^-$	1894.6	1937.5	0.002	6931	5635	98
	Cathode: $4O_2 + 16H^+ + 16e^- \rightarrow 8H_2O$			1.229			
	Overall: $(CH_3O)_2CH_2 + 4O_2 \rightarrow 3CO_2 + 4H_2O$			1.227			
Dimethyl ether	Anode: $(CH_3)_2O + 3H_2O \rightarrow 2CO_2 + 12H^+ + 12e^-$	1387.2	1460.3	0.031	8377	6981	95
	Cathode: $3O_2 + 12H^+ + 12e^- \rightarrow 6H_2O$			1.229			
	Overall: $(CH_3)_2O + 3O_2 \rightarrow 2CO_2 + 3H_2O$			1.198			

(Continued)

TABLE 18.2 (Continued)
Liquid Fuels for Direct Liquid Fuel Cells, Their Electrochemical Reactions, and Their Thermodynamic Features at 25°C and 1 atm

Fuel	Electrode Reactions	$-\Delta G°$ (kJ mol⁻¹)	$-\Delta H°$ (kJ mol⁻¹)	Standard Theoretical Potential (V)	Specific Energy (W h kg⁻¹)	Pure Compound Capacity (A h kg⁻¹)	Theoretical Energy Conversion Efficiency (%)
Ethanol	Anode: $C_2H_5OH + 3H_2O \rightarrow 2CO_2 + 12H^+ + 12e^-$ Cathode: $3O_2 + 12H^+ + 12e^- \rightarrow 6H_2O$ Overall: $C_2H_5OH + 3O_2 \rightarrow 2CO_2 + 3H_2O$	1325	1367	0.084 1.229 1.145	8028	6981	97
Ethylene glycol	Anode: $C_2H_6O_2 + 2H_2O \rightarrow 2CO_2 + 10H^+ + 10e^-$ Cathode: $5/2O_2 + 10H^+ + 10e^- \rightarrow 5H_2O$ Overall: $C_2H_6O_2 + 5/2O_2 \rightarrow 2CO_2 + 3H_2O$	1176.7	1189.5	0.009 1.229 1.220	5268	4318	99
Formic acid	Anode: $HCOOH \rightarrow CO_2 + 2H^+ + 2e^-$ Cathode: $1/2O_2 + 2H^+ + 2e^- \rightarrow H_2O$ Overall: $HCOOH + 1/2O_2 \rightarrow CO_2 + H_2O$	270	254.3	−0.171 1.229 1.4	1630	1165	106
Hydrazine	Anode: $N_2H_4 \rightarrow N_2 + 4H^+ + 4e^-$ Cathode: $O_2 + 4H^+ + 4e^- \rightarrow 2H_2O$ Overall: $N_2H_4 + O_2 \rightarrow N_2 + 2H_2O$	623.4	622.2	−0.386 1.229 1.615	5419	3345	100

Borohydride ion oxidation leads to the formation of metaborate, which is slightly harmful. Methanol and the other alcohols produce hazardous by-products. Table 18.3[3] summarizes the known main hazards of the liquid fuels and the by-products of the DLFCs. It can be seen that all of the liquid fuels are more or less hazardous. Based on Table 18.3, Demirci suggested that hydrazine should be avoided because it is carcinogenic, hazardous toward health and the environment, and unstable.[3] Ethanol is viewed as the *perfect* fuel because it is easily produced, less harmful, and energetically self-sufficient.

18.2.1　ORGANIC LIQUID FUEL

18.2.1.1　Methanol

The most common and studied fuel is methanol.[3] It is considerably more convenient and less dangerous than gaseous hydrogen, which makes it promising for mobile applications. Methanol is predominantly produced by steam reforming of natural gas, although both coal and biomass can also be used.[4]

Methanol oxidation requires active multiple sites.[5] At the first step, methanol molecules undergo dehydrogenation to generate chemisorbed species, COH_{ads}:

$$CH_3OH \rightarrow COH_{ads} + 3H_{ads} \tag{18.1}$$

Next, COH_{ads} are oxidized by chemical interaction with oxygen-containing species, OH_{ads}, adsorbed on neighboring sites of the catalyst (e.g., platinum) surface:

$$COH_{ads} + 3OH_{ads} \rightarrow CO_2 + 2H_2O \tag{18.2}$$

Ionization of the adsorbed hydrogen atoms and the formation of OH_{ads} species from water molecules are the steps producing the current:

$$H_{ads} \rightarrow H^+ + e^- \tag{18.3}$$

$$H_2O \rightarrow OH_{ads} + H^+ + e^- \tag{18.4}$$

Under certain conditions, when the formation of OH_{ads} species is not possible, COH_{ads} species change to CO_{ads} species that are not readily oxidizable and remain strongly adsorbed, preventing fresh methanol from adsorbing and undergoing further reaction. In addition, transient production of small amounts of other oxidation products, such as formaldehyde and formic acid, is seen.[6]

18.2.1.2　Ethanol

Ethanol is similar to methanol in terms of chemical properties, and yet it is much less toxic than methanol. Ethanol is produced by hydration of acetylene or biologically by fermenting agricultural biomasses. Transformations of ethanol from biomass are safer and more renewable. Electrochemical oxidation of ethanol occurs less readily than that of methanol and thus a higher temperature (ranging up to 200°C) is needed.

TABLE 18.3
Hazards of Liquid Fuels and Their Byproducts

	Health Effects	Environmental Hazards	Fire Hazard	Others
(Sodium) borohydride	Toxic, causes severe burns	Polluting	Highly flammable	Corrosive, liberates toxic and highly flammable gas
Dimethoxymethane	Irritant	Slightly polluting	Highly flammable	—
Dimethyl ether	—	—	Extremely flammable	—
Ethanol	Irritant	Slightly polluting	Highly flammable	—
Ethylene glycol	Irritant, harmful if swallowed	Slightly polluting	Flammable	Corrosive
Formic acid	Irritant, harmful, causes severe burns	Slightly polluting	Flammable	Corrosive
Hydrazine	Irritant, harmful, toxic, causes severe burns, sensitizing	Very toxic to aquatic organisms, dangerous for the envi onment	Highly flammable	Carcinogenic, unstable
Methanol	Toxic	Slightly polluting	Highly flammable	—
1-Methoxy-2-propanol	Irritant	—	Flammable	—
1-Propanol	Irritant	Toxic, slightly polluting	Highly flammable	Carcinogenic
2-Propanol	Irritant	Slightly polluting	Highly flammable	—
Tetramethyl orthocarbonate	Irritant	—	Flammable	Sensitive to humidity
Trimethoxymethane	Irritant	Slightly polluting	Highly flammable	—

(Continued)

TABLE 18.3 (Continued)
Hazards of Liquid Fuels and Their Byproducts

		Health Effects	Environmental Hazards	Fire Hazard	Others
Trioxane		Irritant, Toxic for reproduction	Slightly polluting	Highly flammable	—
Ammonia	in DHFC	Irritant, harmful, toxic, causes severe burns	Very toxic to aquatic organisms, dangerous for the environment	Nonflammable	Corrosive
Acetaldehyde	in DEFC	Highly irritant, toxic, harmful	Marine pollutant	Extremely flammable	Unstable, carcinogenic
Acetic acid	in DEFC	Irritant, harmful. causes severe burns	Slightly polluting	Flammable	Corrosive
Formic acid	in DMFC	Irritant, harmful, causes severe burns	Slightly polluting	Flammable	Corrosive
Formaldehyde	in DMFC, DDEFC	Highly irritant, toxic, sensitizing, harmful	Toxic	Flammable	Severely corrosive, carcinogenic
Glycolic acid	in DEGFC	Harmful, causes burns	Slightly polluting	Flammable	Corrosive
Methanol	in DDEFC, DDMFC, DTMFC, DTFC	Toxic	Toxic, slightly polluting	Highly flammable	—
Oxalic acid	in DEGFC	Harmful	Slightly polluting	Flammable	Corrosive
Propanal	in DP1FC	Irritant	Slightly polluting	Highly flammable	—
Propanone	in DP2FC	Irritant	Slightly polluting	Highly flammable	—
Metaborate	in DBFC	Irritant	Slightly polluting	—	—

Source: Adapted from *J. Power Sources*, 169, Demirci, U.B., 239, Copyright 2007, with permission from Elsevier.

Conditions have not been found that would reliably provide CO_2 yields approaching 100%. The reaction of ethanol electrooxidation yields by-products, such as acetic acid and acetaldehyde. The formation of acetic acid by ethanol oxidation is a four-electron process:

$$C_2H_5OH + H_2O \rightarrow CH_3COOH + 4H^+ + 4e^- \qquad (18.5)$$

The formation of acetaldehyde is a two-electron process:

$$C_2H_5OH \rightarrow CH_3CHO + 2H^+ + 2e^- \qquad (18.6)$$

The generation of these side products reduces the number of electrons involved in the reaction and thus leads to a considerable decrease in the specific energy of ethanol. In addition, a number of problems of removal and disposal arise when aldehyde and acetic acid are formed in the reaction. The formation of acetaldehyde and acetic acid is due to the fact that at relatively low temperature (below 200°C), the breaking of C–C bonds (required for forming CO_2) is very difficult, whereas the rupture of C–H bonds occurs much more readily, thus leading to side products.[6] Thus, research efforts need to be put in developing polyfunctional catalysts that have high catalytic activity toward reactions involving C–C bond rupture in addition to C–H bond rupture at relatively low temperature.

18.2.1.3 Formic Acid

Formic acid has a number of attractive properties. First, this substance is ecologically harmless. Its only oxidation products are CO_2 and water without side products or intermediates. Formic acid dissociates into $HCOO^-$ ions in aqueous solutions that have electrostatic repulsion with proton conducting groups in membrane electrolytes, eliminating the crossover issue associated with using other fuels, such as menthol and ethanol. The theoretical electrode potential for the oxidation of formic acid is −0.171 V, more negative than that for liquid fuels. However the disadvantage is that the theoretical specific energy of formic acid (1.6 kW h kg^{-1}) is much lower than for other liquid fuels.

The electrochemical oxidation of formic acid is given as follows:

$$HCOOH \rightarrow CO_2 + 2H^+ + 2e^- \qquad (18.7)$$

This reaction occurs in one of two possible pathways:

According to the first path or the dehydrogenation path, formic acid is catalytically dehydrogenated to give CO_2

$$HCOOH \rightarrow CO_2 + H_2 (\text{or } 2\text{Metal-H}_{ads}) \qquad (18.8)$$

followed by hydrogen ionization

$$H_2 \rightarrow 2H^+ + 2e^- \qquad (18.9)$$

According to the second path or the dehydration path, formic acid is dehydrated into CO, which poisons the electrode or is further oxidized to produce CO_2:

$$HCOOH \rightarrow Metal\text{-}CO_{ads} + H_2O \qquad (18.10)$$

$$H_2O + Metal \rightarrow Metal\text{-}OH_{ads} + H^+ + e^- \qquad (18.11)$$

$$Metal\text{-}CO_{ads} + Metal\text{-}OH_{ads} \rightarrow CO_2 + H^+ + e^- \qquad (18.12)$$

18.2.1.4   Other Organic Fuels

Other organic fuels, such as higher alcohols, like 1-propanol and 2-propanol, and other similar organic compounds (e.g., ethylene glycol, dimethyl ether) have also been investigated. Despite the high theoretical specific energy of these compounds, they undergo more complicated paths of electrochemical oxidation that lead to the formation of more by-products than in the case of methanol, ethanol, and formic acid as mentioned earlier.

18.2.1.5   Alkaline Alcohol

It is known that for many reactions, the reaction kinetics in alkaline medium is better than in acidic medium, which leads to a reduction in catalyst loadings, as well as less expensive, non-noble electrocatalyst. Table 18.4[7] lists the electrochemical oxidation of various alcohols in alkaline medium. The mechanism of electrochemical oxidation of methanol in alkaline systems takes place through a series of reaction steps involving successive electron transfer[7]:

$$Pt + OH^- \rightarrow Pt\text{-}(OH)_{ads} + e^- \qquad (18.13)$$

$$Pt + (CH_3OH)_{sol} \rightarrow Pt\text{-}(CH_3OH)_{ads} \qquad (18.14)$$

$$Pt\text{-}(CH_3OH)_{ads} + OH^- \rightarrow Pt\text{-}(CH_3O)_{ads} + H_2O + e^- \qquad (18.15)$$

$$Pt\text{-}(CH_3O)_{ads} + OH^- \rightarrow Pt\text{-}(CH_2O)_{ads} + H_2O + e^- \qquad (18.16)$$

$$Pt\text{-}(CH_2O)_{ads} + OH^- \rightarrow Pt\text{-}(CHO)_{ads} + H_2O + e^- \qquad (18.17)$$

$$Pt\text{-}(CHO)_{ads} + OH^- \rightarrow Pt\text{-}(CO)_{ads} + H_2O + e^- \qquad (18.18)$$

$$Pt\text{-}(CHO)_{ads} + Pt\text{-}(OH)_{ads} + 2OH^- \rightarrow 2Pt + CO_2 + 2H_2O + 2e^- \qquad (18.19)$$

$$Pt\text{-}(CHO)_{ads} + Pt\text{-}(OH)_{ads} + OH^- \rightarrow Pt + Pt\text{-}(COOH)_{ads} + H_2O + e^- \qquad (18.20)$$

$$Pt\text{-}(CO)_{ads} + Pt\text{-}(OH)_{ads} + OH^- \rightarrow 2Pt + CO_2 + H_2O + e^- \qquad (18.21)$$

$$Pt\text{-}(CO)_{ads} + Pt\text{-}(OH)_{ads} \rightarrow Pt + Pt\text{-}(COOH)_{ads} \qquad (18.22)$$

$$Pt\text{-}(COOH)_{ads} + OH^- \rightarrow Pt\text{-}(OH)_{ads} + HCOO^- \qquad (18.23)$$

$$Pt\text{-}(COOH)_{ads} + Pt\text{-}(OH)_{ads} \rightarrow 2Pt + CO_2 + H_2O \qquad (18.24)$$

TABLE 18.4
Electrochemical Oxidation of Various Alcohols in Alkaline Medium

Fuel	Anode Reactions	E° (V/SHE)	Energy Density (W h kg^{-1})
Methanol	$CH_3OH + 6OH^- \rightarrow CO_2 + 5H_2O + 6e^-$	−0.81	6100
Ethanol	$CH_3CH_2OH + 2OH^- \rightarrow CH_3CHO + 2H_2O + 2e^-$	−0.77	8030
	$CH_3CH_2OH + 4OH^- \rightarrow CH_3COOH + 3H_2O + 4e^-$		
	$CH_3CH_2OH + 12OH^- \rightarrow 2CO_2 + 9H_2O + 12e^-$		
Propanol	$CH_3CHOHCH_3 + 2OH^- \rightarrow CH_3COCH_3 + 2H_2O + 2e^-$	−0.67	8600
	$CH_3COCH_3 + 16OH^- \rightarrow 3CO_2 + 11H_2O + 16e^-$		
Ethylene glycol	$(CH_2OH)_2 + 14OH^- \rightarrow 2CO_3^{2-} + 10H_2O + 10e^-$ or	−0.72	5200
	$(CH_2OH)_2 + 10OH^- \rightarrow (CO_2)_2^{2-} + 8H_2O + 8e^-$		
Glycerol	$HOCH_2CHOHCH_2OH + 20OH^- \rightarrow 3CO_3^{2-} + 14H_2O + 14e^-$	−0.69	5000
	or $HOCH_2CHOHCH_2OH + 12OH^- \rightarrow (COO^--COH-COO^-) + 10H_2O + 10e^-$		

Source: Adapted from Yu, E.H. et al., *Energies*, 3, 1499, Copyright 2010, Authors. With permission.

Some research indicates that the rate-determining step is likely the oxidation of the active intermediate –CHO.[8]

18.2.2 INORGANIC LIQUID FUEL

18.2.2.1 Hydrazine

Hydrazine as a fuel for alkaline fuel cells has been studied since the 1970s.[9,10] The reasons for using hydrazine fuel are as follows: its electrooxidation produces no greenhouse gases, or species that may poison the electrocatalysts (e.g., CO and products of incomplete carbon-based molecule oxidation), and the theoretical electromotive force is relatively high, which results in high power density. However as mentioned earlier, the high toxicity of hydrazine should be taken into account during the design of complete fuel cell systems.[11]

Hydrazine is used in aqueous solution as liquid fuel. Due to its strong alkaline properties, in aqueous solution hydrazine dissociates into $N_2H_5^+$ and OH^- ions:

$$N_2H_4 + H_2O \rightarrow N_2H_5^+ + OH^- \qquad (18.25)$$

The anodic oxidation of hydrazine can be written as

$$N_2H_5^+ + OH^- \rightarrow N_2 + 4H^+ + 4e^- \qquad (18.26)$$

A problem associated with using hydrazine is the decomposition of hydrazine that leads to lower fuel utilization

$$N_2H_4 \rightarrow N_2 + 2H_2 \tag{18.27}$$

and/or

$$3N_2H_4 \rightarrow 3N_2 + 4NH_3 \tag{18.28}$$

18.2.2.2 Borohydride

Borohydride ion, in aqueous alkaline medium, can be oxidized directly on a large variety of electrode materials liberating a maximum of eight electrons. A big problem associated with the anodic reaction in borohydride fuel cells is that BH_4^- hydrolyzes quasi-spontaneously to generate a hydroxyl borohydride intermediate and hydrogen on various electrode materials. Hydrolysis of BH_4^- takes place through the formation of trihydrohydroxy borate ion intermediate to generate hydrogen[12] as shown in the following equations:

$$BH_4^- + H_2O \rightarrow BH_3(OH)^- + H_2 \tag{18.29}$$

$$BH_3(OH)^- + H_2O \rightarrow BO_2^- + 3H_2 \tag{18.30}$$

The presence of atomic hydrogen on the borohydride fuel cell anode makes the anode potential a mixed potential between -1.24 and -0.828 V vs. SHE.[13]

The detailed mechanism of BH_4^- electrooxidation is not yet fully understood. However, a possible reaction pathway for the electrooxidation of BH_4^- on the platinum electrode is reported in literature.[14,15] Mirkin et al.[16] reported that BH_4^- electrooxidation on gold electrode took place by an electrochemical–chemical–electrochemical reaction mechanism involving unstable intermediates:

$$BH_4^- \leftrightarrow BH_4^{\cdot} + e^- \tag{18.31}$$

$$BH_4^{\cdot} + OH^- \leftrightarrow BH_3^- + H_2O \tag{18.32}$$

$$BH_3^- \leftrightarrow BH_3 + e^- \tag{18.33}$$

The monoborane (BH_3) intermediate then undergoes further reaction to produce a total of eight electrons.

18.3 MATERIALS AND PERFORMANCE OF DIRECT LIQUID FUEL CELLS

The success of the DLFCs largely depends on two key materials, the membrane (as separator and ion-conducting electrolyte) and the electrodes. An electrode consists of a electrocatalyst with a binder loaded on the electrode substrate or the diffusion layer. Key components of DLFCs and their functions are summarized in Table 18.5.[17]

TABLE 18.5
Membrane Electrode Assembly Components and Their Tasks

MEA Component	Task/Effect
Anode substrate	Fuel supply and distribution (hydrogen/fuel gas)
	Electron conduction
	Heat removal from reaction zone
	Water supply (vapor) into electrocatalyst
Anode catalyst layer	Catalysis of anode reaction
	Proton conduction into membrane
	Electron conduction into substrate
	Water transport
	Heat transport
Proton exchange membrane	Proton conduction
	Water transport
	Electronic insulation
Cathode catalyst layer	Catalysis of cathode reaction
	Oxygen transport to reaction sites
	Proton conduction from the membrane to reaction sites
	Electron conduction from the substrate to the reaction zone
	Water removal from the reaction zone into the substrate
	Heat generation/removal
Cathode substrate	Oxidant supply and distribution (air/oxygen)
	Electron conduction toward the reaction zone
	Heat removal
	Water transport (liquid/vapor)

Source: Adapted from Zainoodin, A.M. et al., *Int. J. Hydrogen Energy*, 35, 4606, Copyright 2010. With permission.

18.3.1 ELECTRODE

The most active catalysts for DLFCs are based on platinum or platinum alloys. Platinum is the most active metallic catalyst for the dissociative adsorption of methanol, but it is readily poisoned by CO. Therefore, Pt-based bimetallic alloys are developed. It has been shown that the platinum–ruthenium bimetallic catalyst has a significantly higher catalytic activity for methanol electrooxidation reaction than pure platinum. There is a consensus about the fact that Pt–Ru is the best material for methanol oxidation. Such improvement of Pt–Ru over pure Pt is explained by the bifunctional mechanism, according to which the methanol is adsorb primarily on the platinum sites while ruthenium sites facilitate adsorption of the OH_{ads} species needed for oxidation of the organic species.[18] The optimum ruthenium content depends on the temperature, and some studies suggested that an alloy with 50 at% ruthenium represented the optimum at a temperature of 60°C.[19,20]

For the ethanol oxidation, the electrocatalysts must be active toward dehydrogenation, C–O and C–C bond cleavages. Unlike methanol oxidation, the best binary catalyst for ethanol oxidation in an acid environment is not Pt–Ru, but Pt–Sn since it promotes the cleavage of the C–C bond and improves the removal of the CO_{ads} species.[21] The optimum Sn content in the catalyst depends on the ratio of alloyed and nonalloyed Sn and also on the cell temperature.[22]

Platinum-based and palladium-based catalysts are commonly used for the oxidation of formic acid. A series of Pt-based bimetallic systems including PtRu, PtBi, PtPd, PtAu, PtPb, and PtIr have been reported as effective catalysts for formic acid fuel cells.[23] Pd-based catalysts generally have higher activity for formic acid oxidation but tend to become less active with cell operation time. Many research efforts have been put in the development of carbon-supported palladium-based catalysts, and a variety of synthesis routes for Pd/C (or Pd–M/C) catalysts have been reported.[24]

On the cathode side, it is important to deal with the negative effect of fuel crossover on catalyst activity. The key performance-limiting factors related to the anode and cathode electrodes that inhibit commercialization of the DLFC (methanol as fuel) are presented in Figure 18.2. In both cases, poor kinetics and mass transport limitations result in low electrode performance.[17]

18.3.2 MEMBRANE ELECTROLYTE

Membranes in fuel cells have the essential functions of an ion-conducting medium as well as a separator to prevent direct contact between the anode and cathode chamber. The commercial Nafion® membrane by DuPont has been widely used as a proton-conductive membrane in DLFCs. The Nafion membrane has high chemical stability, high mechanical strength, and high proton conductivity. However, Nafion membranes also have disadvantages, such as high cost, high alcohol permeability, and an operational temperature limited to 100°C.[25] Some characteristics of Nafion are listed in Table 18.6.[26] Besides Nafion, there have been several engineering thermoplastic polymers, such as poly(etheretherketone) (PEEK), polysulfone (PSF), and polybenzimidazole (PBI) (Figure 18.3),[26] which are used as alternative membranes due to their lower cost and high mechanical and thermal stability in high-temperature operations. The proton conductivity of these polymers can be increased by the addition of the sulfonic group.

18.3.3 ALKALINE FUEL CELL

Direct alkaline alcohol fuel cells can benefit from the more active reaction in alkaline media. Unlike conventional proton exchange membrane fuel cells, anion exchange membranes are used to transfer OH^- from the cathode to the anode chamber. Various alcohol fuels have been investigated in alkaline fuel cells. However, the power output of alkaline alcohol fuel cell is still lower than that using proton exchange membranes. In addition to the widely used precious metal alloys, such as Pt and Pt alloys, other lower-cost metal catalysts, such as Pd and Ni, and metal alloys have exhibited potential to be used as alternatives. Lanthanum, strontium oxides, and perovskite-type oxides have also been investigated for alcohol oxidation.[7] Alkaline alcohol fuel cell performances reported for different alcohols, catalysts, membranes, and operating parameters are summarized in Table 18.7.

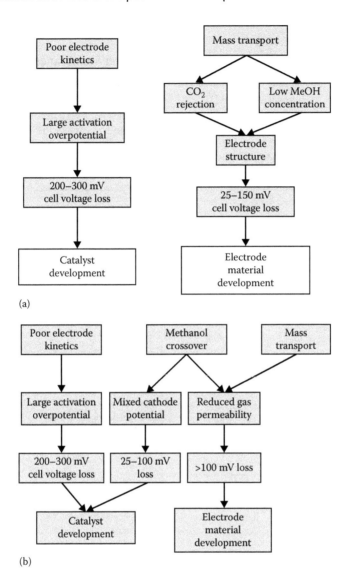

FIGURE 18.2 Technological limitations for (a) anode, and (b) cathode, in a methanol fuel cell. (Adapted from Zainoodin, A.M. et al., *Int. J. Hydrogen Energy*, 35, 4606, Copyright 2010. With permission.)

It should be noted that inorganic fuels such as hydrazine and borohydride are preferably operated in alkaline medium. Several metals were found as possible candidates for direct hydrazine fuel cells, such as Ag, Pd, and Pt. The use of several noble metals and Ni–Zr alloys was investigated under membrane electrode assembly (MEA) operational conditions. The current cation exchange membranes do not lead to good performance in hydrazine fuel cells. The development and commercialization

TABLE 18.6

Characteristics of Commercial Nafion Membranes

Membrane	Nafion 112	Nafion 115	Nafion 117
Thickness (mm)	51	127	178
J (mol s^{-1} cm^{-2})	15.3×10^{-8}	9.3×10^{-8}	7.2×10^{-8}
D (s^{-1} cm^{-2})	7.8×10^{-7}	11.8×10^{-7}	12.8×10^{-7}

Source: Adapted from Ahmad, H. et al., *Int. J. Hydrogen Energy*, 35, 2160, Copyright 2010. With permission.

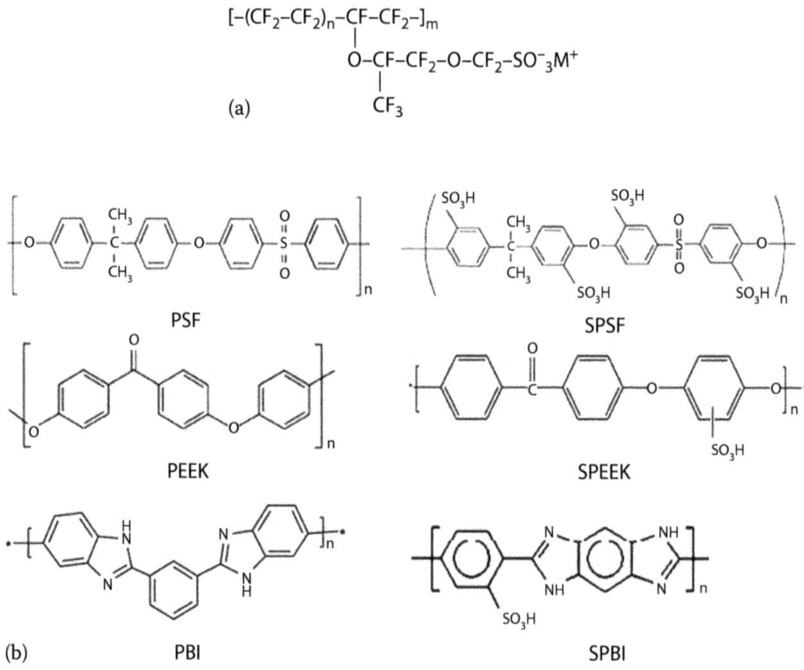

FIGURE 18.3 Structures of Nafion, PSF, PEEK, and PBI and their sulfonated counterparts. (Adapted from Ahmad, H. et al., *Int. J. Hydrogen Energy*, 35, 2160, Copyright 2010. With permission.)

of an anion exchange membrane is the crucial step for a breakthrough in hydrazine fuel cell.[11]

For borohydride fuel cells, electrocatalysts, such as Ni, Pt, and Pd, have good catalytic activity toward both the electrochemical oxidation reaction and the hydrolysis reaction. Therefore, these metals as anode give a high power density but low faradic efficiencies, for example, 50% for nickel. Gold and silver exhibit slow kinetics toward electrooxidation of BH_4^- but higher efficiencies.

TABLE 18.7
Summary of Alkaline Alcohol Fuel Cell Characteristics and Performance

Fuel/Oxidant	Anode Catalysts	Cathode Catalysts	Electrolyte Membrane	T/°C	OCV/V	P_{max}/mW cm^{-2}
2 M methanol/O$_2$	Pt	Pt	Quaternized, radiation	50	0.46	1.5
No backpressure	4 mg cm^{-2}	4 mg cm^{-2}	Grafted ethylene	60	0.52	2.4
No backpressure			Tetrafluoroethylene	60	0.58	4.2
2.5 bar back pressure			Alkaline anion Exchange membrane	80	0.63	8.5
2.5 bar back pressure						
1 M methanol/O$_2$	Pt/Ru	Pt/C	Tokuyama	80	0.57	2.6
	1 mg cm^{-2}	Pd/C	A201		0.64	2.6
		0.5 mg cm^{-2}				
2 M methanol/air	Pt/Ru	Pt 1 mg cm^{-2}	ADP	30	0.6	5.9
	1 mg cm^{-2}			40	0.62	6.9
				50	0.65	7.6
				60	0.65	9.0
16% methanol in N$_2$/air	PtRu	Pt	PVA+10M KOH	40	0.9	22
			PVA+10M KOH/ Ni-LDH			35
1 M methanol/O$_2$	Pt/C	Pt/C	ADP	RT		
no NaOH	2 mg cm^{-2}	2 mg cm^{-2}			0.48	0.2
0.25 M NaOH					0.52	1
1 M NaOH					0.6	8
4 M NaOH (2 M methanol)					0.6	18
2 M methanol, 1 M NaOH/air	Pt/C	Pt/C	ADP	60	0.7	18
	2 mg cm^{-2}	2 mg cm^{-2}				
2 M methanol, 1 M NaOH/ air	Pt/C	Pt/C	Nafion	60	0.80	4.5
	2 mg cm^{-2}	2 mg cm^{-2}				
7 M methanol, 1 M KOH/ air (passive)	PtRu	Pt/C	Tokuyama	RT	0.71	12.8
	4 mg cm^{-2}	1 mg cm^{-2}				
2 M methanol, 2 M KOH/ O$_2$	PtRu/C	Pt/C	PBI/KOH	90	1.0	30
	2 mg cm^{-2}	1 mg cm^{-2}				
2 M methanol, 2 M KOH/ O$_2$	PtRu/C	Pt/C	PBI/KOH	75	0.92	49
	2 mg cm^{-2}	1 mg cm^{-2}		90	0.98	61
1 M methanol, 1 M KOH/saturated O$_2$ in 1 NH$_2$SO$_4$	PtRu 2 mg cm^{-2}	Pt 2 mg cm^{-2}	Laminar flow– based micro fuel cells	RT	1.4	12
4 M methanol, 4 M KOH/ air	Pt/Ru 4 mg cm^{-2}	MnO$_2$/C 4 mg cm^{-2}	QPVA/Al2O3	RT	0.88	36

(Continued)

TABLE 18.7 (*Continued*)
Summary of Alkaline Alcohol Fuel Cell Characteristics and Performance

Fuel/Oxidant	Anode Catalysts	Cathode Catalysts	Electrolyte Membrane	T/°C	OCV/V	P_{max}/mW cm^{-2}
1 M KOH / humidified O$_2$	Pt/C	Ag/C	AHA	50	0.80	
Methanol	1 mg cm^{-2}	1 mg cm^{-2}				6.0
Ethylene glycol		or PtRu/C				9.0
Glycerol		4 mg cm^{-2}				6.8
Erythritol						5.5
Xylitol						4.0
1 M ethanol, KOH/ humidified O$_2$	PtRu	Pt	AHA	RT		
0.1 M KOH	3 mg cm^{-2}	3 mg cm^{-2}			1.17	18
0.5 M KOH					0.83	58
1.0 M KOH					0.84	58
4 M KOH/air	Pt/Ru	MnO$_2$/C	PVA/TiO$_2$ composite	RT	0.80	
2 M methanol	3.6 mg cm^{-2}		membrane			9.3
2 M ethanol						8.0
2 M isopropanol						5.5
2 M glycerol, 4 M NaOH/O$_2$	Pt/C	Pt/C	ADP	RT	0.68	4.2
	Pd/C	2 mg cm^{-2}			0.59	2.4
	Au/C				0.60	1.0
	AuPd/C				0.49	0.3
2 M EG, 4 M NaOH/O$_2$	Pt	Pt	ADP	20	0.66	19
	PtBi	2 mg cm^{-2}			0.83	22
	PtPdBi				0.81	28
	2 mg cm^{-2}					
2 M EG, NaOH/O$_2$	Pt	Pt	ADP	20		
1 M NaOH	2 mg cm^{-2}	2 mg cm^{-2}			0.54	4
2 M NaOH					0.63	13
4 M NaOH					0.66	19

Source: Adapted from Yu, E.H. et al., *Energies*, 3, 1499, Copyright 2010, Authors. With permission.

18.4 SUMMARY

Liquid fuels have higher energy densities than hydrogen, and they are easier to handle. These features make DLFCs a promising alternative as a low-cost power supply for both mobile and stationary applications. This chapter reviews the theoretical energy properties, the environmental concerns, as well as the materials and performance of liquid chemicals used as hydrogen carrier fuels for the PEFC-type system. Special emphasis is placed on methanol, ethanol, and formic acid as well as inorganic liquid fuels. Alcohol electrooxidation still relies on a noble metal as a catalyst. The alkaline medium opens up the possibility of using some of the less expensive metals as electrocatalysts. The boost of alkaline fuel cells is needed for the development of a robust and effective anion exchange membrane.

REFERENCES

1. M. Baldauf and W. Preidel, *J. Power Sources*, 1999, **84**, 161.
2. W. Qian, D. P. Wilkinson, J. Shen, H. Wang and J. Zhang, *J. Power Sources*, 2006, **154**, 202.
3. U. B. Demirci, *J. Power Sources*, 2007, **169**, 239.
4. U.S. Department of Energy, Energy efficiency and renewable energy, http://www.eere. energy.gov. Accessed on 2013.
5. E. Antolini, J. R. C. Salgado and E. R. Gonzalez, *Appl. Catal. B*, 2006, **63**, 137.
6. C. Lamy, A. Lima, V. Le Rhun, C. Coutanceau and J. M. Léger, *J. Power Sources*, 2002, **105**, 283.
7. E. H. Yu, U. Krewer and K. Scott, *Energies*, 2010, **3**, 1499.
8. A. V. Tripkovic, K. D. Popovic, J. D. Momcilovic and D. M. Draic, *J. Electroanal. Chem.*, 1996, **418**, 9.
9. M. R. Andrew, W. J. Gressler, J. K. Johnson, R. T. Short and K. R. Williams, *J. Appl. Electrochem.*, 1972, **2**, 327.
10. K. Tamura and T. Kahara, *J. Electrochem. Soc.*, 1976, **123**, 776.
11. A. Serov and C. Kwak, *Appl. Catal. B*, 2010, **98**, 1.
12. J. A. Gardiner and J. W. Collat, *J. Am. Chem. Soc.*, 1965, **87**, 1692.
13. B. Liu and S. Suda, *J. Alloy Compd.*, 2008, **454**, 280.
14. J. P. Elder and A. Hickling, *Trans. Faraday Soc.*, 1962, **58**, 1852.
15. J. H. Morris, H. J. Gysing and D. Reed, *Chem. Rev.*, 1985, **85**, 51.
16. M. V. Mirkin, H. Yang and A. J. Bard, *J. Electrochem. Soc.*, 1992, **139**, 2212.
17. A. M. Zainoodin, S. K. Kamarudin and W. R. W. Daud, *Int. J. Hydrogen Energy*, 2010, **35**, 4606.
18. A. S. Arico, P. Creti, H. Kim, R. Mantegna, N. Giordano and V. Antonucci, *J. Electrochem. Soc.*, 1996, **143**, 3950.
19. D. Chu and S. Gilman, *J. Electrochem. Soc.*, 1996, **143**, 1685.
20. H. A. Gasteiger, N. Markovic, P. N. Ross and E. J. Cairns, *J. Electrochem. Soc.*, 1994, **141**, 1795.
21. U. B. Demirci, *J. Power Sources*, 2007, **173**, 11.
22. E. Antolini, *J. Power Sources*, 2007, **170**, 1.
23. X. Yu and P. G. Pickup, *J. Power Sources*, 2008, **182**, 124.
24. H. Gregor, *Fuel Cell Technology Handbook*, CRC Press LLC, Boca Raton, FL, 2003.
25. C. Coutanceau, R. K. Koffi, J. M. Léger, K. Marestin, R. Mercier, C. Nayoze and P. Capron, *J. Power Sources*, 2006, **160**, 334.
26. H. Ahmad, S. K. Kamarudin, U. A. Hasran and W. R. W. Daud, *Int. J. Hydrogen Energy*, 2010, **35**, 2160.

19 Developments in Electrodes, Membranes, and Electrolytes for Direct Borohydride Fuel Cells

Irene Merino-Jiménez, Carlos Ponce de León, and Frank C. Walsh

CONTENTS

19.1 PEM FUEL CELLS FOR ENERGY CONVERSION

While lithium-ion battery technology might contribute to energy storage in a sustainable energy economy, their application in lightweight vehicles, for example, is limited to short distances. There are also concerns about the energy requirements

589

and the power grid infrastructure needed for a battery-based transport technology. Efficient internal combustion technologies and hybrid systems have been currently proposed and are likely to satisfy the transport energy needs in the near future.[1] Higher-power and higher-charge-capacity fuel cell devices could emerge as a sustainable energy alternative if the fuel cell technology and hydrogen supply infrastructure can be implemented.

As one of the most promising technologies for transport and other niche applications, the polymer electrolyte membrane (PEM) fuel cell has remained as a practical demonstrator and applications beyond buses (Ballard) are now beginning to emerge such as cars (e.g., the Honda Clarity) and motorcycles (e.g., the Intelligent Energy demonstrator). However, serious questions remain concerning the relatively poor electrochemical performance; the high cost of the fuel cell systems that include the perfluorosulfonate PEMs, bipolar plates and auxiliary components, and noble metal catalysts; and their long-term stability.

Fuel cell performance is affected by water management, loss of active surface area of the catalyst, membrane corrosion, contamination, and structural and chemical changes in the membrane electrode assembly (MEA) and in the gas diffusion layer. Due to its excellent proton conductivity and thermal, mechanical, and high chemical and electrochemical stability, Nafion® (DuPont) membranes are the most used PEM in fuel cells. However, the performance of Nafion membranes decreases over time due to degradation arising via hydrogen peroxide formation and contamination by hydrogen sulfide.[2] Currently, platinum-based metals are the most commonly used catalysts since they present high activity toward oxygen reduction and hydrogen oxidation. Since the catalyst cost of the fuel cell stack could represent around 50% of the cost, considerable efforts have been made to reduce its content and a large number of research groups dedicate their efforts to the study of nanoparticulate catalysts supported on different materials such as nanotubes or nanofibers.[3]

Alternative and readily available fuels for fuel cell energy conversion need to be safe, easily transported, and rapidly oxidized and have a high weight density and volumetric energy density, methanol[4–6] and, most recently, sodium borohydride[7–9] have been potential contenders. Direct borohydride fuel cell (DBFC) is a high-energy system with high specific energy storage capacity that has attracted academic and industrial interest for its potential applications in portable electronic devices, transportation, and underwater vehicles. Many challenges still need to be addressed to improve cell performance including the following: electrocatalysts, hydrolysis reaction, stabilization of electrolytes, materials for fuel cell construction, membranes, oxidation mechanism, and recycling process.[10]

Borohydride is the most promising fuel compared to hydrogen and methanol, and its predicted specific energy is 9.3 kW h kg^{-1} as a solid, while methanol is 6.2 kW h kg^{-1} and hydrogen 0.45 kW h kg^{-1} at 4500 psi. These values cannot be achieved, as a number of inefficiencies and limitations need to be resolved; rather, they are a guide to the most promising systems.[11,12] The lack of an effective catalyst and the use of an appropriate membrane are two of the most urgent aspects to be resolved; most metals, except gold, catalyze borohydride decomposition to hydrogen gas (hydrolysis) and borates, impeding the direct oxidation and thus the extraction of the maximum energy, while anionic membranes that keep the chemistry of the borohydride fuel cell in balance are unstable in strong alkaline solutions where sodium borohydride is stable.[7] This chapter focuses on the borohydride fuel cell components, its performance, and the major challenges facing its development.

19.2   CASE FOR LIQUID FUEL CELLS AND DBFCs

The predicted energy density of solid sodium borohydride is higher than pure methanol. However, sodium borohydride must be used in solution generally up to 30 wt.%, making its energy density lower, but the advantage is that unlike hydrogen or alcohol, which is nonflammable, it is highly soluble in concentrated sodium hydroxide. Direct oxidation of BH_4^- ions in alkaline medium transfers eight electrons at an electrode potential of -1.24 V vs. the standard hydrogen electrode (SHE):

$$BH_4^- + 8OH^- \rightarrow BO_2^- + 6H_2O + 8e^- \quad E^\circ = -1.24 \text{ V vs. SHE} \qquad (19.1)$$

Combined with the reduction of O_2, the total reaction and its thermodynamic standard cell potential is 1.64 V:

$$2O_2 + 4H_2O + 8e^- \rightarrow 8OH^- \quad E^\circ = 0.401 \text{ V vs. SHE} \qquad (19.2)$$

$$BH_4^- + 2O_2 \rightarrow BO_2^- + 2H_2O \quad E^\circ_{CELL} = 1.64 \text{ V} \qquad (19.3)$$

Hydrogen peroxide can be used for anaerobic applications (specific energy 17 kW h kg^{-1}):

$$4H_2O_2 + 8H^+ + 8e^- \rightarrow 8H_2O \quad E^\circ = -1.77 \text{ V vs. SHE} \qquad (19.4)$$

The major inefficiency is the loss of borohydride ions through the hydrolysis in acid or neutral solutions:

$$BH_4^- + 2H_2O \rightarrow BO_2^- + 4H_2 \qquad (19.5)$$

This reaction occurs together with reaction (19.1) and can be minimized in strong alkaline solutions. The oxidation of BH_4^- (Equation 19.1) is also >400 mV more negative than the evolution of hydrogen from H_2O, making it thermodynamically more favorable:

$$2H_2O + 2e^- \rightarrow H_2 + 2OH^- \quad E^\circ = -0.8277 \text{ V vs. SHE} \qquad (19.6)$$

This will decrease the number of electrons and the cell potential. The relationship between these two reactions needs more research to establish the conditions to favor the direct oxidation of borohydride. Another aspect to be considered is that the products of the partial hydrolysis of BH_4^-, for example, BH_3OH^-, oxidizes at more negative potentials than BH_4^- ions, increasing the open-circuit cell potential:

$$BH_4^- + H_2O \rightarrow BH_3OH^- + \frac{1}{2}H_2. \qquad (19.7)$$

19.3 CELL DESIGN AND COMPONENTS OF DBFCs

19.3.1 CELL DESIGN

The majority of borohydride fuel cells proposed in the literature use the parallel plate configuration design with a membrane separating the two electrodes. The membrane could be permeable to cations or anions; each choice has advantages and limitations. Figure 19.1a and b shows each type of membrane immersed in sodium hydroxide electrolyte solution. The figure shows the chemical changes during energy generation from the oxidation of borohydride reaction (19.1) in the absence of the hydrolysis of borohydride. When the cation-permeable membrane is used, eight moles of sodium ions migrate through the membrane per mole of borohydride oxidized and the concentration of sodium hydroxide on the borohydride side will decrease over time making borohydride less stable. When an anionic membrane is used, the alkalinity of the borohydride compartment is maintained as hydroxide ions migrate through the membrane from the catholyte.

Cell designs are complicated by the use of membranes and some attempts have been made to design undivided cells by using cathodes that are not active for borohydride

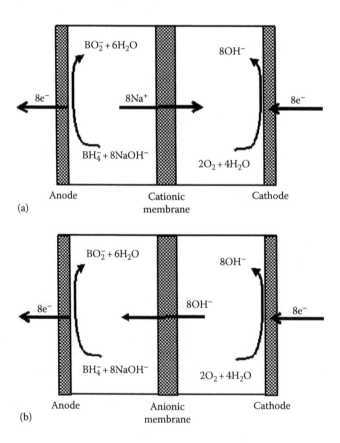

FIGURE 19.1   DBFCs using (a) a cationic membrane and (b) an anionic membrane.

reduction. For example, Ma et al.[13] developed a simple, cost-effective DBFC stack of four unit cells with a metal phthalocyanine catalyst for oxygen reduction that does not react with borohydride obtaining an open-circuit voltage (OCV) of 3.6 V and 400 mW power at 1.5 V at ambient temperature. The cell ran for over 3 h with 1 mol dm^{-3} NaBH$_4$ solution. The performance is comparable with some cell designs divided by a membrane.

19.3.2 ANODES FOR BOROHYDRIDE OXIDATION

Early work on Hg electrodes recognized that the eight electrons predicted during oxidation of borohydride occurred only in concentrated KOH or NaOH solutions where NaBH$_4^-$ ions are stable while less than eight electrons indicated partial oxidation and hydrolysis at the anode. Nickel,[14] platinum,[15,16] palladium,[16] and gold[17] at ≈0.1 A cm^{-2} in strong NaOH solutions have been used and the number of electrons transferred is 4 at Ni, <4 at Pt, and ≈8 on Au. Nickel boride seems to be significantly better than other metals.[16] The open-circuit potential is more negative on the high-surface-area Pt than on Pd when using Ni and C substrates, but sometimes Pt on sintered porous Ni plate yielded 6e$^-$ at 50–200 mA cm^{-2}.[18] A complete 8e$^-$ oxidation of borohydride, which became controlled by mass transport at high overpotentials, was reported using Au ultramicroelectrodes, and the identification of metastable intermediates was possible by cyclic voltammetry at 200 V s^{-1}.[19,20] A voltammetric study of the oxidation of borohydride on Pt electrodes using thiourea to inhibit hydrogen hydrolysis found low n values.[21] The oxidation of borohydride on nickel electrodes seems to deliver four electrons (half the theoretical energy value) at several hundreds of mA per cm^2 at an acceptable overpotential.[22]

Other approaches include catalysts for H$_2$ storage where the role of borohydride is to replenish the H$_2$ within the lattice of the alloys, Zr$_{0.9}$Ti$_{0.1}$Mn$_{0.6}$V$_{0.2}$Co$_{0.1}$Ni$_{1.1}$,[23] ZrCr$_{0.8}$Ni$_{1.2}$,[24] and LmNi$_{4.78}$Mn$_{0.22}$, where Lm is a lanthanum-rich mischmetal, and achieving up to 300 mA cm^{-2} can be at −0.7 V vs. SHE at <50% (i.e., n = 4).[25,26] High-surface-area alloy electrodes (Au 97 wt.%/Pt 3 wt.%) achieved n values of ≈7 at current densities above 0.1 A cm^{-2} at 343 K.[27] High-surface-area Pt electrodes at 10–60 mA cm^{-2} showed n values of 5–5.5.[28] Other catalysts include Ni$_2$B and Pd–Ni [16], Pt,[28] Au,[21] colloidal Au and Au alloys with Pt and Pd,[29] MnO$_2$,[30] mischmetals,[30] AB$_5$-type hydrogen storage alloys,[25,30] Ni including Raney Ni, Cu[31] colloidal Os, and Os alloys.[32]

Except for Au, none of these electrode materials realizes an eight-electron oxidation. While these studies are interesting, it is difficult to assess their impact on fuel cell technology because the oxidation waves/peaks reported occur at a potential significantly more positive to the open-circuit potentials in fuel cell conditions. Figure 19.2 shows a comparison of the power density obtained in a BH$_4^-$/O$_2$ fuel cell using different anode materials with the same concentration of borohydride and sodium hydroxide. It can be seen that Pt–Ni was the anode material that gave the highest power density, 250 mW cm^{-2}, even when the temperature used was lower than in all the other systems.

19.3.3 CATHODES FOR OXYGEN OR PEROXIDE REDUCTION

Air cathodes made of highly dispersed platinum particles supported on high-surface-area carbon (Johnson Matthey) are the most common electrodes used;

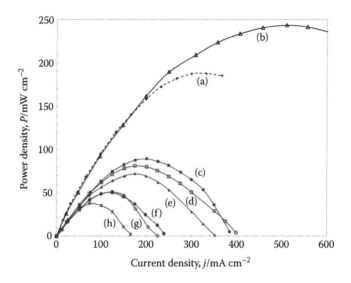

FIGURE 19.2 Power density (P) vs. current density in a BH_4^-/O_2 DBFC with various anodes of 4 cm^2 active area: (a) $Zr_{0.9}Ti_{0.1}Mn_{0.6}V_{0.2}Co_{0.1}Ni_{1.1}$, (b) Pt–Ni, (c) Pd/C, (d) Au/Ti, (e) Au/C, (f) Ag/Ti, (g) Pt/C, and (h) Ni/C. Cathode: Pt/C (2 mg Pt cm^{-2}). Membrane: Nafion 117. Temperature, 85°C, except (b) 60°C. Fuel, (a) 10 wt.% (2.64 mol dm^{-3}) NaBH$_4$ in 20 wt.% (5 mol dm^{-3}) NaOH[23] and (b–h) 5 wt.% (1.32 mol dm^{-3}) NaBH$_4$ in 10 wt.% (2.5 mol dm^{-3}) NaOH. (Adapted from Li, Z.P. et al., *J. Electrochem. Soc.*, 150, A868, 2003. With permission from The Electrochemical Society; Bessette, R.R. et al., *J. Power Sources*, 96, 240, 2001; Cheng, H. and Scott, K., *J. Electroanal. Chem.*, 596, 117, 2006; Adapted from *Electrochem. Commun.*, 10, Ma, J., Liu, Y., Zhang, P., and Wang, J., 100–102, Copyright 2008, with permission from Elsevier.)

however if there is no membrane to separate the anolyte and catholyte compartments, the catalyst can deplete the concentration of borohydride through hydrolysis.[25] The supported catalyst is mixed with Nafion and additives to prepare the ink.[33] Other catalysts include manganese dioxide and the standard Pt/Ni anode (Electro-Chem-Technic, United Kingdom) that has been reported being used with no membrane.[34] The cell voltage and power density of a Ni electrode shows higher performance than Au/Pt, Pt inks, and MnO$_2$ electrodes. These experiments cannot be closely compared as the operational conditions and membranes are different.

Due to sluggish oxygen reduction, the choice of a cathode catalyst has traditionally been limited to Pt, but other alloys such as Pt/C, Au/C, Ag/C, MnOx/C, and MnOx–Mg/C have been reported.[35] Pt/C, Ag/C, and Au/C are catalytic to borohydride oxidation and should not be used as DBFC cathodes, while MnOx/C and MnOx–Mg/C are more appropriate cathodes for BH$_4^-$/O$_2$ fuel cells. Cathodes made of Pt, Au, Pd, Ag, Raney Ag, Pd/Ir, Pd/Ru, and Pd/Ag have been used in BH$_4^-$/H$_2$O$_2$ fuel cells.[35–37] A BH$_4^-$/H$_2$O$_2$ fuel cell showed up to 680 mW cm^{-2} power density at 60°C with a Pd anode catalyst combined with a Au cathode catalyst (0.5 mg cm^{-2}) and 18 wt.% NaBH$_4$ in 17 wt.% NaOH.[38] The performance of a BH$_4^-$/O$_2$ fuel cell with a Au/C (2 mg cm^{-2}) anode in combination with different cathode materials is shown

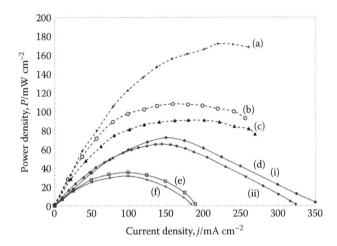

FIGURE 19.3 Power density (P) vs. current density curves measured in a BH_4/H_2O_2 fuel cell at 25°C using a Pd/C anode of 25 cm² active area and different cathode materials,[36] (a) Pd, (b) Pt, and (c) Ag, and in a BH_4/O_2 fuel cell at 85°C, using a Au/C anode of 4 cm² active area (2 mg Au cm²) and different cathode materials (2 mg cm²), (d) (i) Pt/C, (ii) FeTMPP, (e) Ni/C, and (f) Ag/C. Membrane: Nafion 117. Fuel, 5 wt.% (1.32 mol dm³) NaBH₄ in 10 wt.% (2.5 mol dm⁻³) NaOH. (From Cheng, H. et al., *J. Electroanal. Chem.*, 596, 117, 2006; Ma, J. et al., *Electrochem. Commun.*, 10, 100, 2008; Adapted from *J. Power Sources*, 219, Merino-Jiménez, I., Ponce de León, C., Shah, A.A., and Walsh, F.C., 339, Copyright 2012, with permission from Elsevier.)

in Figure 19.3 together with the comparison of the same cathodes but in a BH_4/H_2O_2 fuel cell with Pd/C as the anode. The curves in the figure show that the peak power density varied from 33 to 35 mW cm⁻², when Ag/C and Ni/C were used, respectively, and 65–72 mW cm⁻² when iron tetramethoxyphenyl prophyrin (FeTMPP) and Pt/C, respectively, were used.[39]

Other cathode materials based on cobalt phthalocyanine have shown considerable activity for oxygen reduction without being influenced by the concentration of sodium borohydride giving up to 90 mW cm⁻² at a discharge current density of 175 mA cm⁻² in a simple DBFC.[40] The figure also shows that higher current densities can be obtained by using H_2O_2 as the catholyte, varying the power density from 90 to 171 mW cm⁻² for Ag and Pd, respectively, at a discharge current density of 180 mA cm⁻².[38]

19.3.4 ELECTROLYTES

In order to avoid the decomposition of borohydride ions, the electrolyte is restricted to a strong alkaline aqueous sodium or potassium hydroxide at concentrations in the range of 10–40 wt.%. Potassium hydroxide is more conductive but more expensive than sodium hydroxide and may provide better fuel cell performance. The physical properties of borohydride fuel cell electrolytes have been reported.[22] A polymer gel electrolyte with 0.270 S cm⁻¹ conductivity has been proposed; giving an open-circuit cell potential of about 0.9 V and 6 mol dm^{-3} KOH and 2 mol dm^{-3} NaBH$_4$, the maximum power density obtained was 8.818 mW cm^{-2} at ~230 mA h discharge during 23 h.[41]

19.3.5 SEPARATORS AND ION EXCHANGE MEMBRANES

Membranes are necessary to prevent contact between the borohydride aqueous solutions with the cathode catalyst. Borohydride ions will decompose in a typical cathode used in DBFC and can form a passive layer of borate on its surface. When a high concentration of borohydride is used, a high degree of crossover would be expected and it is important to use an efficient ion-permeable membrane to prevent it.[42] In order to maintain the alkalinity of the borohydride solution, an anionic ion exchange membrane will be the ideal separator in a DBFC to transport the hydroxyl ions from the catholyte to the anolyte compartment and ensure the stability of the borohydride ions. However, an anionic membrane will also transport borohydride ions, and as a result the majority of the DBFCs use cation exchange membranes to limit the borohydride crossover and provide mechanical, chemical, and thermal stability in high sodium or potassium borohydride concentration.[7]

Thin membranes exhibit higher rates of borohydride ion crossover, but Nafion-961 membrane helps prevent the crossover of BH$_4$.[43] Other types of noncommercial membranes that have been used with a reasonable performance compared with systems using Nafion are polyethylenetetrafluoroethylene,[44] polyvinyl alcohol hydrogel,[45] and multivalent phosphate cross-linked chitosan biopolymer membrane.[46] Membraneless DBFCs have been reported using the MnO$_2$ cathode[29] giving a reasonable cell potential of 0.6 V at 5 mA cm^{-2} current density.

19.4 CHALLENGES TO BE ADDRESSED

DBFCs offer a promising alternative to incumbent electrical power generation technologies (primarily based on fossil fuels), for large-scale applications (i.e., remote or backup power), as well as for small-scale applications (i.e., laptops or mobile phones). DBFC are potentially sustainable, clean, safe, and efficient sources of energy. However, there several challenges that must be addressed to increase the energy density, which remains far below the theoretical maximum value.

The number of publications on borohydride oxidation/DBFC and borohydride hydrolysis has been consistently increased since 1960, when Indig and Snyder[14] and Elder and Hickling[15] first reported the use of Pt and Ni as anode materials for borohydride oxidation. This can be appreciated in the histogram of the number of papers published in Figure 19.4.[47] In the past 2 years the number of publications on borohydride hydrolysis was slightly higher than those on the DBFC, reflecting its relevance to both direct and indirect borohydride fuel cells.

Despite 30 years of progress in the design and operation of DBFCs and our increasing knowledge of the scientific principles governing their operation and increased experience of their design, scale-up, and operation, a number of critical challenges remain to realize satisfactory performance:

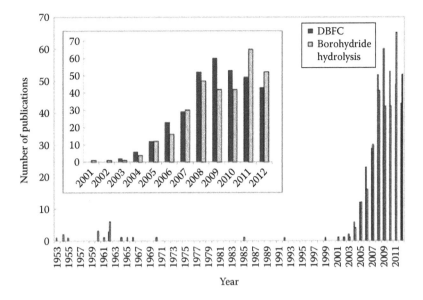

FIGURE 19.4   Evolution of the number of publications related to the DBFC and the hydrolysis of borohydride. The inset shows an expanded view of years 2001–2012. (Adapted from *Renew. Sustain. Energy Rev.*, 15, Santos, D.M.F. and Sequeira, C.A.C., 3980, Copyright 2011, with permission from Elsevier.)

1. *Improved anodes*: The lack of a selective catalyst for borohydride oxidation that releases eight electrons while minimizing hydrolysis is still a remaining problem and the need for quantitative studies such as combinatorial techniques allied to molecular modeling is plausible. The use of molecular modeling, first principles, and density functional theory (DFT) has already been applied to the characterization and design of a suitable catalyst, together with the elucidation of the oxidation mechanism. Further studies on different catalyst structures and configurations could be useful to improve electrode design. Surface spectroscopy would help corroborate the intermediates of borohydride oxidation and analyze the effect of electrolyte additives to suppress borohydride hydrolysis. Anode structures and materials[9] are needed that have moderate costs and high stability over a long period of time and are electrocatalytic for borohydride oxidation while avoiding significant rates of side reactions or borohydride hydrolysis. Cathode structures and materials must have moderate costs and high stability over a long period of time while being electrocatalytic for oxygen or peroxide reduction and avoiding significant side reactions or borohydride hydrolysis. They must also have a suitable structure for efficient operation over long periods; gas diffusion for oxygen and 3D electrodes, for example, meshes, foams, and fibers, is viable, but it is important to realize the limitations of carbon supports in very alkaline electrolytes at high current

densities, for example, nickel substrates being a more sensible but poorly studied area since carbon substrates are so widely used in other proton exchange membrane fuel cells.

2. *Ion exchange membranes* that are moderate in cost yet have high selectivity and longevity as well as high ionic conductivity. Stable hydroxyl membranes remain scarce.

3. *More stable, ionically conductive electrolytes* that offer minimum degradation to cell components while being easy to procure, store, circulate, and transport. Suitable surfactants capable of minimizing hydrolysis while maintaining good anodic kinetics are important as are compatible half-cell electrolytes in divided cells.

4. *Improved cell designs* that offer practical, modular, and easily scaled use. Cell design and construction are also key aspects requiring attention. Fuel flow plate designs with appropriate flow fields for liquid solutions should be optimized for best performance while avoiding mass transport issues such as channel blocking. Gas generation due to the hydrolysis reaction, which decreases the ionic conductivity of the electrolyte and the effective diffusion coefficients of the reactants, should be considered in the cell design. By, for example, leaving a gap between the anode and the membrane through which the hydrogen bubbles can easily escape from the anode, the dead zone formed in the anode as a consequence of the generation of hydrogen can be alleviated.[48,49] Operational conditions such as pressure drops and concentration, temperature, and potential distributions within the cell components in localized areas of the cell should be used to characterize the reaction environments in DBFCs. The use of further developed mathematical models and simulation tools, together with experimental approaches, would increase the understanding and development of the catalyst design, cell design, optimum operation conditions, and hydrogen inhibition necessary to increase the DBFC performance. As with other electrochemical reactors, it is important to maintain an efficient reaction environment with the control of concentration, current, potential, and flow distributions.[50–52]

5. *Multiphysics mathematical models* are needed that not only describe and rationalize the performance of existing cells and stacks but also predict the behavior of design changes and new materials for cell components. Recent attempts[53] should be used to build models that are easy to use, are fast to run, require modest computing power, and utilize all appropriate operational variables, including cell design, materials selection, and electrode type. Gas evolution, side reactions, and time-dependent performance are critical areas that are of utmost importance. The early nature of DBFC is apparent when the scarce literature on borohydride fuel cell modeling is contrasted with advances in PEM fuel cell modeling for H_2–O_2 and CH_3OH–O_2 fuel cells.[53]

19.5   EXAMPLES OF RECENT DEVELOPMENTS

In this section, we illustrate some recent developments in materials for DBFCs using examples from our laboratory and others.

19.5.1 Nanostructured Electrocatalysts

The most common catalyst supports for nanostructured materials[21,54] that provide a large active area for high power densities include all carbon-related products such as carbon powder, carbon paper, and carbon cloth. Electrode performance also depends on the selectivity, structure, particle size, and dispersion during the manufacture of the catalyst layer.[38,55] It has been reported that the number of electrons during the oxidation of borohydride can vary from 8 to 2 on a Pt/C nanoparticulate catalyst when the thickness of the active layer changes from 3 to <1 μm, respectively.[56] A thick layer allows enough residence time to complete the hydrogen oxidation reaction and/or the borohydride oxidation reaction increasing the apparent number of electrons. Yi et al.[57] designed a low-cost nanostructured catalyst consisting of hollow Pd/C nanospheres increasing the power density up to 25% compared to that obtained on Pd solid nanoparticles.

The catalyst cost and selectivity can be modified by bimetallic metal alloys resulting in less expensive and more selective catalysts with larger current densities compared to that of pure metal. For example, low-cost Au–Cu and $Au_{(1-x)}Zn_x$ ($0<x<1$) alloys showed increased activity compared to a pure gold electrode while maintaining the selectivity to borohydride oxidation.[58,59] Nanocatalysts such as Pd–Ir, Pd–Ni, Pd–Au, and Pd–Au have also been found to provide up to 30% more current density compared to pure Pd.[60]

Copper and titanium have also been used as supports for nanostructured electrodes for borohydride oxidation; gold–cobalt catalyst deposited onto a titanium sheet via galvanic displacement resulted in higher current peaks of BH_4^- and H_2 (from BH_4^- hydrolysis) oxidations.[61] Other nanostructured catalysts such as Ni/Zn–Ni prepared by electrodeposition and further dissolution of Zn^{62} showed a higher current peak (125.8 mA cm^{-2}) than Ni/C (18 mA cm^{-2}), Ni_{37}–Pt_3/C (26 mA cm^{-2}),[63] and nanoporous gold wire array (73.6 mA cm^{-2}).[64]

19.5.2 Porous, 3D Electrode Supports

Three-dimensional electrode supports increase the space-time yield and space velocity in the electrochemical reaction, leading to a high rate of fuel conversion per unit volume.[65] Using reticulated vitreous carbon (RVC), reticulated nickel (RN), nanotubes, and silver sponge[66,67] as a 3D electrode/support for borohydride oxidation shows that the kinetic rate constants and the current generated increase with the porosity grade. Figure 19.5a displays a PVD gold-coated RVC electrode (80 ppi) and the inset shows the nodular particles uniformly distributed on top of the RVC substrate. Figure 19.5b shows titanium oxide nanotubes, prepared by anodization at a cell voltage of 60 V, with gold sputtered to form a layer of 20–100 nm.[66] The nanotubes formed a pseudo-regular hexagonal array of the titanium dioxide nanotube of ≈200 nm with centers separated by 200 nm. At longer sputtering times (30 min), the electrode surface was almost completely covered by the gold deposit. Cyclic voltammograms (CVs) showed that the deposition time and thickness did not increase the peak current or the electrical charge per unit mass of the gold surface area, with a deposition time of 10 min being sufficient to utilize the high surface area created on the titanium sheet.[66]

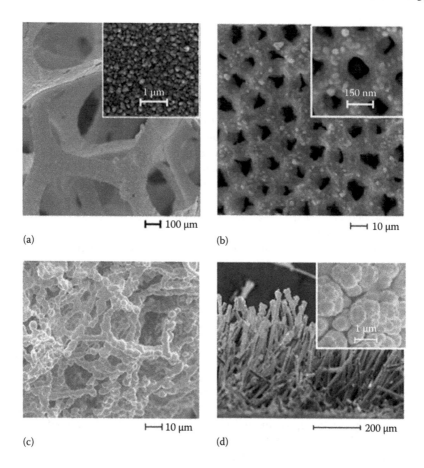

FIGURE 19.5 Scanning electron microscopy (SEM) images of four 3D electrodes used for the oxidation of BH_4^- ions: (a) Au-coated 80 ppi RVC prepared by sputtering; the inset shows the Au distribution, (b) an Au-coated Ti nanotube array produced by anodizing, (c) a silver sponge, and (d) a Pd/Ir alloy coating on carbon fibers. (a and c: From Ponce de León, C. et al., *Catal. Today*, 170, 148, 2011; b: From Low, C.T.J. et al., *Electrochem. Commun.*, 22, 166, 2012; d: Adapted from *Electrochem. Commun.*, 10, Ponce de León, C., Walsh, F.C., Patrissi, C.J., Medeiros, M.G., Bessette, R.R., Reeve, R.W., Lakeman, J.B., Rose, A., and Browning, D., 1610, Copyright 2008, with permission from Elsevier; Adapted from Patrissi, C.J. et al., *J. Electrochem. Soc.*, 155, B558, 2008. With permission from The Electrochemical Society.)

CV studies on 3D supported electrodes suggested that the energy barriers for borohydride oxidation reaction are higher than on a 2D electrode, but this can be compensated by the higher current density (per geometric area) provided by the 3D structure.[9,67] Silver sponge electrodes shown in Figure 19.5c were prepared from the calcination of a polymer matrix and silver nitrate mixtures and showed reasonable activity toward the oxidation of borohydride ions at positive potentials.[67] Figure 19.5d shows a carbon fiber support coated with a Pd/Ir alloy manufactured by direct charging electrostatic flocking (DCEF) used as a cathode in a direct borohydride hydrogen peroxide fuel cell.[37,68] This electrode was also used as an anode for borohydride oxidation showing high catalytic activity, giving current densities between 100 and 200 mA cm^{-2} in a solution containing 0.5 mol dm^{-3} NaBH$_4$ in 3 mol dm^{-3}, with low activity toward the hydrolysis reaction (<0.1 cm^3 min^{-1}) at potentials between −0.8 and −0.1 V vs. Hg/HgO. Other 3D electrodes including Pt(Ni)/TiO$_2$ nanotubes, prepared via galvanic displacement reactions, exhibited a higher electrocatalytic activity toward the oxidation of BH$_4^-$ ions than that of pure Pt electrode.[61]

19.5.3 SURFACTANT ADDITIONS TO THE ELECTROLYTE TO SUPPRESS BOROHYDRIDE HYDROLYSIS

The use of additives to inhibit borohydride hydrolysis has been barely investigated. One of the few studies reported includes the effect of thiourea on the cyclic voltammetry of borohydride ions in order to suppress the hydrolysis reaction.[69] This study demonstrated that only the peak corresponding to the direct oxidation of borohydride ions can be observed suggesting that there is no hydrogen generation that leads to a H$_2$ oxidation peak or formation of intermediates. However, there is disagreement that thiourea can poison the metallic catalysts causing deterioration of the catalyst and decreasing fuel cell performance.[70] The presence of particular surfactants in solution can significantly alter the voltammetric response of an electrooxidation reaction as has been demonstrated for other electrochemical reactions such as the oxidation of estradiol on a nano-Al$_2$O$_3$-modified glassy carbon electrode by adding cetyltrimethylammonium bromide (CTAB) to the solution. The peak current increased with the concentration of surfactant, until micelles are formed and can block access to the electrode, decreasing the current density and lowering the rate of electron transfer.[71] The presence of a surfactant shows a complex dependence on the peak current with the concentration of surfactant depending on the concentration at which the micelles are formed, which can block access to the electrode and lower the rate of electron transfer. Polymer carboxymethyl cellulose (CMetC) at 0.25 wt.% helped increase hydrogen generation from 6 wt.% sodium borohydride.[72]

The addition of a nonionic surfactant such as Triton X-100 at 0.001 wt.% concentration decreases the hydrogen generation rate from the hydrolysis of borohydride by 23% during the electrolysis at constant potential of 1 mol dm^{-3} NaBH$_4$ in 3 mol dm^{-3} NaOH when an Au/C electrode was used. Figure 19.6 shows the cyclic voltammogram at 0 and 2500 rpm of a borohydride solution containing 0.02 mol dm^{-3} NaBH$_4$ in 3 mol dm^{-3} NaOH in the absence (straight line) and in the presence of 0.001 wt.% Triton X-100 (dashed line). The curves demonstrated that the presence of the surfactant did not affect the current density at low rotation rates but slightly increased it at high rotation rates. The use of surfactants might play an important role in the inhibition of borohydride hydrolysis and further investigations should be encouraged.

19.5.4 MODELING AND SIMULATION OF DBFCS

Simulation and modeling can play an important role in the improvement of the design and performance of the DBFC, accelerating experimental studies and reducing the cost and timescale associated with laboratory evaluation.[9] Verma and Basu[73]

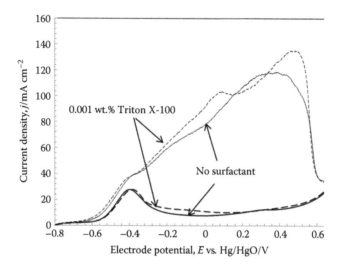

FIGURE 19.6 Cyclic voltammogram of the oxidation of BH_4^- ions at an Au/C RDE (at 0 rpm and at 2500 rpm) in the absence (black lines) and in the presence (broken lines) of 0.001 wt.% Triton X-100 at 25°C. (After Merino-Jiménez, I. et al., unpublished work.)

predicted the cell voltage at a given current density of BH_4^-/O_2. The activation, ohmic, and concentration overpotentials were considered together with the assumption of fuel cell operation under steady state and isothermal conditions. The model captures the influence of experimental conditions, such as temperature, on current–voltage characteristics but disregards the diffusion layer in the catalyst, which has a large influence on mass transfer limitations.[9] A similar model includes the activation overpotential of hydrogen peroxide in a BH_4^-/H_2O_2 DBFC.[74] A reasonable fit to the experimental data was achieved; however, the current density range considered was narrow, up to 0.02 A cm^{-2}.

A recent study calculated the cell voltage from a known constant applied current considering borohydride oxidation, borohydride hydrolysis, water reduction, ohmic drop, concentration polarizations, and hydrogen oxidation.[75] The model suggests that borohydride oxidation is followed by hydrogen evolution from the reduction of water (Tafel–Volmer–Heyrovsky mechanism), which is thermodynamically favorable (−0.828 V vs. SHE). The model showed reasonable comparison with experimental results from the literature; however, it could be further developed by considering NaOH conductivity, oxidation mechanism, membrane water content, and temperature.

19.5.5 IMPROVED MATERIALS AND CELL DESIGN

The catalyst activity and selectivity of a material toward borohydride oxidation can be predicted through mathematical modeling, calculating the adsorption free energy of the species involved in the reaction. The interest of using mathematical modeling to design a more efficient catalyst for the DBFC has been recently increased, after Rostamikia and Janik[76] applied DFT methods to evaluate the reaction mechanism of borohydride oxidation on Au(111) and Pt(111) crystalline surfaces.[77,78] The preferred oxidation mechanism was elected presuming that the reaction follows the minimum energy path, from the free energy of adsorption of borohydride ions and all the possible intermediates. The activation energy for each step of borohydride oxidation was calculated from the adsorbed initial (reactant) and final (product) states, finding the saddle point between reactant and products and assuming the lowest energy configuration.[75–80] The non-potential-dependent activation barriers were extrapolated, using Butler–Volmer formalism, and an approach has been recently made to make them potential dependent.[79] The results obtained at Au(111) showed that at −1 V vs. SHE, initial BH_4^- adsorption is not favorable and B–H bond dissociation steps present high barriers, but from 0.05 V vs. SHE, the reaction starts to be thermodynamically favorable. At −0.2 V vs. SHE, all barriers were accessible and the overall oxidation reaction proceeds to lead to $B(OH)_4^-$ (or hydrated BO_2) as a final product. At a Pt(111) surface, the B–H bond breaking is very favorable at any potential between −1 and 1 V vs. SHE, leading to a dissociated adsorption of BH_4 to $BH^* + 3H^*$ and hydrogen evolution. On the contrary, the dissociation of the O–H bond has higher barriers, finding that boric acid $(B(OH)_3)$ was the final product.[78] The first principles of a microkinetic model were also used to simulate the kinetic constants and the linear sweep voltammograms, suggesting that species containing B–H bonds are stable surface intermediates at electrode potential where an oxidation current is observed. The presence of BH_3^* as a stable intermediate compound was confirmed by surface-enhanced Raman spectroscopy.[79] Although these results give a good approximation of the catalyst behavior, it must be noted that the electrode was represented by a single-crystal surface and any subsequent solution-based reactions were not considered in the model.[9]

Bimetallic alloys have also been designed and evaluated using DFT studies. Arevalo et al.[81] used first principles calculations based on spin-polarized DFT to study the adsorption of borohydride ions on Au and $Au_3M(111)$ (M = Cr, Mn, Fe, Co and Ni), obtaining more favorable configurations and larger adsorption energies on the alloys than on pure Au, demonstrating, one more time, the prefeasibility of using bimetallic alloys for borohydride oxidation.

DFT studies can save the time and cost of preparation and experimental testing of several possible structures of different materials that might not be the most selective or active toward borohydride oxidation. It is, however, always important to validate such models using practical experimental data.

19.6 SUMMARY AND FURTHER NEEDS

Over the last 30 years, DBFCs have risen from conceptual to practical devices. They are fast becoming a potential alternative to H_2–O_2 and CH_3OH–air fuel cells in some applications involving small- to medium-scale energy conversion and specialized or strategic power supplies. Examples of 50 kW DBFC cells have been reported.

Further developments to face ever-increasing demands for efficient, competitive, and strategic energy conversion in DBFCs face continuing challenges:

1. Improved anode structures and materials that have moderate costs, have high stability over a long period time, and are electrocatalytic for borohydride

oxidation while avoiding significant rates of side reactions or borohydride hydrolysis. Molecular modeling studies of electrocatalysis need to be validated by experimental studies of borohydride oxidation.

2. Cathode structures and materials that have moderate costs and high stability over a long period of time are electrocatalytic for oxygen or peroxide reduction while avoiding significant side reactions or borohydride hydrolysis and have a suitable structure enabling efficient operation over long periods; gas diffusion for both oxygen and 3D electrodes, for example, meshes, foams, and fibers, is viable, but it is important to realize the limitations of carbon supports in very alkaline electrolytes at high current densities, for example, nickel substrates being a more sensible but poorly studied area since carbon substrates are so widely used in other proton exchange membrane fuel cells.

3. Ion exchange membranes that are moderate in cost yet have high selectivity and longevity as well as high ionic conductivity together with more stable hydroxyl membranes.

4. More stable, ionically conductive electrolytes that offer minimum degradation to cell components while being easy to procure, store, circulate, and transport, including suitable surfactants that minimize hydrolysis while maintaining good anodic kinetics, which are important here as are compatible half-cell electrolytes in divided cells.

5. Improved cell designs that offer practical, modular use while being easy to scale-up. As with other electrochemical reactors, it is important to maintain an efficient reaction environment with the control of concentration, current, potential, and flow distributions.

6. Multiphysics mathematical models are needed that not only describe and rationalize performance of existing cells and stacks but also predict the behavior of design changes and new materials of construction. Future models should be easy to use and fast to run, require modest computing power, and utilize all appropriate operational variables, including cell design, materials selection, and electrode type. Gas evolution, side reactions, and time-dependent performance are critical areas that are of utmost importance. The maturity of DBFC modeling and operational experience needs to aspire to that of PEM fuel cells for H_2–O_2 and CH_3OH–O_2.

REFERENCES

1. S. G. Li, S. M. Sharkh, F. C. Walsh and C. N. Zhang, *IEEE Trans. Veh. Technol.*, 2011, **60**, 3571.
2. A. A. Shah and F. C. Walsh, *J. Power Sources*, 2008, **185**, 287.
3. M. S. Saha, V. Neburchilov, D. Ghosh and J. Zhang, *Adv. Rev.*, 2013, **2**, 31.
4. T. S. Zhao, W. W. Yang, R. Chen and Q. X. Wu, *J. Power Sources*, 2010, **195**, 3451.
5. S. Wasmus and A. Küver, *J. Electroanal. Chem.*, 1999, **461**, 14.
6. F. Lufrano, V. Baglio, P. Staiti, V. Antonucci and A. S. Arico, *J. Power Sources*, 2013, **243**, 519.
7. C. Ponce de León, F. C. Walsh, A. Rose, J. B. Lakeman, D. J. Browning and R. W. Reeve, *J. Power Sources*, 2007, **164**, 441.
8. M. Jia, N. A. Choudhury and S. Yogeshwar, *Renew. Sustain. Energy Rev.*, 2010, **14**, 183.

9. I. Merino-Jiménez, C. Ponce de León, A. A. Shah and F. C. Walsh, *J. Power Sources*, 2012, **219**, 339–357.
10. T. Kemmitt and G. J. Gainsford, *Int. J. Hydrogen Energy*, 2009, **34**, 5726.
11. B. H. Liu, Z. P. Li and S. Suda, *J. Alloy Compd.*, 2009, **468**, 493.
12. G. Hoogers (ed.), *Fuel Cell Technology Handbook*, CRC Press, Boca Raton, FL, 2003.
13. J. Ma and Y. Liu, *J. Fuel Cell Sci. Technol.*, 2012, **9**, 011004.
14. M. E. Indig and R. N. Snyder, *J. Electrochem. Soc.*, 1962, **109**, 1104.
15. J. P. Elder and A. Hickling, *Trans. Faraday Soc.*, 1962, **58**, 1852.
16. R. Jasinski, *Electrochem. Technol.*, 1965, **3**, 40.
17. Y. Okinsaka, *J. Electrochem. Soc.*, 1973, **120**, 739.
18. M. Kubokawa, M. Yamashita and K. Abe, *Denki Kagaku*, 1968, **36**, 788.
19. M. V. Mirkin, H. Yang and A. J. Bard, *J. Electrochem. Soc.*, 1992, **139**, 2212.
20. M. V. Mirkin and A. J. Bard, *Anal. Chem.*, 1991, **63**, 532.
21. E. Gyenge, *Electrochim. Acta*, 2004, **49**, 965.
22. B. H. Liu, Z. P. Li and S. Suda, *J. Electrochem. Soc.*, 2003, **150**, A398.
23. Z. P. Li, B. H. Liu, K. Arai and S. Suda, *J. Electrochem. Soc.*, 2003, **150**, A868–A872.
24. S.-M. Lee, J.-H. Kim, H.-H. Lee, P. S. Lee and J.-Y. Lee, *J. Electrochem. Soc.*, 2002, **149**, A603.
25. L. Wang, C. Ma, Y. Sun and S. Suda, *J. Alloy Compd.*, 2005, **391**, 318.
26. L. Wang, C.-A. Ma and X. Mao, *J. Alloy Compd.*, 2005, **397**, 313.
27. S. C. Amendola, P. Onnerud, P. T. Kelly, P. J. Petillo, S. L. Sharp-Goldman and M. Binder, *J. Power Sources*, 1999, **84**, 130.
28. J.-H. Kim, H.-S. Kim, Y.-M. Kang, M.-S. Song, S. Rajendran, S.-C. Han, D.-H. Jung and J.-Y. Lee, *J. Electrochem. Soc.*, 2004, **151**, A1039.
29. M. H. Atwan, C. L. B. Macdonald, D. O. Northwood and E. L. Gyenge, *J. Power Sources*, 2006, **158**, 36.
30. R. X. Feng, H. Dong, Y. D. Wang, X. P. Ai, Y. L. Cao and H. X. Yang, *Electrochem. Commun.*, 2005, **7**, 449.
31. B. H. Liu, Z. P. Li and S. Suda, *Electrochim. Acta*, 2004, **49**, 3097.
32. M. H. Atwan, D. O. Northwood and E. L. Gyenge, *Int. J. Hydrogen Energy*, 2005, **30**, 1323.
33. B. H. Liu, Z. P. Li, K. Arai and S. Suda, *Electrochim. Acta*, 2005, **50**, 3719.
34. A. Verma, A. K. Jha and S. Basu, *J. Power Sources*, 2005, **141**, 30.
35. M. Chatenet, F. Micoud, I. Roche, E. Chainet and J. Vondrák, *Electrochim. Acta*, 2006, **51**, 5452.
36. R. R. Bessette, M. G. Medeiros, C. J. Patrissi, C. M. Deschenes and C. N. LaFratta, *J. Power Sources*, 2001, **96**, 240–244.
37. C. Ponce de León, F. C. Walsh, C. J. Patrissi, M. G. Medeiros, R. R. Bessette, R. W. Reeve, J. B. Lakeman, A. Rose and D. Browning, *Electrochem. Commun.*, 2008, **10**, 1610–1613.
38. L. Gu, N. Luo and G. H. Miley, *J. Power Sources*, 2007, **173**, 77.
39. H. Cheng and K. Scott, *J. Electroanal. Chem.*, 2006, **596**, 117–123.
40. J. Ma, Y. Liu, P. Zhang and J. Wang, *Electrochem. Commun.*, 2008, **10**, 100–102.
41. A. Jamaludin, Z. Ahmad, Z. A. Ahmad and A. A. Mohamad, *Int. J. Hydrogen Energy*, 2010, **35**, 11229.
42. H. Cheng and K. Scott, *J. Power Sources*, 2006, **160**, 407.
43. R. K. Raman, S. K. Prashant and A. K. Shukla *J. Power Sources*, 2006, **162**, 1073.
44. H. Cheng, K. Scott, K. V. Lovell, J. A. Horsfall and S. C. Waring, *J. Membr. Sci.*, 2007, **288**, 168.
45. J. Ma, N. A. Choudhury, Y. Sahai and R. G. Buchheit, *Fuel Cells*, 2011, **11**, 603.
46. J. Ma, Y. Sahai and R. G. Buchheit, *J. Power Sources*, 2012, **202**, 18.
47. D. M. F. Santos and C. A. C. Sequeira, *Renew. Sustain. Energy Rev.*, 2011, **15**, 3980–4001.

48. C. Kim, K.-J. Kim and M. Y. Ha, *J. Power Sources*, 2008, **180**, 154.
49. K. T. Park, U. H. Jung, S. U. Jeong and S. H. Kim, *J. Power Sources*, 2006, **162**, 192.
50. D. Pletcher and F. C. Walsh, *Industrial Electrochemistry*, Chapman & Hall, London, 2nd edn., 1990.
51. F. C. Walsh, *A First Course in Electrochemical Engineering*, The Electrochemical Consultancy, Romsey, 1993.
52. F. C. Walsh, *Pure Appl. Chem.*, 2001, **73**, 1819–1837.
53. A. A. Shah, K. Luo, T. R. Ralph and F. C. Walsh, *Electrochim. Acta*, 2011, **58**, 3731–3757.
54. V. Kiran, T. Ravikumar, N. T. Kalyanasundaram, S. Krishnakumar, A. K. Shukla and S. Sampath, *J. Electrochem. Soc.*, 2010, **157**, B1201.
55. V. Ganesh, D. Vijayaraghavan and V. Lakshminarayanan, *Appl. Surf. Sci.*, 2005, **240**, 286.
56. C. B. Molina and M. Chatenet, *Electrochim. Acta*, 2009, **54**, 6130.
57. L. Yi, Y. Song, X. Wang, L. Yi, J. Hu, G. Su, W. Yi and H. Yan, *J. Power Sources*, 2012, **205**, 63.
58. P. He, X. Wang, P. Fu, H. Wang and L. Yi, *Int. J. Hydrogen Energy*, 2011, **36**, 8857.
59. L. Yi, Y. Song, X. Liu, X. Wang, G. Zou, P. He and W. Yi, *Int. J. Hydrogen Energy*, 2011, **36**, 15775.
60. M. H. Atwan, E. Gyenge and D. O. Northwood, *J. New Mater. Electrochem. Syst.*, 2010, **13**, 21.
61. L. Tamašauskaitė-Tamašiūnaitė, A. Balčiūnaitė, A. Zabielaitė, J. Vaičiūnienė, A. Selskis, V. Pakštas and E. Norkus, *J. Electroanal. Chem.*, 2013, **707**, 31.
62. M. G. Hosseini, M. Abdolmaleki and S. Ashrafpoor, *Chin. J. Catal.*, 2012, **33**, 1817.
63. X. Geng, H. Zhang, W. Ye, Y. Ma and H. Zhong, *J. Power Sources*, 2008, **185**, 627.
64. L. C. Nagle and J. F. Rohan, *Int. J. Hydrogen Energy*, 2011, **36**, 10319.
65. A. Tentorio and U. Casolo-Ginelli, *J. App. Electrochem.*, 1978, **8**, 195.
66. C. T. J. Low, C. Ponce-de-León and F. C. Walsh, *Electrochem. Commun.*, 2012, **22**, 166–169.
67. C. Ponce de León, A. Kulak, S. Williams, I. Merino-Jiménez and F. C. Walsh, *Catal. Today*, 2011, **170**, 148–154.
68. C. J. Patrissi, R. R. Bessette, Y. K. Kim and C. R. Schumacher, *J. Electrochem. Soc.*, 2008, **155**, B558–B562.
69. J. I. Martins, M. C. Nunes, R. Koch, L. Martins and M. Bazzaoui, *Electrochim. Acta*, 2007, **52**, 6443.
70. Ü. B. Demirci, *Electrochim. Acta*, 2007, **52**, 5119.
71. Q. He, S. Yuan, C. Chen and S. Hu, *Mater. Sci. Eng. C Mater. Biol. Appl.*, 2003, **23**, 621.
72. M. J. F. Ferreira, V. R. Fernandes, L. Gales, C. M. Rangel and A. M. F. R. Pinto, *Int. J. Hydrogen Energy*, 2010, **35**, 11456.
73. A. Verma and S. Basu, *J. Power Sources*, 2007, **168**, 200.
74. A. E. Sanli, M. L. Aksu and B. Z. Uysal, *Int. J. Hydrogen Energy*, 2011, **36**, 8542.
75. A. A. Shah, R. Singh, C. Ponce de León, R. G. Wills and F. C. Walsh, *J. Power Sources*, 2013, **221**, 157.
76. G. Rostamikia and M. J. Janik, *Energy Environ. Sci.*, 2010, **3**, 1262.
77. G. Rostamikia and M. J. Janik, *J. Electrochem. Soc.*, 2009, **156**, B86.
78. G. Rostamikia and M. J. Janik, *Electrochim. Acta*, 2010, **55**, 1175.
79. G. Rostamikia, A. J. Mendoza, M. A. Hickner and M. J. Janik, *J. Power Sources*, 2011, **196**, 9228.
80. G. Helkeman, B. P. Uberuaga and H. Jonsson, *J. Chem. Phys.*, 2000, **113**, 9901.
81. R. L. Arevalo, M. C. S. Escanio, A. Y. Wang and H. Kasai, *Dalton Trans.*, 2012, **42**, 770.
82. I. Merino-Jiménez, C. Ponce de León and F. C. Walsh, unpublished work.

Index

9 780367 575816